D1751334

Handbook of Chemical Glycosylation

Edited by
Alexei V. Demchenko

Further Reading

Seeberger, P. H., Werz, D. (eds.)

Automated Carbohydrate Synthesis

2008
ISBN: 978-3-527-31875-9

Kollár, L. (ed.)

Modern Carbonylation Methods

2008
ISBN: 978-3-527-31896-4

Lindhorst, T. K.(ed.)

Essentials of Carbohydrate Chemistry and Biochemistry

2007
ISBN: 978-3-527-31528-4

Kinzel, T., Major, F., Raith, C., Redert, T. Stecker, F. Tölle, N., Zinngrebe, J.

Organic Synthesis Workbook III

2007
ISBN: 978-3-527-31665-6

Yudin, A. K. (ed.)

Aziridines and Epoxides in Organic Synthesis

2006
ISBN: 978-3-527-31213-9

Handbook of Chemical Glycosylation

Advances in Stereoselectivity and
Therapeutic Relevance

Edited by
Alexei V. Demchenko

WILEY-VCH Verlag GmbH & Co. KGaA

The Editor

Prof. Dr. Alexei V. Demchenko
University of Missouri
434 Benton Hall (MC27)
One University Boulevard
St. Louis, MO 63121-4499
USA

All books published by **Wiley-VCH** are carefully produced. Nevertheless, authors, editors, and publisher do not warrant the information contained in these books, including this book, to be free of errors. Readers are advised to keep in mind that statements, data, illustrations, procedural details or other items may inadvertently be inaccurate.

Library of Congress Card No.: applied for

British Library Cataloguing-in-Publication Data
A catalogue record for this book is available from the British Library.

Bibliographic information published by the Deutsche Nationalbibliothek
Die Deutsche Nationalbibliothek lists this publication in the Deutsche Nationalbibliografie; detailed bibliographic data are available on the Internet at <http://dnb.d-nb.de>.

© 2008 WILEY-VCH Verlag GmbH & Co. KGaA, Weinheim

All rights reserved (including those of translation into other languages). No part of this book may be reproduced in any form – by photoprinting, microfilm, or any other means – nor transmitted or translated into a machine language without written permission from the publishers. Registered names, trademarks, etc. used in this book, even when not specifically marked as such, are not to be considered unprotected by law.

Composition   Thomson Digital, Noida, India
Printing   Strauss GmbH, Mörlenbach
Bookbinding   Litges & Dopf GmbH, Heppenheim
Cover Design   Adam Design, Weinheim

Printed in the Federal Republic of Germany
Printed on acid-free paper

ISBN: 978-3-527-31780-6

Contents

Preface *XV*
List of Contributors *XIX*

1	**General Aspects of the Glycosidic Bond Formation**	*1*
	Alexei V. Demchenko	
1.1	Introduction *1*	
1.2	Major Types of *O*-Glycosidic Linkages *1*	
1.3	Historical Development: Classes of Glycosyl Donors *2*	
1.4	General Reaction Mechanism *4*	
1.5	Anomeric Effects *7*	
1.6	Stereoselectivity of Glycosylation *8*	
1.6.1	Structure of the Glycosyl Donor *8*	
1.6.1.1	Protecting Groups *8*	
1.6.1.2	Leaving Group *9*	
1.6.2	Structure of the Glycosyl Acceptor *9*	
1.6.2.1	Position of the Hydroxyl *9*	
1.6.2.2	Protecting Groups *10*	
1.6.3	Reaction Conditions *10*	
1.6.3.1	Solvent Effect *10*	
1.6.3.2	Promoter (Catalyst), Additions *11*	
1.6.3.3	Temperature and Pressure *11*	
1.6.4	Other Factors *11*	
1.7	Special Cases of Glycosylation *12*	
1.7.1	Aminosugars *12*	
1.7.2	Sialosides *13*	
1.7.3	Synthesis of 2-Deoxyglycosides *15*	
1.7.4	Synthesis of β-Mannosides *15*	
1.7.5	Synthesis of Furanosides *16*	
1.8	Glycosylation and Oligosaccharide Sequencing *16*	
1.8.1	Leaving-Group-Based Strategies *17*	
1.8.2	Two-Step Activation and Preactivation Strategies *18*

Handbook of Chemical Glycosylation: Advances in Stereoselectivity and Therapeutic Relevance.
Edited by Alexei V. Demchenko.
Copyright © 2008 WILEY-VCH Verlag GmbH & Co.KGaA. All rights reserved.
ISBN: 978-3-527-31780-6

1.8.3	Protecting-Group-Based Strategies 19
1.9	Conclusions and Outlook 21
	References 21

2	**Glycoside Synthesis from Anomeric Halides** 29
2.1	Glycosyl Fluorides 29
	Shin-ichiro Shoda
2.1.1	Background 29
2.1.2	Synthesis of Glycosyl Fluoride Donors 31
2.1.2.1	Fluorinating Reagents 31
2.1.2.2	Glycosyl Fluorides from Hemiacetals 32
2.1.2.3	Glycosyl Fluorides from Glycosyl Esters 33
2.1.2.4	Glycosyl from Glycosyl Halides 34
2.1.2.5	Glycosyl Fluorides from S-Glycosides 35
2.1.2.6	Glycosyl Fluorides from Other Anomeric Moieties 35
2.1.3	Glycosylation Using Glycosyl Fluorides as Glycosyl Donors 36
2.1.3.1	A Weak Lewis Acid Cleaves the C–F Bond. How Was the Glycosyl Fluoride Method Discovered? 36
2.1.3.2	Various Promoters Employed in Glycosylation by the Glycosyl Fluoride Method 38
2.1.3.3	Glycosylations Promoted by Various Promoters 38
2.1.3.4	Glycosylation of Silylated Compounds as Glycosyl Acceptors 41
2.1.3.5	Two-Stage Activation Procedure 42
2.1.3.6	Protecting-Group-Based Strategy 44
2.1.4	Application to Natural Product Synthesis 44
2.1.5	Special Topics 51
2.1.5.1	C-Glycoside Synthesis via O-Glycosylation 51
2.1.5.2	Glycosyl Fluorides for the Synthesis of a Combinatorial Library 51
2.1.5.3	Glycosyl Fluorides as Glycosyl Donors for Chemoenzymatic Synthesis 52
2.1.6	Conclusions and Future Directions 53
2.1.7	Typical Experimental Procedures 53
2.1.7.1	Preparation of the Glycosyl Donors 53
2.1.7.2	Glycosylation Using Glycosyl Fluorides as Glycosyl Donors 54
	References 56
2.2	Glycosyl Chlorides, Bromides and Iodides 59
	Suvarn S. Kulkarni, Jacquelyn Gervay-Hague
2.2.1	Background 59
2.2.2	Glycosyl Chlorides 60
2.2.2.1	Preparation of Glycosyl Chlorides 60
2.2.2.2	Reactions of Glycosyl Chlorides 62
2.2.3	Glycosyl Bromides 66
2.2.3.1	Preparation of Glycosyl Bromides 66
2.2.3.2	Reactivity Patterns and Some Useful Reactions of Glycosyl Bromides 68

2.2.3.3	Stereoselective Glycosylations Employing Glycosyl Bromides and Applications	69
2.2.4	Glycosyl Iodides	74
2.2.4.1	Preparation of Glycosyl Iodides	75
2.2.4.2	Reactions of Glycosyl Iodides	77
2.2.5	Conclusions	89
2.2.5.1	General Procedure for One-Pot Glycosylation Using Glycosyl Iodides	90
	References	90

3 Glycoside Synthesis from 1-Oxygen Substituted Glycosyl Donors 95

3.1	Hemiacetals and O-Acyl/Carbonyl Derivatives	95
	Daniel A. Ryan, David Y. Gin	
3.1.1	Introduction	95
3.1.2	Dehydrative Glycosylation via Electrophilic Activation of C1-Hemiacetals	95
3.1.3	Acid Activation of C1-Hemiacetals	96
3.1.4	Hemiacetal Activation with Silicon Electrophiles	100
3.1.5	Hemiacetal Activation with Phosphorus Electrophiles	103
3.1.6	Hemiacetal Activation with Sulfur Electrophiles	107
3.1.7	Hemiacetal Activation with Carbon Electrophiles	111
3.1.8	Other Methods	114
3.1.9	Glycosylation with Anomeric Esters	116
3.1.9.1	Glycosyl Acetate and Glycosyl Benzoate Donors	117
3.1.10	Activation of O-Carbonyl Derivatives	122
3.1.11	Conclusion	128
3.1.12	Representative Experimental Procedures	128
3.1.12.1	Representative Procedure for Preparation of C1-Hemiacetal Donors Through a Peracylation-Selective Anomeric Deacylation Sequence	128
3.1.12.2	Representative Procedure for Brønsted Acid Promoted Glycosylation with C1-Hemiacetal Donors Using Methoxyacetic Acid	128
3.1.12.3	Representative Procedure for Lewis Acid Promoted Glycosylation with C1-Hemiacetal Donors Using $Sn(OTf)_2$ and $LiClO_4$	129
3.1.12.4	Representative Procedure for Silicon Promoted Glycosylation with C1-Hemiacetal Donors Using Me_3SiBr and $CoBr_2$	129
3.1.12.5	Representative Procedure for Mitsunobu-Type Glycosylation with C1-Hemiacetal Donors and Phenol Glycosyl Acceptors	129
3.1.12.6	Representative Procedure for Appel-Type Glycosylation with C1-Hemiacetal Donors	129
3.1.12.7	Representative Procedure for Nosyl Chloride Promoted Glycosylation with C1-Hemiacetal Donors	130
3.1.12.8	Representative Procedure for Diphenyl Sulfoxide and Triflic Anhydride Promoted Glycosylation with C1-Hemiacetal Donors	130

3.1.12.9	Representative Procedure for Carbodiimide Promoted Glycosylation with C1-Hemiacetal Donors *130*
3.1.12.10	Representative Procedure for Carbonyl Promoted Glycosylation with C1-Hemiacetal Donors Using Trichloroacetic Anhydride *131*
3.1.12.11	Representative Procedure for Lewis Acid Promoted Glycosylation with Glycosyl Acetate Donors Using $SnCl_4$ *131*
3.1.12.12	Representative Procedure for Iodotrimethylsilane and Phosphine Oxide Promoted Glycosylation with Glycosyl Acetate Donors *131*
3.1.12.13	Representative Procedure for Lewis Acid Promoted Glycosylation with TOPCAT Glycosyl Donor Using Silver Triflate *131*
3.1.12.14	Representative Procedure for TMS Triflate Promoted Glycosylation with Glycosyl N-Tosyl Carbamate Donors *132*
3.1.12.15	Representative Procedure for Trityl Salt Promoted Glycosylation with Glycosyl Phenyl Carbonate Donors *132*
	References *132*
3.2	Glycoside Synthesis from 1-Oxygen-Substituted Glycosyl Imidates *143*
	Xiangming Zhu, Richard R. Schmidt
3.2.1	Introduction *143*
3.2.2	Methodological Aspects *144*
3.2.2.1	Preparation of Anomeric *O*-Trichloroacetimidates *144*
3.2.2.2	Glycosidation of *O*-Glycosyl Trichloroacetimidates *145*
3.2.3	Synthesis of Oligosaccharides *146*
3.2.3.1	β-Glucosides, β-Galactosides, α-Mannosides and Others *146*
3.2.3.2	Aminosugar-Containing Oligosaccharides *149*
3.2.3.3	1,2-*cis* Glycosides *155*
3.2.3.4	Miscellaneous Oligosaccharides *156*
3.2.4	Synthesis of Glycoconjugates *160*
3.2.4.1	Glycosphingolipids and Mimics *160*
3.2.4.2	Glycosyl Phosphatidyl Inositol Anchors *162*
3.2.4.3	Glycosyl Amino Acids and Glycopeptides *163*
3.2.4.4	Saponins *166*
3.2.4.5	Other Natural Products and Derivatives *168*
3.2.4.6	Miscellaneous Glycoconjugates *171*
3.2.5	Solid-Phase Oligosaccharide Synthesis *171*
3.2.6	Trifluoroacetimidates *174*
3.2.6.1	Preparation and Activation *174*
3.2.6.2	Application to Target Synthesis *176*
3.2.7	Conclusions and Outlook *178*
3.2.8	Experimental Procedures *178*
3.2.8.1	Typical Procedure for the Preparation of *O*-Glycosyl Trichloroacetimidates *178*
3.2.8.2	Typical Procedure for the Glycosylation with *O*-Glycosyl Trichloroacetimidates *179*
3.2.8.3	Typical Procedure for the Preparation of *O*-Glycosyl *N*-Phenyl Trifluoroacetimidates *179*

3.2.8.4	Typical Procedure for the Glycosylation with O-Glycosyl N-Phenyl Trifluoroacetmidates *179*	
	References *179*	
3.3	Anomeric Transglycosylation *185*	
	Kwan-Soo Kim, Heung-Bae Jeon	
3.3.1	Introduction *185*	
3.3.2	Alkyl Glycosides *187*	
3.3.3	Silyl Glycosides *187*	
3.3.4	Heteroaryl Glycosides *190*	
3.3.5	2-Hydroxy-3,5-Dinitrobenzoate (DISAL) Glycosides *193*	
3.3.6	Vinyl Glycosides *194*	
3.3.7	n-Pentenyl Glycosides *200*	
3.3.8	2′-Carboxybenzyl Glycosides *212*	
3.3.9	Conclusions and Outlook *217*	
3.3.10	Experimental Procedures *218*	
3.3.10.1	Glycosylation Employing Vinyl Glycosides *218*	
3.3.10.2	Glycosylation Employing n-Pentenyl Glycosides with NIS/TESOTf *219*	
3.3.10.3	Glycosylation Employing n-Pentenyl Glycosides with IDCP *219*	
3.3.10.4	Preparation of n-Pentenyl Glycosides from Glycosyl Bromides *219*	
3.3.10.5	Glycosylation Employing CB Glycosides with Tf$_2$O *219*	
3.3.10.6	Preparation of BCB Glycosides from Glycosyl Bromides *220*	
3.3.10.7	Preparation of CB Glycosides from BCB Glycosides *220*	
	References *220*	
3.4	Phosphates, Phosphites and Other O–P Derivatives *223*	
	Seiichi Nakamura, Hisanori Nambu, Shunichi Hashimoto	
3.4.1	Introduction *223*	
3.4.2	Glycosyl Phosphates *224*	
3.4.2.1	Preparation of Glycosyl Phosphates *224*	
3.4.2.2	Glycosidation Using Glycosyl Phosphates *228*	
3.4.2.3	Mechanism of Glycosidation Reaction with Glycosyl Phosphates *231*	
3.4.3	Glycosyl Phosphites *232*	
3.4.3.1	Preparation of Glycosyl Phosphites *232*	
3.4.3.2	Glycosidation Using Glycosyl Phosphites *233*	
3.4.3.3	Mechanism of Glycosidation Reaction with Glycosyl Phosphites *237*	
3.4.4	Glycosyl Donors Carrying Other Phosphorus-Containing Leaving Groups *238*	
3.4.4.1	Glycosyl Dimethylphosphinothioates *238*	
3.4.4.2	Glycosyl Phosphinimidates and Other N=P Derivatives *238*	
3.4.4.3	Glycosyl N,N,N′,N′-Tetramethylphosphorodiamidates *239*	
3.4.4.4	Miscellaneous O–P Derivatives *240*	
3.4.5	Construction of Other Types of Glycosidic Linkages *241*	
3.4.5.1	Construction of the β-Mannosidic Linkage *241*	
3.4.5.2	Construction of 2-Acetamido-2-deoxyglycosidic Linkages *241*	
3.4.5.3	Construction of 2-Deoxyglycosidic Linkages *243*	
3.4.5.4	Construction of α-Sialosidic Linkages *244*

3.4.6	Chemoselective Glycosidation Strategies 246
3.4.7	Application to the Synthesis of Natural Products 248
3.4.8	Conclusion 249
3.4.9	Experimental Procedures 249
3.4.9.1	Preparation of the Glycosyl Donors 249
3.4.9.2	Glycosidation 252
	References 254

4 Glycoside Synthesis from 1-Sulfur/Selenium-Substituted Derivatives 261

4.1	Thioglycosides in Oligosaccharide Synthesis 261
	Wei Zhong, Geert-Jan Boons
4.1.1	Preparation and O-Glycosidation of Thioglycosides 261
4.1.2	Preparation of Thioglycosides 261
4.1.3	Indirect Use of Thioglycosides in Glycosidations 263
4.1.4	Direct Use of Thioglycosides in Glycosidations 264
4.1.5	Anomeric Control in Glycosidations of Thioglycosides 267
4.1.6	Glycosylation Strategies Using Thioglycosides 274
4.1.6.1	Chemoselective Glycosylations 274
4.1.6.2	Orthogonal and Semiorthogonal Glycosylations 282
4.1.6.3	Two-Directional Glycosylation Strategies 288
4.1.7	Aglycon Transfer 292
4.1.8	General Procedure for Synthesis of Thioglycosides from Peracetylated Hexapyranosides Promoted by BF_3-Etherate 292
4.1.9	General Procedure for Synthesis of Thioglycosides by Displacement of Acylated Glycosyl Bromide with Thiolate Anion 293
4.1.10	General Procedure for Synthesis of Sialyl Thioglycosides Using TMSSMe and TMSOTf 293
4.1.11	General Procedure for Activation of Thioglycosides with Ph_2SO/Tf_2O 293
4.1.12	General Procedure for Activation of Thioglycosides with $BSP/TTBP/Tf_2O$ 294
4.1.13	General Procedure for Activation of Sialyl Thioglycosides with NIS/TfOH 294
	References 294
4.2	Sulfoxides, Sulfimides and Sulfones 303
	David Crich, Albert A. Bowers
4.2.1	Introduction 303
4.2.2	Donor Preparation 303
4.2.2.1	Sulfoxides 303
4.2.2.2	Sulfimides 306
4.2.2.3	Sulfones 306
4.2.2.4	Other Oxidized Derivatives of Thioglycosides 307
4.2.2.5	1,2-Cyclic Sulfites 307
4.2.3	Glycosylation 307

4.2.3.1	Sulfoxides	307
4.2.3.2	Sulfimides	315
4.2.3.3	Sulfones	316
4.2.3.4	Cyclic Sulfites	316
4.2.4	Applications in Total Synthesis	317
4.2.5	Special Topics	319
4.2.5.1	Intramolecular Aglycone Delivery (IAD)	319
4.2.5.2	Polymer-Supported Synthesis	321
4.2.5.3	Ring Closing and Glycosylation	321
4.2.5.4	Activation of Thioglycosides by Sulfoxides and Related Reagents	323
4.2.6	Experimental Procedures	324
4.2.6.1	General Procedure for the Preparation of Glycosyl Sulfoxides	324
4.2.6.2	General Procedure for Sulfoxide Glycosidation	325
4.2.7	Conclusion	325
	References	325
4.3	Xanthates, Thioimidates and Other Thio Derivatives	329
	Wiesław Szeja, Grzegorz Grynkiewicz	
4.3.1	Introduction	329
4.3.2	Dithiocarbonates – Preparation and Application as Glycosyl Donors	330
4.3.3	Glycosyl Thioimidates – Preparation and Application as Glycosyl Donors	335
4.3.4	Glycosyl Thiocyanates as Glycosyl Donors	349
4.3.5	Glycosyl Dithiophosphates as Glycosyl Donors	350
4.3.6	Conclusions	352
4.3.7	Typical Experimental Procedures	353
4.3.7.1	Preparation of Xanthates	353
4.3.7.2	Glycosidation of Xanthates	353
4.3.7.3	Preparation of Thioimidates	356
4.3.7.4	Synthesis of Glycosyl Thiocyanates	356
4.3.7.5	Glycosidation of Thiocyanates	357
4.3.7.6	Synthesis of *S*-(2-Deoxyglycosyl) Phosphorodithioates	357
4.3.7.7	Glycosidation of Glycosyl Phosphorodithioates	357
	References	357
4.4	Selenoglycosides	361
	Robert A. Field	
4.4.1	Background	361
4.4.2	Selenoglycoside Preparation	362
4.4.3	Selenides as Donors	365
4.4.3.1	Promoters for Selenoglycoside Activation	365
4.4.4	Selenoglycosides as Acceptors	371
4.4.5	Exploiting Selenoglycoside Relative Reactivity in Oligosaccharide Synthesis	372
4.4.6	Summary	375

4.4.7	Examples of Experimental Procedures 376
4.4.7.1	Typical Procedure for the Preparation of Selenoglycosides from Glycosyl Bromides 376
4.4.7.2	Typical Procedure for the Preparation of Selenoglycosides from Glycals 376
4.4.7.3	Typical Procedure for NIS/TfOH-Promoted Glycosylation with Selenoglycosides 376
4.4.7.4	Typical Procedure for BAHA-Promoted Glycosylation with Selenoglycosides 377
	References 377
5	**Other Methods for Glycoside Synthesis** 381
5.1	Orthoesters and Related Derivatives 381
	Bert Fraser-Reid, J. Cristóbal López
5.1.1	Introduction 381
5.1.2	Sugar 1,2-Orthoesters 382
5.1.2.1	1,2-O-Alkyl Orthoesters as Glycosyl Donors – Early Developments 384
5.1.2.2	1,2-O-Cyanoethylidene Derivatives 385
5.1.2.3	1,2-Thioorthoester Derivatives 387
5.1.2.4	Internal Orthoesters 388
5.1.2.5	Miscellaneous Orthoesters 389
5.1.3	Orthoester to Glycoside Rearrangement – The Two-Stage Glycosylation Method Revisited 390
5.1.3.1	Self-Condensation of Mannose 1,2-Orthoesters: Ready Access to (1 → 2)-Linked Mannose Oligosaccharides 394
5.1.3.2	Rearrangement of Sugar–Sugar Orthoesters Leading to 1,2-*cis*-Glycosidic Linkages 394
5.1.4	*n*-Pentenyl-1,2-Orthoesters: Glycosyl Donors with Novel Implications 394
5.1.4.1	Divergent–Convergent Synthesis of Glycosylaminoglycan 120 from Glycosyl Donors and Acceptors Ensuing from NPOEs 396
5.1.4.2	From NPOEs to the 1,2-β-Linked Oligomannans of *Candida albicans* 398
5.1.4.3	From NPOEs to the Synthesis of a Malaria Candidate Glycosylphosphatidylinositol (GPI) 398
5.1.4.4	From NPOEs to the Preparation of Glycolipids for Multivalent Presentation 399
5.1.4.5	The Lipoarabinomannan Components of the Cell Wall Complex of *Mycobacterium tuberculosis*: NPOEs in Chemoselective, Regioselective and Three-Component Double Differential Glycosidations 401
5.1.4.6	Relevance of NPOEs to the Regioselectivity in the Glycosylation of Primary Versus Secondary Hydroxyls 405
5.1.4.7	Iterative Regioselective Glycosylations of Unprotected Glycosyl Donors and Acceptors 407

5.1.4.8	NPOEs of Furanoses: Key Intermediates in the Elaboration of the Arabino Fragment of LAM	*408*
5.1.5	Conclusions and Future Directions	*410*
5.1.6	Typical Experimental Procedures	*411*
5.1.6.1	General Procedure for the Preparation of Orthoesters	*411*
5.1.6.2	General Procedure for Glycosidation with *n*-Pentenyl Orthoesters	*411*
	References	*412*
5.2	Other Methods for Glycoside Synthesis: Dehydro and Anhydro Derivatives	*416*
	David W. Gammon, Bert F. Sels	
5.2.1	Introduction	*416*
5.2.2	Glycals in Glycoside Synthesis	*417*
5.2.2.1	Preparation of Glycals	*417*
5.2.2.2	Glycals as Glycosyl Donors	*420*
5.2.3	Anhydro Sugars as Glycosyl Donors	*436*
5.2.3.1	1,2-Anhydro Sugars	*436*
5.2.3.2	1,6-Anhydro Sugars as Glycosyl Donors	*441*
5.2.4	Conclusion	*443*
5.2.5	General Experimental Procedures	*444*
5.2.5.1	General Method for the Preparation of 2-Deoxy-2-Iodoglycosides from Glycals	*444*
5.2.5.2	Preparation of 1,2-Anhydro-tri-*O*-Benzyl-α-D-Glucose and General Method for Its Use as a Glycosyl Donor in the Formation of β-Glycosides	*444*
5.2.5.3	General Method for the Preparation of 2-Deoxy-2-Iodoglycosylbenzenesulfonamides from Glycals and Its Use as Glycosyl Donors in the Synthesis of 2-Benzenesulfonamido-2-Deoxy-β-Glycosides	*444*
	References	*445*
5.3	Miscellaneous Glycosyl Donors	*449*
	Kazunobu Toshima	
5.3.1	Introduction	*449*
5.3.2	1-*O*-Silyl Glycoside	*449*
5.3.3	Diazirine	*450*
5.3.4	Telluroglycoside	*452*
5.3.5	Carbamate	*452*
5.3.6	2-Iodosulfonamide	*453*
5.3.7	*N*-Glycosyl Triazole	*453*
5.3.8	*N*-Glycosyl Tetrazole	*454*
5.3.9	*N*-Glycosyl Amide	*456*
5.3.10	DNA and RNA Nucleosides	*457*
5.3.11	Oxazoline	*457*
5.3.12	Oxathiine	*458*
5.3.13	1,6-Lactone	*459*

5.3.14	Sulfate	460
5.3.15	1,2-Cyclic Sulfite	461
5.3.16	1,2-Cyclopropane	461
5.3.17	1,2-O-Stannylene Acetal	462
5.3.18	6-Acyl-2H-Pyran-3(6H)-One	463
5.3.19	exo-Methylene	464
5.3.20	Concluding Remarks	465
5.3.21	Typical Experimental Procedure	465
5.3.21.1	General Procedure for the Preparation of Diazirines from Glycosyl Sulfonates	465
5.3.21.2	General Procedure for the Glycosylation of Diazirines	465
5.3.21.3	General Procedure for the Preparation of Glycosyl Sulfonylcarbamates from Hemiacetals	465
5.3.21.4	General Procedure for the Glycosylation of Glycosyl Sulfonylcarbamates	466
5.3.21.5	General Procedure for the Preparation of 1,2-O-Stannyl Acetals from Hemiacetals and the Glycosylation	466
5.3.21.6	General Procedure for the Preparation of 6-Acyl-2H-Pyran-3(6H)-Ones from1-(2′-Furyl)-2-tert-Butyldimethylsilanyloxyethan-1-Ols	466
5.3.21.7	General Procedure for the Glycosylation of 6-Acyl-2H-Pyran-3(6H)-Ones	467
	References	467
5.4	The Twenty First Century View of Chemical O-Glycosylation	469
	Thomas Ziegler	
5.4.1	Indirect and Special Methods	469
5.4.1.1	Intramolecular O-Glycosylation	469
5.4.1.2	Leaving-Group-Based Concept	469
5.4.1.3	Prearranged Glycoside Concept	479
5.4.2	Other Indirect and Special Methods	488
5.4.2.1	[4 + 2] Cycloadditions of Glycals	488
5.4.2.2	1,2-Cyclopropanated Sugars	492
	References	494

Index 497

Preface

Carbohydrates are the most abundant biomolecules on Earth. Although information about these fascinating natural compounds is not yet complete, we have already learned about some crucial aspects of the carbohydrate involvement in damaging cellular processes such as bacterial and viral infections, development and growth of tumors, metastasis, septic shock that are directly associated with deadly diseases of the twenty-first century, such as AIDS, cancer, meningitis and septicemia. The tremendous medicinal potential of glycostructures has already been acknowledged by the development of synthetic carbohydrate-based vaccines and therapeutics. The elucidation of the mechanisms of carbohydrate involvement in disease progression would be further improved if we could rely on the detailed knowledge of the structure, conformation and properties of the carbohydrate molecules. Therefore, the development of effective methods for the isolation and synthesis of complex carbohydrates has become critical for the field of glycosciences. Although significant improvements of the glycoside and oligosaccharide synthesis have already emerged, a variety of synthetic targets containing challenging glycosidic linkages cannot yet be directly accessed.

A vast majority of biologically and therapeutically active carbohydrates exist as polysaccharides (cellulose, chitin, starch, glycogen) or complex glycoconjugates (glycolipids, glycopeptides, glycoproteins) in which monosaccharide units are joined via glycosidic bonds. This linkage is formed by a glycosylation reaction, most commonly a promoter-assisted nucleophilic displacement of the leaving group (LG) of the glycosyl donor with the hydroxyl moiety of the glycosyl acceptor. Other functional groups on both the donor and the acceptor are temporarily masked with protecting groups (P). These reactions are most commonly performed in the presence of an activator: promoter or catalyst. As the new glycosidic linkage creates a chirality center, particular care has to be taken with regard to the stereoselectivity. Although in the natural environment specificity and selectivity of an enzyme ensure the stereoselectivity of glycosylation, synthesis of synthetic carbohydrate faces a major challenge in comparison to the synthesis of other natural biopolymers, that is proteins and nucleic acids.

Handbook of Chemical Glycosylation: Advances in Stereoselectivity and Therapeutic Relevance.
Edited by Alexei V. Demchenko.
Copyright © 2008 WILEY-VCH Verlag GmbH & Co.KGaA. All rights reserved.
ISBN: 978-3-527-31780-6

PO–[sugar]–LG + HO–[sugar]–OP →(activator)→ PO–[sugar]–O–[sugar]–OP

glycosyl donor + glycosyl acceptor → 1,2-cis and/or 1,2-trans-linked glycosides

Although mechanistic studies of the glycosylation reaction are scarce, certain conventions have already been established. Pioneering mechanistic work of Lemieux was enriched by recent studies by Bols, Boons, Crich, Gin, Kochetkov, Schmidt, Whitfield and others. 1,2-*trans* Glycosides are often stereoselectively obtained with the assistance of the 2-acyl neighboring participating group. In case of ether-type nonparticipating substituents, the glycosylation proceeds with poorer stereocontrol that results in mixtures of diastereomers, which makes the synthesis of 1,2-*cis* glycosides a notable challenge.

Since the first attempts at the turn of the twentieth century, enormous progress has been made in the area of the chemical O-glycoside synthesis. However, it is only in the past two–three decades that the scientific world has witnessed a dramatic improvement in the methods used for glycosylation. Recently, an abundance of glycosyl donors that can be synthesized under mild reaction conditions and that are sufficiently stable toward purification, modification and storage have been developed. Convergent synthetic strategies enabling convenient and expeditious assembly of oligosaccharides from properly protected building blocks with the minimum synthetic steps have also become available.

As it stands, many of the recent developments in the area of chemical glycosylation still remain compromised when applied to the stereoselective synthesis of difficult glycosidic linkages. These special cases include the synthesis of 1,2-*cis* glycosides, especially β-mannosides and *cis*-furanosides, 2-amino-2-deoxyglycosides, 2-deoxyglycosides and α-sialosides. In spite of the considerable progress and the extensive effort in this field, no universal method for the synthesis of targets containing these types of linkages has yet emerged. Therefore, these difficult cases will be discussed individually.

This book summarizes the recent advances in the area of chemical glycosylation and provides updated information regarding the current standing in the field of synthetic carbohydrate chemistry. An expansive array of methods and strategies available to a modern synthetic carbohydrate chemist is discussed. The first chapter (Chapter 1) discusses major principles of chemical glycosylation, reaction mechanisms, survey methods for glycosylation and factors influencing the reaction outcome, as well as describes the strategies for expeditious synthesis of oligosaccharide. Each subsequent chapter discusses a certain class of glycosyl donors. Methodologies developed to date are classified and discussed based on the type of the anomeric leaving group: halogens (Chapter 2), oxygen-based derivatives (Chapter 3) and sulfur/selenium-based derivatives (Chapter 4). Bicyclic compounds, 1,2-dehydro derivatives, miscellaneous glycosyl donors and indirect synthetic methods are discussed in Chapter 5. Each chapter will discuss the following aspects of a particular methodology or approach, wherever it is applicable:

(1) Introduction (relevant to this class of glycosyl donors/methods)
(2) Synthesis of glycosyl donor
(3) Glycosylation (major activators/promoters, particulars of the reaction mechanism, examples of both 1,2-*cis* and 1,2-*trans* glycosylations)
(4) Application to target/total synthesis (oligosaccharides, glycoconjugates, natural products)
(5) Special topics (synthesis of β-mannosides, furanosides, sialosides, glycosides of aminosugars and deoxysugars, if applicable)
(6) Conclusions and future directions
(7) Typical experimental procedures
(8) References.

Alexei V. Demchenko
University of Missouri – St. Louis
USA
January, 2008

List of Contributors

Geert-Jan Boons
University of Georgia
Complex Carbohydrate Research Center
315 Riverbend Road
Athens, GA 30606
USA

Albert A. Bowers
Colorado State University
Department of Chemistry
Fort Collins, CO 80523
USA

Jadwiga Bogusiak
Silesian Medical School
Faculty of Pharmacy
41-200 Sosnowiec
Poland

David Crich
Wayne State University
Department of Chemistry
Detroit, MI 48202
USA

J. Cristóbal López
Instituto de Química Orgánica General
CSIC, Juan de la Cierva 3
28006 Madrid
Spain

Alexei V. Demchenko
University of Missouri, St. Louis
Department of Chemistry and
Biochemistry
434 Benton Hall (MC27)
One University Boulevard
St. Louis, MO 63121-4499
USA

Robert A. Field
John Innes Centre
Department of Biological Chemistry
Colney Lane
Norwich NR4 7UH
UK

Bert Fraser-Reid
Natural Products and Glycotechnology
Research Institute
595 F Weathers Field Road
Fearrington 595 F
Pittsboro, NC 27312
USA

David W. Gammon
University of Cape Town
Department of Chemistry
7701 Rondebosch
South Africa

Handbook of Chemical Glycosylation: Advances in Stereoselectivity and Therapeutic Relevance.
Edited by Alexei V. Demchenko.
Copyright © 2008 WILEY-VCH Verlag GmbH & Co.KGaA. All rights reserved.
ISBN: 978-3-527-31780-6

Jacquelyn Gervay-Hague
University of California, Davis
Department of Chemistry
One Shields Avenue
Davis, CA 95616
USA

David Y. Gin
Memorial Sloan-
Kettering Cancer Center
1275 York Avenue, Mailbox 379
New York, NY 10065
USA

Grzegorz Grynkiewicz
Pharmaceutical Research Institute
Warszawa
Poland

Shunichi Hashimoto
Hokkaido University
Faculty of Pharmaceutical Sciences
Kita 12 Nishi 6, Kita-Ku
Sapporo 060-0812
Japan

Heung-Bae Jeon
Kwangwoon University
Department of Chemistry
Seoul 139-701
Korea

Kwan-Soo Kim
Yonsei University
Center for Bioactive Molecular Hybrids
and Department of Chemistry
Yonsei University
Seoul 120-749
Korea

J. Cristóbal López
Natural Products and Glycotechnology
Research Institute
595 F Weathers Field Road
Fearrington 595 F
Pittsboro, NC 27315
USA

Suvarn S. Kulkarni
University of California, Davis
Department of Chemistry
One Shields Avenue
Davis, CA 95616
USA

Seiichi Nakamura
Hokkaido University
Faculty of Pharmaceutical Sciences
Kita 12 Nishi 6, Kita-Ku
Sapporo 060-0812
Japan

Hisanori Nambu
Hokkaido University
Faculty of Pharmaceutical Sciences
Kita 12 Nishi 6, Kita-Ku
Sapporo 060-0812
Japan

Daniel A. Ryan
Memorial Sloan-
Kettering Cancer Center
1275 York Avenue, Mailbox 379
New York, NY 10065
USA

Richard R. Schmidt
Universität Konstanz
Fachbereich Chemie
Fach M 725
78457 Konstanz
Germany

Bert F. Sels
Katholieke Universiteit Leuven
Centrum Oppervlaktechemie en
Katalyse
Kasteelpark Arenberg 23
3001 Heverlee (Leuven)
Belgium

Shin-ichiro Shoda
Tohoku University
School of Engineering
6-6-04 Aramaki Aza Aoba
Aoba-ku, Sendai
Miyagi 980-8579
Japan

Wiesław Szeja
Silesian Technical University
Department of Chemistry
44-100 Gliwice
Poland

Kazunobu Toshima
Keio University
Faculty of Science and Technology
Department of Applied Chemistry
3-14-1 Hiyoshi, Kohoku-ku
Yokohama 223-8522
Japan

Wei Zhong
University of Georgia
Complex Carbohydrate Research Center
315 Riverbend Road
Athens, GA 30606
USA

Xiangming Zhu
University College Dublin
School of Chemistry
and Chemical Biology
Belfield, Dublin 4
Ireland

Thomas Ziegler
University of Tübingen
Institute of Organic Chemistry
Auf der Morgenstelle 18
72076 Tübingen
Germany

1
General Aspects of the Glycosidic Bond Formation
Alexei V. Demchenko

1.1
Introduction

Since the first attempts at the turn of the twentieth century, enormous progress has been made in the area of the chemical synthesis of *O*-glycosides. However, it was only in the past two decades that the scientific world had witnessed a dramatic improvement the methods used for chemical glycosylation. The development of new classes of glycosyl donors has not only allowed accessing novel types of glycosidic linkages but also led to the discovery of rapid and convergent strategies for expeditious oligosaccharide synthesis. This chapter summarizes major principles of the glycosidic bond formation and strategies to obtain certain classes of compounds, ranging from glycosides of uncommon sugars to complex oligosaccharide sequences.

1.2
Major Types of *O*-Glycosidic Linkages

There are two major types of *O*-glycosides, which are, depending on nomenclature, most commonly defined as α- and β-, or 1,2-*cis* and 1,2-*trans* glycosides. The 1,2-*cis* glycosyl residues, α-glycosides for D-glucose, D-galactose, L-fucose, D-xylose or β-glycosides for D-mannose, L-arabinose, as well as their 1,2-*trans* counterparts (β-glycosides for D-glucose, D-galactose, α-glycosides for D-mannose, etc.), are equally important components in a variety of natural compounds. Representative examples of common glycosides are shown in Figure 1.1. Some other types of glycosides, in particular 2-deoxyglycosides and sialosides, can be defined neither as 1,2-*cis* nor as 1,2-*trans* derivatives, yet are important targets because of their common occurrence as components of many classes of natural glycostructures.

Handbook of Chemical Glycosylation: Advances in Stereoselectivity and Therapeutic Relevance.
Edited by Alexei V. Demchenko.
Copyright © 2008 WILEY-VCH Verlag GmbH & Co. KGaA. All rights reserved.
ISBN: 978-3-527-31780-6

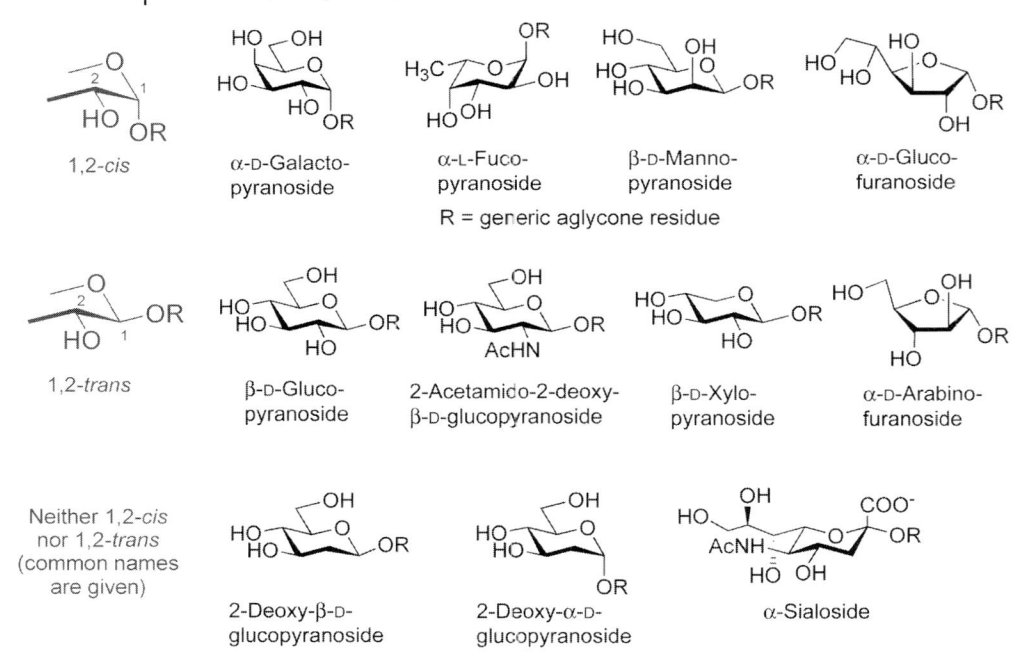

Figure 1.1 Common examples of O-glycosides.

1.3
Historical Development: Classes of Glycosyl Donors

The first reactions performed by Michael (synthesis of aryl glycosides from glycosyl halides) [1] and Fischer (synthesis of alkyl glycosides from hemiacetals) [2] at the end of the nineteenth century showed the complexity of the glycosylation process. The discovery of the first controlled, general glycosylation procedure involving the nucleophilic displacement of chlorine or bromine at the anomeric center is credited to Koenigs and Knorr [3]. The glycosylations were performed in the presence of Ag_2CO_3, which primarily acted as an acid (HCl or HBr) scavenger. At that early stage, glycosylations of poorly nucleophilic acceptors such as sugar hydroxyls were sluggish and inefficient; hence, even the synthesis of disaccharides represented a notable challenge. The first attempts to solve this problem gave rise to the development of new catalytic systems that were thought to be actively involved in the glycosylation process [4]. Thus, Zemplen and Gerecs [5] and, subsequently, Helferich and Wedermeyer [6] assumed that the complexation of the anomeric bromides or chlorides with more reactive, heavy-metal-based catalysts would significantly improve their leaving-group ability. This approach that has become a valuable expansion of the classic Koenigs–Knorr method made it possible to replace Ag_2CO_3 or Ag_2O by more active mercury(II) salt catalysts. The early attempts to improve the glycosylation process have revealed the necessity to find a delicate balance between the reactivity and stereoselectivity [7,8]. Indeed, it was noted that faster reactions often result in a decreased stereoselectivity. At around the same time, the first attempts to involve other classes of anomeric leaving groups (LGs) resulted in the investigation of peracetates as glycosyl donors [9].

Seminal work of Lemieux [10] and Fletcher and coworkers [11,12] has led to the appreciation that the reactivity of the glycosyl halides and the stereoselectivity of glycosylation are directly correlated to the nature of the protecting groups, especially at the neighboring C-2 position. From early days, it has been acknowledged that peracylated halides often allow stereoselective formation of 1,2-*trans* glycosides. Later, this phenomenon was rationalized by the so-called participatory effect of the neighboring acyl substituent at C-2. Although occasionally substantial amounts of 1,2-*cis* glycosides were obtained even with 2-acylated glycosyl donors, the purposeful 1,2-*cis* glycosylations were best achieved with a nonparticipating ether group at C-2, such as methyl or benzyl. Further search for suitable promoters for the activation of glycosyl halides led to the discovery of Ag-silicate that proved to be very efficient for direct β-mannosylation, as these reactions often proceed via a concerted S_N2 mechanism [13,14].

For many decades classic methods, in which anomeric bromides, chlorides, acetates or hemiacetals were used as glycosyl donors, had been the only procedure for the synthesis of a variety of synthetic targets ranging from simple glycosides to relatively complex oligosaccharides (Figure 1.2). Deeper understanding of the reaction mechanism, driving forces and principles of glycosylation have stimulated the development of other methods for glycosylation, with the main effort focusing on the development of new anomeric leaving groups [15,16]. During the 1970s to early 1980s, a few new classes of glycosyl donors were developed. The following compounds are only the most representative examples of the first wave of the leaving-group development: thioglycosides by Ferrier *et al.* [17], Nicolaou *et al.* [18], Garegg *et al.* [19] and others [20]; cyanoethylidene and orthoester derivatives by Kochetkov and coworkers [21,22]; *O*-imidates by Sinay and coworkers [23] and Schmidt and Michel [24]; thioimidates including *S*-benzothiazolyl derivatives by Mukaiyama *et al.* [25]; thiopyridyl derivatives by Hanessian *et al.* [26] and Woodward *et al.* [27] and glycosyl fluorides by Mukaiyama *et al.* [28] (Figure 1.2). Many glycosyl donors introduced during that period gave rise to excellent complimentary glycosylation methodologies. Arguably, trichloroacetimidates [29,30], thioglycosides [31–33] and fluorides [34,35] have become the most common glycosyl donors nowadays.

A new wave of methods arose in the end of the 1980s, among which were glycosyl donors such as glycosyl acyl/carbonates [36–38], thiocyanates [39], diazirines [40], xanthates [41], glycals [42,43], phosphites [44,45], sulfoxides [46], sulfones [47], selenium glycosides [48], alkenyl glycosides [49–51] and heteroaryl glycosides [52] (Figure 1.2). These developments were followed by a variety of more recent methodologies and improvements, among which are glycosyl iodides [53], phosphates [54], Te-glycosides [55], sulfonylcarbamates [56], disulfides [57], 2-(hydroxycarbonyl) benzyl glycosides [58] and novel thio- [59,60] and *O*-imidates [61,62] (Figure 1.2). In

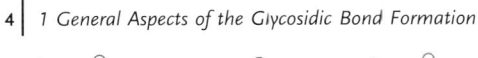

Figure 1.2 Survey of glycosyl donors.

addition, a variety of new recent methodologies bring the use of classic glycosyl donors such as hemiacetals to entirely different level of flexibility and usefulness [63]. These innovative concepts will be discussed in the subsequent chapters dealing with particular classes of clycosyl donors..

1.4
General Reaction Mechanism

Detailed glycosylation mechanism has not been elucidated as yet; therefore, speculations and diagrams presented herein are a commonly accepted prototype of the glycosylation mechanism. Most commonly, the glycosylation reaction involves nucleophilic displacement at the anomeric center. As the reaction takes place at the secondary carbon atom with the use of weak nucleophiles (sugar acceptors), it often follows a unimolecular S_N1 mechanism. Glycosyl donors bearing a nonparticipating and a participating group will be discussed separately (Scheme 1.1a and b, respectively). In most cases, an activator (promoter or catalyst) assisted departure of the anomeric leaving group results in the formation of the glycosyl cation. The only

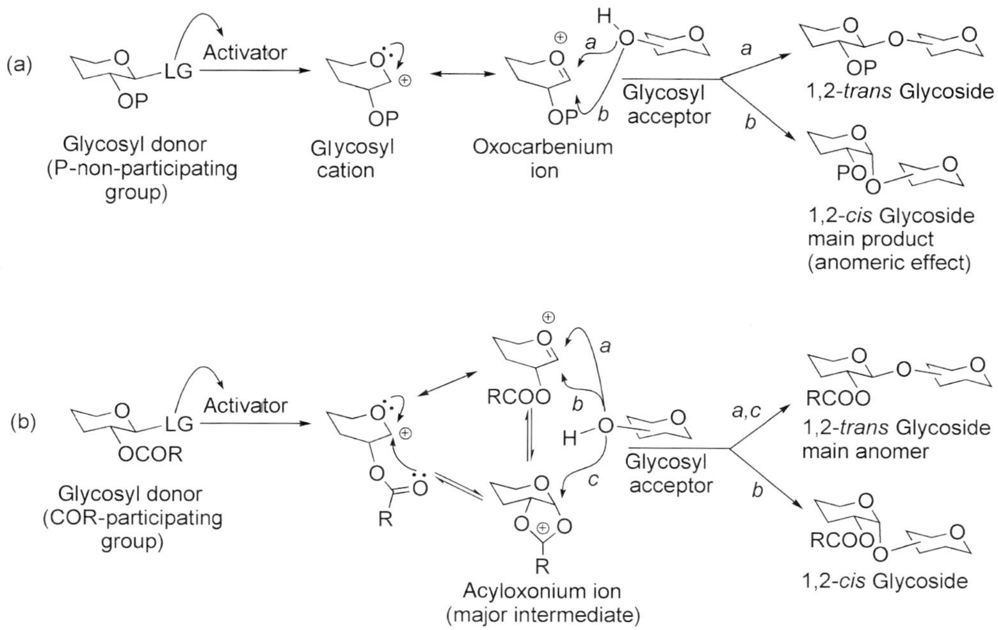

1.4 General Reaction Mechanism

Scheme 1.1

possibility to intramolecularly stabilize glycosyl cation formed from the glycosyl donor bearing a non-participating group is by resonance from O-5 that results in oxocarbenium ion (Scheme 1.1a). The most commonly applied nonparticipating groups are benzyl (OBn) for neutral sugars and azide (N_3) for 2-amino-2-deoxy sugars; however, other moieties have also been occasionally used. The anomeric carbon of either resonance contributors is sp^2 hybridized; hence, the nucleophilic attack would be almost equally possible from either the top (trans, β- for the D-gluco series) or the bottom face (cis, α-) of the ring. Even though the α-product is thermodynamically favored because of the so-called anomeric effect (discussed in the subsequent section) [64], a substantial amount of the kinetic β-linked product is often obtained owing to the irreversible character of glycosylation of complex aglycones. Various factors such as temperature, protecting groups, conformation, solvent, promoter, steric hindrance or leaving groups may influence the glycosylation outcome (discussed below) [65,66].

1,2-trans Glycosidic linkage can be stereoselectively formed with the use of anchimeric assistance of a neighboring participating group, generally an acyl moiety such as O-acetyl (Ac), O-benzoyl (Bz), 2-phthalimido (NPhth) and so on [67–69]. These glycosylations proceed primarily via a bicyclic intermediate, the acyloxonium ion (Scheme 1.1b), formed as a result of the activator-assisted departure of the leaving group followed by the intramolecular stabilization of the glycosyl cation. In this case, the attack of a nucleophile (alcohol, glycosyl acceptor) is only possible from the top face of the ring (pathway c), therefore allowing stereoselective formation of a 1,2-trans glycoside. Occasionally, substantial amounts of 1,2-cis-linked products are also

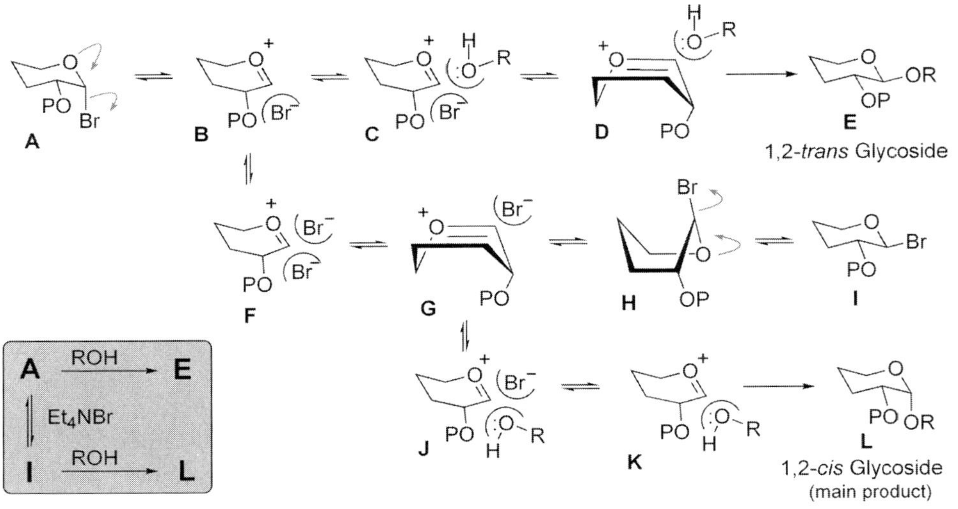

Scheme 1.2

formed, most often when unreactive alcohols are used as the substrates and/or poorly nucleophilic participatory substituents are present at C-2. In these cases, glycosylation assumingly proceeds via oxocarbenium ion, via pathways a and b (Scheme 1.1b), resulting in the formation of 1,2-*trans* and 1,2-*cis* glycosides, respectively, or most commonly mixtures thereof.

Seminal work by Lemieux on the halide-ion-catalyzed glycosidation reaction involved extensive theoretical studies that gave rise to a more detailed understanding of the reaction mechanism [70]. Thus, it was postulated that a rapid equilibrium could be established between a relatively stable α-halide **A** and its far more reactive β-counterpart **I** by the addition of tetraalkylammonium bromide (Et$_4$NBr, Scheme 1.2). In this case, a glycosyl acceptor (ROH) would preferentially react with the more reactive glycosyl donor (**I**) in an S$_N$2 fashion, possibly via the tight ion-pair complex **K**, providing the α-glycoside **L**. It is likely that the energy barrier for a nucleophilic substitution **I** → **L** (formation of the α-glycoside) is marginally lower than that for the reaction **A** → **E** (formation of a β-glycoside). If the difference in the energy barrier were sufficient, it should be possible to direct the reaction toward the exclusive formation of α-anomers.

Therefore, to obtain complete stereoselectivity, the entire glycosylation process has to be performed in a highly controlled manner. In this particular case, the control is achieved by the use of extremely mild catalyst (R$_4$NBr), although very reactive substrates and prolonged reaction at times are required.

Other common approaches to control the stereoselectivity of glycosylation will be discussed in the subsequent sections. In addition to the apparent complexity of the glycosidation process, there are other competing processes that cannot be disregarded. These reactions often cause the compromised yields of the glycosylation products and further complicate the studies of the reaction mechanism.

Elimination, substitution (formation of unexpected substitution products or hydrolysis at the anomeric center), cyclization (inter- and intramolecular orthoesterification), migration and redox are only a few to mention [71].

1.5
Anomeric Effects

A basic rule of conformational analysis known from the introductory organic chemistry is that an *equatorial* substituent of cyclic six-membered hydrocarbons is energetically favored. Hence, it is more stable owing to 1,3-diaxial interactions that would have occurred if a large substituent were placed in the *axial* position (Figure 1.3). For sugars, this rule is only applicable to hemiacetals (1-hydroxy derivatives) that are stabilized in β-orientation via intramolecular hydrogen-bond formation with O-5. Other polar substituents such as halide, OR or SR attached to the anomeric center of pyranoses/pyranosides prefer the axial orientation, which would be exclusive if the equilibrium at the anomeric center could be achieved. This phenomenon, which was first observed by Edward [72] and defined as *anomeric effect* by Lemieux [73], is partially responsible for the stereochemical outcome of processes taking place at the anomeric center of sugars [64,74,75].

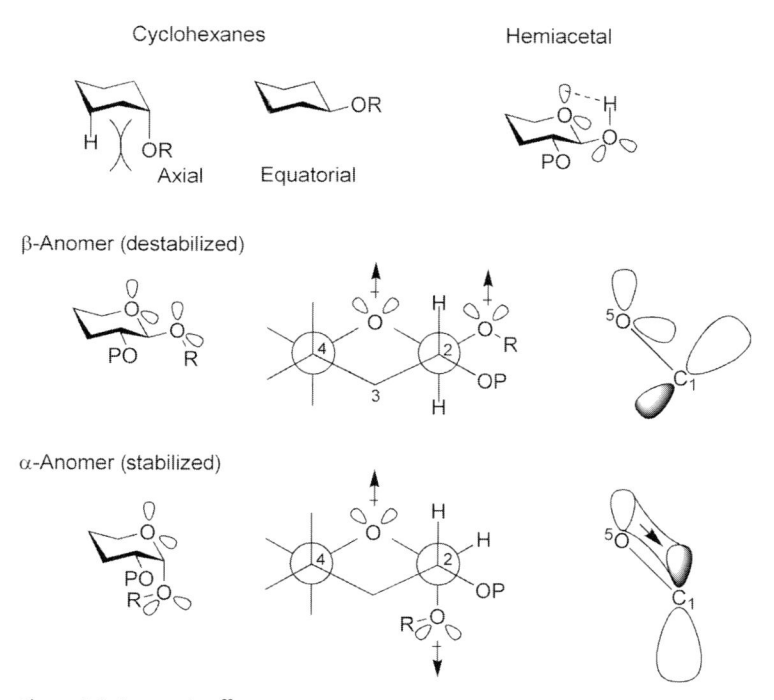

Figure 1.3 Anomeric effect.

What are the origins of the anomeric effect, sometimes referred to as *endo*-anomeric effect? In this context, the so-called *exo*-anomeric effect dealing with the stabilization of the β-anomer is of somewhat lesser influence on the overall process and will not be discussed herein [64]. One factor is that the substituent on the atom bonded to the ring at C-1 has lone-pair electrons, which would have repulsive interactions with those of the ring oxygen (O-5) if the anomeric substituent is in the equatorial position (β-position for D-sugars in $^4C_1$ conformation) but not if it is in the axial position (Figure 1.3). In addition, an electron-withdrawing axial substituent (α-anomer for D-sugars in $^4C_1$ conformation) is stabilized via hyperconjugation owing to the periplanar orientation of both nonbonding orbital of O-5 and antibonding orbital of C-1. This does not occur with the β-anomer, as the nonbonding orbital of O-5 and antibonding orbital of C-1 are in different planes and therefore are unable to interact.

1.6
Stereoselectivity of Glycosylation

As noted above, it is a general experience of carbohydrate synthesis that stereoselective preparation of 1,2-*cis* glycosides is more demanding than that of 1,2-*trans* glycosides. The formation of 1,2-*trans* glycosides is strongly favored by the neighboring-group participation (generation of intermediate acyloxonium ion). Typically, the use of a participating substituent at C-2 is sufficient to warrant stereoselective 1,2-*trans* glycosylation.

One of the factors affecting the stereochemical outcome of glycosidation of glycosyl donors bearing a nonparticipating substituent at C-2 is the anomeric effect, which favors α-glycoside formation (1,2-*cis* for the D-gluco series). However, because of the irreversible character of glycosylation, the role of the anomeric effect is diminished and other factors affecting the orientation of the new glycosidic bond (discussed below) often come to the fore. Although variation of reaction conditions or structural elements of the reactants may lead to excellent 1,2-*cis* stereoselectivity, no successful comprehensive method for 1,2-*cis* glycosylation has emerged as yet.

1.6.1
Structure of the Glycosyl Donor

1.6.1.1 Protecting Groups
The most powerful impact on the stereoselectivity is produced from the neighboring group at C-2. Neighboring-group participation is one of the most powerful tools to direct stereoselectivity toward the formation of a 1,2-*trans*-linked product. The neighboring substituent at C-2 is also responsible for the 'armed–disarmed' chemoselective glycosylation strategy [76]. The effects of the remote substituents are of lesser importance; however, there is strong evidence that a substituent at C-6 position may influence the stereochemical outcome of glycosylation dramatically. Although experimental proof has not emerged as yet, a possibility for the long-range 6-*O*-acyl or carbonate

group assistance resulting in the preferential formation of α-glucosides cannot be overruled [77–81]. It was also found that the steric bulkiness or strong electron-withdrawing properties of a substituent at C-6 are beneficial for 1,2-*cis* glycosylation, most likely because of shielding (sterically or electronically) the top face of the ring and, therefore, favoring the nucleophilic attack from the opposite side [14,82–88].

Although the effect of the C-6 substituent was found to be of minor importance for the derivatives of the D-galacto series [89], a remote effect is sufficiently strong when a participating moiety is present at C-4 [90,91]. Thus, the use of *p*-methoxybenzoyl (anisoyl) [91] and diethylthiocarbamoyl [81] groups was found to be exceptionally beneficial for the formation of α-galactosides. Similar effects (including C-3 participation) were also detected for the derivatives of the L-fuco [92,93], L-rhamno [94], D-manno and D-gluco [14,82,95] series [96]. It was noted that when the unprotected hydroxyl is present at C-4 of the sulfamidogalacto donor, the expected β-glycosyl formation occurs. However, when the hydroxyl group is blocked with benzyl or acyl, the process unexpectedly favors α-glycoside formation. This phenomenon was rationalized via the formation of the intramolecular hydrogen bonding (C4−O−H ··· O−C5), destabilizing oxocarbenium ion contribution to the reaction mechanism that favors α-glycosylation (pathway b, Scheme 1.1a). Torsional effects induced by the cyclic acetal protecting groups may also strongly affect the stereoselectivity at the anomeric center; however, these effects remain unpredictable at this stage [88,97–99].

1.6.1.2 Leaving Group

There are a large number of publications describing the comparison of various glycosylation methods applied for particular targets. However, only few principles could be reliably outlined. It has been unambiguously demonstrated that halides activated in the presence of a halide ion (from, e.g. Bu_4NBr) often provide the highest ratios of α-/β-glycosides [100–104]. Since in most cases the glycosylation reactions proceed via unimolecular S_N1 mechanism, the orientation of the leaving group at the anomeric center is of lesser importance. However, the glycosylation reactions occasionally proceed via bimolecular S_N2 mechanism with the inversion of the anomeric configuration. In this context, glycosyl donors with 1,2-*cis* orientation form 1,2-*trans* glycosides: for example glycosyl halides with insoluble catalysts (also used for β-mannosylation) [105], α-imidates in the presence of boron trifluoride etherate (BF_3–Et_2O) at low temperature [106] and 1,2-anhydro sugars [107]. Conversely, 1,2-*trans*-oriented glycosyl donors stereospecifically afford 1,2-*cis* glycosides, for example highly reactive β-glucosyl halides [70], glycosyl thiocyanates [39,108] and anomeric triflates formed *in situ* were found superior for the synthesis of β-mannosides [109,110].

1.6.2
Structure of the Glycosyl Acceptor

1.6.2.1 Position of the Hydroxyl

Alcohol reactivity is typically inversely correlated with the 1,2-*cis* stereoselectivity – the most reactive hydroxyls give the lowest α-/β-ratios – the stronger the nucleophile, the faster the reaction, and hence the more difficult it is to control. Regarding the sugar or aliphatic glycosyl acceptors, the general rule normally states that glycosylation of more reactive primary hydroxyl provides poorer stereoselectivity in comparison to that when the secondary hydroxyls are involved [111]. The same principles are applicable for the synthesis of glycopeptides; thus, the glycosylation of the secondary hydroxyl of threonine typically gives higher α-stereoselectivity than when primary hydroxyl group of serine is glycosylated with 2-azido-2-deoxy-galactosyl bromide or trichloroacetimidates [112,113]. Occasionally, primary hydroxyls provide somewhat higher 1,2-*cis* stereoselectivity in comparison to that of the secondary hydroxyl groups. This can serve as an indirect evidence of the glycosylation reaction proceeding via the bimolecular mechanism, at least partially.

1.6.2.2 Protecting Groups

It is well established that ester-electron-withdrawing substituents reduce electron density of the neighboring hydroxyl group by lowering its nucleophilicity [88,105,114]. This may improve stereoselectivity, as the reaction can be carried out in a more controlled manner. As an example, glycosylation of axial 4-OH of galactose often gives excellent 1,2-*cis* stereoselectivity, especially in combination with electron-withdrawing substituents (e.g. *O*-benzoyl, OBz) [115]. However, poorly reactive hydroxyls can lose their marginal reactivity completely when surrounded by the deactivating species, resulting in lower glycosylation yields.

1.6.3
Reaction Conditions

1.6.3.1 Solvent Effect

Another important factor that influences the stereoselectivity at the anomeric center is the effect of the reaction solvent. In general, polar reaction solvents increase the rate of the β-glycoside formation via charge separation between O-5 and β-O-1. If the synthesis of α-glycosides is desired, CH_2Cl_2, $ClCH_2CH_2Cl$ or toluene would be suitable candidates as reaction solvents. However, there are more powerful forces than simple solvation that have to be taken into consideration. The so-called participating solvents, such as acetonitrile and diethyl ether, were found to be the limiting cases for the preferential formation of β-D- and α-D-glucosides, respectively [78]. These observations were rationalized as follows: if the reactions are performed in acetonitrile, the nitrilium cation formed *in situ* exclusively adopts axial orientation, allowing stereoselective formation of equatorially substituted glycosides (Scheme 1.3). This approach allows obtaining 1,2-*trans* glucosides with good stereoselectivity even with glycosyl donors bearing a nonparticipating substituent. On the contrary, ether-type reaction solvents such as diethyl ether, tetrahydrofuran [116] or dioxane [117] can also participate in the glycosylation process. Differently, in these cases the equatorial intermediate is preferentially formed, leading toward the axial glycosidic bond formation [85,86,118–120]. Nitroethane was also employed as a suitable solvent for 1,2-*cis* glycosylation [121].

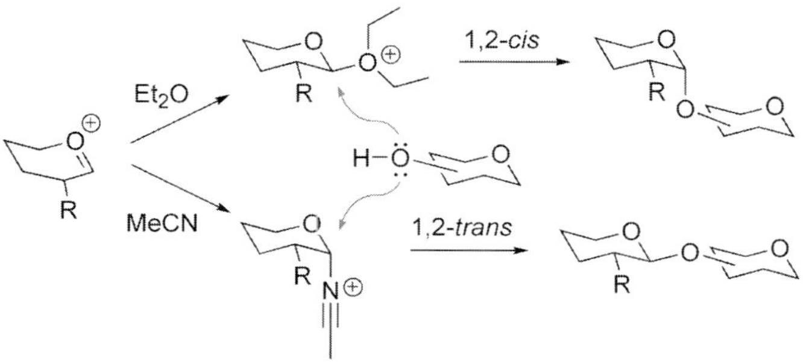

Scheme 1.3

1.6.3.2 Promoter (Catalyst), Additions

Milder activating conditions are generally beneficial for 1,2-*cis* glycosylation. Thus, halide-ion-catalyzed reactions give the best results for the glycosylation with glycosyl halides [70]; thioglycosides perform better when activated with a mild promoter, such as iodonium dicollidine perchlorate (IDCP) [122,123]; whereas trichloroacetimidates are best activated with the strong acidic catalysts, such as trimethylsilyl trifluoromethanesulfonate (TMS-triflate, TMSOTf) or triflouromethanesulfonic acid (triflic acid, TfOH) [106]. Various additions to the promoter systems often influence the stereochemical outcome of the glycosylation. Among the most remarkable examples is the use of perchlorate ion additive that was found to be very influential in 1,2-*cis* glycosylations [118,124].

1.6.3.3 Temperature and Pressure

High pressure applied to the reactions with participating glycosyl donors further enhances 1,2-*trans* selectivity [125]; when the high-pressure conditions were applied for glycosylations with a nonparticipating glycosyl donor, remarkable increase in the reaction yield was noted with only marginal changes in stereoselectivity [126]. Kinetically controlled glycosylations at lower temperatures generally favor 1,2-*trans* glycoside formation [100,120,127–130], although converse observations have also been reported [131,132].

1.6.4
Other Factors

Unfavorable steric interactions that occur between glycosyl donor and acceptor in the transition state or other factors or conditions may unexpectedly govern the course and outcome of the glycosylation process. One of the most remarkable effects, the so-called 'double stereodifferentiation' takes place when stereochemical interactions between bulky substituents in glycosyl donor and glycosyl acceptor prevail the stereodirecting effect of a neighboring participating group. The pair of reagents where these interactions occur is called a 'mismatched pair'. Thus, only α-linked product was unexpectedly formed with 2-phthalimido glycosyl acceptor (Scheme 1.4). [133]. A coupling of the same glycosyl donor with conformationally

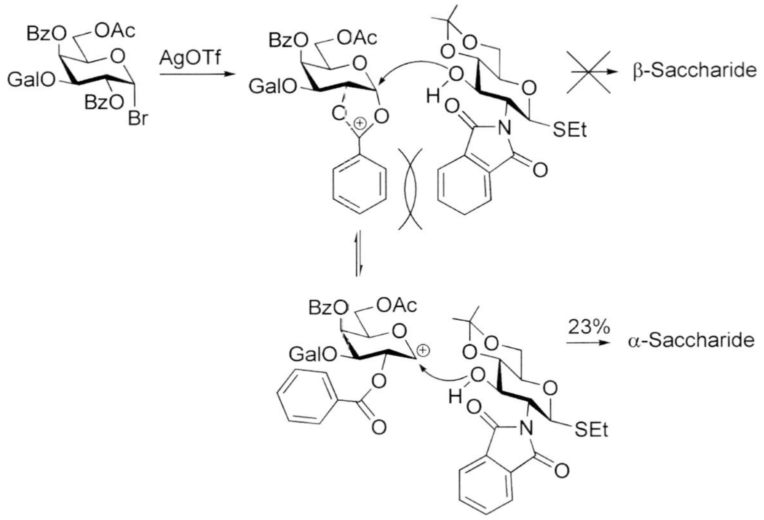

Scheme 1.4

modified 1,6-anhydro acceptor afforded β-linked oligosaccharide with 75% yield. It was also demonstrated that if this effect takes place with a sugar of the D-series, its L-enantiomer forms a matched pair with the same glycosyl acceptor.

1.7
Special Cases of Glycosylation

This section outlines special cases of glycosylation, not necessarily uncommon, which do not follow general conventions discussed above. It is not unusual when general glycosylation methods do not work or cannot be applied to the synthesis of glycosides described herein. The synthesis of each of these classes of compounds requires careful selection of techniques, their modification or design of conceptually new approaches. Sometimes special indirect or total synthesis based technologies have been developed and applied specifically to the synthesis of these targets.

1.7.1
Aminosugars

Glycosides of 2-amino-2-deoxy sugars are present in the most important classes of glycoconjugates and naturally occurring oligosaccharides, in which they are connected to other residues via either 1,2-*cis* or, more frequently, 1,2-*trans* glycosidic linkage [134–136]. In particular, 2-acetamido-2-deoxyglycosides, most common of the D-gluco and D-galacto series, are widely distributed in living organisms as glycoconjugates (glycolipids, lipopolysaccharides, glycoproteins) [134], glycosaminoglycans (heparin,

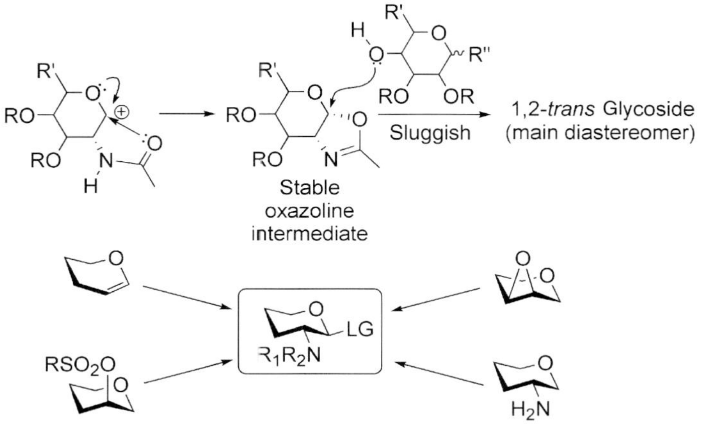

Scheme 1.5

heparin sulfate, dermatan sulfate, chondroitin sulfate, hyaluronic acid) [137] and so on [138,139]. Special efforts for the synthesis of glycosyl donors of 2-amino-2-deoxy sugars have been focusing on the development of simple, efficient, regio- and stereoselective procedures.

As a vast majority of naturally occurring 2-amino-2-deoxy sugars are N-acetylated, from the synthetic point of view, a 2-acetamido-2-deoxy substituted glycosyl donor would be desirable to minimize protecting-group manipulations. For this type of glycosyl donors, however, the oxocarbenium ion rearranges rapidly into an oxazoline intermediate (Scheme 1.5). Even under harsh Lewis acid catalysis, this highly stable oxazoline intermediate does not exert strong glycosyl-donor properties. Although the synthesis of 1,2-*trans* glycosides is possible with the use of this type of glycosyl donors, the synthesis of 1,2-*cis* glycosides is a burden. As a matter of fact, the participating nature of the N-acetyl moiety presents an obvious hindrance when the formation of the α-linkage is desired. A minimal requirement for the synthesis of 1,2-*cis* glycosides would be the use of a C-2 nonparticipating moiety.

Nowadays, a variety of synthetic approaches to the synthesis of 2-amino-2-deoxyglycosides have been developed, and progress in this area has been reviewed [140–142]. These syntheses are started either from a glycosamine directly or by the introduction of nitrogen functionality to glycose or glycal derivatives. To this end, various glycosamine donors with modified functionalities have been investigated; in particular, those bearing an N-2 substituent capable of either efficient participation via acyloxonium, but not (2-methyl) oxazoline, intermediate for 1,2-*trans* glycosylation or a nonparticipating moiety for 1,2-*cis* glycosylation.

1.7.2
Sialosides

Sialic acids are nine-carbon monosaccharides involved in a wide range of biological phenomena. Their unique structure is characterized by the presence of a carboxylic group (ionized at physiological pH), deoxygenated C-3, glycerol chain at C-6 and differently functionalized C-5. Among the 50 derivatives reported so far, N-acetylneuraminic acid (5-acetamido-3, 5-dideoxy-D-*glycero*-D-*galacto*-non-2-ulopyranosonic acid, Neu5Ac) is the most widespread. The natural equatorial glycosides and their unnatural axial counterparts are classified as α- and β-glycosides, respectively. In spite of extensive efforts and notable progress, the chemical synthesis of sialosides remains a significant challenge [143–146]. The presence of a destabilizing electron-withdrawing carboxylic group and the lack of a participating auxiliary often drive glycosylation reactions toward competitive elimination reactions, resulting in the formation of a 2,3-dehydro derivative and in poor stereoselectivity (β-anomer). To overcome these problems, a variety of leaving groups and activation conditions for direct sialylations have been developed. It was also demonstrated that the N-substituent at C-5 plays an influential role in both stereoselectivity of sialylation and the reactivity of sialyl donors [147].

Along with these studies, a variety of indirect methods for chemical sialylation have been developed. Several glycosyl donors derived from Neu5Ac have been prepared that possess an auxiliary at C-3. This auxiliary should control the anomeric selectivity of glycosylation by neighboring-group participation, leading to the formation of 2,3-*trans*-glycosides [143]. Thus, α-glycosides are favored in the case of equatorial auxiliaries (Scheme 1.6), whereas β-glycosides are preferentially formed when the participating auxiliary is axial. The auxiliaries also help in preventing 2,3-elimination that often constitutes a major side reaction in the direct O-sialylations. One of the most important requirements is that an auxiliary should be easily installed prior to, and removed after, glycosylation. Usually, the auxiliaries are introduced by a chemical modification of the readily accessible 2,3-dehydro derivative of Neu5Ac [148]. These transformations can be performed

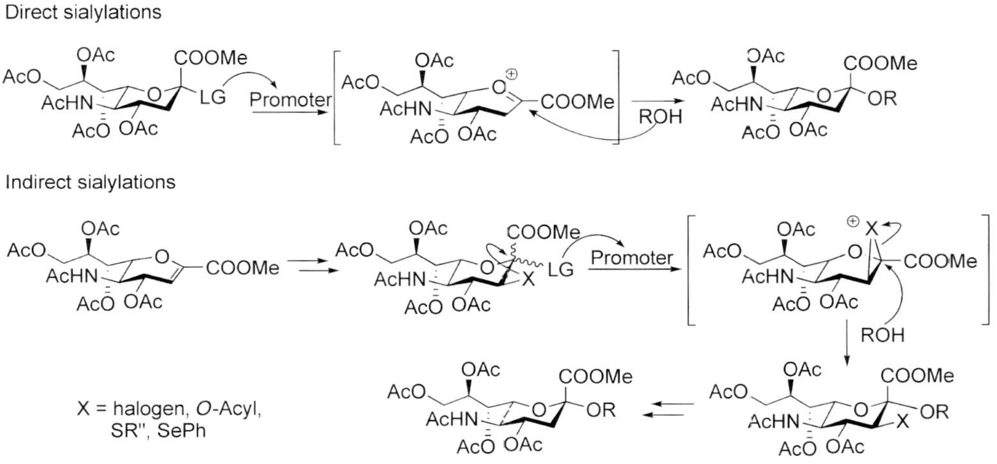

Scheme 1.6

either through a 2,3-oxirane derivative or by a direct addition reaction to the double bond.

1.7.3
Synthesis of 2-Deoxyglycosides

2-Deoxyglycosides are important constituents of many classes of antibiotics. The development of reliable methods for stereoselective synthesis of both α- and β-2-deoxyglycosides has become an important area of research and development of new classes of drugs and glycomimetics [149,150]. It should be noted that because of the lack of anchimeric assistance of the substituent at C-2, the synthesis of both types of linkages represents a notable challenge. On one hand, the direct glycosylation of 2-deoxy glycosyl donors often results in the formation of anomeric mixtures. Similar to that of conventional glycosylation, the solvent and promoter effects play important stereodirecting roles in the synthesis. On the other hand, similar to that discussed for the sialosides, a participating auxiliary can be used to add to the stereoselectivity of glycosylation. Usually this moiety is introduced through 1,2-dehydro derivatives with concomitant or sequential introduction of the anomeric leaving group. The methods employing both axial and equatorial substituents are known to result in the formation of 1,2-*trans* glycosides, which upon 2-deoxygenation can be converted into the respective targets. Although this latter approach requires additional synthetic steps, it is often preferred because it provides higher level of stereocontrol.

1.7.4
Synthesis of β-Mannosides

β-Mannosyl residues are frequently found in glycoproteins. The chemical synthesis of 1,2-*cis*-β-mannosides cannot be achieved by relying on the anomeric effect that would favor axial α-mannosides at the equilibrium. In addition, it is further disfavored by the repulsive interactions that would have occurred between the axial C-2 substituent and the nucleophile approaching from the top face of the ring. For many years the only direct procedure applicable to β-mannosylation – Ag-silicate promoted glycosidation of α-halides – was assumed to follow bimolecular S_N2 mechanism [13,14]. The difficulty of the direct β-mannosylation was addressed by developing a variety of indirect approaches such as C-2 oxidation-reduction, C-2 inversion, anomeric alkylation and intramolecular aglycone delivery (Scheme 1.7) [151–155]. This was the standing in this field before Crich and coworkers discovered that 4,6-*O*-benzylidene protected sulfoxide [109] or thiomannoside [110] glycosyl donors provide excellent β-manno stereoselectivity. Mechanistic and spectroscopic studies showed that anomeric α-*O*-triflates generated *in situ* as reactive intermediates can be stereospecifically substituted. On a similar note, 2-(hydroxycarbonyl)benzyl glycosides have proven to be versatile glycosyl donors for the synthesis of β-mannosides via anomeric triflate intermediates [58].

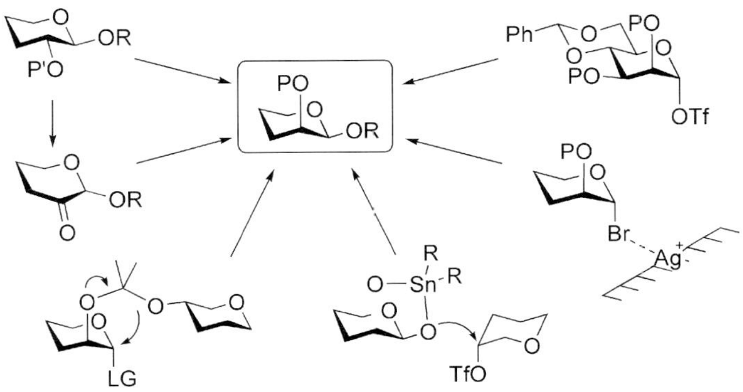

Scheme 1.7

1.7.5
Synthesis of Furanosides

In comparison to their six-membered ring counterparts, furanosides are relatively rare. Nevertheless, their presence in a variety of glycostructures from bacteria, parasites and fungi makes this type of glycosidic linkage an important synthetic target [156,157]. The synthesis of 1,2-*trans* furanosides is relatively straightforward and, similar to that of pyranosides, can be reliably achieved with the use of glycosyl donors bearing a participating group at C-2. In contrast, the construction of 1,2-*cis*-glycofuranosidic linkage is difficult, even more so than with pyranosides, because the stereocontrol in glycofuranosylation cannot be added by the anomeric effect owing to the conformational flexibility of the five-membered ring. In fact, both stereoelectronic and steric effects favor the formation of 1,2-*trans* glycofuranosides. Despite some recent progress, stereoselective synthesis of 1,2-*cis* glycofuranosides has been one of the major challenges of synthetic chemistry. General glycosylation methods, involving glycosyl fluorides [158], trichloroacetimidates [159], and thioglycosides [156,160] along with less common and indirect techniques [161–164], were applied to 1,2-*cis* furanosylation. More recently, a notable improvement in stereoselectivity of 1,2-*cis* furanosylation was achieved by using glycosyl donors in which the ring has been locked into a single conformation. These examples include 2,3-anhydro [165–169], 3,5-*O*-(di-*tert*-butylsilylene) [170,171] and 3,5-*O*-tetraisopropyldisiloxanylidene [172] protected bicyclic glycosyl donors.

1.8
Glycosylation and Oligosaccharide Sequencing

Stereoselective glycosylation is only a part of the challenge that synthetic chemists confront during the synthesis of oligosaccharides. Regardless of the efficiency of a single glycosylation, a traditional stepwise approach requires subsequent conversion of the disaccharide derivative into the second-generation glycosyl acceptor or glycosyl

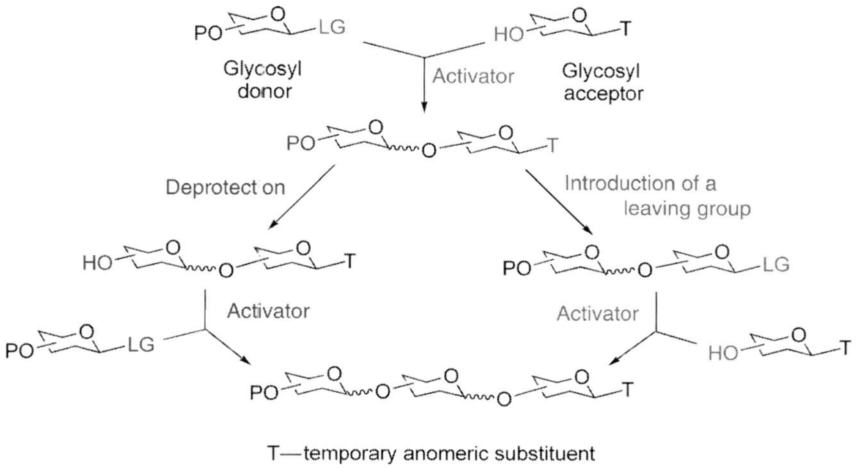

1.8 Glycosylation and Oligosaccharide Sequencing

Scheme 1.8 Linear oligosaccharide synthesis.

donor (see Scheme 1.8). The modified disaccharides are then coupled with a glycosyl donor (or acceptor) to obtain a trisaccharide. This reaction sequence is then repeated again until oligosaccharide of the desired chain length is obtained.

It has become apparent that the linear approach is too inefficient and lengthy, especially when applied to the synthesis of large oligosaccharides. As a result, the past two decades have witnessed a dramatic improvement in the methods and strategies used for oligosaccharide synthesis. The first attempt to address challenges associated with the linear stepwise approach was the development of a convergent approach [105,173,174]. According to this approach, oligosaccharide building blocks are obtained separately and then coupled together, which, in comparison to the linear approach, allows formation of larger saccharides faster. It is especially advantageous if the synthesis of two or more sequential repeat units is desired for the introduction of a 'difficult' linkage, such as 1,2-*cis*, at an earlier stage of the saccharide assembly to avoid complicated diastereomer separation at the later stage.

When the arsenal of the glycosylation techniques was limited to Fischer and Koenigs–Knorr approaches (or their variations), the oligosaccharide assembly was limited to the linear and block techniques. However, when stable glycosyl donors such as thioglycosides or *O*-alkenyl glycosides emerged, a question of selective or chemoselective activation arose. In brief, most recently developed strategies fit into three broad categories: leaving-group-based (selective or orthogonal activation), preactivation-based and protecting-group-based (armed–disarmed) strategies [175].

1.8.1
Leaving-Group-Based Strategies

The schematic outline of the stepwise selective activation is shown in Scheme 1.9; thus, a glycosyl donor bearing a reactive LG^a, is coupled with a glycosyl acceptor,

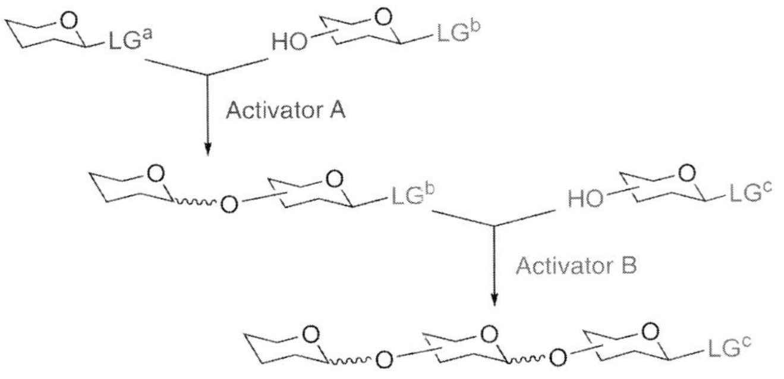

Scheme 1.9 Stepwise selective activation.

bearing a relatively stable LG^b at the anomeric center. The major requirement for this reaction to take place is the use of a suitable activator that would selectively activate LG^a but not LG^b (Activator A). Early studies involved the activation of alkyl halides over S-alkyl/aryl glycosides [173,176–179]. Subsequently, other examples of suitable LG^a for selective activations have become available: trichloroacetimidates, phosphite, phosphate, thioimidate, hemiacetal, sulfoxide, selenoglycoside, orthoester and so on [175]. Many of these couplings were performed with the use of either S-alkyl/aryl or O-alkenyl/hetaryl moieties as LG^b. In principle, the activation sequence can be continued providing that there is an LG^c that would withstand reaction conditions for the LG^b activation (Activator B). However, these elongated multistep sequences are not yet routinely available. Few available examples include the following three-step activation sequences: bromide as LG^a, Se-phenyl as LG^b, and S-ethyl as LG^c [180]; S-benzoxazolyl (SBox) as LG^a, S-ethyl as LG^b, and O-pentenyl as LG^c [181]; SBox as LG^a, S-thiazolinyl (STaz) as LG^b, S-ethyl as LG^c [182] and so on [183].

The combination of two chemically distinct glycosylation reactions, in which one of the leaving groups is activated while the other one remains intact and vice versa, has led to the discovery of the orthogonal glycosylation strategy [184]. This unique and virtually one of the most advanced techniques for oligosaccharide synthesis requires the use of orthogonal glycosyl donors [185]. Typically, phenyl thioglycosides are selectively activated over glycosyl fluorides and vice versa [184,186,187]. Recently, it has been discovered that a combination of S-ethyl and S-thiazolinyl glycosides also allows orthogonal activation [188].

A one-pot technique, combining two or more glycosylation steps based on activation of one donor over another, have also been developed [182,189–191]. This one-pot technique is virtually a variation of a simple stepwise selective activation strategy, which allows further improvement in the efficiency of the synthesis by avoiding the necessity for isolation (and purification) of the intermediates.

1.8.2
Two-Step Activation and Preactivation Strategies

According to the two-stage activation (or preactivation) approach, both glycosyl donor and glycosyl acceptor initially bear the same type of a leaving group (LG^a).

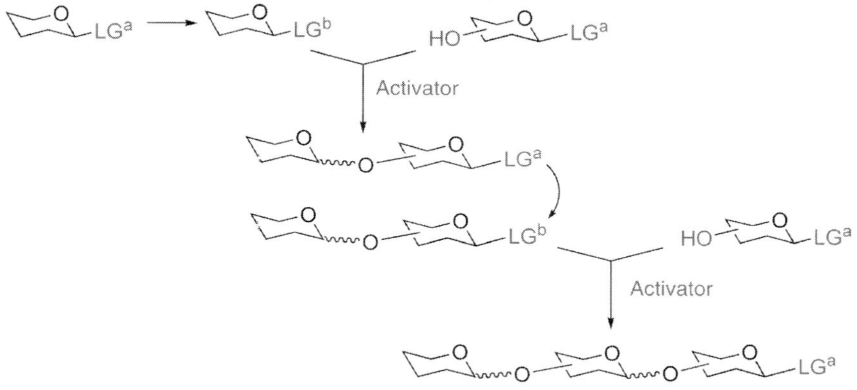

Scheme 1.10 Two-step activation concept.

However, to couple these two reactants, the LG^a of the potential glycosyl donor unit is first converted into an LG^b, which is then selectively activated over LG^a of the glycosyl acceptor in the presence of a selective activator (Activator B, Scheme 1.10). This two-step activation sequence can be reiterated; for this purpose the leaving group of the obtained disaccharide LG^a is again converted into LG^b and so on. This concept was discovered for S-ethyl (LG^a) and bromo (LG^b) moieties [173] and further explored for other systems [176,192–194]. The same principle was applied by Danishefsky in the reiterative glycal assembly approach [43,195,196]. This technique involves the transformation of glycals into 1,2-anhydro sugars, which can be easily glycosidated with partially protected glycal-based glycosyl acceptors. Recently, the versatility of the two-step activation was demonstrated by a one-pot preactivation procedure [197,198], according to which S-tolyl glycosides are converted *in situ* into a reactive intermediate.

Similar principle is executed in the active–latent approach that has been applied to a number of systems [51,193,199,200]. It is important to note that the conversion between LG^a and LG^b in the active–latent approach does not involve substitution at the anomeric center; instead, a simple modification of the leaving group itself is executed. Recently, the application of this technique was enhanced by the discovery that 2-(benzyloxycarbonyl)benzyl glycosides are perfectly stable compounds, whereas the corresponding 2-(hydroxycarbonyl)benzyl moiety, obtained by the selective hydrogenolysis is an excellent leaving group that can be readily activated [58]. This and other conceptually similar recent discoveries [56,201] open new exciting perspectives for further development of the active–latent strategy.

1.8.3
Protecting-Group-Based Strategies

Another efficient strategy, the armed–disarmed approach, developed by Fraser-Reid is based on the chemoselectivity principle [193,202]. According to this

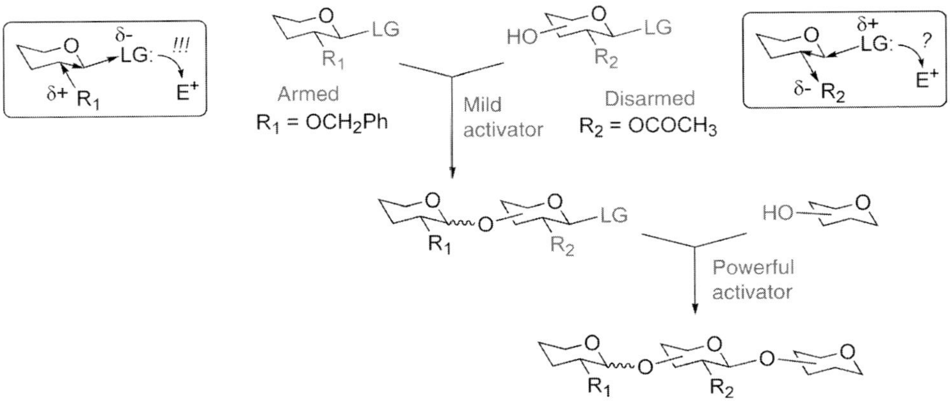

Scheme 1.11 Armed–disarmed strategy.

technique, a benzylated (electronically activated, armed) glycosyl donor is chemoselectively activated over the acylated (electronically deactivated, disarmed) derivative bearing the same type of LG in the presence of a mild promoter (Scheme 1.11). At the first step a 1,2-*cis*-linked disaccharide is preferentially obtained because of the use of the ether-type arming substituent, a nonparticipating group (*O*-benzyl). The obtained disaccharide can then be used for 1,2-*trans* glycosylation directly in the presence of a more potent promoter, capable of activating the disarmed LG.

Initially developed for *O*-pentenyl glycosides, this concept was further explored for the chemoselective glycosidations of ethyl thioglycosides, selenoglycosides, fluorides, phosphoroamidates, substituted thioformimidates, glycals, benzoxazolyl and thiazolinyl glycosides [175]. The chemoselective activation has become the basis of programmable multistep reactivity-based technique, including highly efficient one-pot approaches [203,204]. Further insights into the armed–disarmed approach came with the observation of the unusual reactivity pattern of the SBox glycosides. Thus, it was demonstrated that glycosyl donors with 2-*O*-benzyl-3,4,6-tri-*O*-acyl protecting-group pattern are even more deactivated than the corresponding 'disarmed' peracyl derivatives [205]. Although the exact nature of this so-called *O*-2/*O*-5 cooperative interesting effect is not yet understood, it has been postulated that the intermediate carbocation stabilization via the acyloxonium ion of the peracylated donor is favored over the oxocarbenium ion stabilization of the 2-*O*-benzyl-3,4,6-tri-*O*-acylated donors. Another interesting concept is the use of 2-*O*-picolyl substituent as an arming participating group. According to this so-called inverse armed–disarmed method, the 1,2-*trans* glycosidic linkage can be chemoselectively introduced at the first glycosylation step [206].

1.9
Conclusions and Outlook

The progress in the area of chemical glycosylation has significantly improved our ability to synthesize various glycosidic linkages with impressive yields and stereoselectivity. But, can we conclude that we have entirely solved the problem of chemical glycosylation? Unfortunately not, and hopefully this chapter introduced the reader to the challenge of chemical glycosylation, a variety of factors, conditions, and driving forces influencing all aspects of this complex chemical reaction as well as prepared for studying more specialized material dedicated to particular methods and strategies employed in modern carbohydrate chemistry. Recent progress made in the area of development of new coupling methods and highly efficient strategies for expeditious oligosaccharide synthesis will ultimately provide an efficient and trouble-free access to complex saccharides. This goal cannot be achieved without the comprehensive knowledge of the glycosylation mechanism and driving forces of glycosylation and competing side processes. It is likely that the consecutive scientific development in this field will be focusing on studying the fundamental mechanistic aspects of glycosylation rather than developing additional anomeric leaving groups.

References

1 Michael, A. (1879) *American Chemical Journal*, **1**, 305–312.
2 Fischer, E. (1893) *Chemische Berichte*, **26**, 2400–2412.
3 Koenigs, W. and Knorr, E. (1901) *Chemische Berichte*, **34**, 957–981.
4 Igarashi, K. (1977) *Advances in Carbohydrate Chemistry and Biochemistry*, **34**, 243–283.
5 Zemplen, G. and Gerecs, A. (1930) *Chemische Berichte*, **63B**, 2720–2729.
6 Helferich, B. and Wedemeyer, K.F. (1949) *Liebigs Annalien der Chemie*, **563**, 139–145.
7 Helferich, B. and Zirner, J. (1962) *Chemische Berichte*, **95**, 2604–2611.
8 Conrow, R.B. and Bernstein, S. (1971) *Journal of Organic Chemistry*, **36**, 863–870.
9 Helferich, B. and Schmitz-Hillebrecht, E. (1933) *Chemische Berichte*, **66B**, 378–383.
10 Lemieux, R.U. (1954) *Advances in Carbohydrate Chemistry and Biochemistry*, **9**, 1–57, and references therein.
11 Ness, R.K., Fletcher, H.G. and Hudson, C.S. (1951) *Journal of the American Chemical Society*, **73**, 959–963.
12 Ness, R.K. and Fletcher, H.G. (1956) *Journal of the American Chemical Society*, **78**, 4710–4714.
13 Paulsen, H. and Lockhoff, O. (1981) *Chemische Berichte*, **114**, 3102–3114.
14 van Boeckel, C.A.A., Beetz, T. and van Aelst, S.F. (1984) *Tetrahedron*, **40**, 4097–4107, and references therein.
15 Toshima, K. and Tatsuta, K. (1993) *Chemical Reviews*, **93**, 1503–1531.
16 Davis, B.G. (2000) *Journal of the Chemical Society, Perkin Transactions*, **1**, 2137–2160.
17 Ferrier, R.J., Hay, R.W. and Vethaviyasar, N. (1973) *Carbohydrate Research*, **27**, 55–61.
18 Nicolaou, K.C., Seitz, S.P. and Papahatjis, D.P. (1983) *Journal of the American Chemical Society*, **105**, 2430–2434.
19 Garegg, P.J., Henrichson, C. and Norberg, T. (1983) *Carbohydrate Research*, **116**, 162–165.
20 Oscarson, S. (2001) *Glycoscience: Chemistry and Chemical Biology*, vol. **1** (eds B. Fraser-Reid, K. Tatsuta and J. Thiem), Springer, Berlin, Heidelberg, New York, pp. 643–671.

21 Bochkov, A.F. and Kochetkov, N.K. (1975) *Carbohydrate Research*, **39**, 355–357.
22 Kochetkov, N.K., Backinowsky, L.V. and Tsvetkov, Y.E. (1977) *Tetrahedron Letters*, **41**, 3681–3684.
23 Pougny, J.R., Jacquinet, J.C., Nassr, M., Duchet, D., Milat, M.L. and Sinay, P. (1977) *Journal of the American Chemical Society*, **99**, 6762–6763.
24 Schmidt, R.R. and Michel, J. (1980) *Angewandte Chemie (International Edition in English)*, **19**, 731–732.
25 Mukaiyama, T., Nakatsuka, T. and Shoda, S.I. (1979) *Chemistry Letters*, 487–490.
26 Hanessian, S., Bacquet, C. and Lehong, N. (1980) *Carbohydrate Research*, **80**, c17–c22.
27 Woodward, R.B., Logusch, E., Nambiar, K.P., Sakan, K., Ward, D.E., Au-Yeung, B.W., Balaram, P., Browne, L.J., Card, P.J. and Chen, C.H. (1981) *Journal of the American Chemical Society*, **103**, 3215–3217.
28 Mukaiyama, T., Murai, Y. and Shoda, S. (1981) *Chemistry Letters*, 431–432.
29 Schmidt, R.R., Castro-Palomino, J.C. and Retz, O. (1999) *Pure and Applied Chemistry*, **71**, 729–744.
30 Schmidt, R.R. and Jung, K.H. (2000) *Carbohydrates in Chemistry and Biology*, vol. 1 (eds B. Ernst, G.W. Hart and P. Sinay), Wiley-VCH Verlag GmbH, Weinheim, New York, pp. 5–59.
31 Garegg, P.J. (1997) *Advances in Carbohydrate Chemistry and Biochemistry*, **52**, 179–205.
32 Oscarson, S. (2000) *Carbohydrates in Chemistry and Biology*, vol. 1 (eds B. Ernst, G.W. Hart and P. Sinay), Wiley-VCH Verlag GmbH, Weinheim, New York, pp. 93–116.
33 Codee, J.D.C., Litjens, R.E.J.N., van den Bos, L.J., Overkleeft, H.S. and van der Marel, G.A. (2005) *Chemical Society Reviews*, **34**, 769–782.
34 Nicolaou, K.C. and Ueno, H. (1997) *Preparative Carbohydrate Chemistry* (ed. S. Hanessian), Marcel Dekker, Inc., New York, pp. 313–338.
35 Mukaiyama, T. (2004) *Angewandte Chemie (International Edition)*, **43**, 5590–5614.
36 Boursier, M. and Descotes, G. (1989) *Comptes Rendus de l'Academie des Sciences Serie II*, **308**, 919–921.
37 Koide, K., Ohno, M. and Kobayashi, S. (1991) *Tetrahedron Letters*, **32**, 7065–7068.
38 Kunz, H. and Zimmer, J. (1993) *Tetrahedron Letters*, **34**, 2907–2910.
39 Kochetkov, N.K., Klimov, E.M. and Malysheva, N.N. (1989) *Tetrahedron Letters*, **30**, 5459–5462.
40 Briner, K. and Vasella, A. (1989) *Helvetica Chimica Acta*, **72**, 1371–1382.
41 Marra, A. and Sinay, P. (1990) *Carbohydrate Research*, **195**, 303–308.
42 Friesen, R.W. and Danishefsky, S.J. (1989) *Journal of the American Chemical Society*, **111**, 6656–6660.
43 Halcomb, R.L. and Danishefsky, S.J. (1989) *Journal of the American Chemical Society*, **111**, 6661–6666.
44 Kondo, H., Ichikawa, Y. and Wong, C.H. (1992) *Journal of the American Chemical Society*, **114**, 8748–8750.
45 Martin, T.J. and Schmidt, R.R. (1992) *Tetrahedron Letters*, **33**, 6123–6126.
46 Kahne, D., Walker, S., Cheng, Y. and van Engen, D. (1989) *Journal of the American Chemical Society*, **111**, 6881–6882.
47 Brown, D.S., Ley, S.V. and Vile, S. (1988) *Tetrahedron Letters*, **29**, 4873–4876.
48 Mehta, S. and Pinto, B.M. (1991) *Tetrahedron Letters*, **32**, 4435–4438.
49 Fraser-Reid, B., Konradsson, P., Mootoo, D.R. and Udodong, U. (1988) *Journal of the Chemical Society: Chemical Communications*, 823–825.
50 Marra, A., Esnault, J., Veyrieres, A. and Sinay, P. (1992) *Journal of the American Chemical Society*, **114**, 6354–6360.
51 Boons, G.J. and Isles, S. (1994) *Tetrahedron Letters*, **35**, 3593–3596.
52 Huchel, U., Schmidt, C. and Schmidt, R.R. (1998) *European Journal of Organic Chemistry*, 1353–1360, and references therein.
53 Gervay, J. and Hadd, M.J. (1997) *Journal of Organic Chemistry*, **62**, 6961–6967.

54 Plante, O.J., Andrade, R.B. and Seeberger, P.H. (1999) *Organic Letters*, **1**, 211–214.

55 Stick, R.V., Tilbrook, D.M.G. and Williams, S.J. (1997) *Australian Journal of Chemistry*, **50**, 237–240.

56 Hinklin, R.J. and Kiessling, L.L. (2001) *Journal of the American Chemical Society*, **123**, 3379–3380.

57 Davis, B.G., Ward, S.J. and Rendle, P.M. (2001) *Chemical Communications*, 189–190.

58 Kim, K.S., Kim, J.H., Lee, Y.J., Lee, Y.J. and Park, J. (2001) *Journal of the American Chemical Society*, **123**, 8477–8481.

59 Demchenko, A.V., Malysheva, N.N. and De Meo, C. (2003) *Organic Letters*, **5**, 455–458.

60 Demchenko, A.V., Pornsuriyasak, P., De Meo, C. and Malysheva, N.N. (2004) *Angewandte Chemie (International Edition)*, **43**, 3069–3072.

61 Adinolfi, M., Barone, G., Iadonisi, A. and Schiattarella, M. (2002) *Synlett*, 269–270.

62 Adinolfi, M., Barone, G., Iadonisi, A. and Schiattarella, M. (2002) *Tetrahedron Letters*, **43**, 5573–5577.

63 Gin, D. (2002) *Journal of Carbohydrate Chemistry*, **21**, 645–665.

64 Tvaroska, I. and Bleha, T. (1989) *Advances in Carbohydrate Chemistry and Biochemistry*, **47**, 45–123.

65 Demchenko, A.V. (2003) *Synlett*, 1225–1240.

66 Demchenko, A.V. (2003) *Current Organic Chemistry*, **7**, 35–79.

67 Goodman, L. (1967) *Advances in Carbohydrate Chemistry and Biochemistry*, **22**, 109–175.

68 Fife, T.H., Bembi, R. and Natarajan, R. (1996) *Journal of the American Chemical Society*, **118**, 12956–12963.

69 Nukada, T., Berces, A., Zgierski, M.Z. and Whitfield, D.M. (1998) *Journal of the American Chemical Society*, **120**, 13291–13295.

70 Lemieux, R.U., Hendriks, K.B., Stick, R.V. and James, K. (1975) *Journal of the American Chemical Society*, **97**, 4056–4062, and references therein.

71 Bochkov, A.F. and Zaikov, G.E. (1979) *Chemistry of the O-Glycosidic Bond: Formation and Cleavage*, Pergamon Press.

72 Edward, J.T. (1955) *Chemistry & Industry*, 1102–1104.

73 Lemieux, R.U. (1971) *Pure and Applied Chemistry*, **25**, 527–548, and references therein.

74 Wolfe, S., Whangbo, M.H. and Mitchell, D.J. (1979) *Carbohydrate Research*, **69**, 1–26.

75 Box, V.G.S. (1990) *Heterocycles*, **31**, 1157–1181.

76 Fraser-Reid, B., Wu, Z., Udodong, U.E. and Ottosson, H. (1990) *Journal of Organic Chemistry*, **55**, 6068–6070.

77 Ishikawa, T. and Fletcher, H.G. (1969) *Journal of Organic Chemistry*, **34**, 563–571.

78 Eby, R. and Schuerch, C. (1974) *Carbohydrate Research*, **34**, 79–90, and references therein.

79 Koto, S., Yago, K., Zen, S., Tomonaga, F. and Shimada, S. (1986) *Bulletin of the Chemical Society of Japan*, **59**, 411–414.

80 Dasgupta, F. and Garegg, P.J. (1990) *Carbohydrate Research*, **202**, 225–238.

81 Mukaiyama, T., Suenaga, M., Chiba, H. and Jona, H. (2002) *Chemistry Letters*, 56–57.

82 van Boeckel, C.A.A. and Beetz, T. (1985) *Recueil Des Travaux Chimiques Des Pays-Bas*, **104**, 171–173.

83 Fei, C.P. and Chan, T.H. (1987) *Tetrahedron Letters*, **28**, 849–852.

84 Houdier, S. and Vottero, P.J.A. (1992) *Carbohydrate Research*, **232**, 349–352.

85 Fukase, K., Kinoshita, I., Kanoh, T., Nakai, Y., Hasuoka, A. and Kusumoto, S. (1996) *Tetrahedron*, **52**, 3897–3904.

86 Fukase, K., Nakai, Y., Kanoh, T. and Kusumoto, S. (1998) *Synlett*, 84–86.

87 Fukase, K., Nakai, Y., Egusa, K., Porco, J.A. and Kusumoto, S. (1999) *Synlett*, 1074–1078.

88 Green, L.G. and Ley, S.V. (2000) *Carbohydrates in Chemistry and Biology*, vol. **1** (eds B. Ernst, G.W. Hart and

P. Sinay),Wiley-VCH Verlag GmbH, Weinheim, New York, pp. 427–448.
89 Lucas, T.J. and Schuerch, C. (1975) *Carbohydrate Research*, **39**, 39–45.
90 Nakahara, Y. and Ogawa, T. (1987) *Tetrahedron Letters*, **28**, 2731–2734.
91 Demchenko, A.V., Rousson, E. and Boons, G.J. (1999) *Tetrahedron Letters*, **40**, 6523–6526.
92 Zuurmond, H.M., van der Laan, S.C., van der Marel, G.A. and van Boom, J.H. (1991) *Carbohydrate Research*, **215**, c1–c3.
93 Gerbst, A.G., Ustuzhanina, N.E., Grachev, A.A., Tsvetkov, D.E., Khatuntseva, E.A. and Nifant'ev, N.E. (1999) *Mendeleev Communications*, 114–116.
94 Yamanoi, T., Nakamura, K., Takeyama, H., Yanagihara, K. and Inazu, T. (1994) *Bulletin of the Chemical Society of Japan*, **67**, 1359–1366.
95 Ustyuzhanina, N., Komarova, B., Zlotina, N., Krylov, V., Gerbst, A.G., Tsvetkov, Y. and Nifantiev, N.E. (2006) *Synlett*, **6**, 921–923.
96 Kwon, O. and Danishefsky, S.J. (1998) *Journal of the American Chemical Society*, **120**, 1588–1599.
97 Crich, D. and Sun, S. (1998) *Journal of the American Chemical Society*, **120**, 435–436.
98 Anilkumar, G., Nair, L.G., Olsson, L., Daniels, J.K. and Fraser-Reid, B. (2000) *Tetrahedron Letters*, **41**, 7605–7608.
99 Crich, D., Cai, W. and Dai, Z. (2000) *Journal of Organic Chemistry*, **65**, 1291–1297.
100 Andersson, F., Fugedi, P., Garegg, P.J. and Nashed, M. (1986) *Tetrahedron Letters*, **27**, 3919–3922.
101 Kihlberg, J.O., Leigh, D.A. and Bundle, D.R. (1990) *Journal of Organic Chemistry*, **55**, 2860–2863.
102 Garegg, P.J., Oscarson, S. and Szonyi, M. (1990) *Carbohydrate Research*, **205**, 125–132.
103 Vermeer, H.J., van Dijk, C.M., Kamerling, J.P. and Vliegenthart, J.F.G. (2001) *European Journal of Organic Chemistry*, 193–203.
104 Hadd, M.J. and Gervay, J. (1999) *Carbohydrate Research*, **320**, 61–69.
105 Paulsen, H. (1982) *Angewandte Chemie (International Edition in English)*, **21**, 155–173.
106 Schmidt, R.R. and Jung, K.H. (1997) *Preparative Carbohydrate Chemistry* (ed. S. Hanessian), Marcel Dekker, Inc., New York, pp. 283–312.
107 Williams, L.J., Garbaccio, R.M. and Danishefsky, S.J. (2000) *Carbohydrates in Chemistry and Biology*, **vol. 1** (eds B. Ernst, G.W. Hart and P. Sinay), Wiley-VCH Verlag GmbH, Weinheim, New York, pp. 61–92.
108 Kochetkov, N.K., Klimov, E.M., Malysheva, N.N. and Demchenko, A.V. (1991) *Carbohydrate Research*, **212**, 77–91.
109 Crich, D. and Sun, S. (1996) *Journal of Organic Chemistry*, **61**, 4506–4507.
110 Crich, D. and Smith, M. (2000) *Organic Letters*, **2**, 4067–4069.
111 Chen, Q. and Kong, F. (1995) *Carbohydrate Research*, **272**, 149–157.
112 Sames, D., Chen, X.T. and Danishefsky, S.J. (1997) *Nature*, **389**, 587–591.
113 Schwartz, J.B., Kuduk, S.D., Chen, X.T., Sames, D., Glunz, P.W. and Danishefsky, S.J. (1999) *Journal of the American Chemical Society*, **121**, 2662–2673.
114 Sinay, P. (1978) *Pure and Applied Chemistry*, **50**, 1437–1452.
115 Hashimoto, S.I., Sakamoto, H., Honda, T., Abe, H., Nakamura, S.I. and Ikegami, S. (1997) *Tetrahedron Letters*, **38**, 8969–8972.
116 Wulff, G. and Rohle, G. (1974) *Angewandte Chemie (International Edition in English)*, **13**, 157–170.
117 Demchenko, A., Stauch, T. and Boons, G.J. (1997) *Synlett*, 818–820.
118 Fukase, K., Hasuoka, A., Kinoshita, I., Aoki, Y. and Kusumoto, S. (1995) *Tetrahedron*, **51**, 4923–4932.
119 Hashimoto, S., Hayashi, M. and Noyori, R. (1984) *Tetrahedron Letters*, **25**, 1379–1382.
120 Manabe, S., Ito, Y. and Ogawa, T. (1998) *Synlett*, 628–630.

121 Uchiro, H. and Mukaiyama, T. (1996) *Chemistry Letters*, 271–272.
122 Veeneman, G.H., van Leeuwen, S.H. and van Boom, J.H. (1990) *Tetrahedron Letters*, **31**, 1331–1334.
123 Veeneman, G.H. and van Boom, J.H. (1990) *Tetrahedron Letters*, **31**, 275–278.
124 Mukaiyama, T. and Matsubara, K. (1992) *Chemistry Letters*, 1041–1044.
125 Klimov, E.M., Malysheva, N.N., Demchenko, A.V., Makarova, Z.G., Zhulin, V.M. and Kochetkov, N.K. (1989) *Doklady Akademii Nauk*, **309**, 110–114, and references therein.
126 Sasaki, M., Gama, Y., Yasumoto, M. and Ishigami, Y. (1990) *Tetrahedron Letters*, **31**, 6549–6552.
127 Wegmann, B. and Schmidt, R.R. (1987) *Journal of Carbohydrate Chemistry*, **6**, 357–375.
128 Nishizawa, M., Shimomoto, W., Momii, F. and Yamada, H. (1992) *Tetrahedron Letters*, **33**, 1907–1908.
129 Shimizu, H., Ito, Y. and Ogawa, T. (1994) *Synlett*, 535–536.
130 Chenault, H.K., Castro, A., Chafin, L.F. and Yang, J. (1996) *Journal of Organic Chemistry*, **61**, 5024–5031.
131 Schmidt, R.R. and Rucker, E. (1980) *Tetrahedron Letters*, **21**, 1421–1424.
132 Dohi, H., Nishida, Y., Tanaka, H. and Kobayashi, K. (2001) *Synlett*, 1446–1448.
133 Spijker, N.M. and van Boeckel, C.A.A. (1991) *Angewandte Chemie (International Edition in English)*, **30**, 180–183.
134 Dwek, R.A. (1996) *Chemical Reviews*, **96**, 683–720.
135 Herzner, H., Reipen, T., Schultz, M. and Kunz, H. (2000) *Chemical Reviews*, **100**, 4495–4537.
136 Rai, R., McAlexander, I. and Chang, C.W.T. (2005) *Organic Preparations and Procedures International*, **37**, 337–375.
137 Casu, B. (1985) *Advances in Carbohydrate Chemistry and Biochemistry*, **43**, 51–134.
138 Varki, A. (1993) *Glycobiology*, **3**, 97–130.
139 Varki, A., Cummings, R., Esko, J., Freeze, H., Hart, G. and Marth, J. (1999) *Essentials of Glycobiology*, Cold Spring Harbor Laboratory Press, Cold Spring Harbor, NY.
140 Banoub, J., Boullanger, P. and Lafont, D. (1992) *Chemical Reviews*, **92**, 1167–1195.
141 Debenham, J., Rodebaugh, R. and Fraser-Reid, B. (1997) *Liebigs Annalen-Recueil*, 791–802.
142 Bongat, A.F.G. and Demchenko, A.V. (2007) *Carbohydrate Research*, **342**, 374–406.
143 Boons, G.J. and Demchenko, A.V. (2000) *Chemical Reviews*, **100**, 4539–4565.
144 Kiso, M., Ishida, H. and Ito, H. (2000) *Carbohydrates in Chemistry and Biology*, vol. 1 (eds B. Ernst, G.W. Hart and P. Sinay), Wiley-VCH Verlag GmbH, Weinheim, New York, pp. 345–366.
145 von Itzstein, M. and Thomson, R.J. (1997) *Topics in Current Chemistry*, **186**, 119–170.
146 Ress, D.K. and Linhardt, R.J. (2004) *Current Organic Synthesis*, **1**, 31–46.
147 De Meo, C. (2007) *American Chemical Society Symposium Series*, **960**, 118–131.
148 Okamoto, K., Kondo, T. and Goto, T. (1987) *Bulletin of the Chemical Society of Japan*, **60**, 631–636.
149 Veyrieres, A. (2000) *Carbohydrates in Chemistry and Biology*, vol. 1 (eds B. Ernst, G.W. Hart and P. Sinay), Wiley-VCH Verlag GmbH, Weinheim, New York, pp. 367–406.
150 Marzabadi, C.H. and Franck, R.W. (2000) *Tetrahedron*, **56**, 8385–8417.
151 El Ashry, E.S.H., Rashed, N. and Ibrahim, E.S.I. (2005) *Current Organic Synthesis*, **2**, 175–213.
152 Crich, D. (2002) *Journal of Carbohydrate Chemistry*, **21**, 667–690.
153 Ito, Y. and Ohnishi, Y. (2001) *Glycoscience: Chemistry and Chemical Biology*, vol. 2 (eds B. Fraser-Reid, K. Tatsuta and J. Thiem), Springer, Berlin, Heidelberg, New York, pp. 1589–1620.
154 Gridley, J.J. and Osborn, H.M.I. (2000) *Journal of the Chemical Society Perkin Transactions*, **1**, 1471–1491.
155 Pozsgay, V. (2000) *Carbohydrates in Chemistry and Biology*, vol. 1 (eds B. Ernst, G.W. Hart and P. Sinay), Wiley-VCH

156 Gelin, M., Ferrieres, V. and Plusquellec, D. (2000) *European Journal of Organic Chemistry*, 1423–1431.
157 Lowary, T.L. (2002) *Journal of Carbohydrate Chemistry*, **21**, 691–722.
158 Mukaiyama, T., Hashimoto, Y. and Shoda, S. (1983) *Chemistry Letters*, 935–938.
159 Gelin, M., Ferrières, V. and Plusquellec, D. (1997) *Carbohydrate Letters*, **2**, 381–388.
160 Yin, H., D'Souza, F.W. and Lowary, T.L. (2002) *Journal of Organic Chemistry*, **67**, 892–903.
161 Bogusiak, J. and Szeja, W. (2001) *Carbohydrate Research*, **330**, 141–144.
162 Gadikota, R.R., Callam, C.S. and Lowary, T.L. (2001) *Organic Letters*, **3**, 607–610.
163 Marotte, K., Sanchez, S., Bamhaoud, T. and Prandi, J. (2003) *European Journal of Organic Chemistry*, 3587–3598.
164 Lee, Y.J., Lee, K., Jung, E.H., Jeon, H.B. and Kim, K.S. (2005) *Organic Letters*, **7**, 3263–3266.
165 Gadikota, R.R., Callam, C.S. and Lowary, T.L. (2001) *Organic Letters*, **3**, 607–610.
166 Gadikota, R.R., Callam, C.S., Wagner, T., Del Fraino, B. and Lowary, T.L. (2003) *Journal of the American Chemical Society*, **125**, 4155–4165.
167 Callam, C.S., Gadikota, R.R., Krein, D.M. and Lowary, T.L. (2003) *Journal of the American Chemical Society*, **125**, 13112–13119.
168 Cociorva, O.M. and Lowary, T.L. (2004) *Tetrahedron*, **60**, 1481–1489.
169 Bai, Y. and Lowary, T.L. (2006) *Journal of Organic Chemistry*, **71**, 9658–9671.
170 Zhu, X.M., Kawatkar, S., Rao, Y. and Boons, G.J. (2006) *Journal of the American Chemical Society*, **128**, 11948–11957.
171 Crich, D., Pedersen, C.M., Bowers, A.A. and Wink, D.J. (2007) *Journal of Organic Chemistry*, **72**, 1553–1565.
172 Ishiwata, A., Akao, H. and Ito, Y. (2006) *Organic Letters*, **8**, 5525–5528.
173 Koto, S., Uchida, T. and Zen, S. (1973) *Bulletin of the Chemical Society of Japan*, **46**, 2520–2523.
174 Ogawa, T., Yamamoto, H., Nukada, T., Kitajima, T. and Sugimoto, M. (1984) *Pure and Applied Chemistry*, **56**, 779–795.
175 Demchenko, A.V. (2005) *Letters in Organic Chemistry*, **2**, 580–589.
176 Nicolaou, K.C., Dolle, R.E., Papahatjis, D.P. and Randall, J.L. (1984) *Journal of the American Chemical Society*, **106**, 4189–4192.
177 Lonn, H. (1985) *Carbohydrate Research*, **139**, 105–113.
178 Garegg, P.J. and Oscarson, S. (1985) *Carbohydrate Research*, **136**, 207–213.
179 Fugedi, P., Garegg, P.J., Lonn, H. and Norberg, T. (1987) *Glycoconjugate Journal*, **4**, 97–108.
180 Baeschlin, D.K., Chaperon, A.R., Charbonneau, V., Green, L.G., Ley, S.V., Lucking, U. and Walther, E. (1998) *Angewandte Chemie (International Edition)*, **37**, 3423–3428.
181 Demchenko, A.V., Kamat, M.N. and De Meo, C. (2003) *Synlett*, 1287–1290.
182 Pornsuriyasak, P. and Demchenko, A.V. (2005) *Tetrahedron: Asymmetry*, **16**, 433–439.
183 Tanaka, H., Adachi, M., Tsukamoto, H., Ikeda, T., Yamada, H. and Takahashi, T. (2002) *Organic Letters*, **4**, 4213–4216.
184 Kanie, O., Ito, Y. and Ogawa, T. (1994) *Journal of the American Chemical Society*, **116**, 12073–12074.
185 Kanie, O. (2000) *Carbohydrates in Chemistry and Biology*, vol. 1 (eds B. Ernst, G.W. Hart and P. Sinay), Wiley-VCH Verlag GmbH, Weinheim, New York, pp. 407–426.
186 Ito, Y., Kanie, O. and Ogawa, T. (1996) *Angewandte Chemie (International Edition)*, **35**, 2510–2512.
187 Ferguson, J. and Marzabadi, C. (2003) *Tetrahedron Letters*, **44**, 3573–3577.
188 Pornsuriyasak, P. and Demchenko, A.V. (2006) *Chemistry – A European Journal*, **12**, 6630–6646.
189 Yamada, H., Harada, T., Miyazaki, K. and Takahashi, T. (1994) *Tetrahedron Letters*, **35**, 3979–3982.

190 Yamada, H., Harada, T. and Takahashi, T. (1994) *Journal of the American Chemical Society*, **116**, 7919–7920.

191 Mukaiyama, T., Ikegai, K., Jona, H., Hashihayata, T. and Takeuchi, K. (2001) *Chemistry Letters*, 840–841.

192 Nicolaou, K.C., Caulfield, T., Kataoka, H. and Kumazawa, T. (1988) *Journal of the American Chemical Society*, **110**, 7910–7912.

193 Fraser-Reid, B., Udodong, U.E., Wu, Z.F., Ottosson, H., Merritt, J.R., Rao, C.S., Roberts, C. and Madsen, R. (1992) *Synlett*, 927–942, and references therein.

194 Yamago, S., Yamada, T. Hara, O., Ito, H., Mino, Y. and Yoshida, J. (2001) *Organic Letters*, **3**, 3867–3870.

195 Danishefsky, S.J. and Bilodeau, M.T. (1996) *Angewandte Chemie (International Edition in English)*, **35**, 1380–1419.

196 Deshpande, P.P., Kim, H.M., Zatorski, A., Park, T.K., Ragupathi, G., Livingston, P.O., Live, D. and Danishefsky, S.J. (1998) *Journal of the American Chemical Society*, **120**, 1600–1614.

197 Huang, X., Huang, L., Wang, H. and Ye, X.S. (2004) *Angewandte Chemie (International Edition)*, **43**, 5221–5224.

198 Wang, Y., Yan, Q., Wu, J., Zhang, L.H. and Ye, X.S. (2005) *Tetrahedron*, **61**, 4313–4321.

199 Roy, R., Andersson, F.O. and Letellier, M. (1992) *Tetrahedron Letters*, **33**, 6053–6056.

200 Cao, S., Hernandez-Mateo, F. and Roy, R. (1998) *Journal of Carbohydrate Chemistry*, **17**, 609–631.

201 Huang, L., Wang, Z. and Huang, X. (2004) *Chemical Communications*, 1960–1961.

202 Mootoo, D.R., Konradsson, P., Udodong, U. and Fraser-Reid, B. (1988) *Journal of the American Chemical Society*, **110**, 5583–5584.

203 Douglas, N.L., Ley, S.V., Lucking, U. and Warriner, S.L. (1998) *Journal of the Chemical Society Perkin Transactions*, **1**, 51–65.

204 Zhang, Z., Ollmann, I.R., Ye, X.S., Wischnat, R., Baasov, T. and Wong, C.H. (1999) *Journal of the American Chemical Society*, **121**, 734–753.

205 Kamat, M.N. and Demchenko, A.V. (2005) *Organic Letters*, **7**, 3215–3218.

206 Smoot, J.T., Pornsuriyasak, P. and Demchenko, A.V. (2005) *Angewandte Chemie (International Edition)*, **44**, 7123–7126.

2
Glycoside Synthesis from Anomeric Halides

2.1
Glycosyl Fluorides
Shin-ichiro Shoda

2.1.1
Background

Selection of an appropriate glycosyl donor is of great importance in planning strategies for glycoside syntheses (see Section 1.3). Factors to be considered regarding the structure of glycosyl donors used in glycosylations are the chemical character and stereochemistry of leaving groups, the stability of protecting groups among others. The following two characteristics are preferred for the glycosyl donors to be employed in practical glycoside syntheses (Figure 2.1). First, the covalent bond between the leaving group (X) and the anomeric center should possess a moderate thermal stability, because the use of an unstable glycosyl donor often results in difficulty in its handling as well as low reaction yields. Second, the Lewis acid (LA) that promotes glycosylation by cleaving the C–X bond must have a weak acidity so that acid-sensitive functional groups in the glycosyl donors are not damaged during the glycosylation reactions. Glycosyl donors should fulfill these requirements to be employed practically for glycosylation reactions.

Recently, glycosyl fluorides (X = F, Figure 2.1) [1–4] have become one of the most useful glycosyl donors that fulfill the above-mentioned requirements. Fluorine has the smallest covalent radius among the halogens and the largest electronegativity among all elements [5] (Table 2.1). Because of the large bond-dissociation energy of the C–F bond (552 kJ mol^{-1}), it had been believed that glycosyl fluorides were too stable to be used for glycosidic bond formations. These beliefs influenced chemists for a long time and resulted in glycosyl fluorides playing a less significant role than other glycosyl halides in carbohydrate chemistry, until the first publication of the glycosyl fluoride method appeared in 1981 [6]. In fact, glycosyl halides employed in glycoside synthesis had been restricted to relatively unstable glycosyl chlorides and

Handbook of Chemical Glycosylation: Advances in Stereoselectivity and Therapeutic Relevance.
Edited by Alexei V. Demchenko.
Copyright © 2008 WILEY-VCH Verlag GmbH & Co. KGaA. All rights reserved.
ISBN: 978-3-527-31780-6

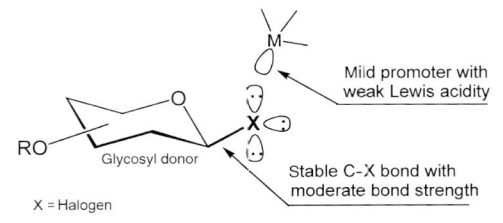

Figure 2.1 Preferred characteristics of glycosyl donors: stability of C–X bond and mildness of promoter.

bromides, and successful results could not always be expected with regard to either yields and/or stereoselectvity.

The first glycosylation reaction by using a glycosyl fluoride was based on the hypothesis that if the chlorine or bromine atom at the anomeric center of glycosyl donors were replaced by fluorine, it would make the glycosyl donor more stable, which would potentially lead to higher yields. Once it was disclosed that the C–F bond of glycosyl fluorides could be activated by a weak Lewis acid, stannous chloride ($SnCl_2$), the large bond-dissociation energy of the C–F bond became inconsequential. On the contrary, the following merits of using glycosyl fluorides as glycosyl donors have emerged: (a) the ease of preparation under a variety of mild reaction conditions, (b) relatively high stability toward chromatography on silica gel, (c) a variety of suitable promoters available for coupling with glycosyl acceptors, (d) orthogonality of the conditions required for the activation of glycosyl fluorides over other glycosyl donors, for example thioglycosides, (e) application to 'armed–disarmed concept' for convergent synthesis of oligosaccharides (see Sections 1.8.3 and 2.1.3.6), (f) possibility to be used in chemoenzymatic glycosylations and (g) ease of monitoring the reaction of glycosyl fluorides by NMR spectroscopy as the spin number of the naturally occurring isotope of ^{19}F is 1/2.

Table 2.1 Physical data for elements that can be used to change the chemical character of the leaving groups on the anomeric center of glycosyl donors.

	Covalent radius (nm)	Electronegativity (Pauling)	C–halogen bond-dissociation energy (kJ mol^{-1}) [5]	H–halogen bond-dissociation energy (kJ mol^{-1}) [5]
F	0.064	3.98	552	570
Cl	0.099	3.16	397	432
Br	0.114	2.96	280	366

The bond-dissociation energy is defined as the standard enthalpy change of the reaction in which the bond is broken: $R:X \rightarrow R + X$.

As a result of the extensive research on glycosylations using glycosyl fluorides during the past 25 years, an enormous amount of experience has been gained concerning reactions and manipulations of glycosyl fluorides [2–4]. This chapter deals with advances in the synthesis of glycosyl fluorides, glycosylation reactions with glycosyl fluoride donors promoted by various activators and applications to natural product synthesis. Several special topics relevant to glycosidic bond-forming reactions that use glycosyl fluorides will also be discussed.

2.1.2
Synthesis of Glycosyl Fluoride Donors

2.1.2.1 Fluorinating Reagents

The number of publications on the synthesis of glycosyl fluorides dramatically increased as many researchers started to use glycosyl fluorides for glycosylation in the early 1980s. The synthesis can be achieved by substitution or addition reactions with many types of fluorinating reagents (Figure 2.2), such as 2-fluoro-1-methylpyridinium tosylate **1** [7], diethylaminosulfur trifluoride (DAST) **2** [8,9], CF_3ZnBr–TiF_4 [10], diethyl azodicarboxylate (DEAD)–PPh_3 (Mitsunobu reagent)/$Et_3O^+BF_4^-$ [11], 1-chloromethyl-4-fluoro-1,4-diazoniabicyclo[2.2.2]octane bis(tetrafluoroborate) (Selectfluor) **3** [12], bis(2-methoxyethyl)aminosulfur trifluoride (Deoxo-Fluor) **4** [13], *N*,*N*-diisopropyl(1-fluoro-2-methyl-1-propenyl)amine **5** [14], HF [15], HF-pyridine [16], HF–MeCN–Ac_2O [17], AgF [1,18], $AgBF_4$ [19], KHF_2 [20], ZnF_2 [21], Et_3N–HF [22], CF_3ZnBr [10], DAST *N*-bromosuccinimide (NBS) or *N*-iodosuccinimide (NIS) [23,24], 4-methyl(difluoroiodo)benzene **6** [25,26], tetrabutylammonium fluoride (TBAF) [27], $PhI(OAc)_2$–SiF_4 [28], XeF_2 [29], *N*,*N*-diethyl-α,α-difluoro-(*m*-methylbenzyl)amine (DFMBA) **7** [30], Bu_4NBF_4 (electrochemical) [31] and BF_3–OEt_2 [32].

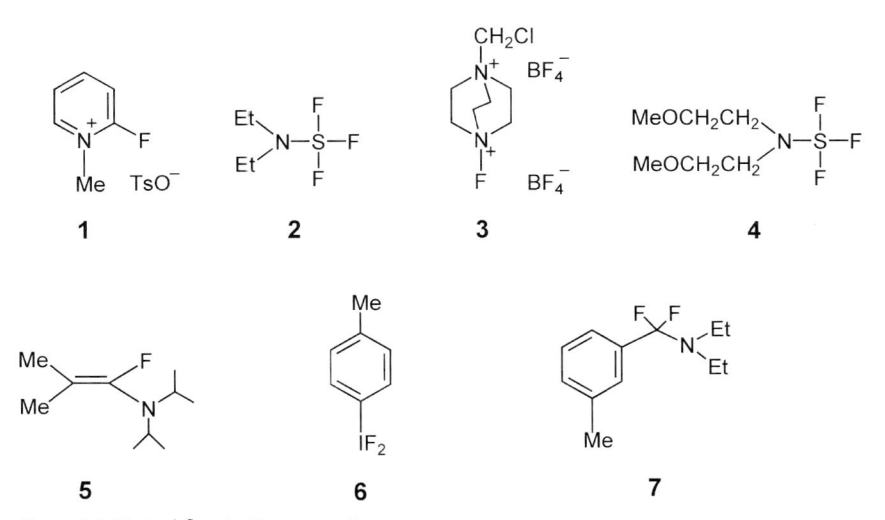

Figure 2.2 Typical fluorinating reagents.

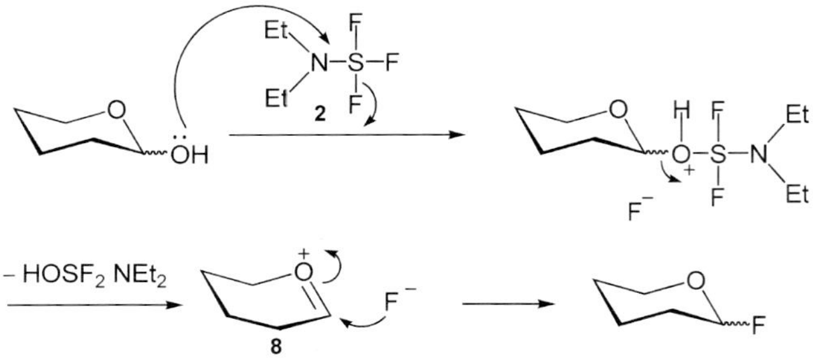

Scheme 2.1

In general, a synthetic route that involves glycosylation processes strongly depends on the availability of the precursors. Taking this into consideration, fluorination reactions are classified according to synthetic precursors as follows: hemiacetals, glycosyl halides, glycosyl esters, S-glycosides and other derivatives. Comprehensive reviews on the synthesis of glycosyl fluorides are available and are highly recommended to the reader [33,34].

2.1.2.2 Glycosyl Fluorides from Hemiacetals

Various hemiacetals can be converted into the corresponding glycosyl fluorides with diethylaminosulfur trifluoride **2** as the fluorinating reagent. The reaction proceeds presumably via a process that involves an oxocarbenium ion intermediate **8** (Scheme 2.1). Commonly used hydroxyl-protecting groups such as benzyl, benzoyl and acetonide functionalities do not interfere with the fluorinations. The α/β ratio of the resulting glycosyl fluorides depends on the solvent. For example, when 2,3,5-tri-O-benzyl-D-ribose is treated with DAST in THF, the α/β ratio of the resulting fluoride is 1/9.9, which becomes 1/2.0 when CH_2Cl_2 is used [8]. A nonexplosive reagent Deoxo-Fluor (($MeOCH_2CH_2$)$_2NSF_3$) **4** was developed for the kilogram scale usage [13].

Trifluoromethylzinc bromide (CF_3ZnBr)–TiF_4 is also an effective reagent that has been used to replace the anomeric hydroxyl group by fluorine. For example, 2,3,4,6-tetra-O-acetyl-D-glucopyranose can be converted into the corresponding glycosyl fluoride in 83% yield (α/β = 40/60) [10]. N,N-Diisopropyl(1-fluoro-2-methyl-1-propenyl)amine **5** is an effective reagent for the conversion of 1-hydroxy furanose and pyranose derivatives into the corresponding glycosyl fluorides. For example, 2,3,4,6-tetra-O-benzyl-D-glucopyranosyl fluoride is obtained from the corresponding 1-hydroxy sugar as a mixture of anomers (α/β = 7/18) [14]. As fluorination occurs under neutral conditions, this reagent does not interfere with protecting groups such as benzyl, benzoyl, acetyl, acetonide or silyl functionalities.

The transformation of 1-hydroxy sugars to the corresponding glycosyl fluorides can be achieved by a mixture of Selectfluor **3** (Figure 2.2) and methyl sulfide presumably through a fluorosulfonium ion **9**, which then reacts with the anomeric

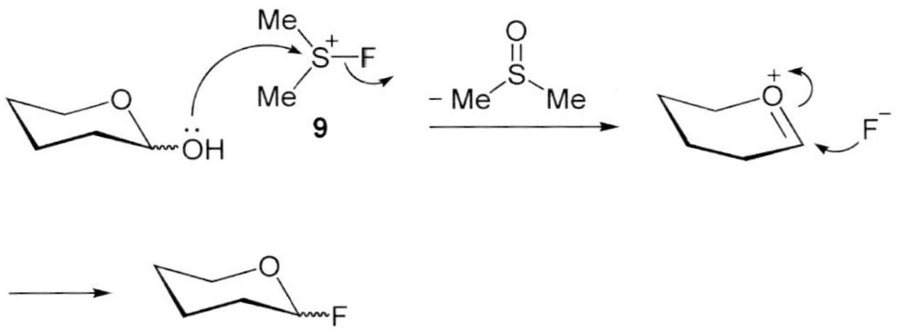

Scheme 2.2

hydroxyl group of the 1-hydroxy sugar followed by the displacement of the intermediate sulfoxide moiety by fluoride ion (Scheme 2.2) [12].

Some dehydrating reagents are also effective for the fluorination of 1-hydroxy sugars. For example, 2,3:5,6-di-O-isopropylidene-α-D-mannofuranose can be converted into the corresponding fluoride under modified Mitsunobu conditions by using diethyl azodicarboxylate–PPh$_3$/Et$_3$O$^+$BF$_4^−$ reagent [11]. 2-Fluoro-1-methylpyridinium tosylate **1** was found to be effective for the synthesis of glycofuranosyl fluoride **11** starting from the corresponding 1-hydroxy sugar **10** (Scheme 2.3) [7].

2.1.2.3 Glycosyl Fluorides from Glycosyl Esters

The classical method for introducing the fluorine atom to the anomeric center is to treat per-O-acetylated sugars with hydrogen fluoride [15], using which α-glycosyl fluorides can be prepared stereoselectively, however, the procedure is incompatible with any acid-sensitive functionalities present in the molecule. In addition, hydrogen fluoride is highly corrosive, which makes this method unattractive to many

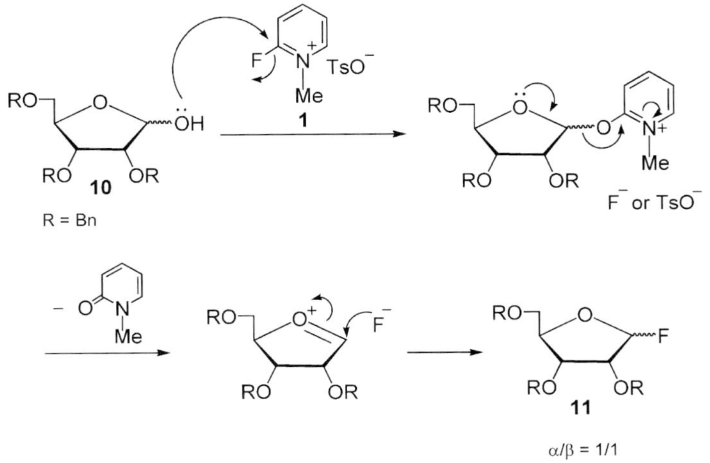

Scheme 2.3

researchers. A much milder procedure for the formation of α-glycosyl fluorides from 1-acyl derivatives uses pyridinium poly(hydrogen fluoride) [35,36].

2.1.2.4 Glycosyl Fluorides from Glycosyl Halides

Isomers of both α-glycosyl fluoride and β-glycosyl fluoride in their protected forms are prepared starting from glycosyl chlorides or bromides by the following two procedures. β-Glycosyl fluorides are prepared by the displacement of the corresponding per-O-acylated glycosyl chlorides or bromides with silver fluoride via an S_N2-type reaction (Scheme 2.4a). The synthetic route to β-fluorides can be replaced by another route that involves a nucleophilic substitution of α-glycosyl bromide with potassium hydrogen fluoride (KHF_2) (Scheme 2.4b) [20]. The opposite anomer (α-glycosyl fluoride) can be prepared via the anomerization of the β-fluoride by the action of BF_3–OEt_2 (Scheme 2.4b) [7]. For laboratory scale experiments, the use of KHF_2 and BF_3–OEt_2 is recommended instead of silver fluoride, because the experimental procedures for the reactions without the involvement of a silver salt or hydrogen fluoride are much simpler and more environmentally benign.

Trifluorozincbromide reagent (CF_3ZnBr) is employed for the synthesis of glycosyl fluorides from glycosyl bromides. When 2,3,4,6-tetra-O-acetyl-α-D-glycopyranosyl

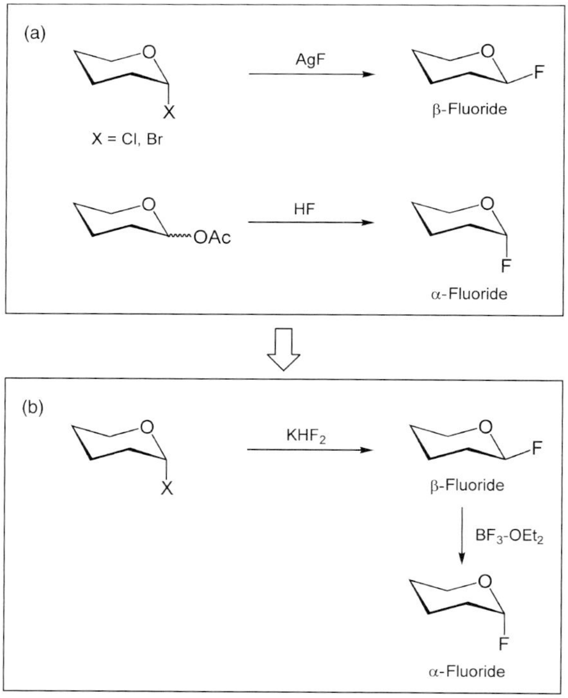

Scheme 2.4

bromide was treated with this reagent, the corresponding fluoride was obtained in a β-selective manner. This result can be explained by assuming an acyloxonium ion intermediate resulted from the neighboring-group participation at C-2 [10].

2.1.2.5 Glycosyl Fluorides from S-Glycosides

Some of the most significant progress concerning glycosyl fluoride synthesis is the development of conversion methods starting from thioglycoside derivatives. Phenyl thioglycosides can be converted into the corresponding glycosyl fluorides by the combined use of DAST **2** and NBS [23]. Selectfluor **3** is also able to convert thioglycosides into glycosyl fluorides [12]. The fact that the anomeric alkylthio group can be converted into the fluorine atom enables one to design an efficient synthetic route for oligosaccharides based on a two-step activation concept (see Figure 2.7 in Section 2.1.3.5).

2.1.2.6 Glycosyl Fluorides from Other Anomeric Moieties

The triazole derivative **12**, prepared from the corresponding glycosyl azide, can be converted into the corresponding glycosyl fluoride by the action of HF-pyridine (Figure 2.3a) [37]. The [1-phenyl-1H-tetrazol-5-yl]glycosides **13** were also used to produce the corresponding fluorides with HF-pyridine (Figure 2.3b) [38].

The per-O-benzylated 1,2-anhydro-α-D-hexopyranose **14**, prepared by epoxidation of a glycal, reacts with tetrabutylammonium fluoride to give the corresponding glycosyl fluoride with β-configuration (Figure 2.4a) [27]. The hydroxyfluorination of glycal derivative **15** using PhI(OAc)$_2$–SiF$_4$ proved to be a stereoselective route to α-glycosyl fluorides (Figure 2.4b) [28].

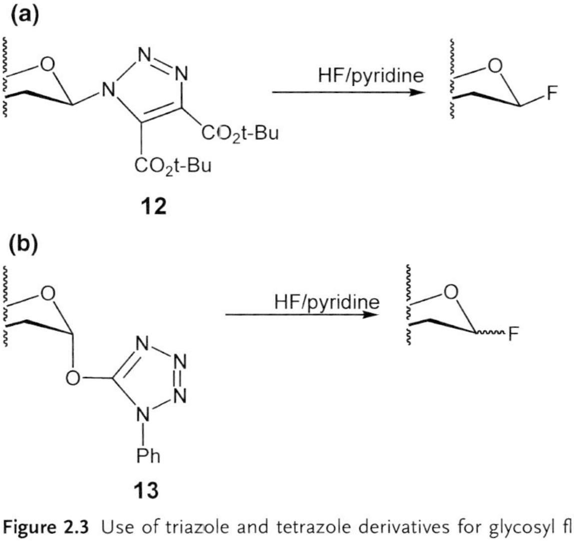

Figure 2.3 Use of triazole and tetrazole derivatives for glycosyl fluoride synthesis.

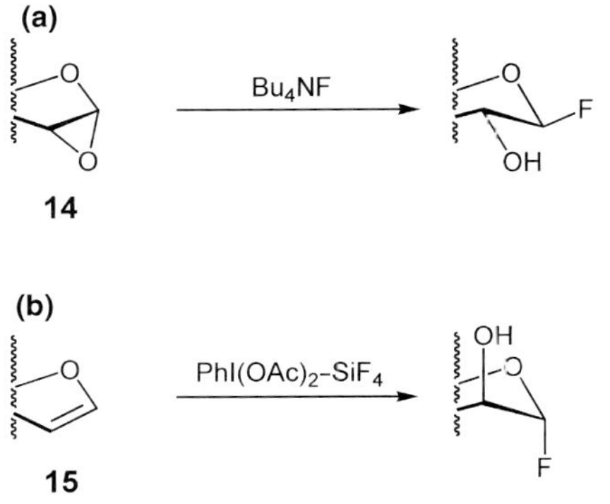

Figure 2.4 Synthesis of glycosyl fluorides from 1,2-anhydro sugars and glycal derivatives.

2.1.3
Glycosylation Using Glycosyl Fluorides as Glycosyl Donors

2.1.3.1 A Weak Lewis Acid Cleaves the C—F Bond. How Was the Glycosyl Fluoride Method Discovered?

The discovery of the glycosyl fluoride method originates from two unexpected findings made during early experiments. The first finding was the realization of a new synthetic route to glycosyl fluorides by using a dehydrating reagent, 2-fluoro-1-methylpyridinium tosylate **1** (Scheme 2.3). In the course of investigating nucleoside synthesis using the dehydrating reagent **1**, glycosyl fluoride **11** was formed as the by-product in good yield starting from the hemiacetal **10**. In the early 1980s, the synthesis of glycosyl fluorides was a problem because preparative methods usually included the use of hazardous anhydrous hydrogen fluoride. As the fluorination reaction using **1** required no hazardous reagents, this method triggered extensive investigations of glycosyl fluorides as glycosyl donors.

The second unexpected finding occurred during the course of examining various Lewis acids as potential promoters for the coupling reaction of glycosyl fluorides and alcohols. Interestingly, the desired glycosylation reaction proceeded smoothly with SnCl$_2$, a fairly weak Lewis acid at room temperature, giving rise to 1,2-*cis* glycosides that are difficult to prepare compared to the 1,2-*trans* glycosides [6]. When the reaction was carried out in the presence of silver perchlorate as a copromoter in diethyl ether, the 1,2-*cis* selectivity further increased. Various 1,2-*cis* glucosides were prepared in a stereoselective manner by the reaction of 2,3,4,6-tetra-*O*-benzyl-β-D-glucopyranosyl fluoride **16** with alcohols including sterically hindered ones, such as methyl 2,3,6-tri-*O*-benzyl-α-D-glucopyranoside and *t*-butyl alcohol (Scheme 2.5, Table 2.2) [6]. The tendency of predominant formation of 1,2-*cis* glycosides in diethyl ether can be explained by the preferential formation of the equatorial intermediate because of solvent participation as described in Section 1.6.3.

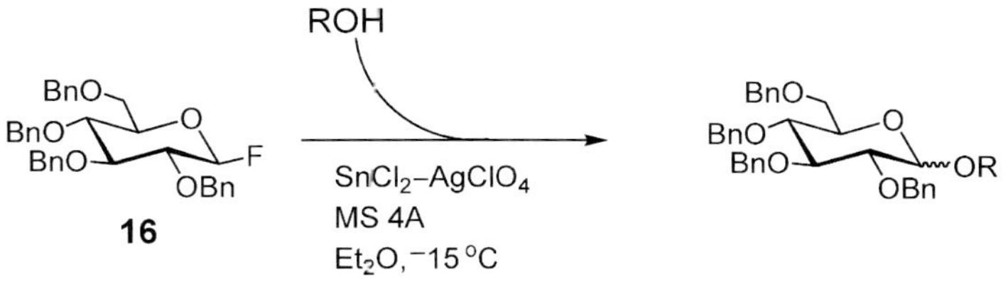

2.1 Glycosyl Fluorides

Scheme 2.5

This method was also found to be applicable to the synthesis of 1,2-*cis* glycofuranosides. 2,3,5-Tri-*O*-benzyl-β-glucofuranosyl fluoride **17** was found to react with various alcohols, including sterically hindered *t*-butyl alcohol, in the presence of $SnCl_2$–$TrClO_4$ (Tr = Ph_3C), thereby affording a good route to 1,2-*cis* furanosides that are difficult to obtain by other means (Scheme 2.6, Table 2.2) [7].

The characteristic feature of the divalent tin compounds is that they have both a vacant orbital and a lone pair of electrons (Figure 2.5). In this glycosylation reaction, it is assumed that $SnCl_2$ behaves as a Lewis acid, where the vacant orbital accepts one of the three lone pairs in the fluorine atom of the glycosyl fluoride. As a result of this interaction, the C–F bond cleaves to give the oxocarbenium ion intermediate that is then attacked by an alcohol to give the glycoside.

It is noteworthy that the addition of strong Lewis acids is not necessary to promote the glycosylation; the use of relatively weak Lewis acids like $SnCl_2$ is sufficient to activate the considerably strong C–F bond of glycosyl fluorides. All of these results

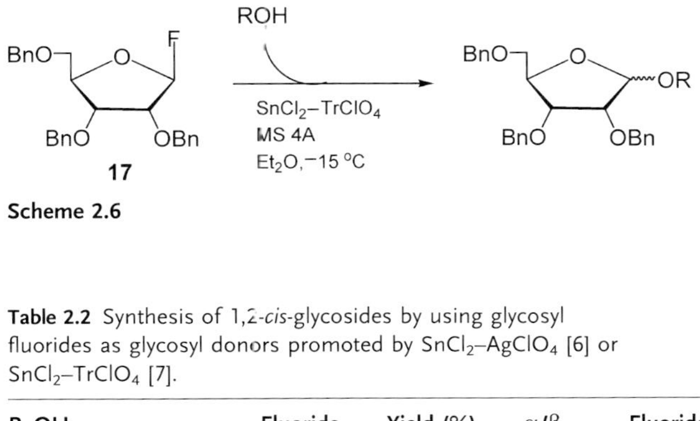

Scheme 2.6

Table 2.2 Synthesis of 1,2-*cis*-glycosides by using glycosyl fluorides as glycosyl donors promoted by $SnCl_2$–$AgClO_4$ [6] or $SnCl_2$–$TrClO_4$ [7].

R–OH	Fluoride	Yield (%)	α/β	Fluoride	Yield (%)	α/β
3β-Cholestanol	16	96	92/8	17	88	81/19
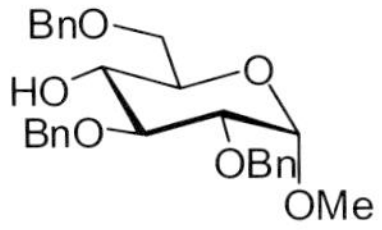 BnO, HO, BnO, OBn, OMe	16	91	80/20	17	96	85/15
t-BuOH	16	87	81/19	17	90	85/15

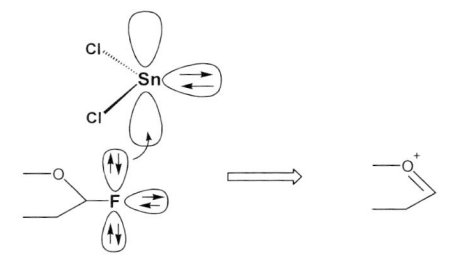

Figure 2.5 Activation of the C—F bond of glycosyl fluoride by the divalent tin species, giving rise to oxocarbenium ion intermediate.

clearly show that the fluorophilic character of Lewis acids is much more important than their Lewis acidities [39]. These basic findings have greatly stimulated the development of a series of new glycosylation reactions in these years, where a variety of combinations of glycosyl fluorides and Lewis acids have been designed on the basis of the fluorophilicity of the promoters [39].

2.1.3.2 Various Promoters Employed in Glycosylation by the Glycosyl Fluoride Method

The promoters found for the activation of glycosyl fluorides are as follows: $SnCl_2$–$AgClO_4$ [6], $SnCl_2$–$TrClO_4$ (Tr = Ph_3C) [7], $SnCl_2$–AgOTf [40,41], SiF_4 [42], Me_3SiOTf (Tf = trifluoromethanesulfonyl) [42], BF_3–OEt_2 [43–46], TiF_4 [47], SnF_4 [48], Cp_2MCl_2–$AgClO_4$ (M = Ti, Zr, Hf; Cp = cyclopentadiene) [49–53], Cp_2ZrCl_2–$AgBF_4$ [54], Cp_2HfCl_2–AgOTf [54–57], $Bu_2Sn(ClO_4)_2$ [58], Me_2GaCl [59], Tf_2O [60,61], $LiClO_4$ [62–64], $Yb(OTf)_3$ [65], $La(ClO_4)_3$–nH_2O [66], $La(ClO_4)_3$–nH_2O/$Sn(OTf)_2$ [67], Yb-Amberlyst 15 [68], sulfated zirconia [69,70], $TrB(C_6F_5)_4$ [71], TfOH [72–77], $HB(C_6F_5)_4$ [73–77], carbocationic species paired with $B(C_6F_5)_4^-$ and TfO^- [78], $SnCl_2$–$AgB(C_6F_5)_4$ [79], $SnCl_4$–$AgB(C_6F_5)_4$ [80], $Ce(ClO_4)_3$ [81], ytterbium(III)tris[bis(perfluorobutylsulfonyl)amide] ($Yb[N(SO_2CF_2CF_2CF_2CF_3)_2]_3$) [82] and $Cu(OTf)_2$ [83]. Among these promoters, tin(II) species (SnX_2), bis(cyclopentadienyl)metal derivatives (Cp_2MCl_2), BF_3–OEt_2 and protic acids are the most frequently employed. Several examples using these promoters will be described in the following section.

2.1.3.3 Glycosylations Promoted by Various Promoters

$SnCl_2$–AgX The L-fucosyl fluoride derivative **18** reacts with glycosyl acceptor **19** in the presence of $SnCl_2$–$AgClO_4$ to give the corresponding disaccharide **20** as α-anomer only (Scheme 2.7) [84].

The coupling reaction of glycosyl fluoride of N-acetylneuraminic acid derivative **21**, possessing an equatorial auxiliary at C-3, and a sugar alcohol **22** promoted by $SnCl_2$–AgOTf leads to the corresponding α-linked disaccharide **23** as the major product (Scheme 2.8) [85]. The phenylthio auxiliary acts as a participating group to control the stereochemical course of the glycosylation, giving rise to a naturally occurring sialoside derivative.

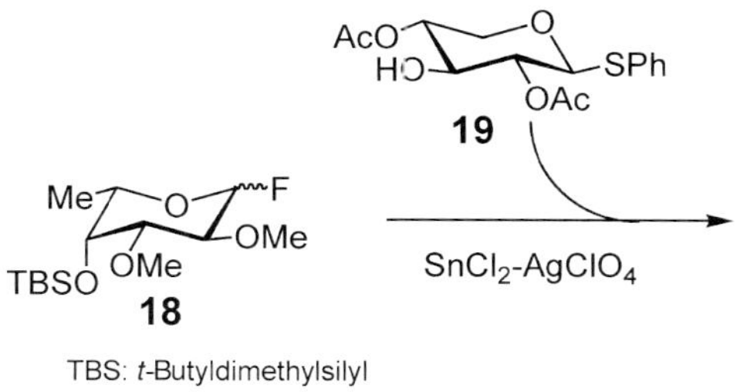

Scheme 2.7

The β-selective mannosylation is a challenging topic in glycoside synthesis because the formation of β-glycosidic linkage is stereoelectronically disfavored (see Section 1.7.4). Several approaches so far have been tried aiming at the preferential formation of β-mannnosides by changing the structure of promoters. The principle of the intramolecular aglycon delivery [86–89] has been applied to glycosyl fluoride 24 by the use of $SnCl_2$–AgOTf in dichloroethane, giving rise to a 1,2-*cis*-mannoside derivative 25 (Scheme 2.9) [90].

Bis(cyclopentadienyl)metal-Based Promoters (Cp_2MCl_2)–AgX A combination of bis(cyclopentadienyl) metal dichloride and silver perchlorate acts as a powerful pro-

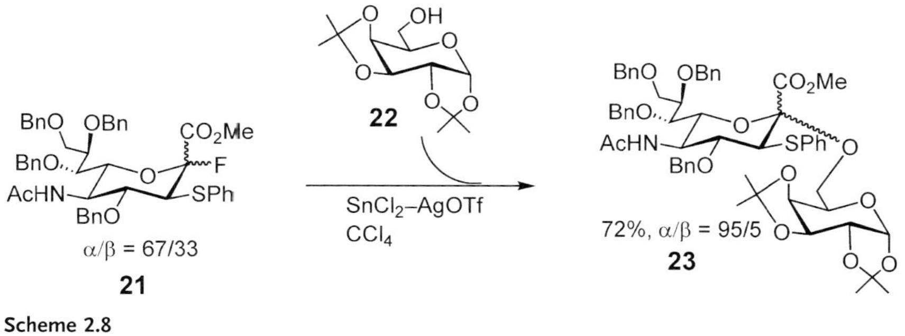

Scheme 2.8

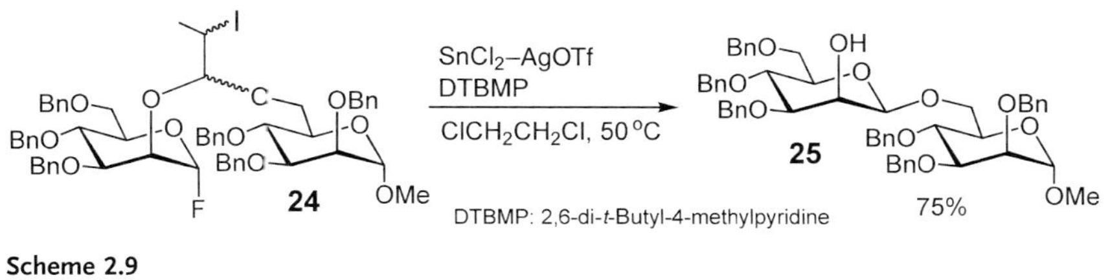

Scheme 2.9

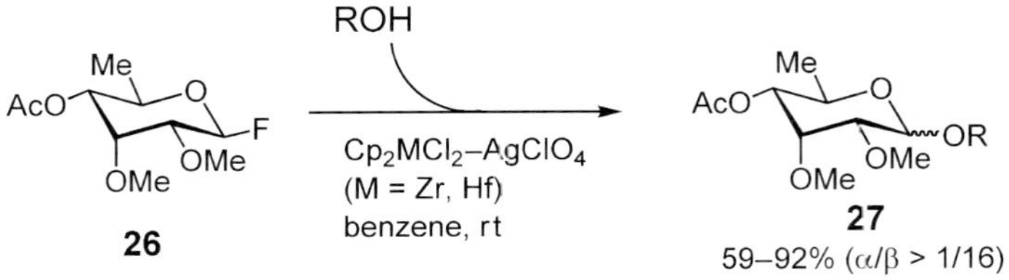

Scheme 2.10

moter for the glycosylation of glycosyl fluoride. For example, the promoter system of Cp_2MCl_2–$AgClO_4$ (M = Zr, Hf) is effective for the activation of glycosyl fluoride **26**, which enables highly β-selective formation of D-mycinoside **27** in benzene (Scheme 2.10) [49]. The use of Cp_2HfCl_2–$AgClO_4$ in 1:2 molar ratio rather than 1:1 molar ratio provides much higher reactivity for the activation of glycosyl fluoride, where some implication for the involvement of $Cp_2Hf(ClO_4)_2$ is obtained [52].

BF$_3$–OEt$_2$ Unnatural β-sialoside derivatives are produced preferentially by using O-acetylated glycosyl fluoride donor and BF$_3$–OEt$_2$ as the promoter. When β-sialyl fluoride derivative **28** was reacted with acceptor **22** in the presence of BF$_3$–OEt$_2$ in dichloromethane, the β-sialoside **28** was obtained in a mixture of anomers (α/β 17/83) (Scheme 2.11) [44].

Protic Acids According to hard and soft acids and bases (HSAB) principle, proton is considered to be fluorophilic because of its hard character and because it has higher dissociation energy for H–F bond than for H–Cl and H–Br (Table 2.1) [91]. Various protic acid catalysts can promote stereoselective glycosylation of alcohols using glycosyl fluoride donors in the presence of MS 5 Å in an appropriate solvent [72–77]. For example, when glycosyl fluoride was reacted with various alcohols in CH_2Cl_2 at room temperature in the presence of a catalytic amount of trifluoromethanesulfonic acid (TfOH), the corresponding glycoside is obtained in good yields [73–75].

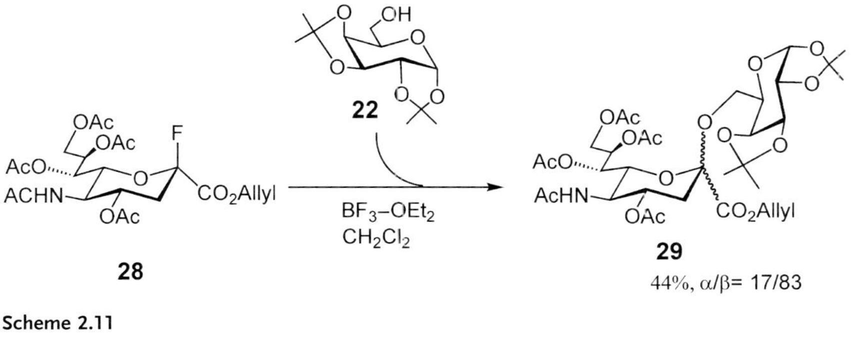

Scheme 2.11

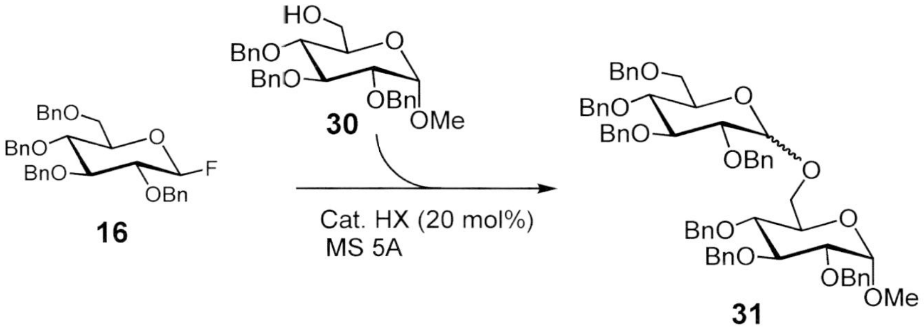

2.1 Glycosyl Fluorides | 41

Scheme 2.12

Donors possessing other halogens as leaving groups, such as glycosyl chloride and bromide, are not effectively activated in contrast to glycosyl fluoride.

When strong acid catalysts are generated *in situ* (supernatant) and used for the coupling of **16** with the glycosyl acceptor **30** either in Et$_2$O at room temperature or in benzotrifluoride (BTF)/^{t}BuCN (5:1) at 0 °C, the corresponding α- or β-linked disaccharides **31**, respectively, is obtained in high yields (Scheme 2.12 and Table 2.3). Interestingly, the stereoselectivity varies as the combination of catalyst and solvent is switched.

2.1.3.4 Glycosylation of Silylated Compounds as Glycosyl Acceptors

Glycosyl fluorides can be coupled with silylated glycosyl acceptors in place of compounds with a free hydroxyl by using a catalytic amount of Lewis acid as the promoter. In this reaction, an oxocarbenium ion **32** is first generated from a glycosyl fluoride in the presence of a Lewis acid (Figure 2.6) [92]. The resulting ion is then attacked by the silylated oxygen atom of the acceptor **33** to form the *O*-glycoside by liberating R$_3$Si–F and LA. As no acid is generated during the glycosylation, it is not necessary to add a base to the reaction mixture, which often favors the formation of

Table 2.3 Effect of solvent and counteranion of protic acids on stereoselectivity of formation of **31** [76].

Catalyst	Yield (%) (Et$_2$O, rt)	α/β	Yield (%) (BTF–^{t}BuCN(5:1)0 °C)	α/β
HOTf	98	88/12	Quant	47/53
HOTf	96	88/12	99	49/51
HClO$_4$	98	92/8	Quant	60/40
HOSO$_2$C$_4$F$_9$	99	88/12	96	47/53
HNTf$_2$	Quant	49/51	99	9/91
HNTf$_2$	Quant	50/50	Quant	9/91
HSbF$_6$	99	56/44	Quant	12/88
HB(C$_6$F$_5$)$_4$	95	55/45	99	7/93

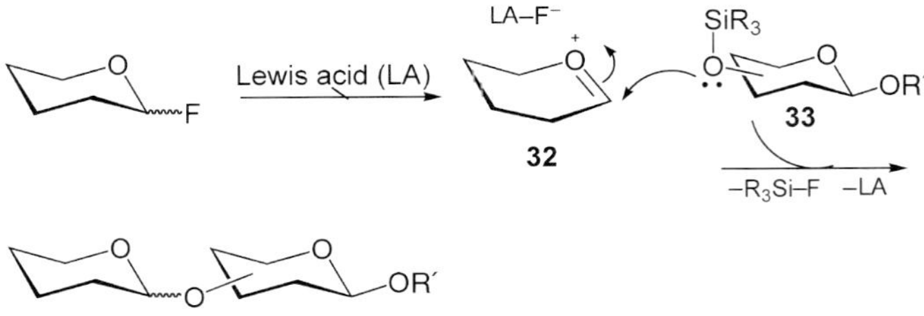

Figure 2.6 Glycosyltion of silylated acceptor by glycosyl fluoride.

orthoesters in the classical Koenigs–Knorr reactions. Trimethylsilyl triflate (TMSOTf) effectively promotes the coupling reaction of various silyl ethers and glycosyl fluorides [42]. When the glycosyl fluoride **34** is reacted with the 4,6-TIPS-protected methyl glucoside **35** in the presence of a catalytic amount of BF$_3$–OEt$_2$, a regioselective β(1–6) glycosylation is observed, affording the gentiobioside **36** (Scheme 2.13) [93].

2.1.3.5 Two-Stage Activation Procedure

The strategy of combining the chemistry of glycosyl fluorides with that of thioglycosides (see Section 4.1) for the synthesis of oligosaccharides was first suggested in 1984 (see Section 2.7) (Figure 2.7) [23]. In the first activation stage, the phenylthio group of compound **37** is converted into the corresponding glycosyl fluoride **38** by treating it with DAST and NBS (see Section 2.1.2.5). In the second activation stage, the resulting glycosyl fluoride **38** is coupled to a glycosyl acceptor **39**, which may carry a phenylthio group at the reducing end for further propagation of the oligosaccharide chain. The disaccharide **40** obtained may then be deprotected at the desired position to give an acceptor **42** or converted into a glycosyl fluoride **41** that may be used as a new glycosyl donor. Reiteration of the process can produce the desired target oligosaccharide.

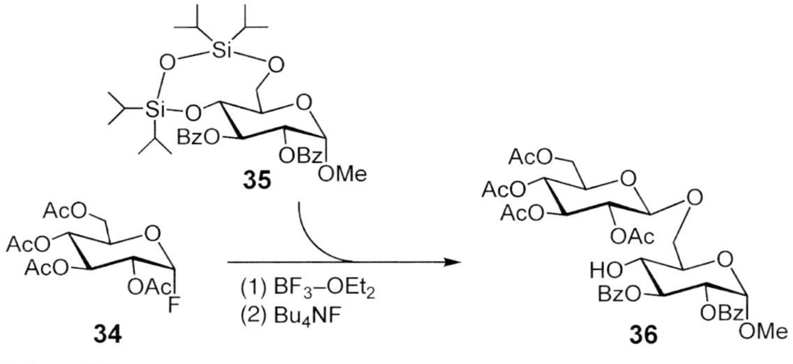

Scheme 2.13

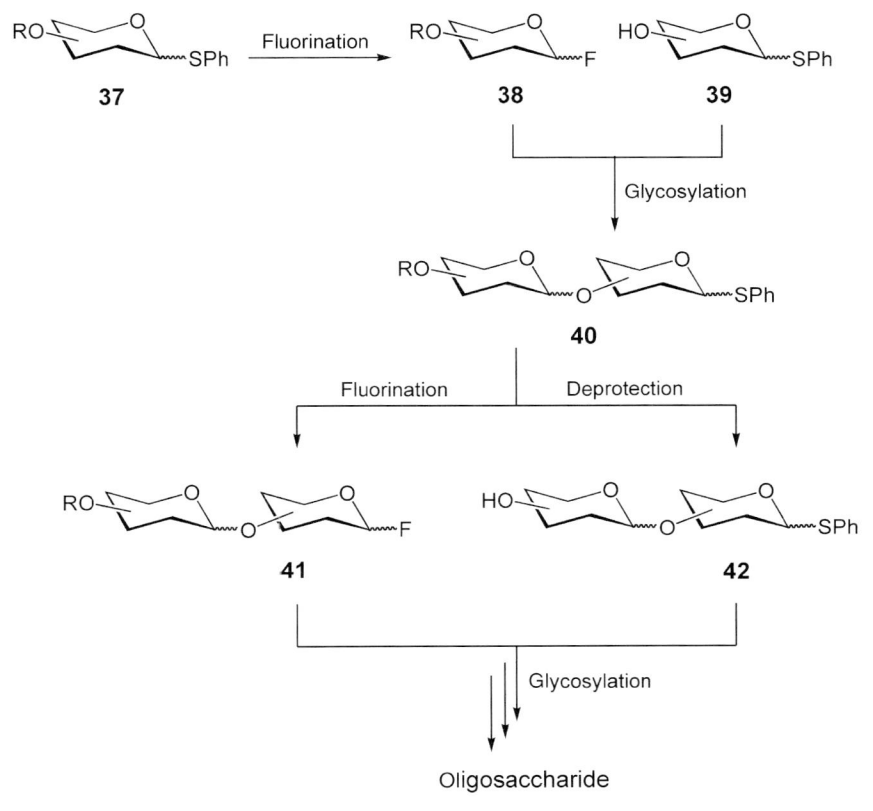

Figure 2.7 Two-stage activation procedure using glycosyl fluoride and thioglycoside as glycosyl donor.

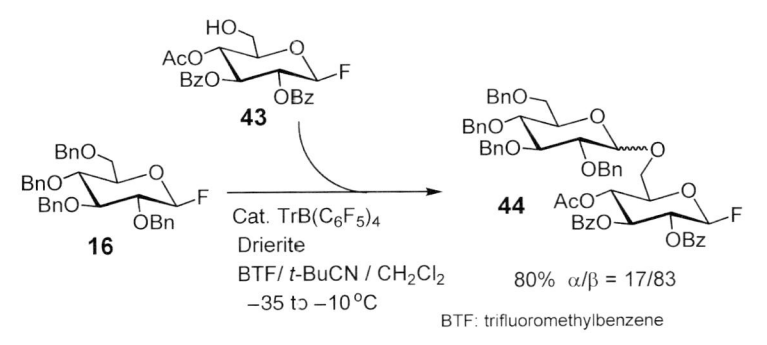

Scheme 2.14

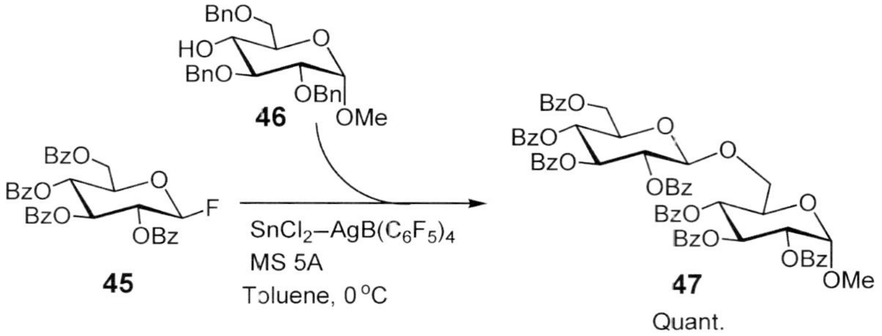

Scheme 2.15

2.1.3.6 Protecting-Group-Based Strategy

The chemoselective activation of glycosyl fluorides based on the 'armed–disarmed concept' (see Section 1.8.4) has been described (Scheme 2.14) [94]. The catalytic glycosylation of the disarmed glycosyl fluoride 43 with the armed glycosyl fluoride 16 was promoted by trityl tetrakis(pentafluorophenyl)borate (TrB(C$_6$F$_5$)$_4$) in trifluoromethylbenzene (BTF)–pivalonitorile–CH$_2$Cl$_2$ (5:1:1) to give the corresponding disaccharide 44 in good yield and β-selectivity [72,72].

On the contrary, activation of the disarmed glycosyl fluoride 45 by the combined use of SnCl$_2$ and silver tetrakis(pentafluorophenyl)borate (AgB(C$_6$F$_5$)$_4$) could be achieved to afford β-D-disaccharide 47 in excellent yields even in the cases of using acceptor 46 having secondary alcohol that generally exhibit low nucleophilicity (Scheme 2.15) [79]. These reactions are the first examples of catalytic glycosylation using armed and disarmed glycosyl fluorides with various glycosyl acceptors having free hydroxyl groups.

2.1.4
Application to Natural Product Synthesis

In the previous section, several glycosylation reactions have been described, showing a variety of promoters employed in the glycosyl fluoride method. Applications to natural product synthesis that build on these basic reactions will be described herein. The promoters most frequently employed for natural product synthesis are SnCl$_2$–AgX (X = ClO$_4$, OTf), Cp$_2$MCl$_2$–AgX (M = Hf, Zr, X = ClO$_4$, OTf), BF$_3$–OEt$_2$ and a protic acid (HOTf).

In the synthesis of avermectin B$_{1a}$, the disaccharide fluoride 48 is prepared and coupled with avermectin aglycon 49 in the presence of SnCl$_2$–AgClO$_4$ in dry ether to give protected avermectin B$_{1a}$ 50 in 62% yield (Scheme 2.16) [23].

As the reaction proceeds under mild reaction conditions, this original promoter, SnCl$_2$–AgClO$_4$, can also be employed in the synthesis of complex glycosides such as rhynchosporosides [95], cyclodextrin [41,96], trimeric Le^x [55,97], globotriaosylceramide (Gb$_3$) [98] or Elfamycin [99]. As a representative example, the synthetic route

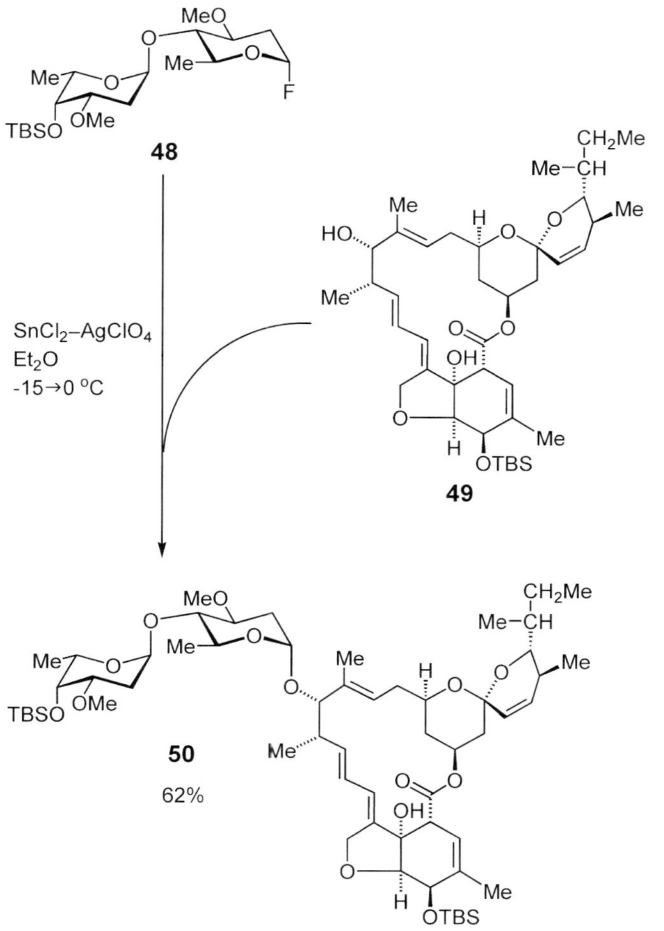

Scheme 2.16

to the trimeric Le^x **61** is illustrated in Schemes 2.17–2.19. The lactosyl sphingosine acceptor **51** is constructed by coupling the disaccharide fluoride donor **52** and the alcohol **53** in the presence of SnCl$_2$–AgClO$_4$, followed by the regioselective deprotection. However, the coupling of galactosyl fluoride derivative **54** with the sugar acceptor **55** affords the corresponding β-glycoside **56** stereoselectively. After selective deprotection, the resulting disaccharide alcohol can be coupled with the glycosyl fluoride **57** using the SnCl$_2$–AgClO$_4$ promoter system, giving rise to the trisaccharide derivative **58**. The trisaccharide **58** is then converted into the corresponding fluoride donor **59** and coupled with the disaccharide sphingosine derivative **51** by using the Cp$_2$HfCl$_2$–AgOTf promoter to afford the pentasaccharide derivative **60**.

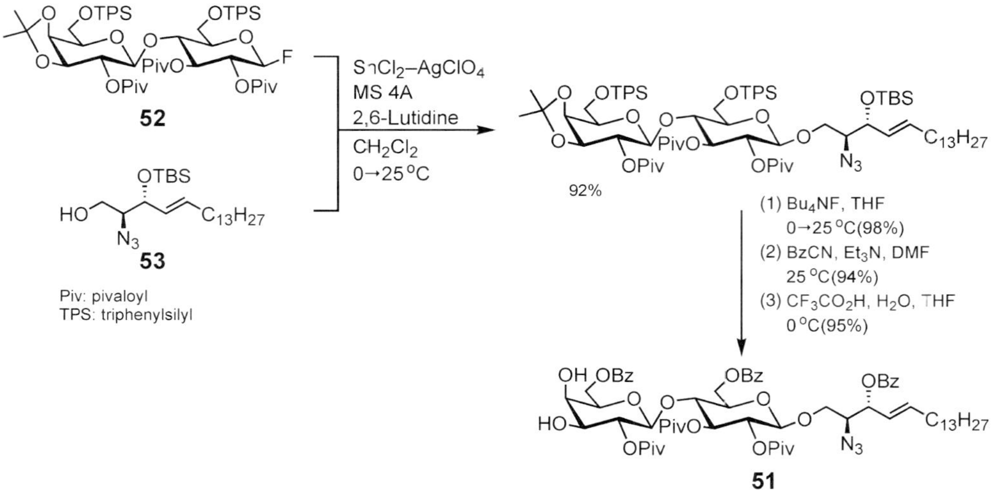

46 | *2 Glycoside Synthesis from Anomeric Halides*

Scheme 2.17

The glycosylation process by using the trisaccharide fluoride is repeated twice to give trimeric Le^x **61** after deprotections.

The total synthesis of agelagalastatins **62a** and **62b**, an antineoplastic glycosphingolipid, has been achieved by using an α-selective glycosylation of the ceramide

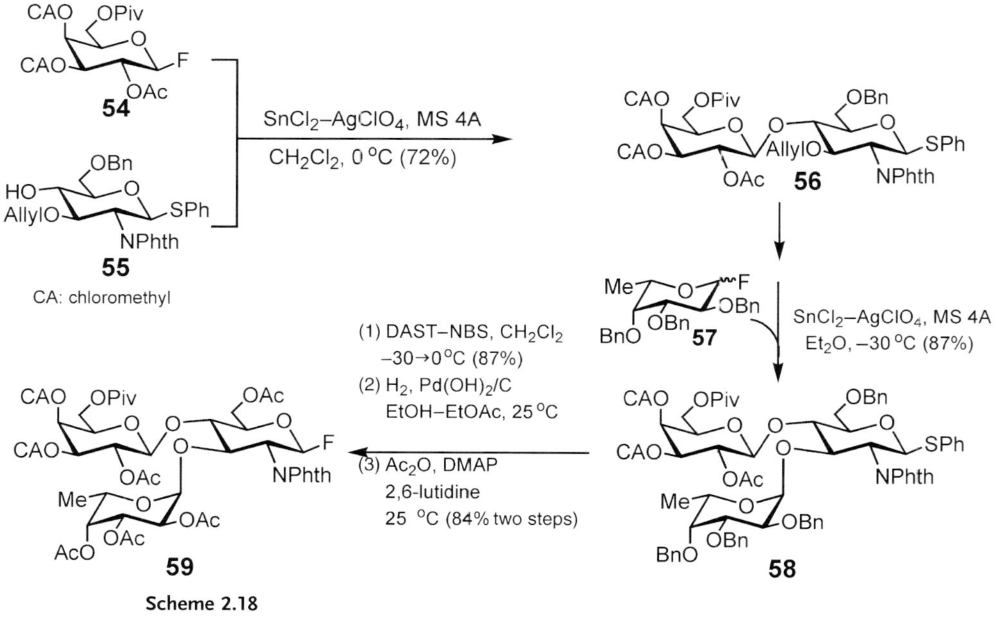

Scheme 2.18

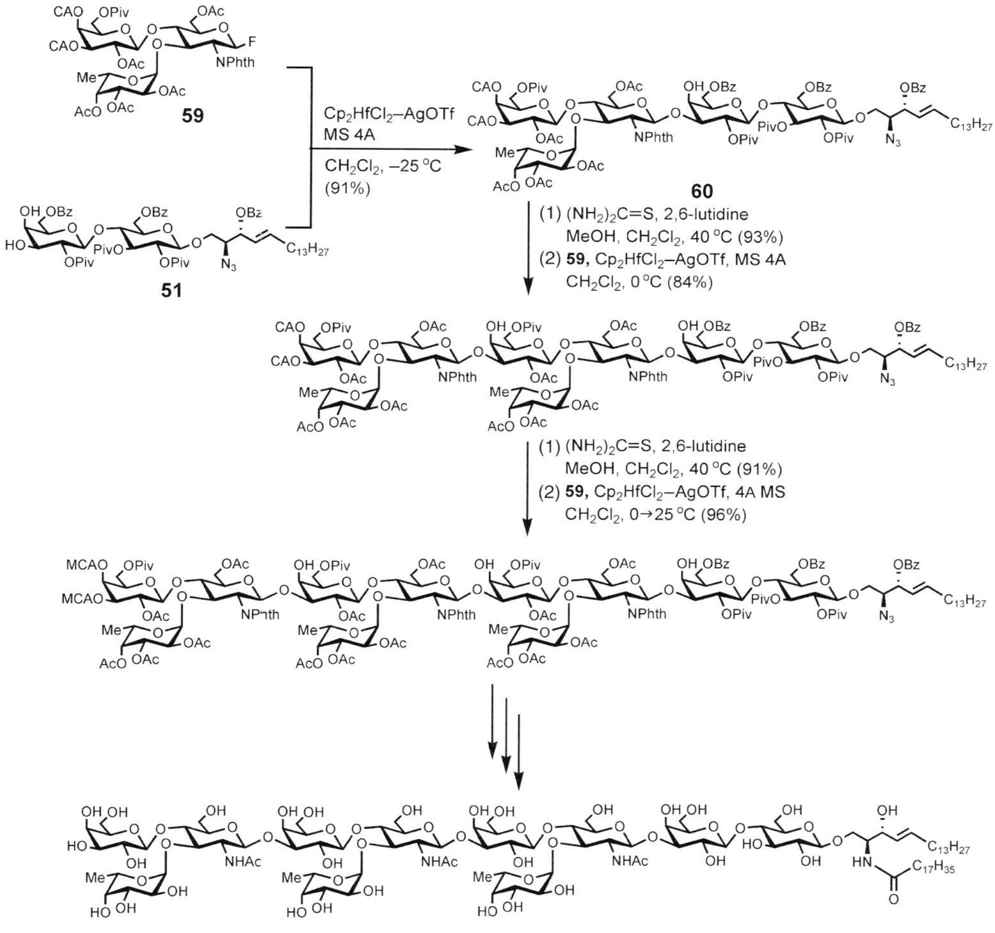

Scheme 2.19

derivatives **63a** and **63b** with the trisaccharide fluoride **64** promoted by SnCl$_2$–AgClO$_4$ (Scheme 2.20) [100].

The usefulness of the Cp$_2$MCl$_2$–AgX system (M = Hf, Zr, X = ClO$_4$, OTf) can be also seen in the total synthesis of various natural products. The [Cp$_2$MCl$_2$]–AgClO$_4$ (M = Zr, Hf) promoters were used for the total synthesis of mycinamicin IV [51]. The first total synthesis of neohancosides A and B, monoterpene diglycosides isolated from *Cynanchum hancockianum*, has been achieved using a glycosyl fluoride as a glycosyl donor. The key step involves the coupling of glycosyl fluoride and linalool in the presence of [Cp$_2$ZrCl$_2$]–AgClO$_4$ [101].

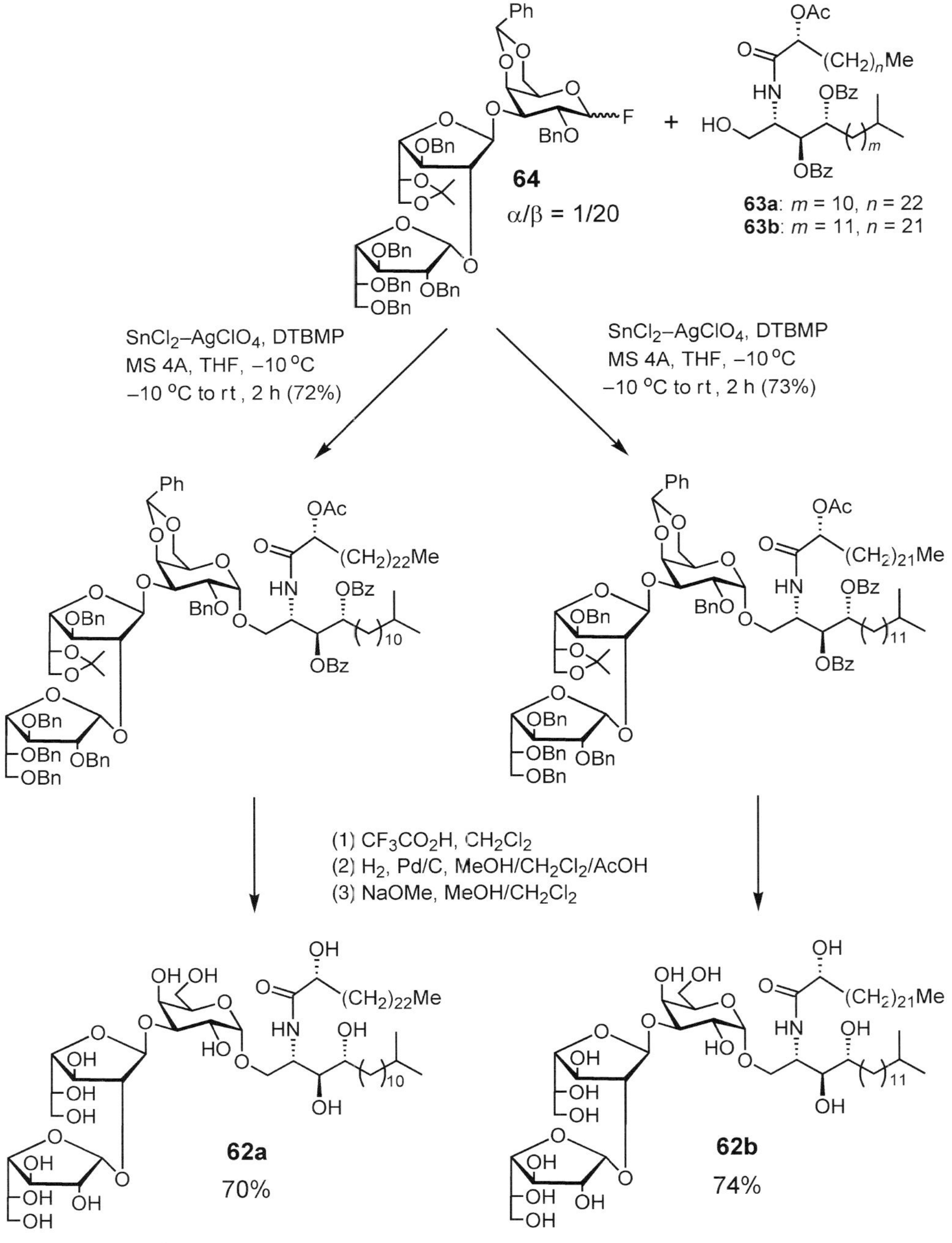

48 | *2 Glycoside Synthesis from Anomeric Halides*

Scheme 2.20

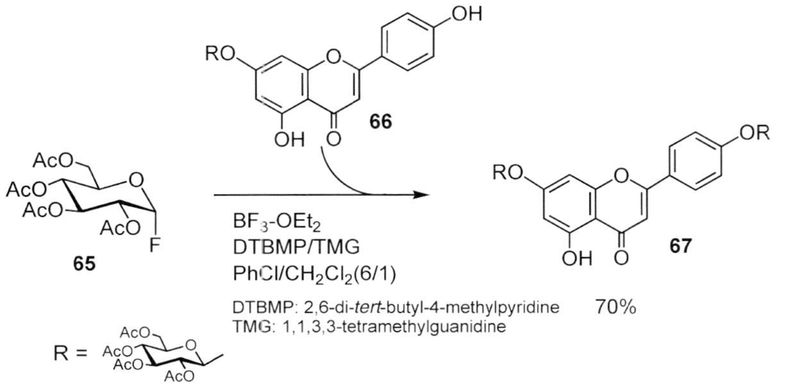

2.1 Glycosyl Fluorides | 49

Scheme 2.21

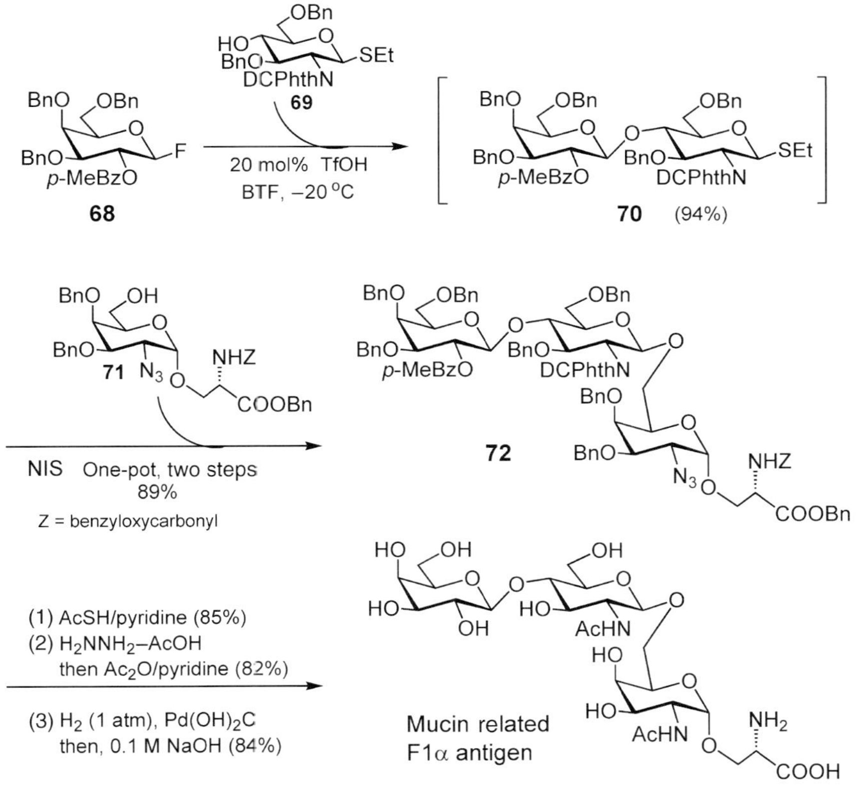

Scheme 2.22

Z = benzyloxycarbonyl

(1) AcSH/pyridine (85%)
(2) H$_2$NNH$_2$–AcOH
 then Ac$_2$O/pyridine (82%)
(3) H$_2$ (1 atm), Pd(OH)$_2$C
 then, 0.1 M NaOH (84%)

Mucin related F1α antigen

DTBMP: 2,6-di-*tert*-butyl-4-methylpyridine
TMG: 1,1,3,3-tetramethylguanidine

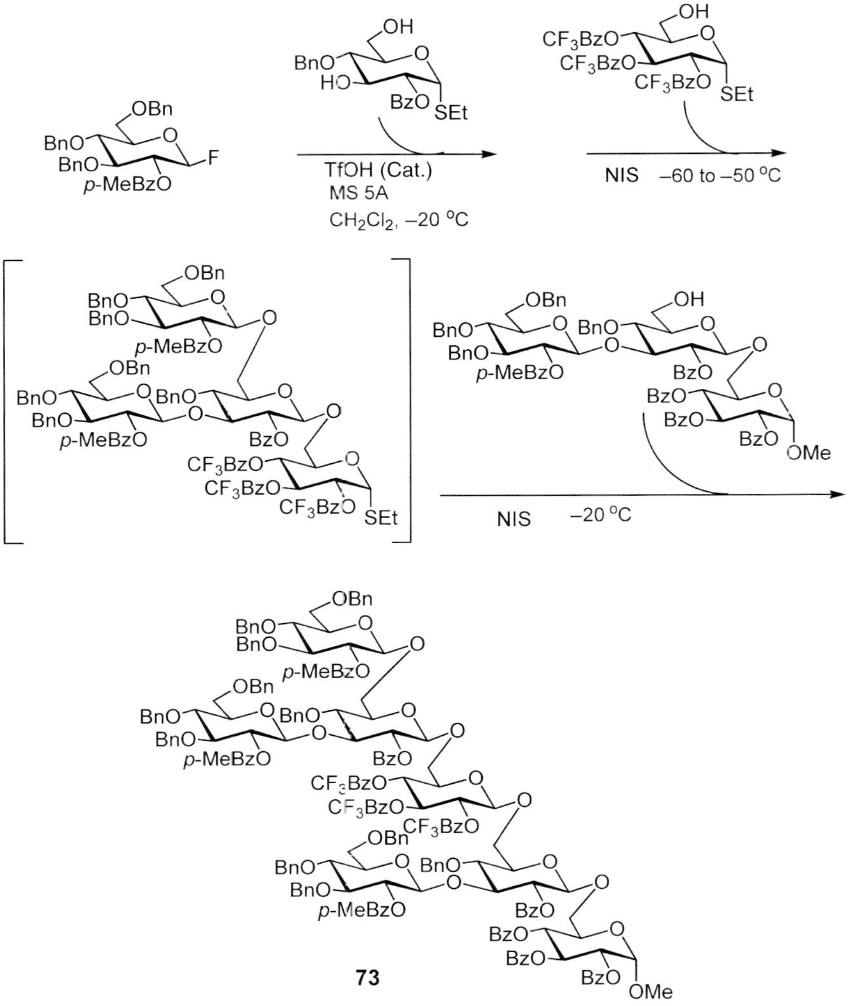

Scheme 2.23

The glycosyl fluoride glycosylation promoted by BF$_3$–OEt$_2$ is applied in the key step of the total synthesis of ipopolysaccharide (LPS) derivative [102] and D-*myo*-inositol monomannoside [103]. Also, glycosylation of 4′-OH of **66** with 2,3,4,6-tetra-O-acetyl-α-D-glucopyranosyl fluoride **65** is achieved using a BF$_3$–OEt$_2$/2,6-di-*tert*-butyl-4-methylpyridine/1,1,3,3-tetramethylguanidine system in 70% yield (Scheme 2.21). The resulting glycoside **67** is converted into apigenin glucopyranoside derivative, a component of blue flower pigment of *Salvia patens* [104].

The one-pot sequential glycosylation strategy is applied to the convergent total synthesis of the F1α antigen (Scheme 2.22) [105]. The glycosyl fluoride **68** is first

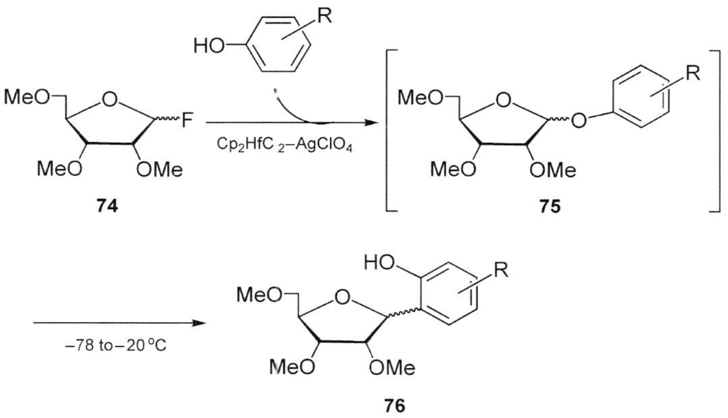

Scheme 2.24

reacted with thioglycoside **69** in the presence of a catalytic amount of TfOH to afford disaccharide **70**. The β-selectivity of the reaction can be controlled by the neighboring-group participation of the *p*-MeBz moiety at the C-2 position. Second glycosylation of glycopeptide **71** with **70** occurs by subsequently adding NIS in a one-pot fashion, and fully protected trisaccharide derivative **72** is obtained stereoselectively in 89% yield.

Moreover, the one-pot sequential glycosylation method has been applied successfully to the rapid assembly of a branched heptasaccharide **73** (Scheme 2.23) [106].

2.1.5
Special Topics

2.1.5.1 C-Glycoside Synthesis via O-Glycosylation
Glycosyl fluorides have also proved useful in the synthesis of *C*-glycosides, *N*-glycosides and *S*-glycosides, so that a broad application is guaranteed for this class of glycosyl donors [107–109]. Here, a *C*-glycosylation reaction that involves a rearrangement of *O*-glycoside is described as a representative example (Scheme 2.24). When glycosyl fluoride donor **74** is reacted with a phenol derivative in the presence of CpHfCl$_2$–AgClO$_4$, the *O*-glycosylation takes place rapidly at −78 °C. After the formation of the *O*-glycoside **75**, gradual warming results in the formation of *C*-glycoside **76** via rearrangement [110]. This reaction was successfully applied to the total synthesis of Veneomycine B$_2$ [111].

2.1.5.2 Glycosyl Fluorides for the Synthesis of a Combinatorial Library
The use of a single glycosylating tool is not enough to accomplish the synthesis of a complex oligosaccharide. It is often necessary to combine a glycosylation reaction with other glycosylation reactions. The principle of the orthogonal strategy [112] by

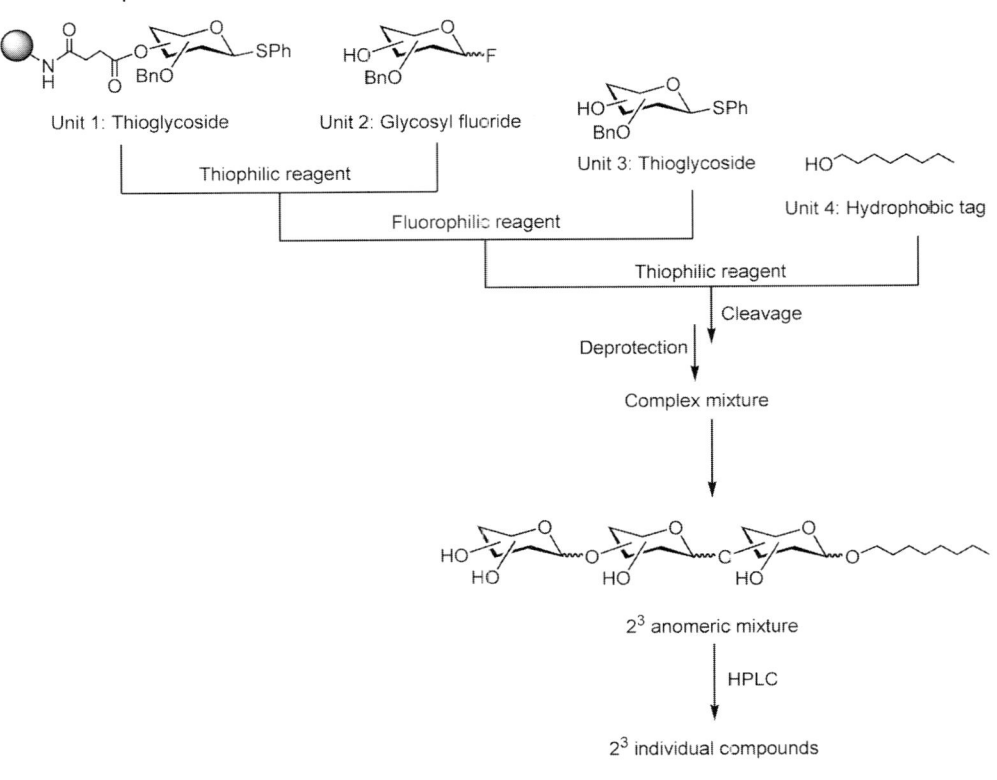

Figure 2.8 Orthogonal synthesis leading to combinatorial library of trisaccharides.

the combination of two types of glycosylation reactions, in which one of the leaving groups is activated while the other one remains intact and vice versa, will be discussed.

A new method of orthogonal oligosaccharide synthesis leading to a combinatorial library of trisaccharides, α/β-L-Fuc-(1–6)-α/β-D-Gal-(1–2/3/4/6)-α/β-D-Glc-octyl, has been developed [113]. This method is based on the combined use of a stationary solid-phase reaction, in which no mechanical mixing is required, an orthogonal glycosylation strategy [112] and solid-phase extraction (Figure 2.8). The glycosyl fluoride method plays the key role in the synthesis. Four individual synthetic equivalents were sequentially coupled by means of the orthogonal-glycosylation strategy. The last component introduced acts as a hydrophobic tag that facilitates rapid isolation of the products after cleavage of the substances accumulated on the support and the deprotection reactions. The trisaccharides obtained were finally isolated by reverse-phase HPLC [113].

2.1.5.3 Glycosyl Fluorides as Glycosyl Donors for Chemoenzymatic Synthesis

Unlike any of the other protected glycosyl halides, glycosyl fluorides can be deprotected without the loss of the halide function. This makes glycosyl fluorides very rare compounds that can be used as glycosyl donors in aqueous solutions as well as in organic solvents. Several efficient glycosylations have been demonstrated catalyzed by glycosidases [114–116] and glycosynthases [117].

2.1.6
Conclusions and Future Directions

The glycosyl fluoride method, which was developed in the early 1980s, has now reached a stage at which it has become possible to produce extremely complex glycosides while controlling their stereochemistry. Although the history of glycosyl fluoride method is much shorter than that of the classical Koenigs–Knorr glycosylation, many applications to natural product synthesis have already been demonstrated, clearly indicating its great utility in the field of carbohydrate chemistry. The reason why glycosyl fluorides have been so frequently employed as glycosyl donors is that they show higher activities in the presence of various promoters as well as ease of handling because of their thermal stability. Almost all elements that can activate the C–F bond of glycosyl fluoride donors are located in the region of hard elements with higher fluorophilicity in the periodic table. It can easily be predicted that other Lewis acids that promote glycosylations using glycosyl fluorides will be found in future on the basis of the concept of fluorophilicity. It should also be noted that various macromolecular catalysts of well-defined structures would be designed and prepared for chemoenzymatic glycosylation on a larger scale based on rapid advances in genetic engineering [118].

2.1.7
Typical Experimental Procedures

2.1.7.1 Preparation of the Glycosyl Donors

Fluorination with Bis(2-Methoxyethyl)aminosulfur trifluoride (Deoxo-Fluor) from 1-Hydroxy Sugar [13] A solution of 2,3,4,6-tetra-O-benzyl-D-glucopyranose (10 mmol) in dry CH_2Cl_2 (3.0 ml) was added at room temperature under nitrogen to a solution of bis(2-methoxyethyl)aminosulfur trifluoride (Deoxo-Fluor) 4 (11 mmol) in CH_2Cl_2 (2.0 ml). The resulting mixture was poured into saturated $NaHCO_3$ (25 ml), extracted with the help of CH_2Cl_2 (3 × 15 ml), dried (Na_2SO_4), filtered and evaporated *in vacuo*. Flash chromatography (silica gel, hexane/ethyl acetate) afforded the pure product of 2,3,4,6-tetra-O-benzyl-D-glucopyranosyl fluoride (98%, α/β = 28:72).

Fluorination with Silver Fluoride from Glycosyl Bromide [17] A mixture of 2,3,4,6-tetra-O-acetyl-α-D-glucopyranosyl bromide (5 g) and anhydrous silver fluoride (5 g) in dry acetonitrile (25 ml) was shaken under argon overnight. The resulting solution was filtered and aqueous sodium chloride was added to precipitate any silver ions from the solution. The mixture was filtered and concentrated to a syrup that was dissolved in chloroform, extracted with water, dried (Na_2SO_4) and evaporated under reduced pressure at 35 °C, with precautions to exclude moisture. The residue was dissolved in boiling diethyl ether. Subsequently, crystallization was completed by the addition of light petroleum (bp 65–110 °C), giving rise to 2,3,4,6-tetra-O-acetyl-β-D-glucopyranosyl fluoride (80%).

Fluorination with Pyridinium Poly(hydrogen fluoride) from Glycosyl Ester [16] Pyridinium poly(hydrogen fluoride) (4 ml) was added dropwise to an ice-cooled solution of 1,2-O-acetyl-3,5-di-O-benzoyl-6-deoxy-D-glucofuranose (2.5 mmol) in dry toluene (5 ml). The mixture was left at 0 °C for 5 h. Ether (10 ml) and saturated KI solution (30 ml) were added to the reaction mixture, which was then extracted with a mixture of ether and hexane (3 : 1, 3 × 30 ml) and the organic layer was washed with brine (30 ml), dried (Na_2SO_4) and filtered and the solvent was removed *in vacuo* to give white solid particles. Recrystallization from ethanol gave 2-O-acetyl-3,5-di-O-benzoyl-6-deoxy-D-glucofuranosyl fluoride as white needles (30%).

Fluorination with DAST–NBS from Thioglycoside [23] Carefully dried thioglycoside (0.3 mmol) was dissolved in CH_2Cl_2 (3 ml) under argon and cooled to −15 °C. The stirred solution was then treated with DAST (60 µl, 0.45 mmol) and allowed to stir for 2 min before NBS (0.39 mmol) was added. After 25 min, the reaction mixture was diluted with CH_2Cl_2 (25 ml) and poured into a cold saturated $NaHCO_3$ solution (3 ml). The organic phase was separated and washed with saturated $NaHCO_3$ solution (3 ml) and brine (3 ml), before being dried ($MgSO_4$) and evaporated. The oily product was subjected to flash column chromatography (silica gel, ether–petroleum ether mixtures) to afford pure fluoride (85%). R_f 0.26 (60% ether in petroleum ether).

2.1.7.2 Glycosylation Using Glycosyl Fluorides as Glycosyl Donors

$SnCl_2$–$AgClO_4$ Promoter [6] A solution of 2,3,4,6-tetra-O-benzyl-β-D-glucopyranosyl fluoride (0.2 mmol) and 3β-chlostanol (0.17 mmol) in ether (4 ml) was added to a mixture of stannous chloride (0.2 mmol), silver perchlorate (0.2 mmol) and 4 Å molecular sieves at −15 °C, and the reaction mixture was stirred at the indicated temperature for 24 h. After filtration, the filtrate was washed with cold saturated $NaHCO_3$ solution and dried (Na_2SO_4). After the removal of the solvent, the residue as purified by preparative TLC (silica gel) to give 3β-chloestanyl 2,3,4,6-tetra-O-benzyl-α-D-glucopyranoside (88%) and the corresponding β-anomer (8%).

$SnCl_2$–$TrClO_4$ Promoter [7] A solution of 3β-cholestanol (1.7 mmol) and 2,3,5-tri-O-benzyl-β-ribofuranosyl fluoride (2.0 mmol) in ether (4 ml) was added to a stirred suspension of stannous chloride (2.0 mmol), trityl perchlorate (2.0 mmol) and 4 Å molecular sieves in ether (1 ml) at −15 °C. After the reaction was completed, saturated $NaHCO_3$ solution was added to the reaction mixture. The mixture was filtered through Celite and extracted with ether. The organic layer was dried (Na_2SO_4), the solvent was removed under reduced pressure and the residue was purified by preparative TLC to give 3β-cholestanyl 2,3,5-tri-O-benzyl-α-ribofuranoside (71%) and the corresponding β-anomer (17%).

Bis(cyclopentadienyl)metal Derivatives (Cp_2ZrCl_2) Promoter [49] Cp_2ZrCl_2 (271 mmol) and $AgClO_4$ (271 mmol) were added to a mixture of D-mycinosyl fluoride **26** (54.2 mmol) and cyclohexylmethanol (108 mmol) and powdered molecular sieves 4 Å (approximately 100 mg) in benzene (2.5 ml), and the mixture was stirred for 10 min at room temperature. After the addition of saturated $NaHCO_3$ solution and filtration through a pad of Celite, the mixture was extracted with ethyl acetate, and washed with saturated aqueous $NaHCO_3$ solution and brine. After drying (Na_2SO_4), the solvent was removed under reduced pressure, and the residue was purified by preparative TLC (hexane : ether = 1 : 1) to give the corresponding glycoside **27** (92%, $\alpha/\beta = 1/16$).

Glycosylation of a Silylated Glycosyl Acceptor Using SiF_4 Promoter [42] An acetonitrile solution of SiF_4 (0.08 M, 7 ml, 0.57 mmol) was added to a mixture of 2,3,4,6-tetra-O-benzyl-α-D-glucopyranosyl fluoride (1.9 mmol) and cyclohexyl trimethylsilyl ether (1.9 mmol) in acetonitrile (3 ml) at 0 °C. The mixture was stirred for 4 h at the same temperature and poured into a mixture of KF (5 g) and 0.1 M phosphate buffer solution (pH 7.4, 30 ml). The resulting mixture was extracted with ether–hexane (2 : 1, 60 ml), and the organic layer was washed with a saturated $NaHCO_3$ solution (30 ml), a mixture of saturated $NaHCO_3$ solution and brine (1 : 5, 30 ml), and dried (Na_2SO_4). After evaporation, the residue was purified by chromatography (silica gel, 15 g, ether : hexane = 1 : 2) to give cyclohexyl 2,3,4,6-tetra-O-benzyl-D-glucopyranosides (90%, $\alpha/\beta = 15/85$). Separation of the anomers was done by medium-pressure column chromatography (silica gel, ethyl acetate : $CHCl_3 = 1 : 6$).

TfOH Promoter [71] TfOH (3.0 mg in toluene, 0.2 ml, 0.020 mol) was added to a stirred suspension of MS 5 Å (300 mg), a glycosyl fluoride (0.12 mmol), and a glycosyl acceptor (0.10 mmol) in ether (2.5 ml) at 0 °C. After completion of the glycosylation reaction by monitoring TLC, the reaction mixture was quenched by the addition of saturated aqueous $NaHCO_3$ (2 ml). Then, the mixture was diluted with ethyl acetate and 1 M HCl, and the aqueous layer was extracted with ethyl acetate. The combined organic layer was washed with water and brine, and was dried ($MgSO_4$). After filtration and evaporation, the resulting residue was purified by preparative TLC (silica gel) to afford the corresponding glycoside.

Acknowledgment

The author thanks Professor Teruaki Mukaiyama, Kitasato Institute, Japan, for helpful discussions.

References

1 Micheel, F. and Klemer, A. (1961) *Advances in Carbohydrate Chemistry*, **16**, 85–103.
2 Tsuchiya, T. (1990) *Advances in Carbohydrate Chemistry and Biochemistry*, **48**, 91–277.
3 Shimizu, M., Togo, H. and Yokoyama, M. (1998) *Synthesis*, 799–822.
4 Toshima, K. (2000) *Carbohydrate Research*, **327**, 15–26.
5 Kerr, J.A. (1995) *CRC Handbook of Chemistry and Physics*, 76th edn, CRC Press, Boca Raton, FL, pp. 9/51–9/62.
6 Mukaiyama, T., Murai, Y. and Shoda, S. (1981) *Chemistry Letters*, **10**, 431–432.
7 Mukaiyama, T., Hashimoto, Y. and Shoda, S. (1983) *Chemistry Letters*, **12**, 935–938.
8 Rosenbrook, W., Jr, Riley, D.A. and Lartey, P.A. (1985) *Tetrahedron Letters*, **26**, 3–4.
9 Posner, G.H. and Haines, S.R. (1985) *Tetrahedron Letters*, **26**, 5–8.
10 Miethchen, R., Hager, C. and Hein, M. (1997) *Synthesis*, 159–161.
11 Kunz, H. and Sanger, W. (1985) *Helvetica Chimica Acta*, **68**, 283–287.
12 Burkart, M.D., Zhang, Z., Hung, S.C. and Wong, C.H. (1997) *Journal of the American Chemical Society*, **119**, 11743–11746.
13 Lal, G.S., Pez, G.P., Pesaresi, R.J. Prozonic, F.M. and Cheng, H. (1999) *Journal of Organic Chemistry*, **64**, 7048–7054.
14 Ernst, B. and Winkler, T. (1989) *Tetrahedron Letters*, **30**, 3081–3084.
15 Brauns, D.H. (1923) *Journal of the American Chemical Society*, **45**, 833–835.
16 Buchanan, J.G., Hill, D.G., Wightman, R.H., Boddy, I.K. and Hewitt, B.D. (1995) *Tetrahedron*, **51**, 6033–6050.
17 Miethchen, R., Gabriel, T. and Kolp, G. (1991) *Synthesis*, 885–888.
18 Hall, L.D., Manville, J.F. and Bhaccha, N.S. (1969) *Canadian Journal of Chemistry*, **47**, 1–17.
19 Igarashi, K., Honma, T. and Irisawa, J. (1969) *Carbohydrate Research*, **11**, 577–587.
20 Thiem, J., Kreuzer, M., Fritsche-Lang, W. and Deger, H.M. Ger. Pat 3626028A1 (1987). (1987) *Chemical Abstract*, **107**, 176407e.
21 Goggin, K.D., Lambert, J.F. and Walinsky, S.W. (1994) *Synlett*, 162–164.
22 Miethchen, R. and Kolp, G. (1993) *Journal of Fluorine Chemistry*, **60**, 49–55.
23 Nicolaou, K.C., Dolle, R.E., Papahatjis, D.P. and Randall, J.L. (1984) *Journal of the American Chemical Society*, **106**, 4189–4192.
24 Horne, G. and Mackie, W. (1999) *Tetrahedron Letters*, **40**, 8697–8700.
25 Caddick, S., Motherwell, W.B. and Wilkinson, J.A. (1991) *Journal of the Chemical Society: Chemical Communications*, 674–675.
26 Caddick, S., Gazzard, L., Motherwell, W.B. and Wilkinson, J.A. (1996) *Tetrahedron*, **52**, 149–156.
27 Gordon, D.M. and Danishefsky, S.J. (1990) *Carbohydrate Research*, **206**, 361–366.
28 Shimizu, M., Nakahara, Y. and Yoshioka, H. (1999) *Journal of Fluorine Chemistry*, **97**, 57–60.
29 Koivula, A., Ruohonen, L., Wohlfahrt, G., Reinikainen, T., Teeri, T., Piens, K., Claeyssens, M., Weber, M., Vasella, A., Becker, D., Sinnott, M.L., Zou, J., Kleyweg, G.J., Szardenings, M., Stahlberg, J. and Jones, T.A. (2002) *Journal of the American Chemical Society*, **124**, 10015–10024.
30 Kobayashi, S., Yoneda, A., Fukuhara, T. and Hara, S. (2004) *Tetrahedron*, **60**, 6923–6930.
31 Suzuki, S., Matsumoto, K., Kawamura, K., Suga, S. and Yoshida, J. (2004) *Organic Letters*, **6**, 3755–3758.
32 Pearson, A.G., Kiefel, M.J. Ferro, V. and von Itzstein, M. (2005) *Carbohydrate Research*, **340**, 2077–2085.
33 Yokoyama, M. (2000) *Carbohydrate Research*, **327**, 5–14.
34 Davis, B.G., Chambers, D., Cumpstey, I., France, R. and Gamblin, D. (2003)

Carbohydrates, (ed. H.M. Osborn), Academic Press, Oxford, pp. 84–91.
35 Hayashi, M., Hashimoto, S.I. and Noyori, R. (1984) *Chemistry Letters*, **13**, 1747–1750.
36 Szarek, W.A., Grynkiewicz, G., Doboszewski, B. and Hay, G.W. (1984) *Chemistry Letters*, **13**, 1751–1754.
37 Broeder, W. and Kunz, H. (1993) *Carbohydrate Research*, **249**, 221–241.
38 Palme, M. and Vasella, A. (1995) *Helvetica Chimica Acta*, **78**, 959–969.
39 Ahrland, S., Chatt, J. and Davies, N.R. (1958) *Quarterly Review*, **12**, 265–276.
40 Ogawa, T. and Takahashi, Y. (1985) *Carbohydrate Research*, **138**, C5–C9.
41 Takahashi, Y. and Ogawa, T. (1987) *Carbohydrate Research*, **164**, 277–296.
42 Hashimoto, S., Hayashi, M. and Noyori, R. (1984) *Tetrahedron Letters*, **25**, 1379–1382.
43 Nicolaou, K.C., Chucholowski, A., Dolle, R.E. and Randall, J.L. (1984) *Journal of the Chemical Society: Chemical Communications*, 1155–1156.
44 Kunz, H. and Waldmann, H. (1985) *Journal of the Chemical Society: Chemical Communications*, 638–640.
45 Kunz, H. and Sager, W. (1985) *Helvetica Chimica Acta*, **68**, 283–287.
46 Yamaguchi, M., Horiguchi, A., Fukuda, A. and Minami, T. (1991) *Journal of the Chemical Society: Chemical Communications*, 1079–1082.
47 Kreuzer, M. and Thiem, J. (1986) *Carbohydrate Research*, **149**, 347–361.
48 Juennermann, J., Lundt, I. and Thiem, J. (1991) *Liebigs Annalen der Chemie*, **1991**, 759–764.
49 Matsumoto, T., Maeta, H., Suzuki, K. and Tsuchihashi, G. (1988) *Tetrahedron Letters*, **29**, 3567–3570.
50 Suzuki, K., Maeta, H., Matsumoto, T. and Tsuchihashi, G. (1988) *Tetrahedron Letters*, **29**, 3571–3574.
51 Matsumoto, T., Maeta, H., Suzuki, K. and Tsuchihashi, G. (1988) *Tetrahedron Letters*, **29**, 3575–3578.
52 Suzuki, K., Maeta, H. and Matsumoto, T. (1989) *Tetrahedron Letters*, **30**, 4853–4856.
53 Suzuki, K. and Matsumoto, T. (1993) *Journal of Synthetic Organic Chemistry Japan*, **51**, 718–732.
54 Suzuki, K., Maeta, H., Suzuki, T. and Matsumoto, T. (1989) *Tetrahedron Letters*, **30**, 6879–6882.
55 Nicolaou, K.C., Caulfield, T.J., Kataoka, H. and Stylianides, N.A. (1990) *Journal of the American Chemical Society*, **112**, 3693–3695.
56 Nicolaou, K.C., Hummel, C.W. and Iwabuchi, Y. (1992) *Journal of the American Chemical Society*, **114**, 3126–3128.
57 Matsuzaki, Y., Ito, Y., Nakahara, Y. and Ogawa, T. (1993) *Tetrahedron Letters*, **34**, 1061–1064.
58 Maeta, H., Matsumoto, T. and Suzuki, K. (1993) *Carbohydrate Research*, **249**, 49–56.
59 Kobayashi, S., Koide, K. and Ohno, M. (1990) *Tetrahedron Letters*, **31**, 2435–2438.
60 Wessel, H.P. (1990) *Tetrahedron Letters*, **31**, 6863–6866.
61 Wessel, H.P. and Ruiz, N. (1991) *Journal of Carbohydrate Chemistry*, **10**, 901–910.
62 Boehm, G. and Waldmann, H. (1995) *Tetrahedron Letters*, **36**, 3843–3846.
63 Boehm, G. and Waldmann, H. (1996) *Liebigs Annalen der Chemie*, **1996**, 613–619.
64 Boehm, G. and Waldmann, H. (1996) *Liebigs Annalen der Chemie*, **1996**, 621–625.
65 Hosono, S., Kim, W.S., Sasai, H. and Shibasaki, M. (1995) *Journal of Organic Chemistry*, **60**, 4–5.
66 Kim, W.S., Hosono, S., Sasai, H. and Shibasaki, M. (1995) *Tetrahedron Letters*, **36**, 4443–4446.
67 Kim, W.S., Sasai, H. and Shibasaki, M. (1996) *Tetrahedron Letters*, **37**, 7797–7800.
68 Yu, L., Chen, D., Li, J. and Wang, P.G. (1997) *Journal of Organic Chemistry*, **62**, 3575–3581.
69 Toshima, K., Kasumi, K. and Matsumura, S. (1998) *Synlett*, 643–645.
70 Toshima, K., Kasumi, K. and Matsumura, S. (1999) *Synlett*, 813–815.
71 Takeuchi, K. and Mukaiyama, T. (1998) *Chemistry Letters*, **27**, 555–556.

72 Mukaiyama, T., Takeuchi, K., Jona, H., Maeshima, H. and Saitoh, T. (2000) *Helvetica Chimica Acta*, **83**, 1901–1918.
73 Mukaiyama, T., Jona, H. and Takeuchi, K. (2000) *Chemistry Letters*, **29**, 696–697.
74 Jona, H., Takeuchi, K. and Mukaiyama, T. (2000) *Chemistry Letters*, **29**, 1278–1279.
75 Jona, H., Mandai, H. and Mukaiyama, T. (2001) *Chemistry Letters*, **30**, 426–427.
76 Jona, H., Mandai, H., Chavasiri, W., Takeuchi, K. and Mukaiyama, T. (2002) *Bulletin of the Chemical Society of Japan*, **75**, 291–309.
77 Mukaiyama, T., Suenaga, M., Chiba, H. and Jona, H. (2002) *Chemistry Letters*, **31**, 56–57.
78 Yanagisawa, M. and Mukaiyama, T. (2001) *Chemistry Letters*, **30**, 224–225.
79 Mukaiyama, T., Maeshima, H. and Jona, H. (2001) *Chemistry Letters*, **30**, 388–389.
80 Jona, H., Maeshima, H. and Mukaiyama, T. (2001) *Chemistry Letters*, **30**, 726–727.
81 Packard, G.K. and Rychnovsky, S.D. (2001) *Organic Letters*, **3**, 3393–3396.
82 Yamanoi, T., Nagayama, S., Ishida, H., Nishikido, J. and Inazu, T. (2001) *Synthetic Communications*, **31**, 899–903.
83 Yamada, H. and Hayashi, T. (2002) *Carbohydrate Research*, **337**, 581–585.
84 Johnson, J.N. and Paquette, L.A. (1995) *Tetrahedron Letters*, **36**, 4341–4344.
85 Ito, Y. and Ogawa, T. (1988) *Tetrahedron Letters*, **29**, 3987–3990.
86 Barresi, F. and Hindsgaul, O. (1991) *Journal of the American Chemical Society*, **113**, 9376–9377.
87 Stork, G. and Kim, G. (1992) *Journal of the American Chemical Society*, **114**, 1087–1088.
88 Bols, M. (1993) *Tetrahedron*, **49**, 10049–10060.
89 Ito, Y. and Ogawa, T. (1994) *Angewandte Chemie (International Edition in English)*, **33**, 1765–1767.
90 Cumpstey, I., Fairbanks, A.J. and Redgrave, A.J. (2001) *Organic Letters*, **3**, 2371–2374.
91 Pearson, R.G. (1973) *Hard and Soft Acids and Bases*, Dowden, Hutchinson, & Ross Inc., Stroudsburg, PA.
92 Ziegler, T. (1998) *Journal Fur Praktische Chemie*, **340**, 204–213.
93 Ziegler, T., Neumann, K., Eckhardt, E., Herold, G. and Pantkowsky, G. (1991) *Synlett*, 699–701.
94 Mootoo, D.R., Konradsson, P., Udodong, U. and Fraser-Reid, B. (1988) *Journal of the American Chemical Society*, **110**, 5583–5584.
95 Nicolaou, K.C., Randall, J.L. and Furst, G.T. (1985) *Journal of the American Chemical Society*, **107**, 5556–5558.
96 Ogawa, T. and Takahashi, Y. (1985) *Carbohydrate Research*, **138**, C5–C9.
97 Nicolaou, K.C., Bockvich, N.J. and Carcanague, D.R. (1993) *Journal of the American Chemical Society*, **115**, 8843–8844.
98 Nicolaou, K.C., Caulfield, T.J. and Kataoka, H. (1990) *Carbohydrate Research*, **202**, 177–191.
99 Dolle, R.E. and Nicolaou, K.C. (1985) *Journal of the American Chemical Society*, **107**, 1695–1698.
100 Lee, Y.J., Lee, B.Y., Jeon, H.B. and Kim, K.S. (2006) *Organic Letters*, **8**, 3971–3974.
101 Konda, Y., Toida, T., Kaji, E., Takeda, K. and Harigaya, Y. (1997) *Carbohydrate Research*, **301**, 123–143.
102 Kusumoto, S., Kusunose, N., Kamikawa, T. and Shiba, T. (1988) *Tetrahedron Letters*, **29**, 6325–6326.
103 Elie, C.J.J., Verduyn, R., Dreef, C.E., Brounts, D.M., van der Marel, G.A. and van Boom, J.H. (1990) *Tetrahedron*, **46**, 8243–8254.
104 Oyama, K. and Kondo, T. (2004) *Tetrahedron*, **60**, 2025–2034.
105 Mukaiyama, T., Ikegai, K., Jona, H., Hashihayata, T. and Takeuchi, K. (2001) *Chemistry Letters*, **30**, 840–841.
106 Hashihayata, T., Ikegai, K., Takeuchi, K., Jona, H. and Mukaiyama, T. (2003) *Bulletin of the Chemical Society of Japan*, **76**, 1829–1848.
107 Posner, G.H. and Haines, S.R. (1985) *Tetrahedron Letters*, **26**, 1823–1826.

108 Araki, Y., Watanabe, K., Kuan, F.H., Itoh, K., Kobayashi, N. and Ishido, Y. (1984) *Carbohydrate Research*, **127**, C5–C9.

109 Nicolaou, K.C., Dolle, R.E., Chucholowski, A. and Randall, J.L. (1984) *Journal of the Chemical Society: Chemical Communications*, 1153–1154.

110 Matsumoto, T., Katsuki, M. and Suzuki, K. (1988) *Tetrahedron Letters*, **29**, 6935–6938.

111 Matsumoto, T., Katsuki, M., Jona, H. and Suzuki, K. (1991) *Journal of the American Chemical Society*, **113**, 6982–6992.

112 Kanie, O., Ito, Y. and Ogawa, T. (1994) *Journal of the American Chemical Society*, **116**, 12073–12074.

113 Ako, T., Daikoku, S., Ohtsuka, I., Kato, R. and Kanie, O. (2006) *Chemistry – An Asian Journal*, **1**, 798–813.

114 Zhang, J., Wu, B., Liu, Z., Kowal, P., Chen, S., Shao, J. and Wang, P.G. (2001) *Current Organic Chemistry*, **5**, 1169–1176.

115 Shoda, S., Fujita, M. and Kobayashi, S. (1998) *Trends in Glycoscience and Glycotechnology*, **54**, 279–289.

116 Shoda, S., Kawasaki, T., Obata, K. and Kobayashi, S. (1993) *Carbohydrate Research*, **249**, 127–137.

117 Mackenzie, F.L., Wang, Q., Warren, R.A.J. and Withers, S.G. (1998) *Journal of the American Chemical Society*, **120**, 5583–5584.

118 Shoda, S. (2001) *Glycoscience*, vol. 2 (eds B.O. Fraser-Reid, K. Tatsuta and J. Thiem), Springer, Heidelberg, pp. 1465–1496.

2.2
Glycosyl Chlorides, Bromides and Iodides

Suvarn S. Kulkarni, Jacquelyn Gervay-Hague

2.2.1
Background

Since the advent of Koenigs–Knorr glycosylation in 1901 [119], glycosyl bromides are among the most popular glycosyl donors that continue to find wide application in oligosaccharide and natural product synthesis [120,121]. Glycosyl chlorides, on the contrary, have been used for specific applications but their general use is limited owing to their reduced activity [121]. In contrast, for much of the twentieth century, glycosyl iodides were thought to be unstable and too reactive for any useful purposes. This notion has been refuted and over the past decade glycosyl iodide chemistry has enjoyed a renaissance, with increasing reports on the preparation and use of glycosyl iodides since last reviewed in 1998 [122]. The chemistry of glycosyl chlorides and bromides was reviewed by Nitz and Bundle [121]. O-Glycosylations of glycosyl bromides and chlorides with applications in total synthesis was also reviewed by Pellissier [120]. Along with these reports, virtually every review article on stereoselective glycosylations includes examples of bromides and chlorides as glycosyl donors. In this chapter, we will first briefly discuss major advances in the twenty-first century pertaining to the preparation and use of glycosyl chlorides and bromides and devote the major part of the discussion to the novel chemistry of glycosyl iodides.

2.2.2
Glycosyl Chlorides

2.2.2.1 Preparation of Glycosyl Chlorides

Anomeric halides follow the typical reactivity order $F < Cl < Br < I$ for nucleophilic substitutions. They have been used in stereoselective *O*-glycosylation, nucleophilic displacement, and carbanion as well as in radical reactions.

There are several procedures for the selective preparation of anomerically pure α- and β-glycosyl chlorides. Thermodynamically unstable 1,2-*trans*-isomers (β-isomers of D-glucose and D-galactose) are usually prepared via neighboring-group participation from 1,2-*trans*-glycosyl acetates using various chlorinating agents such as anhydrous tin tetrachloride ($SnCl_4$), titanium tetrachloride ($TiCl_4$) and aluminum trichloride ($AlCl_3$) in nonpolar solvents or by the action of HCl in dry ether. Combinations of dichloromethyl methyl ether–BF_3OEt_2 and thionyl chloride with acetic acid are also used. The aforementioned methods have recently been compared to Ibatullin's method [123], which utilizes phosphorous pentachloride (PCl_5) in the presence of a catalytic amount of BF_3OEt_2 (Scheme 2.25). This methodology has been shown to give consistently high yields and selectivity [124] in the presence of $AlCl_3$ or in the absence of any catalyst when conducted in polar solvents such as acetonitrile. 1,2-*cis* Acetates do not react with PCl_5. Moreover, cleavage of interglycosidic linkage does not occur under these conditions. An alternative method to generate β-glycosyl chlorides is by the chlorination of alkyl or aryl thioglycosides using iodine monochloride (ICl) [125]. Similarly, the use of IBr generates the corresponding α-glycosyl bromide under mild conditions that tolerate sensitive protecting groups on the sugar substrate [125].

A few methods exist for the formation of 1,2-*cis* isomers. Protic acids such as HCl or AcOH in conjunction with thionyl chloride ($SOCl_2$) furnish β-chlorides from

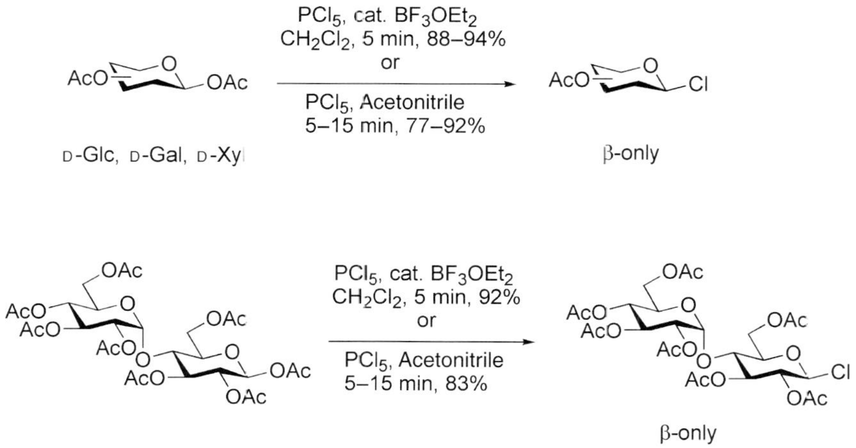

Scheme 2.25

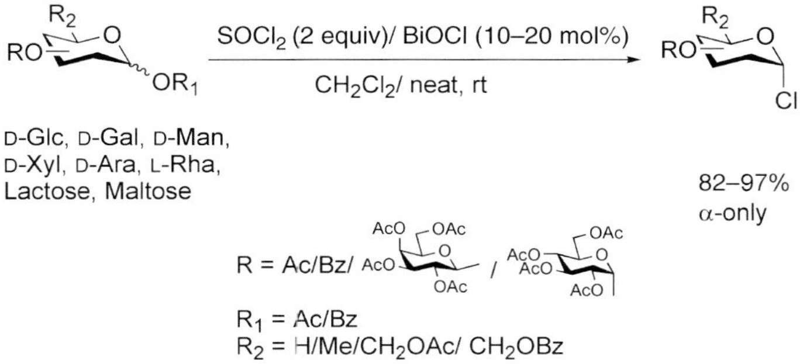

Scheme 2.26

glycosyl acetates, whereas Lewis acids such as $ZnCl_2$ or $BiCl_3$ reverse the selectivity. A recent solvent-free protocol, amenable to large-scale preparation, involves the reaction of glycosyl peracetates (mono- and disaccharides; α or β) with $SOCl_2$ and $BiCl_3$, generated *in situ* from an amount of 10–20 mol% of a procatalyst BiOCl to afford α-anomeric chlorides in high yields and selectivity (Scheme 2.26) [126]. The reaction probably proceeds through a concerted mechanism without neighboring-group participation to give α-chlorides.

Very recently, it has been shown that unprotected reducing sugars can be directly converted into acetylated α-glycosyl chlorides using AcCl and basic Al_2O_3 on solid support (Scheme 2.27) [127]. The reaction conditions are mild and generally the yields and selectivity are high. Even, *N*-acetyl D-glucosamine can be converted into glycosyl chloride using this method.

Yet another method involves the treatment of a hemiacetal with oxalyl chloride in DMF. The protocol allows for an efficient preparation of α-chlorides of 2-deoxy-L-hexopyranosides (Scheme 2.28) [128].

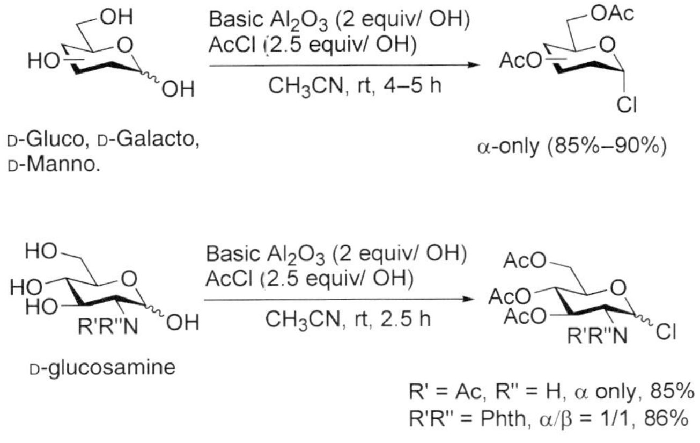

Scheme 2.27

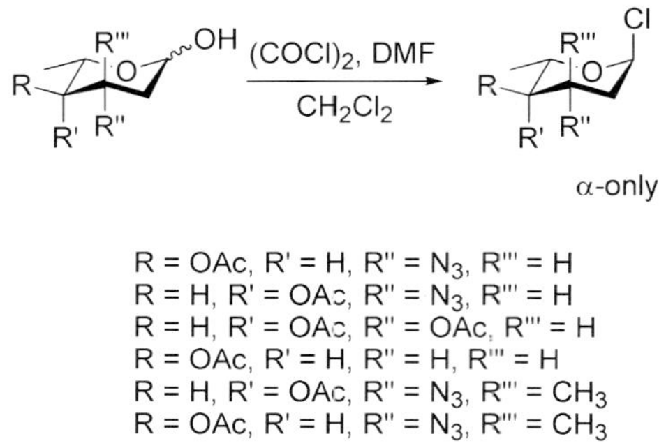

[Scheme 2.28 shown: glycosyl hemiacetal converted to α-glycosyl chloride using (COCl)$_2$, DMF in CH$_2$Cl$_2$, α-only]

R = OAc, R' = H, R'' = N$_3$, R''' = H
R = H, R' = OAc, R'' = N$_3$, R''' = H
R = H, R' = OAc, R'' = OAc, R''' = H
R = OAc, R' = H, R'' = H, R''' = H
R = H, R' = OAc, R'' = N$_3$, R''' = CH$_3$
R = OAc, R' = H, R'' = N$_3$, R''' = CH$_3$

Scheme 2.28

2.2.2.2 Reactions of Glycosyl Chlorides

Glycosylation Classical Koenigs–Knorr reaction involves the coupling of a glycosyl bromide or chloride with glycosyl acceptors using heavy metal ion, typically mercury or silver [119–121]. Over the years, Lewis acid catalysis and phase transfer catalysis (PTC) have been introduced as useful variations of this process. Generally, glycosyl bromides are preferred over chlorides in these glycosylations. Copper(II) trifluoromethanesulfonate in benzotrifluoride (C$_6$H$_5$CF$_3$, BTF) has been shown to promote glycosylation of chlorides along with other glycosyl donors such as glycosyl fluorides, acetates, trichloroacetimidates and hemiacetals, although the yields and the selectivities are moderate (Scheme 2.29) [129].

S$_N$2 Reactions Owing to their inherent stability, glycosyl chlorides are appropriate candidates for S$_N$2 reactions offering complementary stereoselectivity compared to that of bromides and iodides. They are therefore useful precursors for various glycosyl donors. The following examples are noteworthy.

Selenoglycosides Monophasic reactions of acetylated β-glucosyl or galactosyl chlorides with potassium p-methylselenobenzoate in the presence of 18-crown-6 furnish a mixture of α- and β-isomers together with an unidentified product that upon purification affords α-isomer in modest yields (Scheme 2.30) [124]. In contrast, S$_N$2 reactions of the corresponding α-bromides (α-D-Glc, α-D-Gal, α-D-Lac) work well under monophasic as well as biphasic conditions to afford the corresponding β-isomers [130]. These α-and β-p-methylbenzoyl selenoglycosides react rapidly with various electrophiles to produce a diverse array of α-and β-selenoglycosides, respectively.

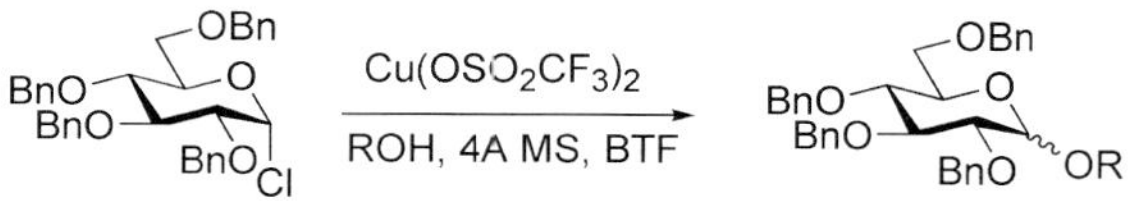

[Scheme 2.29 shown: tetra-O-benzyl-α-glucopyranosyl chloride + ROH, Cu(OSO$_2$CF$_3$)$_2$, 4Å MS, BTF → glycoside product]

Scheme 2.29

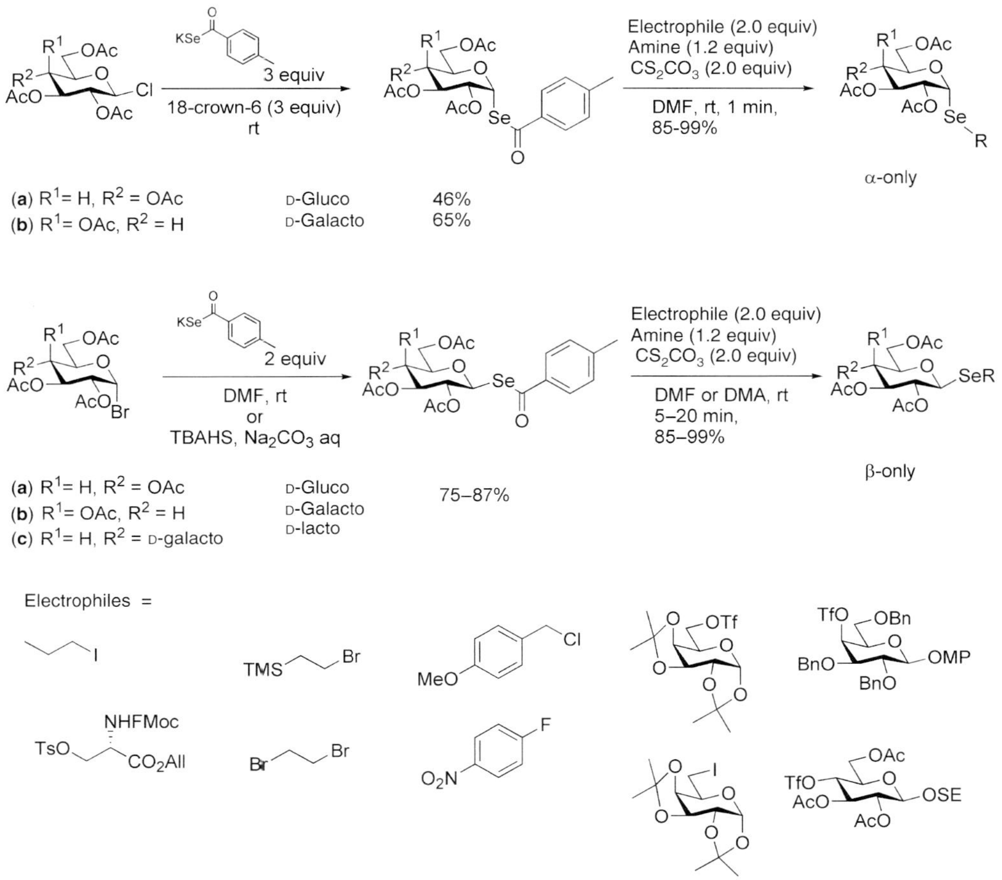

Scheme 2.30

Thioglycosides α-Anomeric chlorides can be displaced smoothly in an S_N2 manner by thioacetate anions to produce β-anomeric thioacetates in high yields. For example, 2-azido lactosyl α-chloride reacts with excess thioacetic acid in the presence of pyridine to yield β-thioacetate with concomitant reductive acetamidation of the azide group [131]. The thioacetate could be reacted further to provide S-glycosides in high yields and subsequently transformed into a new class of glycoclusters (Scheme 2.31). Watt and Boons [132] used this reaction in their convergent synthesis of N-glycan core oligosaccharide thioaldoses (Scheme 2.32). The corresponding per-O-acetylated α-glycosyl chlorides were displaced with thioacetate to afford β-thioacetates in high yields and selectivity, which upon saponification gave the target oligosaccharides to be used for site-specific glycosylation of peptides and proteins bearing free cysteine.

Glycosyl Phosphates Deoxy sugars, which are vital components of various biomolecules such as vancomycin, erythromycin and daunomycin are difficult to be

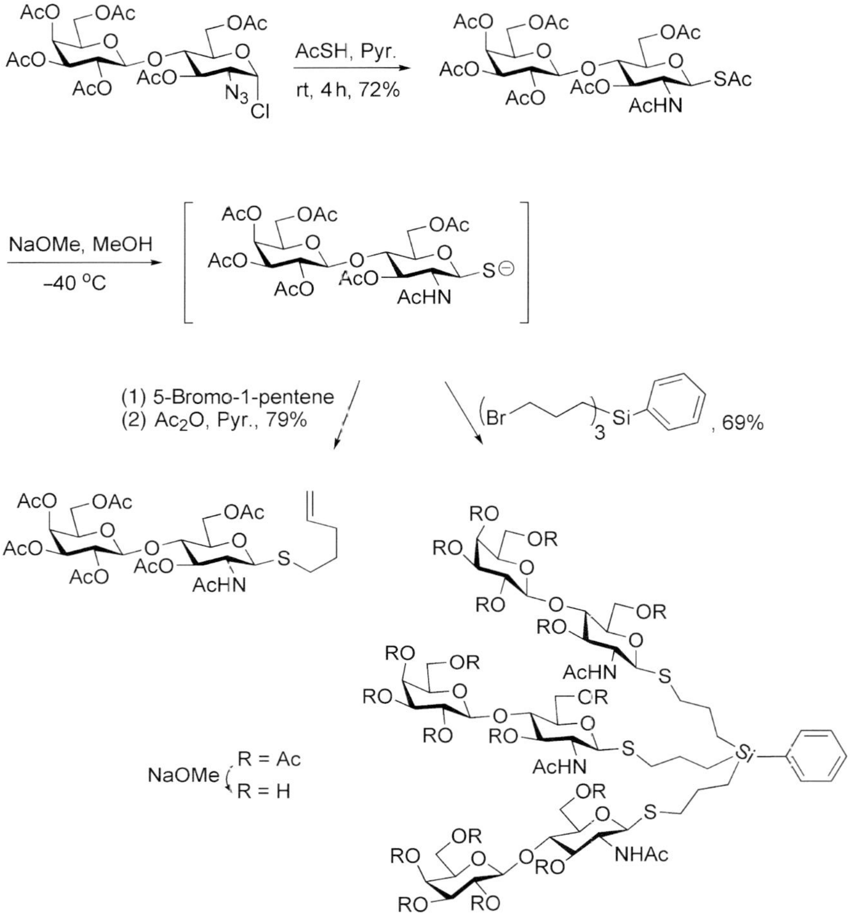

Scheme 2.31

stereoselectively synthesized. β-Selectivity is hard to control without neighboring-group participation as α-linked products are favored. Kahne and coworkers [128] reported a stereoselective method to synthesize 2-deoxy-β-L-glycosyl phosphates from glycosyl chlorides via mainly S_N2 displacement with $Bu_4NH_2PO_4^-$ (α/β = 1 : 5–1 : 9). It should be emphasized that more reactive glycosyl iodides and bromides

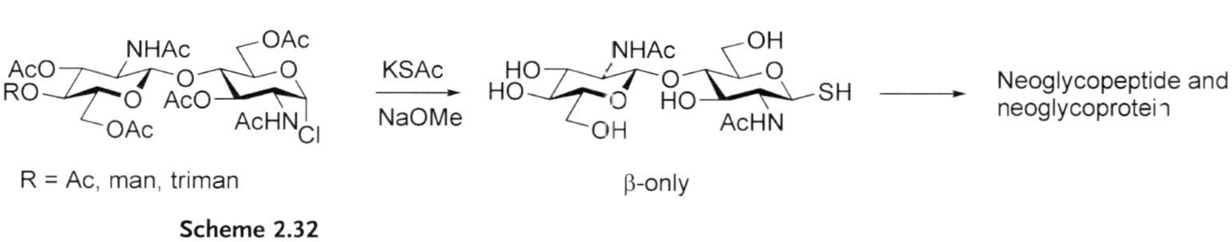

Scheme 2.32

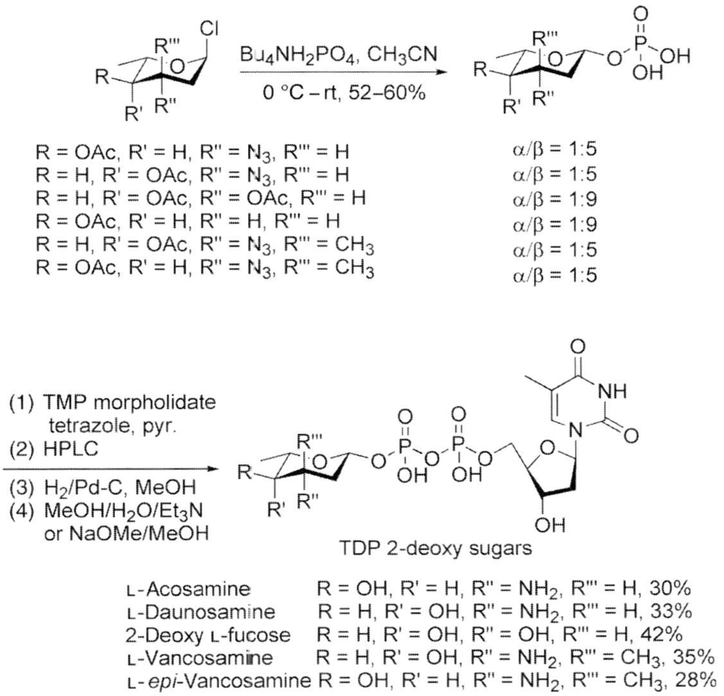
Cl
R'''
R
R'
R''
$Bu_4NH_2PO_4$, CH_3CN
0 °C – rt, 52–60%
R''' O P OH OH
R = OAc, R' = H, R'' = N_3, R''' = H
R = H, R' = OAc, R'' = N_3, R''' = H
R = H, R' = OAc, R'' = OAc, R''' = H
R = OAc, R' = H, R'' = H, R''' = H
R = H, R' = OAc, R'' = N_3, R''' = CH_3
R = OAc, R' = H, R'' = N_3, R''' = CH_3
α/β = 1:5
α/β = 1:5
α/β = 1:9
α/β = 1:9
α/β = 1:5
α/β = 1:5
(1) TMP morpholidate tetrazole, pyr.
(2) HPLC
(3) H_2/Pd-C, MeOH
(4) MeOH/H_2O/Et_3N or NaOMe/MeOH
NH
N
OH
TDP 2-deoxy sugars
L-Acosamine R = OH, R' = H, R'' = NH_2, R''' = H, 30%
L-Daunosamine R = H, R' = OH, R'' = NH_2, R''' = H, 33%
2-Deoxy L-fucose R = H, R' = OH, R'' = OH, R''' = H, 42%
L-Vancosamine R = H, R' = OH, R'' = NH_2, R''' = CH_3, 35%
L-*epi*-Vancosamine R = OH, R' = H, R'' = NH_2, R''' = CH_3, 28%

Scheme 2.33

give α-isomers as major products through S_N1-like reactions (α/β = 2 : 1), whereas stable chlorides shift the mode of the reaction toward S_N2 providing a range of β-deoxy sugar phosphates. Subsequent reaction of the glycosyl phosphates with TMP (thymidine monophosphoryl) morpholidate, followed by a careful HPLC separation from the minor α-isomer, azide reduction and deprotection of acetates gave 2-deoxy-β-TDP sugars in good yields (Scheme 2.33).

Elimination – Glycal Formation Sialic-acid-containing oligosaccharides play vital roles in living systems. Stereoselective sialylations typically require a stereodirecting group at C-3 that can be introduced to a 2,3-glycal (Neu5Ac glycal), which is also a key intermediate in the synthesis of the anti-influenza drug Relenza. Neu5Ac Glycal is obtained by eliminating the corresponding glycosyl chloride under various conditions. According to a recent protocol [133], a treatment of the peracetylated *N*-acetylneuraminic acid glycosyl chloride with anhydrous Na_2HPO_4 in refluxing acetonitrile quantitatively affords glycal (Scheme 2.34). The product can be isolated by simple filtration and evaporation of solvent thus obviating the need for chromatographic purification. Notably, no glycal formation takes place at room temperature.

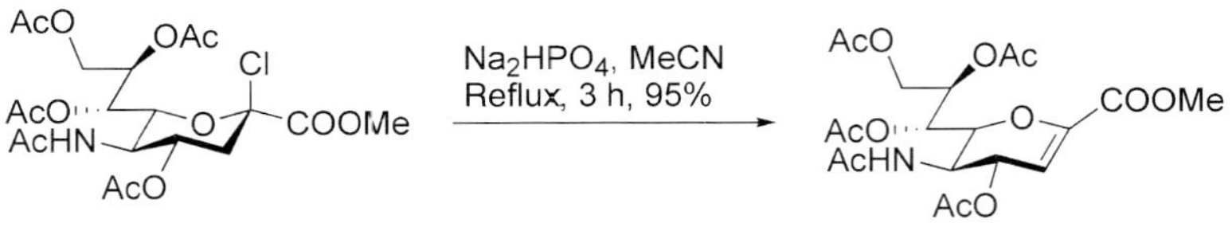

Scheme 2.34

2.2.3
Glycosyl Bromides

As stated earlier, glycosyl bromides possess activity and stability intermediate to that of other halides; they are more reactive than fluorides or chlorides but more stable than iodides. Their popularity is largely attributed to controlled O-glycosylation via Koenigs–Knorr and related methods, phase transfer catalysis, solvolysis and displacement reactions. In this section, new methods of preparing glycosyl bromides and their modes of reactions are discussed. Their recent applications in oligosaccharide, glycoconjugate and natural product synthesis since 2001 are also presented.

2.2.3.1 Preparation of Glycosyl Bromides

Earlier established methods for the generation of glycosyl bromides include the treatment of free reducing sugars with AcBr, AcBr–AcOH, AcBr–MeOH, PBr$_3$, Ac$_2$O–HBr–AcOH, or treatment of peracetylated sugars with HBr–AcOH or BiBr$_3$–TMSBr [134]. A recently reported two-step one-pot procedure workable on a large scale, affords α-anomeric bromides from free sugars (D-glucose, D-mannose, D-lactose, D-cellobiose, D-maltose) almost quantitatively (Scheme 2.35) [135]. Sequential treatment of a free sugar with Ac$_2$O (1.05 equiv per OH) in the presence of LiClO$_4$ (0.1 equiv per OH) followed by bromination using 33% HBr/AcOH solution furnishes acetylated glycosyl bromides.

Hunsen's procedures [134] use a combination of AcBr and MeOH for *in situ* generation of HBr (Scheme 2.36). Thus, free sugars including mono- (D-Glc, D-Man), β-linked disaccharides (D-cellobiose, D-lactose) and α-linked di- and trisaccharides (D-maltose, D-maltotriose) could be effortlessly converted into α-glycosyl bromides by treatment with Ac$_2$O and cat HClO$_4$ for peracetylation, followed by AcBr and MeOH. Alternatively, premixing of AcBr and MeOH in AcOH, to generate HBr, followed by the addition of sugar and Ac$_2$O affords the title compounds. The former protocol works better for galactose and maltotriose.

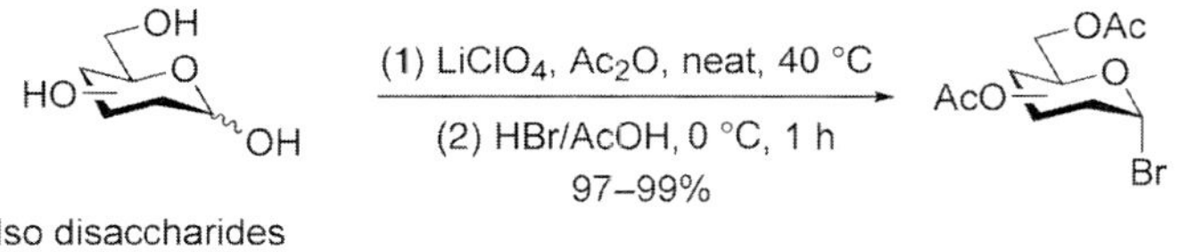

Also disaccharides

Scheme 2.35

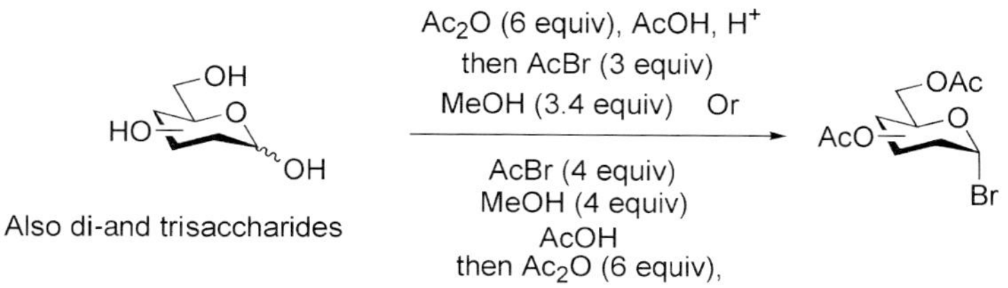

Scheme 2.36

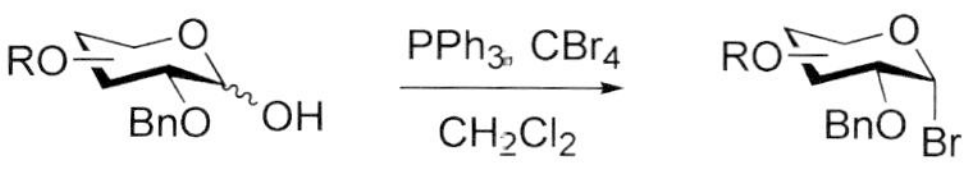

Scheme 2.37

The treatment of 2-O-benzylated hemiacetal sugars with Appel agents, triphenylphosphine/carbon tetrabromide (PPh$_3$ + CBr$_4$), in dichlormethane generates the corresponding 2-OBn α-glycosyl bromides, which are too reactive to purify (Scheme 2.37) [136]. The addition of diethyl ether to the reaction mixture precipitates the side product triphenylphosphine oxide (Ph$_3$PO), which can be filtered and the crude bromide is then obtained by evaporation.

Polat and Linhardt reported a unique reagent combination of zinc triflate and benzoyl bromide for one-pot conversion of benzyl ethers to the corresponding benzoates. In these reactions, methyl or p-methoxyphenyl (OMP) glycosides were converted into perbenzoylated glycosyl bromides in near quantitative yields (Scheme 2.38) [137]. Per-O-benzylated lactose however underwent cleavage of both glycoside bonds generating per-O-benzoylated α-galactosyl and α-glucosyl bromides.

$$\text{R-O-R' + PhCOBr + Zn(OTf)}_2 \longrightarrow \text{PhCOOR' + R-Br}$$

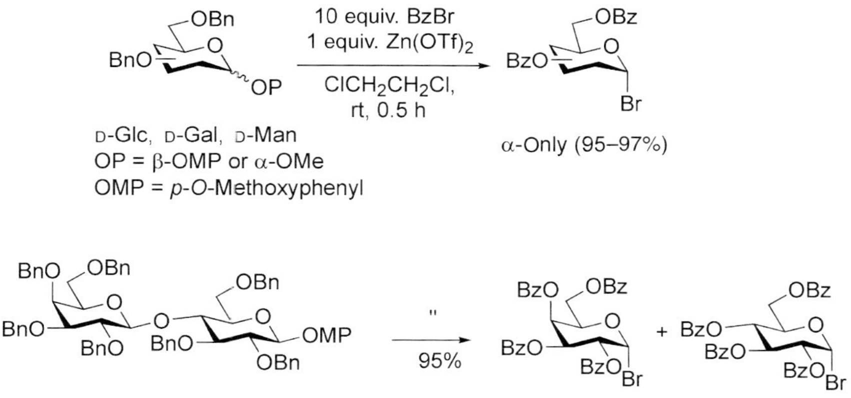

Scheme 2.38

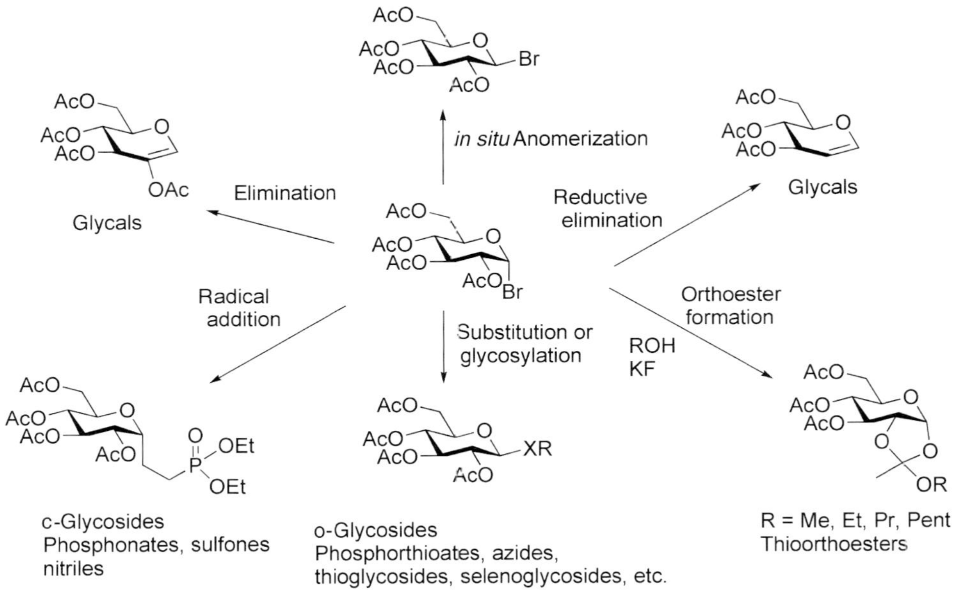

Scheme 2.39

2.2.3.2 Reactivity Patterns and Some Useful Reactions of Glycosyl Bromides

Apart from direct glycosylations, glycosyl bromides can be converted into a panoply of synthons for diverse applications (Scheme 2.39). For example, glycals are useful synthetic precursors for the synthesis of glycosides (via Ferrier glycosylation or epoxides), aminoglycosides and oligosaccharides. Acetylated glycosyl bromides (pyranose, furanose mono- and disaccharides) form glycals via reductive elimination under various conditions [138] including a simple electrochemical setup [139]. Glycosyl bromides are susceptible to halide-assisted anomerization. Alternatively, bromides can be converted into 1,2-orthoesters, yet another synthetically useful entity, generated conventionally in the presence of a sterically hindered base or more recently by using potassium fluoride in acetonitrile at 50 °C [140].

Acetylated α-glycosyl bromides can be converted into β-anomeric azides or thioglycosides using NaN$_3$ or RSH in the presence of tetrabutylammonium hydrogen sulfate (TBAHS) via a one-pot protocol starting from free sugars under phase transfer catalysis [141]. Azidolysis has also been shown to be accelerated under sonication-mediated conditions (5–10 min, 99%) [142]. The 1-azido derivatives can be subsequently transformed into dipeptides by incorporating sugar amino acids [143] and novel β-linked N-glycoside neoglycotrimers employing the Staudinger–aza-Wittig process or click chemistry [144].

Anomeric bromides can be converted into other important common glycosyl donors such as 1,2-*trans* selenoglycosides with indium(I) iodide-mediated cleavage of diselenides [145]. Seleno- and thioglycosides are also obtained from zinc/zinc chloride mediated cleavage of dichalconides [146], through thiophenolysis under phase transfer [141,147] as well as homogeneous conditions [148], and *n*-pentenyl glycosides [149]. Glycosyl bromides have also been transformed into phosphorothioates using microwave techniques [150]. Various S- [151] and N-glycosylated [151,152] heterocycles have been accessed through glycosyl bromides as well.

Tin-catalyzed radical reactions of glycosyl bromides (D-Glc, D-Gal and L-Fuc) with diethyl vinylphosphonate have been shown to proceed in a diastereoselective fashion leading to the formation of corresponding α-linked C-glycosides [153], which were elaborated to the C-glycosyl analogs of natural NDP sugars as glycosyltransferase inhibitors [154]. Radical reactions of glycosyl bromides with benzothiazoyl vinyl sulfone afford α-C-glycoside sulfones with 60–73% yields [155].

2.2.3.3 Stereoselective Glycosylations Employing Glycosyl Bromides and Applications

Since the introduction of the first stereoselective glycosylation protocol more than a century ago, the Koenigs–Knorr reaction [119], glycosyl bromides have remained the most extensively used donors in glycosidation reactions. Anomeric selectivity is mainly controlled by the nature of the C-2 substituent; an ether-type group allows the formation of 1,2-*cis* glycosides owing to the anomeric effect, whereas ester-type groups lead to the formation of 1,2-*trans* glycosides through neighboring-group participation. Other factors such as solvent participation, temperature and metal chelation also play an important role.

Over the years, several modifications including the use of Hg(II) salts and AgOTf as catalysts have been incorporated into the protocol that originally involved Ag$_2$CO$_3$ as an acid scavenger. These catalysts are especially suitable for solid-phase synthesis as exemplified by the synthesis of O-linked glycopeptide analogs of Enkephalin (Scheme 2.40) [156]. The synthesis of 18 N-α-Fmoc-amino acid glycosides, for solid-phase glycopeptide assembly, was carried out from either the corresponding O'Donnell Schiff bases or the N-α-Fmoc-amino-protected serine or threonine and the appropriate glycosyl bromide using Hanessian's modification of the Koenigs–Knorr method utilizing AgOTf. The observed differences in the

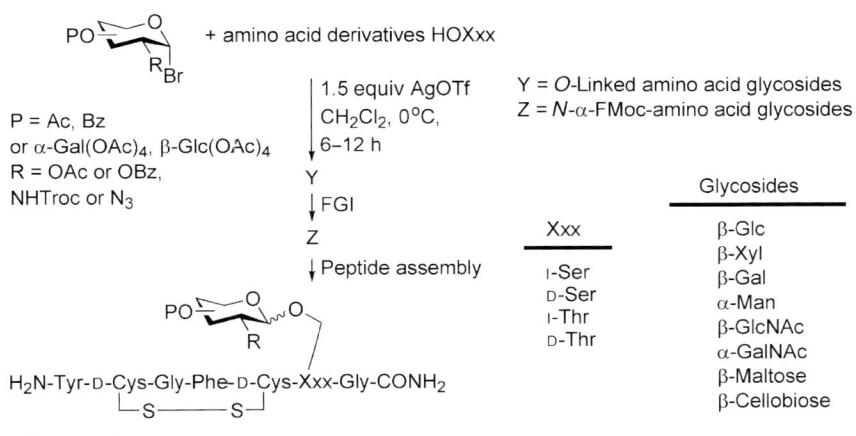

Scheme 2.40

reaction rates of D-glycosyl bromides with the L- and D-forms of serine and threonine were rationalized in terms of the steric interactions within the two types of diastereomeric transition states for the D/L and D/D reactant pairs. The N-α-Fmoc-protected glycosides [monosaccharides Xyl, Glc, Gal, Man, GlcNAc and GalNAc; disaccharides Gal-β-(1–4)-Glc (lactose), Glc-β-(1–4)-Glc (cellobiose) and Gal-α-(1–6)-Glc (melibiose)] were incorporated into 22 enkephalin glycopeptide analogs. Fluorobenzoyl groups have been successfully used as alternatives to benzoyl or acetyl groups in solid-phase synthesis, suppressing β-elimination of base-sensitive O-serine-linked glycopeptides during base-catalyzed deacylation [157]. AgOTf-catalyzed glycosylations also work well in solution, as exemplified by syntheses of the spacer-armed pentasaccharide sialyl lacto-N-neoteraose and sialyl lacto-N-tetraose. In a systematic study of glycosylating N-trichloroacetyl-D-glucosamine derived mono- and disaccharide donors with disarmed acceptors (galactose, lactose, and lactosamine), AgOTf-activated acetylated glycosyl bromide donors displayed the best results among other types of donors [158].

Another modification of the Koenigs–Knorr method that uses HgCN$_2$ has continued to find applications in the synthesis of challenging O-disaccharides including 3-β-D-glucopyranosyl-D-glucitol [159] and more recently in the total synthesis of a bioactive cerebroside (Scheme 2.41) [160]. Hg(CN)$_2$ promoted coupling of acetylated glucosyl bromide with a fully functionalized ceramide acceptor afforded β-linked O-glycoside (50% yield), which upon global deprotection afforded the target cerebroside. Recently, glycosylations of acetylated glycosyl bromides were shown to be promoted with Lewis acid catalysis using InCl$_3$ under essentially neutral conditions, affording 1,2-trans glycosides and disaccharides in good yields [161].

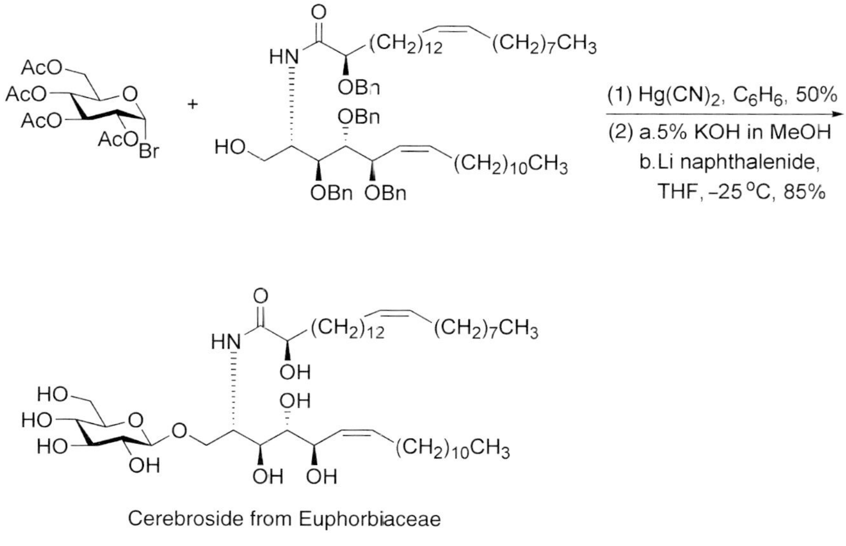

Cerebroside from Euphorbiaceae

Scheme 2.41

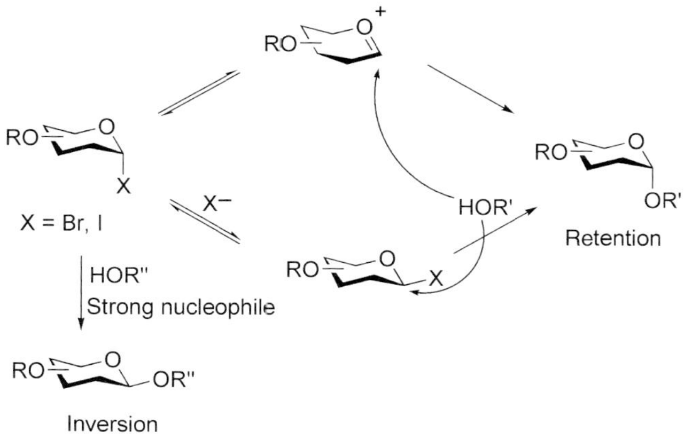

Scheme 2.42

Installation of a 1,2-cis-glycosidic bond is more challenging as compared to the 1,2-trans linkage [162]. Although, a nonparticipating group at C-2 of a glycosyl donor favors the formation of 1,2-cis glycoside by virtue of the anomeric effect, α-selectivity is often only moderate. In principle, S_N2 displacement of a 1,2-trans donor should furnish a 1,2-cis glycoside. However, this is often complicated by partial intervention of S_N1-like transition states giving α/β-mixtures. Pioneering work by Lemieux et al. [163] revealed new pathways for stereoselective 1,2-cis glycosylation through halide-catalyzed in situ anomerization. Accordingly, reactions of thermodynamically stable α-glycosyl bromides with tetrabutylammonium bromide generated the more reactive β-glycosyl bromides, which reacted with various alcohols under neutral conditions to afford α-glycosides, stereoselectively (Scheme 2.42).

Stable thioglycosides can be readily converted into α-glycosyl bromides, which upon in situ anomerization could be coupled with thioglycoside acceptors with high α-selectivity or, conversely, with high β-selectivity using AgOTf via neighboring-group participation. This two-stage activation provides a useful tool for stereoselective orthogonal glycosylation, the merits of which can be gauged from Oscarson's recent syntheses of monodeoxy analogs of an α-linked branched trisaccharide Glcp-(1 → 3)-α-D-Manp(1 → 2)- α-D-ManpOMe [164]. This approach was further exploited in the synthesis of oligosaccharides corresponding to Vibrio cholerae – a spacer equipped tetrasaccharide α-L-Colp-(1 → 2)-β-D-Galp(1 → 3)-[α-L-Colp-(1 → 4)]-β-D-Glcp-NAc (colitose = 3,6-dideoxy-L-xylo-hexose), containing a 4,6-cyclic phosphate in the galactose residue (Scheme 2.43) [165]. Thus, galactose β-thioglycoside was first converted into an α-anomeric bromide and subsequently coupled with a glucosyl thioglycoside acceptor in the presence of AgOTf to obtain a β-linked disaccharide. Selective deacylation and reductive ring opening of the benzylidene acetal at O-4 afforded a diol acceptor that underwent double glycosylation with a L-colitose donor under in situ anomerization conditions affording a protected tetrasaccharide with high α-stereoselectivity and good yields.

Two alternate protocols for the in situ anomerization procedure have been introduced by Kobayashi and Nishida, both involving Appel agents [166–168]. According to

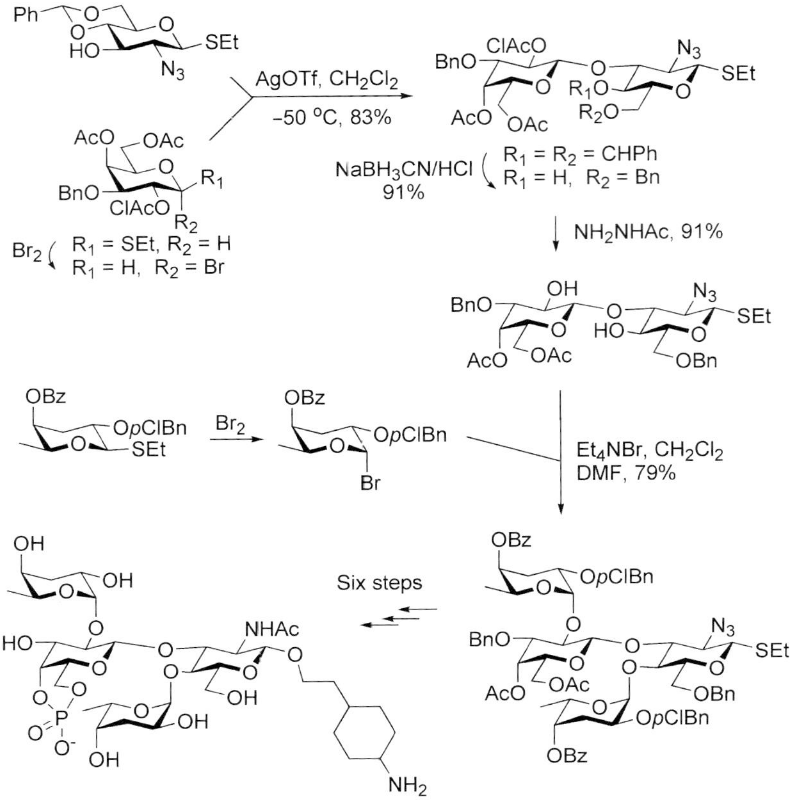

Scheme 2.43

the first pathway (Scheme 2.44), the reaction of a 2-*O*-benzyl-1-hydroxy sugar with CBr$_4$ and PPh$_3$ generates a reactive glycosyl bromide *in situ* [136], which is subsequently coupled with an acceptor in the presence of Br$^-$ and *N*,*N*-tetramethylurea (TMU) at room temperature to afford α-glycoside quantitatively [166]. This reagent

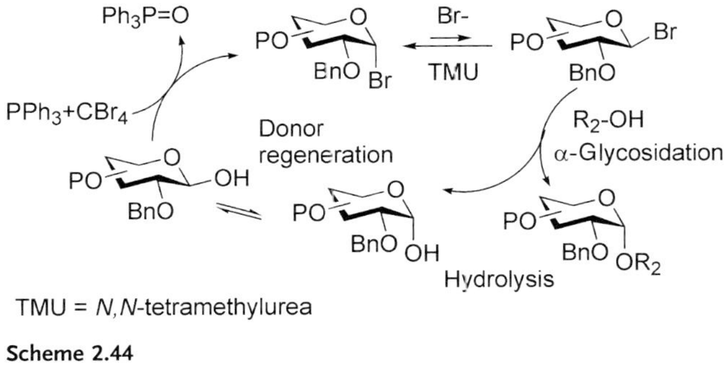

TMU = *N*,*N*-tetramethylurea

Scheme 2.44

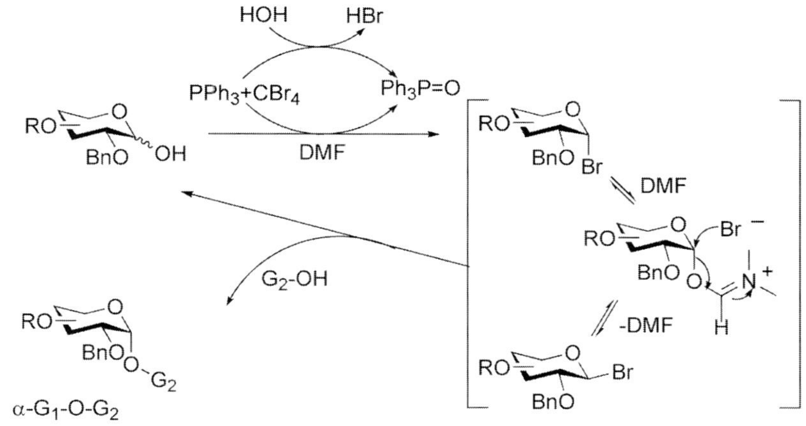

Scheme 2.45

G_1 or G_2 = D-Glc, D-Gal, L-Fuc

combination plays multiple roles including the glycosyl bromide formation, *in situ* anomerization and glycosylation. It also serves to scavenge water allowing the reactions to be performed without special attention to moisture. The reaction works well for D-gluco, D-galacto and L-fuco donors using various acceptors. The second dehydrative protocol employs DMF as a solvent, which obviates the necessity to add TMU (Scheme 2.45) [167]. On the basis of NMR studies, a DMF–glycosyl adduct (cationic α-glycosyl imidate-like [168] intermediate with Br$^-$ counterion) is implicated in this reaction.

Glycosyl bromides have also proven to be excellent donors for highly regioselective glycosylation of flavonols leading to a total synthesis of numerous natural products [120]. For example, peracetylglucosyl bromide was regioselectively coupled with naringenin (Scheme 2.46) under classical Konigs–Knorr conditions resulting in 80% yield [169]. Consecutive coupling with a glucosyl fluoride also proceeded regioselectively at O-4′. On the basis of these results, Kondo and coworkers reported the first total synthesis of apigenin 7,4′-di-O-β-D-glucopyraoside, a component of the blue pigment, protodelphin, along with seven chiral analogs [169,170]. Partially protected quercetin (Scheme 2.47) was regioselectively coupled with a glucosyl bromide at O-3 under basic conditions to afford the 3-O-β-D-glycoside in 54% yield. The glycoside was further transformed into quercetin-3-O-β-D-glucuronide via benzylation, deacetylation, TEMPO oxidation and hydrogenation [171]. Yu [172], and subsequently Linhardt [173], employed phase transfer conditions to effect regioselective O-3 glycosylation of 3,5,3′ and 3,5,4′-triols of the quercetin nucleus using various mono- and disaccharide glycosyl bromides (Scheme 2.48) [172–176]. Phase transfer conditions have also been used for the coupling of glycosyl bromides in solid-phase synthesis [177].

Very recently, direct displacement of acylated glycosyl bromides (D-Man and L-Fuc) with nucleotide 5′-diphosphates has been shown to proceed stereoselectively

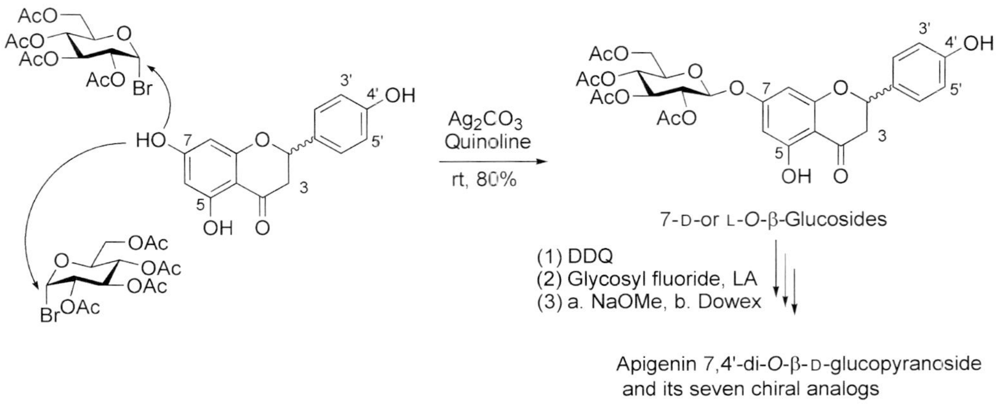

74 | *2 Glycoside Synthesis from Anomeric Halides*

Scheme 2.46

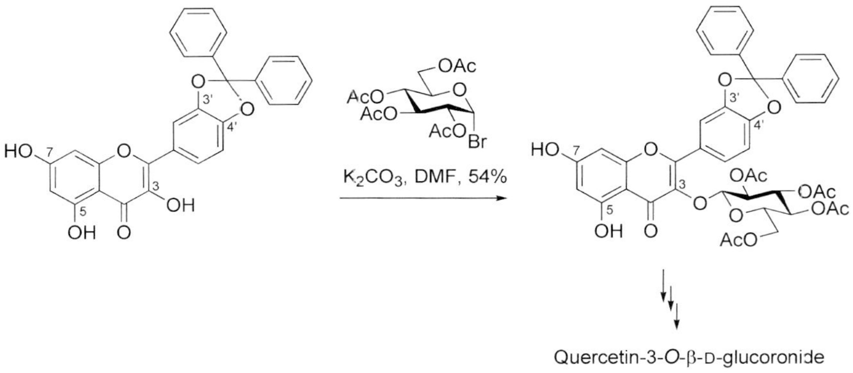

Scheme 2.47

through classical neighboring-group participation. This route is utilized for the preparation of diastereomerically pure α-D-manno and β-L-fuco-linked sugar nucleotide diphosphates UDP and GDP (Scheme 2.49) [178].

2.2.4
Glycosyl Iodides

Over the past decade, glycosyl iodides [122] have clearly become an important reactive intermediate in carbohydrate synthesis. Their reactivity and stability can be tuned by altering the protecting-group pattern. Thus, per-*O*-silylated iodides, which are usually generated *in situ*, are on the extreme high side of the reactivity scale, partially benzylated iodides possess intermediate reactivity and can be stored for longer times at subzero temperatures, whereas per-*O*-acylated glycosyl iodides are stable crystalline solids with long shelf life. The unique reactivity profile of

Scheme 2.48

glycosyl iodides can be advantageously exploited for solvolysis and stereospecific S_N2 glycosylations.

2.2.4.1 Preparation of Glycosyl Iodides

Several methods are available to access glycosyl iodides (Scheme 2.50). Anomeric hemiacetals bearing diverse protecting groups (Bn, Bz, Ac, N_3, CMe_2) upon treatment with a polymer-bound triphenylphosphine–iodine complex and imidazole can be converted into α-glycosyl iodides [179]. The precipitated by-products,

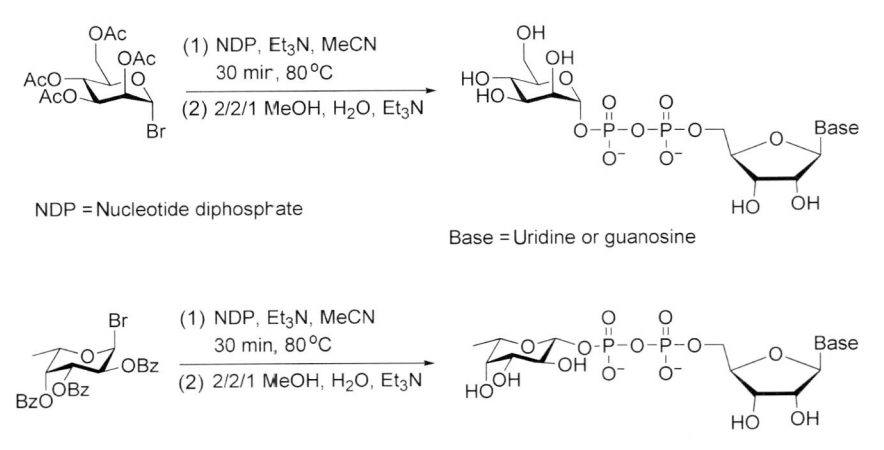

Scheme 2.49

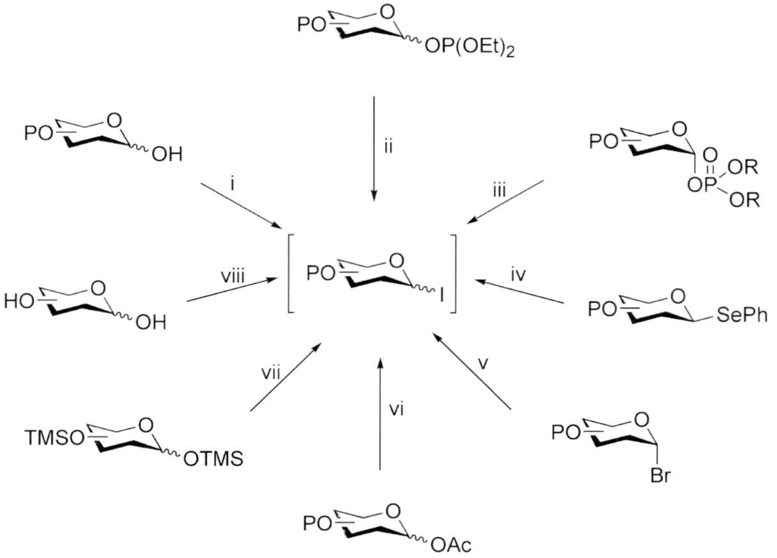

(i) -C$_6$H$_4$-PPh$_2$-I$_2$, ImH or PPh$_3$-I$_2$, ImH
(ii) DTBPI, 4 Å MS, CH$_2$Cl$_2$
(iii) 1M LiClO$_4$, Li(Na)I, 4 Å MS, CH$_2$Cl$_2$
(iv) I$_2$, CDCl$_3$
(v) I$_2$ or IBr or ICl or NIS
(vi) TMSI or I$_2$/Et$_3$SiH or HI (I$_2$ + RSH)
(vii) TMSI, CH$_2$Cl$_2$
(viii) a. Ac$_2$O, cat. I$_2$; b. I$_2$, HMDS (TMSI)

Scheme 2.50

that is excess imidazole and solid polymer-bound phosphine oxide, can be removed by filtration through Celite yielding iodides that are pure enough to be used for further reaction. Per-O-benzylated glycosyl diethylphosphites (D-Glc, D-Gal, L-Fuc) have also been used to generate glycosyl iodides using 2,6-di-*tert*-butyl pyridinium iodide (DTBPI) in CH$_2$Cl$_2$ at ambient temperature [180]. Waldman's method [181] employs per-O-benzylated glycosyl phosphates as precursors to glycosyl iodides generated by the reaction of LiI in 1 M solution of LiClO$_4$ in organic solvents, such as CH$_2$Cl$_2$ or CH$_3$CN. Selenoglycosides provide the corresponding iodides upon treatment with molecular I$_2$. NMR monitoring experiments revealed that per-O-benzylated (armed) selenoglucosides are rapidly converted (5 min) into the corresponding iodides, whereas the disarmed counterpart takes 4 days [182]. Acylated bromosugars can be converted into glycosyl iodides by its reaction with iodine [183] or other reagents such as IBr, ICl and NIS [184].

One of the most reliable and commonly used methods to prepare glycosyl iodides involves the treatment of anomeric acetates with TMSI. Thiem and Meyer introduced this method to generate iodides from anomeric acetates, acetals, methyl glycosides or anhydro sugars [185]. In this reaction, the acetate is first activated by silylation and concomitantly undergoes displacement to generate anomeric iodide. In the first mechanistic studies with glycosyl iodides [186], it was shown that α-iodides are stereoselectively formed from β-anomeric acetates. Conversely, β-glycosyl iodides are the initial products derived from α-acetates but β-iodide readily converts into the thermodynamically more stable α-anomer. Per-O-acetates of mono- and disaccharides can also be transformed into α-glycosyl iodides upon treatment with HI, generated *in situ* by the reaction of solid iodine and thiol [187], or by I_2/triethylsilane [188]. Per-O-trimethylsilylated mono- [189,190] and disaccharides [191] also undergo conversion to the corresponding glycosyl iodides upon treatment with TMSI. A recent procedure for one-pot preparation of glycosyl iodides from free hexoses involves per-O-acetylation followed by the treatment with I_2/hexamethyl disilane (TMSI generated *in situ*) [192]. Among all these procedures, TMSI is often the reagent of choice due to the ease of removing volatile by-products (TMSOMe, TMSOTMS or TMSOAc), which can compete as acceptors if left in the reaction mixture.

2.2.4.2 Reactions of Glycosyl Iodides

Nucleophilic Anionic Substitutions Classical Koenigs–Knorr glycosylation and variants thereof involves metal chelation for halide activation and often proceeds through oxonium ion formation, allowing the stereochemical outcome to be dictated by the C-2 substituent on the donor sugar. Alternatively, glycosyl iodides undergo direct displacement through an S_N2-like mechanism. In an attempt to develop efficient nonmetal-catalyzed glycosylations, anionic additions to glycosyl iodides were studied [193]. These reactions proceeded with inversion of configuration at the anomeric center to give β-glycosides even in the absence of a C-2 participating group, with the following order of reactivity –2,3,4,6-tetra-O-benzyl-α-D-galactosyl iodide > 2,3,4,6-tetra-O-benzyl- α-D-glucosyl iodide > 2,3,4,6-tetra-O-benzyl- α-D-mannosyl iodide. Glycal formation was observed with glucosyl and galactosyl iodides when highly basic anions were employed whereas no elimination took place with mannosyl iodides. A variety of nucleophiles such as malonate, CN^-, N_3, phthalimide, phenoxide and acetate anions were stereoselectively added to glycosyl iodides to afford β-linked C-, N- and O-glycosides in good yields. The formation of the β-mannosyl cyanate was particularly noteworthy (Scheme 2.51).

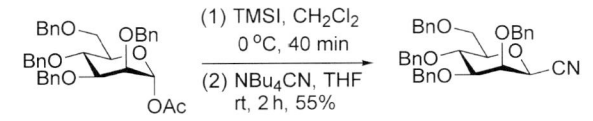

Scheme 2.51

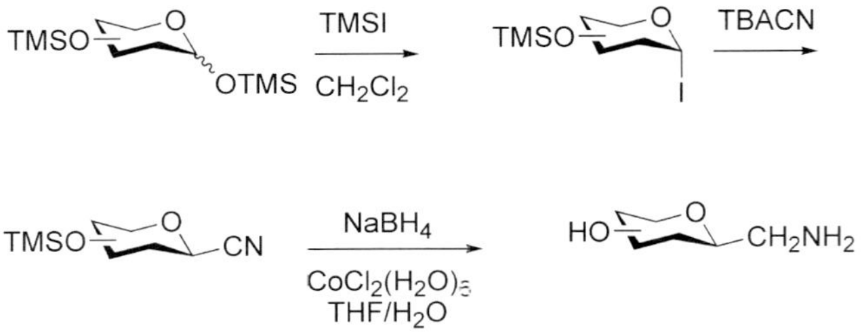

Scheme 2.52

Similarly, per-O-trimethylsilylated mono- and β-linked disaccharides (lactose and cellobiose, not melibiose) could be converted into the corresponding α-glycosyl iodides, which upon S_N2 displacement with CN^- using TBACN mainly afforded β-cyano derivatives in good overall yields [191]. The cyanoglycosides were transformed into aminomethyl glycosides via reduction under mild conditions (Scheme 2.52).

These methods were extended to include disarmed glycosyl iodides as a general method for the synthesis of glycopyranosyluronic acid azides (Scheme 2.53) [194]. Peracetylated mono- (D-Glc, D-Gal and D-Man) and disaccharides (cellobiose, lactose, melibiose) were first treated with TMSI to generate the corresponding glycosyl iodides, which were then reacted with $TBAN_3$ or tetramethylguanidium azide (TMGA) in CH_2Cl_2 to afford β-anomeric azides in good yields. These azides were transformed into the corresponding uronic acids after deacetylation and low-temperature TEMPO oxidation. Along similar lines, glycosyl iodides of dimethylmaleolyl (DMM) or phthaloyl (Phth)-protected D-glucosamine were generated and coupled with various nucleophiles (alcohols – MeOH, AllOH, i-PrOH and BnOH without promoter, sugar-OH with AgOTf, PhSH, and allyl-TMS as well as $TMSN_3$ with $BF_3 \cdot OEt_2$) to obtain β-linked O-, S-, C-, and N-glycosides in good yields (Scheme 2.54) [195]. This reaction goes through the intermediacy of an unstable β-iodide as evidenced by NMR studies. The stereochemical outcome of the glycosylation is believed to be controlled by neighboring-group participation of the C-2 functionality.

S_N2 reactions of glycosyl iodides have proven especially advantageous in the synthesis of 2-deoxy β-O-aryl-D-glycosides. This is a challenging linkage to make, as there is no neighboring group to participate. Sometimes, stereochemistry is

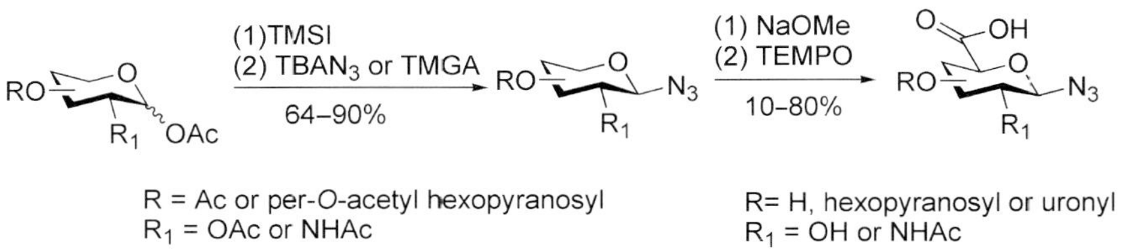

R = Ac or per-O-acetyl hexopyranosyl
R_1 = OAc or NHAc

R = H, hexopyranosyl or uronyl
R_1 = OH or NHAc

Scheme 2.53

2.2 Glycosyl Chlorides, Bromides and Iodides | 79

Scheme 2.54

R$_1$ = Ac, R$_2$ = DMM
R$_1$ = TMS, R$_2$ = DMM
R$_1$ = Ac, R$_2$ = Phth

R$_1$ = Ac, R$_2$ = DMM
R$_1$ = TMS, R$_2$ = DMM
R$_1$ = Ac, R$_2$ = Phth

R$_2$ = DMM or Phth

β-Glycosides

Nu = MeOH, *i*-PrOH, AllOH, BnOH,
Sugar–OH, PhSH, AllylTMS, TMSN$_3$

controlled by temporary introduction of a stereo-directing functionality at C-2, and the C-2 functional group is removed after glycosylation. Direct displacement of α-glycosyl iodides obviated the necessity for a C-2 directing group. The conversion of glycals to 2-deoxy glycosyl acetates followed by the reaction with TMSI readily afforded the corresponding α-glycosyl iodides, which underwent facile S$_N$2 reactions with aryl alkoxy anions (*o*-crysol or 2-naphthol) to provide aryl β-2-deoxy-glycosides in good yields (Scheme 2.55) [196].

Stereoselective Glycosylations of Glycosyl Iodides General methods for α-selective glycosylation of glycosyl iodides via *in situ* anomerization have been established [197]. Armed glycosyl iodides undergo the reaction in the presence of tetrabutylammonium iodide (TBAI) and Hünig's base (diisopropyl ethylamine DIPEA) with

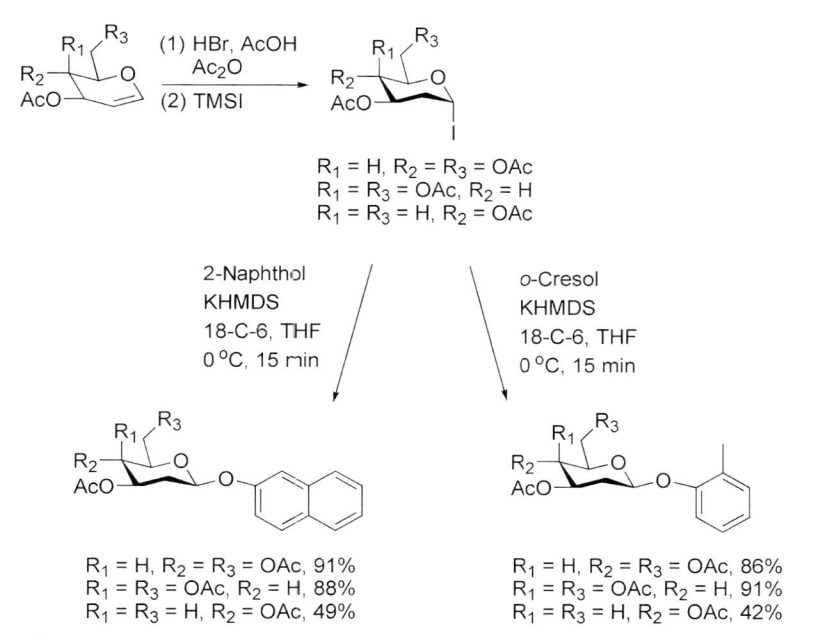

R$_1$ = H, R$_2$ = R$_3$ = OAc
R$_1$ = R$_3$ = OAc, R$_2$ = H
R$_1$ = R$_3$ = H, R$_2$ = OAc

2-Naphthol
KHMDS
18-C-6, THF
0 °C, 15 min

o-Cresol
KHMDS
18-C-6, THF
0 °C, 15 min

R$_1$ = H, R$_2$ = R$_3$ = OAc, 91%
R$_1$ = R$_3$ = OAc, R$_2$ = H, 88%
R$_1$ = R$_3$ = H, R$_2$ = OAc, 49%

R$_1$ = H, R$_2$ = R$_3$ = OAc, 86%
R$_1$ = R$_3$ = OAc, R$_2$ = H, 91%
R$_1$ = R$_3$ = H, R$_2$ = OAc, 42%

Scheme 2.55

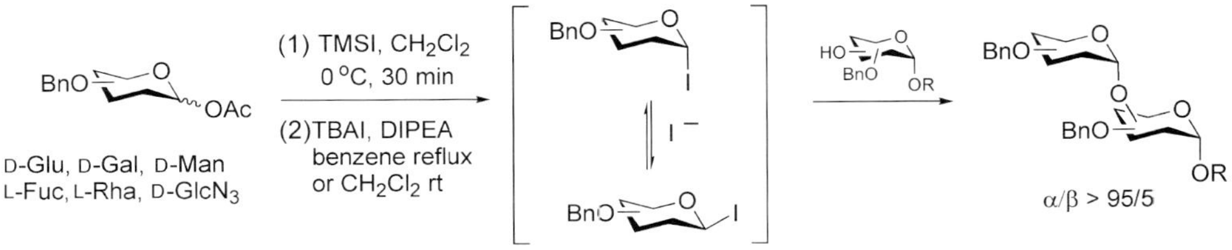

Scheme 2.56

various acceptors (Scheme 2.56), including hemiacetals, 6-OH and sterically hindered secondary sugar acceptors. Under these conditions, the first formed α-iodide undergoes attack by I⁻ to generate the thermodynamically unstable β-iodide, which being orders of magnitude more reactive than the corresponding α-iodide instantaneously reacts with nucleophilic acceptors to form α-glycosidic linkages. Such glycosylations involving anomeric iodides were found to offer advantages over the corresponding bromides in terms of time, yield and overall efficiency. The donor iodides follow the following order of reactivity: L-Fuc > D-Gal > D-Man > D-Glc. Thus, glucosyl iodides show higher reactivity than mannosyl iodides in direct nucleophilic displacement reactions, whereas this order is reversed in TBAI-promoted glycosylations. Solvent effects have also been observed in these glycosylations when using acetonitrile as a participating solvent and allyl alcohol as an acceptor. Intriguingly, per-O-benzylated glucosyl iodide afforded the corresponding β-allyl glucoside as the major isomer (α/β = 1 : 10), whereas per-O-benzylated galactosyl iodide rapidly and exclusively generated α-galactoside in acetonitrile. The mannosyl iodide furnished 1/1 α/β mixture under identical conditions [197]. These observations could be rationalized on the basis of the relative rate of formation of oxonium ion versus nitrilium ion intermediates.

Analogous to the *in situ* anomerization method, α-selectivity is also achieved by adding triphenylphosphine oxide in place of TBAI as a promoter, as first established with glycosyl iodides [198,199] (Scheme 2.57) and subsequently with glycosyl bromides [200]. Upon treatment with Ph₃PO, benzylated glycosyl iodides are believed to generate transient glycosyl phosphonium iodides. Glycosyl acceptors predominantly react with the β-form to afford α-disaccharides in very high yields and with high

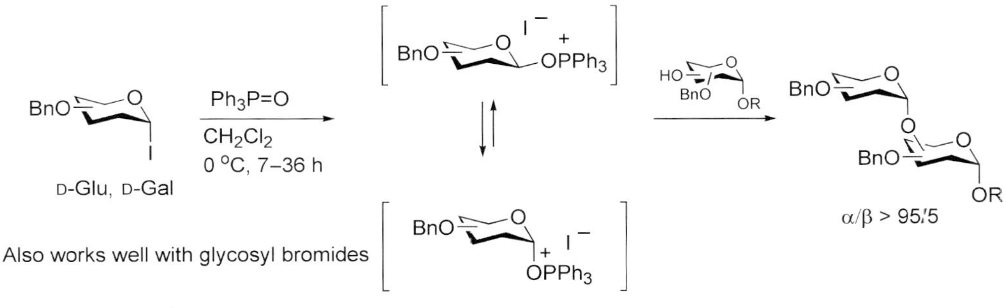

Scheme 2.57

stereoselectivity. This procedure also works well with glycosyl bromides but requires longer reaction times.

Field and coworkers [201] used iodine for the activation of per-O-acetylated glycosyl iodides. Although disarmed glycosyl iodides are activated by I_2, the stereochemical outcome is dominated by the nature of the O-2 protecting group and the reactivity of the acceptor. For example, glycosylations between acetate protected donors and reactive acceptors like methanol gave exclusively α-product, whereas per-O-benzoylated iodides gave only β-linked products. Good α-selectivity was also observed with 2-deoxy-2-azido donors using serine and threonine acceptors (Scheme 2.58). The α-linked products in per-O-acetate sugars presumably arise through S_N2 displacement of an β-iodide existing at equilibrium in the reaction mixture (Scheme 2.59). Per-O-benzoylated donors have more effective neighboring-group participation yielding β-isomers. Donors bearing nonparticipating azides at C-2 mainly give α-isomers, and the selectivity decreases with decreasing reactivity of the acceptor.

Per-O-acylated glycosyl iodides are stable at room temperature and can be purified on a silica gel column and stored at 0 °C. Stachulski and coworkers [202] synthesized methyl 2,3,4-tri-O-pivaloyl-glucopyranuroate iodide, which is a stable solid at 20 °C and can be stored for months at room temperature or for more than a year at 0 °C. The X-ray crystal structure of this compound, the first one of this class, shows a typical chair structure. Importantly, such a disarmed and stable iodide can be coupled with primary and secondary steroidal alcohols using I_2 as a promoter, as demonstrated by the synthesis of morphine-6-glucuronide, an analgesic [202]. The glycosyl donor ability

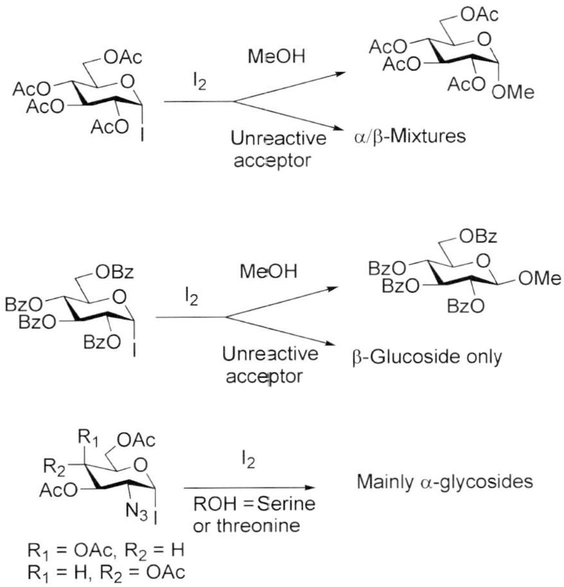

Scheme 2.58

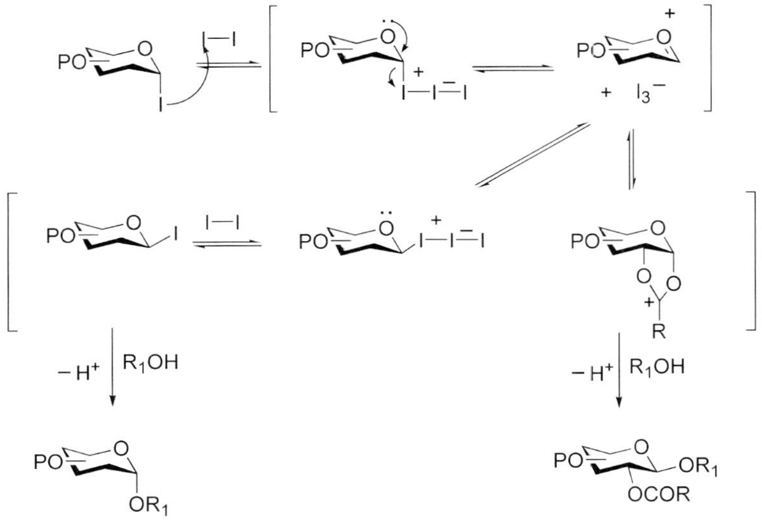

Scheme 2.59

of the iodide is contrasted with the corresponding bromide, which gives a yield of only 20% for the coupling with 3-O-pivaloyl morphine acceptor compared to 55% yield obtained with the iodide. This iodide donor could also be coupled with several other steroidal alcohols using NIS or metal salts as promoters [203]. In a similar fashion, the iodide glycosyl donor can be coupled with disarmed sugar acceptors in the presence of NIS/I_2/TMSOTf, FeCl$_3$/I_2 or CuCl/I_2 to obtain β-linked disaccharides in good yields (Scheme 2.60) [204]. Disarmed glycosyl iodides (and bromides) have also been used to

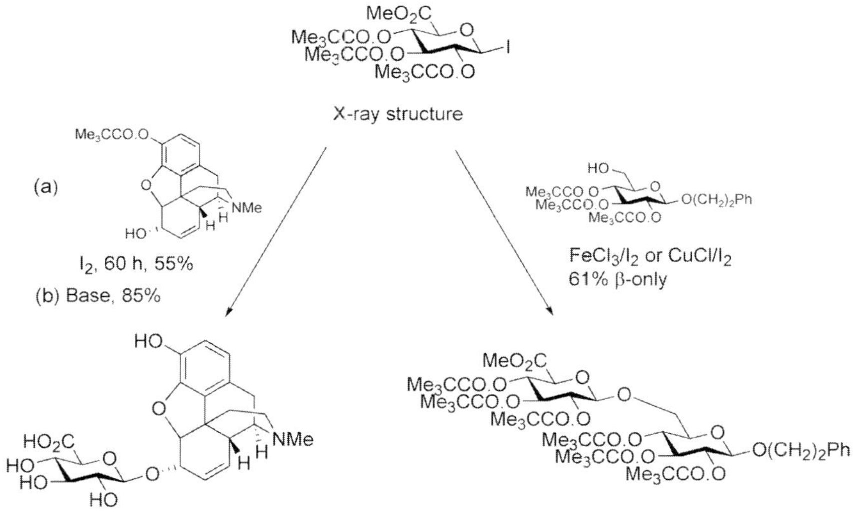

Scheme 2.60

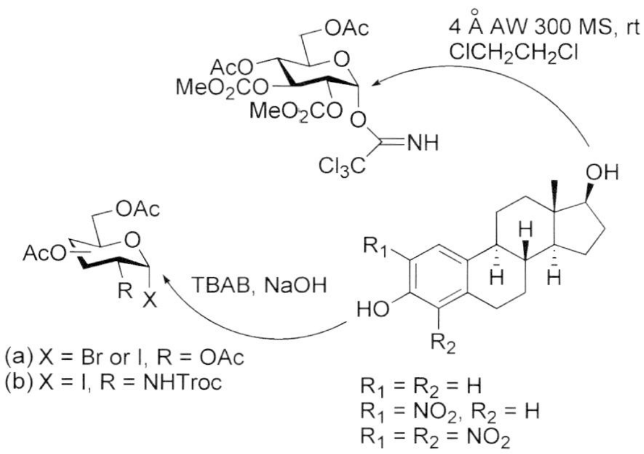

Scheme 2.61

(a) X = Br or I, R = OAc
(b) X = I, R = NHTroc

$R_1 = R_2 = H$
$R_1 = NO_2, R_2 = H$
$R_1 = R_2 = NO_2$

achieve regioselectivity in the glycosylation of 17β-estradiol and its derivatives [205]. Glycosyl iodides undergo regioselective glycosylation of the phenolic alcohol under phase transfer catalysis conditions, whereas trichloroacetimidates selectively couple with the carbinol under mild activation with 4 Å acid-washed molecular sieves (Scheme 2.61).

α-Selective Glycosylation: Applications to Oligosaccharide and Glycolipid Synthesis   *In situ* anomerization has been successfully applied in α-glycosylations of orthogonally protected armed glycosyl iodides for oligosaccharide synthesis under solution- [206–209] and solid-phase [207] conditions. α-(1→6)-Linked glucosyl homooligomers (isomaltobiose) were synthesized with high yields (84–94% for each coupling), giving α-glycoside as the only product in each step. A 1 + 1 + 1 iterative coupling strategy utilizing a 1,6-di-O-acetyl-2,3,4-tri-O-benzyl glucopyranoside monomeric unit and convergent 2 + 2 + 2 and 2 + 4 couplings was equally successful [206,207]. Notably, the corresponding glycosyl iodide could be stored under argon in benzene solution in the refrigerator for a month without significant degradation. Solution-phase synthesis via 1 + 1 + 1 strategy began with the glucosyl iodide donor that was first coupled with a 6-OH thioglucoside acceptor to afford α-1→6-linked disaccharide, which upon selective de-O-acetylation generated the corresponding 6-OH disaccharide acceptor. Iterative coupling with the same iodide donor furnished tri- and tetrasaccharides in very high yields (Scheme 2.62) [206]. Under solid-phase conditions, the 6-OH thioglucoside acceptor was first linked to tentagel NH$_2$ resin via amide bond formation with the anomeric thioglycolic acid and then the coupling–deacetylation sequence was repeated [207].

In the 2 + 2 + 2 strategy, the α-1→6-inked disaccharide was first assembled from the glucosyl iodide donor and 6-OH glucoside acceptor bearing an anomeric acetate. This disaccharide was then used to generate iodide, which was employed in the coupling with the disaccharide thioglycoside acceptor. Repetitive coupling and deactylation sequences on the tetrasaccharide afforded hexasaccharide as only the

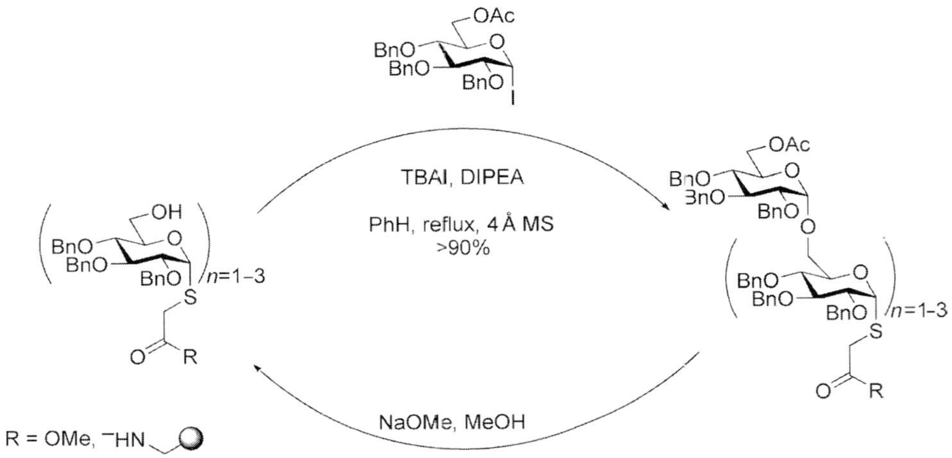

Scheme 2.62

α-anomer in high yield (Scheme 2.63). Similarly, the 2 + 4 strategy furnished the corresponding hexasaccharide in high yields.

It should be emphasized that under these conditions, neither cleavage of inter-glycoside bond by TMSI was observed nor β-isomer was detected in any of the couplings. Although the solid-phase strategy was advantageous in terms of ease of purification, it required the use of excess donor (7.5 equiv per coupling) and longer reaction times (12 h for each coupling) [207]. In contrast, solution-phase reactions

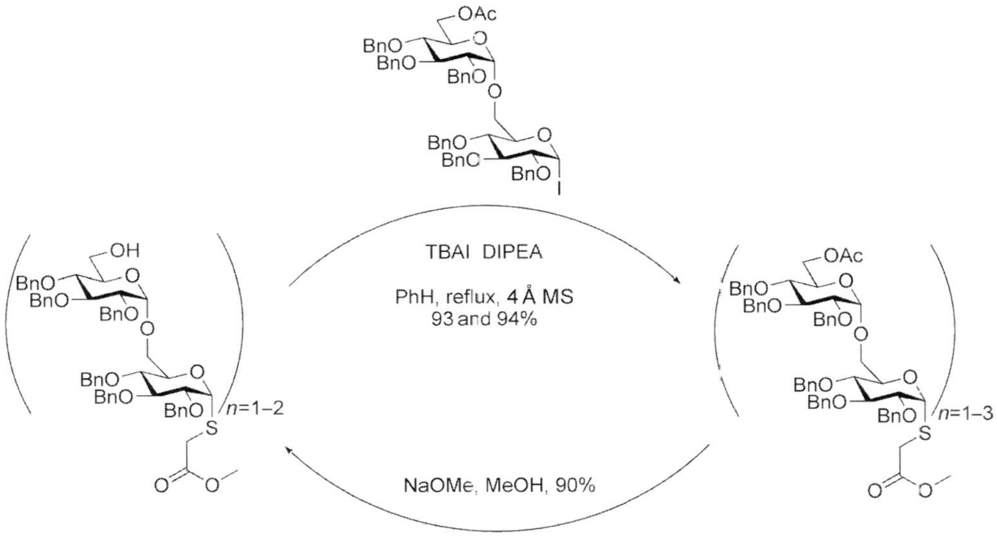

Scheme 2.63

utilized only 2.5 equiv of donor and required 2–3 h for the completion of each glycosylation reaction, making solution phase the preferred strategy for oligomer synthesis.

These studies were extended to develop a highly efficient synthesis of HIV-1-associated glycoprotein (gp120) mannose di-, tri- and pentasaccharides (Man-3 and Man-5). The α-(1 → 6)-linked disaccharide constructs could be prepared in solution from glycosyl iodide precursors with only a slight excess of the iodide donor [208], and this process offers advantages over solid-phase methods that require more than 5 equiv of donor. During the TBAI-assisted reaction, excess glycosyl iodide is converted into a glycal that is not easily separable from the desired disaccharide. This problem could be overcome using a scavenging protocol involving selective epoxidation of the intervening glycal followed by nucleophilic attack (Scheme 2.64) [208]. Alternatively [209], glycosylation of a O-2-acetyl mannosyl iodide donor in the presence of silver triflate at −40 °C furnished the desired disaccharide along with the orthoester, which could be rearranged to the disaccharide by simply warming the reaction to room temperature. The methodology was applied in the synthesis of pentasaccharide (Man-5). Through double glycosylation of a 3,6-dihydroxy acceptor, high mannose sugars were readily obtained in nearly quantitative yields (Scheme 2.65).

The glycosyl iodide methodology has worked especially well in more challenging α-galactosylations [210–212], as demonstrated by the synthesis of a potent immunostimulator α-galactosyl ceramide KRN7000 (Scheme 2.66) [210]. Commonly employed donors such as fluorides, trichloroacetimidates, phosphites and hemiacetals usually furnish difficult-to-separate α-/β-mixtures with yields typically ranging from 30 to 70%. In contrast, (2S,3S)-2-azido-3-para-methoxybenzyl sphingosine and (2S,3S,4R)-2-azido-3,4-para-methoxybenzyl phytosphingosine react with per-O-benzylated galactosyl iodide affording only α-O-glycosidic linkages with yields over 90%. Subsequent conversion of azido groups to an amine, followed by fatty acid coupling and debenzylation along with the reduction of double bond under hydrogenolysis conditions afforded pure KRN7000 and 4-deoxy-KRN7000.

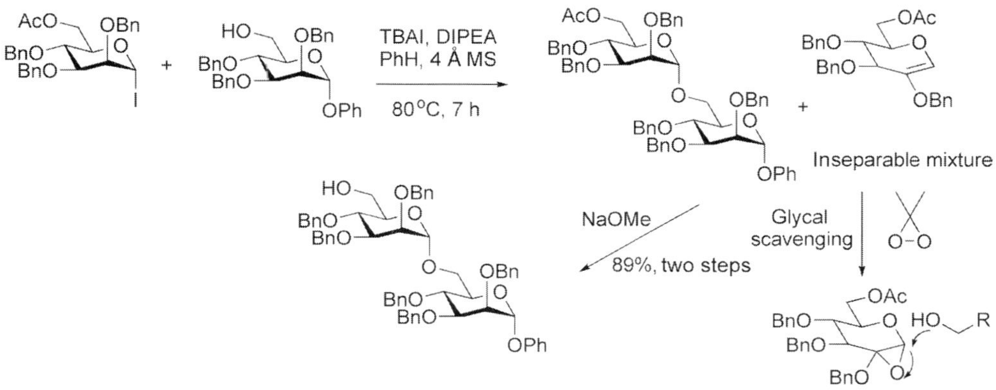

Scheme 2.64

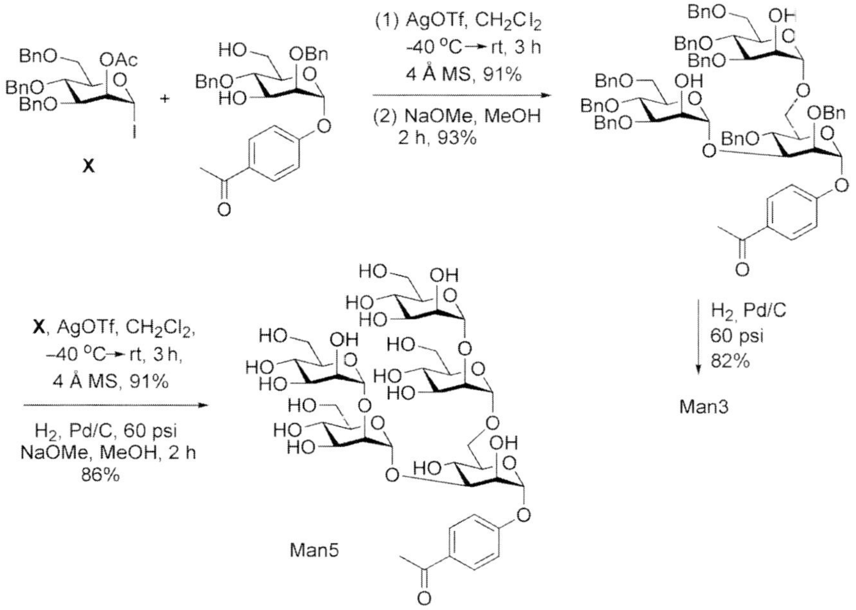

Scheme 2.65

Although this method offered advantages over existing technology, it required several steps to prepare the glycolipid for coupling. A higher degree of efficiency and simplicity was achieved by using per-O-trimethylsilylated (O-TMS) sugars as precursors to glycosyl iodides in a one-pot endeavor [211]. Under these conditions, per-O-TMS galactosyl iodides underwent α-glycosidation with fully functionalized glyceride and ceramide acceptors producing α-linked glycolipids (Schemes 2.42 and 2.43). The treatment of the crude product in the same reaction vessel with acidic resin in methanol afforded biologically relevant biomolecules in high yields and with high stereoselectivity. This mild one-pot protocol allows the synthesis of pure α-anomeric glycolipids bearing various sensitive functionalities such as esters, amides and double bonds (Scheme 2.67) [212]. Microwave radiation has proven useful in these reactions when utilizing lipids having limited solubility.

Hindsgaul and Uchiyama also used *in situ*-generated per-O-TMS fucosyl iodides for α-L-fucosylation [189]. Beau and coworkers employed anomeric-TMS sugars for the synthesis of 1,2-*trans*-C-glycosyl compounds via reductive samariation of glycosyl iodides [190]. Very recently, the first synthesis of indigo N-glycosides (blue sugars) was reported from the reaction of dehydroindigo with *in situ*-generated per-O-TMS L-rhamnosyl, D-glucosyl and D-mannosyl iodides [213].

A useful extension of the *in situ* anomerization process involves the employment of C-nucleophiles such as vinyl and allyl magnesium bromides. Grignard reactions to per-O-benzylated glycosyl iodides proceed stereoselectively when a strong nucleophile like allyl magnesium bromide is used, giving β-C-allyl fucosides (95% β-only)

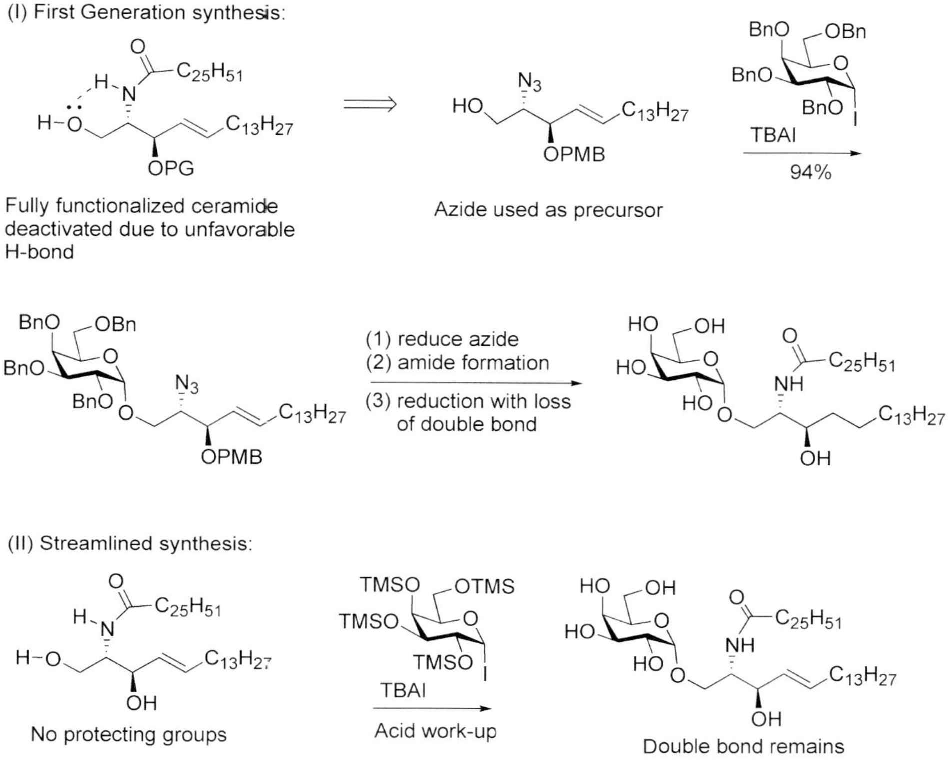

2.2 Glycosyl Chlorides, Bromides and Iodides

(I) First Generation synthesis:

Fully functionalized ceramide deactivated due to unfavorable H-bond

Azide used as precursor

Scheme 2.66

(II) Streamlined synthesis:

No protecting groups

Acid work-up

Double bond remains

and galactosides (85% β-only) in high yields [214]. In contrast, reactions of benzylated α-D-galactosyl iodides with vinyl magnesium bromide generate an α/β mixture favoring the α-isomer. The scenario is reversed when the reaction is carried out under *in situ* anomerization conditions using TBAI in toluene at reflux, in which case the α-isomer is formed in high yields (79%, α/β = 12/1) [215]. This methodology proved useful in the first synthesis of a α-linked C-glycolipid corresponding to the immunoreactive bacterial glycolipid BbGL2 (Scheme 2.68).

β-Mannosylation Using Glycosyl Iodides The unique reactivity of glycosyl iodides was further revealed when glucosyl, galactosyl and mannosyl iodide donors were treated with strained oxacycloalkane acceptors to afford O-glycosides with high β-selectivity (Scheme 2.69) [216]. These reactions proceed without donor activation in CH_2Cl_2 and are highly β-selective with reactive acceptors, such as propylene oxide and trimethylene oxide. Glycosyl iodides are unique in this respect, as analogous reactions with the corresponding bromides failed. These reactions were used for the synthesis of β-thiomannosides from thiocycloalkane acceptors. In the absence of neighboring-group participation, β-selectivity is thought to arise from direct nucleophilic displacement of the α-iodide, whereas the minor α-product may result from

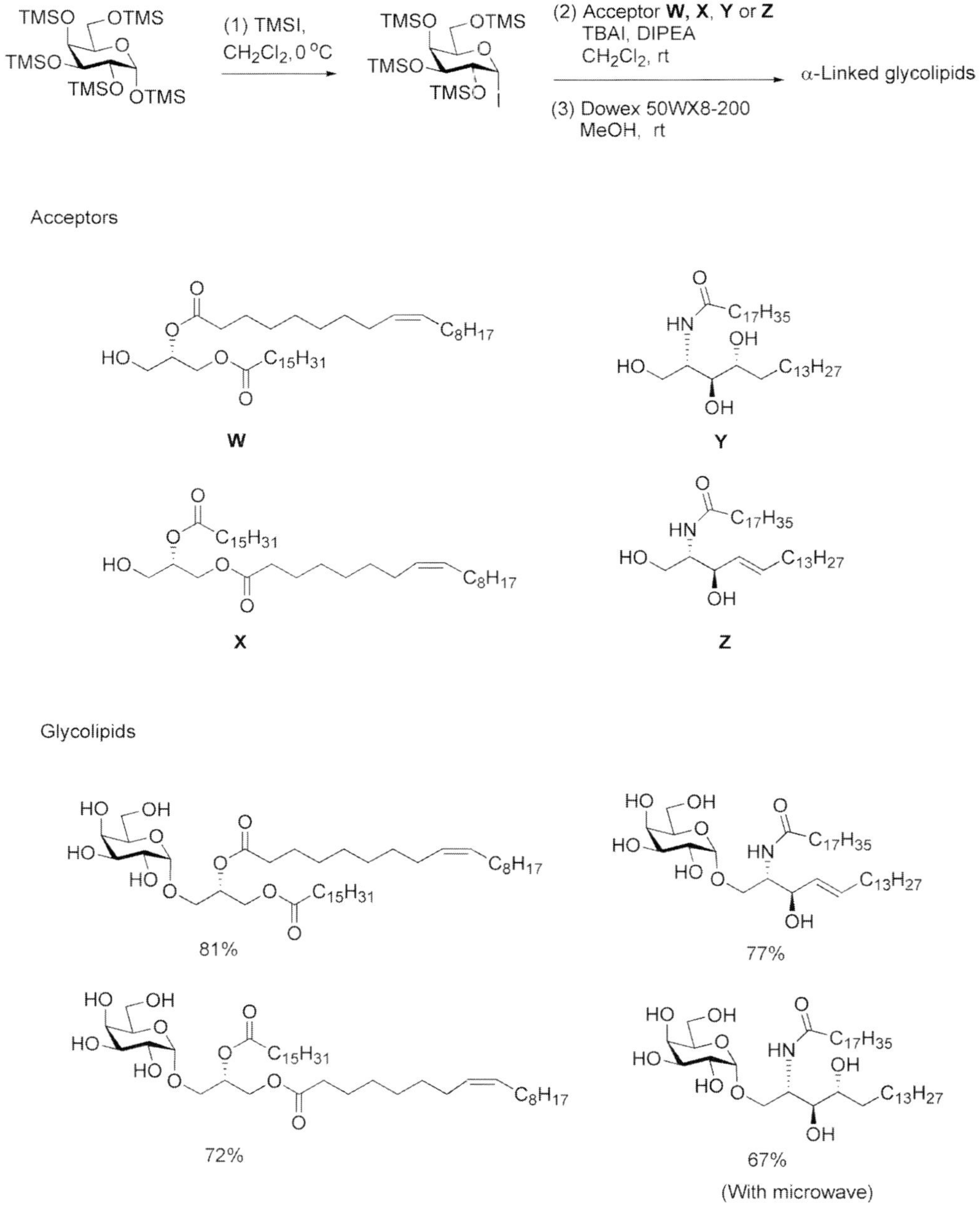

2 Glycoside Synthesis from Anomeric Halides

Scheme 2.67

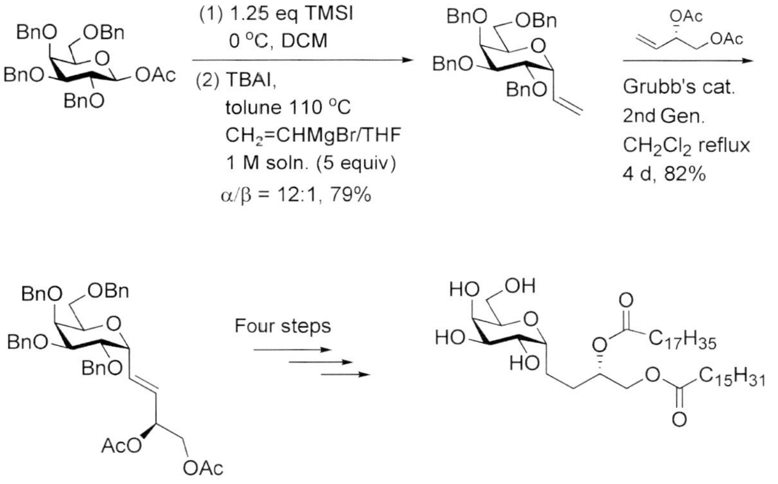

Scheme 2.68

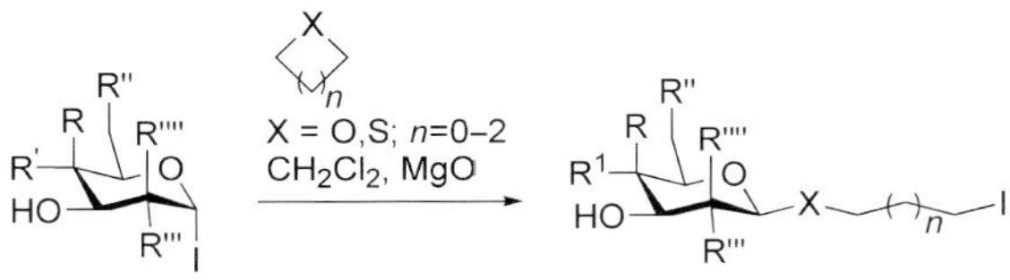

Scheme 2.69

Glc: R, R'''' = H; R', R'', R''' = OBn
Gal: R', R'''' = H; R, R'', R''' = OBn
Man: R, R''' = H; R', R'''' = OBn, R'' = OAc

nucleophilic attack on the β-iodide formed by *in situ* anomerization by the action of liberated I⁻. Limiting the *in situ* anomerization is required to drive the reaction mechanism toward the exclusive formation of β-glycoside [217]. Achieving β-selectivity is particularly difficult in D-mannosides as the anomeric effect as well as the C2-acyl directing group favor the formation of the α-isomer. Studies further indicated that β-selectivity could be improved using the reverse thermal effect [218].

2.2.5
Conclusions

Over the years, glycosyl halides have been the most utilized donors in stereoselective glycosylations. *In situ* anomerization is a powerful way of introducing 1,2-*cis* glycosides under neutral conditions, whereas direct displacement of anomeric halides typically leads to 1,2-*trans* glycosylations in the absence of neighboring-group participation. Glycosyl iodide donors offer several advantages over previously reported chloride or bromide donors, as reactions employing iodides are faster, highly stereoselective, and high yielding. In many cases, solution-phase reactions and purification can be carried out faster than solid-phase reactions. Current literature dispels the notion that glycosyl iodides are too reactive to be synthetically utilized. Instead, glycosyl iodides have emerged as important players in stereoselective glycoconjugate synthesis. Glycosylations using iodides, *in situ* generated from per-*O*-TMS sugars, are even more advantageous, as the final target molecules can be accessed in a one-pot manner after a single-column chromatography purification. This feature is especially attractive for rapidly synthesizing diverse analogs of oligosaccharides, glycoconjugates and glycolipids for structure–activity relationships (SARs) studies. Recent developments streamline complex oligosaccharide assembly providing powerful tools for drug discovery.

2.2.5.1 General Procedure for One-Pot Glycosylation Using Glycosyl Iodides

TMSI (30 mg, 0.15 mmol) is added to a solution of 1,2,3,4,6-penta-*O*-trimethylsilyl-D-galactopyranose (81 mg, 0.15 mmol) in CH_2Cl_2 (2 ml) at 0 °C and the reaction is stirred for 20 min. Anhydrous benzene (5 ml) is added and solvents are azeotroped twice on rotary evaporator. The yellowish oil is dissolved in CH_2Cl_2 (2 ml) and kept under argon. In a separate flask, molecular sieves (MS, 4 Å, 100 mg), TBAI (165 mg, 0.45 mmol), acceptor (0.05 mmol) and DIPEA (58 mg, 0.45 mmol) are added to CH_2Cl_2 (2 ml). The mixture is stirred under argon at room temperature. The glycosyl iodide solution is cannulated into the reaction mixture and stirring is continued at room temperature. After the completion of the reaction, as indicated by TLC (12–48 h), the solvent is evaporated and EtOAc is added. Precipitated TBAI and other solid materials are filtrated through Celite and the solvent is evaporated. MeOH (10 ml) and Dowex 50WX8-200 ion-exchange resin (0.5 g) are added and the reaction is stirred at ambient temperature for 4 h. The resin is filtered and the solvent is removed *in vacuo*. The resulting residue is purified by column chromatography on silica gel (gradient MeOH–CH_2Cl_2) to obtain the product (typical range 65–90%) as a white solid.

References

119 Koenigs, W. and Knorr, E. (1901) *Chemische Berichte*, **34**, 957–981.
120 Pellissier, H. (2005) *Tetrahedron*, **61**, 2947–2993.
121 Nitz, M. and Bundle, D.R. (2001) *Glycoscience: Chemistry and Biology*, **vol. 2** (eds B. Fraser-Reid, K. Tastuta and J. Thiem), Springer, Heildelberg, pp. 1497–1542.
122 Gervay, J. (1998) *Organic Synthesis: Theory and Applications*, JAI Press, Greenwich, pp. 121–153.
123 Ibatullin, F.M. and Selivanov, S.I. (2002) *Tetrahedron Letters*, **43**, 9577–9580.
124 Nanami, M., Ando, H., Kawai, Y., Koketsu, M. and Ishihara, H. (2007) *Tetrahedron Letters*, **48**, 1113–1116.
125 Kartha, K.P.R., Cura, P., Aloui, M., Readman, K., Rutherford, T.J. and Field, R.A. (2000) *Tetrahedron: Asymmetry*, **11**, 581–593.
126 Ghosh, R., Chakraborty, A. and Maiti, S. (2004) *Tetrahedron Letters*, **45**, 9631–9634.

127 Tiwari, P. and Misra, A.K. (2006) *Carbohydrate Research*, **341**, 339–350.
128 Oberthuer, M., Leimkuhler, C. and Kahne, D. (2004) *Organic Letters*, **6**, 2873–2876.
129 Yamada, H. and Hayashi, T. (2002) *Carbohydrate Research*, **337**, 581–585.
130 Kawai, Y., Ando, H., Ozeki, H., Koketsu, M. and Ishihara, H. (2005) *Organic Letters*, **7**, 4653–4656.
131 Matsuoka, K., Ohtawa, T., Hinou, H., Koyama, T., Esumi, Y., Nishimura, S., Hatano, K. and Terunuma, D. (2003) *Tetrahedron Letters*, **44**, 3617–3620.
132 Watt, G.M. and Boons, G.-J. (2004) *Carbohydrate Research*, **339**, 181–193.
133 Kulikova, N.Y., Shpirt, A.M. and Kononov, L.O. (2006) *Synthesis*, **24**, 4113–4114.
134 Hunsen, M., Long, D.A., D'Ardenne, C.R. and Smith, A.L. (2005) *Carbohydrate Research*, **340**, 2670–2674, and references therein.
135 Lin, C.-C., Huang, L.-C., Liang, P.-H., Liu, C.-Y. and Lin, C.-C. (2006) *Journal of Carbohydrate Chemistry*, **25**, 303–313.
136 Shingu, Y., Nishida, Y., Dohi, H., Matsuda, K. and Kobayashi, K. (2002) *Journal of Carbohydrate Chemistry*, **21**, 605–611.
137 Polat, T. and Linhardt, R.J. (2003) *Carbohydrate Research*, **338**, 447–449.
138 Stick, R.V., Stubbs, K.A., Tilbrook, D.M.G. and Watts, A.G. (2002) *Australian Journal of Chemistry*, **55**, 83–85.
139 Parrish, J.D. and Little, R.D. (2001) *Tetrahedron Letters*, **42**, 7371–7374.
140 Shoda, S.-i., Moteki, M., Izumi, R. and Noguchi, M. (2004) *Tetrahedron Letters*, **45**, 8847–8848.
141 Kumar, R., Tiwari, P., Maulik, P.R. and Misra, A.K. (2006) *European Journal of Organic Chemistry*, 74–79.
142 Deng, S., Gangadharmath, U. and Chang, C.-W.T. (2006) *Journal of Organic Chemistry*, **71**, 5179–5185.
143 Czifrák, K., Szilágyi, P. and Somsák, L. (2005) *Tetrahedron: Asymmetry*, **16**, 127–141.
144 Temelkoff, D.P., Zeller, M. and Norris, P. (2006) *Carbohydrate Research*, **341**, 1081–1090.
145 Tiwari, P. and Misra, A.K. (2006) *Tetrahedron Letters*, **47**, 2345–2348.
146 Mukherjee, C., Tiwari, P. and Misra, A.K. (2006) *Tetrahedron Letters*, **47**, 441–445.
147 Larsen, K., Olsen, C.E. and Motawia, M.S. (2003) *Carbohydrate Research*, **338**, 199–202.
148 Asnani, A. and Auzanneau, F.-I. (2003) *Carbohydrate Research*, **338**, 1045–1054.
149 Yamada, A., Hatano, K., Koyama, T., Matsuoka, K., Takahishi, N., Hidari, J., Suzuki, T., Suzuku, Y. and Terunuma, D. (2007) *Bioorganic and Medicinal Chemistry*, **15**, 1606–1614.
150 Cipolla, L., Redaelli, C., Faria, I. and Francesco, N. (2006) *Journal of Carbohydrate Chemistry*, **25**, 163–171.
151 Attia, A.M.E. (2002) *Nucleosides, Nucleotides and Nucleic Acids*, **21**, 207–216.
152 Saleh, M.A. and Abdel-Megeed, M.F. (2003) *Journal of Carbohydrate Chemistry*, **22**, 79–94.
153 Praly, J.-P., Ardakani, A.S., Bruyère, I., Marie-Luce, C. and Qin, B.B. (2002) *Carbohydrate Research*, **337**, 1623–1632.
154 Vidal, S., Bruyère, I., Maleron, A., Augé, C. and Praly, J.-P. (2006) *Bioorganic and Medicinal Chemistry*, **14**, 7293–7301.
155 Radha Krishna, P., Lavanya, B., Jyothi, Y. and Sharma, G.V.M. (2003) *Journal of Carbohydrate Chemistry*, **22**, 423–431.
156 Mitchell, S.A., Pratt, M.R., Hruby, V.J. and Polt, R. (2001) *Journal of Organic Chemistry*, **66**, 2327–2342.
157 Sjöelin, P. and Kihlberg, J. (2001) *Journal of Organic Chemistry*, **66**, 2957–2965.
158 Sherman, A.A., Yudina, O.N., Mironov, Y.V., Sukhova, E.V., Shashkov, A.S., Menshov, V.M. and Nifantiev, N.E. (2001) *Carbohydrate Research*, **336**, 13–46.
159 Kuszmann, J., Medgyes, G. and Boros, S. (2004) *Carbohydrate Research*, **339**, 2407–2414.
160 Cateni, F., Zacchigna, M., Zilic, J. and Di Luca, G. (2007) *Helvetica Chimica Acta*, **90**, 282–290.

161 Mukherjee, D., Ray, P.K. and Choudhary, U.S. (2001) *Tetrahedron*, **57**, 7701–7704.
162 Demchenko, A.V. (2003) *Synlett*, 1225–1240.
163 Lemieux, R.U., Hendriks, K.B., Stick, R.V. and James, K. (1975) *Journal of the American Chemical Society*, **97**, 4056–4062.
164 Gemma, E., Lahmann, M. and Oscarson, S. (2006) *Carbohydrate Research*, **341**, 1533–1542.
165 Turek, D., Sundgren, A., Lahmann, M. and Oscarson, S. (2006) *Organic and Biomolecular Chemistry*, **4**, 1236–1241.
166 Shingu, Y., Nishida, Y., Dohi, H. and Kobayashi, K. (2003) *Organic and Biomolecular Chemistry*, **1**, 2518–2521.
167 Nishida, Y., Shingu, Y., Dohi, H. and Kobayashi, K. (2003) *Organic Letters*, **5**, 2377–2380.
168 Shingu, Y., Miyachi, A., Miura, Y., Kobayashi, K. and Nishida, Y. (2005) *Carbohydrate Research*, **340**, 2236–2244.
169 Kondo, T., Oyama, K.-i. and Yoshida, K. (2001) *Angewandte Chemie (International Edition)*, **40**, 894–897.
170 Oyama, K.-i. and Kondo, T. (2004) *Tetrahedron*, **60**, 2025–2034.
171 Bouktaib, M., Atmani, M. and Rolando, C. (2002) *Tetrahedron Letters*, **43**, 6263–6266.
172 Li, M., Han, X. and Yu, B. (2002) *Tetrahedron Letters*, **43**, 9467–9470.
173 Du, Y., Wei, G. and Linhardt, R.J. (2003) *Tetrahedron Letters*, **44**, 6887–6890.
174 Du, Y., Wei, G. and Linhardt, R.J. (2004) *Journal of Organic Chemistry*, **69**, 2206–2209.
175 Peng, W., Li, Y., Zhu, C., Han, X. and Yu, B. (2005) *Carbohydrate Research*, **340**, 1682–1688.
176 Zhu, C., Peng, W., Li, Y., Han, X. and Yu, B. (2006) *Carbohydrate Research*, **341**, 1047–1051.
177 Tanaka, H., Zenkoh, T., Setoi, H. and Takahashi, T. (2002) *Synlett*, 1427–1430.
178 Timmons, S.C. and Jakeman, D.L. (2007) *Organic Letters*, **9**, 1227–1230.
179 Caputo, R., Kunz, H., Mastroianni, D., Palumbo, G., Pedatella, S. and Solla, F. (1999) *European Journal of Organic Chemistry*, 3147–3150.
180 Tanaka, H., Sakamoto, H., Sano, A., Nakamura, S., Nakajima, M. and Hashimoto, S. (1999) *Chemical Communications*, 1259–1260.
181 Schmid, U. and Waldmann, H. (1996) *Tetrahedron Letters*, **37**, 3837–3840.
182 Van Well, R.M., Kärkkäinen, K.P., Kartha, K.P.R. and Field, R.A. (2006) *Carbohydrate Research*, **341**, 1391–1397.
183 Kartha, K.P.R., Ballell, L., Bilke, J., McNeil, M. and Field, R.A. (2001) *Journal of the Chemical Society Perkin Transactions*, **1**, 770–772.
184 Stachulski, A.V. (2001) *Tetrahedron Letters*, **42**, 6611–6613.
185 Thiem, J. and Meyer, B. (1980) *Chemische Berichte*, **113**, 3075–3085.
186 Gervay, J., Nguyen, T.N. and Hadd, M.J. (1997) *Carbohydrate Research*, **300**, 119–125.
187 Chervin, S.M., Abada, P. and Koreeda, M. (2000) *Organic Letters*, **2**, 369–372.
188 Adinolfi, M., Iadonisi, A., Ravidá A. and Schiattarella, M. (2003) *Tetrahedron Letters*, **44**, 7863–7866.
189 Uchiyama, T. and Hindsgaul, O. (1996) *Synlett*, 499–501.
190 Miquel, N., Doisneau, G. and Beau, J.-M. (2000) *Chemical Communications*, 2347–2348.
191 Bhat, A.S. and Gervay-Hague, J. (2001) *Organic Letters*, **3**, 2081–2084.
192 Mukhopadhyay, B., Kartha, K.P.R., Russell, D.A. and Field, R.A. (2004) *Journal of Organic Chemistry*, **69**, 7758–7760.
193 Gervay, J. and Hadd, M.J. (1997) *Journal of Organic Chemistry*, **62**, 6961–6967.
194 Ying, L. and Gervay-Hague, J. (2003) *Carbohydrate Research*, **338**, 835–841.
195 Miquel, N., Vignando, S., Russo, G. and Lay, L. (2004) *Synlett*, 341–343.
196 Lam, S.N. and Gervay-Hague, J. (2003) *Organic Letters*, **5**, 4219–4222.

197 Hadd, M.J. and Gervay, J. (1999) *Carbohydrate Research*, **320**, 61–69.
198 Mukaiyama, T., Kobashi, Y. and Shintou, T. (2003) *Chemistry Letters*, **32**, 900–901.
199 Kobashi, Y. and Mukaiyama, T. (2004) *Chemistry Letters*, **33**, 874–875.
200 Mukaiyama, T. and Kobashi, Y. (2004) *Chemistry Letters*, **33**, 10–11.
201 Van Well, R.M., Kartha, K.P.R. and Field, R.A. (2005) *Journal of Carbohydrate Chemistry*, **24**, 463–474.
202 Bickley, J., Cottrell, J.A., Ferguson, J.R., Field, R.A., Harding, J.R., Hughes, D.L., Kartha, K.P.R., Law, J.L., Scheinmann, F. and Stachulski, A.V. (2003) *Chemical Communications*, 1266–1267.
203 Harding, J.R., King, C.D., Perrie, J.A., Sinnott, D. and Stachulski, A.V. (2005) *Organic and Biomolecular Chemistry*, **3**, 1501–1507.
204 Perrie, J.A., Harding, J.R., King, C.D., Sinnott, D. and Stachulski, A.V. (2003) *Organic Letters*, **5**, 4545–4548.
205 Adinolfi, M., Iadonisi, A., Pezzella, A. and Ravidà A. (2005) *Synlett*, 1848–1852.
206 Lam, S.N. and Gervay-Hague, J. (2002) *Organic Letters*, **4**, 2039–2042.
207 Lam, S.N. and Gervay-Hague, J. (2002) *Carbohydrate Research*, **337**, 1953–1965.
208 Lam, S.N. and Gervay-Hague, J. (2005) *Journal of Organic Chemistry*, **70**, 2387–2390.
209 Lam, S.N. and Gervay-Hague, J. (2005) *Journal of Organic Chemistry*, **70**, 8772–8779.
210 Du, W. and Gervay-Hague, J. (2005) *Organic Letters*, **7**, 2063–2065.
211 Du, W., Kulkarni, S.S. and Gervay-Hague, J. (2007) *Chemical Communications*, 2336–2339.
212 Kulkarni, S.S. and Gervay-Hague, J. (2007) Chemical glycobiology ACS Symposium Series, in press.
213 Hein, M., Phuong, N.T.B., Michalik, D., Goerls, H., Lalk, M. and Langer, P. (2006) *Tetrahedron Letters*, **47**, 5741–5745.
214 Kulkarni, S.S. and Gervay-Hague, J. unpublished results.
215 Kulkarni, S.S. and Gervay-Hague, J. (2006) *Organic Letters*, **8**, 5765–5768.
216 Dabideen, D.R. and Gervay-Hague, J. (2004) *Organic Letters*, **6**, 973–975.
217 El-Badri, M.H., Willenbring, D., Tantillo, D.J. and Gervay-Hague, J. (2007) *Journal of Organic Chemistry*, **72**, 4663–4672.
218 El-Badry, M.H. and Gervay-Hague, J. (2005) *Tetrahedron Letters*, **46**, 6727–6728.

3
Glycoside Synthesis from 1-Oxygen Substituted Glycosyl Donors

3.1
Hemiacetals and O-Acyl/Carbonyl Derivatives
Daniel A. Ryan, David Y. Gin

3.1.1
Introduction

This chapter outlines the development, achievements and limitations of glycosylation methods that rely on C1-hemiacetal donors and C1-O-acyl donors. These are among the simplest glycosyl donors to prepare in standard O-glycosylation reactions. As such, developments in the use of these donors constitute valuable advances in the field of synthetic carbohydrate chemistry.

3.1.2
Dehydrative Glycosylation via Electrophilic Activation of C1-Hemiacetals

Glycosylation with C1-hemiacetal donors offers a notable variation from most of the other glycosylation strategies, in that it combines the steps of anomeric derivatization and activation/glycosylation into a one-pot procedure (Scheme 3.1). In this process, the hemiacetal donor **1** is exposed to an electrophilic reagent (El^+) that activates the hemiacetal by converting it to a potent leaving group for expulsion from the anomeric center. With the introduction of a glycosyl acceptor (NuH) to the activated intermediate **2**, the desired glycoside **3** is formed directly, wherein the controlled extraction of 1 equiv of water mediates the union of glycosyl donor and acceptor.

An attractive feature of this dehydrative coupling approach is that it avoids the need for isolation of intermediate glycosyl donors. This can be desirable if a glycosyl donor is not stable to isolation or purification. Moreover, the use of a hemiacetal donor reduces the number of synthetic manipulations of the carbohydrate donor by avoiding hemiacetal derivatization to alternative donor types. In this way, the approach has the potential to streamline time and labor-intensive multiglycosylation sequences. Although there increasingly have been reports of these direct dehydrative

Handbook of Chemical Glycosylation: Advances in Stereoselectivity and Therapeutic Relevance.
Edited by Alexei V. Demchenko.
Copyright © 2008 WILEY-VCH Verlag GmbH & Co. KGaA. All rights reserved.
ISBN: 978-3-527-31780-6

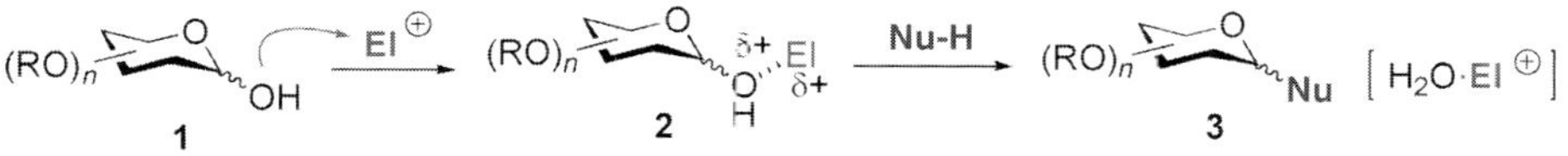

Scheme 3.1 Electrophilic activation of hemiacetals for dehydrative glycosylation.

R = Protective group   El = Electrophilic activator   Nu = Nucleophile

glycosylations having performed with high synthetic utility, the methods that employ derivatized hemiacetal donors are more commonly practiced. This stems from unique challenges posed by hemiacetal donors, given that the hemiacetal can also serve as a nucleophilic acceptor. As a result, any process that generates an activated hemiacetal intermediate **2** in the presence of unreacted hemiacetal **1** is in danger of promoting self-condensation to generate the corresponding 1,1'-linked anhydro dimer as an unwanted side product. Thus, hemiacetal activation relies on either transient activation (i.e. via acid) with thermodynamic control over the acetal exchange process, or electrophiles that efficiently form irreversible complexes with the hemiacetal hydroxyl to initiate the water extraction process.

The preparation of C1-hemiacetal glycosyl donors follows many of the traditional strategies for selective anomeric functionalization. Although many synthetic sequences can be envisioned for the preparation of a selectively protected C1-hemiacetal donor, one of the two general synthetic approaches is employed. The first approach involves a complete and indiscriminate protection of all hydroxyl groups (including the hemiacetal hydroxyl) on a furanose or pyranose substrate, and then reliance on the differential reactivity of the acetal functionality to expose the anomeric hydroxyl in **1**. These synthetic sequences include hydroxyl peralkylation and anomeric hydrolysis [1], hydroxyl perbenzylation and anomeric hydrogenolysis [2], and hydroxyl peracylation and anomeric deacylation [3,4]. The second, perhaps more common, approach involves an initial acid promoted hemiacetal-to-acetal exchange at the anomeric center of a pyranose or furanose to effect selective protection at the C1-position (i.e. Fischer glycosylation, see below). Subsequent orthogonal protection of the periphery hydroxyl groups on the carbohydrate sets the stage for the final, selective anomeric deprotection to C1-hemiacetal donor **1** [5].

3.1.3
Acid Activation of C1-Hemiacetals

The Fischer glycosylation is among the earliest chemical glycosylations and involves acid-catalyzed reaction with an unprotected hemiacetal donor **4** (Scheme 3.2) [6,7]. Because the reaction is under thermodynamic control, it is conducted with a large excess of glycosyl acceptor alcohol (ROH) to drive the equilibrium to glycoside **6**. Consequently, the Fischer glycosylation is best employed with simple alcohol solvents. In addition, the reaction typically requires high temperatures and/or long reaction times, which are not ideal conditions for complex or sensitive substrates. Not surprisingly, anomeric selectivity in the glycosylation process is usually dictated

3.1 Hemiacetals and O-Acyl/Carbonyl Derivatives

$$(HO)_n\text{-sugar-OH} \; \underset{}{\overset{H^+}{\rightleftharpoons}} \; (HO)_n\text{-sugar-}\overset{+}{O}H_2 \; \underset{}{\overset{ROH \,(solvent)}{\rightleftharpoons}} \; (HO)_n\text{-sugar-OR} \; + \; H_2O$$
$$\quad\quad 4 \quad\quad\quad\quad\quad\quad\quad\quad 5 \quad\quad\quad\quad\quad\quad\quad\quad 6$$

R = methyl, ethyl, n-propyl, i-propyl, amyl, allyl, benzyl, etc.

Scheme 3.2 Fischer glycosylation.

by the relative ground-state energies of the product glycoside anomers. Despite these constraints, the Fischer glycosylation remains indispensable, in that it directly provides unprotected glycosidic products [8]. The importance of this reaction is reflected in the application of the Fischer glycosylation in the industrial preparation of surfactants used in consumer goods.

Recent investigations have revealed new acid promoters for the preparation of more complex glycosides from *selectively protected* C1-hemiacetal donors such as **1**. In the presence of a desiccant, several types of acid promote glycosylation typically without the requirements for excess glycosyl acceptor and high temperatures. Because of the reversible nature of acid coordination to hydroxyl groups, this class of reactions generally allows for activation of the hemiacetal donor to occur in the presence of the glycosyl acceptor alcohol.

Classically, Brønsted acids are the promoters of the Fischer glycosylation. Though substrate complexity is still limited in scope, recent developments in these reagents more commonly allows for disaccharide synthesis (Scheme 3.3). For instance, Koto *et al.* have reported that the combination of methanesulfonic acid (30 mol%) and

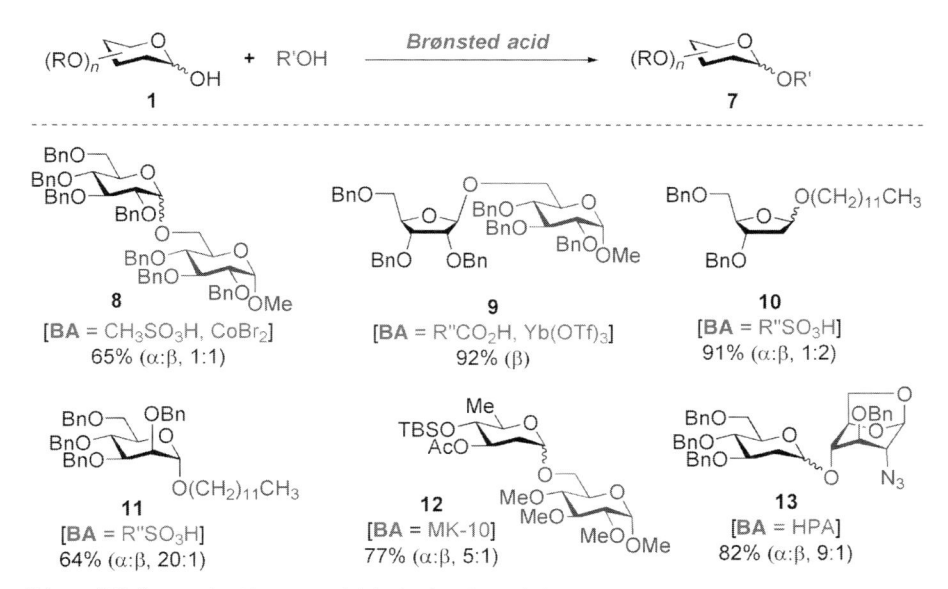

$$(RO)_n\text{-sugar-OH} \; + \; R'OH \; \overset{\text{Brønsted acid}}{\longrightarrow} \; (RO)_n\text{-sugar-OR'}$$
$$\quad\quad\quad 1 \quad\quad\quad\quad\quad\quad\quad\quad\quad\quad\quad\quad\quad 7$$

8
[BA = CH$_3$SO$_3$H, CoBr$_2$]
65% (α:β, 1:1)

9
[BA = R"CO$_2$H, Yb(OTf)$_3$]
92% (β)

10
[BA = R"SO$_3$H]
91% (α:β, 1:2)

11
[BA = R"SO$_3$H]
64% (α:β, 20:1)

12
[BA = MK-10]
77% (α:β, 5:1)

13
[BA = HPA]
82% (α:β, 9:1)

Scheme 3.3 Brønsted acid promoted dehydrative glycosylations.

cobalt(II) bromide (CoBr$_2$, 1 equiv) promotes glycosidic bond formation from hemiacetal donor **1** and acceptor alcohol (1 equiv) in just 2 h at 25 °C. Cobalt(II) bromide functions as both a desiccant and a latent source of hydrobromic acid, allowing for generation of a glycosyl bromide intermediate. From this reaction, disaccharide **8** was prepared in 65% [9,10]. Inanaga et al. reported that the combination of methoxyacetic acid and Yb(OTf)$_3$ (10 mol% each) with 4 Å molecular sieves yields glycosides in good yields [11]. Control experiments indicated that both Lewis and Brønsted acids are necessary. Using this procedure, the ribosyl disaccharide **9** was obtained with high 1,2-*trans* selectivity, though the degree of stereoselectivity was generally substrate dependent.

Kobayashi and coworkers have advanced the use of sulfonic acids in Fischer-type glycosylations with the discovery that surfactant sulfonic acids catalyze the dehydrative glycosylation of hemiacetals using water as the reaction solvent [12]. It was proposed that long-chain acid and alcohol acceptors form emulsions in water with hydrophobic interiors, which promote dehydration within this emulsion. Thus, furanose and pyranose hemiacetals reacted with C-5 to C-12 long-chain alcohol acceptors (1.5 equiv) in the presence of 10 mol% of dodecylbenzenesulfonic acid to afford glycosides such as **10** or **11** with good conversion at elevated temperatures. The method has been extended to the synthesis of aryl *C*-glycosides in water [13].

Toshima and coworkers found that the heterogeneous, layered-silicate acid catalyst Montmorillonite K-10 (MK-10) effectively promotes stereoselective glycosylation with olivoside (2,6-dideoxyglucopyranose) donors [14]. One of the benefits of heterogeneous catalysis is the ability to obtain product by simply filtering the catalyst from the reaction medium, which avoids neutralization steps and salt formation. In the reaction, the dehydrated clay presumably acts as both acid and desiccant. For instance, the treatment of an olivose hemiacetal with a monosaccharide acceptor and MK-10 (150 wt%) at 25 °C provided glycoside **12** in 77% yield. Other reports have appeared wherein 4 Å molecular sieves function as the sole heterogeneous additive to effect *N*-glycoside bond formation [15].

An uncommon approach to in dehydrative glycosylations was detailed by Toshima et al., wherein a heteropoly acid was used to promote the reaction [16]. Heteropoly acids (HPA), such as H$_4$SiW$_{12}$O$_{40}$ used in this study, are strong Brønsted acids with octahedral metal-oxygen core structural units [17]. The researchers comment that H$_4$SiW$_{12}$O$_{40}$ acid is easily dehydrated by heating and acts as both a desiccant and a strong Brønsted acid, thus making it a conspicuous choice for use in dehydrative glycosylations with hemiacetal donors. Using this acid, glycosylation of the secondary alcohol acceptor (1.5 equiv) provided disaccharide **13** in 82% yield after only 1 h at 25 °C. In this investigation, the reaction was applied to various pyranose hemiacetal donors with good results.

In addition to Brønsted acid promoted Fischer-type glycosylations, Lewis acids have been investigated (Scheme 3.4). A variety of Lewis acids promote glycosylation under mild conditions, often in substoichiometric amounts. The earliest examples include ZnCl$_2$ [18] and FeCl$_3$ [19], although these reactions were demonstrated only for preparation of trehalose-type disaccharides. Mukaiyama et al. have very recently developed metal triflate catalysts for the dehydrative glycosylation with

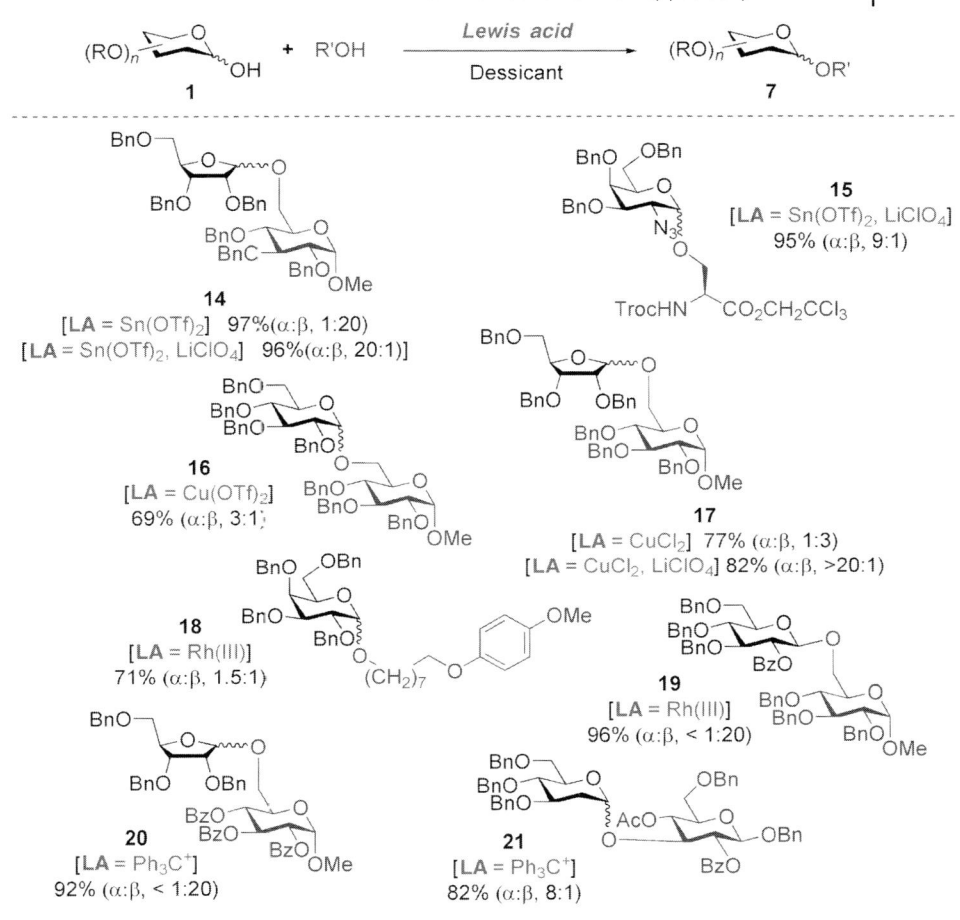

3.1 Hemiacetals and O-Acyl/Carbonyl Derivatives | 99

Scheme 3.4 Lewis acid promoted dehyrative glycosylations.

hemiacetal donors. Tin(II), ytterbium(III) and lanthanum(III) triflates are all viable catalysts at 1 mol% loading. In combination with hexamethyldisiloxane as desiccant, these systems provide glycoside products in a few hours at room temperature [20]. Preliminary investigations suggested that $Sn(OTf)_2$ was the most effective catalyst, which was used in the preparation of disaccharide **14** (97%), favoring the β anomer by 20:1. The anomeric stereoselectivity of this reaction was reversed by addition of lithium perchlorate (1.5 equiv) to the reaction to effect 96% yield of the α anomer, favored by 20:1. The role of lithium perchlorate was proposed to be sterically guided formation of a β-anomeric perchlorate intermediate, which directs nucleophilic addition to the opposite face. Using this modification, the O-linked glycopeptide precursor **15** was achieved in 95% yield (α:β, 9:1). Although the reaction scope has not been extended to secondary alcohols or to more complex substrates, the high yields and stereoselectivity ingrain this catalyst/desiccant combination among the most efficient Lewis acid systems reported to date.

Cu(II) Lewis acids also promote Fischer-type glycosylations. Yamada and Hayashi have reported that Cu(II) triflate (1.1 equiv) in the presence of 4 Å molecular sieves yields disaccharides such as **16** in moderate yield [21]. Benzotrifluoride solvent, introduced by Ogawa and Curran as a less-toxic alternative to dichloromethane [22], gave optimal yields in this reaction. A catalytic Cu(II) system composed of $CuCl_2$, bis(diphenylphosphino)ferrocene and silver perchlorate (1:1:2 mol composition) was effective at 5 mol% loading with calcium sulfate desiccant. The application of this system provided disaccharide **17** with moderate β-selectivity [23]. Again, it was found that the anomeric stereoselectivity could be reversed with the addition of $LiClO_4$, which provided the α anomer of **17** in greater than a 20:1 ratio. The role of lithium perchlorate was suggested to be analogous to that of Mukaiyama's work, which invokes either the intermediacy of a β-anomeric perchlorate [20], or lithium perchlorate affecting *in situ* product anomerization, depending on the activating protocol used [24,25].

Indeed, a variety of Lewis acids have been shown to effect glycosylation with hemiacetal donors. Ernst and coworkers have used 5 mol% of [Rh(III)(MeCN)$_3$(triphos)] tris(triflate) with 4 Å molecular sieves to prepare glycoconjugates **18** and **19** [26]. Mukaiyama's group has used trityl tetrakis(pentafluorophenyl)borate (3–5 mol%) with Drierite in the preparation of disaccharides **20** and **21** [27,28]. In the synthesis of **21**, the α-selectivity was shown to arise from *in situ* anomerization of the β-pyranoside over time.

As illustrated by the above examples, a number of Brønsted and Lewis acids promote Fischer-type glycosylation of hemiacetal donors allowing access to more complex glycosides. The application to oligosaccharide synthesis, or even glycosylation of less-reactive alcohol acceptors, is still uncommon. Beyond these reactions significant contributions have been made to the classic Fischer glycosylation with unprotected glycosyl donors. Some of these innovations include the effect of calcium or strontium cations on product isomer distribution [29], the use of $FeCl_3$ or $BF_3 \cdot OEt_2$ to provide furanoside or pyranoside products using only slight excess of acceptor [30–33], microwave acceleration of the reaction [34] and new developments in the preparation of long-chain alkyl glycosides using heterogeneous acids [35–39].

3.1.4
Hemiacetal Activation with Silicon Electrophiles

Silicon presents an attractive option among electrophilic activating and dehydrating agents of hemiacetals because of the wide commercial availability of electrophilic silicon sources. The two main classes of silicon electrophiles used, namely silyl halides and silyl sulfonates, have been demonstrated to promote a variety of glycosylations including some examples of oligosaccharide synthesis.

One of the earliest reports of silicon-based electrophilic activation comes from the Koto laboratory on the use of silyl halide electrophiles to promote the dehydrative glycosylation with hemiacetal donors [40,41]. In the reaction (Scheme 3.5),

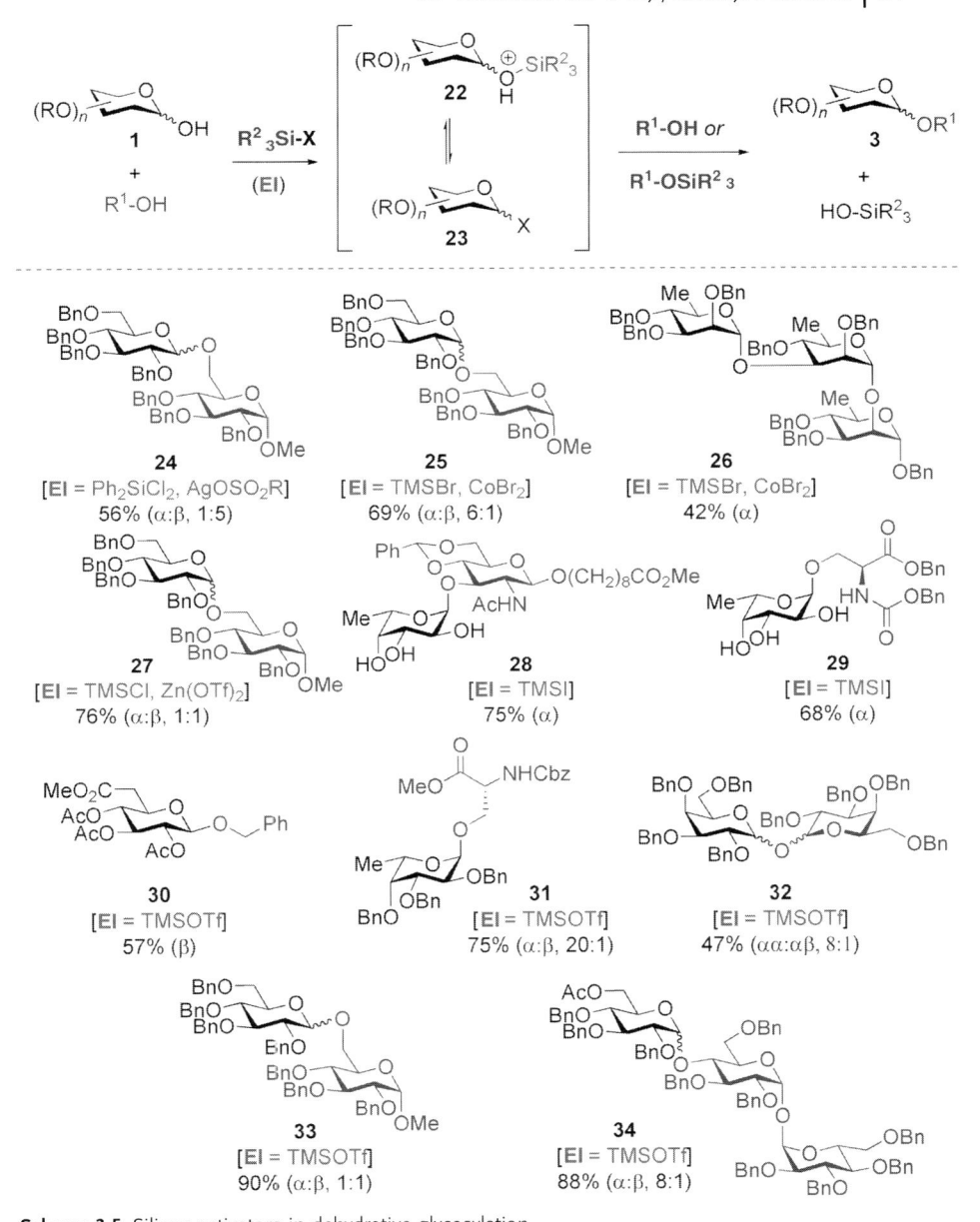

3.1 Hemiacetals and O-Acyl/Carbonyl Derivatives | 101

Scheme 3.5 Silicon activators in dehydrative glycosylation.

diphenyldichlorosilane (1 equiv) and silver sulfonate salts (2 equiv) effect the coupling of glycosyl donor **1** and acceptor (R^{1}OH).

Diphenyldichlorosilane is thought to react with hemiacetal **1** to afford silyl hemiacetal **22** and thereby liberate HCl to perpetuate the reaction. The investigators design that, upon glycosylation of the acceptor, the chlorosilanol by-product polymerizes as an effective mode of dehydration. The role of the silver salt is to facilitate anomeric substitution in the event of anomeric chloride intermediates (23, X = Cl). Specifically, with silver toluenesulfonate and silver triflate additives at 0 °C, this procedure provided disaccharide **24** in 56% yield (α : β 1 : 5) [40]. Although the demonstrated reaction efficiency is only moderate, this early study laid the groundwork for later developments with silyl halide promoters.

The reagent combination of trimethylsilyl bromide and cobalt(II) bromide also promotes dehydrative glycosylation with hemiacetal donors and is notable for its use in oligosaccharide synthesis. The activation of the hemiacetal donor, which proceeds in the presence of the acceptor alcohol (R^1OH), results in glycosyl bromide intermediate **23** (X = Br). Mechanistic pathways may include silylation of the hemiacetal, silylation of the hydroxyl acceptor or both. In control experiments, it was shown that a preformed silyl hemiacetal was capable of transformation to the glycosyl bromide; independently, a silyl ether was shown to be a capable acceptor of glycosyl bromide donors. With the addition of tetrabutylammonium bromide to promote halide-catalyzed glycosidic bond formation, disaccharide **25** was produced in 69% yield (α : β, 6 : 1) over 16 h at 25 °C [42]. The trisaccharide repeat unit **26** of the O-specific polysaccharide of *Pseudomonades* pathogens was synthesized in 42% yield from a disaccharide hemiacetal donor [43]. In related work, Susaki has found that the combination of trimethylsilyl chloride and zinc(II) triflate also effects glycosylation. Using this protocol, disaccharide **27** was formed in 76% as a 1 : 1 anomeric mixture [44].

Trimethylsilyl halides have been used as the sole activator/desiccator in Fischer-type glycosylations. Uchiyama and Hindsgaul reported that the treatment of unprotected L-fucose with excess trimethylsilyl chloride and triethylamine allowed for a quantitative preparation of a tetra-TMS protected fucose donor, which was isolated upon extraction with pentane [45]. A solution of this silylated donor was then treated with a trimethylsilyl iodide (TMSI) promoter and the alcohol acceptor (0.3 equiv) to effect glycosylation in less than 30 min at room temperature. In this way, disaccharide **28** was obtained in 75% yield, exclusively as the α anomer after a work up with methanol. Similarly, the glycopeptide **29** was isolated in 68% yield. Although this reaction involves isolation of the tetra-TMS derivatized donor, it is a notable development using silicon electrophiles to promote Fischer-type glycosylations. Vigorita and coworkers have applied this reaction to unprotected xylopyranose and arabinopyranose donors [46]. Fukase and coworkers have prepared propargyl and allyl glycosides in good yield and selectivity using only TMSCl, a procedure that does not entail isolation of the silylated donor [47].

Trialkylsilyl sulfonates, especially trimethylsilyl triflate (TMSOTf), represent the other broad class of silicon electrophiles used to promote direct dehydrative glycosylations (Scheme 3.5). Among the earliest reports, Nudelman and coworkers found that hemiacetal donor **1** and glycosyl acceptor (R^1OH) can be coupled under the agency of TMSOTf (1 equiv) to provide O-alkyl glycosides in a matter of hours at temperatures below 20 °C [48]. Again, the possibility of nonselective silylation between the glycosyl donor and the acceptor exists, though to no deleterious effect.

In fact, other studies have shown that TMSOTf catalyzes the glycosylation of a silylated acceptor with a silylated hemiacetal donor [49]. Nudelman's procedure was applied to the synthesis β-glucuronide **30**, isolated in 57% yield. Kiyoi and Kondo have applied the TMSOTf activation protocol to protected L-fucose hemiacetal donors for glycopeptide synthesis and obtained glycopeptide fragment **31** in 74% yield (α : β, 20 : 1) [50]. Posner and Bull have developed a procedure that uses excess TMSOTf in the presence of molecular sieves (SYLOSIV A4) to synthesize various 1,1′-linked disaccharides such as the galactopyranose dimer **32** [51,52].

Koto's group reported the use of TMSOTf in one of the rare examples of oligosaccharide synthesis using this class of activators [53]. Activation of the donor and acceptor using TMSOTf (5 equiv) and pyridine (3 equiv) at $-45\,^\circ$C provides glycoside products after a few hours at $0\,^\circ$C. For instance, disaccharide **33** was isolated in 90% yield (α : β, 1 : 1.3). Further, extension of reaction scope was accomplished with the synthesis trisaccharide **34** in 88%, favoring the α anomer (8 : 1). In this case, the C-6 acetate of the glycosyl donor was believed to direct the α-selectivity through long-range participation.

In addition to the silicon-based *in situ* activation of hemiacetal donors, there has been a significant body of work that uses electrophilic silicon activation of preformed C-1 silyl hemiacetal donors [54–67]. However, this work is outside the scope of this discussion.

3.1.5
Hemiacetal Activation with Phosphorus Electrophiles

Phosphorus-based activating agents present an attractive option among electrophilic activation of hemiacetal glycosyl donors. The relatively high bond strength of the phosphorus–oxygen bond provides ample thermodynamic driving force for C1-hydroxyl activation and subsequent dehydration via formation of phosphine oxide in the glycosylation event [68]. Three main modes of electrophilic phosphorus activation exist, and together they exhibit a wide variety of accessible substrate classes in direct dehydrative glycosylations.

Shortly after the discovery of the Mitsunobu reaction in the late 1960s [69], phosphonium activation of hemiacetals was reported using the reagent combination of a phosphine and a dialkyl azodicarboxylate **35** (Scheme 3.6). The reaction between the phosphine and **35** affords *N*-betaine intermediate **37**, which serves as a potent electrophilic activator for hemiacetal **1**. From this activation step, the glycosyl oxophosphonium intermediate **38** is generated along with the liberation of the dialkyl hydrazinedicarboxylate by-product **36**. Subsequent nucleophilic addition of the acceptor occurs to expel phosphine oxide from the anomeric center with formation of the new, anomeric bond. Much like the original Mitsunobu reaction, the successful application of the Mitsunobu protocol to glycosylation typically requires relatively acidic glycosyl acceptors such as imides, hydroxyphthalimides, carboxylic acids and phenols.

Szarek *et al.* were the first to develop this mode of glycosylation, which was demonstrated for ribonucleoside synthesis. In the reaction, a protected mannofuranose

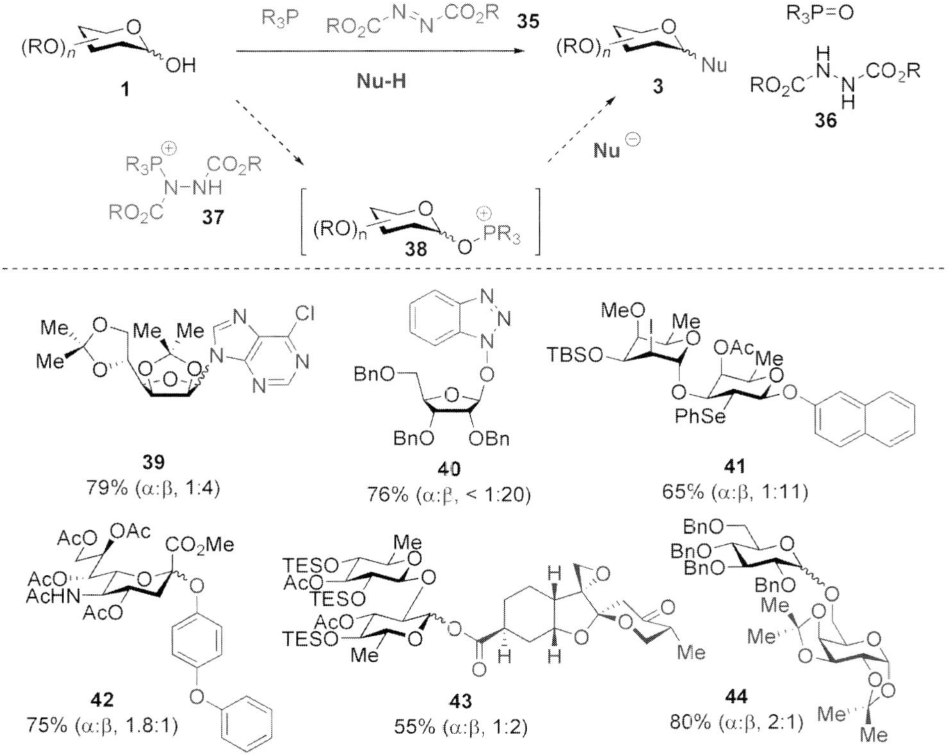

Scheme 3.6 Dehydrative glycosylation via the Mitsunobu protocol.

hemiacetal donor was added to an equimolar solution of methyldiphenylphosphine, diethyl azodicarboxylate, and the 6-chloropurine glycosyl acceptor at ambient temperature. The resulting N-glycoside **39** was isolated in 79% yield, favoring the natural β anomer [70]. This general method has been applied to pyranose hemiacetal donors and other N-acceptors [71–75]. It was later demonstrated that N-hydroxy nucleophiles are efficient glycosyl acceptors in this reaction [76], as evidenced by the formation of riboside **40** in 76% yield with high α-selectivity [77]. The results generally show good stereoselectivity for the 1,2-*trans* riboside, although the stereoselectivity can be reversed when either a trityl or a *tert*-butyldimethylsilyl protective group is used on the C-5 hydroxyl of the ribofuranose donor. Although O-N-glycosides are not a common class of glycosidic bond, this method has been successfully applied to the synthesis of the glycosyl-oxyamine linkage in calicheamicin [78,79].

The glycosylation based on the Mitsunobu reaction has been most commonly directed to the synthesis of O-aryl glycosides, a structural motif found in a variety of natural products [80–82]. Early work by Grynkiewicz [83,84], among others [85–87], established the viability of triphenylphosphine and diethylazodicarboxylate to promote the glycosylation of phenol acceptors at ambient temperature. More recently, Roush and coworkers have discovered that the glycosylation performed well in the stereoselective synthesis of O-aryl glycosides *en route* to 2-deoxy sugars. In synthetic efforts to the antitumor natural products olivomycin and mithramycin, Mitsunobu glycosylation of 2-naphthol provided disaccharide **41** in 65% yield with high β-selectivity [88–90]. Although the hemiacetal donor predominantly favors the α-form and the Mitsunobu reaction generally favors an S_N2 pathway, the participation of the C-2 phenylselenyl group could not be ruled out as the stereocontrolling element. Ernst *et al.* also found that the sialyl glycoside **42** could be obtained in good yield (75%), but with little stereocontrol [91]. This example is notable because of the congested steric environment of sialic acid C-2 hemiketal donors. Not surprisingly, glycosyl esters have also been prepared under similar conditions. In their synthesis of the antitumor natural product phyllanthoside [92,93], Smith and coworkers found that the synthesis of glycoconjugate **43** was not possible by direct esterification of the hemiacetal using an acyl chloride aglycon, as this experiment provided the undesired α-glycoside. Accordingly, the problem was addressed with the Mistunobu glycosylation, thereby capitalizing on a proposed S_N2 pathway via an α-oxophosphonium intermediate. Thus, the glycosyl ester **43** was formed in 55% yield, with increased β-selectivity (α : β, 1 : 2). Advances in the synthesis of O-aryl β-glucuronides have also appeared [94].

Although aliphatic alcohols are typically poor acceptors in the Mitsunobu-type glycosylation, Szarek and coworkers have highlighted one advance to this end [95]. For the triphenylphosphine and diethylazodicarboxylate promoted glycosylation of a monosaccharide acceptor, the addition of mercuric bromide is necessary to promote the reaction. For example, the (1,6)-disaccharide **44** was obtained in 80% yield using this modified Mitsunobu protocol. Unlike previous examples with phenol or N-acceptors, preactivation of the hemiacetal donor was performed for 10 min at room temperature prior to addition of the aliphatic alcohol nucleophile.

In 1975, Hendrickson and Schwartzman reported a different mode of phosphorus-based hydroxyl activation using bis(phosphonium) electrophiles $(R_3P-O-PR_3)^{2+}$. These highly reactive electrophiles are generated from the reagent combination of phosphine oxide and trifluoromethanesulfonic (triflic) anhydride [96]. Mukaiyama

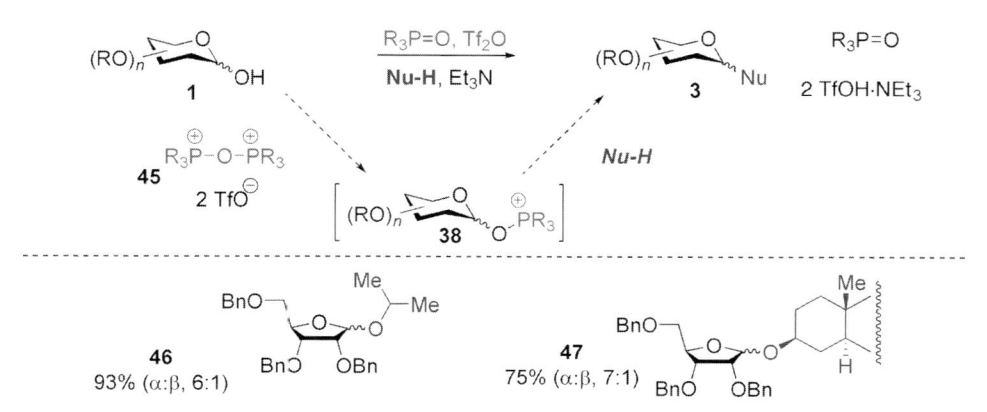

Scheme 3.7 Dehydrative glycosylation with $[R_3PO]_2 \cdot Tf_2O$.

and Sudha later discovered that bisphonium electrophiles effectively promote dehydrative glycosylation with hemiacetal donors (Scheme 3.7) [97]. In this method, a hemiacetal **1** is activated by the bis(phosphonium) ditriflate salt **45** for *in situ* generation of the anomeric oxophosphonium intermediate **38**, a species observed by ^{31}P NMR and ^{13}C NMR. With the introduction of a nucleophile, phosphine oxide is expelled from the anomeric center as the glycosidic bond in **3** is formed.

This method has been used only a few times, despite the high yields reported. In the procedure, the hemiacetal was activated with tributylphosphine oxide (4.5 equiv) and triflic anhydride (2.1 equiv) for 2 h at 0 °C, followed by an addition of the glycosyl acceptor. As a result, the isopropyl riboside **46** was prepared in 93% and the cholestanyl riboside **47** was prepared in 75%, both with α-anomeric selectivity.

The reaction of phosphines and alkyl halides presents an alternative way to generate phosphonium electrophiles (Scheme 3.8). In particular, the combination of a phosphine and carbon tetrabromide (the Appel reaction) allows for *in situ* formation of a phosphonium dibromide salt (**48**, X = Br). Treatment of a hemiacetal donor **1** with the phosphonium halide **48** initially provides the oxophosphonium intermediate **38** (X = Br). However, the oxophosphonium intermediate **38** can react with bromide ion to form the anomeric bromide intermediate **49** (X = Br) with concomitant generation of phosphine oxide. With the aid of bromide ion catalysis (i.e. reversible, catalytic formation of the more reactive β-anomeric bromide **50**) [98], the nucleophile displaces the anomeric bromide to form the desired glycoside product **3**. The hydrobromic acid by-product is typically buffered by the presence of tetramethyl urea (TMU).

In 1980, Gross and coworkers first applied this concept to a direct dehydrative glycosylation using tris(dimethylamino)phosphine and carbon tetrachloride [99].

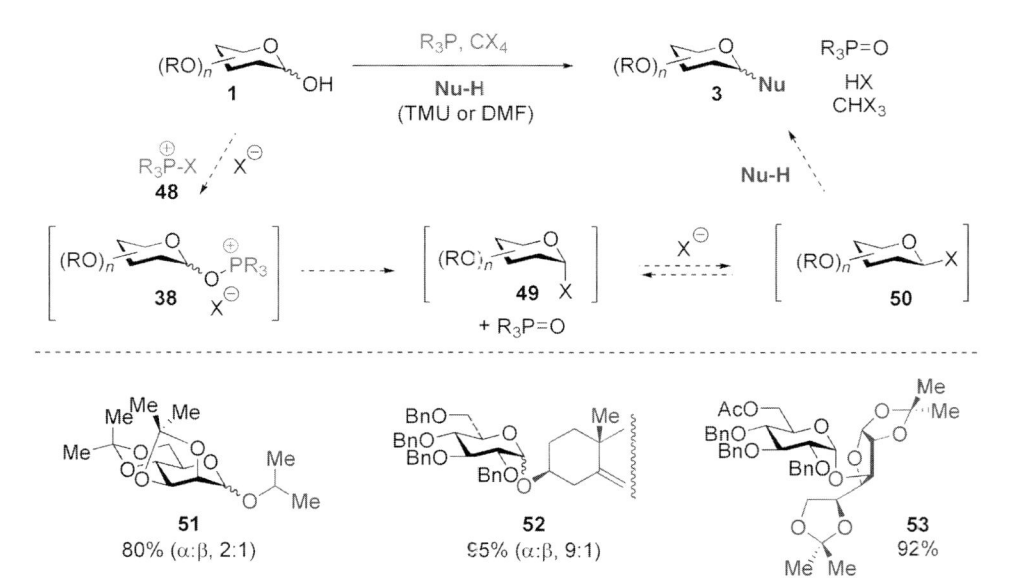

Scheme 3.8 Dehydrative glycosylation with $R_3P\cdot CX_4$.

With these reagents, the corresponding anomeric oxophosphonium intermediate 38 (X = Cl) is formed, allowing for its direct displacement with the nucleophilic acceptor. This reaction afforded the isopropyl glycoside 51 (80%). It was noted that control of the reaction temperature at −40 °C in the presence of silver salts precludes glycosyl chloride formation (49, X = Cl) during hemiacetal activation. More recently, the groups of Nifant'ev [100] and Kobayashi [101] have developed a modified Appel method for glycosidic bond formation. In the early work, the Appel glycosylation involved *in situ* generation of glycosyl bromide 49, which was subjected to halide ion catalysis by addition of tetrabutylammonium bromide and tetramethyl urea. Further reaction insights revealed that the addition of an external halide source is unnecessary, and high yields and selectivities for the 1,2-*cis* α-glycoside are still obtained [102]. In the event, activation of the hemiacetal with triphenylphosphine and carbon tetrabromide for 3 h at 0 °C, followed by addition of the acceptor, provided glycoside products such as cholesteryl glucoside 52 (95%, α : β, 9 : 1) or the (1,3)-disaccharide 53 (92%, α). The higher α selectivity in the formation of 53 was thought to arise from long-range participation of the C-6 acetate. Despite the use of 3 equiv of glycosyl acceptor, the stereoselective formation of 1,2-*cis* α-glycosides from hemiacetal donors with C-2 nonparticipatory protective groups is a notable feature of the method. Interestingly, when the reaction was run in N,N-dimethylformamide (DMF) solvent, glycosyl imidate intermediates were detected by ^{1}H NMR, and shorter reaction times were observed [103,104].

3.1.6
Hemiacetal Activation with Sulfur Electrophiles

The first direct dehydrative glycosylation promoted by sulfur electrophiles was reported by Leroux and Perlin [105]. In this reaction (Scheme 3.9), activation of

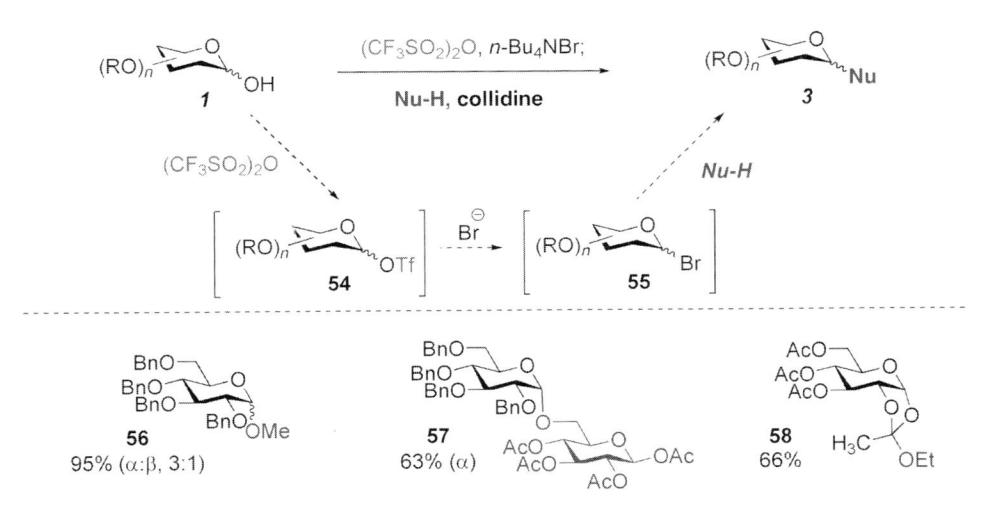

Scheme 3.9 Dehydrative glycosylation with Tf$_2$O.

hemiacetal **1** ensues with triflic anhydride in the presence of the sterically hindered base 2,4,6-collidine at −70 °C to provide the anomeric triflate intermediate **54**. Although the authors report identifying this intermediate by NMR, its exposure to even simple alcohol acceptors led to low yields of product. Increases in reaction efficiency occur with the introduction of tetrabutylammonium bromide, allowing for rapid displacement of the anomeric triflate **54** to form the more stable glycosyl bromide intermediate **55**. Subsequent introduction of a glycosyl acceptor (NuH) provides the desired glycoside **3**.

This method has allowed for the synthesis of methyl glycoside **56** (95%, α : β, 3 : 1) [105] and (1,6)-disaccharide **57** (63%) [106]. The preference for the 1,2-*cis* glycosidic bond in these examples was thought to be the consequence of halide ion catalysis [98]. A limitation of this method was the finding that with a C-2 ester substituted hemiacetal donor, orthoester **58** was isolated as the principal product in the presence of ethanol. Thus, stereoselective formation of β-glycosides appears to be challenging by this approach. Later developments by Pavia and coworkers showed that in the absence of a base and bromide source, self-condensation of the donor occurs to form the 1,1′-linked disaccharide [107–109], although the use of excess acceptor avoids glycosyl donor self-condensation and provides glycoconjugates [110].

Sulfonyl chloride hemiacetal activating agents were first investigated by Leroux and Perlin in the stereoselective synthesis of *O*-alkyl glycosides using methanesulfonyl chloride [106] and later by Szeja using toluenesulfonyl chloride under phase-transfer conditions [111]. These pioneering studies have not yet been extended to the disaccharide synthesis. In 1980, Koto *et al.* reported that a mixture of a hemiacetal glycosyl donor **1** and alcohol acceptor (R^1OH) could be treated with the ternary mixture of *para*-nitrobenzenesulfonyl chloride **59**, triethylamine and silver triflate (AgOTf) chloride scavenger to promote glycosidic bond formation at low temperatures (Scheme 3.10) [112]. The activated hemiacetal donor, in the form of anomeric *p*-nitrobenzenesulfonate **60**, reacts with the acceptor to provide the desired glycoside **3**. Unlike most of the direct dehydrative glycosylations that involve pretreatment of the glycosyl donor with the activating reagents prior to the addition of the alcohol acceptor, in this case, the donor and acceptor are both present when the activating reagents are introduced.

The stereochemistry of the glycosylation was initially found to be highly substrate dependent [113,114]. Advances were made to improve efficiency and generality, whereby addition of *N,N*-dimethylacetamide to the reaction provided optimal α-selectivity without compromising the yield [115]. With this advance, the branched trisaccharide **61** could be obtained in 62%, with α selectivity. Similarly, a C-2 azido galactose hemiacetal, an important class of glycosyl donor in *O*-linked glycopeptide synthesis, reacted to provide disaccharide **62** in 73% [116]. Stereoselectivity in the synthesis of β-glycosides was best achieved by the use of a C-2 participatory group, illustrated by the formation of disaccharide **63** (62%) [117]. Koto and coworkers have applied this stereoselective method to the synthesis of pentosides [118], C-2 aminoglycosides [119] and multiple oligosaccharides [120–125]. The use of common and shelf-stable reagents makes this protocol an attractive choice for glycosylations.

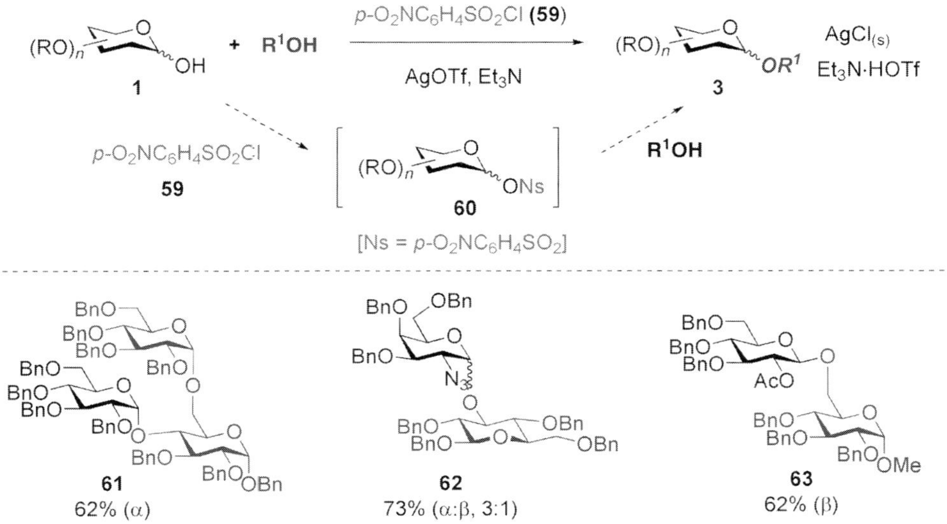

3.1 Hemiacetals and O-Acyl/Carbonyl Derivatives

Scheme 3.10 Dehydrative glycosylation with sulfonyl halides.

Gin and coworkers have developed a sulfonium-based electrophilic activation of hemiacetals (Scheme 3.11). Treatment of diphenyl sulfoxide (2.8 equiv) with triflic anhydride (1.4 equiv) allows for *in situ* formation of the highly reactive sulfonium bistriflate **64** [126]. This putative species engages in efficient activation of hemiacetal **1** over 1 h at −40 °C to provide the activated anomeric oxosulfonium intermediate **65**, which was verified by ^{1}H NMR [127]. With the introduction of a glycosyl acceptor, the glycosidic bond formation occurs to provide glycoside **3**, along with the regeneration of diphenyl sulfoxide. More recent developments have allowed for the use of substoichiometric quantities of sulfoxide reagent, wherein di-(*n*-butyl)sulfoxide (0.2 equiv) can be employed with benzenesulfonic anhydride as the stoichiometric dehydrating reagent [128,129].

Using the diphenyl sulfoxide protocol, a variety of glycoconjugates can be prepared (e.g. **66–69**) [126,127]. The anomeric stereochemistry with nonparticipating C-2 benzyl protective groups is moderate and substrate dependent; however, good 1,2-*trans* β-selectivity is achieved with C-2 ester or C-2 amide directing groups. A number of reports have appeared that extend the method. Seeberger and coworkers have found that glycosylation with a hemiacetal donor incorporating the 4,6-benzylidene directing group [130], allowes for the formation of β-mannosides in moderate to good selectivity [131]. For example, disaccharide **70** was prepared in 79% yield (α : β, 1 : 4). In other developments, (*p*-nitrophenyl)(phenyl) sulfoxide and triflic anhydride were found to be suitable activating reagents for C-2 hemiketal glycosyl donors. Using a C-1 *N,N*-dimethyl glycolamide directing group [132], the sialyl disaccharide **71** was obtained in 63% yield favoring the desired α anomer [133]. The Ph$_2$SO·Tf$_2$O glycosylation has recently been used to establish multiple glycosidic bonds in the synthesis of complex saponin adjuvants for vaccine development [134].

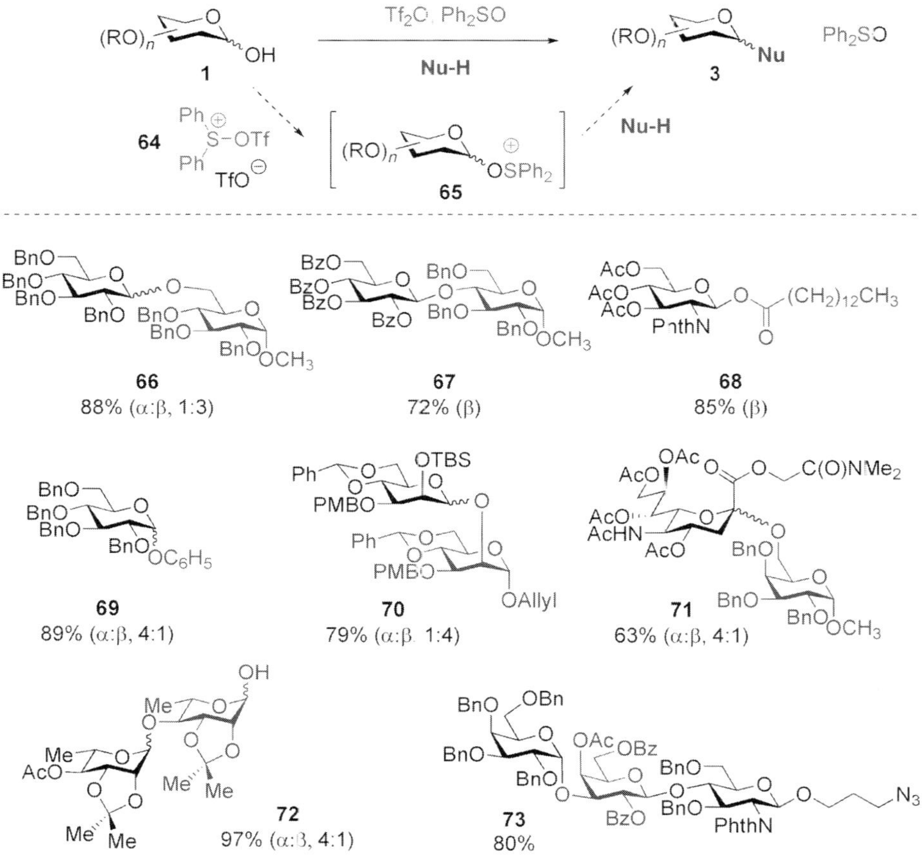

Scheme 3.11 Dehydrative glycosylation with Ph$_2$SO·Tf$_2$O.

Iterative approaches to oligosaccharide formation by this method have also been advanced. A Ph$_2$SO·Tf$_2$O mediated glycosylation of a rhamnopyranose acceptor incorporating both a C-4 hydroxyl and a C-1 hemiacetal was performed, and the C-4 hydroxyl was glycosylated preferentially to form (1,4)-disaccharide **72** in high yield [135]. The disaccharide product **72** presents a hemiacetal that is immediately available for subsequent glycosylation, requiring no additional anomeric protective group manipulations. Recently, a one-pot, three-component trisaccharide synthesis using sulfonium activation of hemiacetals was reported by van der Marel and associates [136]. The Ph$_2$SO·Tf$_2$O dehydration method was used to glycosylate a C-3 hydroxyl acceptor for the in situ formation of an α(1,3)-disaccharide, which contains a latent sulfide donor at the reducing-end. A further addition of triflic anhydride at this time allows for activation of this thioglycoside donor and subsequent glycosylation, forming trisaccharide **73** in 80% yield.

3.1.7
Hemiacetal Activation with Carbon Electrophiles

The earliest examples of carbon-based activation of hemiacetals in a direct dehydrative glycosylation employ hetaryl onium salts, reagents that had previously been used to convert hemiacetals to glycosyl halides [137,138]. In the general approach (Scheme 3.12), the hemiacetal **1** is treated with an *N*-alkyl hetaryl onium salt **74**, which acts as a *C*-electrophile for hemiacetal activation. The resulting *O*-aryl intermediate **76** is susceptible to reaction with a glycosyl acceptor (NuH), leading to the displacement of the anomeric leaving group to form the desired glycoside **3** and the 2-pyridone-derived by-product **75**.

In an early example, Mukaiyama and coworkers used hetaryl onium salts for nucleoside synthesis. The active hetaryl onium salt is generated *in situ* from the reaction of 2-chloro-3-ethylbenzoxazolium tetrafluoroborate **77** and the glycosyl acceptor. With benzimidazole as glycosyl acceptor, the resulting 2-(1-benzimidazoyl)benzoxazolium tetrafluoroborate **78** was obtained. The reaction between the hetaryl onium salt **78** and hemiacetal donor **1** occurs at 60 °C to activate the hemiacetal and thereby reveal the glycosyl acceptor. This procedure led to the formation of nucleoside **80** with exclusive 1,2-*trans* selectivity [139]. The nucleoside **81** was similarly prepared. Alternatively, 2-fluoro-1-methylpyridinium tosylate **79** directly

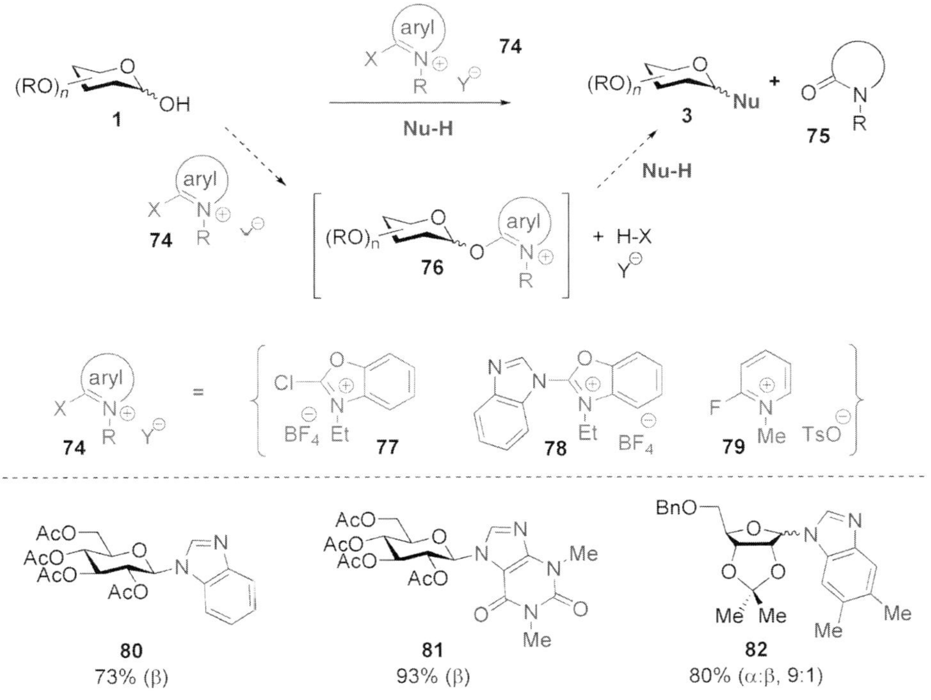

Scheme 3.12 Dehydrative glycosylation with hetaryl onium salts.

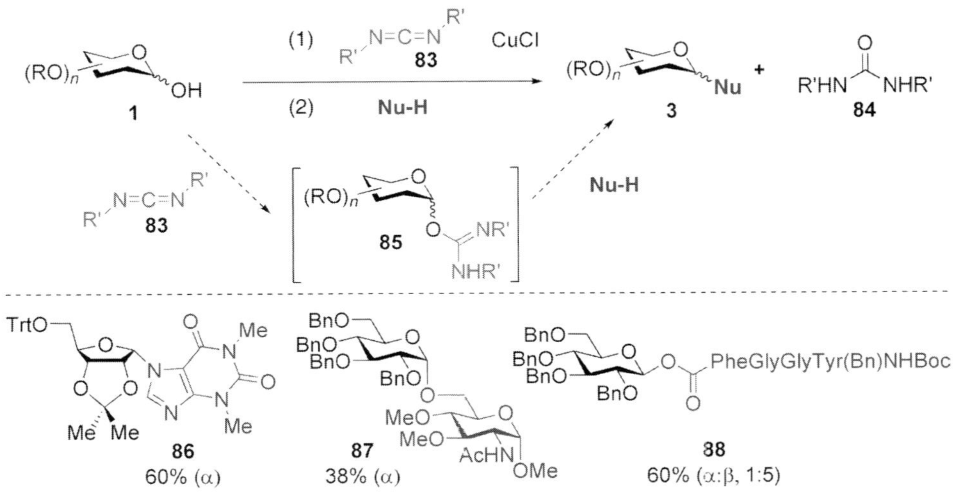

Scheme 3.13 Dehydrative glycosylation with carbodiimides.

promotes dehydrative glycosylations of hemiacetals, providing, for example ribonucleoside **82** in 80% yield [140]. Although research in this area has focused only on the preparation of *N*-glycosides, it nevertheless illustrates an early and intriguing method for dehydrative glycosylation.

In an extension beyond hetaryl onium salt promoted hemiacetal activation, Ishido and coworkers have reported the carbodiimide activation of hemiacetals [141]. In the method (Scheme 3.13), the hemiacetal donor **1** is treated with a carbodiimide electrophile **83** and copper(I) chloride to provide glycosyl isourea intermediate **85**. Highly susceptible to hydrolysis, the isourea **85** was not isolated but could be detected by ^{13}C NMR and IR spectroscopy [142,143]. Accordingly, the reaction between intermediate **85** and the glycosyl acceptor (NuH) provides glycoside product **3**, along with urea by-product **84**.

A typical procedure calls for reaction of the hemiacetal donor with dicyclohexyl carbodiimide and copper(I) chloride (0.1 equiv) at 80 °C, followed by an addition of the acceptor and continued heating. As an early demonstration of this protocol, α-riboside **86** was prepared in moderate yield but with exclusive stereoselectivity [141]. Further measures were required for the glycosylation of monosaccharide acceptors, such as addition of *p*-toluenesulfonic acid (0.1 equiv) to promote the formation of disaccharide **87** [144]. The method was more suitably applied to the synthesis of *O*-acyl glycopeptides, as evidenced by the formation of **88** in 60% yield [143,144]. Various peptides with non-nucleophilic side chains were found to be amenable to this stereoselective reaction. The β-selectivity was suggested to arise from a preponderance of the α-isourea intermediate **85** in the activation step.

Another mode of carbon-based activation of hemiacetals relies on carbonyl-centered electrophiles **89** (Scheme 3.14). These reagents have demonstrated the highest efficiency for disaccharide synthesis among electrophilic carbon activating agents. In the event, the hemiacetal **1** is activated with electrophile **89** for *in situ*

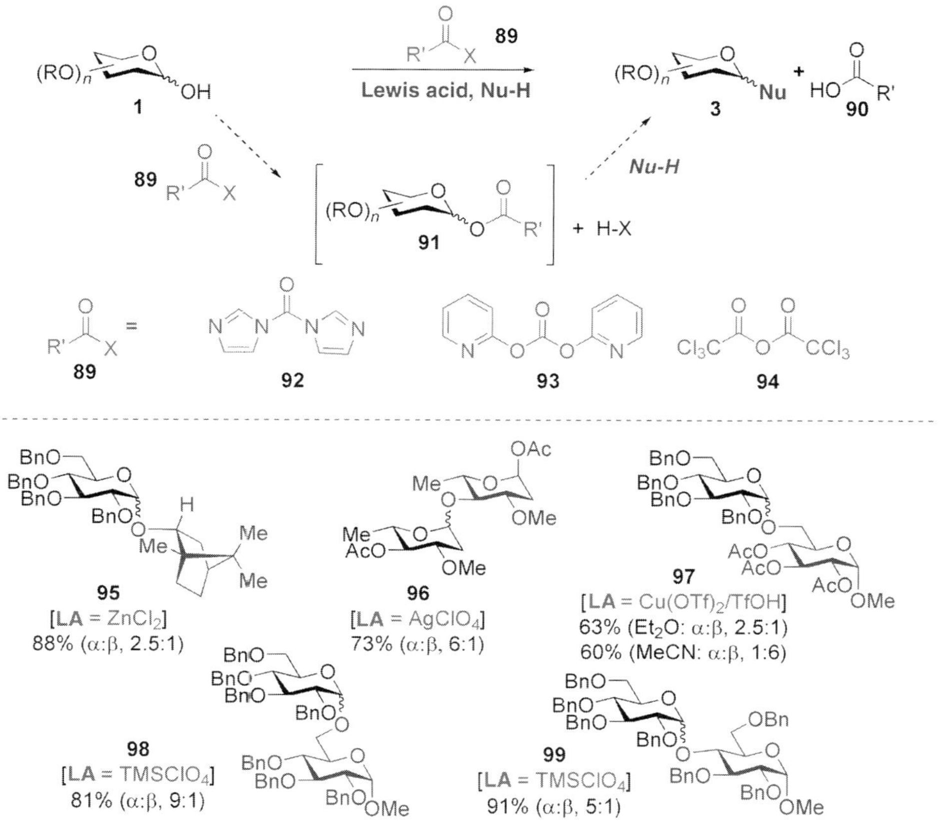

3.1 Hemiacetals and O-Acyl/Carbonyl Derivatives

Scheme 3.14 Dehydrative glycosylation with carbonyl activators.

generation of the glycosyl O-carbonyl intermediate **91**. With Lewis acid assistance, addition of a glycosyl acceptor allows for displacement of the anomeric carboxylate of **91** to give glycoside product **3**. In the case that the active intermediate **91** is a glycosyl carbonate or glycosyl carbamate, carbon dioxide is expelled from the by-product **90** providing added entropic driving force to the reaction.

The earliest examples of one-pot direct dehydrative glycosylations involving this mode of activation was reported by Ford and Ley [145]. Treatment of hemiacetal **1** with 1,1′-carbonyldiimidazole **92** rapidly provides the glycosyl (1-imidazolylcarbonyl) intermediate **91**. The glycosyl carbamate serves as a capable donor with zinc(II) bromide promotion (1 equiv) at elevated temperatures. Through this procedure, the glycosylation of a sterically hindered secondary alcohol afforded glucoside **95** in 88% yield. The method was also applied, with slight modification, in the synthesis of Avermectin B1a, wherein glycoside **96** was formed in 73% yield using silver perchlorate activation [146,147].

Hanessian and coworkers have used a one-pot glycosylation in their development of 2-pyridylcarbonate donors, which were found to be unstable to chromatography

[148]. Hemiacetal **1** is activated with bis(2-pyridyl) carbonate **93** (1.2 equiv) and catalytic *N,N*-dimethyl 4-aminopyridine (DMAP) to allow for *in situ* formation of the corresponding glycosyl carbonate intermediate **91**. This intermediate reacts with alcohol acceptors in the presence of copper(II) triflate (2.5 equiv) and triflic acid (0.5 equiv) to give disaccharides, such as **97**, in moderate yields. The addition of triflic acid was necessary to reduce the amount of unreacted α-glycosyl carbonate, seemingly the less reactive anomeric carbonate intermediate in the glycosylation. The stereoselectivity of this reaction could be reversed by exchanging the reaction solvent from diethyl ether to acetonitrile. In this case, the formation of disaccharide **97** was achieved with a 1:6 α:β ratio in 60% yield. This solvent effect was attributed to the ability of acetonitrile to form an α-glycosyl nitrilium intermediate [149], whereas ether solvents favor β-coordination and the reaction proceeds through the respective intermediate to β or α glycoside product [150,151].

Kusumoto and coworkers have found that the treatment of hemiacetal **1** with trifluoro- or trichloroacetic anhydride **94** (1 equiv) and trimethylsilyl perchlorate (0.2 equiv) selectively provides the corresponding anomeric ester intermediate **91** [152]. Hemiacetal acylation occurs even in the presence of the alcohol acceptor. With Lewis acid assistance, the glycosyl ester intermediate is displaced to provide disaccharide products in good yields. This transformation allowed the synthesis of disaccharides **98** (81%) and **99** (91%). In some cases, acetic anhydride has been used as the electrophilic activator of hemiacetal donors and the reaction with thiol acceptors yields *S*-linked glycosides [153,154].

3.1.8
Other Methods

Activation of hemiacetals with titanium-oxo and tin-sulfide reagents presents an alternative mode of electrophilic activation of hemiacetal donors, though not fundamentally unlike many of the previous methods (Scheme 3.15). Mukaiyama's group has developed much of the chemistry for titanium- and tin-based glycosylation, which has been applied to furanose hemiacetal donors. The original paper reports the use of a [1,2-benzenediolato(2-)-*O,O'*]oxotitanium reagent **103** (catechol titanium oxide, 4 equiv) with triflic anhydride (2 equiv) to provide the postulated titanium bistriflate activating agent **101** [155,156]. The hemiacetal donor **1** reacts with the bis(titanium) complex **101** for over 2 h at −23 °C to form the active glycosyl intermediate **102**. A silyl ether glycosyl acceptor (4 equiv) is then added to the reaction, which in the presence of cesium fluoride, produces riboside products in high yields and anomeric selectivity. For example, the ribosides **105** and **106** were formed in 88 and 94% yields, respectively, favoring the β anomer. It was postulated that coordination of the anomeric oxotitanium moiety of intermediate **102** to the *cis* C-2 oxygen substituent further aided selectivity for the 1,2-*trans* β-glycoside. The reaction was rendered stereoselective upon the introduction of lithium perchlorate (4 equiv) [157], allowing for the generation of disaccharide **106** in 90% yield (α). The lithium perchlorate was proposed to promote dissociation of anomeric oxotitanium species **102**, allowing nucleophiles access to the α-face. Similarly, glycopeptide **107**

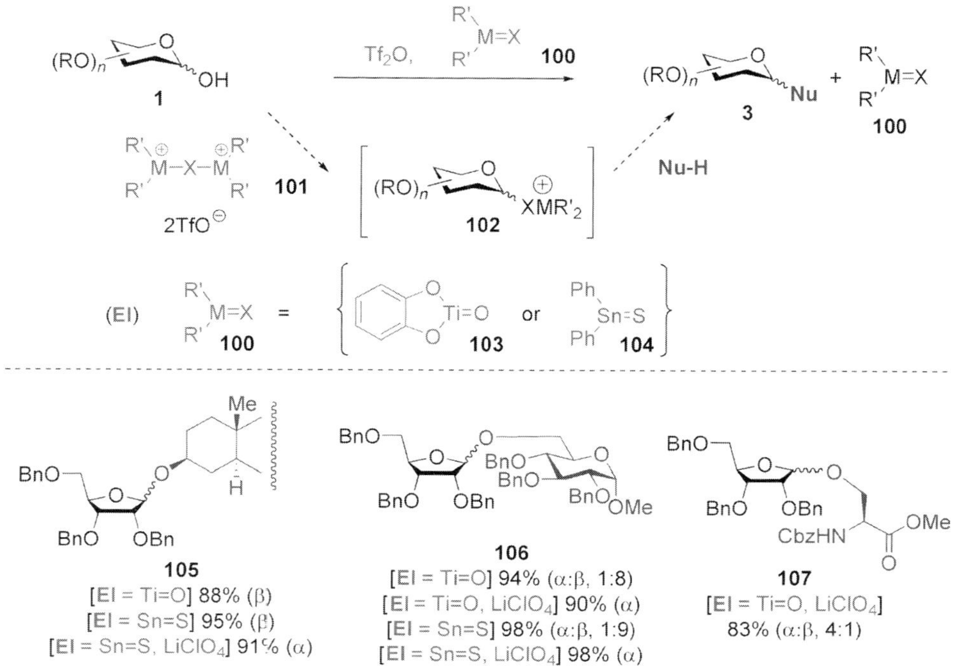

Scheme 3.15 Metal oxide and metal sulfide-promoted glycosylations.

was formed from the glycosylation of the silylated serine acceptor in 83% yield, favoring the α anomer (4:1).

Following from this work, Mukaiyama et al. were able to improve the yields and stereoselectivity using diphenyltin sulfide **104** (1.5 equiv) and triflic anhydride (1.2 equiv) activating agents [158]. Several glycosides were obtained in excellent yield and stereoselectivity, as exemplified by the ribosyl cholesterol **105** (95%, β) and the disaccharide **106** (98%, 1:9 α:β). The α-selective glycosylations with lithium perchlorate additive were equally impressive. The cholesteryl riboside **105** was produced in 91% and the disaccharide **106** in 98% yield. In later work, it was demonstrated that diphenyltin sulfide in combination with silver perchlorate (in the absence of Tf$_2$O) catalyzed the formation of glycosides, albeit with moderate β-selectivity [159]. It was also noted that the combination of Lawesson's reagent and silver salts also effectively catalyzed this transformation. Although these methods have only been demonstrated on ribose hemiacetal donors, the excellent yields and stereoselectivity suggest alluring prospects for the other substrate classes.

Despite the high utility of glycosyl fluorides as stand-alone glycosyl donors, there has been only one example of a direct dehydrative glycosylation whereby hemiacetal activation proceeds through a glycosyl fluoride intermediate. Hirooka and Koto have detailed the use of diethylaminosulfur trifluoride (DAST) for dehydrative glycosylations with hemiacetal donors (Scheme 3.16) [160]. Treatment of a mixture of hemiacetal **1** and alcohol acceptor (R^{1}OH) with DAST **108** (2 equiv) at 0 °C provides the

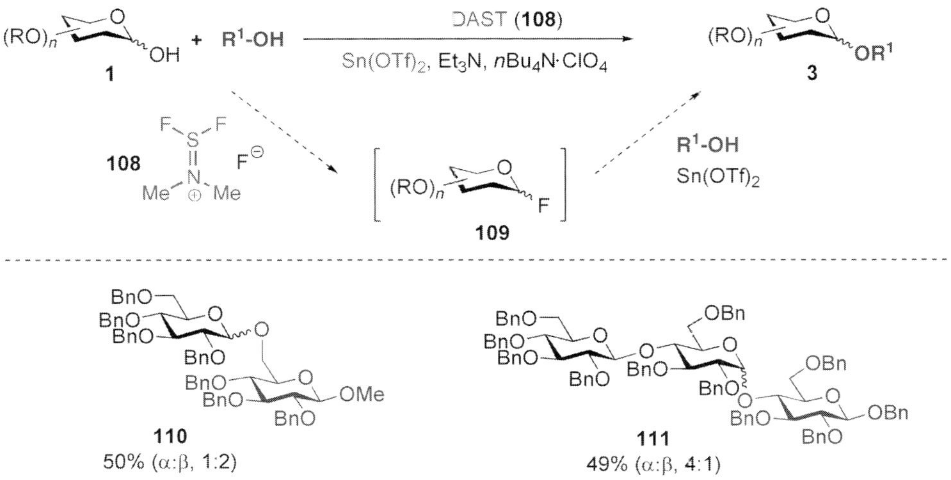

Scheme 3.16 DAST-promoted dehydrative glycosylations.

glycosyl fluoride intermediate **109**. After screening various promoters, an efficient *in situ* conversion of the glycosyl fluoride **109** to glycoside product **3** occurs with Sn(OTf)$_2$. Triethylamine and tetrabutylammonium perchlorate are used in the reaction to minimize the self-condensation of the donor and increase the β-selectivity, respectively.

This procedure has been demonstrated to provide moderate yields and anomeric selectivity in oligosaccharide synthesis. For instance, the disaccharide **110** was obtained in 50% yield as a 1 : 2 α : β ratio. The reaction side products were mainly the self-condensed donor (10–25%) and unreacted hemiacetal (5–10% or higher). Alternatively, the α-linked glycosides were favored with diethyl ether solvent. In this way, trisaccharide **111** was prepared from the disaccharide hemiacetal donor in 49% yield, favoring the α-anomer by 4 : 1.

3.1.9
Glycosylation with Anomeric Esters

Glycosylation with anomeric ester donors is one of the most convenient and simplest approaches to glycosidic bond construction. Advantages of using glycosyl ester donors include their easy preparation and chemical stability, which are characteristics that typically allow these compounds to be prepared in large quantities and stored and handled with relative ease. This presents a practical and convenient option when selecting an approach to glycoside bond formation. In the general reaction (Scheme 3.17), the glycosyl ester **112** is subjected to electrophilic activation of the carbonyl group to provide the reactive intermediate **113**. Subsequent displacement of the anomeric ester by the glycosyl acceptor (NuH) provides the desired glycoside product **3**. A consideration with this class of donor is the anomeric configuration of the starting ester **112**, which can significantly affect the reactivity

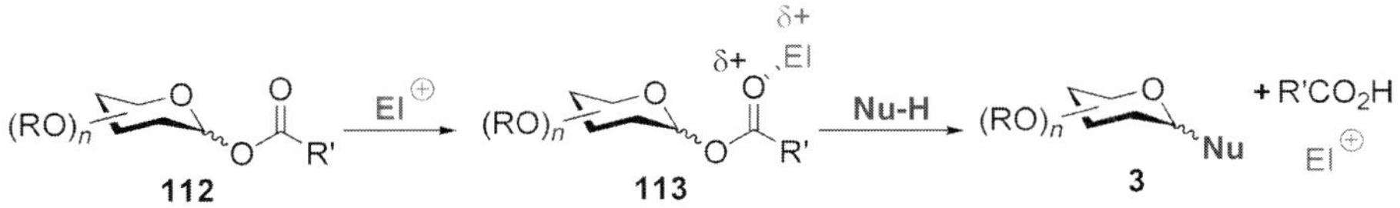

Scheme 3.17 Glycosylations with glycosyl ester donors.

of the glycosyl donor. Indeed, the 1,2-*trans* anomeric ester is most often used in glycosylations, and its higher reactivity is attributed to anchimeric assistance by the C-2 participatory group. Related to this, the nature of the promoter would be expected to play a defining role in the reaction efficiency. Discoveries in the synthesis of glycosides using glycosyl ester donors are discussed below.

3.1.9.1 Glycosyl Acetate and Glycosyl Benzoate Donors

The glycosylation with glycosyl acetate donors was demonstrated by Helferich and Schmitz-Hillerbrecht [151]. In their method (Scheme 3.18), a peracetyl sugar, such as β-D-glucose pentaacetate **114**, is combined with phenol and either zinc(II) chloride or toluenesulfonic acid catalyst. At high temperature and reduced pressure in the absence of solvent, O-aryl glycoside **115** is obtained. The method has been used extensively in the early literature (e.g. [162–166]). For several carbohydrate donors, including glucopyranose, galactopyranose [167], N-acetyl glucosamine, N-acetyl galactosamine [168] and allopyranose [169], the reaction has been reported as being stereoselective, with the outcome depending on the identity of the catalyst and the reaction temperature. Typically, toluenesulfonic acid is used in the reaction at 80–100 °C to favor the 1,2-*trans* β-anomer; however, zinc(II) chloride at higher temperatures leads to the 1,2-*cis* α-anomer even in the presence of a C-2 participatory group. In the latter case, the formation of the α-anomer arises from thermodynamic equilibration under the reaction conditions. A few reports on the preparation of O-aryl disaccharides have been documented [170–175], and the reaction has been used in the synthesis of tyrosine O-aryl glycopeptides [176], ribofuranose nucleosides [177–184] and S-glycosides [185–188]. In these investigations, a diverse set of acid promoters have been successfully employed.

Significant practical advances were made in adapting the process to solution-phase couplings with the appropriate choice of Lewis acid (Scheme 3.19). One of the most commonly used Lewis acids in the solution-phase Helferich glycosylation is tin tetrachloride ($SnCl_4$) [189], which successfully promotes glycosylation with a range of glycosyl acetate donors **116** and acceptors. For instance, the $SnCl_4$ promoted Helferich glycosylation has allowed for the preparation of a variety of O-aryl glycosides [190–192], including the hordenine glycoalkaloid **117** isolated after

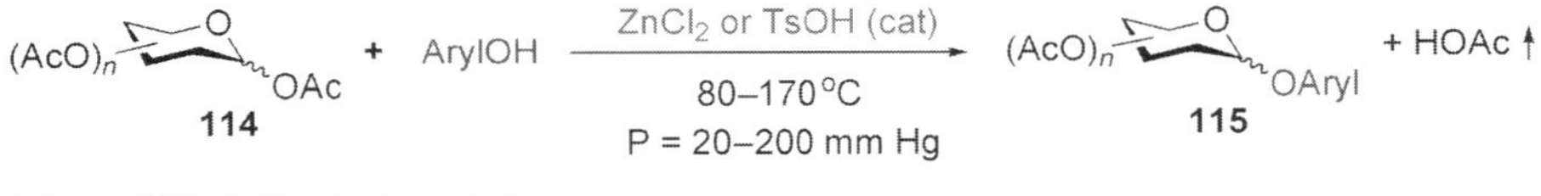

Scheme 3.18 Helferich glycosylation.

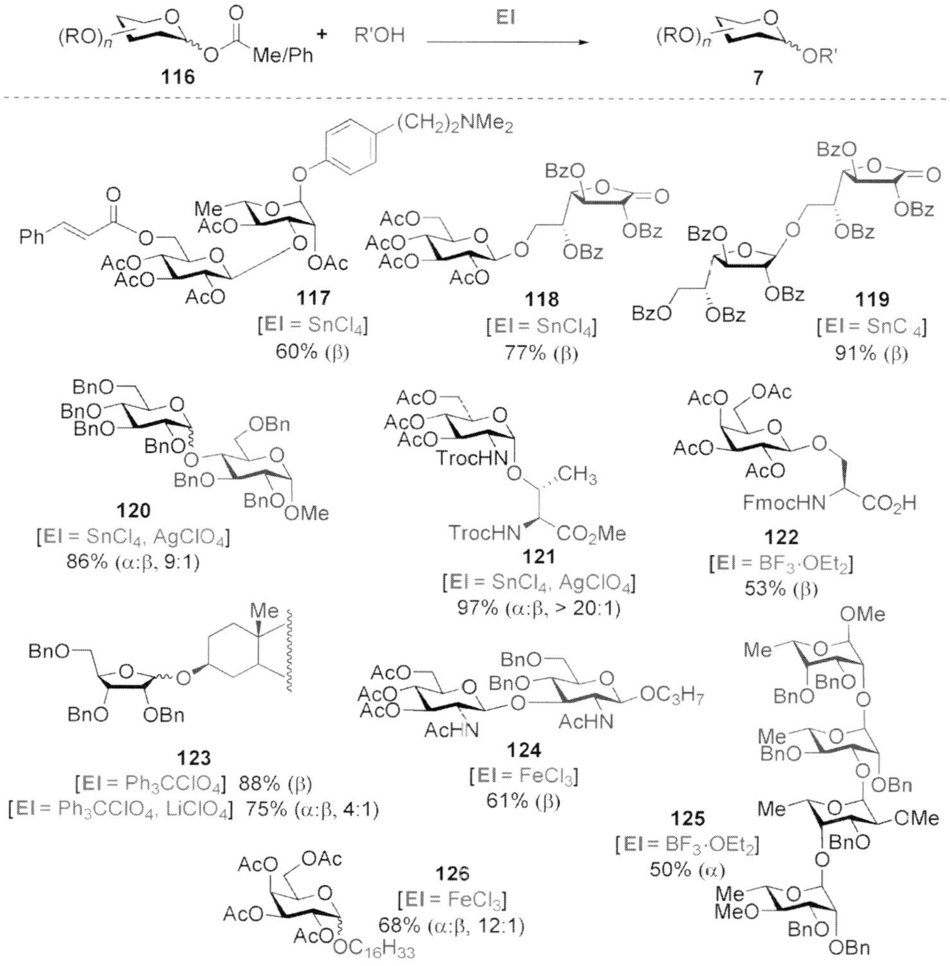

Scheme 3.19 Lewis acid activation of glycosyl ester donors.

24 h at 0 °C [193]. Disaccharides have also been synthesized under similar conditions [194–197], exemplified by the formation of 1,2-*trans* disaccharide **118** in 77% yield [198]. Although less commonly used than anomeric acetate donors, anomeric benzoates have also been activated with stoichiometric SnCl$_4$ [199–201]. With this donor type, disaccharide **119** was obtained in 91% yield after 5 h at ambient temperature [202]. In all of the above examples, the products have been obtained with the expected 1,2-*trans* configurations in high stereoselectivity. Mimicking the thermodynamic control over anomeric stereoselectivity of the original Helferich reaction, it was discovered that employing SnCl$_4$ at elevated temperatures and longer durations can even favor the 1,2-*cis* anomer [190,203–208]. Following this principle, Mukaiyama and coworkers have developed a general protocol to efficiently access 1,2-*cis*

α-glycosides. The combination of SnCl$_4$ and AgOTf (20 mol%) catalyzes the glycosylation with anomeric ester donors and silyl alcohol acceptors to provide good yields of product with high 1,2-*cis* α-stereoselectivity [209–212]. In control experiments, it was shown that the Lewis acid catalyst, postulated to be SnCl$_3^+$ ClO$_4^-$, isomerizes the 1,2-*trans* O-alkyl glycoside to the 1,2-*cis* glycoside. In this way, disaccharide **120** was prepared in 86% yield (α:β, 9:1) over 24 h at 0 °C. Notably, the reaction provides a high 1,2-*cis* α-selectivity even in the presence of C-2 participatory groups. The C2-trichloroethoxycarbonyl (Troc) aminoglycoside **121** was generated in excellent yield and stereoselectivity. It was found that reduction of catalyst loading from 20 to 10% and shortening the reaction time provides the corresponding 1,2-*trans*-linked glycosides in high yield and selectivity. This stereoselective reaction has been adopted for disaccharide and trisaccharide synthesis [213,214]. Taken together, the SnCl$_4$ and SnCl$_4$/AgClO$_4$ mediated glycosylations have demonstrated a significant substrate scope. Apart from the examples described above, various O-alkyl glycosides (e.g. [215–219]) and nucleosides [220–228] have also been achieved by using SnCl$_4$ promotion of the solution-phase Helferich glycosylation.

Boron trifluoride diethyletherate (BF$_3$·OEt$_2$) is another commonly used Lewis acid in solution-phase Helferich glycosylations and was reported as early as 1949 [229,230]. Like SnCl$_4$, BF$_3$·OEt$_2$ promotes glycosylation of phenols with peracetyl donors with high efficiency [231–238]. BF$_3$·OEt$_2$ has also been found to effect C-aryl glycoside formation with armed donors and electron-rich aromatic rings [82,239]. A distinct application of the BF$_3$·OEt$_2$ promoted reaction was demonstrated by Kihlberg and coworkers in a convenient method to glycopeptide fragments by glycosylation of an amino acid alcohol acceptor that incorporates an unprotected C-terminal carboxylate [240,241]. In this reaction, excess BF$_3$·OEt$_2$ (3 equiv) was added to a solution of pentaacetyl galactose and *N*-Fmoc-serine acceptor (1 equiv) to provide the glycoconjugate **122** in 53% yield after 1 h at ambient temperature [242]. By comparison, in the synthesis of L-fucose glycopeptides the use of 6 equiv of BF$_3$·OEt$_2$ over 2 d was demonstrated to reverse the stereoselectivity and provide 1,2-*cis* α-fucosides (35–45% yield) [243]. The 1,2-*cis* stereoselectivity was the result of product anomerization, which has only been reported for L-fucose glycopeptides. A number of other researchers have adopted this general method for glycopeptide bond formation [244–250]. The BF$_3$·OEt$_2$ activation of a glycosyl acetate donors has also been used in oligosaccharide synthesis [251–253]. Gurjar and Viswanadham, in their convergent synthesis of a mycobacterial glycolipid, prepared tetrasaccharide **125** in 50% yield from the coupling of a disaccharide acceptor with a disaccharide acetate donor at 0 °C [254]. Most of the examples with BF$_3$·OEt$_2$ activation result in exclusive formation of the 1,2-*trans* isomer and the formation of 1,2-*cis* glycosides is rare and the efficiency appears to be low [238].

Trityl perchlorate (TrClO$_4$) is a relatively new promoter of the solution phase Helferich glycosylation, though it has shown promising activity for a number of substrates. Early work by Mukaiyama *et al.* established that glycosyl acetate activation with trityl perchlorate (1 equiv) at 0 °C provided ribofuranoside **123** in 88% yield [24]. In this reaction it was observed that the α anomer was first generated and then isomerized to the β anomer under the agency of TrClO$_4$. The authors hypothesized that addition of a mild base would suppress this anomerization without compromising Lewis acid activation of the donor. Interestingly, it was found that addition of $LiClO_4$ and 4 Å molecular sieves provided the corresponding α glycoside of **123** in 75% as a 4 : 1 α : β mixture. The method has been extended to 1,2-*cis* selective glycosylation of glucopyranose acetate donors with C-2 nonparticipatory groups, as well as the synthesis of 1,2-*cis* C-glycosides using a polymer-supported $TrClO_4$ derivative [255]. Trityl perchlorate mediated glycosylation has also been used in the total synthesis of Lepidicin A [256] and Spinosyn A [257].

A number of reports have appeared using $FeCl_3$ activation of anomeric ester donors, though most often with donors incorporating a C-2 amide functionality. Kiso and Anderson advanced the use of $FeCl_3$ for glycosylation with C-2 amide glycosyl acetate donors [258], following from earlier work by Matta and Bahl on the use of $FeCl_3$ for oxazoline synthesis from anomeric acetates [259]. It is widely held that Lewis acid coordinated oxazolines (oxazolinium cations) are the reactive intermediates in glycosylations with these acetate donors [260]. In one example, $FeCl_3$ (1.5 equiv) in combination with $CaSO_4$ and tetramethylurea (TMU) promoted the formation of disaccharide **124** in 61% yield over 2–3 d [261]. The long reaction times are indicative of the low reactivity of C-2 acetamido glycosyl donors, an established challenge with this class of molecules. To aid in the glycosylation of secondary alcohol acceptors, procedures involving excess quantities of both $FeCl_3$ and glycosyl donor have been used [262]. This general method has been adopted by a number of researchers [263–265]. In addition, $FeCl_3$ has been applied to other C-2 amide derivatives, such as *N*-phthaloyl or *N*-chloroacetyl glycosyl acetate donors, wherein electronic tuning of the amide substituent allows for more efficient glycosylations [260,266]. It should be noted that other Lewis acids, such as $SnCl_4$ [267], $BF_3 \cdot OEt_2$ [268], TMSOTf [269–273] and camphor sulfonic acid [274], have all been used to activate C-2 amide glycosyl acetate donors with seemingly comparable efficiency, though TMSOTf was effective at low temperatures. The application of $FeCl_3$ to other, non-C-2 amide donors has provided some interesting results [275,276]. Chatterjee and Nuhn have discovered that the use of $FeCl_3$ at ambient temperature predominately provided the α-galactoside **126** in 68% yield, even with a C-2 ester participatory group in the glycosyl donor.

Trimethylsilyl triflate, a recently developed promoter of glycosyl ester donors, has proven to be generally useful as a consequence of its high electrophilicity [272,277–279]. In these reactions (Scheme 3.20), TMSOTf often can be used in substoichiometric quantities to promote glycosylation with anomeric ester donors (**127**). To this end, the 1,2-*trans*-linked disaccharide **128** was obtained in 75% yield by the action of TMSOTf (0.6 mol equiv) at $-78\,°C$ [280]. This reagent has been further applied to the synthesis of the 2-deoxy galactose segment of the aureolic acid antibiotics. In this work, Durham and Roush reported that the TMSOTf activation (0.2 mol equiv) of a glycosyl acetate donor at $0\,°C$ provided disaccharide **129** in 68% yield (α : β, 1 : 13) [281]. TMSOTf has also been demonstrated in the glycosylation with perbenzoyl glycosyl donors and simple alcohol acceptors [282]. Other silyl triflates have been successfully used as activators, including *tert*-butyldimethylsilyl triflate (TBSOTf) [283], the catalyst system composed of $SiCl_4$ and AgOTf for

Scheme 3.20 Silicon activation of glycosyl ester donors.

nucleoside synthesis [212], and a polymer-supported silyl triflate for deoxyglycoside synthesis [284]. The utility of the silyl triflate protocol is reflected in its use in the synthesis of natural products [82,285,286] and oligosaccharides [287,288].

A variation in electrophilic silicon activation of glycosyl esters includes the use of a nucleophilic promoter to aid in displacement of the anomeric ester (Scheme 3.20). In an early report, Morishima and Mori found that the application of their reagent system used for direct dehydrative glycosylation of hemiacetal donors (Section 3.1.4), namely trimethylsilyl bromide (2 equiv) and cobalt(II) bromide (3 equiv), also promotes glycosylation of ester donors. This protocol first converts the glycosyl ester **127** to a glycosyl bromide intermediate, which in turn reacts with the glycosyl acceptor (NuH) to form glycoside products **3** at room temperature. For example, from this procedure β-glucoside **130** was obtained, but the yield was low (32%) [289]. Gervay-Hague and coworkers have significantly contributed to this approach by intercepting glycosyl iodide intermediates through the action of trimethylsilyl iodide on glycosyl acetate donors. In the reaction, the iodide counterion of the silicon electrophile is sufficiently nucleophilic to displace the activated ester [290]. Thus, the glycosyl iodide intermediate is obtained without purification, and then treated with the acceptor and tetrabutylammonium iodide. From this procedure, glycoside **131** was obtained in 94% after 1.5 h at 65 °C [291]. The efficient formation of the 1,2-*cis* α-glycosides with C-2 nonparticipatory groups results from halide ion catalysis of the intercepted glycosyl iodide intermediate [292]. The method has been extended to solution- and solid-phase oligosaccharide synthesis [293,294], as well as preparation of S-, N- and C-glycosides [67,295,296]. Mukaiyama and coworkers have employed the combination of TMSI and phosphine oxide to effect glycosylation with a glycosyl acetate donor via a proposed anomeric oxophosphonium intermediate [297]. Although the glycosyl oxophosphonium intermediate was not observed by NMR, the phosphine oxide nucleophilic catalysis in glycosidic bond formation was independently demonstrated. The treatment of the glycosyl acetate donor **127** with TMSI, triphenylphosphine oxide and the glycosyl acceptor provides disaccharides after 21 h at room temperature. From this procedure, disaccharides such as **132** or **133** could be obtained in high yield and selectivity. Other researchers have found that in certain cases, silicon activation is not required to access the reactive glycosyl iodide intermediate. Schmid and Waldmann have shown that anomeric trifluoroacetate glycosyl donors undergo displacement by the sole action of lithium iodide (1 equiv) to promote the formation of disaccharides under neutral conditions [298]. In addition, an intriguing report has appeared that uses natural phosphate doped with potassium iodide to promote the reaction of a ribofuranose acetate donor with various silylated nucleobase acceptors [299].

The Helferich glycosylation has been used extensively in carbohydrate chemistry and only the main classes of activating agents are surveyed in this chapter. Besides the acids documented above, many different Brønsted acids (e.g. $HClO_4$ [300], NH_2SO_3H [301], TfOH [302], MK-10 [303]), Lewis acids (e.g. $Sc(OTf)_3$ [304], $Yb(OTf)_3$ [305], $Yb[N(Tf)_2]_3$ [306], $SiCl(OTf)_3$ [307], $AlCl_3$ [220] and Tf_2O [308]) and acid combinations (e.g. $BF_3·OEt_2$ + $Bi(OTf)_3$ [309], TMSCl + $Zn(OTf)_2$ [310], TsOH + $Yb(OTf)_3$ [311]) have been used to similar effect. In terms of variability of the glycosyl ester donor, reports have detailed the use of halogen-substituted acetate donors [24,312], anomeric pivaloate donors [313] and anomeric p-nitrobenzoate donors [314–322] in this reaction.

3.1.10
Activation of O-Carbonyl Derivatives

Apart from glycosyl acetate and benzoate donors, more elaborate O-carbonyl derivatives offer potential for distinct modes of anomeric leaving group activation. Many of these donors (Scheme 3.21) involve activation of an anomeric O-carbonyl derivative **134** at a remote functional group (Y). The activated remote functionality, in turn,

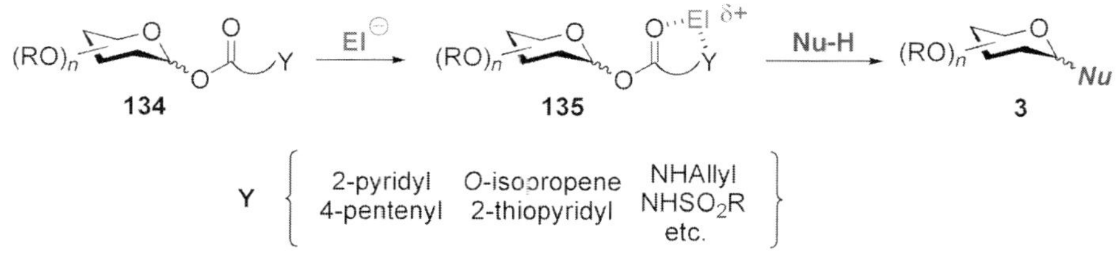

Y	2-pyridyl	O-isopropene	NHAllyl
	4-pentenyl	2-thiopyridyl	$NHSO_2R$
			etc.

Scheme 3.21 Remote activation of C1-O-carbonyl donors.

promotes loss of the anomeric leaving group by an intramolecular reaction with the carbonyl oxygen or through chelated coordination as in intermediate **135**. In the presence of a glycosyl acceptor (NuH), the leaving group is displaced and the glycoside product **3** is formed. This conceptual approach, forwarded by Hanessian and coworkers [323], has the inherent benefit of being able to design a chemoselective pairing of the activating agent and the remote functionality. Consequently, greater possibilities for orthogonal glycosylation are established. In the following section, glycosylation with donors derived from glycosyl esters, O-glycosyl thiocarbonates, glycosyl carbamates and glycosyl carbonates will be summarized.

Glycosyl esters with remote functionality constitute a relatively new class of O-carbonyl glycosyl donors, which fulfill the prospect of mild and chemoselective activation protocols (Scheme 3.22). For example, Kobayashi and coworkers have developed a 2-pyridine carboxylate glycosyl donor **134** (Y = 2-pyridyl), which is activated by the coordination of metal Lewis acid (El$^+$) to the Lewis basic pyridine nitrogen atom and ester carbonyl oxygen atom [324]. In the event, 2-pyridyl(carbonyl) donor **134** and the monosaccharide acceptor were treated with copper(II) triflate (2.2 equiv) in diethyl ether at −50 °C, providing the disaccharide **136** in 70% (α : β,

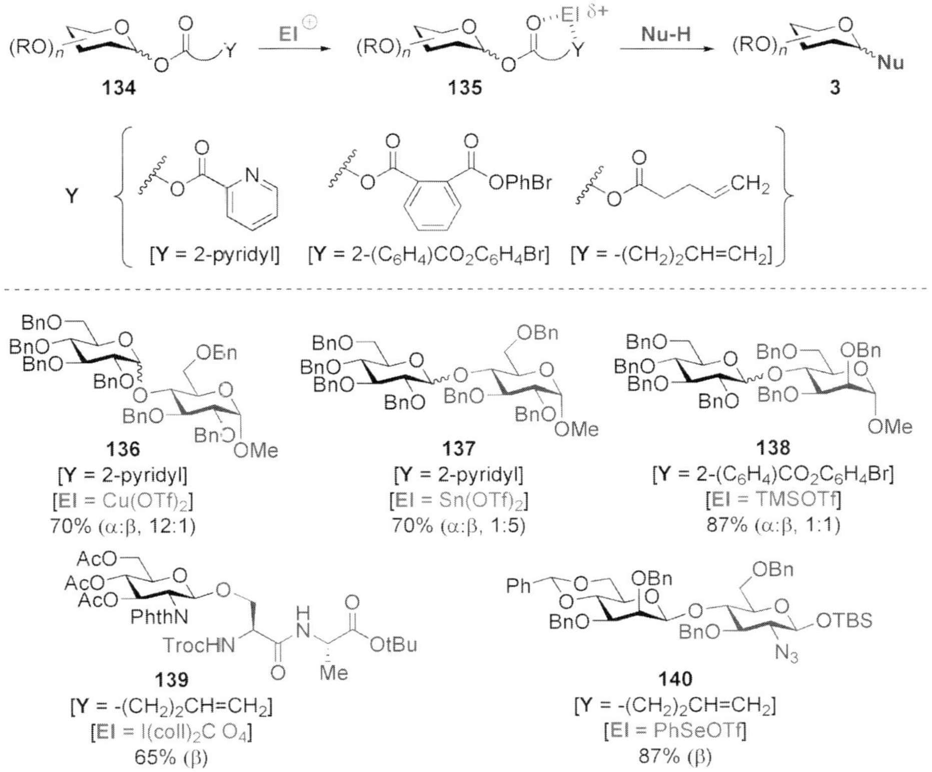

Scheme 3.22 Remote activation of C1-O carbonyl donors.

12 : 1) [325]. Interestingly, it was found that by varying the nature of the Lewis acid and solvent, stereoselectivity could be reversed. Using tin(II) triflate in acetonitrile, 70% yield of disaccharide **137** was isolated, this time favoring the β anomer by 5 : 1. Kim and coworkers have reported glycosylation with glycosyl phthalate donors **134** [Y = 2-(C_6H_4)$CO_2C_6H_4$Br] [326]. The glycosyl phthalate **134** can be activated with TMSOTf (0.5 equiv) at −78 °C to promote the formation of a variety of disaccharides, including **138** (87%, α : β, 1 : 1). Limited stereoselectivity was observed for most of the substrates [327]. Kunz and coworkers have demonstrated the usefulness of anomeric 4-pentenoyl glycosyl donors **134** [Y = −(CH_2)$_2$CH=CH_2] [328], arising from Fraser-Reid's established method using O-pentenyl glycoside donors [329,330]. The 4-pentenoyl glycosyl donors reportedly display increased reactivity relative to O-pentenyl donors. In the reaction, the 4-pentenoyl donor **134** was treated with iodonium di-*sym*-collidine perchlorate at ambient temperature to activate the distal alkene. This remote activation step, in turn, causes γ-lactonization with the anomeric ester carbonyl oxygen of **135** and subsequent expulsion of lactone in the glycosidic bond forming event. From this procedure glycopeptide **139** was obtained in 65% yield with the acid-sensitive *tert*-butyl ester intact. Kim and coworkers have also developed a stereoselective synthesis of β-mannosides by using 4-pentenoyl donor **134** protected with the 4,6-benzylidene stereodirecting group [331,332]. In the reaction, the glycosyl donor is activated with excess phenylselenyl triflate (PhSeOTf) and silver triflate for 15 min at −78 °C, followed by an addition of the acceptor. Glycosylation takes place below 0 °C to provide disaccharides such as **140** in 87% yield, exclusively as the β-anomer. The β-selectivity using the 4-pentenoate glycosyl donors was, in general, high. Innovative methods that use alkynoate ester donors have also been developed to access disaccharides in good yield [333,334].

Through their pioneering efforts in this field, Hanessian's group has devised a number of effective donors in the remote activation approach [8,335]. Covering the topic of remote activation in full is beyond the scope of this chapter, as some of the most useful glycosyl donors include O-heteroaryl glycoside and various S-glycoside donors. However, one relevant class to this discussion is the 2-thiopyridylcarbonate (TOPCAT) donors, compounds such as **141** that are reportedly stable and easily obtainable molecules (Scheme 3.23) [336,337]. It was anticipated that the activation of TOPCAT donor **141** with a Lewis acid (El^+) allows for chelation of the Lewis acid to both the thiopyridyl nitrogen and the carbonyl ester oxygen of intermediate **142**. This two-point coordination aids the displacement of the thiocarbonate moiety to afford the targeted glycoside **3**.

In an early report [335], it was shown that glycosylation with the 2-thiopyridylcarbonate donor **141** and a monosaccharide acceptor proceeds under the promotion of AgOTf (3 equiv) at 0 °C. In diethyl ether cosolvent, this reaction afforded disaccharide **143** in 87%, favoring the α anomer (14 : 1). The TOPCAT donor is also amenable to glycosyl ester protected carbohydrate donors, thereby providing a convenient route to β-linked disaccharides such as **144** with complete anomeric selectivity. The method has been applied to the synthesis of oligosaccharides and sialyl Lewis X mimetics [338–340]. A TOPCAT disaccharide donor was employed with $CuBr_2$ activation to form glycoconjugate **145** in 83%, exclusively as the β anomer

3.1 Hemiacetals and O-Acyl/Carbonyl Derivatives | 125

Scheme 3.23 Remote activation of 2-thiopyridylcarbonate donors.

[341]. The high yields, stereochemical control and mild reaction conditions make this class of O-carbonyl donors an attractive option.

Carbamate glycosyl donors are another useful class of compounds, which present significant variability in chemical structures (Scheme 3.24). In an early example by Kunz and Zimmer, the N-allyl carbamate donors were used in a remote activation protocol to furnish pyranoside and furanoside products. In this reaction, the N-allyl carbamate donor **146** is activated with dimethyl methylthiosulfonium

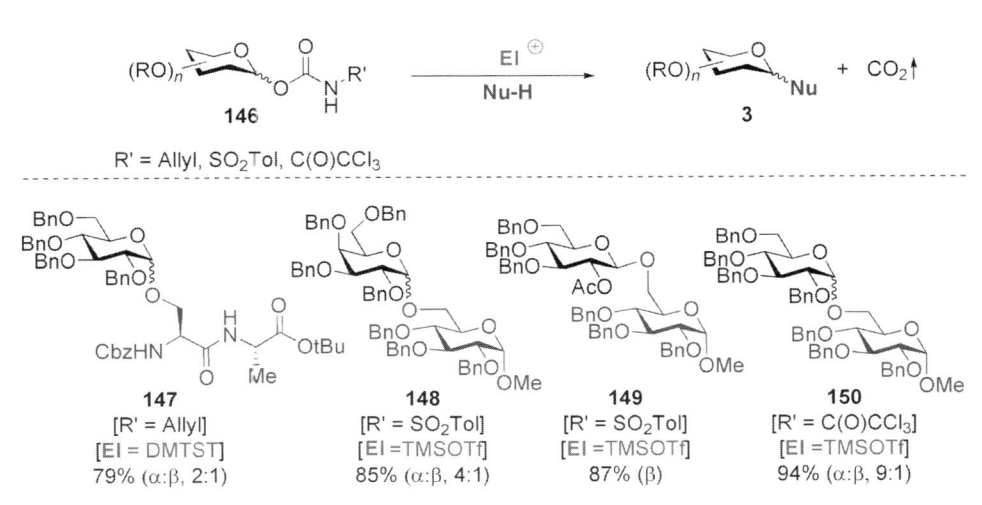

Scheme 3.24 Remote activation of C1-O-carbamate donors.

triflate (DMTST) at room temperature. Subsequent introduction of the glycosyl acceptor affords glycosidic product, such as glycopeptide **147** in 79% yield ($\alpha:\beta$, 2:1) [342]. High stereoselectivity for 1,2-*trans* glycosides was achieved in this reaction using C-2 ester participatory groups. Hinklin and Kiessling have reported on the glycosylation with a variety of N-sulfonylcarbamate glycosyl donors activated by TMSOTf [343]. In the reaction, the N-toluenesulfonyl donor **146** is activated with TMSOTf (1.1 equiv), and upon addition of the monosaccharide acceptor, the desired **3** is obtained. With this protocol, disaccharide **148** was synthesized in 85% yield ($\alpha:\beta$, 4:1). It was also shown that the incorporation of a C-2 ester in the glycosyl donor allows for high yields of the β anomer, illustrated by the preparation of disaccharide **149** in 87% yield. Beyond considerations of anomeric stereoselectivity, further investigations showed that substitution on the sulfonamide nitrogen allowed the reactivity of the glycosyl donor to be tuned. In related work, a variety of substituted carbamate donors and Lewis acid activators were systematically investigated by Redlich's group and others [344,345]. In one example, an N-trichloroacetyl carbamate was activated with TMSOTf (1 equiv) at 0 °C to provide disaccharide **150** in 94% yield ($\alpha:\beta$, 9:1) [346]. Although this example does not react in a remote activation pathway, the ability to tune the electronic character of the glycosyl donor with different substituents again illustrates a beneficial aspect of this donor class.

Glycosyl carbonates constitute another class of glycosyl O-carbonyl donors, which in certain cases show distinct reactivity in the glycosylation event (Scheme 3.25) [347,348]. The carbonate leaving group, upon expulsion of CO_2, has been found to act as the glycosyl acceptor in the reaction. Ishido and coworkers [349–351] first described the heating of the glycosyl phenyl carbonate **151** (R′ = OPh) to 170 °C, which effected decarboxylation to provide the phenyl glycoside **152** in 46% yield along with mixed carbonate by-products. Following the pioneering work by Descotes and coworkers [352], Ikegami and coworkers demonstrated that Lewis acids promote decarboxylation and glycosidic bond formation of bis(mannosyl) carbonates **151** (R′ = Nu = acceptor), even at ambient temperatures [353,354]. Accordingly, disaccharide **153** was prepared in 89% by the treatment of the corresponding mixed carbonate **151** with TMSOTf (1.1 equiv) [355]. A 1,2-*cis* variation of this reaction can be achieved by using the catalyst system composed of $SnCl_4$ and $AgClO_4$ (20 mol %) in diethyl ether solvent. This procedure afforded the α-linked disaccharide **154** in 84% yield as a 9:1 $\alpha:\beta$ ratio [356]. Scheffler and Schmidt have studied the TMSOTf promoted decarboxylative glycosylation through competition experiments between mixed carbonates, and it was concluded that the reaction is an intermolecular process [357].

Simple alkyl carbonates have also been used as glycosyl donors with the addition of an external glycosyl acceptor. Sinay and coworkers have reported that 2-propenyl carbonate donors, such as **151** (R′ = 2-propenyl), are activated with TMSOTf at −25 °C to afford good yields of disaccharide products [358]. Disaccharide **155** was obtained in high yield and α-stereoselectivity after 30 min. Similarly, Mukaiyama *et al.* have used phenyl (R′ = OPh) or methyl (R′ = OMe) carbonates

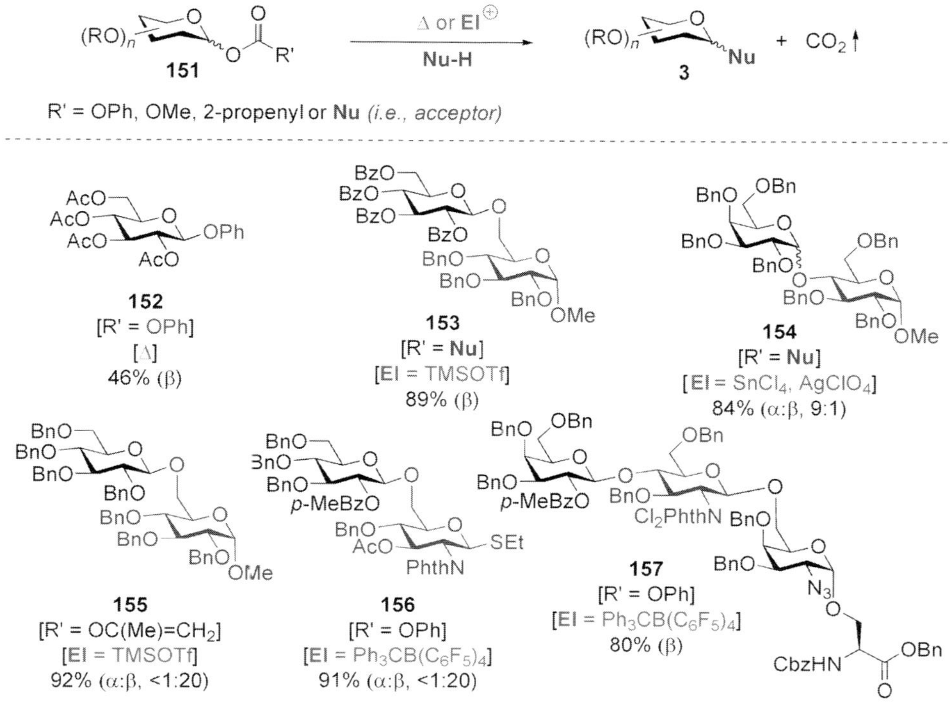

3.1 Hemiacetals and O-Acyl/Carbonyl Derivatives | 127

Scheme 3.25 Glycosylation with C1-O-carbonate donors.

151 as glycosyl donors [359]. By using substoichiometric trityl tetrakis(pentafluorophenyl)borate (10 mol%) as an activator of a phenyl carbonate donor, disaccharide 156 was formed in 91% yield with exclusive β selectivity over 6 h at −20 °C [360]. The stability of the product disaccharide 156, with a latent ethylsulfide leaving group at the reducing end, allowed for the development of one-pot procedures for trisaccharide synthesis [361,362]. In the event, the trisaccharide glycopeptide 157 was prepared by using the trityl perchlorate activation of a galactosyl phenyl carbonate donor to glycosylate a C-4 hydroxyl acceptor, then subsequent *in situ* activation of the latent sulfide donor by addition of N-iodosuccinimide and addition of the C-6 hydroxyl glycopeptide acceptor. This one-pot three-component assembly achieved trisaccharide 157 in 80% yield [363]. Other modes of electrophilic activation of alkyl carbonate donors have been developed, including $SnCl_2$, SbI_3 and TeI_4 [364,365]. In particular, methyl carbonate donors have been activated by the catalyst system composed of either diphenyl tin sulfide and a silver salt or Lawesson's reagent and a silver salt for the preparation of 1,2-*trans*-linked ribonucleosides with good efficiency [366–368]. Trichloromethyl carbonate donors have also been used with TMSOTf or BF_3·OEt_2 activation to effect 1,2-*cis* O-aryl galactopyranosides [369].

3.1.11
Conclusion

There has been a considerable development in glycosylations with C1-hydroxyl donors beyond the classic Fischer glycosylation. These methods employ a wide range of chemistry to effectively deal with the established challenges of the approach, and this is achieved in many cases with good stereocontrol and reaction efficiency. There are many methods that display promising reactivity, but comparatively a few have exhibited expansive substrate scope in the proving ground of oligosaccharide and complex molecule synthesis. It is expected that with continuing developments in this area, the advantages of the direct dehydrative glycosylation will be more fully realized, leading to more efficient reactions with greater contributions to multiglycosylation sequences.

Glycosylations with anomeric acetate and benzoate donors have maintained steady occurrence in the literature. The great advantage of their use is the ease of preparation and stability of the donor. Glycosyl acetate donors have been applied to complex molecule synthesis, though their use in this context has been relatively limited in number. The exploration of anomeric O-carbonyl donors has paved the way for new reactivity in this field. In particular, the remote activation approach allows for mild, chemoselective promotion of various glycosyl donors. The potential of these glycosylations has already been realized in one-pot, iterative oligosaccharide synthesis.

3.1.12
Representative Experimental Procedures

3.1.12.1 Representative Procedure for Preparation of C1-Hemiacetal Donors Through a Peracylation-Selective Anomeric Deacylation Sequence [3,4]

Acetic anhydride (20 equiv) was slowly added to a solution of D-galactose (16.65 mmol, 1 equiv) in dry pyridine (33 ml) at 0 °C. The reaction was stirred at 0 °C for 1 h and then 4-(N,N-dimethylamino)pyridine (0.1 equiv) was added and the reaction was brought to room temperature. After 6 h the reaction mixture was slowly poured into 500 ml of stirring ice water and extracted with ethyl acetate (75 ml). After the evaporation of the ethyl acetate portion and repeated evaporations from toluene, pentaacetyl D-galactose was obtained in 89% yield. The pentaacetyl D-galactose is also widely available from commercial sources. A solution of the peracetyl sugar (12.8 mmol, 1 equiv) and benzylamine (1.5 equiv) in tetrahydrofuran (30 ml) was stirred at room temperature overnight. The mixture was diluted with cold water and extracted with chloroform (3 × 50 ml). The chloroform portion was washed with ice-cold dilute HCl, saturated aqueous $NaHCO_3$, saturated aqueous NaCl, water and then dried (Na_2SO_4), filtered and concentrated. The crude reaction concentrate was purified by silica gel chromatography to provide the acetyl protected C1-hemiacetal.

3.1.12.2 Representative Procedure for Brønsted Acid Promoted Glycosylation with C1-Hemiacetal Donors Using Methoxyacetic Acid [11]

A mixture of hemiacetal donor (1 mmol), glycosyl acceptor (1.2 equiv), methoxyacetic acid (0.1 equiv) and Yb(OTf)$_3$ (0.1 equiv) in dichloromethane (40 ml) was refluxed for 3–5 h under argon through a column of activated 4 Å molecular sieves (28 g).

3.1.12.3 Representative Procedure for Lewis Acid Promoted Glycosylation with C1-Hemiacetal Donors Using Sn(OTf)$_2$ and LiClO$_4$ [20]

The reaction was carried out under an argon atmosphere. Tin(II) trifluoromethanesulfonate and anhydrous calcium sulfate (Drierite) were dried at 100 °C under reduced pressure (0.1 mmHg) for 1 h prior to use. To a stirred suspension of Sn(OTf)$_2$ (0.01 equiv), Drierite (750 mg), LiClO$_4$ (1.5 equiv) and hexamethyldisiloxane (0.10 equiv) in MeNO$_2$ (2 ml) was added a solution of glycosyl donor (0.40 mmol, 1 equiv) and alcohol acceptor (1.2 equiv) in MeNO$_2$ (4 ml) at room temperature. Upon completion of the reaction, as indicated by TLC analysis, saturated aqueous NaHCO$_3$ was added and the mixture was filtered. The organic phase was separated and the aqueous phase was extracted with dichloromethane. The combined organic phases were dried (MgSO$_4$) and concentrated under reduced pressure. The residue was purified by preparative TLC on silica gel.

3.1.12.4 Representative Procedure for Silicon Promoted Glycosylation with C1-Hemiacetal Donors Using Me$_3$SiBr and CoBr$_2$ [42]

The glycosyl donor, glycosyl acceptor, CoBr$_2$, n-Bu$_4$NBr were stored *in vacuo* over P$_2$O$_5$ prior to use. Me$_3$SiBr was used without any pretreatments. Me$_3$SiBr (1 equiv) was stirred into a mixture of glycosyl donor (0.17 mmol, 1 equiv), the glycosyl acceptor (0.8 equiv), CoBr$_2$ (1 equiv), n-Bu$_4$NBr (1 equiv) and 4 Å molecular sieves (180 mg) in CH$_2$Cl$_2$ (0.45 ml). The resulting mixture was stirred at ambient temperature (22–28 °C). Upon completion of the reaction, the reaction mixture was filtered and the filtrate evaporated and purified by silica gel chromatography.

3.1.12.5 Representative Procedure for Mitsunobu-Type Glycosylation with C1-Hemiacetal Donors and Phenol Glycosyl Acceptors [90]

The reaction was conducted in flame-dried glassware under a dry nitrogen atmosphere. A solution of glycosyl donor (0.200 mmol, 1 equiv), phenol (1.4 equiv) and Ph$_3$P (1.5 equiv) in toluene (3 ml) was stirred with 4 Å molecular sieves (~100 mg) for 0.5 h and cooled to 0 °C. DEAD (2 equiv) was added, and the reaction mixture was stirred overnight. The mixture was then diluted with EtOAc and filtered. The purification of the reaction mixture was performed by Kieselgel flash chromatography and preparative TLC.

3.1.12.6 Representative Procedure for Appel-Type Glycosylation with C1-Hemiacetal Donors [102]

Reactions were carried out in a glass vessel closed with a septum cap. Neither molecular sieve nor drying gas was used. The glycosyl donor (0.41 mmol, 1 equiv) in CH$_2$Cl$_2$ (3 ml) was treated with Ph$_3$P (3 equiv) and CBr$_4$ (3 mol equiv) and stirred for 3 h at room temperature. Then, the N,N-tetra-methylurea (300 µl) and the glycosyl acceptor (3 equiv) were added and stirred at room temperature. The reaction was monitored by TLC analysis until the bromide donor was completely consumed. The reaction mixture was then diluted with $CHCl_3$ and washed with saturated aqueous $NaHCO_3$ and aqueous NaCl solution and then dried (Na_2SO_4) and concentrated. The product was purified by silica gel column chromatography.

3.1.12.7 Representative Procedure for Nosyl Chloride Promoted Glycosylation with C1-Hemiacetal Donors [115]

A mixture of a glucosyl acceptor (0.33 mmol, 1 equiv), glycosyl donor (1.3 equiv), 4-nitrobenzenesulfonyl chloride (2.5 equiv), silver trifluoromethanesulfonate (2.5 equiv) and dichloromethane (1.8 ml) was successively treated with N,N-dimethylacetamide (2.5 equiv for secondary alcohol acceptor or 5 equiv for primary alcohol acceptor) and triethylamine (2.5 equiv) under stirring at $-40\,°C$ bath temperature. The bath temperature was gradually raised to $0\,°C$ over 1 h and then the reaction was stirred overnight at $0\,°C$. At this time, solid $NaHCO_3$ was added and the reaction was brought to room temperature with stirring. The reaction mixture was filtered and concentrated, then purified by silica gel chromatography. Contamination of any trace nitrogenous compounds in the glucosides was removed by rechromatography on silica gel in hexane–ethyl acetate.

3.1.12.8 Representative Procedure for Diphenyl Sulfoxide and Triflic Anhydride Promoted Glycosylation with C1-Hemiacetal Donors [127]

The reaction was performed in flame-dried modified Schlenk (Kjeldahl shape) flask fitted with a glass stopper or rubber septum under a positive pressure of argon. Trifluoromethanesulfonic anhydride (1.4 equiv) was added to a solution of glycosyl donor (0.191 mmol, 1 equiv) and diphenyl sulfoxide (2.8 equiv) in a mixture of toluene and dichloromethane (8 ml, 3:1 vol/vol) at $-78\,°C$. The reaction mixture was stirred at this temperature for 5 min and then at $-40\,°C$ for 1 h. At this time, 2-chloropyridine (5.0 equiv) and the glycosyl acceptor (3.0 equiv) were added sequentially at $-40\,°C$. The solution was stirred at this temperature for 1 h, then at $0\,°C$ for 30 min and finally at $23\,°C$ for 1 h before the addition of excess triethylamine (10 equiv). The reaction was diluted with dichloromethane (100 ml) and was washed sequentially with saturated aqueous sodium bicarbonate solution (2×100 ml) and saturated aqueous sodium chloride (100 ml). The organic layer was dried (sodium sulfate) and concentrated. The residue was purified by silica gel flash column chromatography.

3.1.12.9 Representative Procedure for Carbodiimide Promoted Glycosylation with C1-Hemiacetal Donors [144]

A mixture of glycosyl donor (2.0 mmol, 1 equiv), N,N'-dicyclohexylcarbodiimide (1 equiv) and copper(I) chloride (0.01 equiv) was fused for 0.5 h at $80\,°C$. The glycosyl acceptor (0.5 equiv) was added to the resulting melt, which was further heated for 1 h at $80\,°C$. The melt was dissolved in dichloromethane (40 ml) and the N,N'-dicyclohexylurea was removed by filtration. The filtrate was evaporated and the residue was purified by silica gel column chromatography to provide product glycoside.

3.1.12.10 Representative Procedure for Carbonyl Promoted Glycosylation with C1-Hemiacetal Donors Using Trichloroacetic Anhydride [152]

To a suspension of glycosyl donor (1.5 mmol, 1 equiv), glycosyl acceptor (0.7 equiv) and 5 Å molecular sieves (500 mg) in dry diethyl ether (4 ml), AgClO$_4$ (0.2 equiv), (CCl$_3$CO)$_2$O (1 equiv) and TMSCl (0.2 equiv) was added under a N$_2$ atmosphere. The mixture was stirred at room temperature for 8 h. Ethyl acetate and saturated aqueous NaHCO$_3$ solution were added to the mixture and 5 Å molecular sieves were removed by filtration. The organic later was washed with aqueous NaCl, dried (MgSO$_4$) and concentrated *in vacuo*. The residue was purified by silica gel column chromatography.

3.1.12.11 Representative Procedure for Lewis Acid Promoted Glycosylation with Glycosyl Acetate Donors Using SnCl$_4$ [198]

A solution of the glycosyl donor (0.51 mmol, 1 equiv) in MeCN (6 ml) was chilled at 0 °C (ice water bath) and SnCl$_4$ (0.5 mmol, 1 equiv) was added. The solution was stirred for 10 min at 0 °C and the glycosyl acceptor (1 equiv) was added. The reaction was monitored by TLC until no starting material remained. The reaction mixture was diluted with CH$_2$Cl$_2$ (50 ml) and washed with saturated aqueous NaHCO$_3$ (2 × 30 ml) and water, then dried (MgSO$_4$) and evaporated. The residue obtained was purified by column chromatography on silica gel. The product glycoside thus obtained could be further purified by dissolution in Et$_2$O and precipitation by the addition of hexane. The resulting syrup was dissolved in hot EtOH and the product glycoside was crystallized upon cooling with seed crystals.

3.1.12.12 Representative Procedure for Iodotrimethylsilane and Phosphine Oxide Promoted Glycosylation with Glycosyl Acetate Donors [297]

The reaction was carried out under an argon atmosphere in dried glassware. To a stirred suspension of 5 Å molecular sieves (240 mg) and glycosyl acetate donor (0.12 mmol, 1 equiv) in CH$_2$Cl$_2$ (1.2 ml) was added iodotrimethylsilane (1 equiv) at 0 °C. After stirring for 30 min, Ph$_3$P=O (2 equiv) and then glycosyl acceptor (0.7 equiv) were added. The reaction mixture was stirred at room temperature until TLC analysis indicated that the reaction was completed. The reaction mixture was then diluted with EtOAc, filtered through Celite and evaporated. The resulting residue was purified by preparative TLC on silica gel.

3.1.12.13 Representative Procedure for Lewis Acid Promoted Glycosylation with TOPCAT Glycosyl Donor Using Silver Triflate [335]

The reaction was performed under an argon atmosphere. A mixture of the glycosyl donor (0.185 mmol, 1 equiv), glycosyl acceptor (0.7 equiv) and activated powdered 4 Å molecular sieves (200 mg) in 6 ml of Et$_2$O–CH$_2$Cl$_2$ (5 : 1 vol/vol) was stirred overnight at room temperature and then cooled to 0 °C. Silver triflate (3 equiv) was added to the reaction mixture and stirred for 5 h at 0 °C. The resulting suspension was then treated with a few drops of pyridine, filtered through Celite and concentrated. Purification was performed by flash column chromatography on silica gel.

3.1.12.14 Representative Procedure for TMS Triflate Promoted Glycosylation with Glycosyl N-Tosyl Carbamate Donors [343]

Reaction was carried out in an oven-dried glassware under a nitrogen atmosphere. Prior to reaction, the glycosyl donor (0.10 mmol, 1 equiv) and glycosyl acceptor (1.5 equiv) were azeotroped three times with dry toluene. The resulting residue was dissolved in dry Et_2O (1.0 ml) and TMOTf (1.1 equiv) was added dropwise, followed by stirring under nitrogen for 1.5 h. The reaction was quenched by adding solid $NaHCO_3$ and the Et_2O was removed under reduced pressure. The residue was purified using silica gel flash column chromatography.

3.1.12.15 Representative Procedure for Trityl Salt Promoted Glycosylation with Glycosyl Phenyl Carbonate Donors [360]

To a stirred suspension of trityl tetrakis (pentafluorophenyl) borate (4.6 mg, 0.005 mmol) and Drierite (250 mg) in a mixture of pivalonitrile and dichloromethane (0.45 ml, 2 : 1 vol/vol), a solution (pivalonitrile : dichloromethane = 2 : 1, 0.8 ml) of glycosyl acceptor (1.2 equiv) and glycosyl donor (0.05 mmol, 1 equiv) at $-20\,°C$ was successively added. After the reaction mixture was stirred for 6 h at $-20\,°C$, it was quenched by adding saturated aqueous $NaHCO_3$ (10 ml). The mixture was filtered through Celite and extracted with dichloromethane (three times, each of 20 ml). The combined organic layers were washed with brine (5 ml) and the organic layer was dried (Na_2SO_4), filtered and concentrated. The resulting residue was purified by preparative TLC on silica gel.

References

1 Nicolaou, K.C., Snyder, S.A., Nalbandian, A.Z. and Longbottom, D.A. (2004) *Journal of the American Chemical Society*, **126**, 6234–6235.
2 Bieg, T. and Szeja, W. (1990) *Carbohydrate Research*, **205**, C10–C11.
3 Maier, M.A., Yannopoulos, C.G., Mohamed, N., Roland, A., Fritz, H., Mohan, V., Just, G. and Manoharan, M. (2003) *Bioconjugate Chemistry*, **14**, 18–29.
4 Sim, M.M., Kondo, H. and Wong, C.H. (1993) *Journal of the American Chemical Society*, **115**, 2260–2267.
5 Liaigre, J., Dubreuil, D., Pradere, J.P. and Bouhours, J.F. (2000) *Carbohydrate Research*, **325**, 265–277.
6 Fischer, E. (1893) *Chemische Berichte*, **26**, 2400–2412.
7 Capon, B. (1969) *Chemical Reviews*, **69**, 407–498.
8 Hanessian, S. and Lou, B.L. (2000) *Chemical Reviews*, **100**, 4443–4464.
9 Koto, S., Morishima, N. and Zen, S. (1976) *Chemistry Letters*, **5**, 1109–1110.
10 Koto, S., Morishima, N. and Zen, S. (1982) *Bulletin of the Chemical Society of Japan*, **55**, 1543–1547.
11 Inanaga, J., Yokoyama, Y. and Hanamoto, T. (1993) *Journal of the Chemical Society. Chemical communications*, 1090–1091.
12 Aoyama, N. and Kobayashi, S. (2006) *Chemistry Letters*, **35**, 238–239.
13 Shirakawa, S. and Kobayashi, S. (2007) *Organic Letters*, **9**, 311–314.
14 Jyojima, T., Miyamoto, N., Ogawa, Y., Matsumura, S. and Toshima, K. (1999) *Tetrahedron Letters*, **40**, 5023–5026.
15 Chandra, T. and Brown, K.L. (2005) *Tetrahedron Letters*, **46**, 2071–2074.
16 Toshima, K., Nagai, H. and Matsumura, S. (1999) *Synlett*, 1420–1422.

17 Kozhevnikov, I.V. (1998) *Chemical Reviews*, **98**, 171–198.
18 Klemer, A. and Dietzel, B. (1970) *Tetrahedron Letters*, **11**, 275–278.
19 Lubineau, A. and Fischer, J.C. (1991) *Synthetic Communications*, **21**, 815–818.
20 Mukaiyama, T., Matsubara, K. and Hora, M. (1994) *Synthesis*, 1368–1373.
21 Yamada, H. and Hayashi, T. (2002) *Carbohydrate Research*, **337**, 581–585.
22 Ogawa, A. and Curran, D.P. (1997) *Journal of Organic Chemistry*, **62**, 450–451.
23 Suzuki, T., Watanabe, S., Yamada, T. and Hiroi, K. (2003) *Tetrahedron Letters*, **44**, 2561–2563.
24 Mukaiyama, T., Kobayashi, S. and Shoda, S. (1984) *Chemistry Letters*, **13**, 907–910.
25 Uchiro, H. and Mukaiyama, T. (1996) *Chemistry Letters*, **25**, 271–272.
26 Wagner, B., Heneghan, M., Schnabel, G. and Ernst, B. (2003) *Synlett*, 1303–1306.
27 Uchiro, H. and Mukaiyama, T. (1996) *Chemistry Letters*, **25**, 79–80.
28 Takeuchi, K., Higuchi, S. and Mukaiyama, T. (1997) *Chemistry Letters*, **26**, 969–970.
29 Evans, M.E. and Angyal, S.J. (1972) *Carbohydrate Research*, **25**, 43–48.
30 Ferrieres, V., Bertho, J.-N. and Plusquellec, D. (1995) *Tetrahedron Letters*, **36**, 2749–2752.
31 Bertho, J.-N., Ferrieres, V. and Plusquellec, D. (1995) *Journal of the Chemical Society. Chemical communications*, 1391–1393.
32 Velty, R., Benvegnu, T., Gelin, M., Privat, E. and Plusquellec, D. (1997) *Carbohydrate Research*, **299**, 7–14.
33 Regeling, H., Zwanenburg, B. and Chittenden, G.J.F. (1998) *Carbohydrate Research*, **314**, 267–272.
34 Bornaghi, L.F. and Poulsen, S.A. (2005) *Tetrahedron Letters*, **46**, 3485–3488.
35 Corma, A., Iborra, S., Miquel, S. and Primo, J. (1996) *Journal of Catalysis*, **161**, 713–719.
36 Camblor, M.A., Corma, A., Iborra, S., Miquel, S., Primo, J. and Valencia, S. (1997) *Journal of Catalysis*, **172**, 76–84.

37 Corma, A., Iborra, S., Miquel, S. and Primo, J. (1998) *Journal of Catalysis*, **180**, 218–224.
38 Chapat, J.-F., Finiels, A., Joffre, J. and Moreau, C. (1999) *Journal of Catalysis*, **185**, 445–453.
39 van der Heijden, A.M., Lee, T.C., van Rantwijk, F. and van Bekkum, H. (2002) *Carbohydrate Research*, **337**, 1993–1998.
40 Koto, S., Morishima, N. and Zen, S. (1976) *Chemistry Letters*, **5**, 61–64.
41 Koto, S., Morishima, N. and Zen, S. (1979) *Bulletin of the Chemical Society of Japan*, **52**, 784–788.
42 Koto, S., Morishima, N., Kusuhara, C., Sekido, S., Yoshida, T. and Zen, S. (1982) *Bulletin of the Chemical Society of Japan*, **55**, 2995–2999.
43 Hirooka, M., Yoshimura, A., Saito, I., Ikawa, F., Uemoto, Y., Koto, S., Takabatake, A., Taniguchi, A., Shinoda, Y. and Morinaga, A. (2003) *Bulletin of the Chemical Society of Japan*, **76**, 1409–1421.
44 Susaki, H. (1994) *Chemical and Pharmaceutical Bulletin*, **42**, 1917–1918.
45 Uchiyama, T. and Hindsgaul, O. (1996) *Synlett*, 499–501.
46 Alcaro, S., Arena, A., Neri, S., Ottana, R., Ortuso, F., Pavone, B. and Vigorita, M.G. (2004) *Bioorganic and Medicinal Chemistry*, **12**, 1781–1791.
47 Izumi, M., Fukase, K. and Kusumoto, S. (2002) *Bioscience Biotechnology, and Biochemistry*, **66**, 211–214.
48 Fischer, B., Nudelman, A., Ruse, M., Herzig, J., Gottlieb, H.E. and Keinan, E. (1984) *Journal of Organic Chemistry*, **49**, 4988–4993.
49 Tietze, L.F. and Fischer, R. (1981) *Tetrahedron Letters*, **22**, 3239–3242.
50 Kiyoi, T. and Kondo, H. (1998) *Bioorganic and Medicinal Chemistry Letters*, **8**, 2845–2848.
51 Posner, G.H. and Bull, D.S. (1996) *Tetrahedron Letters*, **37**, 6279–6282.
52 Haines, A.H. (2003) *Carbohydrate Research*, **338**, 813–818.

53 Koto, S., Yago, K., Zen, S., Tomonaga, F. and Shimada, S. (1986) *Bulletin of the Chemical Society of Japan*, **59**, 411–414.

54 Tietze, L.F., Fischer, R. and Guder, H.J. (1982) *Tetrahedron Letters*, **23**, 4661–4664.

55 Yoshimura, J., Hara, K., Sato, T. and Hashimoto, H. (1983) *Chemistry Letters*, **12**, 319–320.

56 Tietze, L.F., Fischer, R., Guder, H.J., Goerlach, A., Neumann, M. and Krach, T. (1987) *Carbohydrate Research*, **164**, 177–194.

57 Kolar, C. and Kneissl, G. (1990) *Angewandte Chemie (International Edition in English)*, **29**, 809–811.

58 Kolar, C., Kneissl, G., Wolf, H. and Kampchen, T. (1990) *Carbohydrate Research*, **208**, 111–116.

59 Kolar, C., Kneissl, G., Knodler, U. and Dehmel, K. (1991) *Carbohydrate Research*, **209**, 89–100.

60 Mukaiyama, T. and Matsubara, K. (1992) *Chemistry Letters*, **21**, 1041–1044.

61 Allevi, P., Anastasia, M., Ciuffreda, P., Bigatti, E. and Macdonald, P. (1993) *Journal of Organic Chemistry*, **58**, 4175–4178.

62 Petrakova, E. and Glaudemans, C.P.J. (1995) *Carbohydrate Research*, **268**, 35–46.

63 Noecker, L., Duarte, F. and Giuliano, R.M. (1998) *Journal of Carbohydrate Chemistry*, **17**, 39–48.

64 Daley, L., Guminski, Y., Demerseman, P., Kruczynski, A., Etievant, C., Imbert, T., Hill, B.T. and Monneret, C. (1998) *Journal of Medicinal Chemistry*, **41**, 4475–4485.

65 Daley, L., Mouton, C., Tillequin, F., Seguin, E., Florent, J.-C. and Monneret, C. (1998) *Synthetic Communications*, **28**, 61–73.

66 Noecker, L., Duarte, F., Bolton, S.A., McMahon, W.G., Diaz, M.T. and Giuliano, R.M. (1999) *Journal of Organic Chemistry*, **64**, 6275–6282.

67 Bhat, A.S. and Gervay-Hague, J. (2001) *Organic Letters*, **3**, 2081–2084.

68 Bedford, A.F. and Mortimer, C.T. (1960) *Journal of Chemical Society*, 1622–1625.

69 Mitsunobu, O. (1981) *Synthesis*, 1–28.

70 Szarek, W.A., Depew, C., Jarrell, H.C. and Jones, J.K.N. (1975) *Journal of the Chemical Society. Chemical communications*, 648–649.

71 Nakano, J., Ichiyanagi, T., Ohta, H. and Ito, Y. (2003) *Tetrahedron Letters*, **44**, 2853–2856.

72 Szarek, W.A., Depew, C. and Jones, J.K.N. (1976) *Journal of Heterocyclic Chemistry*, **13**, 1131–1133.

73 Jurczak, J., Grynkiewicz, G. and Zamojski, A. (1975) *Carbohydrate Research*, **39**, 147–150.

74 Peterson, M.L. and Vince, R. (1991) *Journal of Medicinal Chemistry*, **34**, 2787–2797.

75 Seela, F. and Gabler, B. (1994) *Helvetica Chimica Acta*, **77**, 622–630.

76 Grochowski, E. and Jurczak, J. (1976) *Carbohydrate Research*, **50**, C15–C16.

77 Grochowski, E. and Stepowska, H. (1988) *Synthesis*, 795–797.

78 Nicolaou, K.C. and Groneberg, R.D. (1990) *Journal of the American Chemical Society*, **112**, 4085–4086.

79 Groneberg, R.D., Miyazaki, T., Stylianides, N.A., Schulze, T.J., Stahl, W., Schreiner, E.P., Suzuki, T., Iwabuchi, Y., Smith, A.L. and Nicolaou, K.C. (1993) *Journal of the American Chemical Society*, **115**, 7593–7611.

80 Chida, N., Ohtsuka, M. and Ogawa, S. (1988) *Chemistry Letters*, **17**, 969–972.

81 Chida, N., Ohtsuka, M., Nakazawa, K. and Ogawa, S. (1991) *Journal of Organic Chemistry*, **56**, 2976–2983.

82 Duynstee, H.I., De Koning, M.C., Van der Marel, G.A. and Van Boom, J.H. (1999) *Tetrahedron*, **55**, 9881–9898.

83 Grynkiewicz, G. (1977) *Carbohydrate Research*, **53**, C11–C12.

84 Grynkiewicz, G. (1979) *Polish Journal of Chemistry*, **53**, 1571–1579.

85 Lubineau, A., Meyer, E. and Place, P. (1992) *Carbohydrate Research*, **228**, 191–203.

86 Kometani, T., Kondo, H. and Fujimori, Y. (1988) *Synthesis*, 1005–1007.

87 McCleary, B.V. (1988) *Methods in Enzymology*, **160**, 515–518.

88 Roush, W.R. and Lin, X.F. (1993) *Tetrahedron Letters*, **34**, 6829–6832.
89 Roush, W.R. and Lin, X.F. (1991) *Journal of Organic Chemistry*, **56**, 5740–5742.
90 Roush, W.R. and Lin, X.-F. (1995) *Journal of the American Chemical Society*, **117**, 2236–2250.
91 Gao, G., Schwardt, O. and Ernst, B. (2004) *Carbohydrate Research*, **339**, 2835–2840.
92 Smith, A.B., III, Hale, K.J. and Rivero, R.A. (1986) *Tetrahedron Letters*, **27**, 5813–5816.
93 Smith, A.B., III, Rivero, R.A., Hale, K.J. and Vaccaro, H.A. (1991) *Journal of the American Chemical Society*, **113**, 2092–2112.
94 Juteau, H., Gareau, Y. and Labelle, M. (1997) *Tetrahedron Letters*, **38**, 1481–1484.
95 Szarek, W.A., Jarrell, H.C. and Jones, J.K.N. (1977) *Carbohydrate Research*, **57**, C13–C16.
96 Hendrickson, J.B. and Schwartzman, S.M. (1975) *Tetrahedron Letters*, **16**, 277–280.
97 Mukaiyama, T. and Suda, S. (1990) *Chemistry Letters*, **19**, 1143–1146.
98 Lemieux, R.U., Hendriks, K.B., Stick, R.V. and James, K. (1975) *Journal of the American Chemical Society*, **97**, 4056–4062.
99 Chretien, F., Chapleur, Y., Castro, B. and Gross, B. (1980) *Journal of the Chemical Society, Perkin Transactions*, **1**, 381–384.
100 Khatuntseva, E.A., Ustuzhanina, N.E., Zatonskii, G.V., Shashkov, A.S., Usov, A.I. and Nifant'ev, N.E. (2000) *Journal of Carbohydrate Chemistry*, **19**, 1151–1173.
101 Shingu, Y., Nishida, Y., Dohi, H., Matsuda, K. and Kobayashi, K. (2002) *Journal of Carbohydrate Chemistry*, **21**, 605–611.
102 Shingu, Y., Nishida, Y., Dohi, H. and Kobayashi, K. (2003) *Organic and Biomolecular Chemistry*, **1**, 2518–2521.
103 Nishida, Y., Shingu, Y., Dohi, H. and Kobayashi, K. (2003) *Organic Letters*, **5**, 2377–2380.
104 Shingu, Y., Miyachi, A., Miura, Y., Kobayashi, K. and Nishida, Y. (2005) *Carbohydrate Research*, **340**, 2236–2244.
105 Leroux, J. and Perlin, A.S. (1976) *Carbohydrate Research*, **47**, C8–C10.
106 Leroux, J. and Perlin, A.S. (1978) *Carbohydrate Research*, **67**, 163–178.
107 Pavia, A.A., Rocheville, J.M. and Ung, S.N. (1980) *Carbohydrate Research*, **79**, 79–89.
108 Pavia, A.A., Ung-Chhun, S.N. and Durand, J.L. (1981) *Journal of Organic Chemistry*, **46**, 3158–3160.
109 Lacombe, J.M. and Pavia, A.A. (1983) *Journal of Organic Chemistry*, **48**, 2557–2563.
110 Gobbo, M., Biondi, L., Filira, F., Rocchi, R. and Lucchini, V. (1988) *Tetrahedron*, **44**, 887–893.
111 Szeja, W. (1988) *Synthesis*, 223–224.
112 Koto, S., Sato, T., Morishima, N. and Zen, S. (1980) *Bulletin of the Chemical Society of Japan*, **53**, 1761–1762.
113 Koto, S., Morishima, N., Uchino, M., Fukuda, M., Yamazaki, M. and Zen, S. (1988) *Bulletin of the Chemical Society of Japan*, **61**, 3943–3950.
114 Koto, S., Morishima, N., Kihara, Y., Suzuki, H., Kosugi, S. and Zen, S. (1983) *Bulletin of the Chemical Society of Japan*, **56**, 188–191.
115 Koto, S., Morishima, N., Owa, M. and Zen, S. (1984) *Carbohydrate Research*, **130**, 73–83.
116 Koto, S., Asami, K., Hirooka, M., Nagura, K., Takizawa, M., Yamamoto, S., Okamoto, N., Sato, M., Tajima, H., Yoshida, T., Nonaka, N., Sato, T., Zen, S., Yago, K. and Tomonaga, F. (1999) *Bulletin of the Chemical Society of Japan*, **72**, 765–777.
117 Koto, S., Morishima, N., Sato, H., Sato, Y. and Zen, S. (1985) *Bulletin of the Chemical Society of Japan*, **58**, 120–122.
118 Koto, S., Morishima, N., Takenaka, K., Uchida, C. and Zen, S. (1985) *Bulletin of the Chemical Society of Japan*, **58**, 1464–1468.
119 Koto, S., Hirooka, M., Yago, K., Komiya, M., Shimizu, T., Kato, K., Takehara, T., Ikefuji, A., Iwasa, A., Hagino, S., Sekiya, M., Nakase, Y., Zen, S., Tomonaga, F. and

Shimada, S. (2000) *Bulletin of the Chemical Society of Japan*, **73**, 173–183.

120 Koto, S., Haigoh, H., Shichi, S., Hirooka, M., Nakamura, T., Maru, C., Fujita, M., Goto, A., Sato, T., Okada, M., Zen, S., Yago, K. and Tomonaga, F. (1995) *Bulletin of the Chemical Society of Japan*, **68**, 2331–2348.

121 Koto, S., Hirooka, M., Yoshida, T., Takenaka, K., Asai, C., Nagamitsu, T., Sakuma, H., Sakurai, M., Masuzawa, S., Komiya, M., Sato, T., Zen, S., Yago, K. and Tomonaga, F. (2000) *Bulletin of the Chemical Society of Japan*, **73**, 2521–2529.

122 Koto, S., Kusunoki, A. and Hirooka, M. (2000) *Bulletin of the Chemical Society of Japan*, **73**, 967–976.

123 Hirooka, M., Terayama, M., Mitani, E., Koto, S., Miura, A., Chiba, K., Takabatake, A. and Tashiro, T. (2002) *Bulletin of the Chemical Society of Japan*, **75**, 1301–1309.

124 Koto, S., Shinoda, Y., Hirooka, M., Sekino, A., Ishizumi, S., Koma, M., Matuura, C. and Sakata, N. (2003) *Bulletin of the Chemical Society of Japan*, **76**, 1603–1615.

125 Koto, S., Miura, T., Hirooka, M., Tomaru, A., Iida, M., Kanemitsu, M., Takenaka, K., Masuzawa, S., Miyaji, S., Kuroyanagi, N., Yagishita, M., Zen, S., Yago, K. and Tomonaga, F. (1996) *Bulletin of the Chemical Society of Japan*, **69**, 3247–3259.

126 Garcia, B.A., Poole, J.L. and Gin, D.Y. (1997) *Journal of the American Chemical Society*, **119**, 7597–7598.

127 Garcia, B.A. and Gin, D.Y. (2000) *Journal of the American Chemical Society*, **122**, 4269–4279.

128 Boebel, T.A. and Gin, D.Y. (2003) *Angewandte Chemie (International Edition in English)*, **42**, 5874–5877.

129 Boebel, T.A. and Gin, D.Y. (2005) *Journal of Organic Chemistry*, **70**, 5818–5826.

130 Crich, D. and Sun, S.X. (1996) *Journal of Organic Chemistry*, **61**, 4506–4507.

131 Codee, J.D.C., Hossain, L.H. and Seeberger, P.H. (2005) *Organic Letters*, **7**, 3251–3254.

132 Haberman, J.M. and Gin, D.Y. (2001) *Organic Letters*, **3**, 1665–1668.

133 Haberman, J.M. and Gin, D.Y. (2003) *Organic Letters*, **5**, 2539–2541.

134 Kim, Y.-J., Wang, P., Navarro-Villalobos, M., Rohde, B.D., Derryberry, J. and Gin, D.Y. (2006) *Journal of the American Chemical Society*, **128**, 11906–11915.

135 Nguyen, H.M., Poole, J.L. and Gin, D.Y. (2001) *Angewandte Chemie (International Edition in English)*, **40**, 414–417.

136 Codee, J.D.C., van den Bos, L.J., Litjens, R.E.J.N., Overkleeft, H.S., Van Boom, J.H. and van der Marel, G.A. (2003) *Organic Letters*, **5**, 1947–1950.

137 Mukaiyama, T. (1979) *Angewandte Chemie (International Edition in English)*, **18**, 707–721.

138 Mukaiyama, T. (1979) *Pure and Applied Chemistry*, **51**, 1337–1346.

139 Mukaiyama, T., Shoda, S., Nakatsuka, T. and Narasaka, K. (1978) *Chemistry Letters*, **7**, 605–608.

140 Mukaiyama, T., Hashimoto, Y., Hayashi, Y. and Shoda, S. (1984) *Chemistry Letters*, **13**, 557–560.

141 Tsutsumi, H., Kawai, Y. and Ishido, Y. (1978) *Chemistry Letters*, **7**, 629–632.

142 Tsutsumi, H. and Ishido, Y. (1981) *Carbohydrate Research*, **88**, 61–75.

143 Tsutsumi, H. and Ishido, Y. (1982) *Carbohydrate Research*, **111**, 75–84.

144 Horvat, S., Varga, L. and Horvat, J. (1986) *Synthesis*, 209–211.

145 Ford, M.J. and Ley, S.V. (1990) *Synlett*, 255–256.

146 Ford, M.J., Knight, J.G., Ley, S.V. and Vile, S. (1990) *Synlett*, 331–332.

147 Ley, S.V., Armstrong, A., Diezmartin, D., Ford, M.J., Grice, P., Knight, J.G., Kolb, H.C., Madin, A., Marby, C.A., Mukherjee, S., Shaw, A.N., Slawin, A.M.Z., Vile, S., White, A.D., Williams, D.J. and Woods, M. (1991) *Journal of the Chemical Society Perkin Transactions*, **1**, 667–692.

148 Lou, B., Huynh, H.K. and Hanessian, S. (1997) *Prepartive Carbohydrate Chemistry* (ed. S. Hanessian), Marcel Dekker, New York, p. 431.

149 Schmidt, R.R., Behrendt, M. and Toepfer, A. (1990) *Synlett*, 694–696.
150 West, A.C. and Schuerch, C. (1973) *Journal of the American Chemical Society*, **95**, 1333–1335.
151 Wulff, G., Schroder, U. and Wichelhaus, J. (1979) *Carbohydrate Research*, **72**, 280–284.
152 Wakao, M., Nakai, Y., Fukase, K. and Kusumoto, S. (1999) *Chemistry Letters*, **28**, 27–28.
153 Mukhopadhyay, B., Kartha, K.P.R., Russell, D.A. and Field, R.A. (2004) *Journal of Organic Chemistry*, **69**, 7758–7760.
154 Lin, C.-C., Huang, L.-C., Liang, P.-H., Liu, C.-Y. and Lin, C.-C. (2006) *Journal of Carbohydrate Chemistry*, **25**, 303–313.
155 Suda, S. and Mukaiyama, T. (1991) *Chemistry Letters*, **20**, 431–434.
156 Mukaiyama, T. and Hara, R. (1989) *Chemistry Letters*, **18**, 1171–1174.
157 Suda, S. and Mukaiyama, T. (1993) *Bulletin of the Chemical Society of Japan*, **66**, 1211–1215.
158 Mukaiyama, T., Matsubara, K. and Suda, S. (1991) *Chemistry Letters*, **20**, 981–984.
159 Shimomura, N. and Mukaiyama, T. (1993) *Chemistry Letters*, **22**, 1941–1944.
160 Hirooka, M. and Koto, S. (1998) *Bulletin of the Chemical Society of Japan*, **71**, 2893–2902.
161 Helferich, B. and Schmitz-Hillebrecht, E. (1933) *Berichte der Deutschen Chemischen Gesellschaft [Abteilung] B: Abhandlungen*, **66B**, 378–383.
162 Tsou, K.-C. and Seligman, A.M. (1952) *Journal of the American Chemical Society*, **74**, 3066–3069.
163 Jermyn, M.A. (1954) *Australian Journal of Chemistry*, **7**, 202–206.
164 Stellner, K., Westphal, O. and Mayer, H. (1970) *Justus Liebigs Annalen der Chemie*, **738**, 179–191.
165 Wood, H.B., Jr and Fletcher, H.G., Jr (1956) *Journal of the American Chemical Society*, **78**, 207–210.
166 Montgomery, E.M., Richtmyer, N.K. and Hudson, C.S. (1942) *Journal of the American Chemical Society*, **64**, 690–694.
167 Tsuzuki, Y., Koyama, M., Aoki, K., Kato, T. and Tanabe, K. (1969) *Bulletin of the Chemical Society of Japan*, **42**, 1052–1059.
168 Weissman, B. (1966) *Journal of Organic Chemistry*, **31**, 2505–2509.
169 Zissis, E. and Richtmyer, N.K. (1961) *Journal of Organic Chemistry*, **26**, 5244–5245.
170 Hurd, C.D. and Zelinski, R.P. (1947) *Journal of the American Chemical Society*, **69**, 243–246.
171 Jermyn, M.A. (1955) *Australian Journal of Chemistry*, **8**, 403–408.
172 Arita, H., Isemura, M., Ikenaka, T. and Matsushima, Y. (1970) *Bulletin of the Chemical Society of Japan*, **43**, 818–823.
173 McGrath, D., Lee, E.E. and O'Colla, P.S. (1969) *Carbohydrate Research*, **11**, 453–460.
174 Sawai, T., Ohara, S., Ichimi, Y., Okaji, S., Hisada, K. and Fukaya, N. (1981) *Carbohydrate Research*, **89**, 289–299.
175 Yamaha, T. and Cardini, C.E. (1960) *Archives of Biochemistry and Biophysics*, **86**, 133–137.
176 Lu, P.W., Kramer, K.J., Seib, P.A., Mueller, D.D., Ahmed, R. and Hopkins, T.L. (1982) *Insect Biochemistry*, **12**, 377–381.
177 Wolfrom, M.L. and Winkley, M.W. (1966) *Journal of Organic Chemistry*, **31**, 3711–3713.
178 Perini, F., Carey, F.A. and Long, L., Jr (1969) *Carbohydrate Research*, **11**, 159–161.
179 Diekmann, E., Friedrich, K. and Fritz, H.G. (1993) *Journal fur Praktische Chemie – Chemiker Zeitung*, **335**, 415–424.
180 Rokos, H. and Pfleiderer, W. (1971) *Chemische Berichte*, **104**, 748–769.
181 Wright, J.A., Taylor, N.F. and Fox, J.J. (1969) *Journal of Organic Chemistry*, **34**, 2632–2636.
182 Pfleiderer, W. and Robins, R.K. (1965) *Chemische Berichte*, **98**, 1511–1513.
183 Robins, M.J., Bowles, W.A. and Robins, R.K. (1964) *Journal of the American Chemical Society*, **86**, 1251–1252.
184 Ishido, Y. and Sate, T. (1961) *Bulletin of the Chemical Society of Japan*, **34**, 1347–1348.

185 Hurd, C.D. and Bonner, W.A. (1946) *Journal of Organic Chemistry*, **11**, 50–54.
186 Miles, L.W.C. and Owen, L.N. (1950) *Journal of Chemical Society*, 2943–2946.
187 Chawla, M.L. and Bahl, O.P. (1974) *Carbohydrate Research*, **32**, 25–29.
188 Wolfrom, M.L. and Thompson, A. (1934) *Journal of the American Chemical Society*, **56**, 880–882.
189 Lemieux, R.U. and Shyluk, W.P. (1953) *Canadian Journal of Chemistry*, **31**, 528–535.
190 Bellamy, F., Horton, D., Millet, J., Picart, F., Samreth, S. and Chazan, J.B. (1993) *Journal of Medicinal Chemistry*, **36**, 898–903.
191 Clerici, F., Gelmi, M.L. and Mortadelli, S. (1994) *Journal of the Chemical Society Perkin Transactions*, **1**, 985–988.
192 Ingle, T.R. and Bose, J.L. (1970) *Carbohydrate Research*, **12**, 459–462.
193 Chao, L.R., Seguin, E., Skaltsounis, A.L., Tillequin, F. and Koch, M. (1990) *Journal of Natural Products*, **53**, 882–893.
194 Hanessian, S. and Banoub, J. (1977) *Carbohydrate Research*, **59**, 261–267.
195 Barna, P.M. (1971) *Australian Journal of Chemistry*, **24**, 673–674.
196 Hettinger, P. and Schildknecht, H. (1984) *Liebigs Annalen der Chemie*, 1230–1239.
197 Pathak, A.K., El-Kattan, Y.A., Bansal, N., Maddry, J.A. and Reynolds, R.C. (1998) *Tetrahedron Letters*, **39**, 1497–1500.
198 Du Mortier, C., Varela, O. and De Lederkremer, R.M. (1989) *Carbohydrate Research*, **189**, 79–86.
199 De Lederkremer, R.M., Marino, C. and Varela, O. (1990) *Carbohydrate Research*, **200**, 227–235.
200 Gallo-Rodriguez, C., Varela, O. and de Lederkremer, R.M. (1996) *Journal of Organic Chemistry*, **61**, 1886–1889.
201 Gandolfi-Donadio, L., Gallo-Rodriguez, C. and de Lederkremer, R.M. (2006) *Canadian Journal of Chemistry*, **84**, 486–491.
202 Marino, C., Varela, O. and De Lederkremer, R.M. (1989) *Carbohydrate Research*, **190**, 65–76.
203 Apparu, M., Blanc-Muesser, M., Defaye, J. and Driguez, H. (1981) *Canadian Journal of Chemistry*, **59**, 314–320.
204 Banoub, J. and Bundle, D.R. (1979) *Canadian Journal of Chemistry*, **57**, 2085–2090.
205 Konstantinovic, S., Predojevic, J., Pavlovic, V., Gojkovic, S. and Csanadi, J. (2001) *Journal of the Serbian Chemical Society*, **66**, 65–71.
206 Dayal, B., Salen, G., Padia, J., Shefer, S., Tint, G.S., Sasso, G. and Williams, T.H. (1993) *Carbohydrate Research*, **240**, 133–142.
207 Mazur, A.W. and Hiler, G.D., Jr (1987) *Carbohydrate Research*, **168**, 146–150.
208 Honma, K., Nakazima, K., Uematsu, T. and Hamada, A. (1976) *Chemical and Pharmaceutical Bulletin*, **24**, 394–399.
209 Mukaiyama, T., Takashima, T., Katsurada, M. and Aizawa, H. (1991) *Chemistry Letters*, **20**, 533–536.
210 Mukaiyama, T., Shimpuku, T., Takashima, T. and Kobayashi, S. (1989) *Chemistry Letters*, **18**, 145–148.
211 Matsubara, K. and Mukaiyama, T. (1994) *Polish Journal of Chemistry*, **68**, 2365–2382.
212 Mukaiyama, T., Hirano, N., Nishida, M. and Uchiro, H. (1996) *Chemistry Letters*, **25**, 99–100.
213 Petrakova, E. and Glaudemans, C.P.J. (1995) *Carbohydrate Research*, **279**, 133–150.
214 Petrakova, E. and Glaudemans, C.P.J. (1996) *Carbohydrate Research*, **284**, 191–205.
215 Hanessian, S. and Banoub, J. (1976) *Tetrahedron Letters*, **17**, 657–660.
216 Grynkiewicz, G., Barszczak, B. and Zamojski, A. (1979) *Synthesis*, 364–365.
217 Mitaku, S., Skaltsounis, A.L., Tillequin, F. and Koch, M. (1992) *Synthesis – Stuttgart*, 1068–1070.
218 Hisaeda, T., Ohshima, E., Kikuchi, J.-i. and Murakami, Y. (1997) *Tetrahedron Letters*, **38**, 6713–6716.
219 Jacopin, C., Egron, M.-J., Scherman, D. and Herscovici, J. (2002) *Bioorganic and Medicinal Chemistry Letters*, **12**, 1447–1450.

220 Furukawa, Y. and Honjo, M. (1968) *Chemical and Pharmaceutical Bulletin*, **16**, 1076–1080.
221 Niedballa, U. and Vorbruggen, H. (1974) *Journal of Organic Chemistry*, **39**, 3654–3660.
222 Niedballa, U. and Vorbruggen, H. (1974) *Journal of Organic Chemistry*, **39**, 3660–3663.
223 Niedballa, U. and Vorbruggen, H. (1974) *Journal of Organic Chemistry*, **39**, 3664–3667.
224 Niedballa, U. and Vorbruggen, H. (1974) *Journal of Organic Chemistry*, **39**, 3668–3671.
225 Niedballa, U. and Vorbruggen, H. (1974) *Journal of Organic Chemistry*, **39**, 3672–3674.
226 Itoh, T. and Mizuno, Y. (1976) *Heterocycles*, **5**, 285–292.
227 Lynch, B.M. and Sharma, S.C. (1976) *Canadian Journal of Chemistry*, **54**, 1029–1038.
228 Saneyoshi, M. and Satoh, E. (1979) *Chemical and Pharmaceutical Bulletin*, **27**, 2518–2521.
229 Bretschneider, H. and Beran, K. (1949) *Monatshefte fur Chemie*, **80**, 262–270.
230 Magnusson, G., Noori, G., Dahmen, J., Frejd, T. and Lave, T. (1981) *Acta Chemica Scandinavica*, **B35**, 213–216.
231 Nair, V. and Joseph, J.P. (1987) *Heterocycles*, **25**, 337–341.
232 Smits, E., Engberts, J.B.F.N., Kellogg, R.M. and van Doren, H.A. (1996) *Journal of the Chemical Society, Perkin Transactions*, **1**, 2873–2877.
233 Muller, H. and Tschierske, C. (1995) *Journal of the Chemical Society. Chemical Communications*, 645–646.
234 Werschkun, B., Gorziza, K. and Thiem, J. (1999) *Journal of Carbohydrate Chemistry*, **18**, 629–637.
235 Lee, Y.S., Rho, E.S., Min, Y.K., Kim, B.T. and Kim, K.H. (2001) *Journal of Carbohydrate Chemistry*, **20**, 503–506.
236 Sokolov, V.M., Zakharov, V.I. and Studentsov, E.P. (2002) *Russian Journal of General Chemistry*, **72**, 806–811.
237 Erdogan, B., Wilson, J.N. and Bunz, U.H.F. (2002) *Macromolecules*, **35**, 7863–7864.
238 Shie, T.-H., Chiang, Y.-L., Lin, J.-J., Li, Y.-K. and Lo, L.-C. (2006) *Carbohydrate Research*, **341**, 443–456.
239 Andrews, F.L., Larsen, D.S. and Larsen, L. (2000) *Australian Journal of Chemistry*, **53**, 15–24.
240 Elofsson, M., Roy, S., Salvador, L.A. and Kihlberg, J. (1996) *Tetrahedron Letters*, **37**, 7645–7648.
241 Salvador, L.A., Elofsson, M. and Kihlberg, J. (1995) *Tetrahedron*, **51**, 5643–5656.
242 Elofsson, M., Walse, B. and Kihlberg, J. (1991) *Tetrahedron Letters*, **32**, 7613–7616.
243 Kihlberg, J., Elofsson, M. and Salvador, L.A. (1997) *Methods in Enzymology*, **289**, 221–245.
244 Gangadhar, B.P., Jois, S.D.S. and Balasubramaniam, A. (2004) *Tetrahedron Letters*, **45**, 355–358.
245 Arsequell, G., Krippner, L., Dwek, R.A. and Wong, S.Y.C. (1994) *Journal of the Chemical Society. Chemical Communications*, 2383–2384.
246 Steffan, W., Schutkowski, M. and Fischer, G. (1996) *Chemical Communications*, 313–314.
247 Kragol, G. and Otvos, L. (2001) *Tetrahedron*, **57**, 957–966.
248 Palian, M.M. and Polt, R. (2001) *Journal of Organic Chemistry*, **66**, 7178–7183.
249 Arsequell, G., Sarries, N. and Valencia, G. (1995) *Tetrahedron Letters*, **36**, 7323–7326.
250 Maschauer, S., Pischetsrieder, M., Kuwert, T. and Prante, O. (2005) *Journal of Labelled Compounds & Radiopharmaceuticals*, **48**, 701–719.
251 Gurjar, M.K. and Dhar, T.G.M. (1987) *Journal of Carbohydrate Chemistry*, **6**, 313–316.
252 Marino-Albernas, J., Verez-Bencomo, V., Gonzalez-Rodriguez, L., Perez-Martinez, C.S., Castell, E.G. and Gonzalez-Segredo, A. (1988) *Carbohydrate Research*, **183**, 175–182.
253 Chiu-Machado, I., Castro-Palomino, J.C., Madrazo-Alonso, O., Loipetegui-Palacios,

C. and Verez-Bencomo, V. (1995) *Journal of Carbohydrate Chemistry*, **14**, 551–561.
254 Gurjar, M.K. and Viswanadham, G. (1991) *Tetrahedron Letters*, **32**, 6191–6194.
255 Mukaiyama, T. and Kobayashi, S. (1987) *Carbohydrate Research*, **171**, 81–87.
256 Evans, D.A. and Black, W.C. (1993) *Journal of the American Chemical Society*, **115**, 4497–4513.
257 Paquette, L.A., Collado, I. and Purdie, M. (1998) *Journal of the American Chemical Society*, **120**, 2553–2562.
258 Kiso, M. and Anderson, L. (1979) *Carbohydrate Research*, **72**, C12–C14.
259 Matta, K.L. and Bahl, O.P. (1972) *Carbohydrate Research*, **21**, 460–464.
260 Kiso, M. and Anderson, L. (1985) *Carbohydrate Research*, **136**, 309–323.
261 Kiso, M. and Anderson, L. (1979) *Carbohydrate Research*, **72**, C15–C17.
262 Boullanger, P., Descotes, G., Flandrois, J.P. and Marmet, D. (1982) *Carbohydrate Research*, **110**, 153–158.
263 Boldt, P.C., Schumacher-Wandersleb, M.H.M.G. and Peter, M.G. (1991) *Tetrahedron Letters*, **32**, 1413–1416.
264 Akamatsu, S., Ikeda, K. and Achiwa, K. (1991) *Chemical and Pharmaceutical Bulletin*, **39**, 288–296.
265 Eustache, J., Grob, A. and Retscher, H. (1994) *Carbohydrate Research*, **251**, 251–267.
266 Dasgupta, F. and Anderson, L. (1990) *Carbohydrate Research*, **202**, 239–255.
267 Campos-Valdes, M.T., Marino-Albernas, J.R. and Verez-Bencomo, V. (1987) *Journal of Carbohydrate Chemistry*, **6**, 509–513.
268 Dahmen, J., Frejd, T., Magnusson, G. and Noori, G. (1983) *Carbohydrate Research*, **114**, 328–330.
269 Boullanger, P., Banoub, J. and Descotes, G. (1987) *Canadian Journal of Chemistry*, **65**, 1343–1348.
270 Boullanger, P., Jouineau, M., Bouammali, B., Lafont, D. and Descotes, G. (1990) *Carbohydrate Research*, **202**, 151–164.
271 Werner, R.M., Barwick, M. and Davis, J.T. (1995) *Tetrahedron Letters*, **36**, 7395–7398.
272 Ogawa, T., Beppu, K. and Nakabayashi, S. (1981) *Carbohydrate Research*, **93**, C6–C9.
273 Hodosi, G. and Krepinsky, J.J. (1996) *Synlett*, 159–162.
274 Sowell, C.G., Livesay, M.T. and Johnson, D.A. (1996) *Tetrahedron Letters*, **37**, 609–610.
275 Lerner, L.M. (1990) *Carbohydrate Research*, **207**, 138–141.
276 Chatterjee, S.K. and Nuhn, P. (1998) *Chemical Communications*, 1729–1730.
277 Vorbruggen, H. and Krolikiewicz, K. (1975) *Angewandte Chemie (International Edition in English)*, **14**, 421–422.
278 Vorbruggen, H., Krolikiewicz, K. and Bennua, B. (1981) *Chemische Berichte*, **114**, 1234–1255.
279 Paulsen, H. and Brenken, M. (1988) *Liebigs Annalen der Chemie*, 649–654.
280 Rainer, H., Scharf, H.D. and Runsink, J. (1992) *Liebigs Annalen der Chemie*, 103–107.
281 Durham, T.B. and Roush, W.R. (2003) *Organic Letters*, **5**, 1871–1874.
282 Charette, A.B., Marcoux, J.F. and Cote, B. (1991) *Tetrahedron Letters*, **32**, 7215–7218.
283 Abbaci, B., Florent, J.C. and Monneret, C. (1992) *Carbohydrate Research*, **228**, 171–190.
284 Kirschning, A., Jesberger, M. and Schonberger, A. (2001) *Organic Letters*, **3**, 3623–3626.
285 Gargiulo, D., Blizzard, T.A. and Nakanishi, K. (1989) *Tetrahedron*, **45**, 5423–5432.
286 Usui, T. and Umezawa, S. (1988) *Carbohydrate Research*, **174**, 133–143.
287 Olsufyeva, E.N. and Backinowsky, L.V. (1990) *Tetrahedron Letters*, **31**, 4805–4808.
288 Wessel, H.P. and Trumtel, M. (1997) *Journal of Carbohydrate Chemistry*, **16**, 1345–1361.
289 Morishima, N. and Mori, Y. (1996) *Bioorganic and Medicinal Chemistry*, **4**, 1799–1808.

290 Gervay, J., Nguyen, T.N. and Hadd, M.J. (1997) *Carbohydrate Research*, **300**, 119–125.

291 Du, W. and Gervay-Hague, J. (2005) *Organic Letters*, **7**, 2063–2065.

292 Lemieux, R.U. and Hayami, J.I. (1965) *Canadian Journal of Chemistry*, **43**, 2162–2173.

293 Lam, S.N. and Gervay-Hague, J. (2002) *Organic Letters*, **4**, 2039–2042.

294 Lam, S.N. and Gervay-Hague, J. (2002) *Carbohydrate Research*, **337**, 1953–1965.

295 Miquel, N., Vignando, S., Russo, G. and Lay, L. (2004) *Synlett*, 341–343.

296 Gervay, J. and Hadd, M.J. (1997) *Journal of Organic Chemistry*, **62**, 6961–6967.

297 Kobashi, Y. and Mukaiyama, T. (2005) *Bulletin of the Chemical Society of Japan*, **78**, 910–916.

298 Schmid, U. and Waldmann, H. (1997) *Liebigs Annalen – Recueil*, 2573–2577.

299 Lazrek, H.B., Ouzebla, D., Rochdi, A., Redwane, N. and Vasseur, J.-J. (2006) *Letters in Organic Chemistry*, **3**, 313–314.

300 Yokoyama, Y., Hanamoto, T., Jin, X.L., Jin, Y.Z. and Inanaga, J. (2000) *Heterocycles*, **52**, 1203–1206.

301 Ishido, R., Tanaka, H., Yoshino, T., Sekiya, M., Iwabuchi, K. and Sato, T. (1967) *Tetrahedron Letters*, **8**, 5245–5248.

302 Sasaki, K., Matsumura, S. and Toshima, K. (2006) *Tetrahedron Letters*, **47**, 9039–9043.

303 Florent, J.C. and Monneret, C. (1987) *Journal of the Chemical Society, Chemical Communications*, 1171–1172.

304 Yamanoi, T., Oda, Y., Yamazaki, I., Shinbara, M., Morimoto, K. and Matsuda, S. (2005) *Letters in Organic Chemistry*, **2**, 242–246.

305 Yamanoi, T. and Yamazaki, I. (2001) *Tetrahedron Letters*, **42**, 4009–4011.

306 Yamanoi, T., Nagayama, S., Ishida, H.-K., Nishikido, J. and Inazu, T. (2001) *Synthetic Communications*, **31**, 899–903.

307 Mukaiyama, T., Uchiro, H., Hirano, N. and Ishikawa, T. (1996) *Chemistry Letters*, **25**, 629–630.

308 Toshima, K., Nozaki, Y., Mukaiyama, S. and Tatsuta, K. (1992) *Tetrahedron Letters*, **33**, 1491–1494.

309 Ikeda, K., Torisawa, Y., Nishi, T., Minamikawa, J., Tanaka, K. and Sato, M. (2003) *Bioorganic and Medicinal Chemistry*, **11**, 3073–3076.

310 Higashi, K. and Susaki, H. (1992) *Chemical and Pharmaceutical Bulletin*, **40**, 2019–2022.

311 Ponticelli, F., Trendafilova, A., Valoti, M., Saponara, S. and Sgaragli, G. (2001) *Carbohydrate Research*, **330**, 459–468.

312 Mukaiyama, T. and Shimomura, N. (1993) *Chemistry Letters*, **22**, 781–784.

313 Wessel, H.P. and Trumtel, M. (1997) *Carbohydrate Research*, **297**, 163–168.

314 Kimura, Y., Suzuki, M., Matsumoto, T., Abe, R. and Terashima, S. (1984) *Chemistry Letters*, **13**, 501–504.

315 Kimura, Y., Matsumoto, T., Suzuki, M. and Terashima, S. (1985) *Journal of Antibiotics*, **38**, 1277–1279.

316 Kimura, Y., Suzuki, M., Matsumoto, T., Abe, R. and Terashima, S. (1986) *Bulletin of the Chemical Society of Japan*, **59**, 423–431.

317 Csanadi, J., Sztaricskai, F., Batta, G., Dinya, Z. and Bognar, R. (1986) *Carbohydrate Research*, **147**, 211–220.

318 Tamura, Y., Akai, S., Kishimoto, H., Kirihara, M., Sasho, M. and Kita, Y. (1987) *Tetrahedron Letters*, **28**, 4583–4586.

319 Irvine, R.W., Kinloch, S.A., McCormick, A.S., Russell, R.A. and Warrener, R.N. (1988) *Tetrahedron*, **44**, 4591–4604.

320 Kolar, C., Dehmel, K., Knoedler, U., Paal, M., Hermentin, P. and Gerken, M. (1989) *Journal of Carbohydrate Chemistry*, **8**, 295–305.

321 Kolar, C., Dehmel, K. and Moldenhauer, H. (1990) *Carbohydrate Research*, **208**, 67–81.

322 Nicotra, F., Panza, L., Romano, A. and Russo, G. (1992) *Journal of Carbohydrate Chemistry*, **11**, 397–399.

323 Hanessian, S., Bacquet, C. and Lehong, N. (1980) *Carbohydrate Research*, **80**, C17–C22.

324 Furukawa, H., Koide, K., Takao, K. and Kobayashi, S. (1998) *Chemical and Pharmaceutical Bulletin*, **46**, 1244–1247.

325 Koide, K., Ohno, M. and Kobayashi, S. (1991) *Tetrahedron Letters*, **32**, 7065–7068.

326 Kwon, S.Y., Lee, B.-Y., Jeon, H.B. and Kim, K.S. (2005) *Bulletin of the Korean Chemical Society*, **26**, 815–818.

327 Kim, K.S., Lee, Y.J., Kim, H.Y., Kang, S.S. and Kwon, S.Y. (2004) *Organic and Biomolecular Chemistry*, **2**, 2408–2410.

328 Kunz, H., Wernig, P. and Schultz, M. (1990) *Synlett*, 631–632.

329 Fraser-Reid, B., Konradsson, P., Mootoo, D.R. and Udodong, U. (1988) *Journal of the Chemical Society. Chemical Communications*, 823–825.

330 Lopez, J.C., Gomez, A.M., Valverde, S. and Fraser-Reid, B. (1995) *Journal of Organic Chemistry*, **60**, 3851–3858.

331 Baek Ju, Y., Choi Tae, J., Jeon Heung, B. and Kim Kwan, S. (2006) *Angewandte Chemie (International Edition in English)*, **45**, 7436–7440.

332 Choi, T.J., Baek, J.Y., Jeon, H.B. and Kim, K.S. (2006) *Tetrahedron Letters*, **47**, 9191–9194.

333 Imagawa, H., Kinoshita, A., Fukuyama, T., Yamamoto, H. and Nishizawa, M. (2006) *Tetrahedron Letters*, **47**, 4729–4731.

334 Mukai, C., Itoh, T. and Hanaoka, M. (1997) *Tetrahedron Letters*, **38**, 4595–4598.

335 Hanessian, S. (1997) *Preparative Carbohydrate Chemistry*, Marcel Dekker, New York.

336 Hanessian, S., Qiu, D.X., Prabhanjan, H., Reddy, G.V. and Lou, B.L. (1996) *Canadian Journal of Chemistry*, **74**, 1738–1747.

337 Hanessian, S., Conde, J.J. and Lou, B.L. (1995) *Tetrahedron Letters*, **36**, 5865–5868.

338 Hanessian, S., Reddy, G.V., Huynh, H.K., Pan, J.W., Pedatella, S., Ernst, B. and Kolb, H.C. (1997) *Bioorganic and Medicinal Chemistry Letters*, **7**, 2729–2734.

339 Hanessian, S., Huynh, H.K., Reddy, G.V., Duthaler, R.O., Katopodis, A., Streiff, M. B., Kinzy, W. and Oehrlein, R. (2001) *Tetrahedron*, **57**, 3281–3290.

340 Hanessian, S., Mascitti, V. and Rogel, O. (2002) *Journal of Organic Chemistry*, **67**, 3346–3354.

341 Hanessian, S., Huynh, H.K., Reddy, G.V., McNaughton-Smith, G., Ernst, B., Kolb, H.C., Magnani, J. and Sweeley, C. (1998) *Bioorganic and Medicinal Chemistry Letters*, **8**, 2803–2808.

342 Kunz, H. and Zimmer, J. (1993) *Tetrahedron Letters*, **34**, 2907–2910.

343 Hinklin, R.J. and Kiessling, L.L. (2001) *Journal of the American Chemical Society*, **123**, 3379–3380.

344 Knoben, H.-P., Schlueter, U. and Redlich, H. (2004) *Carbohydrate Research*, **339**, 2821–2833.

345 Matsuo, J.-i., Shirahata, T. and Omura, S. (2006) *Tetrahedron Letters*, **47**, 267–271.

346 Jayakanthan, K. and Vankar, Y.D. (2005) *Carbohydrate Research*, **340**, 2688–2692.

347 Jung, K.-H., Mueller, M. and Schmidt, R.R. (2000) *Chemical Reviews*, **100**, 4423–4442.

348 Yu, B., Yang, Z. and Cao, H. (2005) *Current Organic Chemistry*, **9**, 179–194.

349 Inaba, S., Yamada, M., Yoshino, T. and Ishido, Y. (1973) *Journal of the American Chemical Society*, **95**, 2062–2063.

350 Ishido, Y., Inaba, S., Matsuno, A., Yoshino, T. and Umezawa, H. (1977) *Journal of the Chemical Society Perkin Transactions*, **1**, 1382–1390.

351 Ishido, Y., Inaba, S., Komura, H. and Matsuno, A. (1977) *Journal of the Chemical Society. Chemical communications*, 90–91.

352 Boursier, M. and Descotes, G. (1989) *Comptes Rendus de L'Academie Des Sciences Serie Ii*, **308**, 919–921.

353 Iimori, T., Shibazaki, T. and Ikegami, S. (1996) *Tetrahedron Letters*, **37**, 2267–2270.

354 Azumaya, I., Kotani, M. and Ikegami, S. (2004) *Synlett*, 959–962.

355 Azumaya, I., Niwa, T., Kotani, M., Iimori, T. and Ikegami, S. (1999) *Tetrahedron Letters*, **40**, 4683–4686.

356 Iimori, T., Azumaya, I., Shibazaki, T. and Ikegami, S. (1997) *Heterocycles*, **46**, 221–224.

357 Scheffler, G. and Schmidt, R.R. (1997) *Tetrahedron Letters*, **38**, 2943–2946.
358 Marra, A., Esnault, J., Veyrieres, A. and Sinay, P. (1992) *Journal of the American Chemical Society*, **114**, 6354–6360.
359 Mukaiyama, T., Miyazaki, K. and Uchiro, H. (1998) *Chemistry Letters*, **27**, 635–636.
360 Takeuchi, K., Tamura, T. and Mukaiyama, T. (2000) *Chemistry Letters*, **29**, 122–123.
361 Hashihayata, T., Ikegai, K., Takeuchi, K., Jona, H. and Mukaiyama, T. (2003) *Bulletin of the Chemical Society of Japan*, **76**, 1829–1848.
362 Takeuchi, K., Tamura, T. and Mukaiyama, T. (2000) *Chemistry Letters*, **29**, 124–125.
363 Mukaiyama, T., Ikegai, K., Jona, H., Hashihayata, T. and Takeuchi, K. (2001) *Chemistry Letters*, **30**, 840–841.
364 Georgiadis, M.P. and Couladouros, E.A. (1991) *Journal of Heterocyclic Chemistry*, **28**, 1325–1337.
365 Nagai, M., Matsutani, T. and Mukaiyama, T. (1996) *Heterocycles*, **42**, 57–63.
366 Mukaiyama, T., Matsutani, T. and Shimomura, N. (1993) *Chemistry Letters*, **22**, 1627–1630.
367 Shimomura, N., Matsutani, T. and Mukaiyama, T. (1994) *Bulletin of the Chemical Society of Japan*, **67**, 3100–3106.
368 Mukaiyama, T., Matsutani, T. and Shimomura, N. (1994) *Chemistry Letters*, **23**, 2089–2092.
369 Mao, J., Chen, H., Zhang, J. and Cai, M. (1995) *Synthetic Communications*, **25**, 1563–1565.

3.2
Glycoside Synthesis from 1-Oxygen-Substituted Glycosyl Imidates
Xiangming Zhu, Richard R. Schmidt

3.2.1
Introduction

Of the various synthetic strategies developed to date, glycoside syntheses based on glycosyl imidates are probably the most popular. This is because usually only catalytic amounts of a promoter are required to provide very high glycosyl-donor properties of glycosyl imidates whereas other glycosyl donors, such as glycosyl halides, thioglycosides, generally require at least equimolar amounts of a promoter system, which is often associated with disadvantages of various kinds.

In 1980, Schmidt introduced the corresponding trichloroacetimidates [370], which have become one of the most widely used glycosyl donors in contemporary carbohydrate chemistry [371–381]. Glycosyl trichloroacetimidates can be easily prepared by a base-catalyzed addition of an anomeric hydroxyl group to trichloroacetonitrile (Cl_3CCN) using either inorganic or organic bases. This reaction is generally high yielding and, because of its reversibility, high anomeric control can often be achieved. In addition, competing reactions with nonanomeric hydroxyl groups are quite slow. In a typical Schmidt glycosidation reaction, a catalytic amount of Lewis acid, such as trimethylsilyl triflate (TMSOTf) or boron trifluoride etherate ($BF_3 \cdot OEt_2$), is most commonly used as the promoter. Glycosyl trichloroacetimidates exhibit outstanding donor properties in terms of ease of formation, stability, reactivity and general applicability and usually result in high product yields and high anomeric stereocontrol. The anomeric stereochemistry is derived from the anomeric configuration of glycosyl trichloroacetimidates (inversion or retention), anchimeric assistance, the influence of solvents or thermodynamic or kinetic effects. As the O-glycosyl N-methyl acetimidates [382], introduced by Sinay, required lengthy preparation procedures and exhibited low reactivity, they did not gain broad application.

In 1983, Schmidt reported another type of glycosyl imidates, trifluoroacetimidates [383], as glycosyl donors. Afterward, a series of different N-substituted glycosyl trifluoroacetimidates were also prepared from the corresponding glycosyl hemiacetals and N-substituted trifluoroacetimidoyl chlorides [384]. Initial experiments revealed that glycosylations with trifluoroacetimidates were generally less efficient than those with trichloroacetimidates in terms of product yields. Later, Yu and Tao [385] and Iadonisi and coworkers [386] explored the application of glycosyl N-phenyl trifluoroacetimidates and reported particularly good reactivity for some specific glycosylation reactions. On the whole, trifluoroacetimidate donors exhibit reduced reactivity compared to the corresponding trichloroacetimidate donors presumably because of the lower N-basicity or the presence of an N-substituent or smaller trifluoromethyl-group-caused conformational change [387].

This review covers the recent advances in the use of O-glycosyl imidates in oligosaccharide and glycoconjugate synthesis, with emphasis on literature published between 1999 and 2006. However, because of the large volume of work in this area, only the most representative applications will be presented. One can refer to the similar preceding review [381] published in 2000 and another quite comprehensive review [375] on trichloroacetimidate method published in 1994 for earlier application of glycosyl trichloroacetimidates. Trifluoroacetimidate method will be discussed separately in this review in the light of its less popularity in carbohydrate chemistry.

3.2.2
Methodological Aspects

3.2.2.1 Preparation of Anomeric O-Trichloroacetimidates

Conventionally, the use of NaH or Cs_2CO_3 as a base for the reaction of glycosyl hemiacetals with Cl_3CCN often yields the thermodynamically favored α-glycosyl trichloroacetimidates, whereas the use of K_2CO_3 often yields kinetically controlled β-glycosyl trichloroacetimidates (Scheme 3.26). The use of 1,8-diazabicyclo[5.4.0]undec-7-ene (DBU) often provides α/β mixtures, mostly favoring α-products.

Recently, two independent groups reported almost at the same time that polymer-supported DBU [388] and TBD [389] (1,5,7-triazabicyclo[4.4.0]dec-5-ene) were

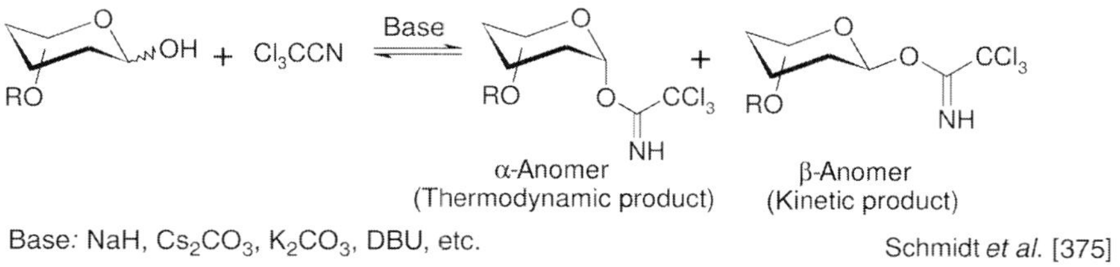

Base: NaH, Cs_2CO_3, K_2CO_3, DBU, etc.

Schmidt et al. [375]

Scheme 3.26

efficient reagents for the preparation of trichloroacetimidates, affording excellent yields of pure products after simple filtration and evaporation. This was found to be particularly useful when the formed trichloroacetimidate donors were highly labile [388]. Another investigation disclosed that the polymer-bound DBU was the most efficient under substoichiometric conditions and was, therefore, the reagent of choice for the general preparation of this important class of glycosyl donors [390].

3.2.2.2 Glycosidation of O-Glycosyl Trichloroacetimidates

As mentioned above, TMSOTf and $BF_3 \cdot OEt_2$ are the most commonly used catalysts for the Schmidt glycosidation. Several new catalysts for the activation of trichloroacetimidate donors have also been reported in the past years. Catalytic amount of $Sm(OTf)_3$ activated armed glycosyl trichloroacetimidates under very mild conditions [391], whereas disarmed trichloroacetimidate could be activated effectively by $Yb(OTf)_3$ [392]. These trivalent lanthanide triflates are generally stable salts that can be easily stored without particular precautions. AgOTf was also reinvestigated recently as a catalyst and proved to be a mild and, in some cases, more efficient catalyst in TMSOTf-sensitive glycosidation reactions [393]. In addition, the nature of the counteranion in catalysts is very influential in controlling the stereoselectivity of the Schmidt glycosidation, as reflected by comparing entries 1 and 2, or 3 and 4 in Table 3.1 [394], but how the anions work has not yet been understood. Appropriately functionalized acyl sulfonamides were also employed as catalysts for trichloroacetimidate glycosidations [395]. More recently, silica-supported perchloric acid ($HClO_4$–SiO_2) has been used as a convenient and efficient promoter in various glycosylation reactions, with trichloroacetimidates as glycosyl donors [396]. Also, the use of $HClO_4$–SiO_2 for 'on-column' glycosylation and subsequent 'in situ' separation provided a novel and robust method for glycoside synthesis [397]. Trichloroacetimidate donors were also promoted by precise microwave heating in the absence of strong Lewis acids, giving the desired products in good yields [398]. A few papers were devoted to the use of ionic liquids as solvents to perform Schmidt glycosylations, in which the reactions proceeded at room temperature under mild conditions and, in some cases, avoided the use of Lewis acid catalyst [399–401].

Table 3.1

Entry	Conditions	Yield (%)	α/β Ratio
1	$HClO_4$, Et_2O	99	91:9
2	$HB(C_6F_5)_4$, Et_2O	97	43:57
3	$HClO_4$, BTF–^{t}BuCN	95	54:46
4	$HB(C_6F_5)_4$, BTF–^{t}BuCN	97	10:90

3.2.3
Synthesis of Oligosaccharides

3.2.3.1 β-Glucosides, β-Galactosides, α-Mannosides and Others

The use of neighboring-group participation of 2-*O*-acyl-protected glycosyl trichloroacetimidates usually allows the synthesis of 1,2-*trans* glycosides such as β-glucosides, β-galactosides, α-mannosides, α-rhamnosides, β-xylosides and so on. A new ether-type protecting group, diphenylmethyl (DPM), was recently introduced into glucose 2-OH position, and interestingly, the resulting glucosyl trichloroacetimidate exclusively gave β-glucosides (Scheme 3.27a) [402]. The steric bulk of this group exerts anchimeric assistance on the anomeric stereocontrol through neighboring-group participation, as an acyl group does. Also, the use of 4-acetoxy-2,2-dimethylbutanoyl (ADMB) protecting group for 2-OH of glucose prevented the orthoester formation during glucosidation reactions, thereby allowing the selective formation of β-glucosides, as shown in Scheme 3.27b [403]. Recently, a glucohexaose was synthesized convergently as its allyl glycoside using trichloroacetimidate method (Scheme 3.28) [404], wherein the glycosylation steps are highly regioselective and high yielding. Similarly, β-(1 → 6)-linked glucooctaose was also synthesized employing glycosyl trichloroacetimidates as donors and partially protected sugars as acceptors [405]. A general approach based on trichloroacetimidate donor for the synthesis of 3,6-branched glucooligosaccharides has also been developed [406]. By this approach, a phytoalexin elicitor glucohexaose was prepared on a 100-g scale [407] and a tetradecasaccharide was also synthesized efficiently [408].

Syntheses of a series of galactans, including a decasaccharide, consisting of a β-(1 → 3)-linked galactosyl backbone and β-(1 → 6)-linked side chains of different size attached at the C-6 were achieved with glycosyl trichloroacetimidates as donors [409]. The β-(1 → 6)-linked galactans branched with α-arabinofuranose, that is arabinogalactan, were also synthesized in the same laboratories [410]. Again, glycosyl trichloroacetimidate was the sole donor used in the whole synthesis (Scheme 3.29). Other arabinogalactan-derived oligosaccharides have also been prepared on the basis of the trichloroacetimidate method [411,412].

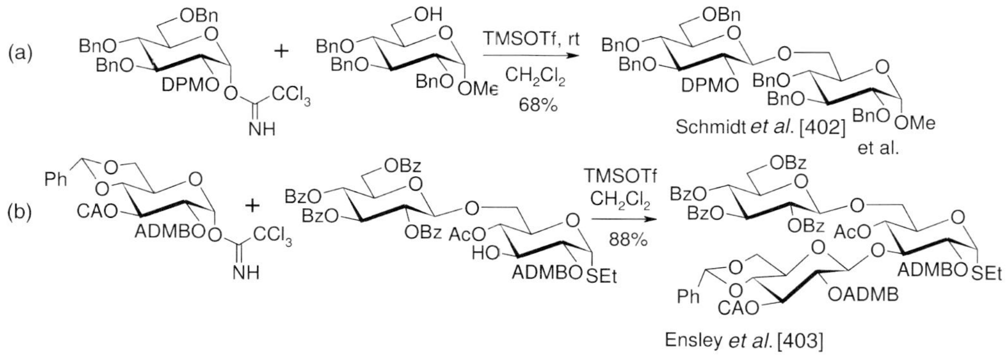

Scheme 3.27

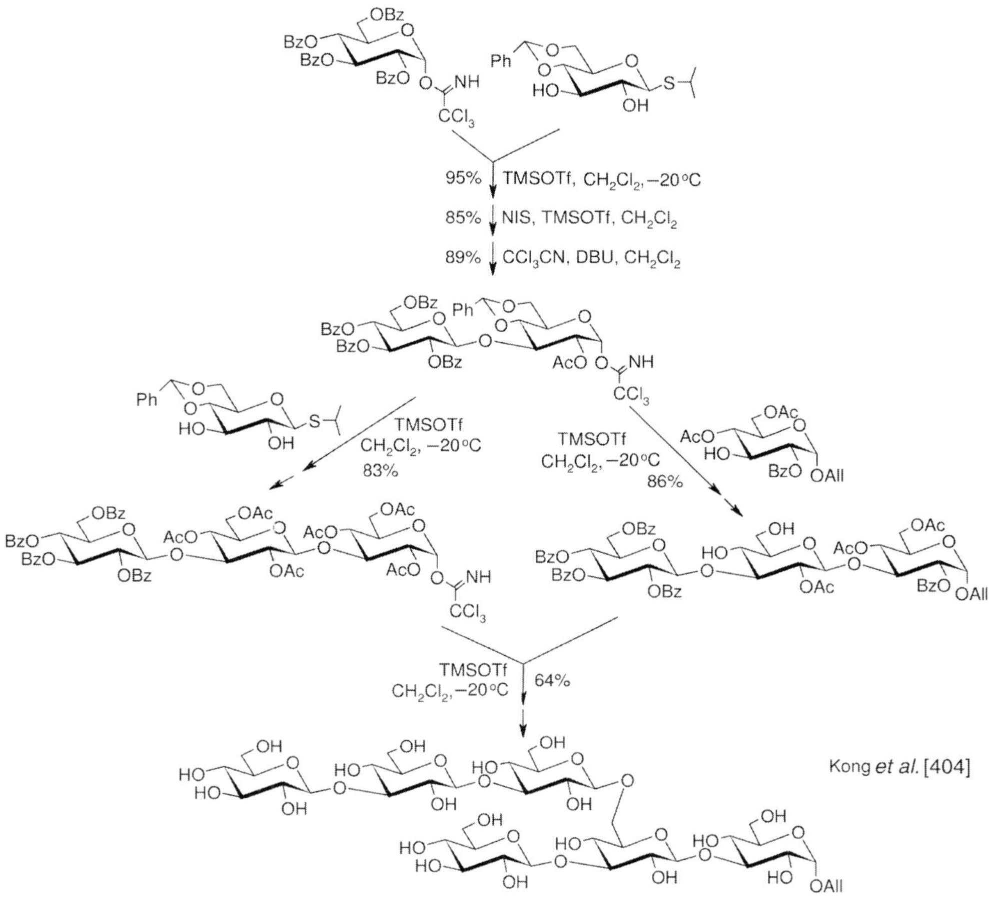

3.2 Glycoside Synthesis from 1-Oxygen-Substituted Glycosyl Imidates | 147

Scheme 3.28

The synthesis of high-mannose-type cell surface glycans, which are found in nature as N-linked glycoconjugates, has been explored for the past two decades. Glycosyl trichloroacetimidates have often been exploited in the synthesis of this type of structures [413–421]. Recently, a linear synthesis of a typical triantennary high-mannose nonasaccharide was accomplished in four high-yielding Schmidt glycosidation events [413], as shown in Scheme 3.30. Also, a convergent procedure has been developed recently for the synthesis of oligosaccharides consisting of α-(1 → 2)- and α-(1 → 3)-linked rhamnan backbones and additional sugar side chains via di- and tetrasaccharides that could be converted either into glycosyl donors by deallylation and transformation into trichloroacetimidates or into acceptors by deacetylation. The efficiency of this procedure was demonstrated by the assembly of a decasaccharide carrying two GlcNAc residues derived from lipopolysaccharides (LPS) of phytopathogenic bacteria (Scheme 3.31) [422]. Trichloroacetimidate method

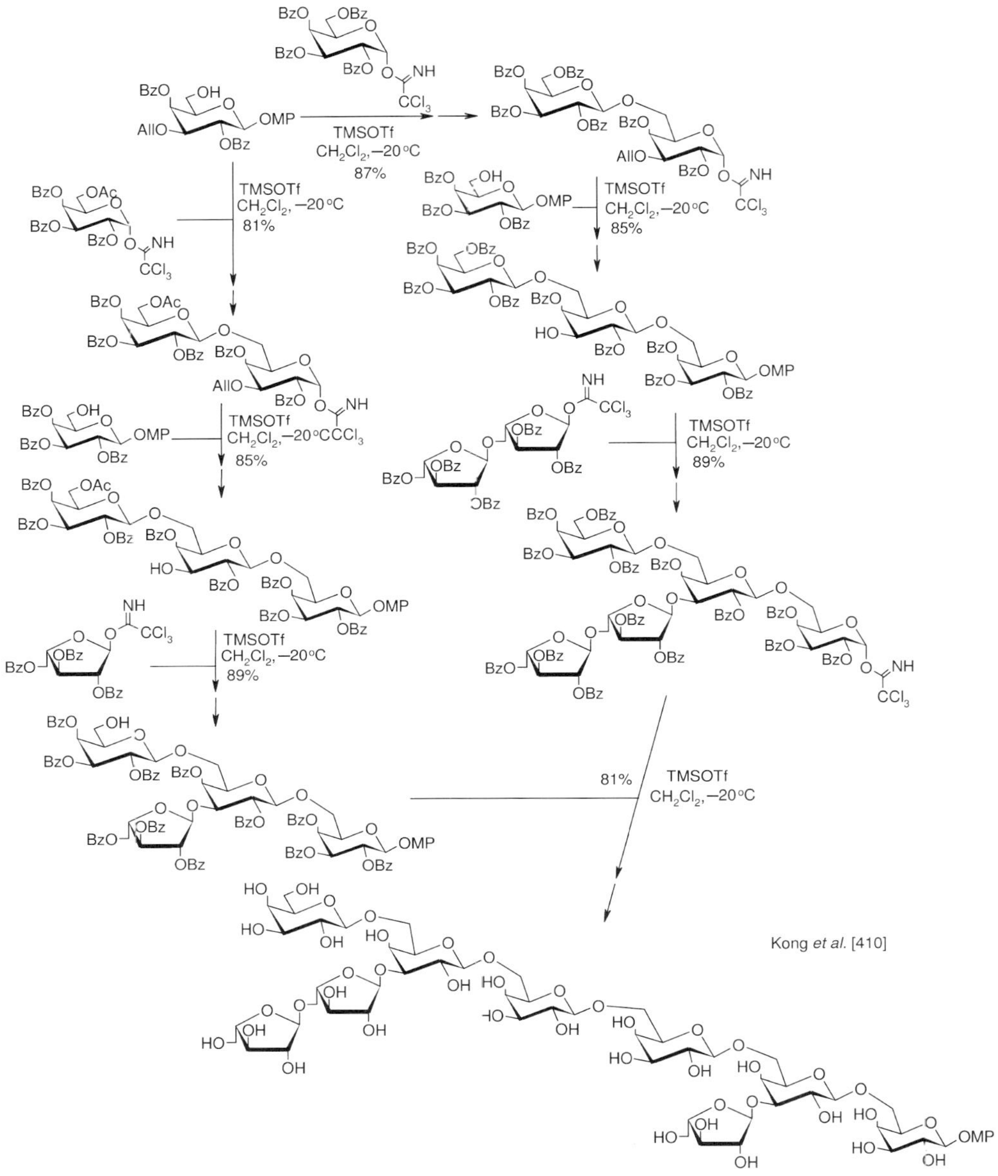

Scheme 3.29

has also been used in the synthesis of other rhamnooligosaccharides branched with different sugars [423–428].

In addition, oligomers of other monosaccharides, such as arabinofuranose [429], galactofuranose [430,431] and xylose [432], were also synthesized efficiently by taking advantage of the Schmidt glycosylation procedure.

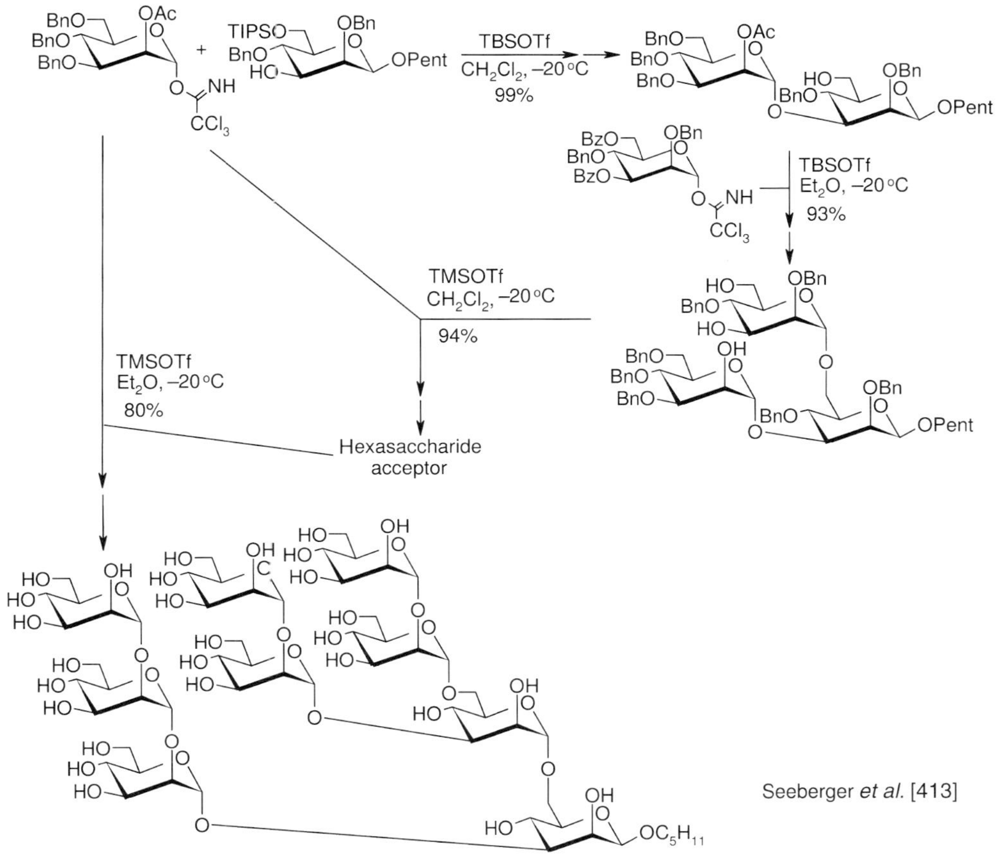

3.2 Glycoside Synthesis from 1-Oxygen-Substituted Glycosyl Imidates | 149

Scheme 3.30

Seeberger et al. [413]

3.2.3.2 Aminosugar-Containing Oligosaccharides

A large number of oligosaccharides of biological significance contain aminosugar units. Therefore, considerable attention has been paid to the synthesis of aminosugar-containing oligosaccharides in the past years. A few new N-protecting groups have been proposed and tested in aminosugar glycoside synthesis [433]. Satisfactory results were obtained with N-diglycolyl (DG) group, which could be easily introduced and removed under basic conditions in high yields, and glycosylation reactions with N-DG-protected trichloroacetimidates gave high yields and β-stereoselectivities (Scheme 3.32) [433].

Chitooligosaccharides, the oligomers of N-acetylglucosamine, have been synthesized recently. In the synthesis, dimethylmaleoyl (DMM) group was used as amino protecting group (Scheme 3.33) [434]. Because of the anchimeric assistance and the electron-withdrawing character of the DMM group, the corresponding trichloroacetimidates were also excellent glycosylating agents to introduce β-aminosugars.

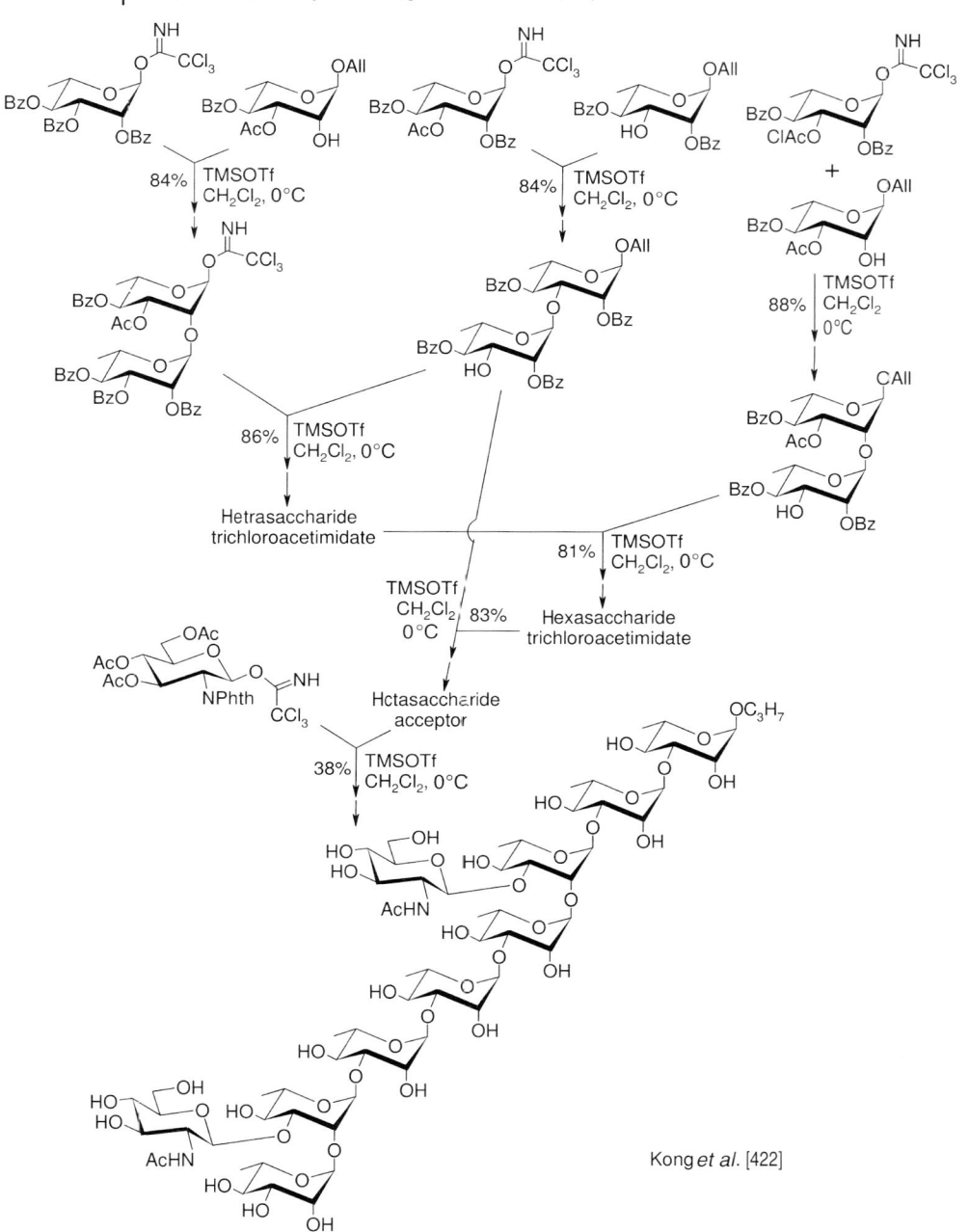

Scheme 3.31

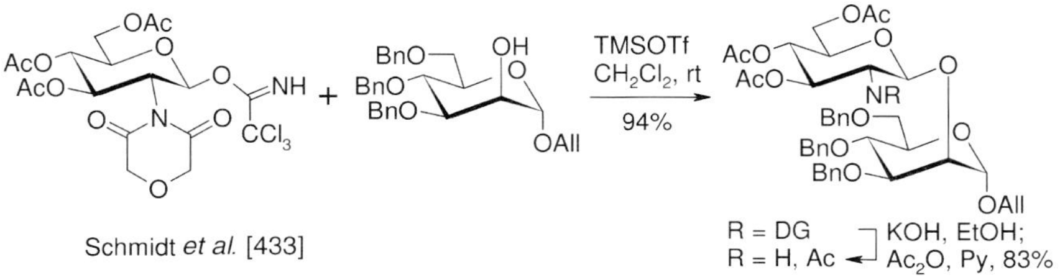

Scheme 3.32

The same approach has also been employed in the synthesis of lacto-N-tetraose and lacto-N-neotetraose [435] that represent core structural elements of more complex oligosaccharides in human milk, glycolipids and glycoproteins. An N-glycan fragment, asparagine-linked heptasaccharide consisting of the pentasaccharide core structure and one N-acetyllactosamine residue, was also assembled successfully using DMM group as the amino protecting group and glycosyl trichloroacetimidates as powerful donors (Scheme 3.34) [436].

The trichloroacetimidate method has also been used to prepare bivalent Le^X oligosaccharides to study the conformational details of carbohydrate clusters by NMR spectroscopy [437]. Two Le^X trisaccharides were covalently linked through the 6-hydroxy group or through the anomeric oxygen to yield the corresponding dimers. The synthesis of anomerically linked dimer was performed with

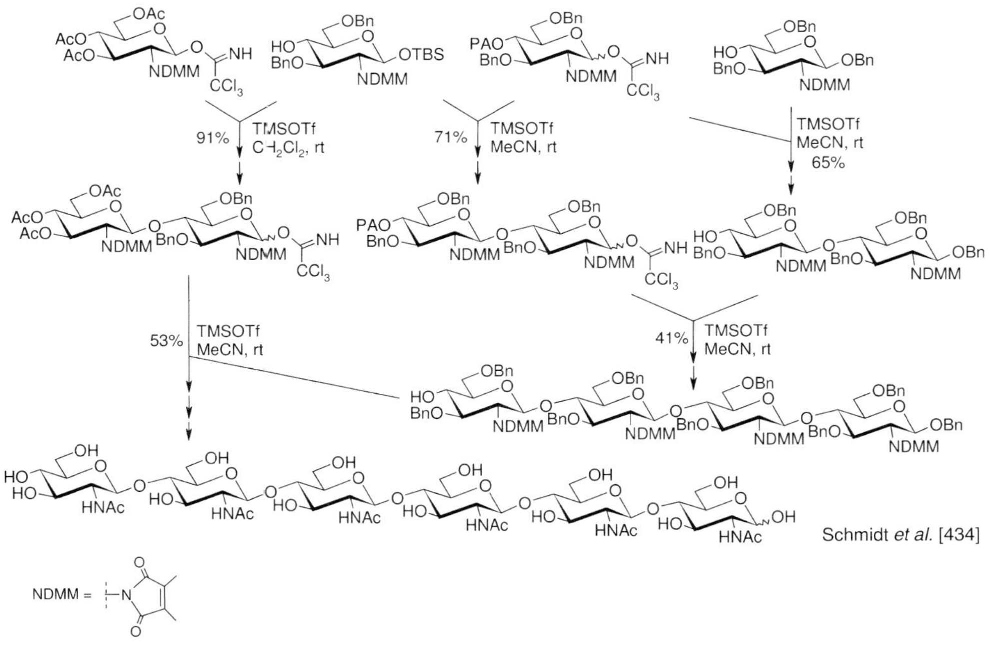

Scheme 3.33

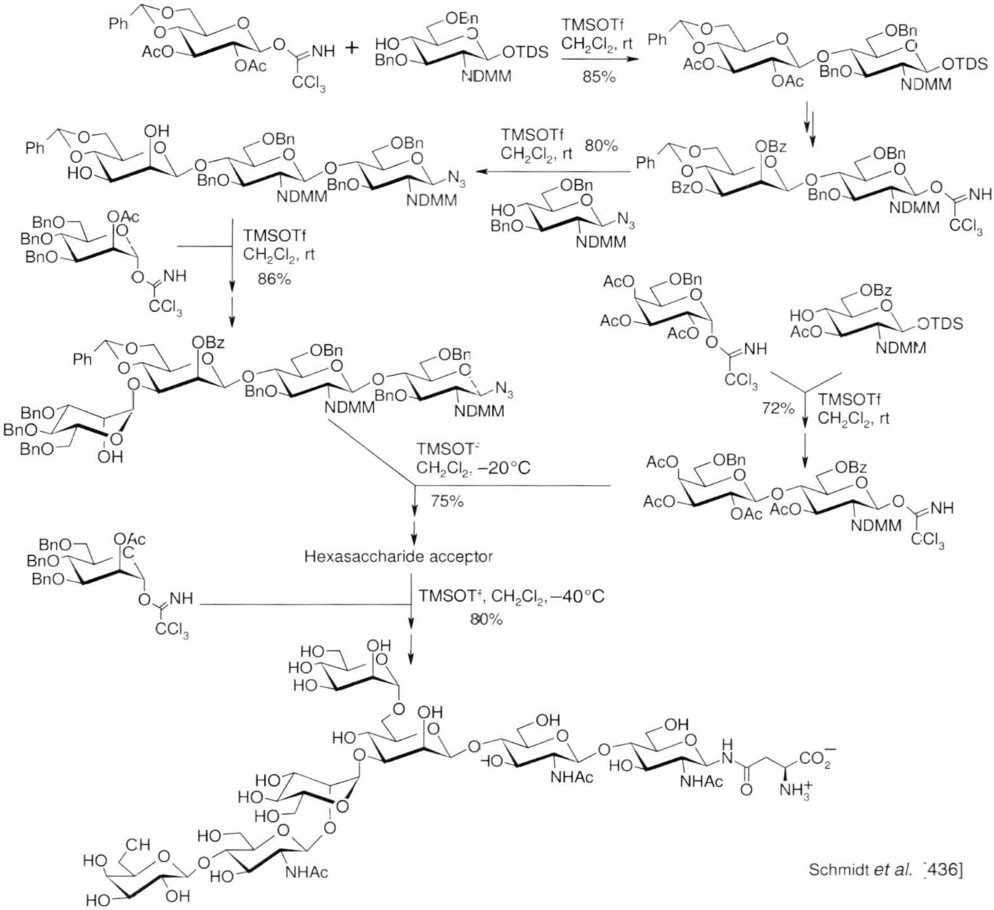

Scheme 3.34

N-trichloroethoxycarbonyl (Troc)-protected aminosugar trichloroacetimidate as the glycosyl donor (Scheme 3.35) [437]. It should be mentioned that the Troc group has often been used in the synthesis of aminosugar glycosides because of its high stability under most reaction conditions, and it could be removed under specific conditions [438,439]. High-yielding and stereoselective glycosylations with N-Troc-protected trichloroacetimidates were also achieved in the synthesis of peptidoglycan fragments, as shown in Scheme 3.35 [440].

In the past years, many syntheses have been described for the glycosaminoglycan (GAG) oligosaccharides [441–450], such as heparin and chondroitin structures, to investigate their biological function in greater detail. One frequently encountered problem in the synthesis of heparin structures is the stereocontrol in the construction of α-glucosamine linkages. Recently, syntheses of a group of heparin oligosaccharides including a nonasaccharide were achieved using the trichloroacetimidate

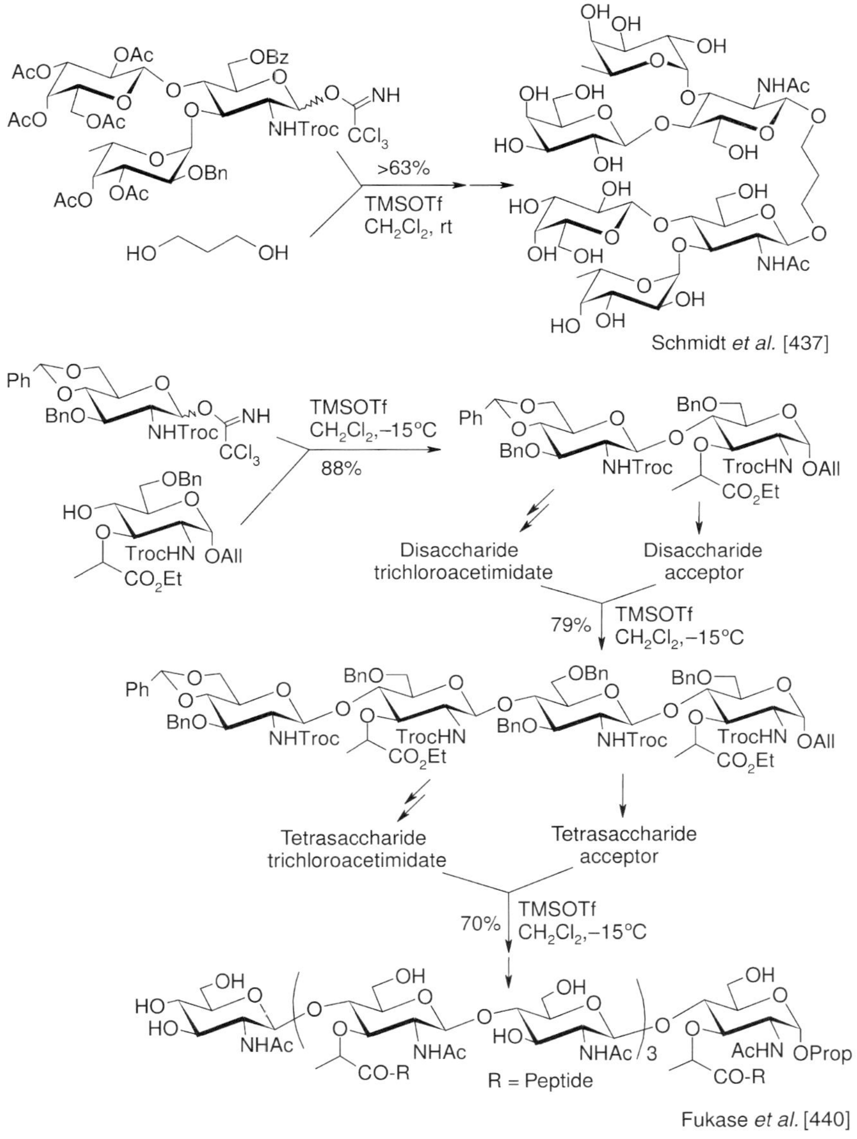

Scheme 3.35

method [441], in which the most commonly used 2-azidoglucosyl trichloroacetimidates were employed to introduce the requisite α-glucosamine linkages (Scheme 3.36). The chain elongation sequence, involving the removal of 2-naphthylmethyl group (NAP) by using DDQ and subsequent glycosylation with the key disaccharide trichloroacetimidate, was repeated to assemble the penta-, hepta- and nonasaccharides, respectively. Also, a group of heparin tetrasaccharides, differing in their sulfation pattern at position C-6 of the glucosamine units, was also synthesized from two common disaccharide precursors by the Schmidt glycosylation procedure [442]. In addition, the stereochemistry of glycosylation reactions with 2-azidoglycosyl trichloroacetimidates has been investigated very recently using a series of *chiro*-inositol derivatives as glycosyl acceptors [443]. The results indicated that the influence of the absolute configuration, the orientation of the acceptor OH group and the conformational constraint of the acceptor on the stereochemical outcome of the reaction are difficult to be assessed. To achieve good stereocontrol of these glycosylations, extensive experimentation is still required.

Chondroitin sulfates (CS) are ubiquitous components of extracellular matrices of all connective tissues such as the artery and tendon and exhibit a variety of biological functions. They are linear copolymers made up of dimeric units composed of glucuronic acid and *N*-acetyl galactosamine. The first biological investigation

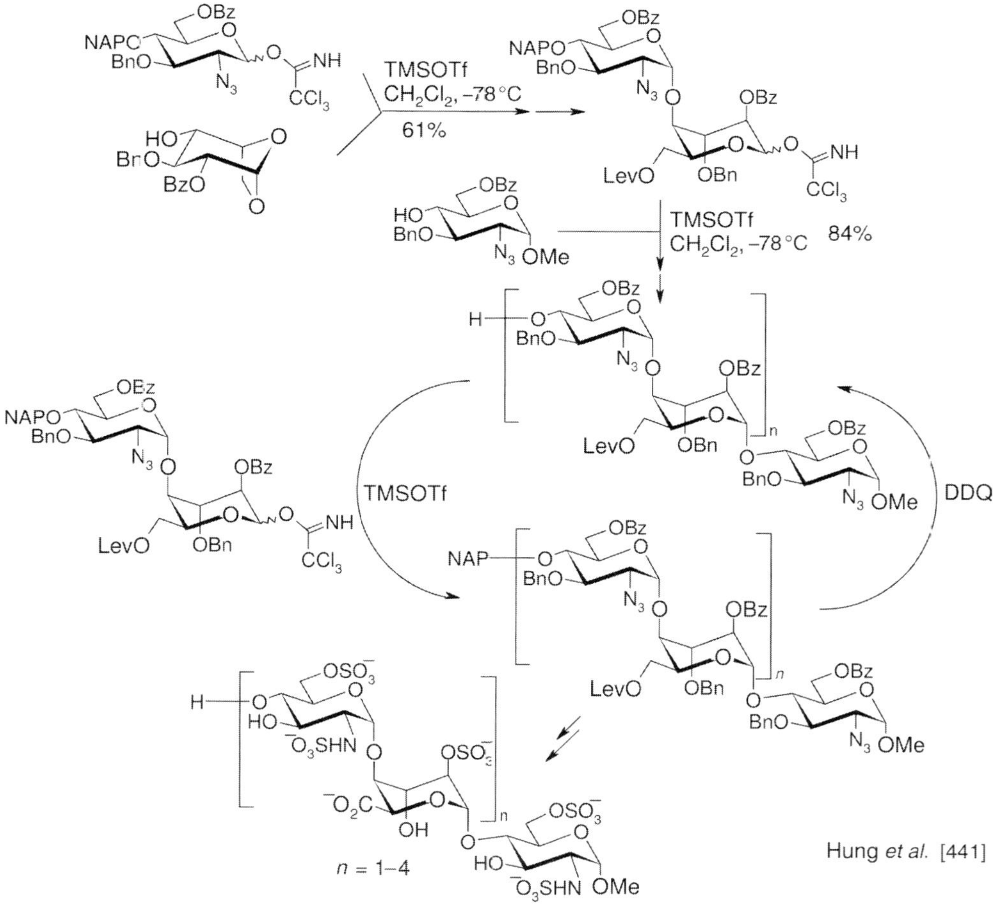

Scheme 3.36

Hung et al. [441]

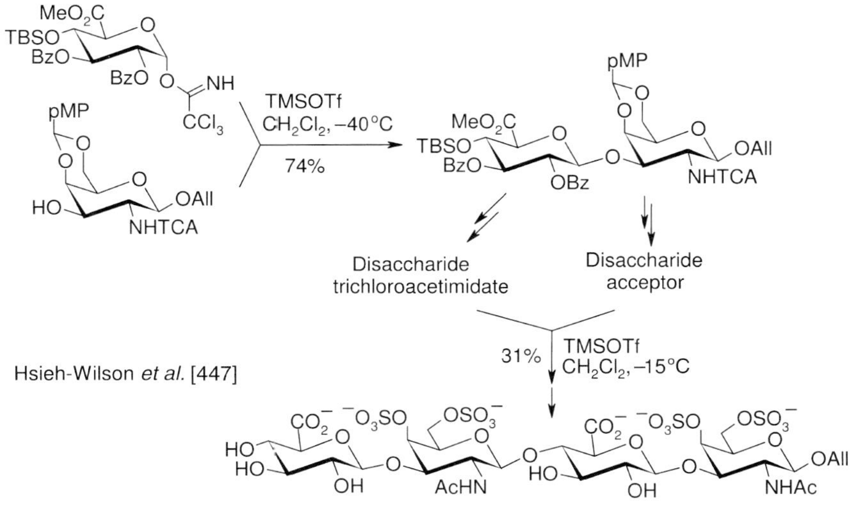

Scheme 3.37

of synthetic chondroitin molecules was carried out very recently, in which a tetrasaccharide fragment was defined as a minimal motif required for activity [447]. A convergent route for the synthesis of this tetrasaccharide, based on the efficient stereocontrolling effect of trichloroacetimido group associated with trichloroacetimidate activation, was developed (Scheme 3.37).

3.2.3.3 1,2-cis Glycosides

The presence of 1,2-cis glycosides in various natural products led to the search for efficient methodologies for constructing this type of glycosidic linkage. Again, trichloroacetimidate chemistry plays a very important role in this context [451–455]. Recently, preferential β-mannoside formation was achieved with 4,6-O-benzylidene-protected mannosyl trichloroacetimidates as glycosyl donors and catalytic amounts of TMSOTf as the promoter [451]; hence, another convenient procedure was available for the preparation of β-mannosides, as shown in Scheme 3.38a. For the reaction course, the intermediacy of a twist-boat-type structure was proposed. β-Mannosides were also prepared conveniently and efficiently with mannosyl trichloroacetimidates possessing a strongly electron-withdrawing benzylsulfonyl group at the O-2 position as glycosyl donors [452]. These donors, upon activation, favor the generation of a flattened twist-boat intermediate conformation because of a strong dipole effect, which is preferentially attacked from the β side to form β-mannosides. Moreover, the benzylsulfonyl group could be easily removed after glycosylation. Biologically interesting cis-(1 → 2)-linked disaccharide derivatives were prepared by a regioselective one-pot benzylation–glycosylation strategy based on the TMSOTf-catalyzed Schmidt glycosylation procedure (Scheme 3.38b) [453]. The high regioselectivity was deemed to be induced not only by the steric hindrance between the

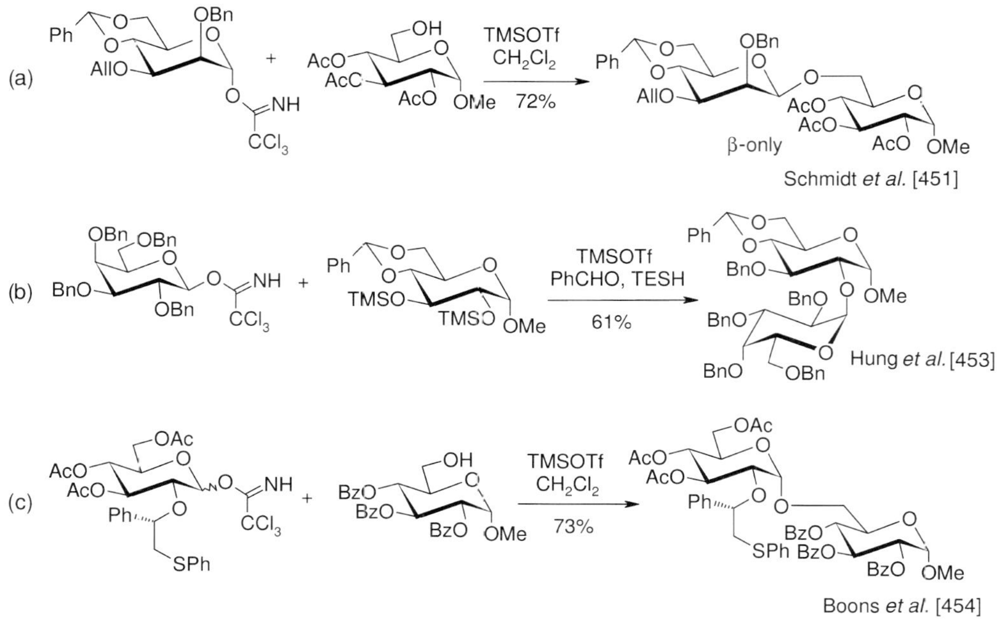

Scheme 3.38

anomeric methoxy group and 2-OTMS group but also by the inductive effect of the two anomeric oxygen atoms that cause a decrease in the nucleophilicity of 2-oxygen.

A new strategy for the stereoselective introduction of 1,2-*cis* glycosidic linkages has also been developed recently based on glycosyl trichloroacetimidates with a (1*S*)-phenyl-2-(phenylsulfanyl)ethyl group at O-2 position (Scheme 3.38c). These donors reacted through an unusual pathway, whereby the phenylsulfanyl moiety of the chiral auxiliary performed the neighboring-group participation to give a quasistable anomeric sulfonium ion with *trans*-decalin conformation. Thus, an acceptor could only approach this sulfonium ion intermediate from the bottom face leading to α-glycosides [454]. Nevertheless, relatively harsh conditions were required to install and cleave this auxiliary.

Besides the applications in the above-mentioned methodologies, trichloroacetimidate chemistry has also been used in the past years to construct various oligosaccharides containing 1,2-*cis* linkages [456–465]. In some examples, notably by Kong and coworkers [460], 1,2-*cis* stereoselectivity was influenced by the glycosidic bonds originally present in either the donor or the acceptor.

3.2.3.4 Miscellaneous Oligosaccharides

The trichloroacetimidate method has also found wide applications in the synthesis of various complex oligosaccharides. In the course of the development of chemically defined glycoconjugate vaccines against shigellosis, a decasaccharide, corresponding to two consecutive repeating units of the O-specific polysaccharide of *Shigella*

3.2 Glycoside Synthesis from 1-Oxygen-Substituted Glycosyl Imidates

flexneri 2a, was synthesized using trichloroacetimidates as glycosylating agents [466], as shown in Scheme 3.39. The convergent route was established by the condensation of two key pentasaccharide building blocks, which were synthesized in a linear fashion. The first glycosylation in the synthesis was conducted in diethyl ether to obtain high α-selectivity, and the β-GlcNAc residue was introduced with *N*-trichloroacetyl-protected glucosaminyl trichloroacetimidate. The whole synthesis reflected the high efficiency and great power of the Schmidt protocol.

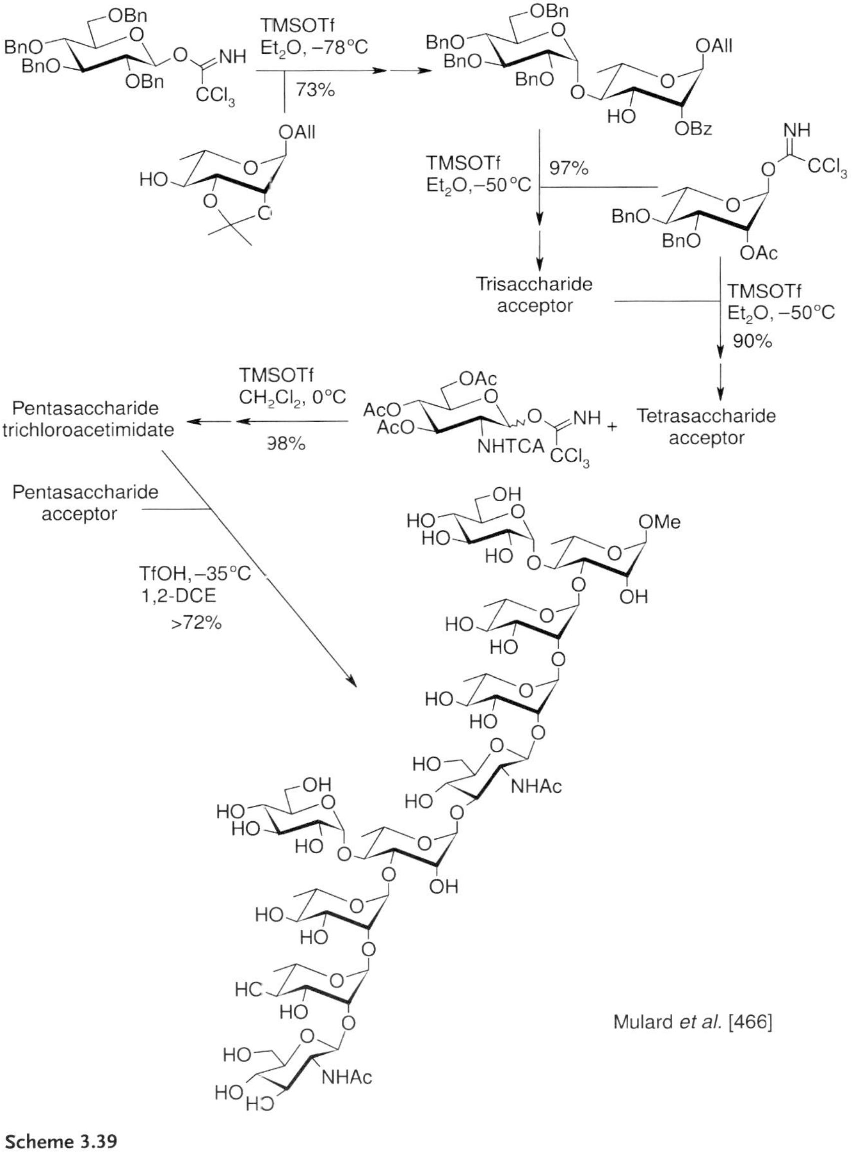

Scheme 3.39

Also, the key step in the synthesis of a mucin oligosaccharide derived from *Trypanosoma cruzi*, the causative agent of Chagas' disease, adopted trichloroacetimidate method in which the solvent effect of acetonitrile was exploited to build up the 1,2-*trans* glycosidic bond in the absence of the neighboring-group participation [467]. A series of branched β-cyclodextrins possessing β-galactose residues at the nonreducing terminal end of the sugar side chains were also prepared using trichloroacetimidate method [468], as enumerated in Scheme 3.40. These types of structures could be useful drug carriers in targeted drug delivery systems, considering the fact that galactose plays important roles in the recognition of receptors on the cell surface.

2-Deoxy-β-glycosides are important structural components of many natural products. Recently, 2-deoxy-2-iodoglycosyl trichloroacetimidates [469] have proved to be

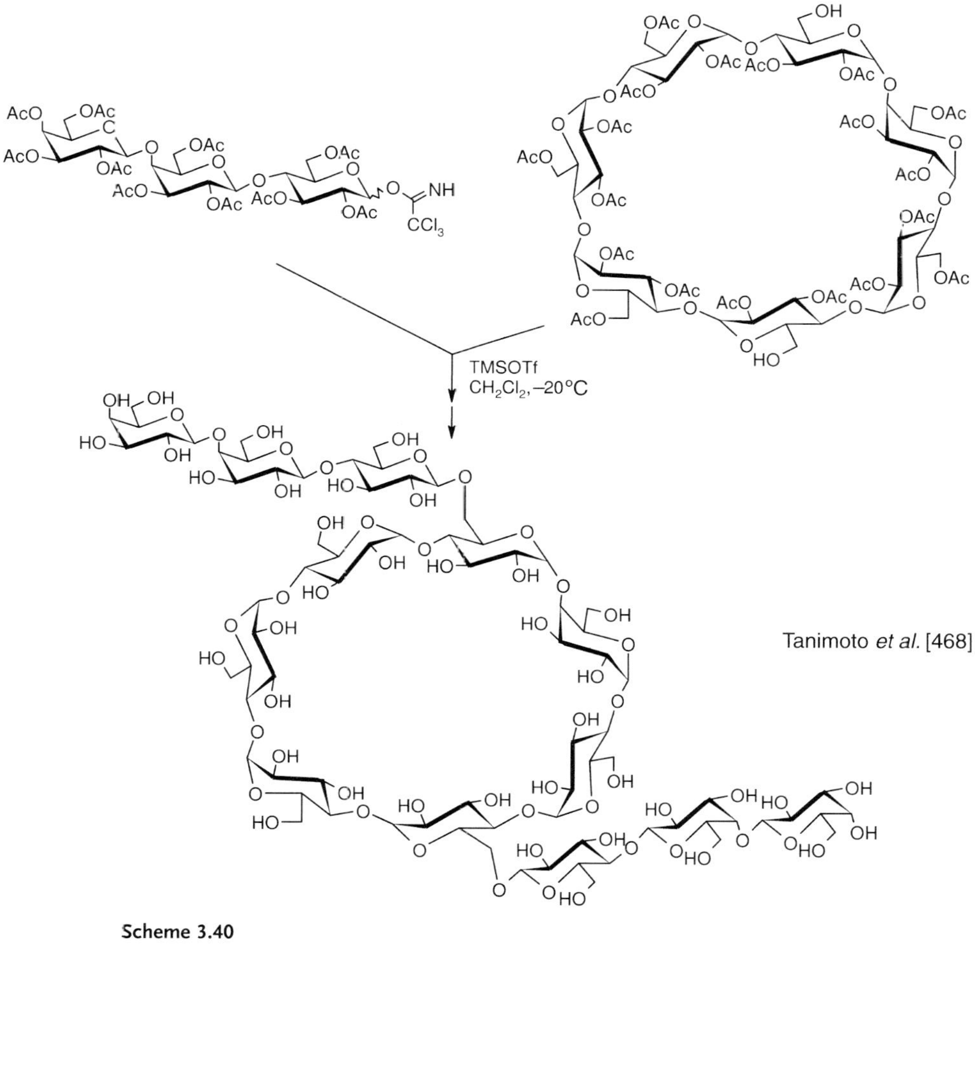

Tanimoto *et al*. [468]

Scheme 3.40

highly reactive glycosyl donors and could undergo highly β-stereoselective glycosylation reactions with various acceptors to form 2-deoxy-2-iodo-β-glycosides, precursors of 2-deoxy-β-glycosides. Investigations have been subsequently carried out on the possible intermediates generated in the reactions using conformationally constrained glycosyl imidates [470], and on the application of this methodology to the synthesis of a complex deoxyhexasaccharide derived from landomycin A, a member of the angucycline antibiotic family (Scheme 3.41) [471].

In the past years, some oligosaccharides containing relatively rare monosaccharides have also been synthesized with glycosyl trichloroacetimidates as donors. For instance, an L-*glycero*-D-*manno*-heptose-containing tetrasaccharide derived from Neisserial lipooligosaccharides was synthesized recently by regioselective glycosylation of mannose derivative with a heptosyl trichloroacetimidate [472]. Also, glycosylations with D-rhamnosyl trichloroacetimidates were performed to construct a branched D-rhamnotetraose [473], a repeating unit of the O-chain from bacterial lipopolysaccharides. In addition, nonnative oligosaccharides containing 5-thiosugars have also been prepared as molecular probes for biological and medicinal studies using the trichloroacetimidate method [474–476]. All the glycosylations proceeded smoothly and gave the desired products in high yields and stereoselectivities.

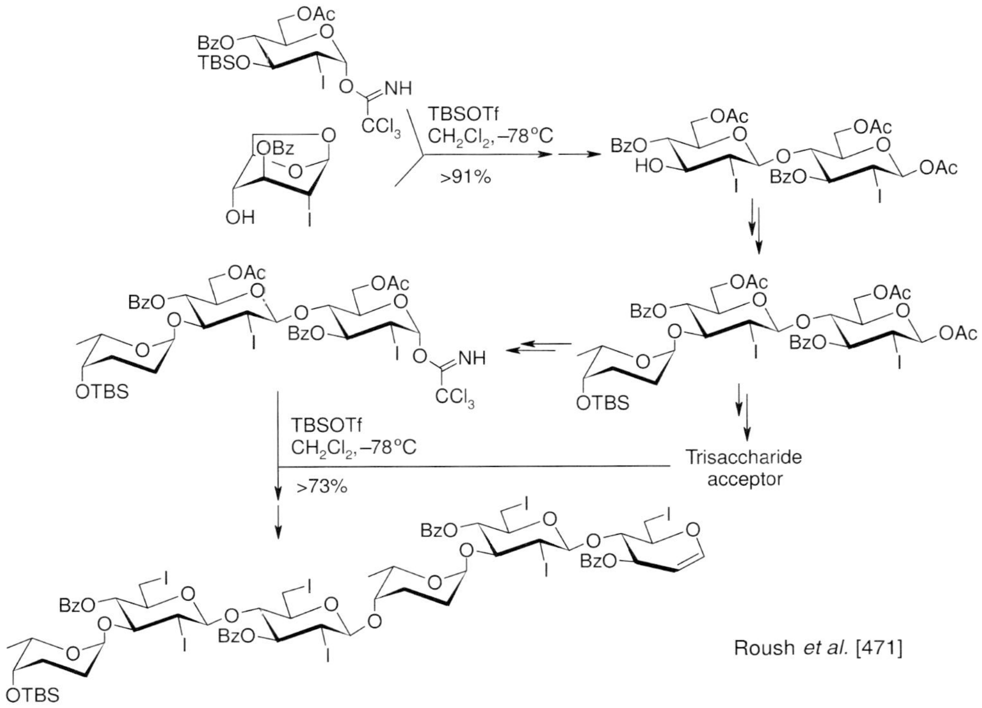

Roush *et al.* [471]

Scheme 3.41

3.2.4
Synthesis of Glycoconjugates

3.2.4.1 Glycosphingolipids and Mimics

Several syntheses of glycosphingolipids (GSLs) based on azidosphingosine glycosylation strategy have been reported in the last few years. Among these, disialoganglioside GD3, a human-melanoma-associated antigen, was synthesized in overall high yield by glycosylation of an azidosphingosine derivative with a tetrasaccharide trichloroacetimidate [477] (Scheme 3.42a). The same strategy was also applied to

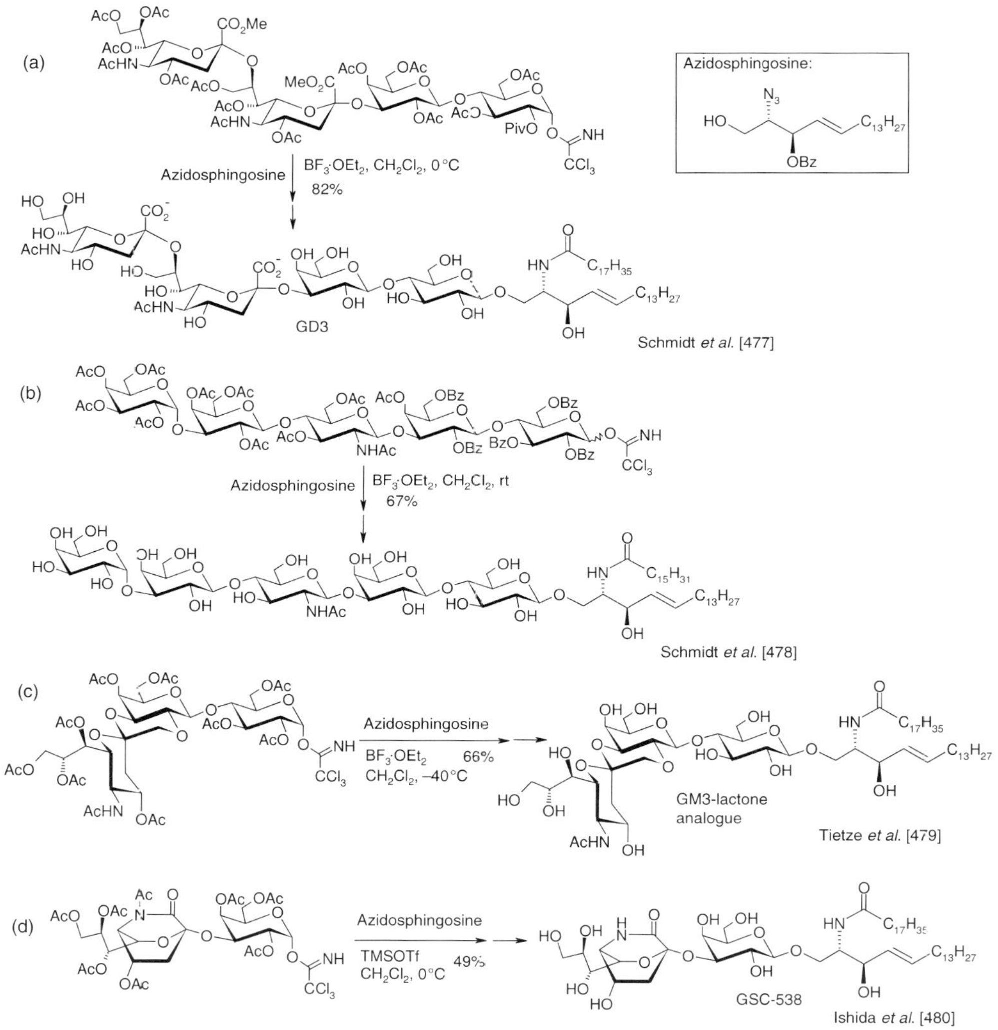

Scheme 3.42

synthesize the natural antigen involved in the hyperacute rejection response to xenotransplants, which consists of a pentasaccharide and a ceramide moiety (Scheme 3.42b). The pentasaccharide itself was also prepared by the Schmidt glycosylation protocol [478]. In view of the low hydrolytic stability of ganglioside lactones, an ether-bridged analog of ganglioside GM3-lactone has been constructed recently as a target for an antibody-based cancer therapy, in which the final ligation between the sugar and azidosphingosine was also achieved using the trichloroacetimidate method [479] (Scheme 3.42c). Also, systematic syntheses of novel lactamized gangliosides, such as GSC-538 in Scheme 3.42d, have been reported recently, in which the key steps were also glycosylations of azidosphingosine with trichloroacetimidates [480].

The sialyl Lewis X (sLe^X) epitope has become a prominent target for biological studies because of its participation in the inflammation process that takes place through binding to selectins. This epitope is located at the terminal end in GSLs, and the lactose unit serves as a spacer to the ceramide moiety. Recently, the influence of the spacer structure and length in regard to the mobility of sLe^X epitope has been investigated with synthetic neoglycolipids [481]. Successive glycosylations of a dialkylglycerol with a lactosyl trichloroacetimidate followed by the attachment of an sLe^X oligosaccharide provided a series of neoglycolipids with one to three lactose units as spacer. The sLe^X oligosaccharide itself was also assembled from the corresponding sugar building blocks using trichloroacetimidate method (Scheme 3.43), wherein the fucosylation of the N-Troc-protected glucosamine trisaccharide exclusively afforded the desired α-linked tetrasaccharide.

Another series of GSL mimics with oligo-ethylene glycol as spacer have also been obtained successfully using trichloroacetimidate method [482]. In addition, fluorescence-labeled sLe^X glycosphingolipids have also been chemically synthesized as targets for investigating microdomain formation in membranes [483].

α-Galactosphingolipids have been found to have interesting immunomodulating activities. They can specifically activate CD1d-restricted natural killer T cells, which are primed to produce and release an array of cytokines such as interferon IFN-γ and interleukin IL-4. These cytokines are recognized subsequently by other cells of the immune system and may have a widespread influence on immune responses, including protection against autoimmune diseases, the host responses to parasites and bacteria and antitumor responses. Many efforts have thus been devoted in the past decade to synthesize this type of compounds to explore further structure–function relationship of individual α-galactosylceramides (α-GalCers). Recently, the total synthesis of α-GalCer was successfully achieved in which the key step was the highly regio- and stereoselective galactosylation of phytosphingosine acceptor with the galactosyl trichloroacetimidate donor [484] (Scheme 3.44a). Also, glucosamine–glycerophospholipid conjugates have been prepared for the investigation of their structure–function relationships (Scheme 3.44b), wherein N-DMM-protected glucosamine trichloroacetimidate was used as the glycosylating agent [485].

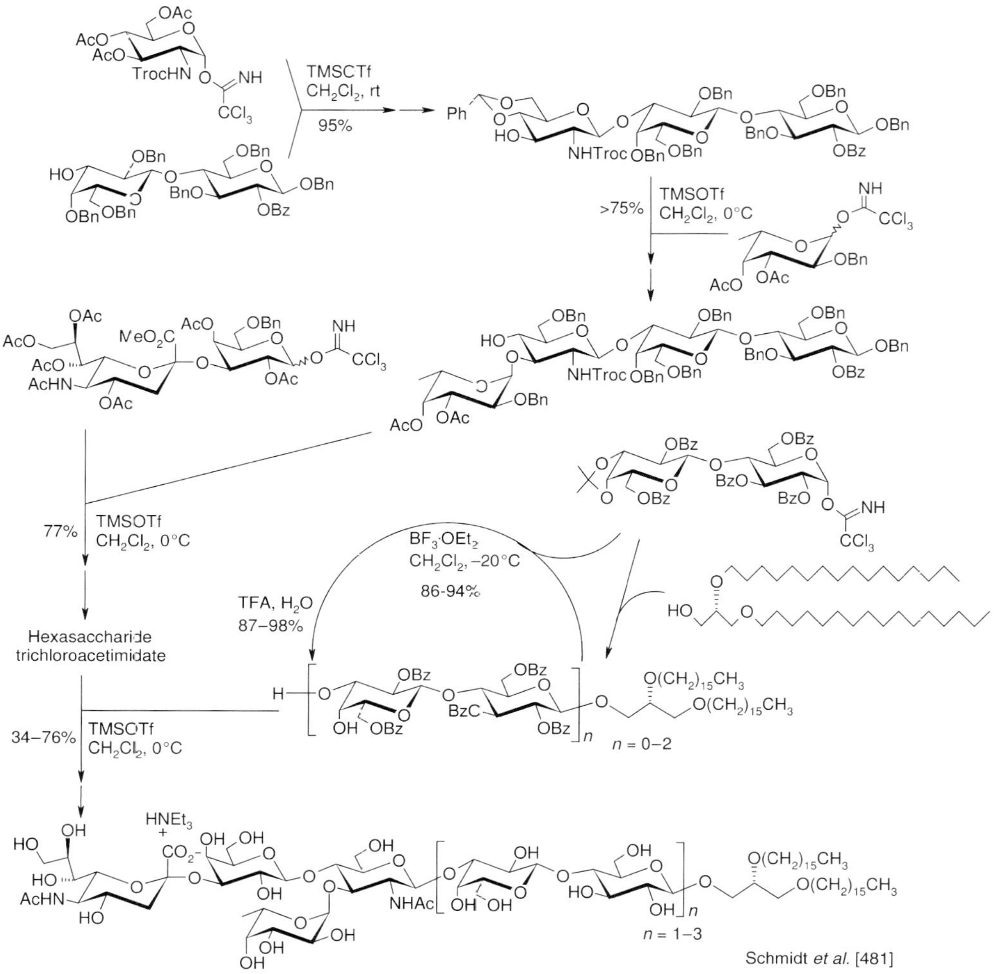

Scheme 3.43

3.2.4.2 Glycosyl Phosphatidyl Inositol Anchors

Glycosyl phosphatidyl inositol (GPI) anchors are a class of naturally occurring glycolipids that serve as anchors for proteins and glycoproteins in membranes. They consist of many variants in both the carbohydrate and the lipid moieties. On the basis of versatile building blocks, a highly variable concept for the synthesis of branched GPI anchors has been established recently, and the building blocks were readily accessible and could be transformed into products in high regio- and stereo-selectivity in all reaction steps [486]. The efficiency of this concept was demonstrated through the synthesis of the 4,6-branched GPI anchor of *rat brain* Thy-1 and *scrapie* prion protein in which a group of trichloroacetimidates were used as glycosylating

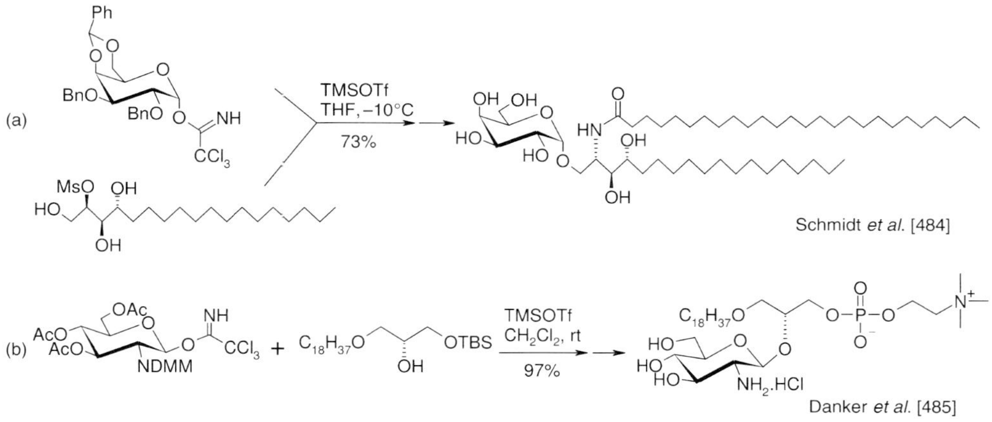

Scheme 3.44

agents (Scheme 3.45). The key intermediate, pentasaccharide trichloroacetimidate, was first built up by a consecutive glycosylation of the mannose acceptor with three suitably protected monosaccharide trichloroacetimidates. With the pentasaccharide donor in hand, the carbohydrate backbone was then assembled by glycosylation of the pseudodisaccharide acceptor. All the glycosylations proceeded stereoselectively and gave the products in high yields. The total synthesis was finally completed by attaching various phosphate residues at the proper positions. Another fully phosphorylated pseudohexasaccharide has also been synthesized efficiently using the trichloroacetimidate method [487]. In addition, synthesis of an inositol-containing pseudohexasaccharide derived from Type A inositolphosphoglycans (IPGs), structurally related to GPIs, has been described [488].

3.2.4.3 Glycosyl Amino Acids and Glycopeptides

The trichloroacetimidate method has been used frequently in the synthesis of glycosyl amino acids and glycopeptides. Glycosyl amino acids carrying tumor-associated antigens were prepared recently by glycosylation of Fmoc-Thr/Ser-OPfp with sialyl-T antigen trisaccharide trichloroacetimidate [489], which gave mainly α-product because of the neighboring nonparticipating azido group (Scheme 3.46a). T-Antigen-containing glycosyl amino acid was also prepared in a similar manner and α-selectivity was obtained again with 2-azidogalactosyl trichloroacetimidate [490]. Both the glycosyl amino acids obtained could be used directly in automated solid-phase glycopeptide synthesis. Glucogalactosyl hydroxylysine, an important biological indicator of collagen turnover, was synthesized relying on the trichloroacetimidate method [491], as shown in Scheme 3.46b. The glucosylation was performed in Et$_2$O as the solvent, and its participation from β-face ensured excellent α-selectivity. A strategy to obtain N-linked glucosyl tryptophan has also been developed recently on the basis of Schmidt glycosylation protocol (Scheme 3.46c). The key steps involved the introduction of a 2-pivaloyl group to the donor to suppress the formation of a tryptophan-1-yl amide acetal by-product. The use of an

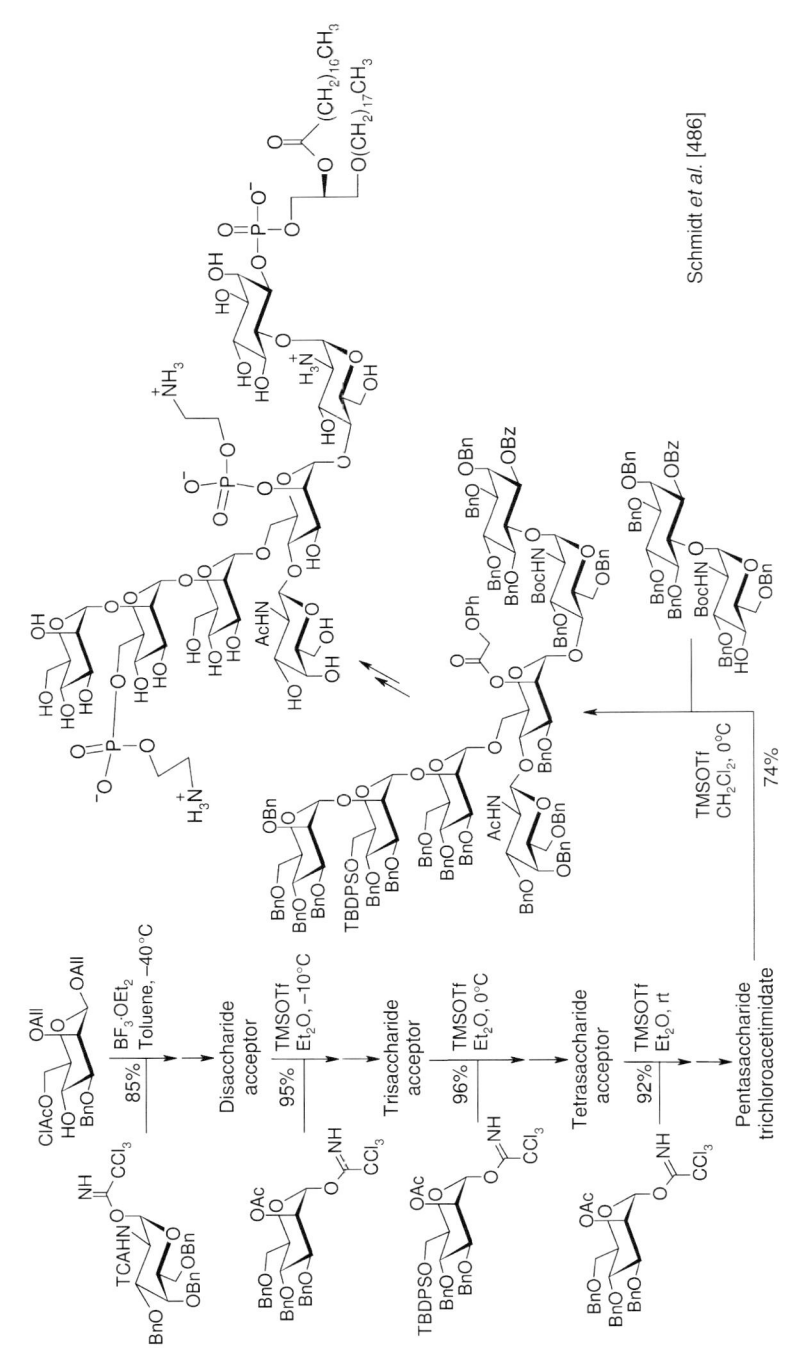

164 | *3 Glycoside Synthesis from 1-Oxygen Substituted Glycosyl Donors*

Scheme 3.45

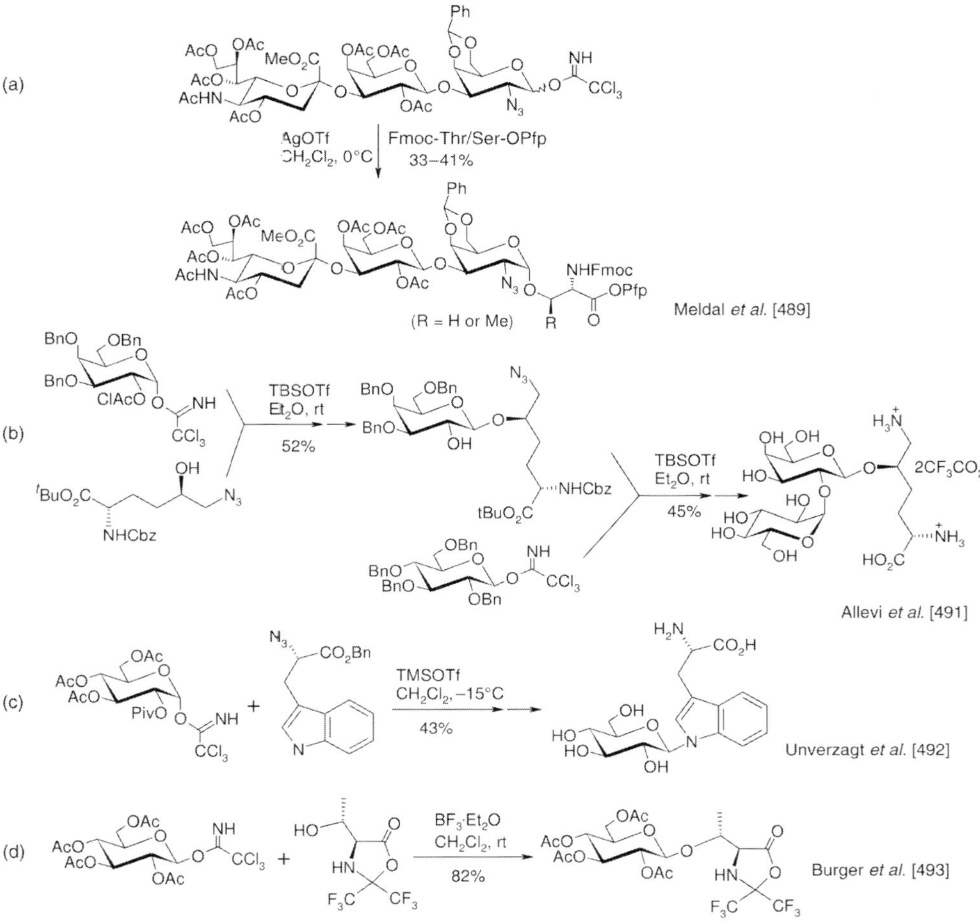

Scheme 3.46

α-azido tryptophan derivative was beneficial to improve the yield [492]. The orthogonal protection of α-amino and carboxylic groups is usually required in the conventional synthesis of glycosyl amino acids to ensure that the amino or carboxylic group can be selectively deprotected and used in subsequent glycopeptide synthesis. Recently, a new protecting-group/activation concept has been developed so that the glycosyl amino acids prepared therein could be directly used to synthesize glycopeptides, and glycodipeptide and tripeptide fragments could be prepared from hydroxyl amino acids by this concept in only three or four synthetic steps [493]. The trichloroacetimidate method exhibited advantages over other glycosylation procedures in the preparation of this type of glycosyl amino acids (Scheme 3.46d).

In general, a convergent coupling between a sugar and a peptide to form an O-glycopeptide is problematic because of the generally poor solubility of peptides under glycosylation conditions and also because of regio- and stereochemical

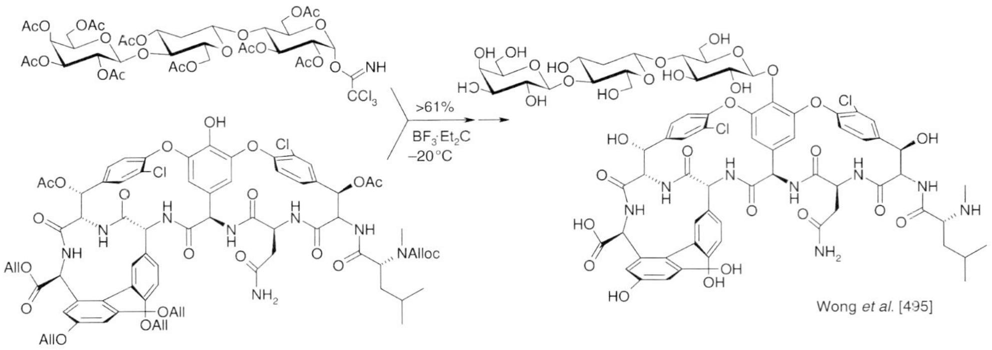

Scheme 3.47

aspects. However, the efficient solid-phase glycosylation of amino acid side chains (Ser, Thr and Tyr) in peptides was accomplished recently with a variety of glycosyl trichloroacetimidates in high yields and purities [494]. Also, as enumerated in Scheme 3.47, direct glycosylations of vancomycin aglycone with different glycosyl trichloroacetimidates were also achieved [495–497], allowing rapid creation of libraries of vancomycin derivatives bearing unnatural sugar substituents.

3.2.4.4 Saponins

Saponins are steroid or triterpenoid glycosides possessing various biological and pharmacological activities. A highly efficient procedure has been reported recently for the glycosylation of sapogenins [498] in which TMSOTf-catalyzed glycosylation with benzoylated glycosyl trichloroacetimidates is the key to success (Scheme 3.48a). The solvent effect of propionitrile has been exploited to control the stereochemistry in the glycosylation of hederagenin derivatives with a (1 → 2)-linked disaccharide trichloroacetimidate to synthesize kalopanaxsaponin A [499], as shown in Scheme 3.48b. Other hederagenin saponins have also been synthesized using a series of disaccharide trichloroacetimidates as donors to investigate the structure–activity relationship between triterpenoid saponins and hemolytic activity [500]. Lycotetraose, one of the major oligosaccharides in steroid saponins, has been installed onto cholesterol via its trichloroacetimidate intermediate to verify its antitumor property (Scheme 3.48c). Unexpectedly, α-lycotetraosyl cholesterol was formed instead of the β-isomer in a stereoselective manner despite the presence of neighboring participating group [501]. To verify the biological role of chacotriose, several chacotriosides of cholesterol, diosgenin and glycyrrhetic acid were also synthesized by a similar *trans*-glycosylation strategy [502].

In addition, the fulvestrant could be glycosylated effectively at its 17-OH position with pivaloylated glycosyl trichloroacetimidates, which suppressed the competing transacylation side reaction and led to improved yields of the desired glycosides (Scheme 3.48d) [503]. In this synthesis, the inverse procedure (i.e. addition of a trichloroacetimidate donor to a mixture of an acceptor and a promoter) was found to be superior for glycosylations. Very recently, a stepwise synthesis of branched

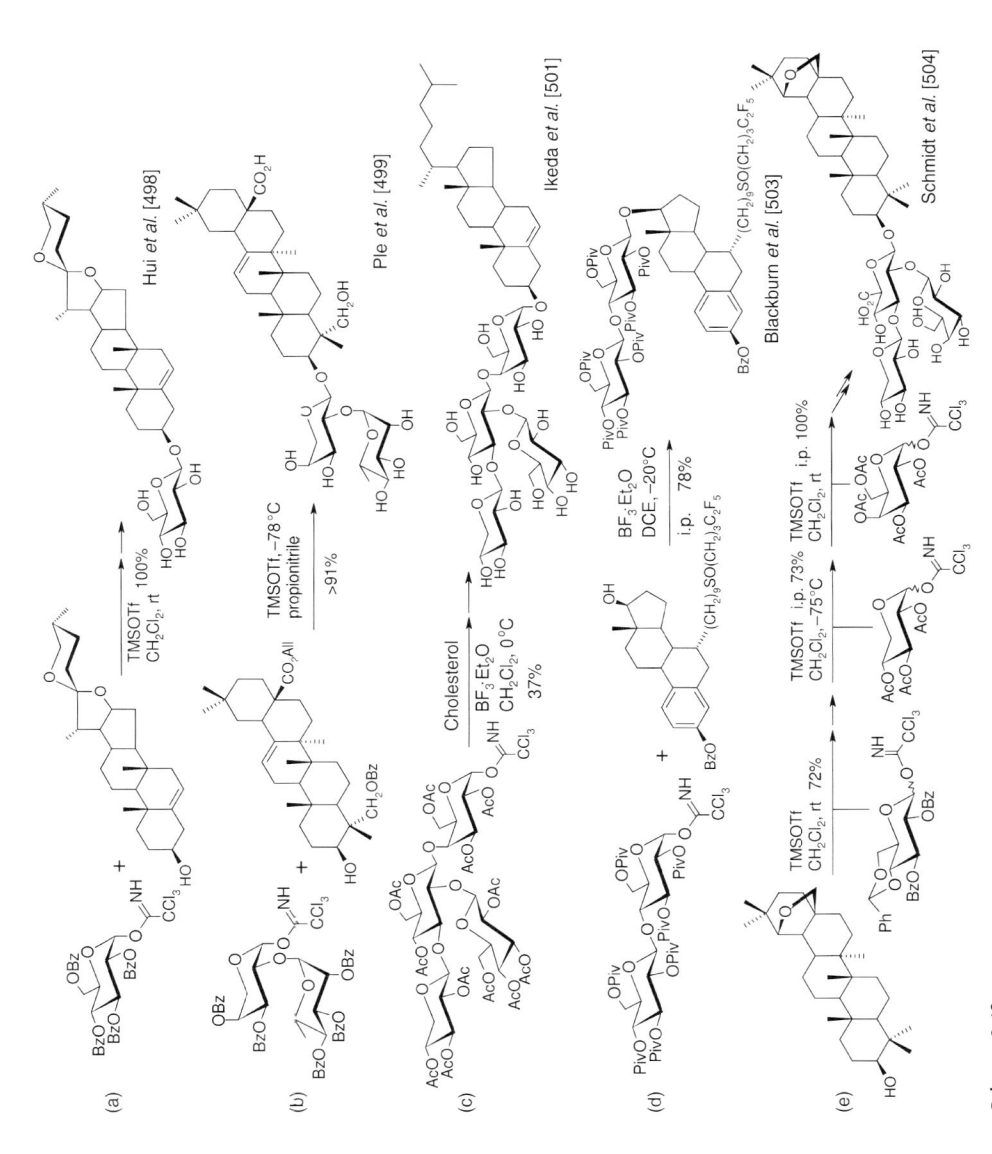

3.2 Glycoside Synthesis from 1-Oxygen-Substituted Glycosyl Imidates | 167

Scheme 3.48

glucuronic-acid-containing trisaccharides of allobetulin and glycyrrhetic acid has been described (Scheme 3.48e). All the glycosylations proceeded smoothly either under normal conditions or by an inverse procedure and gave the corresponding products in high yields [504].

3.2.4.5 **Other Natural Products and Derivatives**

Glycosyl trichloroacetimidates have also been widely used in the synthesis of many other classes of natural products and their derivatives. Conandroside, a 'bitter glycoside' isolated from Conandron ramoidioides, was synthesized using trichloroacetimidates as building blocks [505] (Scheme 3.49a). The first total synthesis of the naturally occurring dimeric ellagitannin, Coriariin A, has also been achieved recently [506], wherein the critical bis-glucosylation step was performed with a trichloroacetimidate donor in the absence of a Lewis acid activator. Strictly speaking, the glycosylation in the synthesis is not a glycosidation process, but we list it in this chapter as an interesting application of trichloroacetimidate donors. As shown in Scheme 3.49b, simply refluxing the glycosyl donor and the acidic acceptor in benzene provided the requisite anomerically pure diglucosyl dehydrodigalloyl diester in good yield, which indicated the excellent reactivity and high stability of the trichloroacetimidate donor.

Resin glycoside has a very widespread occurrence in the plant kingdom and possesses various biological activities, such as cytotoxicity against human cancer cell lines, antibacterial activity, purgative properties and plant-growth-regulating capacity. Structurally, resin glycoside often contains (11*S*)-hydroxyhexadecanoic acid (jalapinolic acid) as a common aglycone, which is usually tied back to form a characteristic macrolide ring that spans two or more sugar units of its oligosaccharide backbone. Following the previous work on resin glycoside synthesis [507], Tricolorin F has been synthesized efficiently using the Schmidt glycosylation protocol [508], as shown in Scheme 3.50. Consecutive glycosylations with three trichloroacetimidate donors, followed by Yamaguchi lactonization, furnished the resin glycoside in an overall good yield after deprotection.

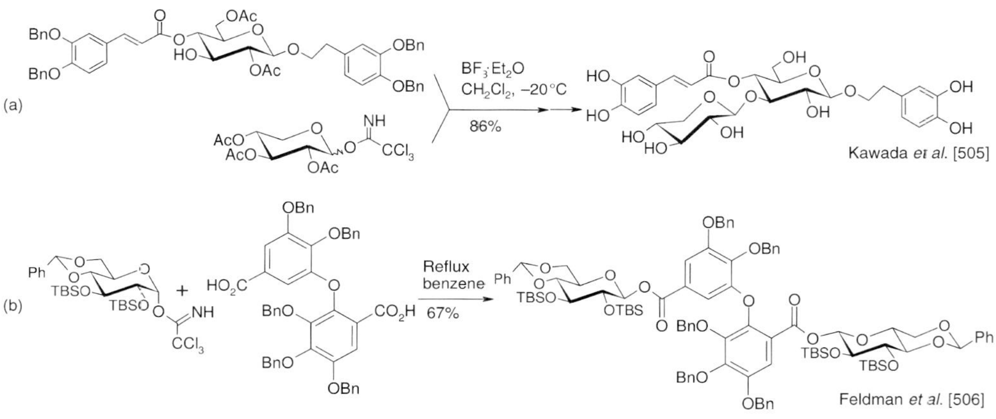

Scheme 3.49

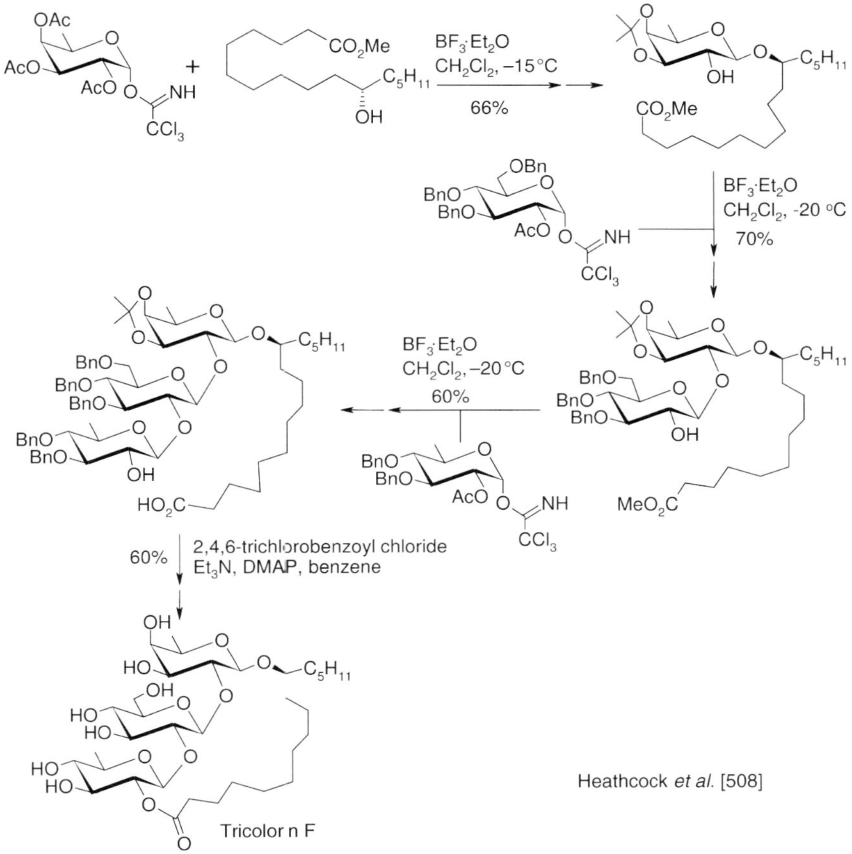

Scheme 3.50

Heathcock et al. [508]

The total synthesis of Woodrosin I, one of the most complex resin glycosides, was also reported recently [509]. The executed approach clearly demonstrated again the great power of the trichloroacetimidate method. The whole synthesis was graced with two regioselectiveglycosylations and the final inverse addition procedure, which dramatically simplified the synthetic route (Scheme 3.51). In addition, cycloviracin B [510] and glucolipsin A [511] have also been synthesized successfully by the same laboratories, taking advantage of the trichloroacetimidate method.

As a drug candidate advances through clinical development, the synthesis of its glucuronide often becomes necessary to provide an analytical standard for use in quantification of metabolite levels in clinical samples and to provide material for further pharmacological evaluation. In the past years, several glucuronides have been prepared by the Schmidt glycosylation protocol [512–516]. For example, morphine-3,6-di-glucuronide could be prepared in a very good yield by direct glycosylation of morphine with isobutyryl-protected glucuronic acid trichloroacetimidates

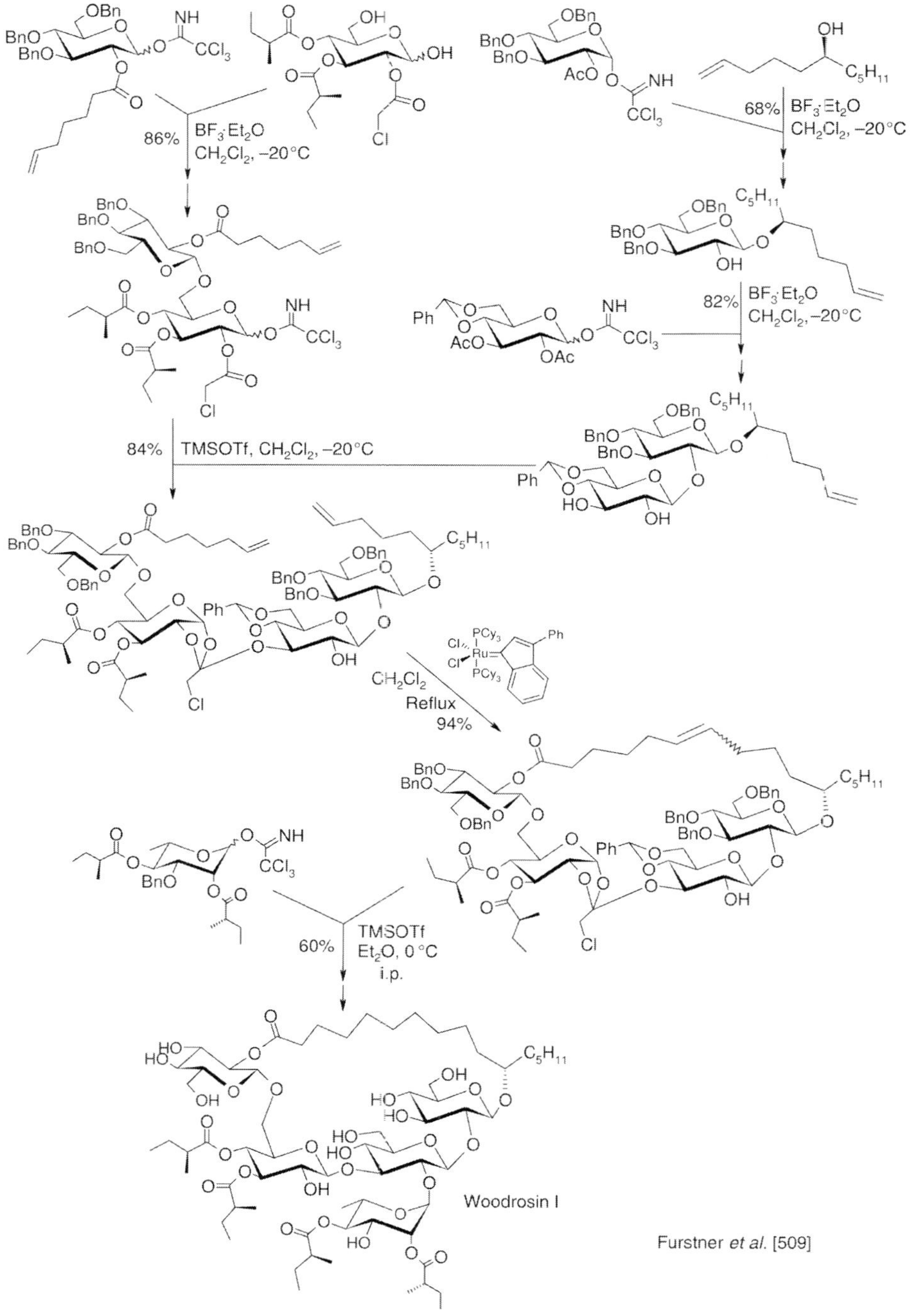

3 Glycoside Synthesis from 1-Oxygen Substituted Glycosyl Donors

Woodrosin I

Furstner et al. [509]

Scheme 3.51

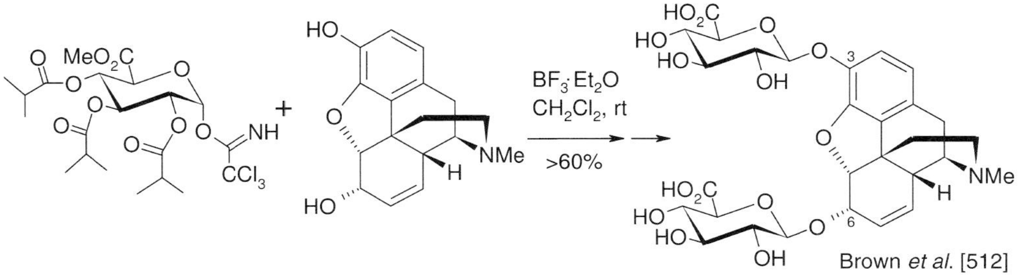

3.2 Glycoside Synthesis from 1-Oxygen-Substituted Glycosyl Imidates

Scheme 3.52

(Scheme 3.52), whereas the corresponding acetyl-protected donor gave little product under the same reaction conditions [512]. Additionally, the trichloroacetimidate method has also been applied to synthesize other natural products, such as buprestin A and B [517], macrophylloside D [518] and neomycin mimetics [519].

3.2.4.6 Miscellaneous Glycoconjugates

The simultaneous presentation of sugars on a macromolecular scaffold can create a multivalent display that amplifies the affinity of glycoside-mediated receptor targeting. Dendrimer-like poly(ethylene oxide) glycopolymers bearing sulfated β-lactose have been prepared recently as potential L-selectin inhibitors, in which protecting-group manipulations were minimized by the use of lactose trichloroacetimidate donors [520]. Glycosyl trichloroacetimidate has also been used to prepare a modified nucleoside, which was then incorporated into oligonucleotides of biological interest by automated solid-phase synthesis [521]. Also, a direct glycosylation of oligonucleotides with trichloroacetimidate donors has been reported recently [522,523]. Glycosyl trichloroacetimidates are also suitable donors for the synthesis of C-/N-glycosides [524–526].

3.2.5
Solid-Phase Oligosaccharide Synthesis

In recent years, a notable progress has been made on solid-phase oligosaccharide synthesis based on the trichloroacetimidate method [527–538]. One important advance is that glycosyl trichloroacetimidates, bearing the O-Fmoc protecting group, have been successfully prepared and proved to be suitable for oligosaccharide synthesis on solid support [527]. Very recently, a series of N-glycan oligosaccharides have been synthesized on Merrifield resin with a hydroxymethyl-benzyl benzoate spacer–linker system [528]. As enumerated in Scheme 3.53, the glycosylations were stereospecifically performed with three types of trichloroacetimidate donors, which allowed chain extension, branching and chain termination, respectively. For chain-branching donors, Fmoc and phenoxyacetyl (PA) were used as temporary protecting groups with Ac, Bz, Bn and/or N-DMM as permanent protecting groups. The crude products released from the resin were of high purity after all glycosylation and

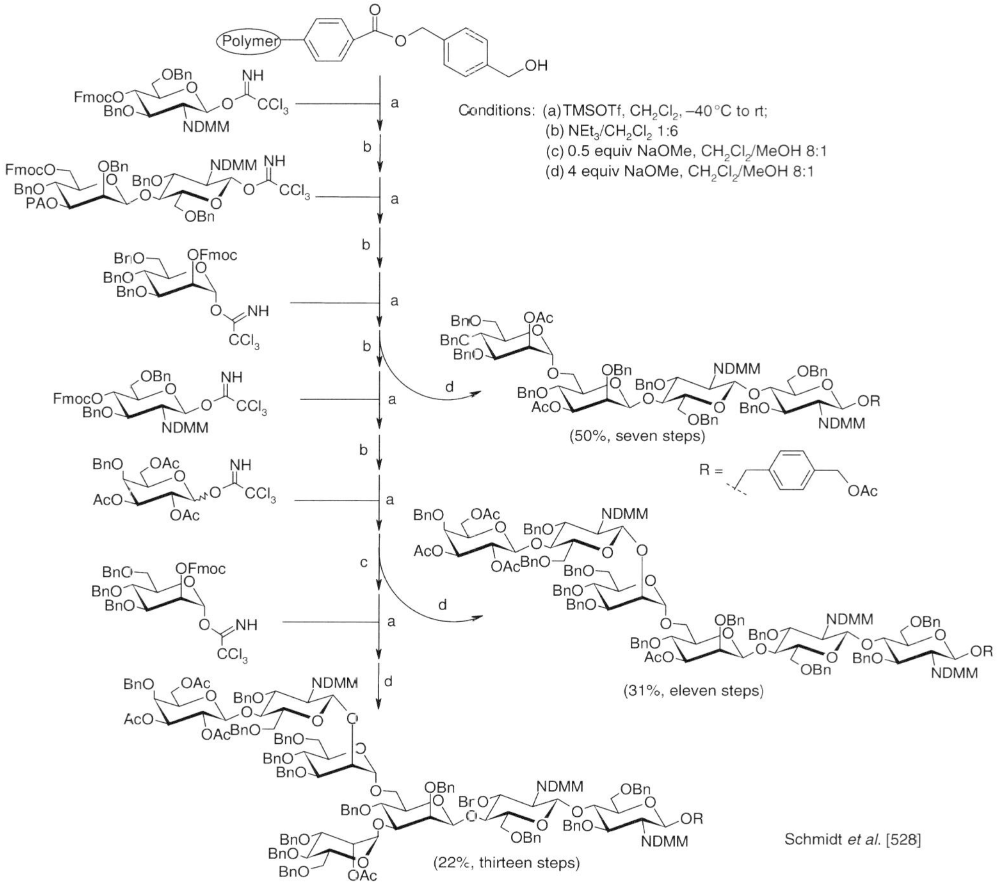

Scheme 3.53

protecting-group manipulation steps. The simplicity and efficiency of the whole synthesis provided a basis for the development of a general approach to the synthesis of oligosaccharides having different glycosidic linkages. For example, a similar strategy has been applied to synthesize a branched lacto-*N*-neohexasaccharide occurring in human milk (Scheme 3.54) [529], and the product was also released from the resin as a benzylic glycoside, which made further deprotection easy. The key building block, lactose trichloroacetimidate, was protected orthogonally with Fmoc and Lev groups, allowing for selective derivatization at both positions. All glycosylations on the solid support were highly stereoselective and high yielding, and the hexasaccharide was furnished in an excellent overall yield of 42%. The great utility of Fmoc-protected glycosyl trichloroacetimidates has also been demonstrated in the synthesis of other oligosaccharides, such as oligomannosides [530], lactosamine- and lactose-containing oligosaccharides [531].

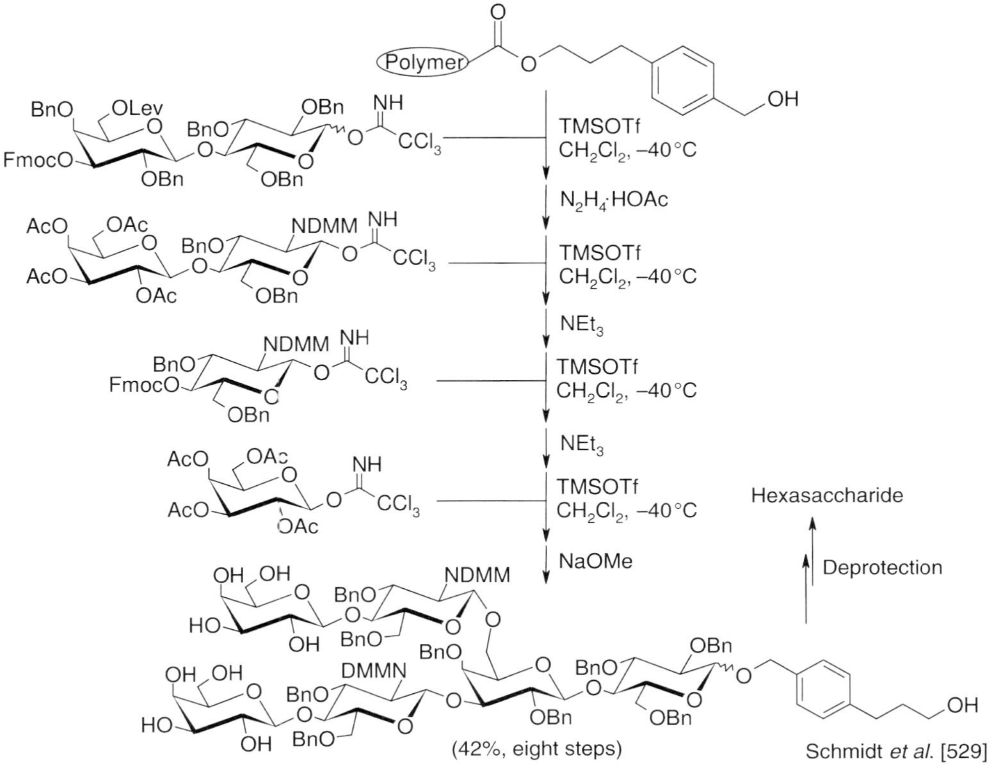

3.2 Glycoside Synthesis from 1-Oxygen-Substituted Glycosyl Imidates | 173

Scheme 3.54

In addition, some other techniques have been developed for solid-phase oligosaccharide synthesis in combination with the Schmidt glycosylation protocol in the past years, including real-time reaction-monitoring method [532] and novel capping reagents [533]. Recently, the automated synthesis of oligosaccharides has been achieved by using a solid-phase synthesizer with trichloroacetimidates as glycosylating agents [534], and N-glycan core pentasaccharide was successfully assembled within 3 days after three consecutive glycosylation reactions (Scheme 3.55) [535]. The final release of the pentasaccharide as its n-pentenyl glycoside from the octenediol-functionalized Merrifield resin was performed with Grubbs catalyst. A rapid synthesis of a tetrasaccharide fragment of malarial toxin has also been accomplished on this synthesizer using trichloroacetimidate method [536].

Heparin-like oligosaccharides have been synthesized on soluble polymer support, polyethylene glycol monomethyl ether, in which the acceptor was bound to the polymer through the carboxylic group of the uronic acid unit by a glycol–succinic ester linkage [539,540]. By this protocol, an octasaccharide fragment, containing the structural motif of the regular region of heparins, has been synthesized using

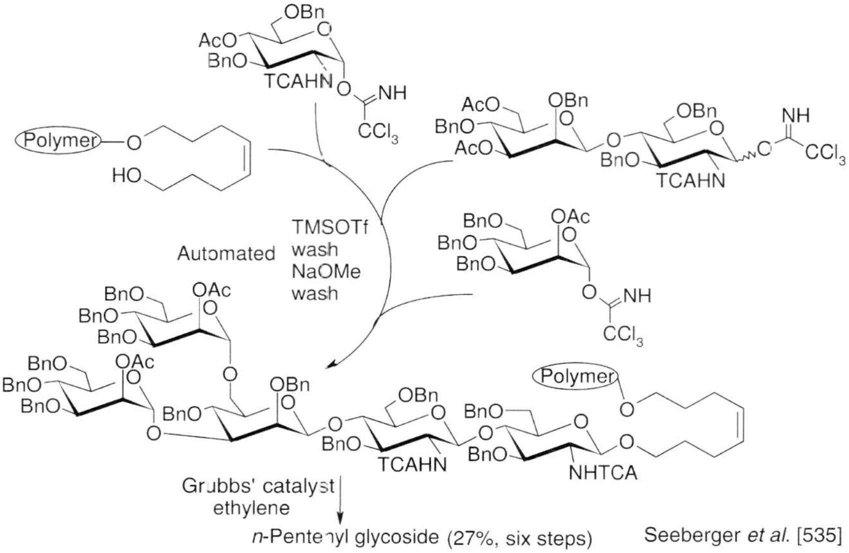

Scheme 3.55 Seeberger *et al.* [535]

n-Pentenyl glycoside (27%, six steps)

trichloroacetimidates as glycosylating agents and a functionalized Merrifield resin as capping agent, as shown in Scheme 3.56 [539].

In addition, a novel fluorous support has been developed recently as an alternative to traditional polymer supports and applied successfully to oligosaccharide synthesis in combination with the trichloroacetimidate method [541]. Each intermediate in the fluorous oligosaccharide synthesis [542,543] could be obtained by simple fluorous-organic solvent extraction, and the reactions could be monitored by TLC, NMR and MS, in contrast to solid-phase reactions. Moreover, the new liquid-phase technique is anticipated to be easily applicable to the large-scale synthesis.

3.2.6
Trifluoroacetimidates

3.2.6.1 Preparation and Activation

As trichloroacetimidate analogs, glycosyl trifluoroacetimidates have received interest many years ago [383]. However, unlike trichloroacetimidates, the N-unsubstituted trifluoroacetimidates are difficult to prepare because the corresponding trifluoroacetonitrile is gaseous (bp $-64\,°C$) and toxic. Nakajima reported a one-pot preparation of glycosyl trifluoroacetimidates [544], wherein volatile trifluoroacetonitrile was generated from trifluoroacetamide with an 'activated' DMSO species at low temperature. In this section, emphasis is placed on glycosyl N-phenyl trifluoroacetimidates (PTFA), the most common and widely investigated trifluoroacetimidates [384,385]. PTFA donors are usually prepared from anomeric hemiacetals by

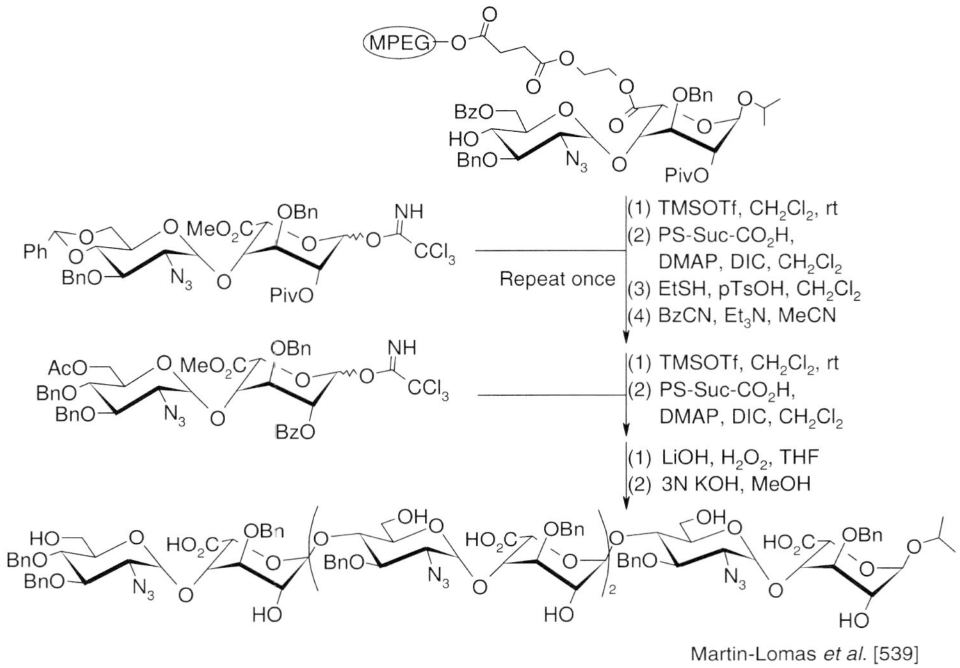

3.2 Glycoside Synthesis from 1-Oxygen-Substituted Glycosyl Imidates | 175

Scheme 3.56

Martin-Lomas et al. [539]

treatment with *N*-phenyl trifluoroacetimidoyl chloride in the presence of a stoichiometric amount of base (Scheme 3.57). In contrast to the trichloroacetimidate formation, the use of K_2CO_3 as the base generally favors α-PTFA [545], whereas the use of NaH [384] or DIPEA [546] mainly yields β-products; more commonly, α-/β-mixtures are produced.

PTFAs are generally less reactive than the corresponding trichloroacetimidate donors presumably because of the lower *N*-basicity or the presence of an *N*-substituent. Although most trichloroacetimidate activators could also be used to promote PTFA glycosidations, such as TMSOTf [384], $BF_3 \cdot Et_2O$ [384,547], TBSOTf [548], $Yb(OTf)_3$ [549,550] and acid-washed molecular sieves [551], the activation of PTFA usually requires more forceful conditions. Several representative Lewis-acid-catalyzed PTFA glycosidation reactions are listed in Scheme 3.58. It is worth

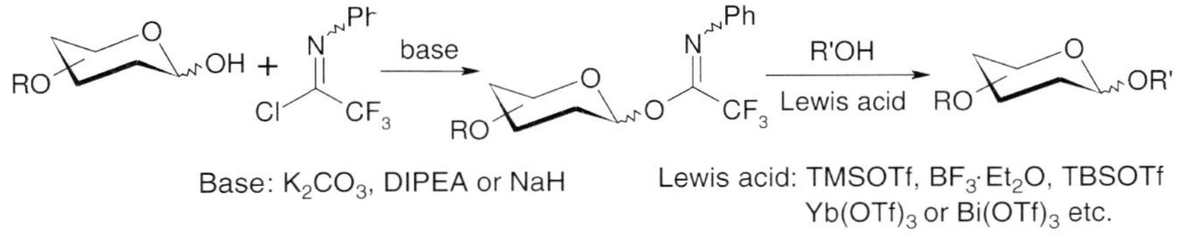

Base: K_2CO_3, DIPEA or NaH

Lewis acid: TMSOTf, $BF_3 \cdot Et_2O$, TBSOTf $Yb(OTf)_3$ or $Bi(OTf)_3$ etc.

Scheme 3.57

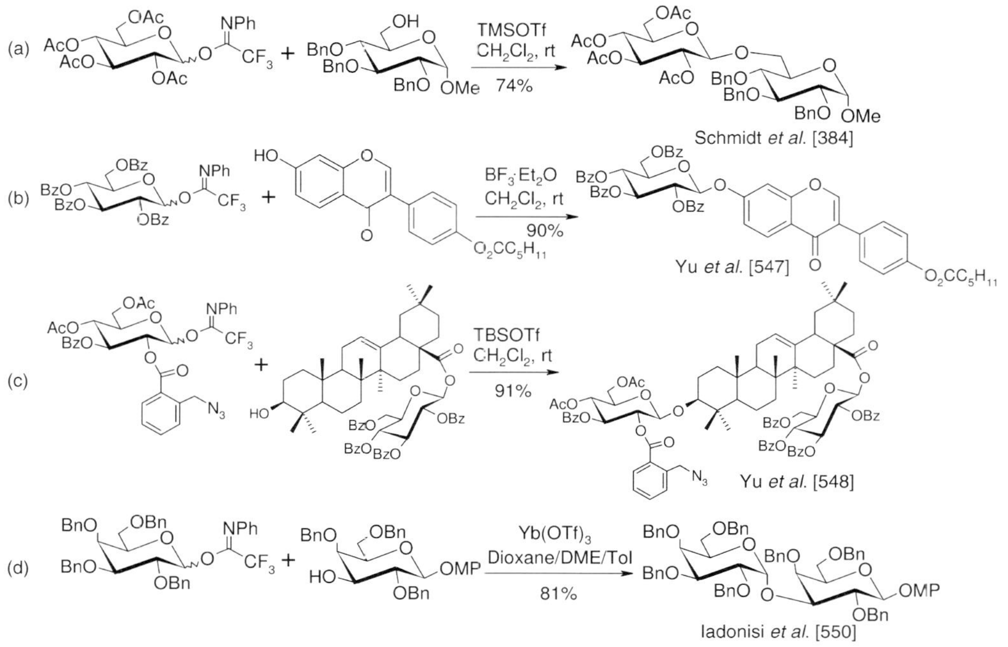

Scheme 3.58

mentioning that glycosyl trifluoroacetimidates bearing 2-(azidomethyl)benzoyl (AZMB) group at O-2 position have been used as efficient glycosylating agents in the synthesis of triterpenoid saponins [548], as enumerated in Scheme 3.58c. The AZMB group ensured 1,2-*trans*-glycosylation, but more importantly, it could be removed selectively in the presence of other acyl protecting groups. Also, some other activation systems such as I_2–Et_3SiH [386], $Bi(OTf)_3$ [552] and $TMSB(C_6F_5)_4$ [553] have been used to promote PTFA glycosidations. The different reactivities of PTFA and trichloroacetimidate donors have been exploited to develop a one-pot multistep procedure featuring selective activation of a trichloroacetimidate donor in the presence of a PTFA moiety [554], in which the PTFA derivative was partially protected to serve as a glycosyl acceptor in the first glycosidation step (Scheme 3.59).

3.2.6.2 Application to Target Synthesis

Trifluoroacetimidate donors have shown advantages over trichloroacetimidates in the synthesis of β-mannosides [553] because of their lower propensity to undergo side reactions during glycosidations. In the course of trichloroacetimidate glycosidation, a certain amount of an *N*-glycoside by-product is occasionally produced by the glycosylation of trichloroacetamide liberated from the donor. Particularly, this side reaction takes place when the acceptor is of low nucleophilicity or sterically hindered, whereas it is diminished in PTFA glycosidation because of the increased steric hindrance of the *N*-phenyl group. PTFA donors have thus gained some applications in oligosaccharide and glycoconjugate synthesis. Also, direct sialylation

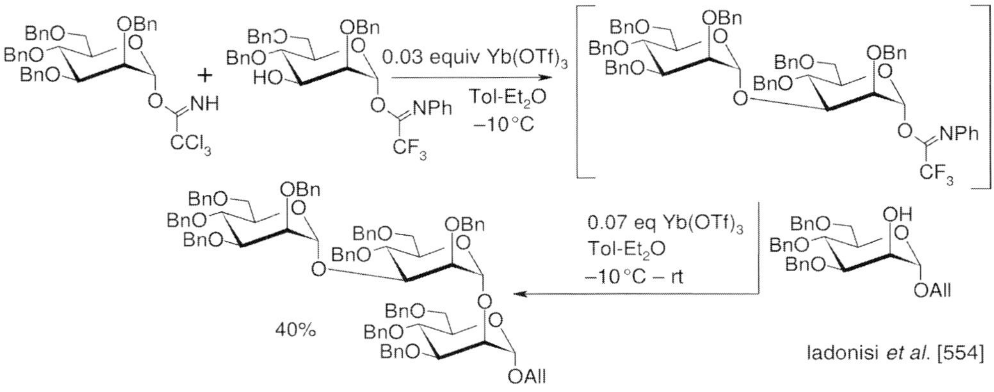

3.2 Glycoside Synthesis from 1-Oxygen-Substituted Glycosyl Imidates | 177

Scheme 3.59

has been achieved with PTFA as the glycosyl donor [555], whereas the corresponding trichloroacetimidate donors are not suitable for sialylation. A linear synthesis of a biologically relevant tetrasaccharide fragment of Globo H antigen has been described recently using PTFA as glycosylating agents (Scheme 3.60) [546]. Neighboring-group participation and solvent effect were exploited to stereoselectively introduce the three glycosidic bonds. The trifluoroacetimidate method has also been used in the synthesis of Fucp3NAc-containing oligosaccharides [556], found exclusively in phytopathogenic bacterial O-antigens, and a tetrasaccharide fragment of clarhamnoside [557], a GSL isolated from marine sponge. A few other applications of PTFA donors in oligosaccharide synthesis have been reported [558–560].

The total synthesis of caminoside A, an antimicrobial glycolipid from the marine sponge, has been achieved with PTFA as the key building blocks (Scheme 3.61) [561]. In this synthesis, the disaccharide trifluoroacetimidate donor was formed

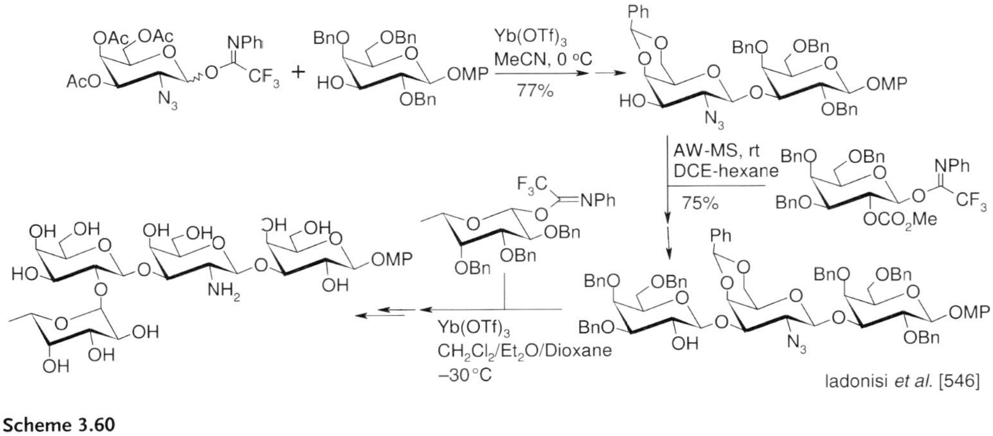

Scheme 3.60

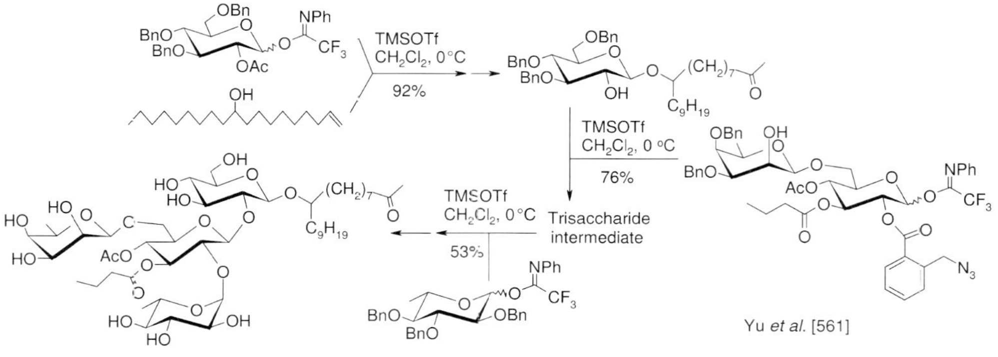

Scheme 3.61

regioselectively in the presence of a nonanomeric hydroxyl group. Toralactone tetraglucoside possessing strong antiallergic activity has also been synthesized using PTFA as glycosyl donors [562]. In addition, trifluoroacetimidate donors have been employed to synthesize glycoconjugates of (E)-resveratrol [563], saponins [564], isoflavone glucuronide [565], C-glycosides [566] and β-lactam glycoconjugates [567].

3.2.7
Conclusions and Outlook

In this review, some important efforts made in the past 8 years in the synthesis of oligosaccharides and glycoconjugates by glycosyl trichloro- and trifluoroimidates have been summarized. It is needless to mention that it is a difficult task to cover all aspects within this brief review. On the whole, the glycosyl trichloroacetimidate protocol has again proven to be an extremely powerful method for carbohydrate synthesis, which has often been featured as the key step in the synthesis of complex sugar-containing natural products. Undoubtedly, this method will continue to contribute tremendously to the future development of the glycoscience. On the contrary, in the recent years some applications of the related glycosyl trifluoroacetimidate method in the carbohydrate synthesis have been observed, and continued advances in this area can surely be expected.

3.2.8
Experimental Procedures

3.2.8.1 Typical Procedure for the Preparation of O-Glycosyl Trichloroacetimidates
Successively excess Cl_3CCN (generally 6.0 mmol) and a catalytic amount of DBU (<0.1 mmol) are added to a solution of an anomeric O-unprotected sugar (1.0 mmol) in dry CH_2Cl_2 (5 ml), and the resulting mixture is stirred at room temperature for 30 min and then concentrated *in vacuo*. The residue is purified by short-column chromatography on silica gel (gradient petroleum ether–EtOAc) to afford the corresponding trichloroacetimidate.

3.2.8.2 Typical Procedure for the Glycosylation with O-Glycosyl Trichloroacetimidates

A solution of a glycosyl trichloroacetimidate (1.2 mmol) and a glycosyl acceptor (1.0 mmol) in dry CH_2Cl_2 (10 ml) is treated at about -40 to $-50\,^\circ C$ with a solution of TMSOTf in CH_2Cl_2 (0.01–0.05 equiv). When TLC analysis indicates completion of the reaction (typically 10–30 min), the reaction is quenched with $NaHCO_3$ or Et_3N. The mixture is filtered and/or directly concentrated *in vacuo* to give a residue, which is purified by flash column chromatography (gradient petroleum ether/EtOAc) to furnish the glycosidation product.

3.2.8.3 Typical Procedure for the Preparation of O-Glycosyl N-Phenyl Trifluoroacetimidates

N-phenyl trifluoroacetimidoyl chloride (1.2 mmol) is added to a stirred mixture of an anomeric O-unprotected sugar (1.0 mmol) and K_2CO_3 (3.0 mmol) in acetone (20 ml) at room temperature. After being stirred overnight, the mixture is filtered and concentrated. The residue is purified by flash column chromatography (gradient petroleum ether/EtOAc) to produce the corresponding trifluoroacetimidate.

3.2.8.4 Typical Procedure for the Glycosylation with O-Glycosyl N-Phenyl Trifluoroacetimidates

a solution of TMSOTf in CH_2Cl_2 (0.1 equiv) is slowly added to a mixture of a glycosyl trifluoroacetimidate (1.2–2.0 mmol), a glycosyl acceptor (1.0 mmol) and 4 Å molecular sieves in dry CH_2Cl_2 (10–20 ml) at room temperature. Upon completion, as indicated by TLC (typically >3 h), the reaction is quenched with $NaHCO_3$ or Et_3N and then filtered and concentrated. The residue is purified by flash column chromatography (gradient petroleum ether/EtOAc) to afford the glycosidation product.

References

370 Schmidt, R.R. and Michel, J. (1980) *Angewandte Chemie (International Edition in English)*, **19**, 731–733.

371 Schmidt, R.R. (1986) *Angewandte Chemie (International Edition in English)*, **25**, 212–235.

372 Schmidt, R.R. (1986) *Pure and Applied Chemistry*, **61**, 1257–1270.

373 Schmidt, R.R. (1991) *Comprehensive Organic Synthesis* (eds B.M. Trost, I. Fleming and E. Winterfeld), vol. 6, Pergamon Press, Oxford, pp. 33–64.

374 Schmidt, R.R. (1992) *Carbohydrates – Synthetic Methods and Application in Medicinal Chemistry* (eds A. Hasegawa, H. Ogura and T. Suami), Kodanasha Scientific, Tokyo, pp. 66–88.

375 Schmidt, R.R. and Kinzy, W. (1994) *Advances in Carbohydrate Chemistry and Biochemistry*, **50**, 21–123.

376 Schmidt, R.R. (1996) *Modern Methods in Carbohydrate Synthesis* (eds S.H. Khan and R.A. O'Neill), Harwood Academic Publisher, Chur, Schweiz, pp. 20–54.

377 Schmidt, R.R. and Jung, K.H. (1997) *Preparative Carbohydrate Chemistry*, (ed. S. Hannessian), Marcel Dekker, New York, pp. 283–312.

378 Schmidt, R.R. (1997) *Glycosciences: Status and Perspectives* (eds H.J. Gabius and S. Gabius), Chapman and Hall, Weinheim, pp. 31–53.

379 Schmidt, R.R. (1998) *Pure and Applied Chemistry*, **70**, 397–402.

380 Schmidt, R.R., Castro-Palomino, J.C. and Retz, O. (1999) *Pure and Applied Chemistry*, **71**, 729–744.

381 Schmidt, R.R. and Jung, K.H. (2000) *Carbohydrates in Chemistry and Biology Part I: Chemistry of Saccharides*, (eds B. Ernst, G.W. Hart and P. Sinay), vol. 1, Wiley-VCH Verlag GmbH, Weinheim, pp. 5–59.

382 Pougny, J.R. and Sinay, P. (1976) *Tetrahedron Letters*, **17**, 4073–4076.

383 Schmidt, R.R., Michel, J. and Roos, M. (1984) *Liebigs Annalen der Chemie*, 1343–1357.

384 Huchel, U. (1998) Dissertation, Universität Konstanz, Papierflieger, Clausthal-Zellerfeld (ISBN 3-89720-221-2).

385 Yu, B. and Tao, H. (2001) *Tetrahedron Letters*, **42**, 2405–2407.

386 Adinolfi, M., Barone, G., Iadonisi, A. and Schiattarella, M. (2002) *Synlett*, 269–270.

387 Schmidt, R.R., Gaden, H. and Jatzke, H. (1990) *Tetrahedron Letters*, **31**, 327–330.

388 Ohashi, I., Lear, M.J., Yoshimura, F. and Hirama, M. (2004) *Organic Letters*, **6**, 719–722.

389 Oikawa, M., Tanaka, T., Fukuda, N. and Kusumoto, S. (2004) *Tetrahedron Letters*, **45**, 4039–4042.

390 Chiara, J.L., Encinas, L. and Díaz B. (2005) *Tetrahedron Letters*, **46**, 2445–2448.

391 Adinolfi, M., Barone, G., Guariniello, L. and Iadonisi, A. (2000) *Tetrahedron Letters*, **41**, 9005–9008.

392 Adinolfi, M., Barone, G., Iadonisi, A., Mangoni, L. and Schiattarella, M. (2001) *Tetrahedron Letters*, **42**, 5967–5969.

393 Wei, G., Gu, G. and Du, Y. (2003) *Journal of Carbohydrate Chemistry*, **22**, 385–393.

394 Jona, H., Mandai, H. and Mukaiyama, T. (2001) *Chemistry Letters*, 426–427.

395 Griswold, K.S., Horstmann, T.E. and Miller, S.J. (2003) *Synlett*, 1923–1926.

396 Du, Y., Wei, G., Cheng, S., Hua, Y. and Linhardt, R.J. (2006) *Tetrahedron Letters*, **47**, 307–310.

397 Mukhopadhyay, B., Maurer, S.V. Rudolph, N., van Well, R.M., Russell, D.A. and Field, R.A. (2005) *Journal of Organic Chemistry*, **70**, 9059–9062.

398 Larsen, K., Worm-Leonhard, K., Olsen, P., Hoel, A. and Jensen, K.J. (2005) *Organic and Biomolecular Chemistry*, **3**, 3966–3970.

399 Pakulski, Z. (2003) *Synthesis*, 2074–2078.

400 Poletti, L., Rencurosi, A., Lay, L. and Russo, G. (2003) *Synlett*, 2297–2300.

401 Rencurosi, A., Lay, L., Russo, G., Caneva, E. and Poletti, L. (2005) *Journal of Organic Chemistry*, **70**, 7765–7768.

402 Ali, I.A.I., Ashry, E.S.H.E. and Schmidt, R.R. (2003) *European Journal of Organic Chemistry*, 4121–4131.

403 Yu, H., Williams, D.L. and Ensley, H.E. (2005) *Tetrahedron Letters*, **46**, 3417–3421.

404 Zeng, Y. and Kong, F. (2003) *Carbohydrate Research*, **338**, 2359–2366.

405 Zhu, Y. and Kong, F. (2000) *Synlett*, 663–667.

406 Yi, Y., Zhou, Z., Ning, J., Kong, F. and Li, J. (2003) *Synthesis*, 491–496.

407 Ning, J., Kong, F., Lin, B. and Lei, H. (2003) *Journal of Agricultural and Food Chemistry*, **51**, 987–991.

408 Ning, J., Yi, Y. and Kong, F. (2002) *Tetrahedron Letters*, **43**, 5545–5549.

409 Li, A. and Kong, F. (2005) *Carbohydrate Research*, **340**, 1949–1962.

410 Li, A. and Kong, F. (2004) *Carbohydrate Research*, **339**, 1847–1856.

411 Ning, J., Yi, Y. and Yao, Z. (2003) *Synlett*, 2208–2212.

412 Li, A., Zeng, Y. and Kong, F. (2004) *Carbohydrate Research*, **339**, 673–681.

413 Ratner, D.M., Plante, O.J. and Seeberger, P.H. (2002) *European Journal of Organic Chemistry*, 826–833.

414 Zeng, Y., Zhang, J. and Kong, F. (2002) *Carbohydrate Research*, **337**, 1367–1371.

415 Ning, J., Heng, L. and Kong, F. (2002) *Carbohydrate Research*, **337**, 1159–1164.

416 Zhu, Y., Chen, L. and Kong, F. (2002) *Carbohydrate Research*, **337**, 207–215.

417 Xing, Y. and Ning, J. (2003) *Tetrahedron: Asymmetry*, **14**, 1275–1283.

418 Smiljanic, N., Halila, S., Moreau, V. and Djedaïni-Pilard, F. (2003) *Tetrahedron Letters*, **44**, 8999–9002.

419 Zeng, Y., Zhang, J., Ning, J. and Kong, F. (2003) *Carbohydrate Research*, **338**, 5–9.
420 Ma, Z., Zhang, J. and Kong, F. (2004) *Carbohydrate Research*, **339**, 29–35.
421 Gu, G., Wei, G. and Du, Y. (2004) *Carbohydrate Research*, **339**, 1155–1162.
422 Zhang, J. and Kong, F. (2003) *Tetrahedron*, **59**, 1429–1441.
423 Zhang, J. and Kong, F. (2002) *Journal of Carbohydrate Chemistry*, **21**, 579–589.
424 Zhang, J. and Kong, F. (2002) *Carbohydrate Research*, **337**, 391–396.
425 Zhang, J. and Kong, F. (2003) *Carbohydrate Research*, **338**, 19–27.
426 Zhang, J., Ning, J. and Kong, F. (2003) *Carbohydrate Research*, **338**, 1023–1031.
427 Ma, Z., Zhang, J. and Kong, F. (2004) *Carbohydrate Research*, **339**, 43–49.
428 Hua, Y., Xiao, J., Huang, Y. and Du, Y. (2006) *Carbohydrate Research*, **341**, 191–197.
429 Du, Y., Pan, Q. and Kong, F. (2000) *Carbohydrate Research*, **329**, 17–24.
430 Gandolfi-Donadío, L., Gallo-Rodriguez, C. and de Lederkremer, R.M. (2003) *Journal of Organic Chemistry*, **68**, 6928–6934.
431 Zhang, G., Fu, M. and Ning, J. (2005) *Carbohydrate Research*, **340**, 155–159.
432 Chen, L. and Kong, F. (2002) *Carbohydrate Research*, **337**, 2335–2341.
433 Aly, M.R.E. and Schmidt, R.R. (2005) *European Journal of Organic Chemistry*, 4382–4392.
434 Aly, M.R.E., Ibrahim, E.S.I., El-Ashry, E.S.H. and Schmidt, R.R. (2001) *Carbohydrate Research*, **331**, 129–142.
435 Aly, M.R.E., Ibrahim, E.S.I., El-Ashry, E.S.H. and Schmidt, R.R. (1999) *Carbohydrate Research*, **316**, 121–132.
436 Chiesa, M.V. and Schmidt, R.R. (2000) *European Journal of Organic Chemistry*, 3541–3554.
437 Gege, C., Geyer, A. and Schmidt, R.R. (2002) *European Journal of Organic Chemistry*, 2475–2485.
438 Xue, J., Khaja, S.D., Locke, R.D. and Matta, K.L. (2004) *Synlett*, 861–865.

439 Sun, B., Pukin, A.V., Visser, G.M. and Zuilhof, H. (2006) *Tetrahedron Letters*, **47**, 7371–7374.
440 (a) Inamura, S., Fukase, K. and Kusumoto, S. (2001) *Tetrahedron Letters*, **42**, 7613–7616. (b) Inamura, S., Fujimoto, Y., Kawasaki, A., Shiokawa, Z., Woelk, E., Heine, H., Lindner, B., Inohara, N., Kusumoto, S. and Fukase, K. (2006) *Organic and Biomolecular Chemistry*, **4**, 232–242.
441 Lee, J.C., Lu, X.A., Kulkarni, S.S., Wen, Y.S. and Hung, S.C. (2004) *Journal of the American Chemical Society*, **126**, 476–477.
442 Poletti, L., Fleischer, M., Vogel, C., Guerrini, M., Torri, G. and Lay, L. (2001) *European Journal of Organic Chemistry*, 2727–2734.
443 Cid, M.B., Alfonso, F. and Martín-Lomas, M. (2005) *Chemistry A European Journal*, **11**, 928–938.
444 de Paz, J.L., Ojeda, R., Reichardt, N. and Martín-Lomas, M. (2003) *European Journal of Organic Chemistry*, 3308–3324.
445 Lohman, G.J.S. and Seeberger, P.H. (2004) *Journal of Organic Chemistry*, **69**, 4081–4093.
446 Lucas, R., Hamza, D., Lubineau, A. and Bonnaffé, D. (2004) *European Journal of Organic Chemistry*, 2107–2117.
447 Tully, S.E., Mabon, R., Gama, C.I., Tsai, S.M., Liu, X. and Hsieh-Wilson, L.C. (2004) *Journal of the American Chemical Society*, **126**, 7736–7737.
448 Karst, N. and Jacquinet, J.C. (2000) *Journal of the Chemical Society Perkin Transactions*, **1**, 2709–2717.
449 Tamura, J.I. and Tokuyoshi, M. (2004) *Bioscience Biotechnology, and Biochemistry*, **68**, 2436–2443.
450 Karst, N. and Jacquinet, J.C. (2002) *European Journal of Organic Chemistry*, 815–825.
451 Weingart, R. and Schmidt, R.R. (2000) *Tetrahedron Letters*, **41**, 8753–8758.
452 Abdel-Rahman, A.A.H., Jonke, S., El-Ashry, E.S.H. and Schmidt, R.R. (2002) *Angewandte Chemie (International Edition)*, **41**, 2972–2974.

453 Wang, C.C., Lee, J.C., Luo, S.Y., Fan, H.F., Pai, C.L., Yang, W.C., Lu, L.D. and Hung, S.C. (2002) *Angewandte Chemie (International Edition)*, **41**, 2360–2362.

454 Kim, J.H., Yang, H., Park, J. and Boons, G.J. (2005) *Journal of the American Chemical Society*, **127**, 12090–12097.

455 Ikeda, T. and Yamada, H. (2000) *Carbohydrate Research*, **329**, 889–893.

456 Twaddle, G.W.J., Yashunsky, D.V. and Nikolaev, A.V. (2003) *Organic and Biomolecular Chemistry*, **1**, 623–628.

457 Damager, I., Olsen, C.E., Møller, B.L. and Motawia, M.S. (1999) *Carbohydrate Research*, **320**, 19–30.

458 Maruyama, M., Takeda, T., Shimizu, N., Hada, N. and Yamada, H. (2000) *Carbohydrate Research*, **325**, 83–92.

459 Gandolfi-Donadio, L., Gallo-Rodriguez, C. and de Lederkremer, R.M. (2002) *Journal of Organic Chemistry*, **67**, 4430–4435.

460 Zeng, Y., Ning, J. and Kong, F. (2003) *Carbohydrate Research*, **338**, 307–311.

461 Chen, L., Zhao, X.E., Lai, D., Song, Z. and Kong, F. (2006) *Carbohydrate Research*, **341**, 1174–1180.

462 Ning, J., Zhang, W., Yi, Y., Yang, G., Wu, Z., Yi, J. and Kong, F. (2003) *Bioorganic and Medicinal Chemistry*, **11**, 2193–2203.

463 Wu, Z. and Kong, F. (2004) *Synlett*, 2594–2596.

464 Yang, F., He, H., Du, Y. and Lü, M. (2002) *Carbohydrate Research*, **337**, 1165–1169.

465 Zhang, G., Fu, M. and Ning, J. (2005) *Carbohydrate Research*, **340**, 597–602.

466 Bélot, F., Wright, K., Costachel, C. Phalipon, A. and Mulard, L.A. (2004) *Journal of Organic Chemistry*, **69**, 1060–1074.

467 Mendoza, V.M., Agusti, R., Gallo-Rodriguez, C. and de Lederkremer, R.M. (2006) *Carbohydrate Research*, **341**, 1488–1497.

468 Ikuta, A., Mizuta, N., Kitahata, S., Murata, T., Usui, T., Koizumi, K. and Tanimoto, T. (2004) *Chemical and Pharmaceutical Bulletin*, **52**, 51–56.

469 Roush, W.R., Gung, B.W. and Bennett, C.E. (1999) *Organic Letters*, **1**, 891–893.

470 Chong, P.Y. and Roush, W.R. (2002) *Organic Letters*, **4**, 4523–4526.

471 Roush, W.R. and Bennett, C.E. (2000) *Journal of the American Chemical Society*, **122**, 6124–6125.

472 Kubo, H., Ishii, K., Koshino, H., Toubetto, K., Naruchi, K. and Yamasaki, R. (2004) *European Journal of Organic Chemistry*, 1202–1213.

473 Bedini, E., Carabellese, A., Corsaro, M.M., Castro, C.D. and Parrilli, M. (2004) *Carbohydrate Research*, **339**, 1907–1915.

474 Izumi, M., Tsuruta, O., Kajihara, Y., Yazawa, S., Yuasa, H. and Hashimoto, H. (2005) *Chemistry – A European Journal*, **11**, 3032–3038.

475 Morii, Y., Matsuda, H., Ohara, K., Hashimoto, M., Miyairi, K. and Okuno, T. (2005) *Bioorganic and Medicinal Chemistry*, **13**, 5113–5144.

476 Ohara, K., Matsuda, H., Hashimoto, M., Miyairi, K. and Okuno, T. (2002) *Chemistry Letters*, 626–627.

477 Castro-Palomino, J.C., Simon, B., Speer, O., Leist, M. and Schmidt, R.R. (2001) *Chemistry – A European Journal*, **7**, 2178–2184.

478 Gege, C., Kinzy, W. and Schmidt, R.R. (2000) *Carbohydrate Research*, **328**, 459–466.

479 Tietze, L.F., Keim, H., Janßen, C.O., Tappertzhofen, C. and Olschimke, J. (2000) *Chemistry – A European Journal.*, **6**, 2801–2808.

480 Yamaguchi, M., Ishida, H., Kanamori, A., Kannagi, R. and Kiso, M. (2003) *Carbohydrate Research*, **338**, 2793–2812.

481 Gege, C., Geyer, A. and Schmidt, R.R. (2002) *Chemistry – A European Journal*, **8**, 2454–2463.

482 Gege, C., Vogel, J., Bendas, G., Rothe, U. and Schmidt, R.R. (2000) *Chemistry – A European Journal.*, **6**, 111–122.

483 Gege, C., Oscarson, S. and Schmidt, R.R. (2001) *Tetrahedron Letters*, **42**, 377–380.

484 Figueroa-Pérez, S. and Schmidt, R.R. (2000) *Carbohydrate Research*, **328**, 95–102.

485 Bartolmäs, T., Heyn, T., Mickeleit, M., Fischer, A., Reutter, W. and Danker, K. (2005) *Journal of Medicinal Chemistry*, **48**, 6750–6755.

486 Pekari, K. and Schmidt, R.R. (2003) *Journal of Organic Chemistry*, **68**, 1295–1308.

487 Pekari, K., Tailler, D., Weingart, R. and Schmidt, R.R. (2001) *Journal of Organic Chemistry*, **66**, 7432–7442.

488 Martín-Lomas, M., Flores-Mosquera, M. and Chiara, J.L. (2000) *European Journal of Organic Chemistry*, 1547–1562.

489 Komba, S., Meldal, M., Werdelin, O., Jensen, T. and Bock, K. (1999) *Journal of the Chemical Society Perkin Transactions*, **1**, 415–419.

490 St Hilaire, P.M., Cipolla, L., Franco, A., Tedebark, U., Tilly, D.A. and Meldal, M. (1999) *Journal of the Chemical Society Perkin Transactions*, **1**, 3559–3564.

491 Allevi, P., Anastasia, M., Paroni, R. and Ragusa, A. (2004) *Bioorganic and Medicinal Chemistry Letters*, **14**, 3319–2231.

492 Schnabel, M., Römpp, B. Ruckdeschel, D. and Unverzagt, C. (2004) *Tetrahedron Letters*, **45**, 295–297.

493 Burger, K., Kluge, M., Fehn, S., Koksch, B., Henning, L. and Müller, G. (1999) *Angewandte Chemie (International Edition)*, **38**, 1414–1416.

494 Halkes, K.M., Gotfredsen, C.H., Grøtli, M., Miranda, L.P., Duus, J.Ø. and Meldal, M. (2001) *Chemistry – A European Journal.*, **7**, 3584–3591.

495 Ritter, T.K., Mong, K.K.T., Liu, H., Nakatani, T. and Wong, C.H. (2003) *Angewandte Chemie (International Edition)*, **42**, 4657–4660.

496 Nicolaou, K.C., Mitchell, H.J., Jain, N.F., Bando, T., Hughes, R., Winssinger, N., Natarajan, S. and Koumbis, A.E. (1999) *Chemistry – A European Journal*, **5**, 2648–2667.

497 Nicolaou, K.C., Cho, S.Y., Hughes, R., Winssinger, N., Smethurst, C., Labischinski, H. and Endermann, R. (2001) *Chemistry – A European Journal*, **7**, 3798–3823.

498 Deng, S., Yu, B., Xie, J. and Hui, Y. (1999) *Journal of Organic Chemistry*, **64**, 7265–7266.

499 Plé, K., Chwalek, M. and Voutquenne-Nazabadioko, L. (2004) *European Journal of Organic Chemistry*, 1588–1603.

500 Chwalek, M., Plé, K. and Voutquenne-Nazabadioko, L. (2004) *Chemical and Pharmaceutical Bulletin*, **52**, 965–971.

501 Ikeda, T., Yamauchi, K., Nakano, D., Nakanishi, K., Miyashita, H., Ito, S.I. and Nohara, T. (2006) *Tetrahedron Letters*, **47**, 4355–4359.

502 Ikeda, T., Miyashita, H., Kajimoto, T. and Nohara, T. (2001) *Tetrahedron Letters*, **42**, 2353–2356.

503 Thompson, M.J., Hutchinson, E.J., Stratford, T.H., Bowler, W.B. and Blackburn, G.M. (2004) *Tetrahedron Letters*, **45**, 1207–1210.

504 Eleutério, M.I.P., Schimmel, J., Ritter, G., Costa, M.D.C. and Schmidt, R.R. (2006) *European Journal of Organic Chemistry*, 5293–5304.

505 Kawada, T., Asano, R., Makino, K. and Sakuno, T. (2000) *European Journal of Organic Chemistry*, 2723–2727.

506 Feldman, K.S. and Lawlor, M.D. (2000) *Journal of the American Chemical Society*, **122**, 7396–7397.

507 Jiang, Z.H., Geyer, A. and Schmidt, R.R. (1995) *Angewandte Chemie (International Edition)*, **34**, 2520–2524.

508 Brito-Arias, M., Pereda-Miranda, R. and Heathcock, C.H. (2004) *Journal of Organic Chemistry*, **69**, 4567–4570.

509 Fürstner, A., Jeanjean, F. and Razon, P. (2002) *Angewandte Chemie (International Edition)*, **41**, 2097–2101.

510 Fürstner, A., Albert, M., Mlynarski, J. and Matheu, M. (2002) *Journal of the American Chemical Society*, **124**, 1168–1169.

511 Fürstner, A., Ruiz-Caro, I., Prinz, H. and Waldmann, H. (2004) *Journal of Organic Chemistry*, **69**, 459–467.

512 Brown, R.T., Carter, N.E., Mayalarp, S.P. and Scheinmann, F. (2000) *Tetrahedron*, **56**, 7591–7594.

513 Engstrom, K.M., Daanen, J.F., Wagaw, S. and Stewart, A.O. (2006) *Journal of Organic Chemistry*, **71**, 8378–8383.

514 Suzuki, T., Mabuchi, K. and Fukazawa, N. (1999) *Bioorganic and Medicinal Chemistry Letters*, **9**, 659–662.

515 Ferguson, J.R., Harding, J.R., Killick, D.A., Lumbard, K.W., Scheinmann, F. and Stachulski, A.V. (2001) *Journal of the Chemical Society Perkin Transactions*, **1**, 3037–3041.

516 Ferguson, J.R., Harding, J.R., Lumbard, K.W., Scheinmann, F. and Stachulski, A.V. (2000) *Tetrahedron Letters*, **41**, 389–392.

517 Schramm, S., Dettner, K. and Unverzagt, C. (2006) *Tetrahedron Letters*, **47**, 7741–7743.

518 Nicolaou, K.C., Pfefferkorn, J.A. and Cao, G.Q. (2000) *Angewandte Chemie (International Edition)*, **39**, 734–739.

519 Rao, Y., Venot, A., Swayze, E.E., Griffey, R.H. and Boons, G.J. (2006) *Organic and Biomolecular Chemistry*, **4**, 1328–1337.

520 Rele, S.M., Cui, W., Wang, L., Hou, S., Barr-Zarse, G., Tatton, D., Gnanou, Y., Esko, J.D. and Chaikof, E.L. (2005) *Journal of the American Chemical Society*, **127**, 10132–10133.

521 de Kort, M., Ebrahimi, E., Wijsman, E.R., van der Marel, G.A. and van Boom, J.H. (1999) *European Journal of Organic Chemistry*, 2337–2344.

522 Adinolfi, M., Barone, G., Napoli, L.D., Guariniello, L., Iadonisi, A. and Piccialli, G. (1999) *Tetrahedron Letters*, **40**, 2607–2610.

523 Adinolfi, M., Napoli, L.D., Fabio, G.D., Guariniello, L., Iadonisi, A., Messere, A., Montesarchio, D. and Piccialli, G. (2001) *Synlett*, **6**, 745–748.

524 Armitt, D.J., Banwell, M.G., Freeman, C. and Parish, C.R. (2002) *Journal of the Chemical Society Perkin Transactions*, **1**, 1743–1745.

525 Furuta, T., Kimura, T., Kondo, S., Mihara, H., Wakimoto, T., Nukaya, H., Tsuji, K. and Tanaka, K. (2004) *Tetrahedron*, **60**, 9375–9379.

526 Hein, M., Michalik, D. and Langer, P. (2005) *Synthesis*, 3531–3534.

527 Roussel, F., Knerr, L., Grathwohl, M. and Schmidt, R.R. (2000) *Organic Letters*, **2**, 3043–3046.

528 Jonke, S., Liu, K.G. and Schmidt, R.R. (2006) *Chemistry – A European Journal*, **12**, 1274–1290.

529 Roussel, F., Takhi, M. and Schmidt, R.R. (2001) *Journal of Organic Chemistry*, **66**, 8540–8548.

530 Grathwohl, M. and Schmidt, R.R. (2001) *Synthesis*, 2263–2272.

531 Roussel, F., Knerr, L. and Schmidt, R.R. (2001) *European Journal of Organic Chemistry*, 2067–2073.

532 Manabe, S. and Ito, Y. (2002) *Journal of the American Chemical Society*, **124**, 12638–12639.

533 Wu, X. and Schmidt, R.R. (2004) *Journal of Organic Chemistry*, **69**, 1853–1857.

534 Plante, O.J., Palmacci, E.R. and Seeberger, P.H. (2001) *Science*, **291**, 1523–1527.

535 Ratner, D.M., Swanson, E.R. and Seeberger, P.H. (2003) *Organic Letters*, **5**, 4717–4720.

536 Hewitt, M.C., Snyder, D.A. and Seeberger, P.H. (2002) *Journal of the American Chemical Society*, **124**, 13434–13436.

537 Wu, X., Grathwohl, M. and Schmidt, R.R. (2002) *Angewandte Chemie (International Edition)*, **41**, 4489–4493.

538 Andrade, R.B., Plante, O.J., Melean, L.G. and Seeberger, P.H. (1999) *Organic Letters*, **1**, 1811–1814.

539 Ojeda, R., Terentí, O., de Paz, J.L. and Martín-Lomas, M. (2004) *Glycoconjugate Journal*, **21**, 179–195.

540 Ojeda, R., de Paz, J.L. and Martín-Lomas, M. (2003) *Chemical Communications*, 2486–2487.

541 Miura, T., Goto, K., Hosaka, D. and Inazu, T. (2003) *Angewandte Chemie (International Edition)*, **42**, 2047–2051.

542 Miura, T., Hirose, Y., Ohmae, M. and Inazu, T. (2001) *Organic Letters*, **3**, 3947–3950.

543 Miura, T. and Inazu, T. (2003) *Tetrahedron Letters*, **44**, 1819–1821.

544 Nakajima, N., Saito, M., Kudo, M. and Ubukata, M. (2002) *Tetrahedron*, **58**, 3579–3588.

545 Yu, B. and Tao, H. (2002) *Journal of Organic Chemistry*, **67**, 9099–9102.

546 Adinolfi, M., Iadonisi, A., Ravidà, A. and Schiattarella, M. (2005) *Journal of Organic Chemistry*, **70**, 5316–5319.

547 Li, M., Han, X. and Yu, B. (2003) *Journal of Organic Chemistry*, **68**, 6842–6845.

548 Peng, W., Sun, J., Lin, F., Han, X. and Yu, B. (2004) *Synlett*, 259–262.

549 Adinolfi, M., Barone, G., Iadonisi, A. and Schiattarella, M. (2002) *Tetrahedron Letters*, **43**, 5573–5577.

550 Adinolfi, M., Iadonisi, A., Ravidà, A. and Schiattarella, M. (2004) *Synlett*, 275–278.

551 Adinolfi, M., Barone, G., Iadonisi, A. and Schiattarella, M. (2003) *Organic Letters*, **5**, 987–989.

552 Adinolfi, M., Iadonisi, A., Ravidà, A. and Valerio, S. (2006) *Tetrahedron Letters*, **47**, 2595–2599.

553 Tanaka, S.I., Takashina, M., Tokimoto, H., Fujimoto, Y., Tanaka K. and Fukase, K. (2005) *Synlett*, 2325–2328.

554 Adinolfi, M., Iadonisi, A. and Ravidà, A. (2006) *Synlett*, 583–586.

555 Cai, S. and Yu, B. (2003) *Organic Letters*, **5**, 3827–3830.

556 Bedini, E., Carabellese, A., Schiattarella, M. and Parrilli, M. (2005) *Tetrahedron*, **61**, 5439–5448.

557 Ding, N., Wang, P., Zhang, Z., Liu, Y. and Li, Y. (2006) *Carbohydrate Research*, **341**, 2769–2776.

558 Komarova, B.S., Tsvetkov, Y.E., Knirel, Y.A., Zähringer, U., Pier, G.B. and Nifantiev, N.E. (2006) *Tetrahedron Letters*, **47**, 3583–3587.

559 Bedini, E., Esposito, D. and Parrilli, M. (2006) *Synlett*, 825–830.

560 Bedini, E., Carabellese, A., Comegna, D., Castro, C.D. and Parrilli, M. (2006) *Tetrahedron*, **62**, 8474–8483.

561 Sun, J., Han, X. and Yu, B. (2005) *Synlett*, 437–440.

562 Zhang, Z. and Yu, B. (2003) *Journal of Organic Chemistry*, **68**, 6309–6313.

563 Zhang, Z., Yu, B. and Schmidt, R.R. (2006) *Synthesis*, 1301–1306.

564 Peng, W., Han, X. and Yu, B. (2004) *Synthesis*, 1641–1647.

565 Al-Maharik, N. and Botting, N.P. (2006) *Tetrahedron Letters*, **47**, 8703–8706.

566 Li, Y., Wei, G. and Yu, B. (2006) *Carbohydrate Research*, **341**, 2717–2722.

567 Adinolfi, M., Galletti, P., Giacomini, D., Iadonisi, A., Quintavalla, A. and Ravidà, A. (2006) *European Journal of Organic Chemistry*, 69–73.

3.3
Anomeric Transglycosylation

Kwan-Soo Kim, Heung-Bae Jeon

3.3.1
Introduction

Enzymatic transglycosylation has often been successfully applied to the synthesis of simple oligosaccharides generally in short reaction steps without or with the minimal use of conventional protecting groups [568]. Construction of complex structures is, however, not easy by the enzymatic procedure. Therefore, chemical transglycosylation is attracting growing interest and has been studied very extensively during the past two decades [569].

The transglycosylation reaction with methyl glycoside, the simplest glycoside, as the glycosyl donor does not generally proceed smoothly in the presence of a Lewis acid because methanol or methoxide generated during the glycosylation is more nucleophilic than the glycosyl acceptor, and it readily recombines with the intermediate oxocarbenium ion to regenerate the original methyl glycoside. So, to overcome the drawback of methyl glycosides, several other glycosides, such as silyl, heteroaryl and methyl 2-hydroxy-3,5-dinitrobenzoate (methyl 3,5-dinitrosalicylate, DISAL) glycosides, have been devised as new glycosyl donors. The advantage of these glycosides is that the departed aglycones are less nucleophilic than the glycosyl acceptor and they cannot compete with the glycosyl acceptor toward the oxocarbenium ion.

Furthermore, vinyl glycosides, n-pentenyl glycosides (NPGs) and 2'-carboxybenzyl (CB) glycosides have been developed as glycosyl donors with increased stability and efficiency for the glycosylation reaction. These glycosides during glycosylation generate nonalcoholic by-products, which are inert toward the glycosylation reaction: ketones are obtained from vinyl glycosides, cyclic ethers from NPGs and lactones from CB glycosides (Figure 3.1). In fact, NPGs and CB glycosides have been successfully utilized for the synthesis of various complex oligosaccharides.

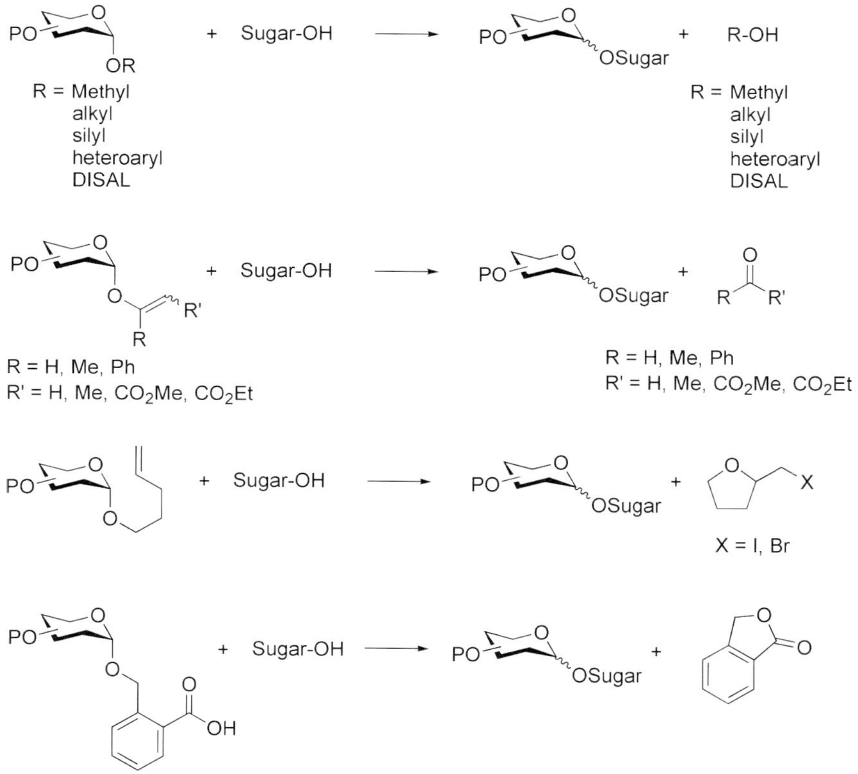

Figure 3.1 Glycosyl donors in chemical transglycosylations.

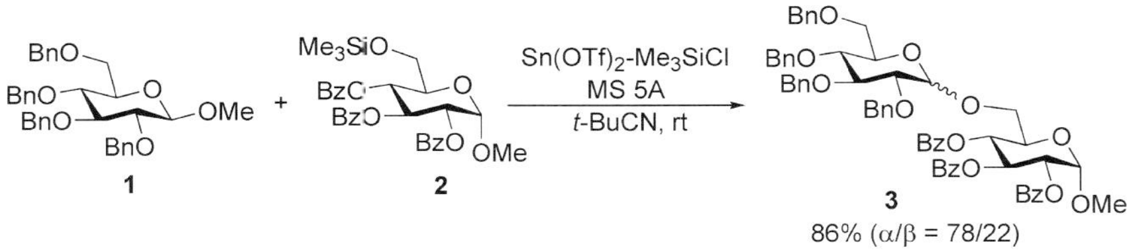

Scheme 3.62

3.3.2
Alkyl Glycosides

In spite of the disadvantage of the methyl glycoside, Mukaiyama and coworkers investigated the effect of solvents, Lewis acids and the reaction temperature on the transglycosylation of various silyl-protected glycosyl acceptors with methyl glycosides. They found that, for example the transglycosylation with methyl glucoside **1** of trimethysilyl-protected glycosyl acceptor **2** using an $Sn(OTf)_2$–Me_3SiCl promoter system in the presence of molecular sieves 5 Å afforded disaccharide **3** in 86% yield (Scheme 3.62) [570]. Recently, Shimizu et al. reported efficient microwave-assisted glycosylations with methyl glycosides. Thus, methyl glucopyranoside **1** could be converted to octyl glucopyranoside **4** using microwave irradiation in the presence of $Yb(OTf)_3$ at 120 °C in high yield (Scheme 3.63). The same reaction without microwave irradiation gave the octyl glucopyranoside **4** in low yield [571].

On the contrary, Higashi and Susaki reported the use of isopropyl glycosides as glycosyl donors. The glycosylation reaction with isopropyl glycoside **5** using TMSBr and $Zn(OTf)_2$ as promoters afforded the corresponding glycoside **6** (Scheme 3.64) [310].

3.3.3
Silyl Glycosides

In the employment of silyl glycoside as a glycosyl donor, trimethylsilyl (TMS) and *tert*-butyldimethylsilyl (TBDMS) groups were preferentially used. Tietze et al. introduced a new glycosylation reaction of aryl trimethylsilyl ether **8** with

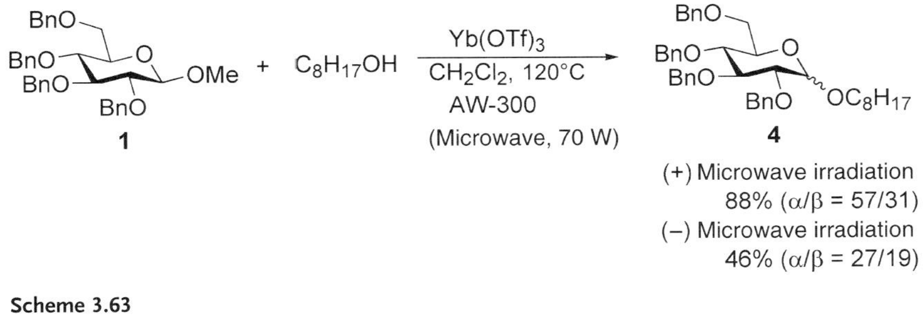

Scheme 3.63

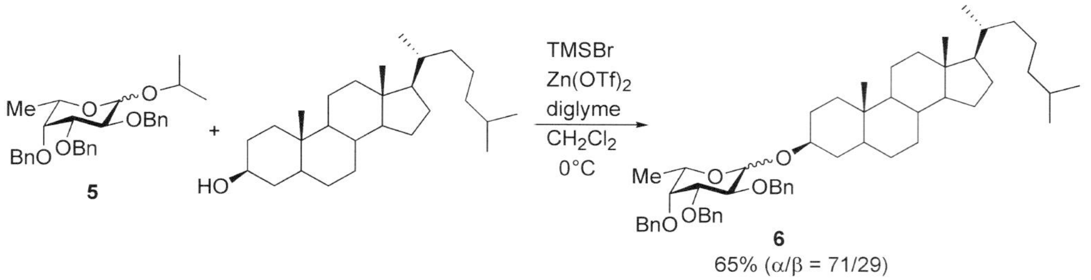

Scheme 3.64

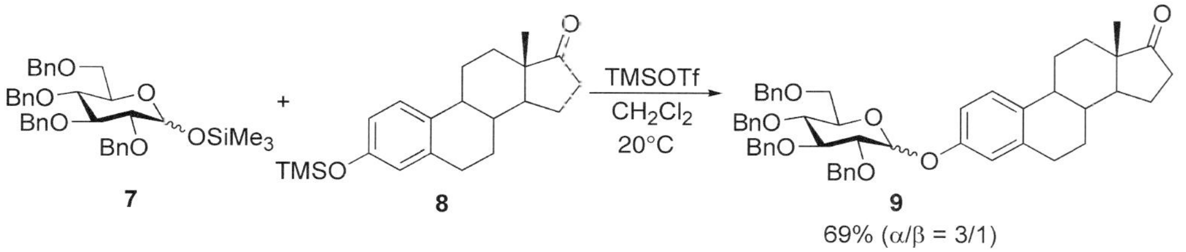

Scheme 3.65

trimethylsilyl glycoside **7** in the presence of a catalytic amount of trimethylsilyl trifluoromethanesulfonate (TMSOTf) as a promoter (Scheme 3.65) [572]. Nashed and Glaudemans reported the formation of the (1→6)-oligosaccharide linkage using a 6-*O*-*tert*-butyldimethylsilyl-protected glycosyl acceptor **11** by the same method (Scheme 3.66) [573]. Cai and coworkers used BF$_3$·Et$_2$O as the activator of trimethylsilyl glycoside **7** for the synthesis of benzyl *O*-glycoside **13** (Scheme 3.67) [574]. Mukaiyama and Matsubara developed stereoselective glycosylation reaction with trimethylsilyl sugars [575]. Thus, 1,2-*trans*-ribofuranosides (**16β**) were predominantly synthesized by the glycosylation of trimethylsilyl ethers (**15**) with 1-*O*-trimethylsilyl ribofuranoses (**14**) in the presence of a catalytic amount of TMSOTf and Ph$_2$Sn=S as an additive, while 1,2-*cis* ribofuranosides (**16α**) and 1,2-*cis* glucopyranosides were selectively prepared using TMSOTf, Ph$_2$Sn=S and LiClO$_4$ as another additive (Scheme 3.68). On the contrary, the *tert*-butyldimethylsilyl glycosyl donors

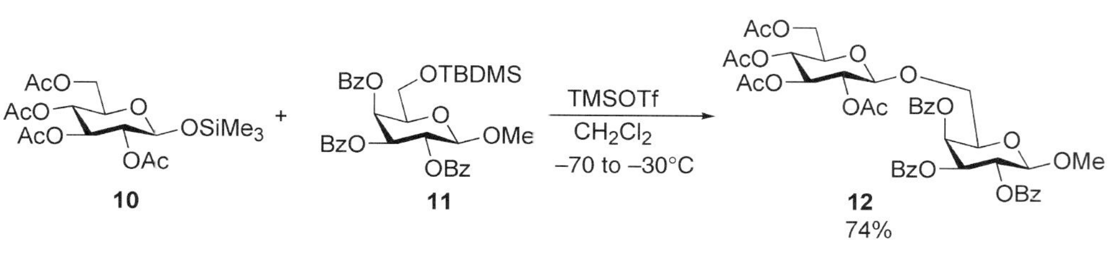

Scheme 3.66

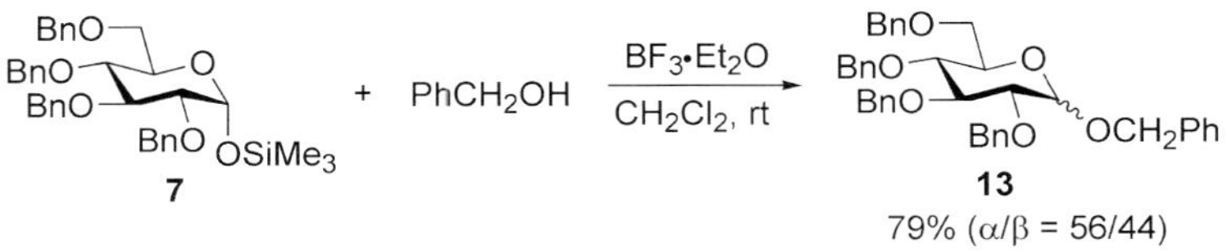

Scheme 3.67

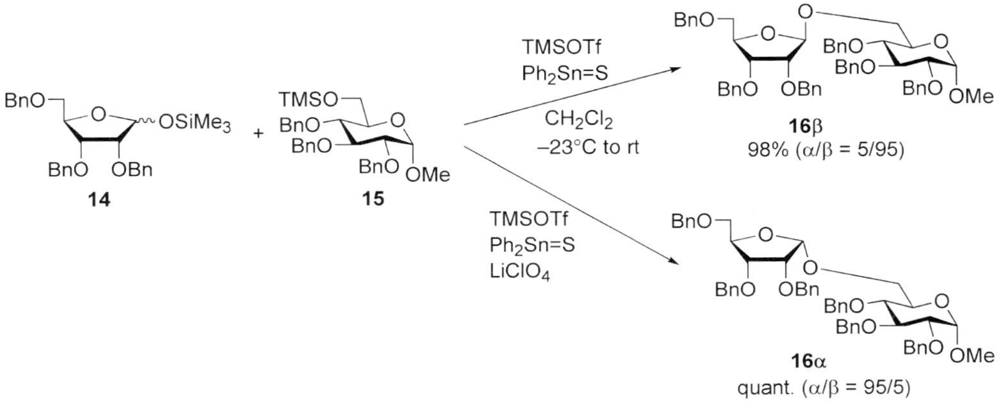

Scheme 3.68

(**17**) were used for the synthesis of 2-deoxy glycosides (**19**) by Priebe and Kolar and coworkers (Scheme 3.69) [576]. A similar glycosyl donor **20** was employed in the anthracycline glycoside (**22**) synthesis by Kolar *et al.* (Scheme 3.70) [577]. Monneret and coworkers reported the use of *tert*-butyldimethylsilyl glycosyl donors **23** and **26** in the synthesis of glycosides of epipodophyllotoxin **25** and **27**, respectively (Scheme 3.71) [578].

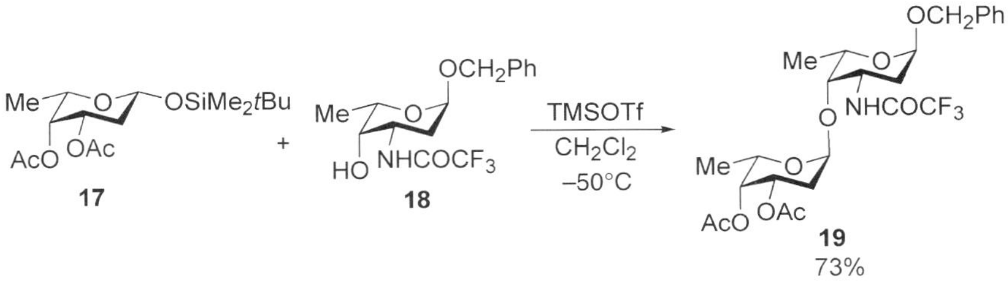

Scheme 3.69

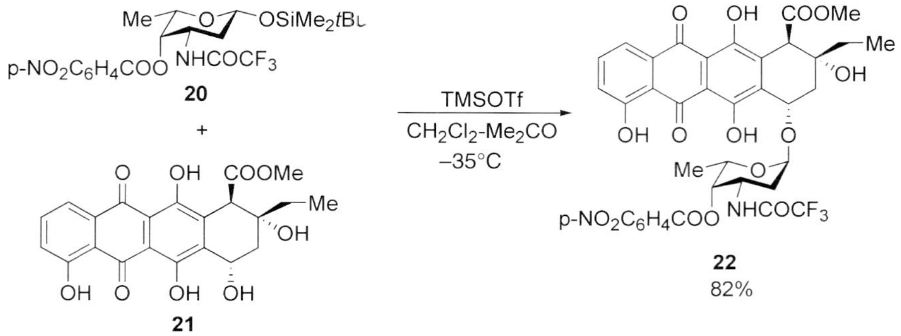

Scheme 3.70

3.3.4
Heteroaryl Glycosides

Mukaiyama and coworkers introduced 3,5-dinitro-2-pyridyl α-glucopyranoside **28**, readily derived from the corresponding α-hemiacetal and 2-chloro-3,5-dinitropyridine, as a new efficient glycosyl donor for the β-stereoselective glycosylation. Thus, the reaction of **28** with alcohol acceptor **29** proceeded smoothly in the

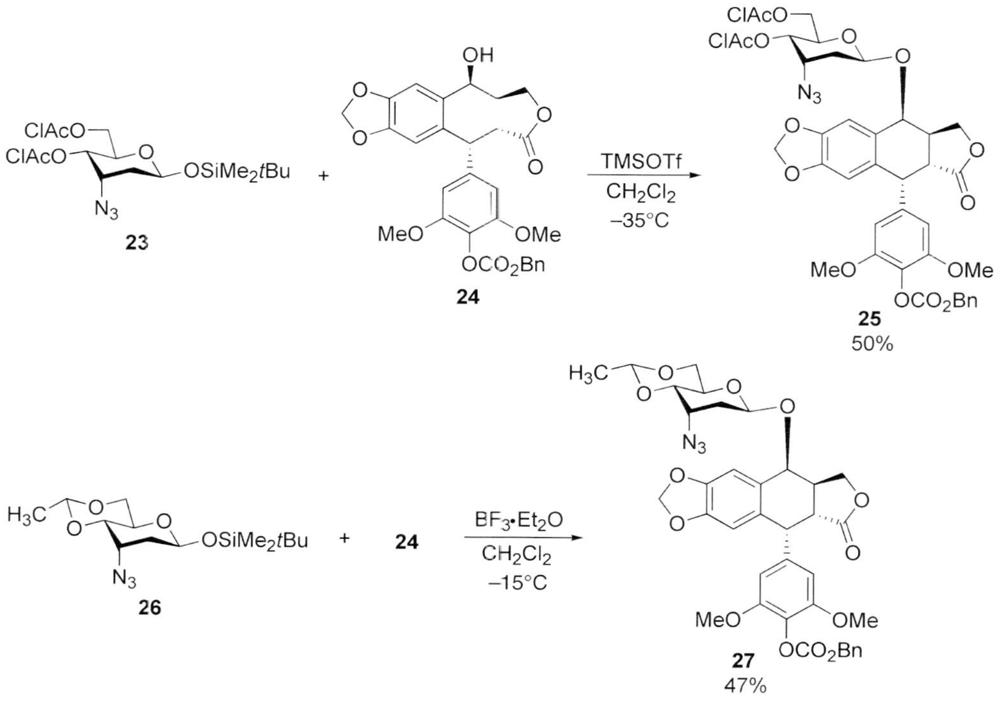

Scheme 3.71

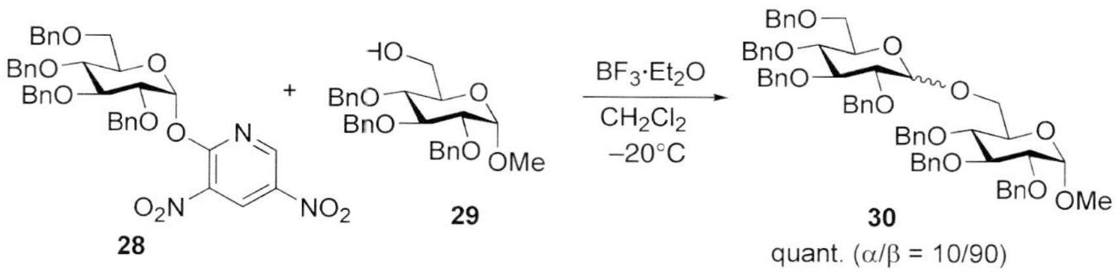

Scheme 3.72

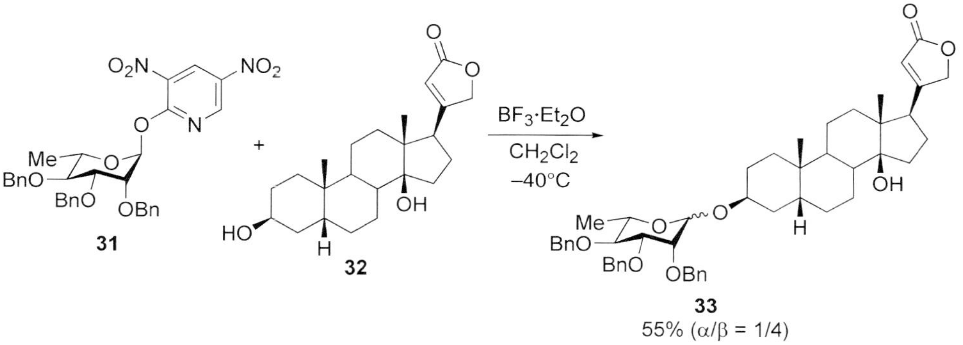

Scheme 3.73

presence of BF$_3$·Et$_2$O or ZnCl$_2$ to afford β-glucoside **30** (Scheme 3.72) [579]. Fullerton and coworkers reported the use of 3,5-dinitro-2-pyridyl α-L-rhamnoside **31** as the glycosyl donor in the presence of BF$_3$·Et$_2$O to obtain digitoxigenin L-rhamnoside **33**, which displayed the Na$^+$, K$^+$-ATPase receptor inhibitory activity (Scheme 3.73) [580]. On a similar note, Kusumoto and coworkers reported the use of 3-nitro-2-pyridyl glycoside **34** as the glycosyl donor in the presence of a catalytic amount of TMSOTf to obtain disaccharide **36** in good yield (Scheme 3.74) [581]. Hanessian *et al.* reported that 2-pyridyl and 3-methoxy-2-pyridyl (MOP) glycosides **37** and **40** could be effectively glycosylated in the presence of TMSOTf, Yb(OTf)$_3$, Cu(OTf)$_2$ or the protic acid HBF$_4$·Et$_2$O to afford α-galactosides **41** and **44** in good yields

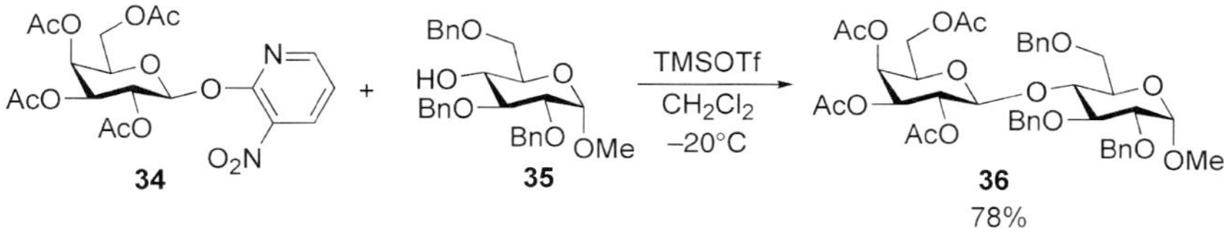

Scheme 3.74

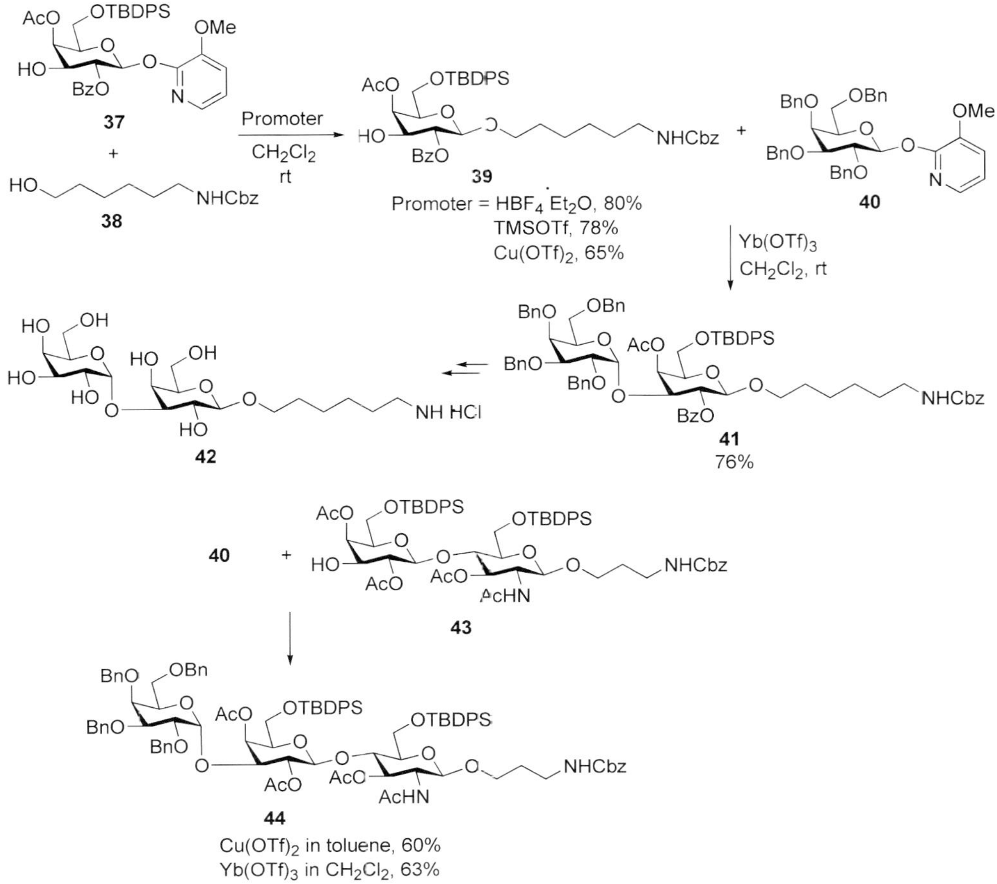

Scheme 3.75

(Scheme 3.75) [582]. So they were able to establish an efficient protocol for the synthesis of hapten-like motifs based on the blood group substance B disaccharide **42** and linear B type 2 trisaccharide **44**, which are nonprimate epitope markers recognized by human anti-α-Gal antibodies causing the hyperacute rejection of xenotransplants.

Schmidt et al. prepared some 1-O-hetaryl-D-glucopyranosides **45** by treating 2,3,4,6-tetra-O-benzyl-D-glucopyranose or 2,3,4,6-tetra-O-acetyl-D-glucopyranose with electron-deficient heteroaromatic compounds (hetaryl chloride or hetaryl fluoride) in the presence of NaH and K_2CO_3 as base. The glycosylation reactions were performed with the 1-O-hetaryl-D-glucopyranosides **45a–f** in the presence of TMSOTf or $BF_3 \cdot Et_2O$ to obtain the corresponding glycosides, such as **30**, in good yields (Scheme 3.76) [583].

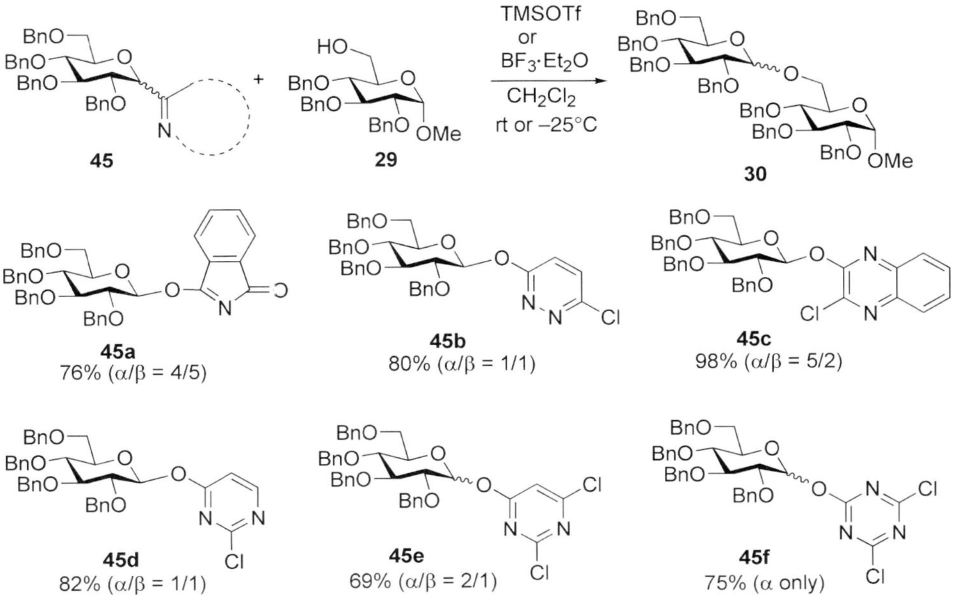

Scheme 3.76

3.3.5
2-Hydroxy-3,5-Dinitrobenzoate (DISAL) Glycosides

Jensen and coworkers reported a glycosylation method employing methyl 2-hydroxy-3,5-dinitrobenzoate (methyl 3,5-dinitrosalicylate, DISAL) glycosides as glycosyl donors. For example, reaction of DISAL glycosyl donor **46** with diacetonegalactose **47** in 1-methylpyrrolidene-2-one (NMP) as the solvent in the absence of a Lewis acid afforded disaccharide **48** in high yield (Scheme 3.77) [584]. Next, they reported the use of DISAL glycosyl donors in Lewis acid-promoted glycosylations. Thus, 2-acetamido acceptor **49** was successfully glycosylated using DISAL donor **46** in CH_2Cl_2 in the presence of $LiClO_4$ (Scheme 3.78) [585]. This approach was extended

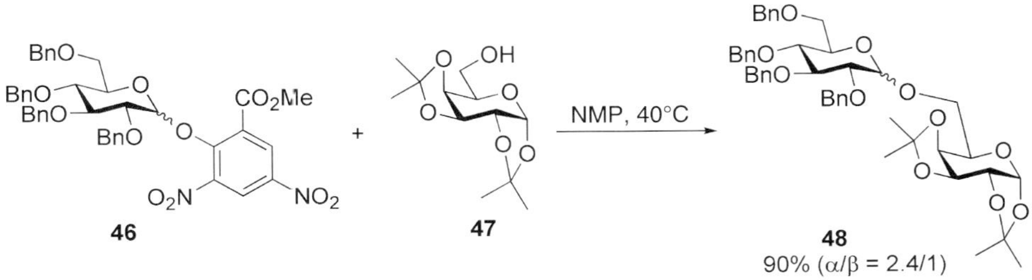

Scheme 3.77

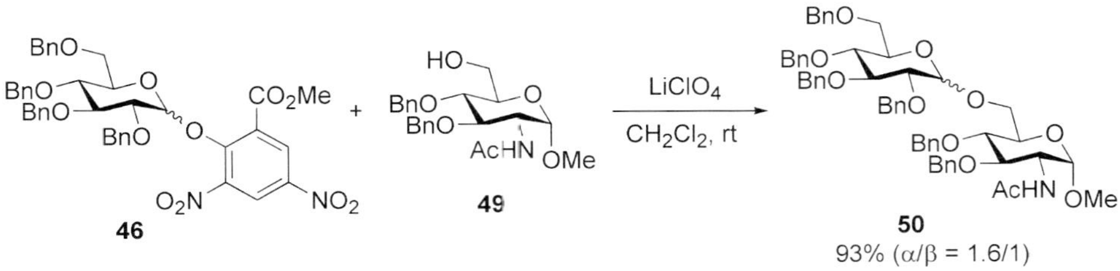

Scheme 3.78

to the solid-phase glycosylation of D-glucosamine derivatives anchored through the 2-amino group by a backbone amide linker (BAL) to a solid support [585]. They also applied the DISAL donor method to the synthesis of a starch-related linear hexasaccharide **55**. Glycosylations proceeded with moderate to high α-stereoselectivity and were compatible with the trityl (Tr) protecting group. Thus, the glycosylation of **52** with 6′-O-tritylated disaccharide DISAL donor **51** in CH_3NO_2 in the presence of $LiClO_4$, Li_2CO_3 and Bu_4NI gave tetrasaccharide **53a** in good yield. Upon detritylation, tetrasaccharide **53b** was glycosylated with disaccharide DISAL donor **54** using $LiClO_4$ and Li_2CO_3 in $(CH_2Cl)_2$ to afford hexasaccharide **55** (Scheme 3.79) [586]. Jensen and coworkers applied the DISAL method to the synthesis of glycosylated phenazine natural products and their analogs (Scheme 3.80) [587]. Recently, the same group reported that the glycosylation of **49** with glucosamine-derived DISAL donor **61** in NMP provided disaccharide **62** (Scheme 3.81) [588]. The DISAL donor concept was developed further to allow intramolecular glycosylations. Thus, dissolving tethered glycoside **63** (α/β = 1.3 : 1) in CH_3NO_2 and warming to 40 °C led to the formation of 1,4-linked disaccharide **64** (α/β = 3.7 : 1) (Scheme 3.82) [589]. The stereoselectivity is low because these reactions, although not concerted, have the tendency to proceed intramolecularly, maybe within the solvent cage.

3.3.6
Vinyl Glycosides

The use of vinyl glycoside donors was independently introduced by Schmidt and Sinay. Schmidt and coworkers reported that glycosyl alkenoate **66** as a glycosyl donor, readily derived from tetra-O-benzyl glucose **65** and ethyl phenyl propiolate using NaH, reacted with acceptor **35** in the presence of TMSOTf in CH_3CN to afford disaccharide **67** (Scheme 3.83) [590]. Takeda and coworkers provided a more efficient method for the preparation of glycosyl alkenoate donors by employing Bu_3P instead of NaH and reported that glycosyl donor **68** reacted with **29** by activation with TMSOTf to afford disaccharide **30** (Scheme 3.84) [591]. Sinay and coworkers initially focused on the use of prop-1-enyl glycoside **70**, prepared by isomerizing allyl glycoside **69** that could be activated with TMSOTf in CH_3CN for the glycosylation reaction (Scheme 3.85) [592]. Their attention then shifted to the use of isopropenyl glycoside **72**, which was prepared from the corresponding

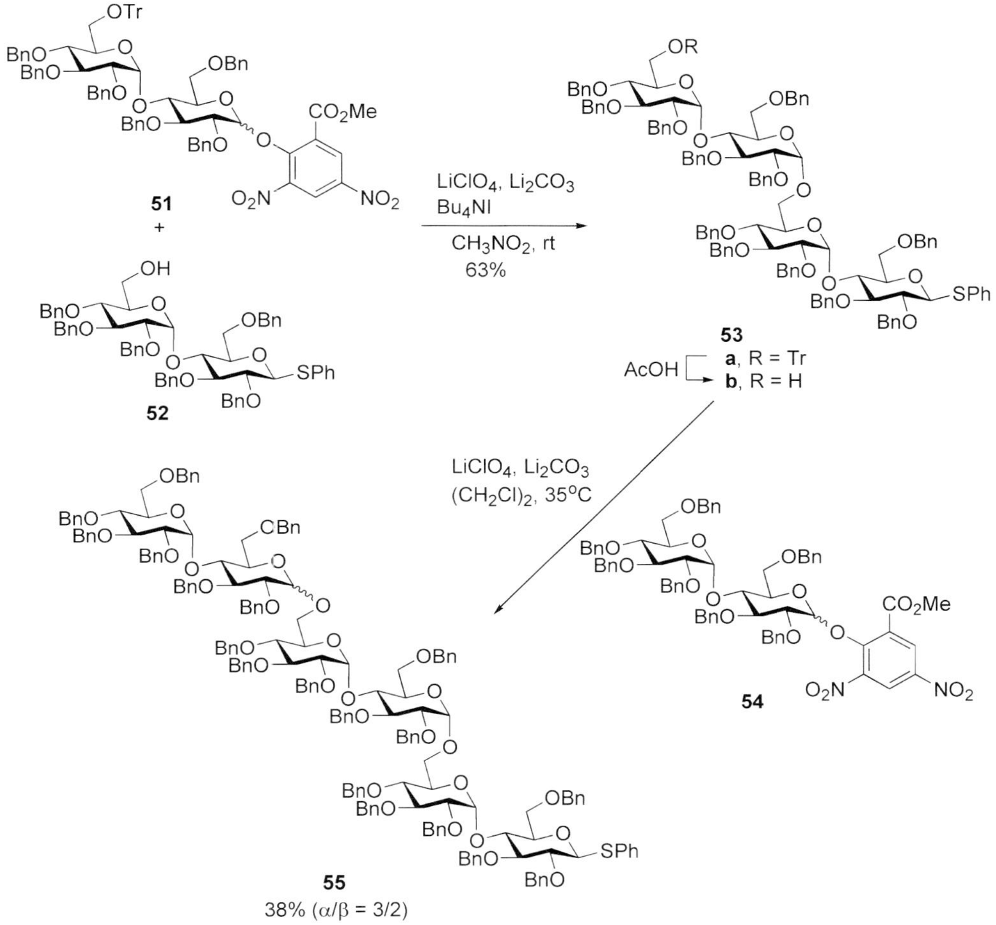

Scheme 3.79

acetate **71** by the Tebbe methylenation. Glycosylation of acceptor **35** with isopropenyl glycoside **72** was carried out in the presence of $BF_3 \cdot Et_2O$ at low temperature to give disaccharide **67** in good yield. Chenault *et al.* reported a new method for the preparation of isopropenyl glycosides from glycosyl halides and bis(acetonyl)mercury (Scheme 3.86) [593]. It was also revealed that strong electrophiles (NIS/TfOH, TMSOTf, Tf_2O or DMTST), polar solvents and electron-donating protecting groups are essential for the transglycosylation with isopropenyl glycosides. For example, the reaction of *O*-isopropenyl glucopyranoside **74** with acceptor **47** in the presence of NIS/TfOH in MeCN afforded the desired disaccharide **75** in good yield, while the same reaction in the presence of iodonium dicollidine perchlorate (IDCP in CH_2Cl_2 gave the mixed iodoacetonide **76**. The same group also showed that selective activation of **74** in the presence of *O-n*-pentenyl glucopyranoside **77** using

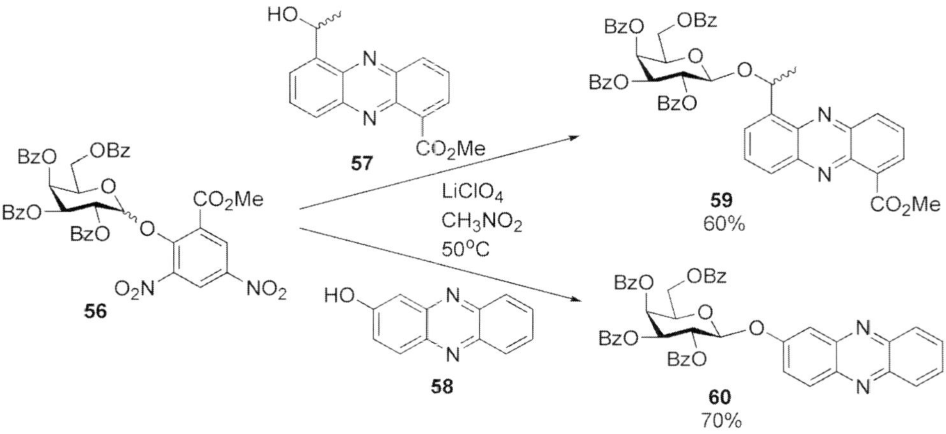

196 | *3 Glycoside Synthesis from 1-Oxygen Substituted Glycosyl Donors*

Scheme 3.80

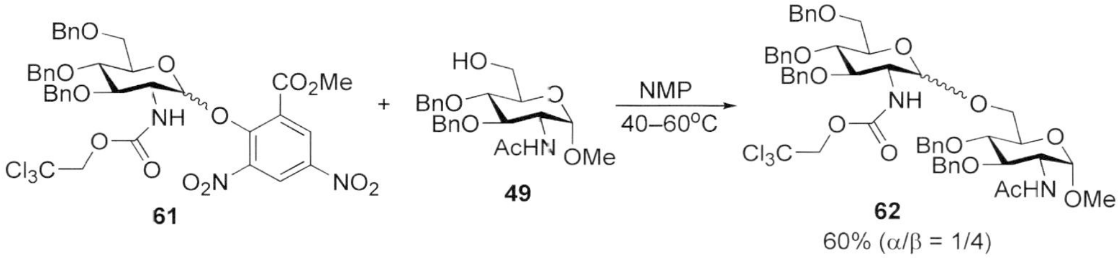

Scheme 3.81

TMSOTf as the promoter allowed two successive glycosylations in one pot to give the corresponding trisaccharide **79** in 25% yield (Scheme 3.87) [594].

Boons and coworkers modified the vinyl glycoside produced by Sinay and reported the use of 2-buten-2-yl glycoside **81**, which could be obtained via isomerization of 3-buten-2-yl glycoside **80** using Wilkinson's catalyst with a hindered base (DABCO)

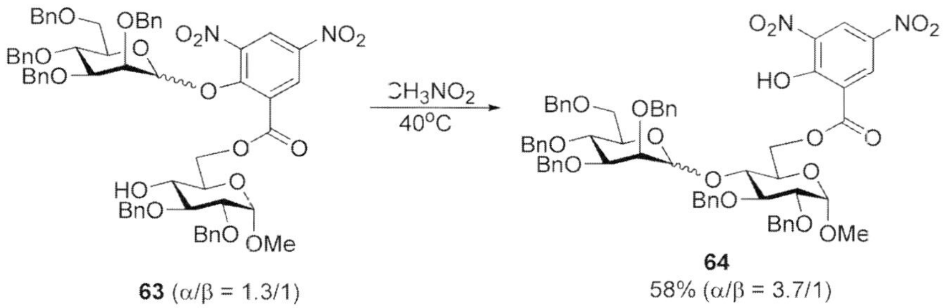

Scheme 3.82

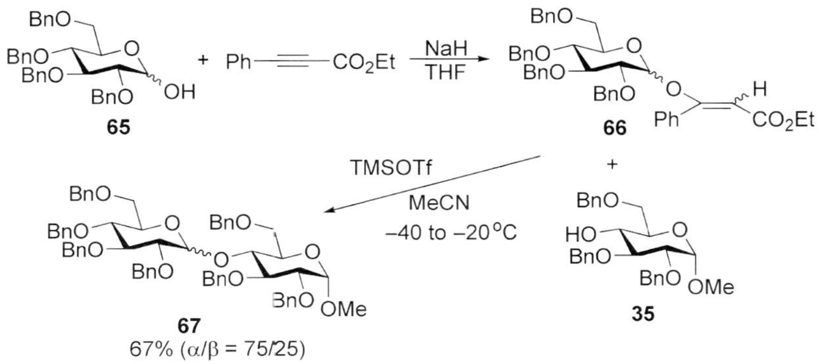

3.3 Anomeric Transglycosylation

Scheme 3.83

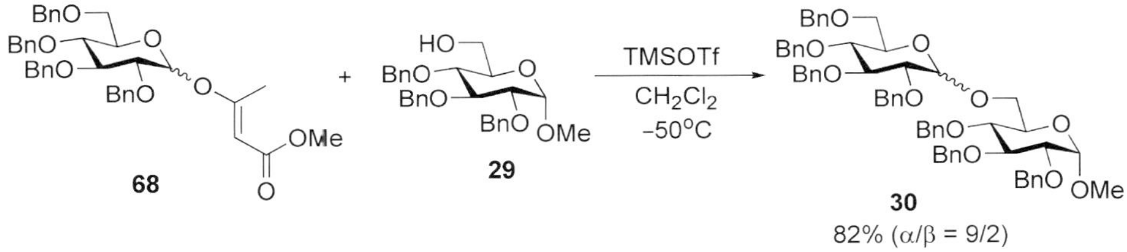

Scheme 3.84

[595], Wilkinson's catalyst with *n*-BuLi- [596] or glycosidase-catalyzed transformation [597]. This approach provided an example of the 'latent–active' concept to the chemoselective glycosylation. Thus, 3-buten-2-yl glycoside **80** is inert toward activation, whereas the isomerized 2-buten-2-yl glycoside **81** can be activated using a catalytic

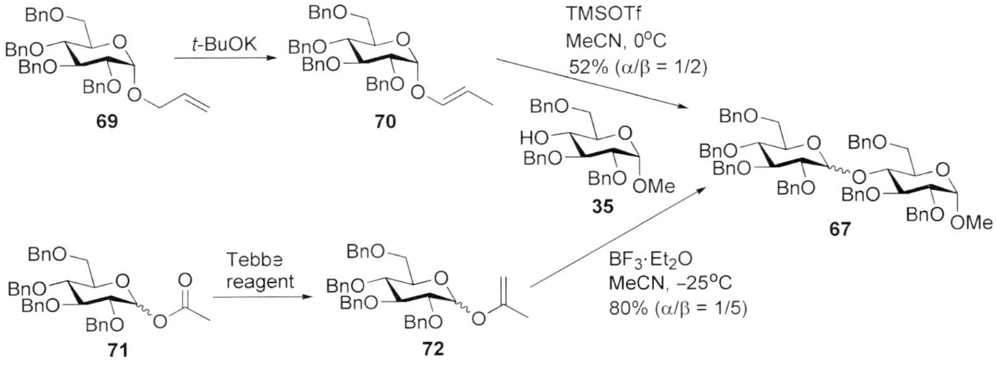

Scheme 3.85

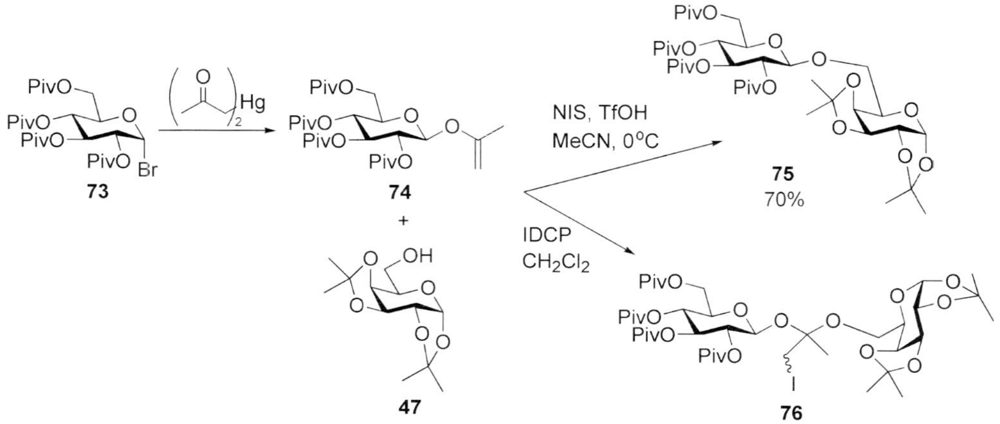

Scheme 3.86

amount of Lewis acid. This approach was applied to the synthesis of latent 3-buten-2-yl disaccharide **83** by the glycosylation of latent 3-buten-2-yl glycoside acceptor **82** with active 2-buten-2-yl glycoside donor **81** in the presence of TMSOTf (Scheme 3.88) [595,598]. Boons *et al.* also reported the synthesis of trisaccharide libraries based on this latent–active glycosylation method (Scheme 3.89) [599]. Thus, vinyl glycoside donors were obtained from allyl glycoside building blocks by isomerization with [(Ph$_3$P)$_3$RhCl]/BuLi and allyl glycoside acceptors by the removal of the acetyl protecting group. Subsequently, several disaccharides were prepared by glycosylations of appropriate acceptors with donors and using a catalytic amount of TMSOTf. Then, the disaccharides were combined and the removal of the acetyl protecting group gave a pool of disaccharide acceptors. In a combinatorial fashion, a vinyl glycosyl donor

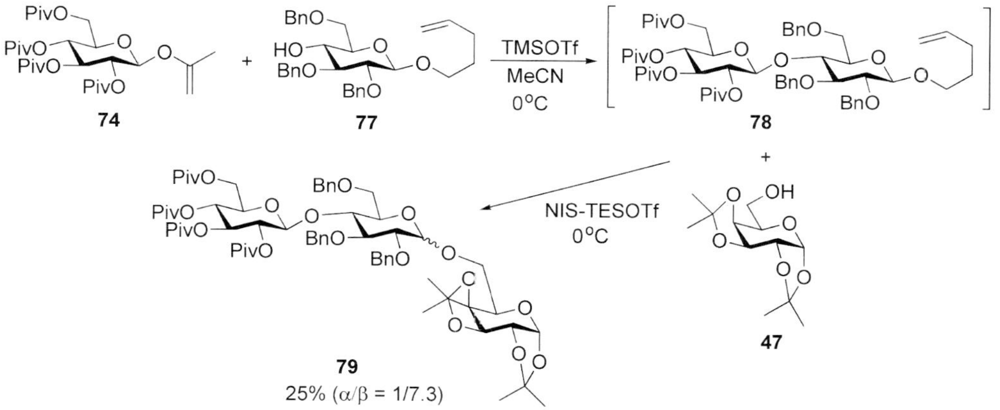

Scheme 3.87

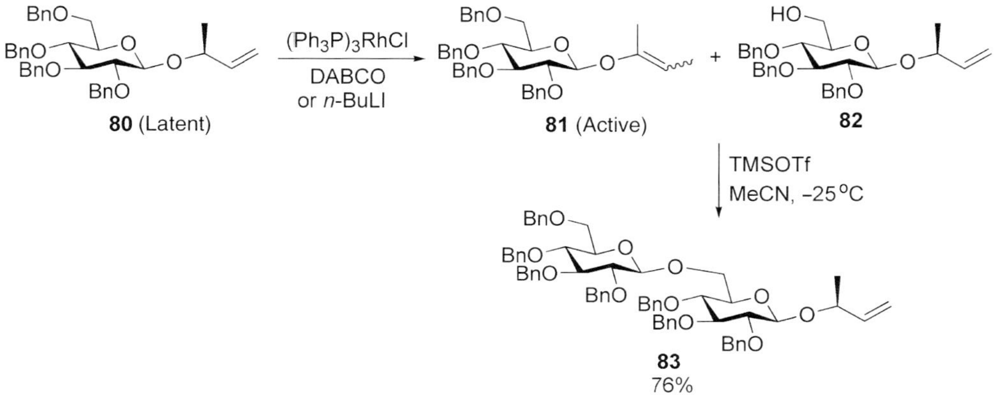

3.3 Anomeric Transglycosylation | 199

Scheme 3.88

was coupled with the acceptor pool to give a library of various trisaccharides. This glycosylation strategy was also extended to the synthesis of disaccharide **85**. 3-Buten-2-yl 2-deoxy-2-azidoglycoside **84a** and 3-buten-2-yl 2-deoxy-2-phthalimidoglycoside **84b** were investigated as glycosyl donors (Scheme 3.90) [600]. 2-Azidoglycoside **84a** gave a mixture of α- and β-disaccharide **85** (α/β = 1:1), whereas 2-phthalimidoglycoside **84b** afforded exclusively β-linked disaccharide.

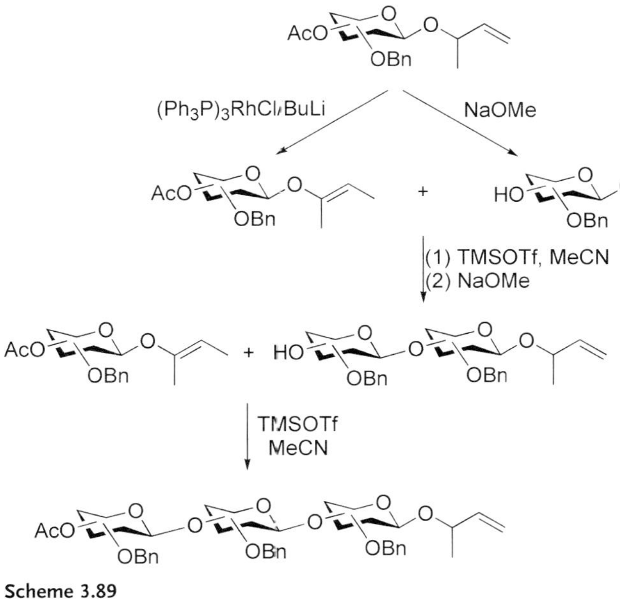

Scheme 3.89

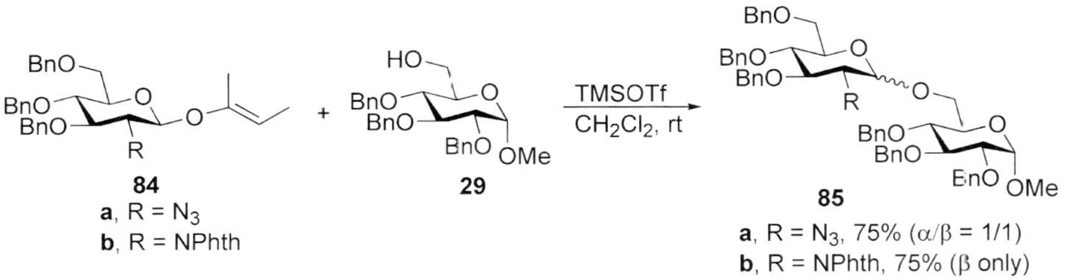

Scheme 3.90

3.3.7
n-Pentenyl Glycosides

Fraser-Reid et al. introduced the n-pentenyl group as the leaving group at the anomeric center of the glycosyl donor in 1988 [601]. It should be noted that n-pentenyl was originally used as the anomeric temporary protective group because it could be selectively deprotected by mild hydrolysis using N-bromosuccinimide (NBS) in CH_3CN-H_2O [602]. The glycosylation reaction probably proceeds via the initial reaction of the electrophile with olefin **A** to give cyclic halonium ion **B**, which is in equilibrium with cyclic oxonium ion **C**. The subsequent release of the halomethyltetrahydrofuran liberates oxocarbenium ion **D**, which is attacked by an alcohol acceptor to afford glycoside **E** (Scheme 3.91).

There is a number of ways by which the n-pentenyl group can be added to the anomeric position (Scheme 3.92). Standard Fischer glycosylation of the unprotected sugar with n-pentenol under acidic condition (**86 → 90**) [603,604] or Koenigs–Knorr glycosylation of the glycosyl bromide (**87 → 90**) with silver(I) salts [605,606] provide access to n-pentenyl glycosides. Glycosylation of n-pentenol with glycosyl acetates using tin(IV) chloride also affords the corresponding n-pentenyl glycosides (**88 → 90**) [605]. Finally, it is interesting to note that NPGs can also be prepared via acid-catalyzed rearrangement of n-pentenyl orthoesters (NPOEs, **89 → 90**). The NPOEs are prepared under basic conditions from 2-O-acyl-protected glycosyl bromides and n-pentenol [605].

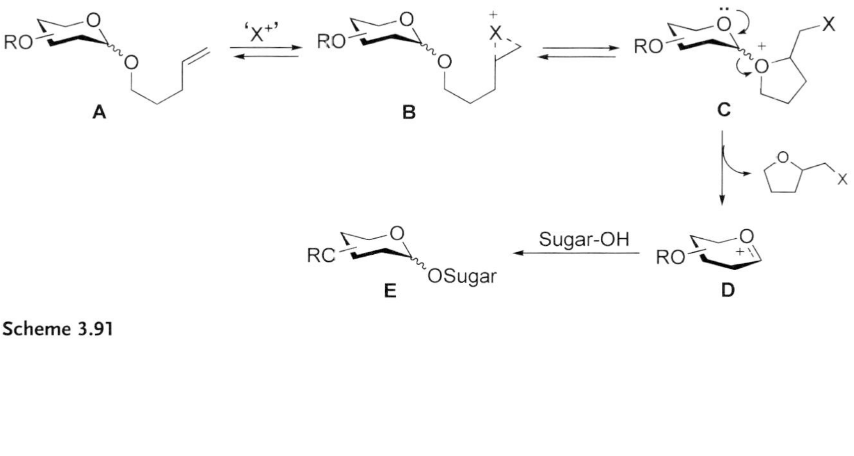

Scheme 3.91

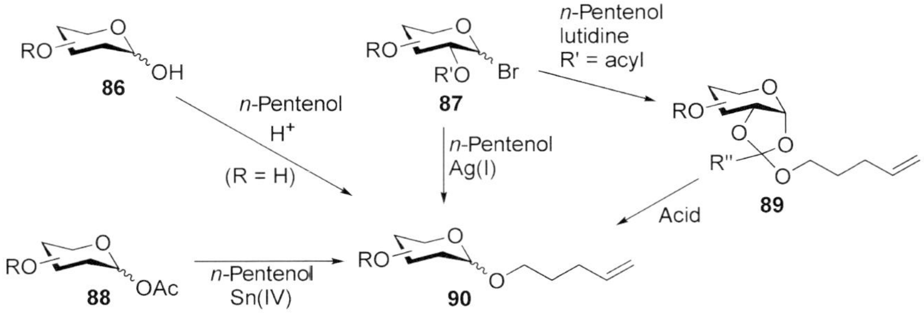

Scheme 3.92

NPGs have been most widely activated by N-bromosuccinimide and N-iodosuccinimide (NIS). The addition of a protic or Lewis acid, such as triflic acid (TfOH), silver triflate (AgOTf) or triethylsilyl triflate (TESOTf), greatly enhances the rate of reaction [607,608]. Iodonium dicollidine perchlorate has also been used as the promoter of intermediate potency for couplings of NPGs [609], being more reactive than NIS alone but less reactive than NIS/TESOTf (Scheme 3.93). It has been successful for the couplings of reactive (electronically armed) NPGs but is not potent enough to be used for less reactive (electronically disarmed) NPGs. This concept is discussed below. Recently, it was also found that armed NPGs can be activated (a) readily with scandium or indium(III) triflates and NIS, (b) moderately with samarium or lanthanide counterparts and NIS and (c) not at all with

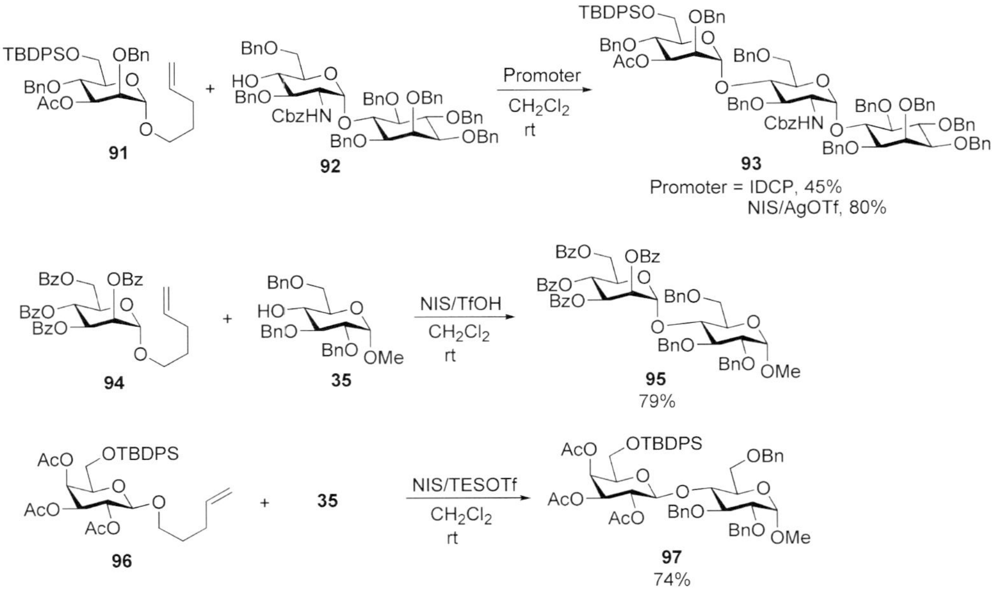

Scheme 3.93

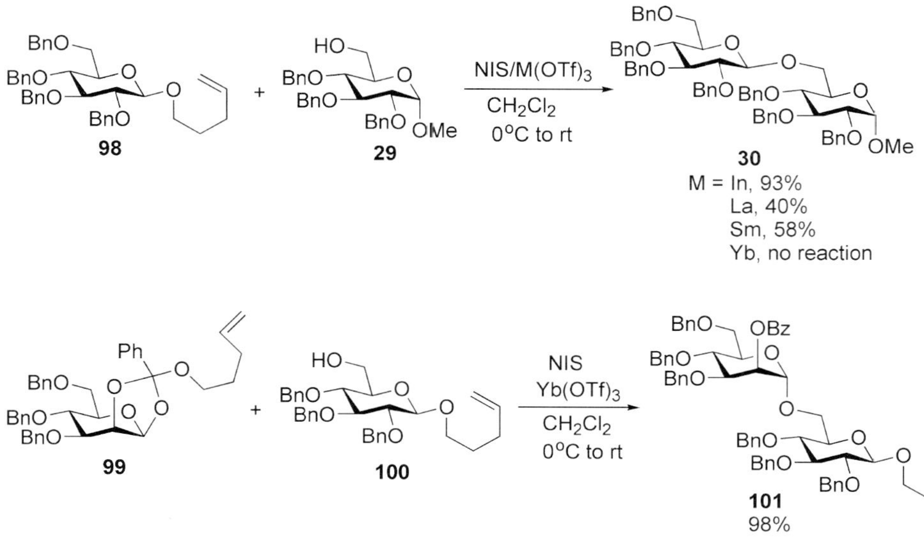

Scheme 3.94

ytterbium counterpart and NIS (Scheme 3.94) [610]. However, the latter combination readily activates NPOE. This chemoselectivity of *n*-pentenyl donors toward lanthanide salts therefore permits armed NPG **100** to function as acceptor toward NPOE donor **99** (Scheme 3.94).

Fraser-Reid and coworkers described a chemoselective glycosylation using the armed–disarmed glycosylation strategy, which depends on the differential reactivity toward halonium ion sources. NPGs with ether-based protecting groups react much more rapidly than those with ester-based protecting groups. Namely, the chemoselectivity relies on the fact that an electron-withdrawing C-2 ester deactivates (disarms) the anomeric center and an electron-donating C-2 ether activates (arms) it. Thus, coupling of an armed donor **98** with a disarmed acceptor **102** in the presence of the mild activator IDCP afforded disaccharide **103** as an anomeric mixture (Scheme 3.95) [609]. No products of self-condensation of **102** have been detected. The disaccharide **103**α can be used either for 1,2-*trans* glycosylation directly with the use of a more powerful promoter such as NIS/TfOH or NIS/TESOTf, or for 1,2-*cis* glycosylation after appropriate protecting-group manipulations. Overall, the armed–disarmed glycosylation strategy offers an efficient way to synthesize oligosaccharides with *cis–trans* or *cis–cis* glycosylation pattern (Scheme 3.95).

Fraser-Reid *et al.* also disclosed that cyclic acetals profoundly affect the glycosyl donor reactivity, thereby paving the way for an armed–disarmed protocol based on torsional effects, complementary to electronic effects [611]. Indeed, it proved possible to chemospecifically couple glucosides **98** and **104** to give the cross-coupling product **105** with no evidence of generation of **106** by the self-condensation of the alcohol acceptor **104** (Scheme 3.96).

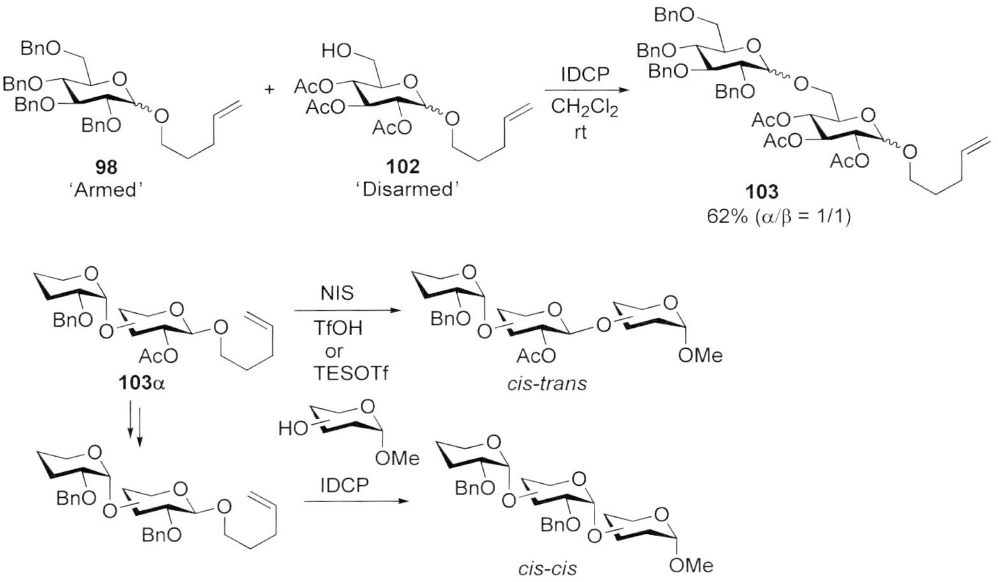

Scheme 3.95

An attractive alternative to the armed–disarmed protocol for chemoselective coupling of NPGs involves dibromination of the double bond to give **G**, which is inert to halonium ion activation. However, depending on how the bromination reaction is carried out, either glycosyl bromide **F** or vicinal dibromide **G** can be obtained (Scheme 3.97). In the absence of other nucleophiles, the ring closure of cyclic bromonium ion **B** occurs to give furanylium ion **C**. Release of 2-bromomethyltetrahydrofuran then leads to the oxocarbenium ion **D**, which is attacked by Br⁻ (the only nucleophile in the reaction mixture) to give glycosyl bromide **F** [612]. However, if excess Br⁻ is assured by the addition of Et$_4$NBr, the bimolecular reaction between **B** and Br⁻ is able to dominate over the intramolecular process, leading to the dibromide **G** [613]. This dibromide **G** can be considered as a latent NPG from which the active form **A** (NPG)

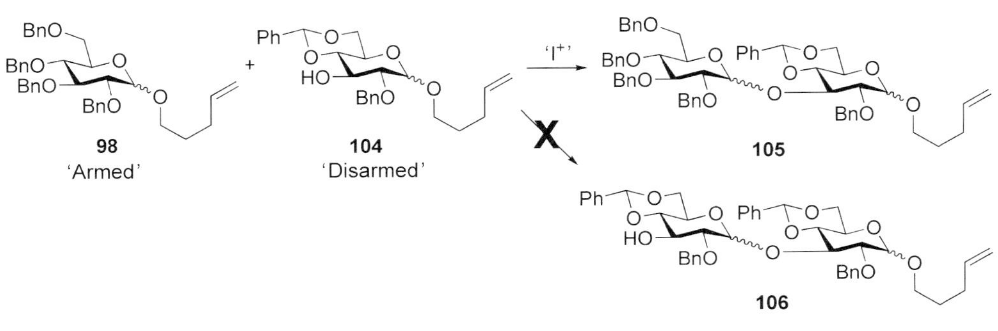

Scheme 3.96

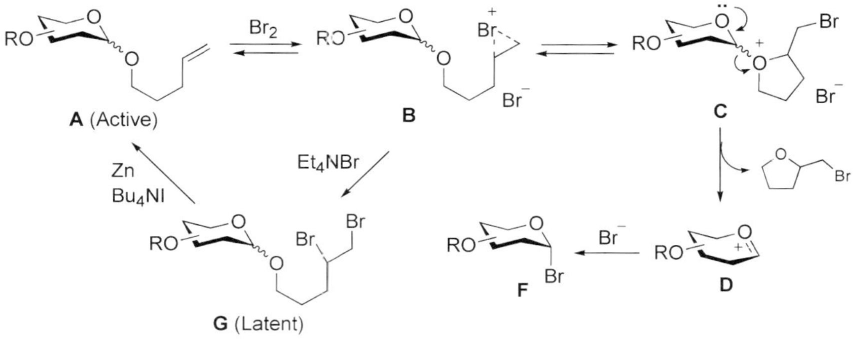

Scheme 3.97

can be regenerated by reductive debromination. The use of this latent/active strategy provides an alternative method for the chemoselective assembly of oligosaccharides (Scheme 3.98).

n-Pentenyl 2-deoxy-2-phthalimido- and 2-anisylimino-2-deoxy-D-glucosides, **107** and **109**, also readily undergo iodonium ion induced coupling with a variety of sugar alcohols to exclusively give β- and α-disaccharides, respectively, in good yields (Scheme 3.99) [614]. On the basis of the latter α-glycoside synthesis, Fraser-Reid *et al.* used NPG **111** in the synthesis of mannan-rich pentasaccharide **118**, the core oligosaccharide of the variant surface glycoprotein found in *Trypanosoma brucei* (Scheme 3.100) [615]. Thus, coupling of *n*-pentenyl 2-anisylimino glycoside **111** and inositol **112** provided the desired α-linked disaccharide **113**. Compound **114**, derived from **113**, coupled with pentenyl mannoside **115** to give an α-disaccharide,

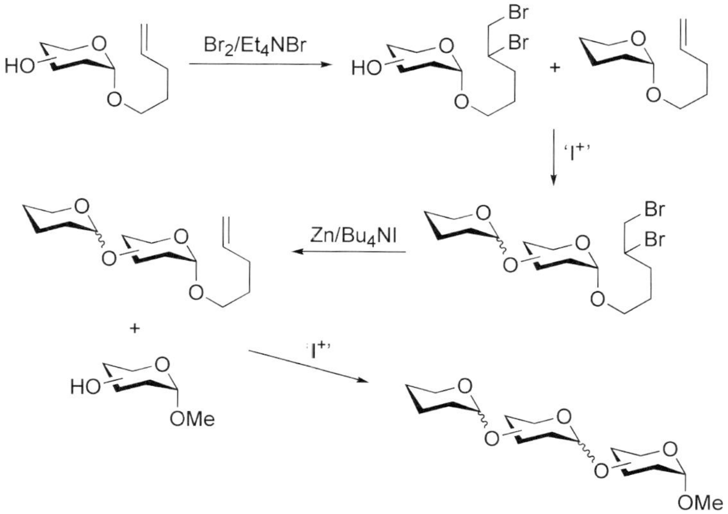

Scheme 3.98

3.3 Anomeric Transglycosylation | 205

Scheme 3.99

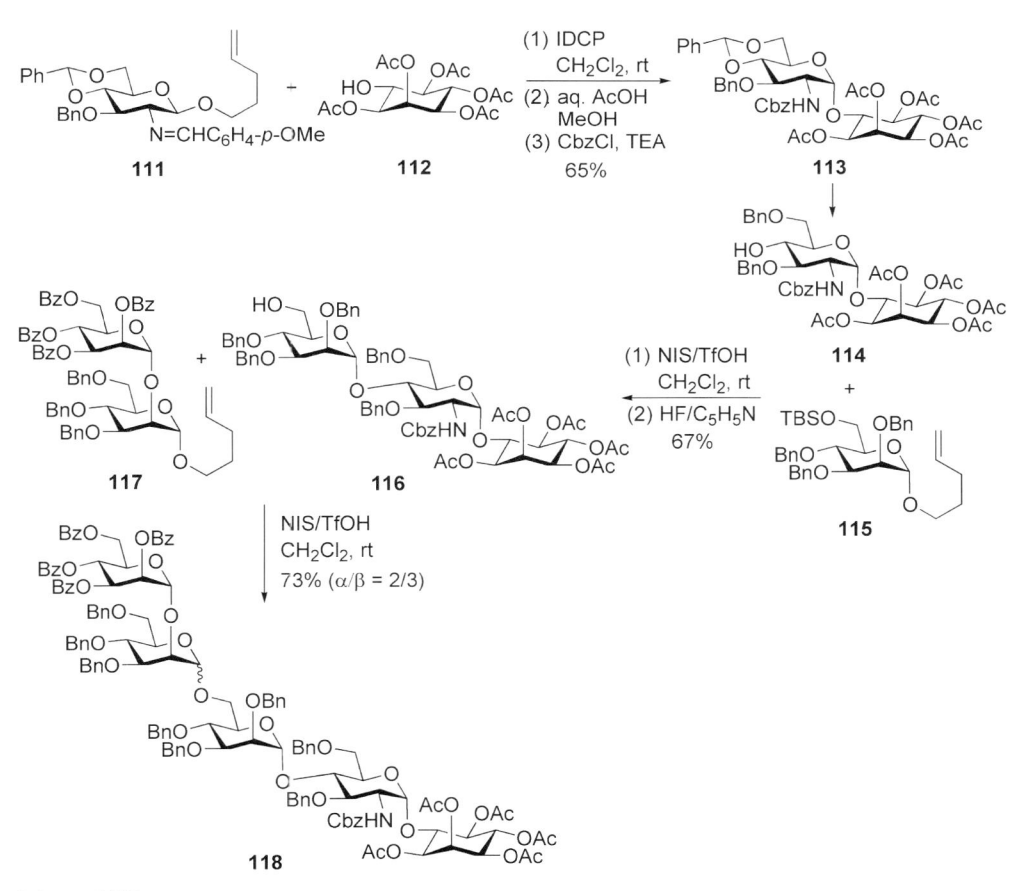

Scheme 3.100

which then converted into trisaccharide acceptor **116**. Coupling of **116** with pentenyl dimannan **117** was mediated by NIS/TfOH to afford pentasaccharide **118** ($\alpha/\beta = 2:3$) in 73% yield in which the poor stereoselectivity might be attributable to the higher reactivity of the primary hydroxyl group of the acceptor **116**.

On a similar note, Matta and coworkers reported the convenient synthesis of 2′-O-α-L-fucopyranosyllactose, found in human milk, using *n*-pentenyl L-fucopyranoside as the glycosyl donor [616]. Fraser-Reid *et al.* reported the synthesis of the blood group substance B tetrasaccharide using NPG method to give, in sequence, the disaccharide **121** (68%), trisaccharide **123** (82%) and tetrasaccharide **125** (91%) (Scheme 3.101) [602]. Moreover, the same group synthesized the nonasaccharide **133**, component of a high-mannan glycoprotein by the latent–active glycosylation strategy of NPGs and their 4,5-dibromo counterpart (Scheme 3.102) [617]. Furthermore, Fraser-Reid reported the synthesis of the complete heptasaccharide sequence **138** of the rat brain Thy-1 membrane glycophosphatidylinositol (GPI) anchor, the first mammalian membrane anchor pseudooligosaccharide characterized (Scheme 3.103) [618]. In the case of the Thy-1 anchor, the synthetic scheme was based on three building blocks comprising *n*-pentenyl galactosamine-mannose **134**, glucosamine-inositol **135** and *n*-pentenyl trimannan **137** residues. The stereoselective construction of the tetrasaccharyl cap portion of Leishmania, Lipophosphoglycan, was also accomplished using NPGs methodology [619].

Houdier and Vottero investigated the I$^+$-promoted glycosylation reaction with *n*-pentenyl 2,3,4-tri-O-benzyl-β-D-glucopyranoside derivatives in which 6-OH was

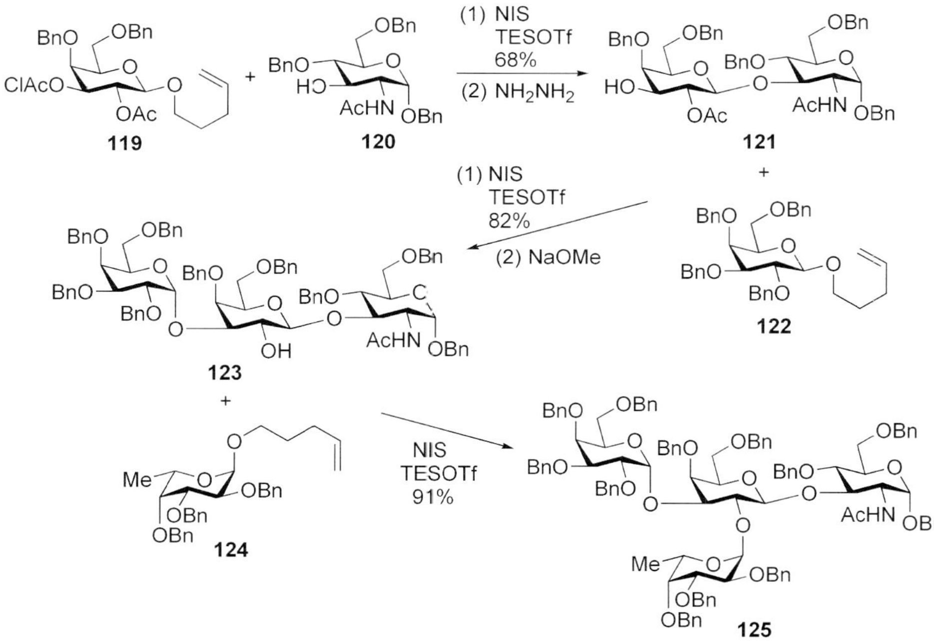

Scheme 3.101

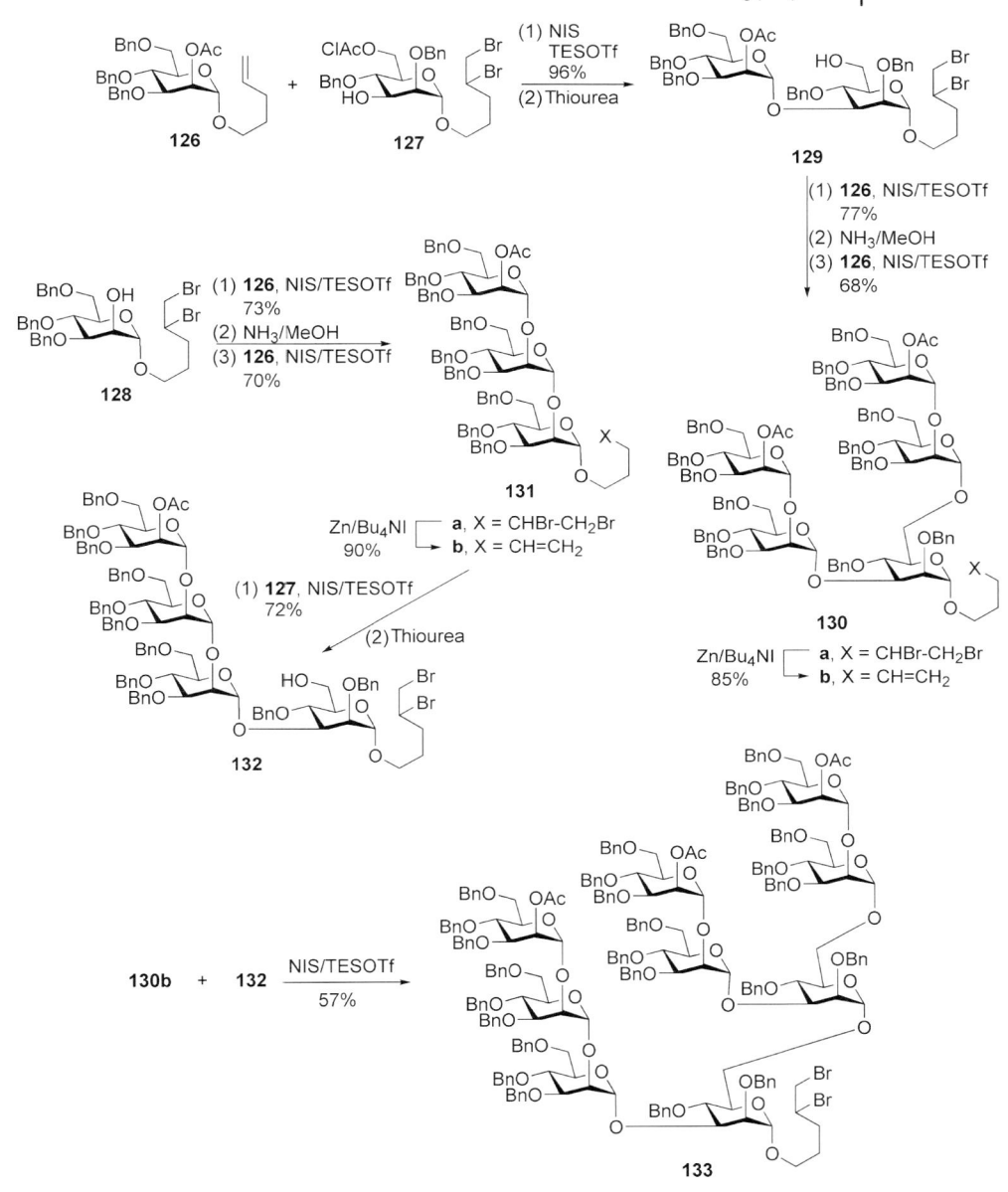

Scheme 3.102

protected by benzyl, trityl (Tr) or TBDMS group to assess the steric effect of bulky substituents at C-6 [620]. Thus, the presence of a bulky substituent at C-6 affects the steric course of the glycosylation reaction: it increases the proportion of α-glycosides but decreases the reaction yield. Lichtenthaler *et al.* reported that *n*-pentenyl

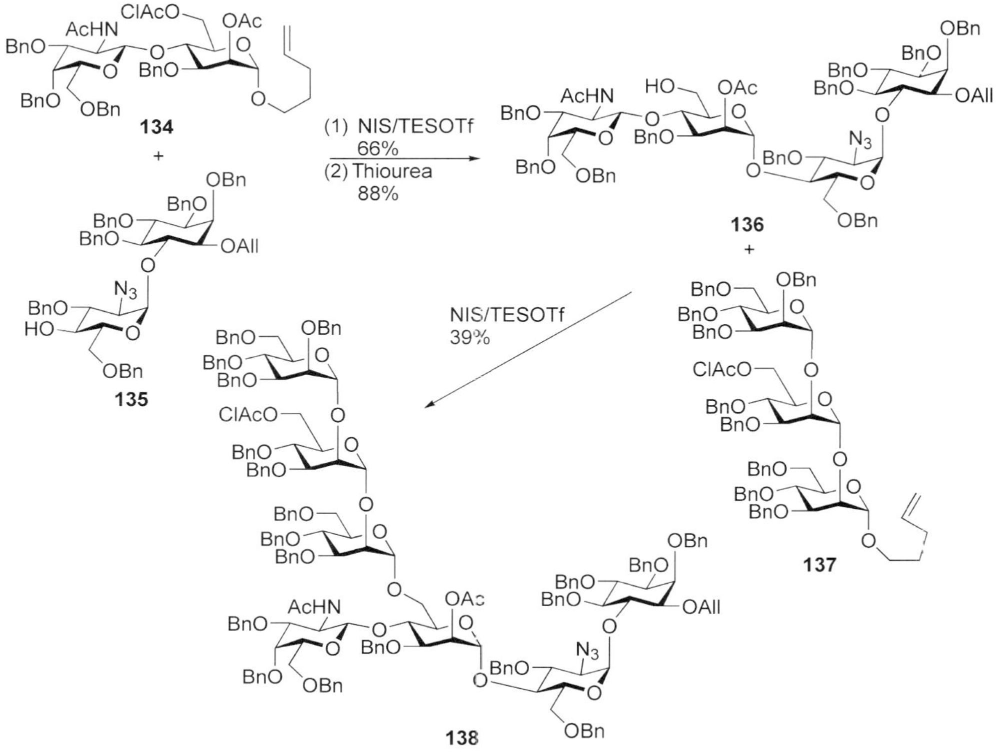

208 | *3 Glycoside Synthesis from 1-Oxygen Substituted Glycosyl Donors*

Scheme 3.103

2-benzoyloxyiminoglycoside **139** readily gave disaccharide **140** using NIS/AgOTf in good yield with exclusive β-selectivity (Scheme 3.104) [621]. Fraser-Reid and coworkers reported that n-pentenyl hex-2-enopyranoside (allylic O-NPG) **141** undergo coupling reaction with **47** in the presence of IDCP to give disaccharide **142** in a moderate yield (Scheme 3.105) [622]. On the contrary, Arasappan and Fraser-Reid [623] and Plusquellec and coworkers [32] reported the use of n-pentenyl furanoside

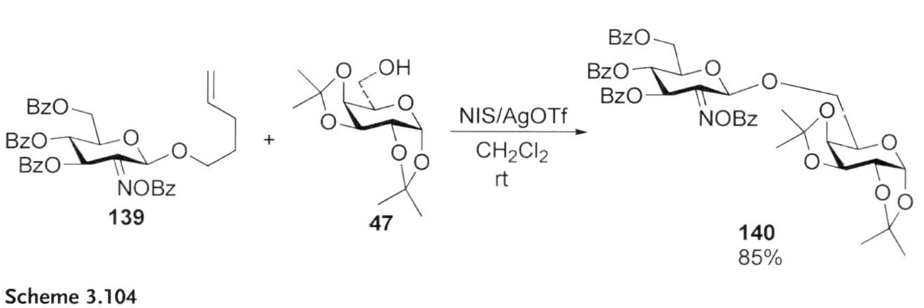

Scheme 3.104

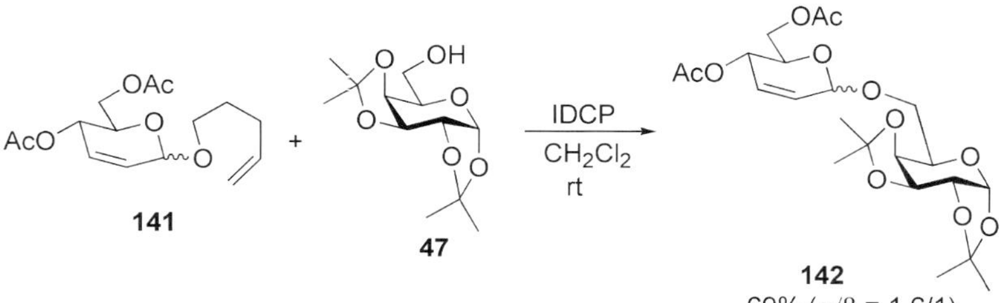

Scheme 3.105

143 with various glycosyl acceptors for the synthesis of the corresponding disaccharides-containing furanosyl moieties (Scheme 3.106).

Demchenko and De Meo reported the semiorthogonal glycosylation strategy using NPGs, **100** and **146**, and thioglycosides, **145** and **147**, for the oligosaccharide synthesis (Scheme 3.107) [624]. Thus, disarmed thioglycoside **145** could be activated with MeOTf in the presence of armed NPG **100** to afford n-pentenyl trisaccharide **146**, which could be subsequently activated with IDCP to give tetrasaccharide **148**. Clausen and Madsen also employed a general protocol with NPGs for the synthesis of the selectively methyl-esterified oligomer of galacturonic acid, **155**, hexasaccharide fragment of pectin (Scheme 3.108) [625]. The methodology was based on the repeated coupling of n-pentenyl galactose mono- and disaccharide donors, **149** and **153**, with galactose acceptor until hexagalactan **154** is obtained. All glycosylations performed with NPGs provided good yields of the desired α-anomers. Heidelberg and Martin utilized the NPG method for the synthesis of the glycopeptidolipid, found in the cell wall of *Mycobacterium avium* Serovar 4 [626].

Svarovsky and Barchi reported that the glycosylation of a serine derivative with β-n-pentenyl 2-azidodisaccharide donor **156β** afforded product **157**, O-glycosylated amino acid containing the tumor-associated T-antigen disaccharide unit in good yield. On the contrary, the glycosylation with α-n-pentenyl donor **156α** gave **157** in a very poor yield (<10%) (Scheme 3.109) [627]. The reason for the low reactivity of **156α** was provided by assuming that the *cis*-1,2 disposition of the n-pentenyl and azido groups in the α-pentenyl glycosides **156α** may facilitate interaction of these two functionalities during the activation step of the glycosylation. Conversion of azidodisaccharide **156β** into **157** is noteworthy because glycosylations of amino acids with n-pentenyl 2-azidoglycosyl donors are rare and usually provide the desired products in poor yields

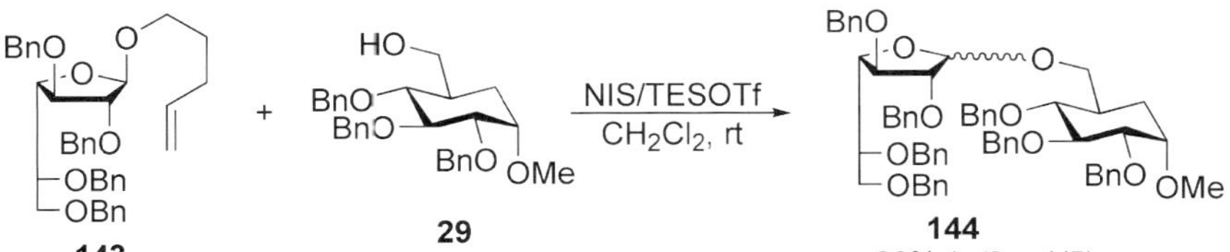

Scheme 3.106

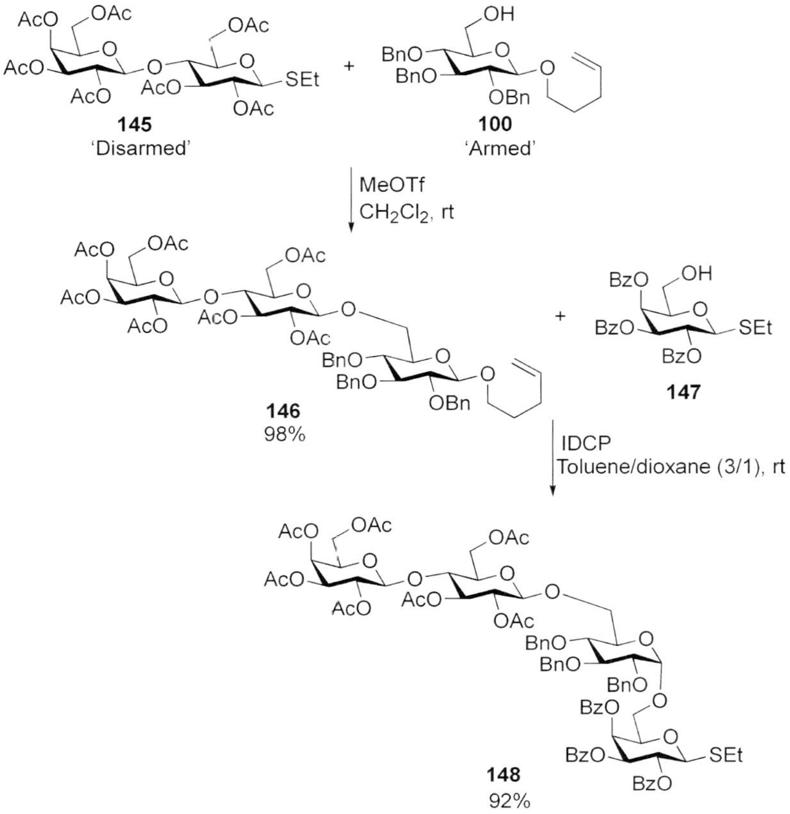

Scheme 3.107

[628]. Ikeda et al. prepared an NPG of N-acetylneuraminic acid **158** that was applied to the glycosylation reaction of various acceptors such as **47** (Scheme 3.110) [629].

Fraser-Reid found that the NPG glycosyl donors preferentially glycosylated the axial-OH in a series of cyclic 1,3-diols [630]. For example, products from the glycosylation reaction of partially protected *myo*-ionsitol acceptor **161** with NPG **160** were mannosides **162a** and **162b** in 3 : 1 (ax. : eq.) ratio. Similarly, reaction of a mannoside diol **164** with NPG **163** gave disaccharide **165** as the major product (Scheme 3.111). When cyclic vicinal diols were examined, the acyl-protected NPG donor frequently reacted with only one of the two hydroxyl groups in good yield, whereas the benzylated NPG donor was less discriminating and frequently gave lower yields and regioselectivity [631]. Furthermore, Fraser-Reid et al. compared NPOEs and NPGs as alternative glycosyl donors through theoretical and experimental studies [632]. They found that with *manno* derivatives, the orthoester (NPOE) is a more reactive donor than the corresponding NPG donor in the presence of NBS and TESOTf, whereas, for *gluco* derivatives, there is no advantage of using one over another.

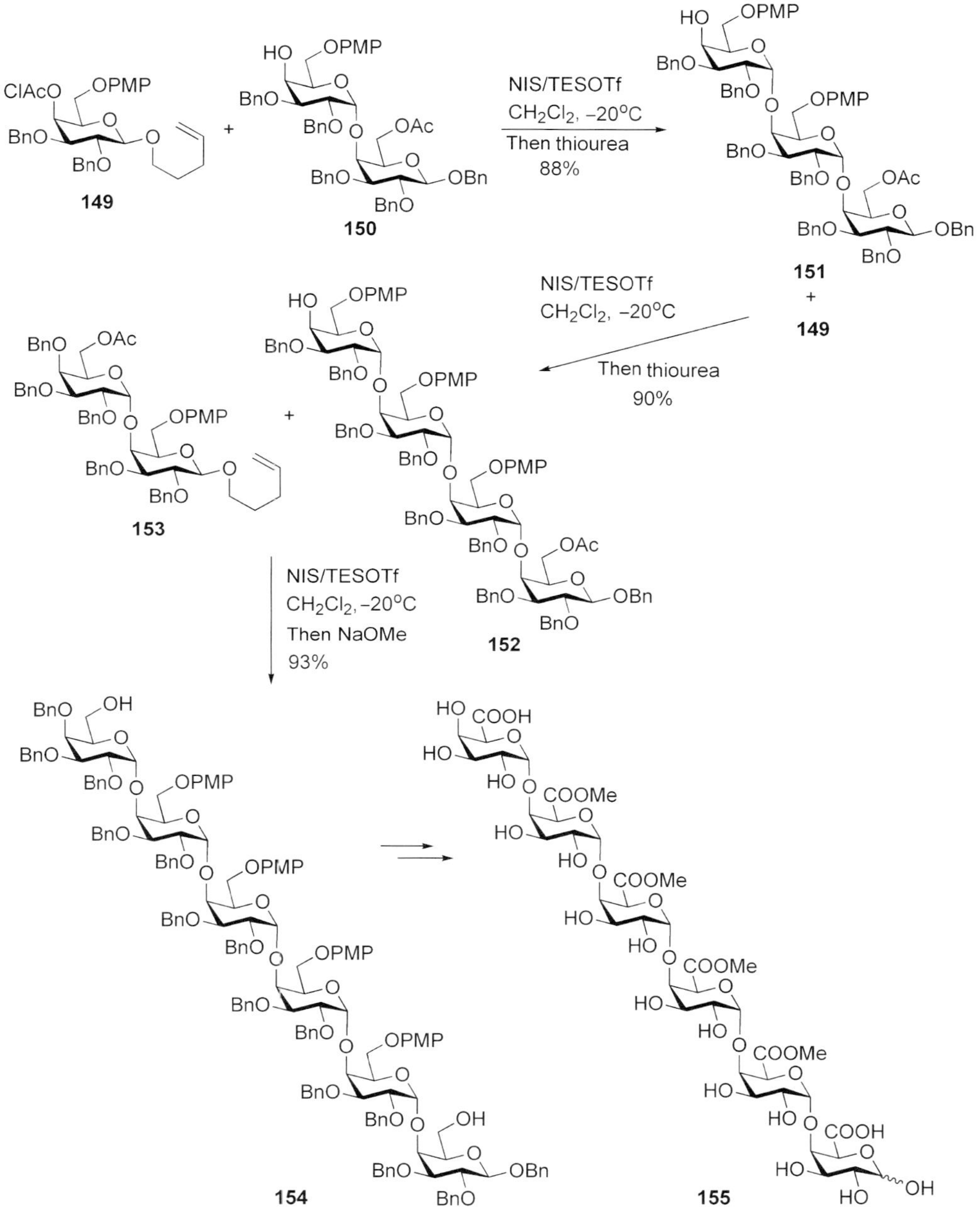
149
150
NIS/TESOTf
CH_2Cl_2, –20°C
Then thiourea
88%
151
+
149
NIS/TESOTf
CH_2Cl_2, –20°C
Then thiourea
90%
153
+
152
NIS/TESOTf
CH_2Cl_2, –20°C
Then NaOMe
93%
154
155

Scheme 3.108

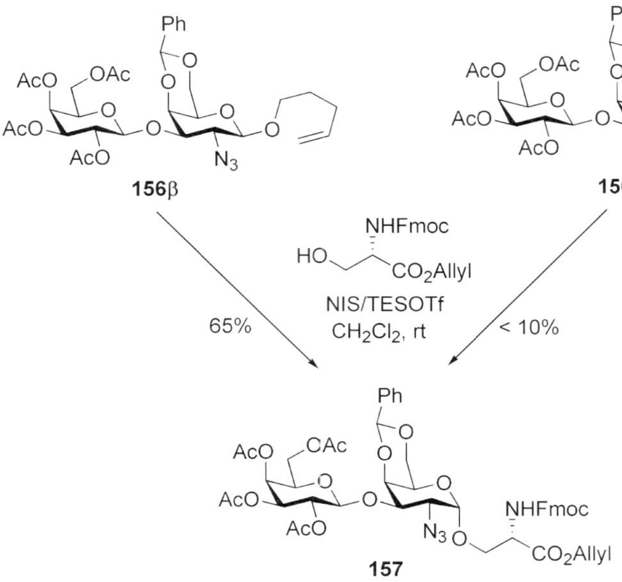

Scheme 3.109

3.3.8
2′-Carboxybenzyl Glycosides

Kim and coworkers introduced a glycosylation method employing CB glycoside **I** as glycosyl donor, which was prepared from 2′-(benzyloxycarbonyl)benzyl (BCB) glycoside **H** by the selective removal of its benzyl ester functionality (Scheme 3.112) [633]. Treatment of **I** with triflic anhydride followed by spontaneous lactonization of the resulting glycosyl triflate **J** afforded oxocarbenium ion **D** by extrusion of stable phthalide. Reaction of **D** with the glycosyl acceptor (Sugar-OH) would give the desired glycoside **E**.

These reactions often proceed in a highly stereocontrolled manner. The CB glycoside was found to be a useful tool for the stereoselective β-mannopyranosylation, using CB 4,6-O-benzylidene-2,3-di-O-benzyl-D-mannoside (**166**), and for the

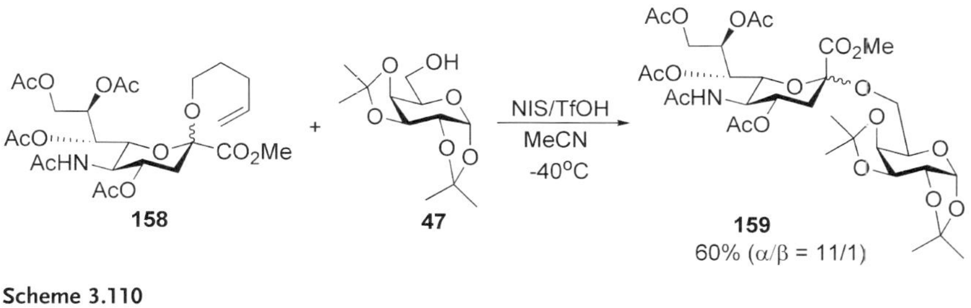

Scheme 3.110

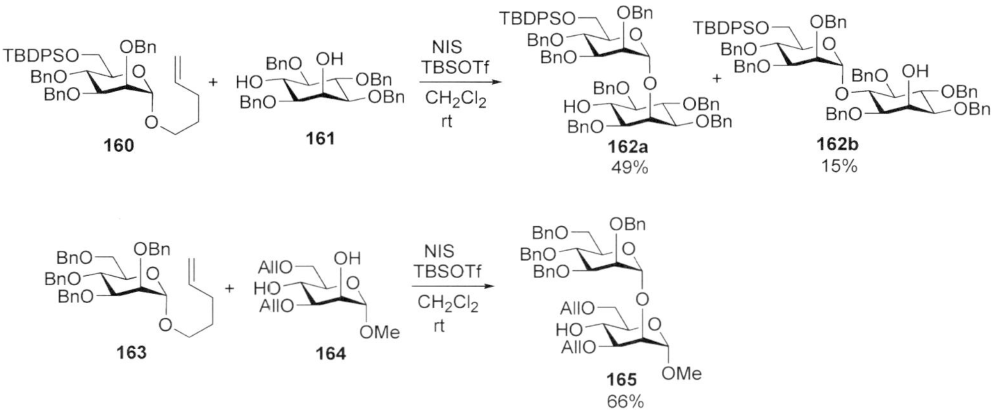

3.3 Anomeric Transglycosylation | 213

Scheme 3.111

stereoselective α-glucopyranosylation, using CB 4,6-*O*-benzylidene-2,3-di-*O*-benzyl-D-glucoside (**168**) as the glycosyl donor and triflic anhydride (Tf$_2$O) as the promoter (Scheme 3.113) [633]. This CB glycoside method was also utilized for the stereoselective construction of 2-deoxyglycosyl linkages. For example, the glycosylation of benzyl-protected CB 2-deoxyglycoside donor **170** with secondary alcohol **35** afforded α-glycoside **172**, whereas the glycosylation of 4,6-*O*-benzylidene-protected CB glycoside donor **171** with **35**-afforded β-glycoside **173** (Scheme 3.114) [634]. They also established a direct method for the stereoselective β-arabinofuranosylation employing CB tri-*O*-benzylarabinoside **174** as the arabinosyl donor, in which the acyl-protective group (**175**) on glycosyl acceptors was essential for the β-stereoselectivity (Scheme 3.115) [635]. Thus, the glycosylation of benzoyl-protected glycosyl acceptor **175** with **174** provided almost exclusively β-disaccharide **176** in high yield, while

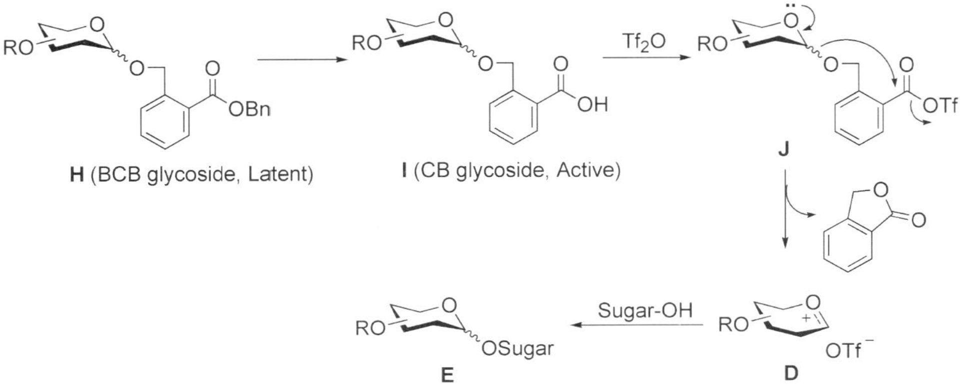

Scheme 3.112

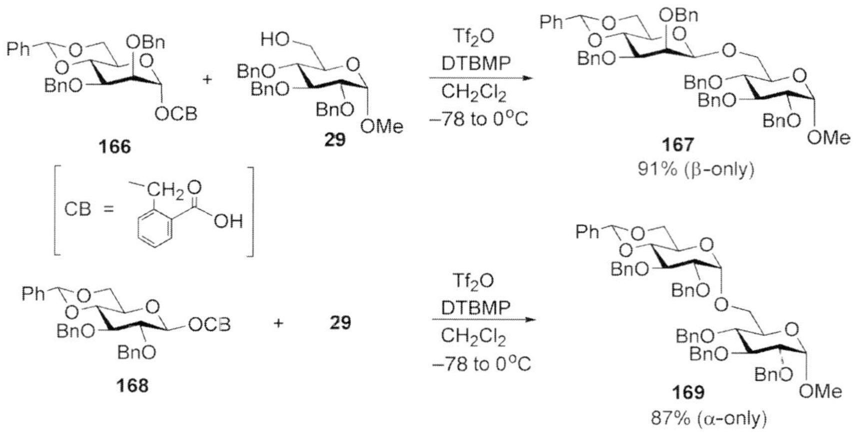

Scheme 3.113

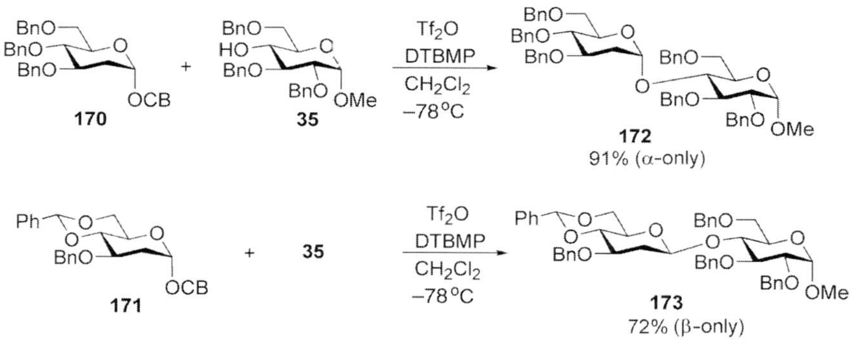

Scheme 3.114

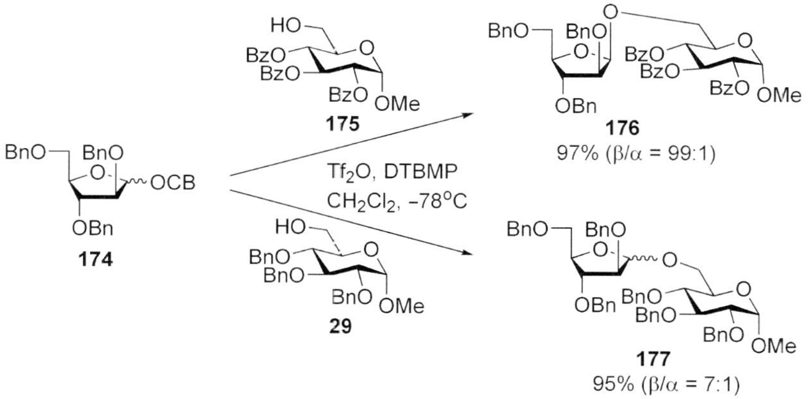

Scheme 3.115

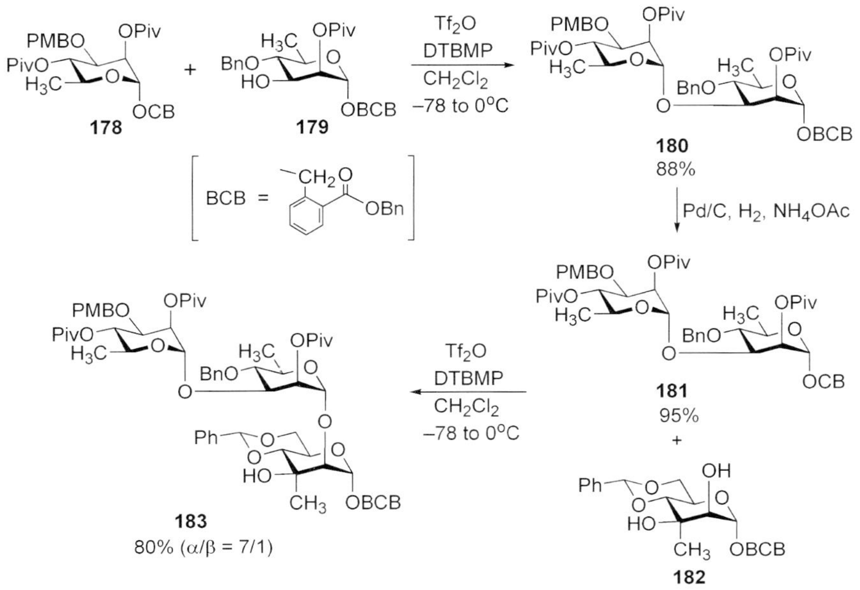

Scheme 3.116

the glycosylation of benzyl-protected glycosyl acceptor **29** afforded a mixture of α- and β-disaccharides **177**.

Furthermore, Kim and coworkers applied the CB glycoside methodology to the synthesis of oligosaccharide repeat units of *O*-antigen polysaccharides of the lipopolysaccharide from gram-negative bacteria. The protected form of the trisaccharide repeat unit of the atypical *O*-polysaccharide from Danish *Helicobacter pylori* strains, **183**, was synthesized from three monosaccharide building blocks **178**, **179** and **182** by the construction of two glycosyl linkages employing CB glycosides **178** and **181** as glycosyl donors (Scheme 3.116) [636]. One of the unusual structural features of the trisaccharide repeat unit **183** is the occurrence of the novel branched sugar, 3-*C*-methyl-D-mannopyranose, which has not been found in nature before. Kim also reported the synthesis of a protected form of the tetrasaccharide repeat unit of the *O*-antigen polysaccharide from the *Escherichia coli* lipopolysaccharide, **191b** (Scheme 3.117) [637]. Thus, β-mannosyl disaccharide **186** was obtained by the glycosylation of 2′-(allyloxylcarbonyl)benzyl (ACB) glycoside **185** with 4,6-*O*-benzylidene CB mannoside **184**. Deprotection of PMB of **186** followed by the glycosylation of the resulting alcohol with CB glycoside **187** provided ACB trisaccharide **188**, which in turn was converted into CB trisaccharide **189** by deallylation. The ACB glycosides were introduced instead of BCB glycosides as a new precursor for CB glycosides, especially useful in the presence of the azide functionality, which could not survive during the conversion of the BCB glycosides into CB glycosides by hydrogenolysis. An

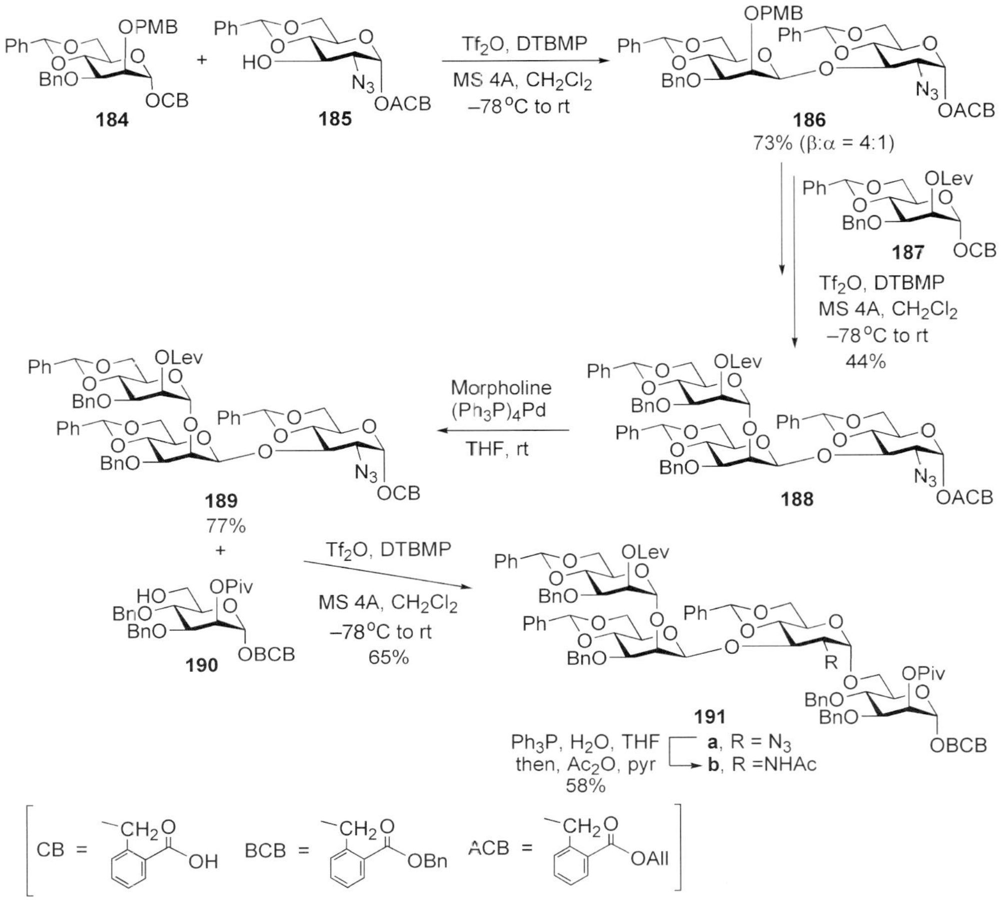

Scheme 3.117

improved synthesis of the trisaccharide repeat unit **191a** was recently reported [638]. Kim and coworkers also reported the synthesis of octaarabinofuranoside **195** in arabinogalactan and lipoarabinomannan, found in mycobacterial cell wall by employing the latent (BCB)–active (CB) glycosylation method (Scheme 3.118) [635]. Thus, completely stereoselective double β-arabinofuranosylation employing benzyl-protected CB arabinoside **174** and benzoyl-protected acceptor diol **192** was reported. Very recently, Kim and coworkers reported the total synthesis of agelagalastatin **201**, which is a trisaccharide sphingolipid and displays significant *in vitro* inhibitory activities against the human cancer cell growth. Among three glycosyl linkages in agelagalastatin, α-galactofuranosyl and β-galactofuranosyl linkages were stereoselectively constructed employing the CB glycoside method (Scheme 3.119) [639].

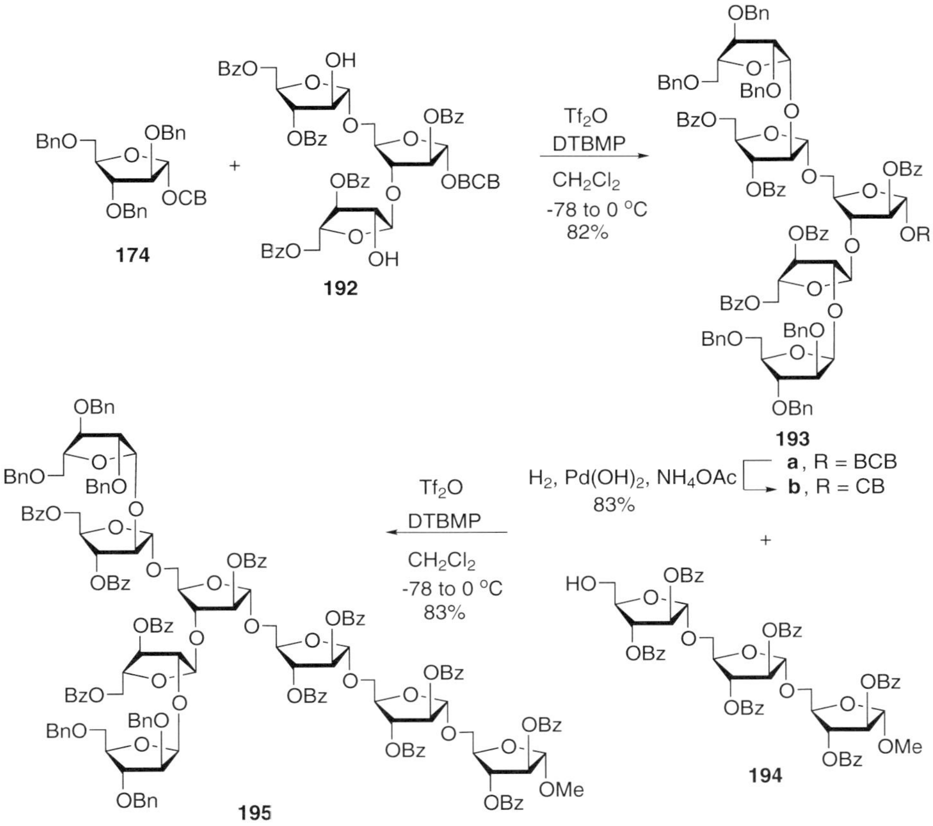

Scheme 3.118

3.3.9
Conclusions and Outlook

The advantages and limitations of anomeric transglycosylation methods employing various glycosyl donors such as alkyl glycosides, silyl glycosides, heteroaryl glycosides, 2-hydroxy-3,5-dinitrobenzoate (DISAL) glycosides, vinyl glycosides, NPGs and CB glycosides have been summarized. Examples for the synthesis of oligosaccharides using all these glycosylation methods have been presented. Among them, vinyl glycosides, n-pentenyl glycosides and 2'-carboxybenzyl glycosides appear to be especially valuable for complex oligosaccharide synthesis because they could be used as active forms in the active–latent glycosylation strategy. The corresponding latent forms, 3-buten-2-yl glycoside, 4,5-dibromopentanyl glycoside and 2'-(benzyloxycarbonyl)benzyl glycoside, are stable enough to survive during the manipulation of protecting groups and glycosylation reactions

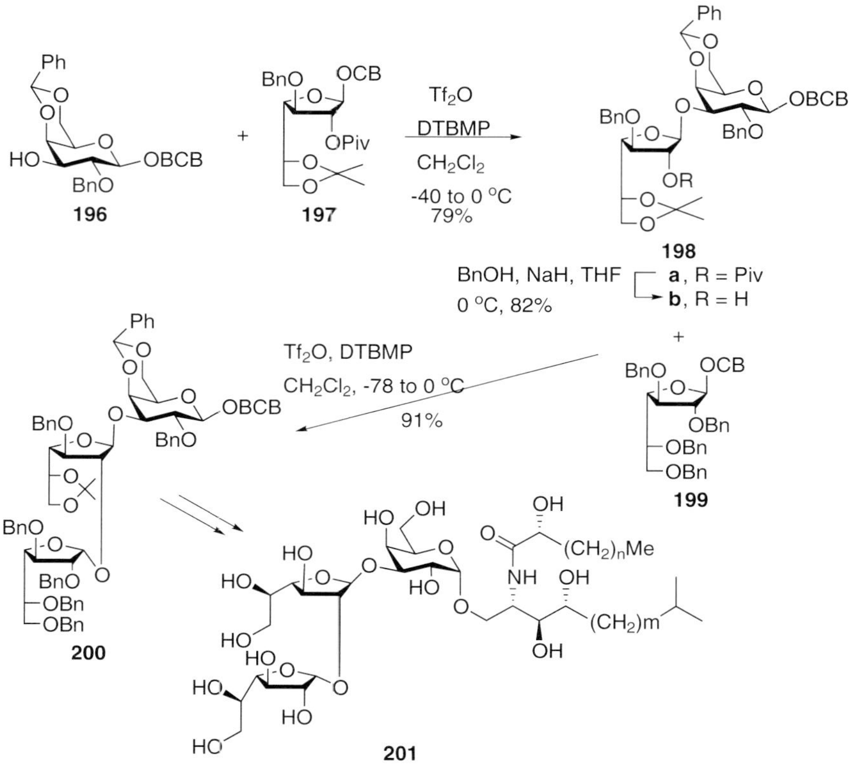

Scheme 3.119

for the oligosaccharide synthesis. In addition, the armed–disarmed protocol for chemoselective glycosylations and its application to the synthesis of oligosaccharides have been described. A novel concept in the glycosylation with NPGs and CB glycosides is that the reaction is mediated by the facile five-membered ring formation, and the departing aglycones, a cyclic ether and a lactone are inert toward oxocarbenium ion generated during glycosylations. Further search for new cyclization-mediated glycosylation methods would help people in this field by providing not only another powerful tool for oligosaccharide synthesis but also mechanistic insights into glycosylations.

3.3.10
Experimental Procedures

3.3.10.1 Glycosylation Employing Vinyl Glycosides
Vinyl glycoside (1.1 equiv) and glycosyl acceptor (1 equiv) were dissolved in the appropriate solvent (0.1 M) and stirred over 4 Å molecular sieves for 1.5 h. After

cooling the resulting solution to the appropriate temperature, a catalytic amount of TMSOTf was added. Upon completion, the reaction mixture was neutralized with TEA, diluted with CH$_2$Cl$_2$ and filtered through a pad of Celite, and the filtrate was washed with saturated aqueous NaHCO$_3$ and brine. The organic phase was dried over MgSO$_4$ and concentrated *in vacuo*. The residue was purified by silica gel flash column chromatography.

3.3.10.2 Glycosylation Employing *n*-Pentenyl Glycosides with NIS/TESOTf
The glycosyl acceptor (1 equiv) was dissolved in CH$_2$Cl$_2$ to give 0.2 M solution. NIS (1.3 equiv) and TESOTf (0.3 equiv) was added to the solution by stirring. The n-pentenyl glycosides (1.3 equiv) that were dissolved in CH$_2$Cl$_2$ to give 0.4 M solution was added dropwise to the glycosyl acceptor solution. Once the whole of the NIS had dissolved, the reaction mixture was quenched with 10% aqueous Na$_2$S$_2$O$_3$ and saturated aqueous NaHCO$_3$. The organic phase was separated, washed with brine, dried over MgSO$_4$ and concentrated *in vacuo*. The residue was purified by silica gel flash column chromatography.

3.3.10.3 Glycosylation Employing *n*-Pentenyl Glycosides with IDCP
A solution of *n*-pentenyl glycoside (1.25 mmol) in Et$_2$O/CH$_2$Cl$_2$ (38 ml, 4:1) was added to a glycosyl acceptor (1.03 mmol), 4A molecular sieves and IDCP (1.65 mmol). After being stirred at room temperature for 10 h, the reaction mixture was quenched with 10% aqueous Na$_2$S$_2$O$_3$ (10 ml) and filtered, and the residue was washed with CH$_2$Cl$_2$. The combined organic phase was washed with saturated aqueous NaHCO$_3$, dried over MgSO$_4$ and concentrated *in vacuo*. The residue was purified by silica gel flash column chromatography.

3.3.10.4 Preparation of *n*-Pentenyl Glycosides from Glycosyl Bromides
A solution of glycosyl bromide (3.79 mmol) in CH$_2$Cl$_2$ (4 ml) was slowly added to a mixture of 4-penten-1-ol, AgOTf (4.67 mmol) and 4A molecular sieves in CH$_2$Cl$_2$ (10 ml) at −30 °C for a period of 10 min. After being stirred at −30 °C for further 2 h, the reaction mixture was quenched with saturated aqueous NaHCO$_3$ and filtered, and the residue was washed with CH$_2$Cl$_2$. The combined organic phase was washed with saturated aqueous NaHCO$_3$, dried over MgSO$_4$ and concentrated *in vacuo*. The residue was purified by silica gel flash column chromatography.

3.3.10.5 Glycosylation Employing CB Glycosides with Tf$_2$O
A solution of the CB glycoside (1 equiv) and 2,6-di-*tert*-butyl-4-methylpyridine (DTBMP, 2 equiv) in CH$_2$Cl$_2$ (0.02 M) was stirred for 30 min at room temperature in the presence of 4 Å molecular sieves and cooled to −78 °C. After the addition of triflic anhydride (Tf$_2$O, 1 equiv), the resulting solution was stirred at −78 °C for 10 min, and then the glycosyl acceptor (1.5 equiv) was added slowly to this solution. The reaction mixture was allowed to warm up over 30 min to room temperature and stirred for further 1.5 h, filtered through Celite, and quenched with saturated aqueous NaHCO$_3$. Collected organic phase was washed with brine, dried over MgSO$_4$ and concentrated *in vacuo*. The residue was purified by silica gel flash column chromatography.

3.3.10.6 Preparation of BCB Glycosides from Glycosyl Bromides

In the presence of 4A molecular sieves in acetonitrile (25 ml) at 0 °C mercury(II) bromide (14.7 mmol, 1.2 equiv), mercury(II) cyanide (14.7 mmol, 1.2 equiv) and finally 2-(hydroxymethyl)benzoate (13.4 mmol, 1.1 equiv) were added to a stirred solution of glycosyl bromide (12.2 mmol). After stirring at 0 °C for further 20 min, the reaction mixture was filtered and the filtrate was concentrated. The resulting oil was dissolved in CH_2Cl_2 (50 ml) and the solution was washed with saturated aqueous $NaHCO_3$ (2×50 ml) and brine (50 ml). The organic phase was dried ($MgSO_4$) and concentrated *in vacuo* and the residue was purified by silica gel flash column chromatography.

3.3.10.7 Preparation of CB Glycosides from BCB Glycosides

BCB glycoside (2.97 mmol) was stirred under hydrogen atmosphere using a balloon in the presence of Pd/C (10%, 0.07 equiv) and ammonium acetate (10.39 mmol, 3.5 equiv) in MeOH (100 ml) at room temperature for 1 h. The reaction mixture was filtered through Celite and the filtrate was concentrated *in vacuo*. The residue was purified by silica gel flash column chromatography to give the pure CB glycosides.

Acknowledgment

Support from the Korea Science and Engineering Foundation through Center for Bioactive Molecular Hybrids is acknowledged.

References

568 (a) Nilsson, K.G.I. (1995) *Modern Methods in Carbohydrate Synthesis* (eds S.H. Khan and R.A. O'Neill), Harwood Academic, Amsterdam, pp. 518–547. (b) Palcic, M.M. (1999) *Current Opinion in Biotechnology*, **10**, 616–624. (c) Wymer, N. and Toone, E.J. (2000) *Current Opinion in Chemical Biology*, **4**, 110–119. (d) Jakeman, D.L. and Withers, S.G. (2002) *Trends in Glycoscience and Glycotechnology*, **14**, 13–25.

569 (a) Boons, G.-J. (2001) *Glycoscience* (eds B. Fraser-Reid, K. Tatsuta and J. Thiem), vol. 1, Springer-Verlag, pp. 551–581. (b) Fairbanks, A.J. and Seward, C.M.P. (2003) *Carbohydrates* (ed. H.M.I. Osborn), Elsevier Science, Oxford, pp. 147–194. (c) Fraser-Reid, B., Anilkumar, G., Gilbert, M.R., Joshi, S. and Kraehmer, R. (2000) *Carbohydrates in Chemistry and Biology* (eds B. Ernst, G.W. Hart and P. Sinay), vol. 1, Wiley-VCH Verlag GmbH, Weinheim, pp. 135–154. (d) Kim, K.S. and Jeon, H.B. (2007) Frontiers in Modern Carbohydrate Chemistry (ed. A.V. Demchenko), ACS Symposium Series 960, American Chemical Society, Washington, DC, pp. 134–149.

570 Uchiro, H., Kurusu, N. and Mukaiyama, T. (1997) *Israel Journal of Chemistry*, **37**, 87–96.

571 Yoshimura, Y., Shimizu, H., Hinou, H. and Nishimura, S.-I. (2005) *Tetrahedron Letters*, **46**, 4701–4705.

572 Tietze, L.-F., Fischer, R. and Guder, H.-J. (1982) *Tetrahedron Letters*, **23**, 4661–4664.

573 Nashed, E.M. and Glaudemans, C.P.J. (1989) *Journal of Organic Chemistry*, **54**, 6116–6118.

574 Qiu, D.-X., Wang, Y.-F. and Cai, M.-S. (1989) *Synthetic Communications*, **19**, 3453–3456.

575 Mukaiyama, T. and Matsubara, K. (1992) *Chemistry Letters*, **21**, 1041–1044.

576 Priebe, W., Grynkiewicz, G. and Neamati, N. (1991) *Tetrahedron Letters*, **32**, 2079–2082.

577 Kolar, C., Kneissl, G., Knödler, U. and Dehmel, K. (1991) *Carbohydrate Research*, **209**, 89–100.

578 Daley, L., Guminski, Y., Demerseman, P., Kruczynski, A., Etiévant, C., Imbert, T., Hill, B.T. and Monneret, C. (1998) *Journal of Medicinal Chemistry*, **41**, 4475–4485.

579 Shoda, S. and Mukaiyama, T. (1979) *Chemistry Letters*, **8**, 847–848.

580 Rathore, H., From, A.H.L., Ahmed, K. and Fullerton, D.S. (1986) *Journal of Medicinal Chemistry*, **29**, 1945–1952.

581 Yasukochi, T., Fukase, K. and Kusumoto, S. (1999) *Tetrahedron Letters*, **40**, 6591–6593.

582 Hanessian, S., Saavedra, O.M., Mascitti, V., Marteter, W., Oehrlein, R. and Mak, C.-P. (2001) *Tetrahedron*, **57**, 3267–3280.

583 Huchel, U., Schmidt, C. and Schmidt, R.R. (1998) *European Journal of Organic Chemistry*, 1353–1360.

584 Petersen, L. and Jensen, K.J. (2001) *Journal of Organic Chemistry*, **66**, 6268–6275.

585 Petersen, L. and Jensen, K.J. (2001) *Journal of the Chemical Society Perkin Transactions*, **1**, 2175–2182.

586 Petersen, L., Laursen, J.B., Larsen, K., Motawia, M.S. and Jensen, K.J. (2003) *Organic Letters*, **5**, 1309–1312.

587 Laursen, J.B., Petersen, L., Jensen, K.J. and Nielsen, J. (2003) *Organic and Biomolecular Chemistry*, **1**, 3147–3153.

588 Grathe, S., Thygesen, M.B., Larsen, K., Petersen, L. and Jensen, K.J. (2005) *Tetrahedron: Asymmetry*, **16**, 1439–1448.

589 Laursen, J.B., Petersen, L. and Jensen, K.J. (2001) *Organic Letters*, **3**, 687–690.

590 Vankar, Y.D., Vankar, P.S., Behrendt, M. and Schmidt, R.R. (1991) *Tetrahedron*, **47**, 9985–9992.

591 Osa, Y., Takeda, K., Sato, T., Kaji, E., Mizuno, Y. and Takayanagi, H. (1999) *Tetrahedron Letters*, **40**, 1531–1534.

592 Marra, A., Esnault, J., Veyrières, A. and Sinay, P. (1992) *Journal of the American Chemical Society*, **114**, 6354–6360.

593 Chenault, H.K., Castro, A., Chafin, L.F. and Yang, J. (1996) *Journal of Organic Chemistry*, **61**, 5024–5031.

594 Chenault, H.K. and Castro, A. (1994) *Tetrahedron Letters*, **35**, 9145–9148.

595 Boons, G.-J. and Isles, S. (1994) *Tetrahedron Letters*, **35**, 3593–3596.

596 Boons, G.-J., Burton, A. and Isles, S. (1996) *Chemical Communications*, 141–142.

597 Gibson, R.R., Dickinson, R.P. and Boons, G.-J. (1997) *Journal of the Chemical Society Perkin Transactions*, **1**, 3357–3360.

598 Boons, G.-J. and Isles, S. (1996) *Journal of Organic Chemistry*, **61**, 4262–4271.

599 (a) Boons, G.-J., Heskamp, B. and Hout, F. (1996) *Angewandte Chemie (International Edition in English)*, **35**, 2845–2847. (b) Johnson, M., Arles, C. and Boons, G.-J. (1998) *Tetrahedron Letters*, **39**, 9801–9804.

600 Bai, Y., Boons, G.-J., Burton, A., Johnson, M. and Haller, M. (2000) *Journal of Carbohydrate Chemistry*, **19**, 939–958.

601 Fraser-Reid, B., Konradsson, P., Mootoo, D.R. and Udodong, U. (1988) *Journal of the Chemical Society. Chemical communications*, 823–825.

602 (a) Mootoo, D.R., Date, V. and Fraser-Reid, B. (1987) *Journal of the Chemical Society. Chemical communications*, 1462–1464. (b) Mootoo, D.R., Date, V. and Fraser-Reid, B. (1988) *Journal of the American Chemical Society*, **110**, 2662–2663.

603 Konradsson, P., Roberts, C. and Fraser-Reid, B. (1991) *Recueil des Travaux Chimiques des Pays-Bas Journal of the Royal Netherlands Chemical Society*, **110**, 23–24.

604 Udodong, U., Rao, C.S. and Fraser-Reid, B. (1992) *Tetrahedron*, **48**, 4713–4724.

605 Fraser-Reid, B., Udodong, U., Wu, Z., Ottosson, H., Merritt, J.R., Rao, C.S., Roberts, C. and Madsen, R. (1992) *Synlett*, 927–942.

606 Legler, G. and Bause, E. (1973) *Carbohydrate Research*, **28**, 45–52.

607 Konradsson, P., Mootoo, D.R., McDevitt, R.E. and Fraser-Reid, B. (1990) *Journal of the Chemical Society, Chemical communications*, 270–272.

608 Konradsson, P., Udodong, U. and Fraser-Reid, B. (1990) *Tetrahedron Letters*, **31**, 4313–4316.

609 Mootoo, D.R., Konradsson, P., Udodong, U. and Fraser-Reid, B. (1988) *Journal of the American Chemical Society*, **110**, 5583–5584.

610 Jayaprakash, K.N. and Fraser-Reid, B. (2004) *Synlett*, 301–305.

611 Fraser-Reid, B., Wu, Z., Andrews, C.W. and Skowronski, E. (1991) *Journal of the American Chemical Society*, **113**, 1434–1435.

612 Konradsson, P. and Fraser-Reid, B. (1989) *Journal of the Chemical Society, Chemical communications*, 1124–1125.

613 Fraser-Reid, B., Wu, Z., Udodong, U.E. and Ottosson, H. (1990) *Journal of Organic Chemistry*, **55**, 6068–6070.

614 Mootoo, D.R. and Fraser-Reid, B. (1989) *Tetrahedron Letters*, **30**, 2363–2366.

615 Mootoo, D.R., Konradsson, P. and Fraser-Reid, B. (1989) *Journal of the American Chemical Society*, **111**, 8540–8542.

616 Jain, R.K., Locke, R.D. and Matta, K.L. (1991) *Carbohydrate Research*, **212**, C1–C3.

617 (a) Merritt, J.R. and Fraser-Reid, B. (1992) *Journal of the American Chemical Society*, **114**, 8334–8336. (b) Merritt, J.R., Naisang, E. and Fraser-Reid, B. (1994) *Journal of Organic Chemistry*, **59**, 4443–4449.

618 (a) Udodong, U.E., Madsen, R., Roberts, C. and Fraser-Reid, B. (1993) *Journal of the American Chemical Society*, **115**, 7886–7887. (b) Roberts, C., Madsen, R. and Fraser-Reid, B. (1995) *Journal of the American Chemical Society*, **117**, 1546–1553. (c) Madsen, R., Udodong, U.E., Roberts, C., Mootoo, D.R., Konradsson, P. and Fraser-Reid, B. (1995) *Journal of the American Chemical Society*, **117**, 1554–1565.

619 Arasappan, A. and Fraser-Reid, B. (1996) *Journal of Organic Chemistry*, **61**, 2401–2406.

620 Houdier, S. and Vottero, P.J.A. (1992) *Carbohydrate Research*, **232**, 349–352.

621 Lichtenthaler, F.W., Kläres, U., Lergenmüller, M. and Schwidetzky, S. (1992) *Synthesis*, 179–184.

622 López, J.C., Gómez, A.M., Valverde, S. and Fraser-Reis, B. (1995) *Journal of Organic Chemistry*, **60**, 3851–3858.

623 Arasappan, A. and Fraser-Reid, B. (1995) *Tetrahedron Letters*, **36**, 7967–7970.

624 Demchenko, A.V. and De Meo, C. (2002) *Tetrahedron Letters*, **43**, 8819–8822.

625 Clausen, M.H. and Madsen, R. (2003) *Chemistry – A European Journal*, **9**, 3821–3832.

626 Heidelberg, T. and Martin, O.R. (2004) *Journal of Organic Chemistry*, **69**, 2290–2301.

627 Svarovsky, S.A. and Barchi, J.J., Jr (2003) *Carbohydrate Research*, **338**, 1925–1935.

628 (a) Anilkumar, G., Nair, L.G., Olsson, L., Daniels, J.K. and Fraser-Reid, B. (2000) *Tetrahedron Letters*, **41**, 7605–7608. (b) Olsson, L., Jia, J.J. and Fraser-Reid, B. (1998) *Journal of Organic Chemistry*, **63**, 3790–3792.

629 Ikeda, K., Fukuyo, J., Sato, K. and Sato, M. (2005) *Chemical and Pharmaceutical Bulletin*, **53**, 1490–1493.

630 (a) Anilkumar, G., Nair, L.G. and Fraser-Reid, B. (2000) *Organic Letters*, **2**, 2587–2589. (b) Fraser-Reid, B., Lopez, J.C., Radhakrishnan, K.V., Mach, M., Schlueter, U., Gomez, A.M. and Uriel, C. (2002) *Journal of the American Chemical Society*, **124**, 3198–3199.

631 Fraser-Reid, B., Anilkumar, G., Nair, L.G., Radhakrishnan, K.V., Lopez, J.C., Gomez, A.M. and Uriel, C. (2002) *Australian Journal of Chemistry*, **55**, 123–130.

632 (a) Mach, M., Schlueter, U., Mathew, F., Fraser-Reid, B. and Hazen, K.C. (2002) *Tetrahedron*, **58**, 7345–7354. (b) Fraser-Reid, B. Crimme, S. Piacenza, M. Mach, M. and Schlueter, U. (2003) *Chemistry – A European Journal*, **9**, 4687–4692.

633 Kim, K.S., Kim, J.H., Lee, Y.J., Lee, Y.J. and Park, J. (2001) *Journal of the American Chemical Society*, **123**, 8477–8481.

634 Kim, K.S., Park, J., Lee, Y.J. and Seo, Y.S. (2003) *Angewandte Chemie (International Edition)*, **42**, 459–462.
635 Lee, Y.J., Lee, K., Jung, E.H., Jeon, H.B. and Kim, K.S. (2005) *Organic Letters*, **7**, 3263–3266.
636 Kwon, Y.T., Lee, Y.J., Lee, K. and Kim, K.S. (2004) *Organic Letters*, **6**, 3901–3904.
637 Lee, B.R., Jeon, J.M., Jung, J.H., Jeon, H.B. and Kim, K.S. (2006) *Canadian Journal of Chemistry*, **84**, 506–515.
638 Lee, B.-Y., Baek, J.Y., Jeon, H.B. and Kim, K.S. (2007) *Bulletin of the Korean Chemical Society*, **28**, 257–262.
639 Lee, Y.J., Lee, B.Y., Jeon, H.B. and Kim, K.S. (2006) *Organic Letters*, **8**, 3971–3974.

3.4
Phosphates, Phosphites and Other O–P Derivatives

Seiichi Nakamura, Hisanori Nambu, Shunichi Hashimoto

3.4.1
Introduction

Owning to the growing biological significance of saccharide residues of carbohydrate-containing biomolecules, the rational design and development of stereocontrolled glycosidation reactions are of paramount importance not only in carbohydrate chemistry but also in medicinal chemistry. Despite recent advances in this field [640], a universal method for glycosidic linkage formation has still not been established because a number of parameters such as anomeric leaving groups, promoters, solvents and protecting groups, as well as the nature of the glycosyl donor and the nucleophilicity of the acceptor alcohols, affect the selectivity and yield of glycosidation reactions. Given that the anomeric leaving group is one of the most fundamental parameters, it is not surprising that a great deal of effort continues to be devoted to the design of the leaving groups of glycosyl donors allowing excellent shelf lives, as well as their activation without resorting to the use of precious, explosive, or toxic heavy-metal salts as promoters.

Glycosyltransferases in the Leloir pathway play a central role in the biosynthesis of glycosidic bonds in mammalian systems. Seven monosaccharides as glycosyl donor substrates are activated by nucleoside diphosphates (NDPs), which are the anomeric leaving groups [641]. These sugars are converted to the corresponding 1-phosphates in several multistep sequences. Specific nucleotide pyrophosphorylases catalyze condensation reactions between sugar 1-phosphates and nucleoside triphosphates (NTPs) to generate NDP sugars, releasing inorganic pyrophosphate. Glycosyltransferases then catalyze the transfer of the NDP sugar to a specific hydroxyl group on the acceptor sugar in a stereospecific manner. In this context, sialyltransferases are unique in that the glycosyl donor is activated by cytidine monophosphate [642]. Because glycosyl phosphates act as structural units in some biologically significant molecules such as glycophospholipids, their synthetic construction was investigated. However, no attention was paid to their synthetic utility as glycosyl donors in chemical glycosidation reactions until Hashimoto *et al.* reported the effective use of diphenyl phosphate as an anomeric leaving group [643]. Three years later, the effectiveness of

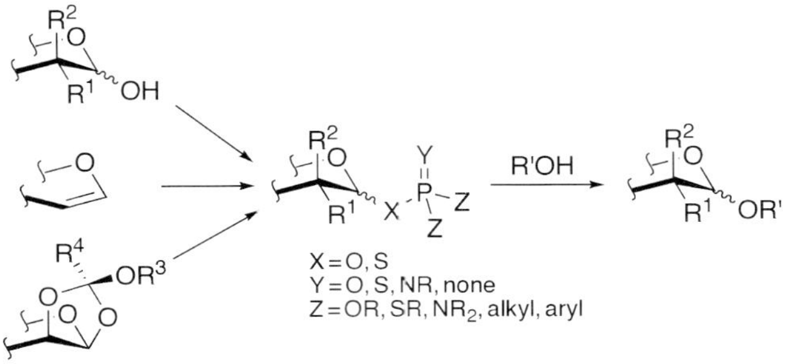

Scheme 3.120 The use of phosphorus-containing leaving groups in glycosidation reactions.

glycosyl phosphites was demonstrated independently by the Schmidt [644] and Wong [645] groups. Because tailor-made glycosyl donors are readily available by varying the substituents (X=O, S; Y=O, S, NR, none; Z=OR, SR, NR_2, alkyl, aryl) on the phosphorus atom (Scheme 3.120) [646], a variety of donors with different phosphorus-containing leaving groups have been developed to date.

This chapter outlines the major progress achieved in glycosidation reactions by capitalizing on phosphorus-containing leaving groups, especially phosphates and phosphites. A number of reviews on this subject have been published previously [647,648]. This chapter is divided into six major sections. The first three sections deal with the synthesis and reactions of glycosyl phosphates (Section 3.4.2), phosphites (Section 3.4.3) and donors with other phosphorus-containing leaving groups (Section 3.4.4). The synthesis of β-mannosides, 2-acetamido-2-deoxy-β-glycosides, 2-deoxy-β-glycosides and α-sialosides are discussed in Section 3.4.5. The next section (Section 3.4.6) describes chemoselective glycosidation strategies based on the difference in anomeric reactivities of the phosphorus-containing leaving groups. The final section (Section 3.4.7) discusses the application of glycosidation methods in natural product synthesis. However, given the limited space, we recommend that more specialized reviews should be consulted for discussion of the solid-phase synthesis [647d,649].

3.4.2
Glycosyl Phosphates

3.4.2.1 Preparation of Glycosyl Phosphates

Diphenyl, dialkyl and propane-1,3-diyl phosphates are introduced at the anomeric carbon and utilized as leaving groups for the subsequent glycosidation reactions. Although a variety of methods have been developed, all approaches broadly fall into two categories. The starting material is either a 1-O-unprotected sugar or another type of glycosyl donor that can also be directly employed in the glycosidation event.

Phosphorylation of 1-O-Unprotected Sugars The reaction of 1-O-unprotected sugars with phosphorochloridate proceeds under basic conditions, yielding the correspond-

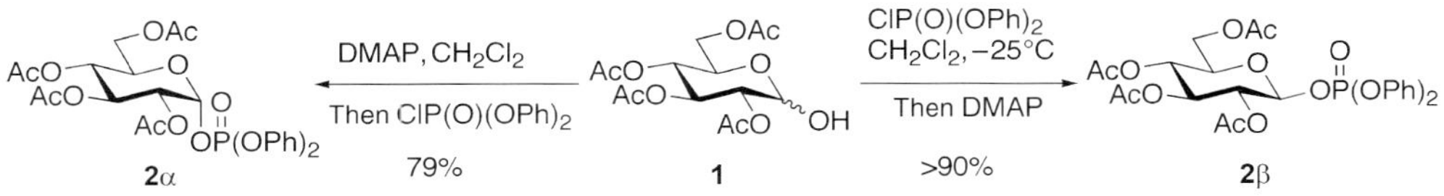

Scheme 3.121 Conversion of hemiacetals into glycosyl phosphates.

ing glycosyl phosphates. 4-(Dimethylamino)pyridine (DMAP) in CH$_2$Cl$_2$ [650] is most frequently used as a base, and N-methylimidazole is an alternative base [651]. Sabesan and Neira reported that the treatment of acylated gluco- and galactopyranoses with DMAP in CH$_2$Cl$_2$ to obtain α-enriched anomer, followed by the addition of diphenyl chlorophosphate, gave predominantly α-glycosyl phosphates (thermodynamic control), whereas phosphorylation of the more reactive β-anomeric hydroxyl group shifted the equilibrium toward β-anomer at low temperature and in the absence of DMAP, eventually leading to the preferential formation of β-glycosyl phosphates (kinetic control) (Scheme 3.121) [650]. However, the stereochemical outcome generally depends on the nature of sugars with various protecting groups, and often using the phosphorylation reaction, it is difficult to obtain the less thermodynamically stable β-phosphates with high selectivity.

Glycosyl phosphates can also be prepared by the lithiation of lactols with BuLi in THF at −78 °C, followed by the addition of phosphorochloridate [652]. Although the use of BuLi does not cause appreciable cleavage or transposition of ester-type protecting groups owning to the short reaction times, rigorous exclusion of moisture is necessary to achieve high yields. O-Thallium(I) salts were shown to promote rapid, high-yielding substitution at O-1 in MeCN or benzene, wherein a preponderance of either α- or β-phosphates was obtained by the choice of solvent [653]. However, the operational inconvenience as well as the toxicity of TlOEt precludes the wide application of this method [654].

An intriguing approach to glycosyl phosphates was introduced by Gin: direct dehydrative coupling of selectively protected pyranose derivatives 3 with dialkyl phosphates. The reaction is mediated by dibenzothiophene-5-oxide (DBTO) and triflic anhydride (Tf$_2$O) in the presence of the acid scavenger 2.4.6-tri-*tert*-butylpyridine (TTBP) (Scheme 3.122) [655]. When hemiacetal donors incorporated O-2 acyl protecting groups, the exclusive formation of 1,2-*trans*-linked glycosyl phosphates was observed as a result of neighboring-group participation. For glycosyl donors

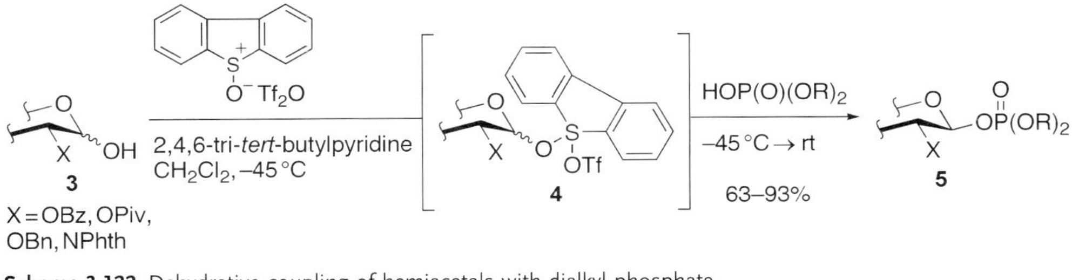

Scheme 3.122 Dehydrative coupling of hemiacetals with dialkyl phosphate.

devoid of O-2 acyl groups, the anomeric product ratio varied with the nature of the coupling partners.

Conversion of Other Glycosyl Donors into Glycosyl Phosphates Although the conversion of one glycosyl donor species into another is not ideal, glycosyl phosphates have been shown to successfully glycosylate hindered or unreactive hydroxyl groups. Thus, the two-step synthesis of glycosyl phosphates is valuable in constructing large oligosaccharides [647d]. Since the seminal work of Zervas on the S_N2-like displacement of tetra-*O*-acetyl-α-glucosyl bromide with silver dibenzyl phosphate [656], a number of glycosyl donors including glycosyl acetates [657], trichloroacetimidates [658] and nitrates [659], thioglycosides [660], 4-pentenyl [661] and 2-buten-2-yl glycosides [662], glycals [663,664], 1,2-orthoesters [665,666] and oxazolines [667] have been used in coupling reactions with phosphonic acid diesters (Scheme 3.123). In general, 1,2-*trans*-linked glycosyl phosphates are exclusively formed when 2-*O*-acyl-protected glycosyl donors, glycals, 1,2-orthoesters or oxazolines are employed for the reaction. In contrast, the protection of the oxygen atom at C-2 as benzyl ether leads to an anomeric mixture of products. The notable exceptions are 2-*O*-benzylated glycosyl trichloroacetimidates. The reaction of these compounds with dialkyl phosphates proceeds, without an addition of any acidic catalyst, via an eight-membered cyclic transition state **9** to provide glycosyl phosphates with inversion of anomeric configuration [658c,d].

It is well documented that the thermodynamically less stable β-phosphates anomerize to the more stable α-phosphates in the presence of acids [655,658b-d]. Thus, the reaction duration under acidic conditions often causes anomerization, resulting in the stereoselective formation of α-phosphates [661,664b,667c,668]. The effectiveness of glycosyl phosphates as glycosylating agents was enhanced by Seeberger and coworkers, who demonstrated the viability of the following one-pot processes. Glycals are attractive starting materials because they possess only three hydroxyl groups that require differentiation by introduction of protecting groups [669]. The conversion of glycals into 1,2-anhydro sugars by epoxydation with dimethyldioxirane

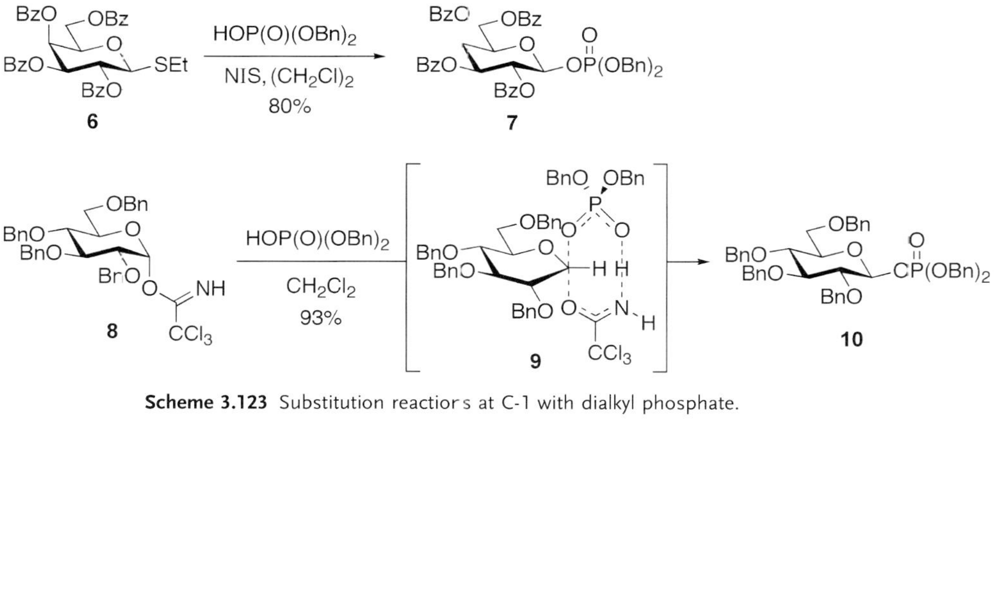

Scheme 3.123 Substitution reactions at C-1 with dialkyl phosphate.

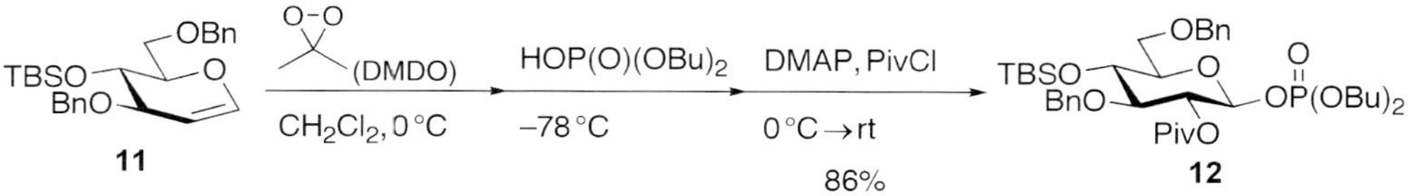

Scheme 3.124 Preparation of glycosyl phosphates from glycals.

(DMDO) was followed by epoxide opening with dibutyl or dibenzyl phosphates at −78 °C and the *in situ* acylation of the newly generated C-2 hydroxyl group to afford fully protected glycosyl phosphates in good yields (Scheme 3.124) [664]. Interestingly, when epoxide opening was carried out in CH$_2$Cl$_2$, β-phosphates were preferentially obtained. Conversely, the use of THF resulted almost exclusively in α-phosphates. Silylation of the C-2 hydroxyl group to install a nonparticipating triethylsilyl (TES) group was also feasible, whereas benzylation was unsuccessful, resulting in phosphate migration to yield 2-*O*-phosphoryl benzyl glycosides. This procedure is efficient, but the use of DMDO, required to be freshly prepared, can be dangerous to handle and make reaction scale-up difficult. Similar to glycals, 1,2-orthoesters possess only three hydroxyl groups requiring differentiation. The treatment of 1,2-orthobenzoates with dibutyl phosphate in CH$_2$Cl$_2$ in the presence of 4-Å molecular sieves (MS) at room temperature produced the corresponding glycosyl phosphates in excellent yields. Also, a one-pot coupling of the *in situ*-generated glycosyl phosphates with acceptor alcohols was achieved by simple addition of trimethylsilyl triflate (TMSOTf) at −30 °C (Scheme 3.125) [666] (see Section 3.4.2.2). It is noteworthy that the formation of methyl glycoside was not observed, probably because the molecular sieves act as a methanol scavenger.

In addition to these methods, oxidation of the corresponding phosphites, the preparation of which will be discussed below, by H$_2$O$_2$ or *tert*-butyl hydroperoxide provides an effective means of preparing glycosyl phosphates (Scheme 3.126) [645,670].

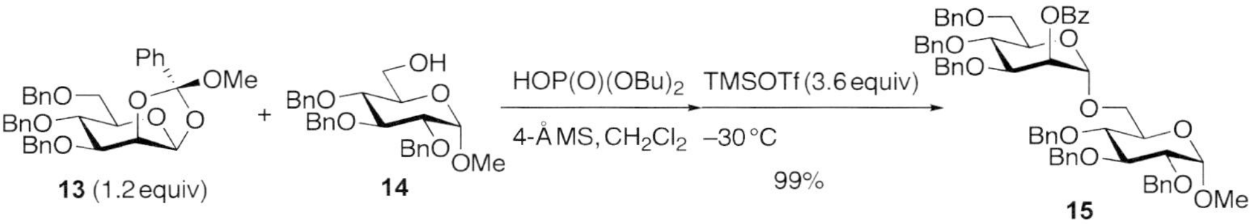

Scheme 3.125 Glycosidation using glycosyl phosphates *in situ* generated from 1,2-orthoesters.

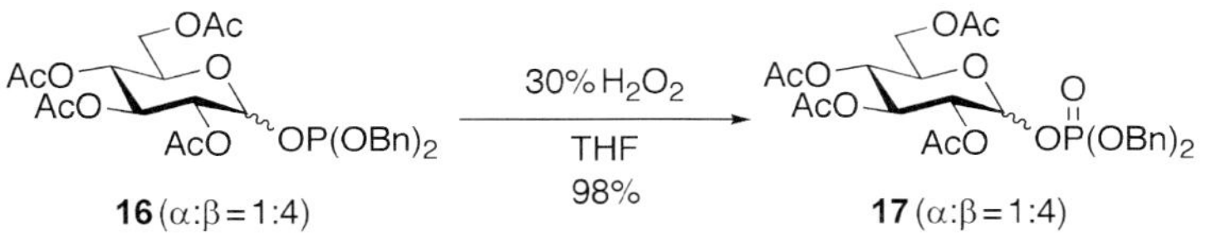

Scheme 3.126 Preparation of glycosyl phosphates from the corresponding phosphites.

3.4.2.2 Glycosidation Using Glycosyl Phosphates

Construction of 1,2-*trans*-β-Glycosidic Linkages with a Participating Group at C-2 The most reliable strategy for achieving stereoselective construction of 1,2-*trans*-glycosidic linkages involves the use of anchimeric assistance by the substituents at C-2. As in the case with other leaving groups, the exclusive formation of 1,2-*trans*-β-glycosides is readily achieved by the reaction of 2-*O*-benzoyl- or 2-*O*-pivaloyl-protected glycosyl phosphates with a range of acceptor alcohols in CH_2Cl_2 in the presence of stoichiometric amounts of TMSOTf (Scheme 3.127) [643,664,671]. In general, an electron-withdrawing ester functionality at C-2 deactivates the anomeric leaving group. Somewhat surprisingly from this point of view, the more reactive β-glycosyl dibutyl phosphates in the glucose and galactose series were rapidly activated at $-78\,°C$, whereas the α-anomeric counterparts required higher temperatures ($-40 \rightarrow -20\,°C$) for activation [664]. Protic acids, such as triflic acid (TfOH) and *para*-toluenesulfonic acid (TsOH), are ineffective for phosphate activation. However, the glycosidation reaction was rendered catalytic by the combined use of 1 mol% of TfOH as a promoter and TMS ether as an acceptor. Thus, TMSOTf was generated *in situ*, along with desilylated alcohol, and it promoted the formation of 1,2-*trans*-glycosides from glycosyl dibutyl phosphate [664b]. The use of 1,1,3,3-tetramethylurea as an acid scavenger was described in the initial report from our laboratory [643] but is not necessary in most cases. As mentioned above, *in situ* generation of glycosyl dibutyl phosphates from 1,2-orthoesters allowed for glycosidation without the intervening step of isolating the donor (Scheme 3.125) [666].

Waldmann and Böhm reported that per-*O*-pivaloylated β-glucosyl dibenzyl phosphate was activated under neutral conditions in 1 M $LiClO_4$ in CH_2Cl_2 or in $CHCl_3$ to react with acceptor alcohols, giving exclusively β-glucosides in moderate yields [672]. It should be noted that neither the β-glucosides nor the orthoesters could be isolated from the reaction of the corresponding trichloroacetimidate with acceptor alcohols under the same conditions.

Construction of 1,2-*trans*-β-Glycosidic Linkages Without a Participating Group at C-2 Without a participating group at C-2, the construction of 1,2-*trans*-β-glycosidic linkages is more difficult to control. However, the benefit of using nitrile as the solvent, first observed by Noyori and coworkers [673], has been well established. Indeed, excellent β-selectivities could be obtained by TMSOTf-promoted glycosidations of benzyl-protected glycosyl diphenyl phosphates **21** of the glucose, galactose

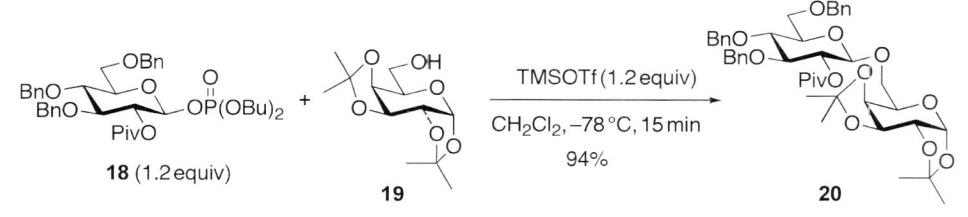

Scheme 3.127 Glycosidation with 2-*O*-acylated glycosyl phosphates.

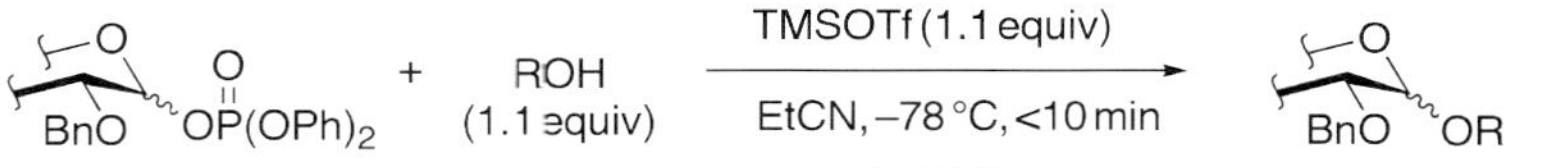

Scheme 3.128 β-Selective glycosidation of glycosyl phosphates without a participating group at C-2.

and glucuronate series with a wide range of acceptor alcohols in EtCN at −78 °C, regardless of the anomeric configuration of the donors (Scheme 3.128) [643]. The so-called nitrile effect plays a pivotal role in this coupling as β-selectivities in the TMSOTf-promoted glycosidation of benzyl-protected glucosyl propane-1,3-diyl phosphate in CH_2Cl_2 were highly dependent on the reactivity of the acceptor alcohols [651]. The examination of the temperature profile of the reaction in EtCN demonstrated that β-selectivity was often increased at lower temperatures. This observation suggests that excellent β-selectivities can be achieved by selecting a highly reactive donor, which drives the reaction at lower temperatures. Donor reactivity is influenced not only by structural effects of monosaccharide cores and protecting groups [674] but also by the substituents on the phosphorus atom. Although the benzyl-protected glycosyl diphenyl phosphates could be activated in EtCN at −78 °C, the temperature limit for an efficient reaction of the corresponding diethyl and propane-1,3-diyl phosphates was −60 °C [675,676]. These results clearly reveal that the more reactive diphenyl phosphate donor, albeit labile in some cases, has a superior leaving group compared to dialkyl phosphates for the construction of 1,2-*trans*-β-glycosidic linkages without the neighboring-group participation. $BF_3 \cdot OEt_2$[658d,677] and $LiClO_4$[672a,678] were also employed in the glycosidation of benzyl-protected glycosyl phosphates in CH_2Cl_2, $CHCl_3$ or Et_2O at room temperature but proved unsatisfactory in terms of reactivity and stereoselectivity.

Seeberger and coworkers found that TMSOTf-promoted glycosidation of 2-*O*-TES-protected β-glucosyl dibutyl phosphate with 6-*O*-unprotected galactoside in CH_2Cl_2 at −78 °C exhibited complete β-selectivity, although the TES group was lost during the reaction [664a,b]. Inspired by this finding, completely β-selective glycosidation was achieved by coupling 2-*O*-unprotected β-glycosyl dibutyl phosphates with a primary alcohol under the same conditions [664b].

Bogusiak and Szeja reported that the S_N2-like displacement of per-*O*-benzylated α-glucosyl diphenyl phosphate with phenols at C-1 proceeded in refluxing benzene in the presence of Bu_4NBr, KOH and K_2CO_3 to afford β-glucosides in good yields [679]. However, the scope of this reaction was limited to the use of phenols as the nucleophile.

Construction of 1,2-*cis*-α-Glycosidic Linkages Despite stereoelectronic preference, the construction of 1,2-*cis*-α-glycosidic linkages with a high degree of stereoselectivity has been an enduring problem in carbohydrate chemistry [648c]. Both the *in situ* anomerization concept introduced by Lemieux *et al.* [98] and the well-known α-directional effect of ethers as solvents [680] have frequently been used to synthesize

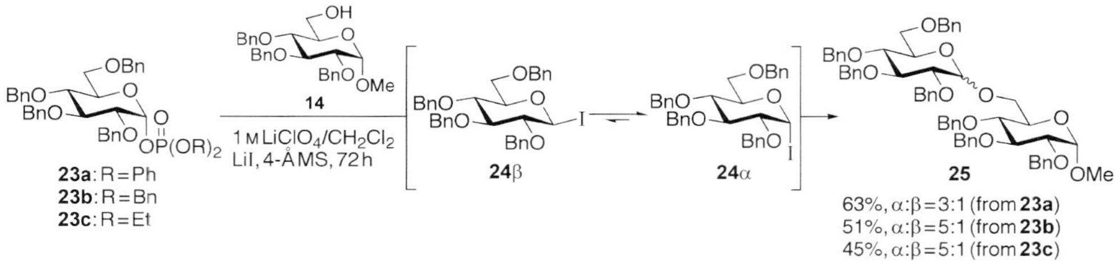

Scheme 3.129 LiClO$_4$-promoted glycosidation of glycosyl phosphates in the presence of LiI.

1,2-*cis*-α-glycosides. To date, there is no generally applicable direct method for the construction of 1,2-*cis*-α-glycosides from glycosyl phosphates because, even without a participating group at C-2, 1,2-*trans*-β-glycosidic linkages are readily formed with these donors, as mentioned above.

Waldmann and Schmid demonstrated that LiClO$_4$-mediated glycosidation of per-*O*-benzylated α-glucosyl phosphates **23** with glycoside alcohols in CH$_2$Cl$_2$ in the presence of LiI afforded di- and trisaccharides in moderate yields with good to excellent α-selectivities (Scheme 3.129) [681]. They assumed that the reaction proceeds by the initial attack of the iodide on the α-phosphate **23** to give the very reactive β-glucosyl iodide **24β**, which is then activated in the LiClO$_4$ solution and attacked by the glycoside alcohols, predominantly from the axial direction (and possibly via the *in situ* anomerization-type process). The formation of the intermediate **24β** as an evidence for the proposed mechanism could be detected only for a few minutes using NMR [681b]. It is interesting to note that evidently more reactive diphenyl phosphate **23a** gave the disaccharide **25** in higher yield, albeit with somewhat lower α-selectivity, than the corresponding dibenzyl phosphate **23b** and diethyl phosphate **23c**.

The remarkable α-directional effect of Et$_2$O was also found in the *tert*-butyldimethylsilyl triflate (TBSOTf)-promoted glycosidation of β-galactosyl dibutyl phosphate **26** (Scheme 3.130) [682]. Excellent α-selectivity (α : β = 20 : 1) was observed at low temperature using the less nucleophilic 4′-*O*-unprotected lactoside **27** in CH$_2$Cl$_2$/Et$_2$O (1 : 4). With regard to the construction of 1,2-*cis*-α-fucosidic linkage, found in many biologically important oligosaccharides, fucosyl dibutyl phosphates bearing an ester functionality at C-4 allowed for completely α-selective fucosylation, with or without the aid of Et$_2$O as a cosolvent [664b,683].

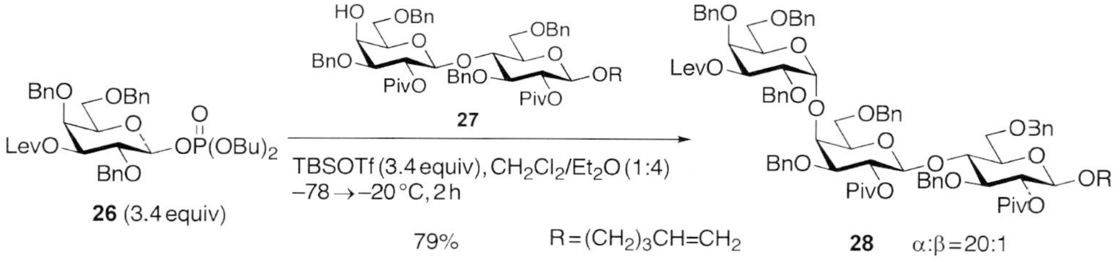

Scheme 3.130 α-Directional effect of Et$_2$O in the glycosidation of glycosyl phosphates.

3.4.2.3 Mechanism of Glycosidation Reaction with Glycosyl Phosphates

Although no detailed mechanistic evidence for the activation of glycosyl phosphates has yet been reported, the effectiveness of silyl triflates, but not protic acids, as promoters strongly suggests that silylation on the phosphoryl oxygen atom triggers the cleavage of the phosphate group (Scheme 3.131). To explain excellent β-selectivities observed with glycosidation of benzyl-protected glycosyl diphenyl phosphates in EtCN, the kinetic formation of an intermediate α-nitrilium ion **32α**, proposed independently by Ratcliffe and Fraser-Reid [684] and Schmidt et al. [685], has been widely invoked. On the basis of their hypothesis, the produced oxocarbenium ion **31** is rapidly trapped by EtCN to form an anomeric mixture of nitrilium ions **32** associated with triflate counterion. In this step, the formation of α-nitrilium ion **32α** is preferred over that of **32β** because of the stereoelectronically favored axial attack of EtCN from the α-face. In addition, **32α** benefits from anomeric stabilization [686]. On kinetic and thermodynamic grounds, the equilibrium between the α- and β-nitrilium ions should heavily favor **32α**. The S_N2-like anomeric displacement of α-nitrilium ion **32α** by acceptor alcohols affords β-glycosides **33β**. The reaction course via a common oxocarbenium ion **31** is consistent with the fact that the stereoselectivities are irrespective of the anomeric configuration of the phosphates **29** employed. In fact, glycosidation of 2-azido-2-deoxy-β-glucosyl diphenyl phosphate competed with anomerization to the α-phosphate via the attack of the departing diphenyl phosphate group on **31** under the above-described reaction conditions [687]. Of particular importance in this system is the isolation of only α-N-imidate by-products like **34**, which could arise from the capture of **32α** by alcohols at the nitrilium carbon. This finding provides substantial evidence for an α-nitrilium ion intermediate in glycosidation reactions.

When CH_2Cl_2 is used as the solvent, glycosidation is assumed to proceed through the intermediacy of the thermodynamically more stable α-glycosyl triflate (or its contact ion pair) [664a,688] or the α-glycosyl phosphate–TMSOTf complex, followed by the backside attack of acceptor alcohols on the intermediate (Scheme 3.132). High to

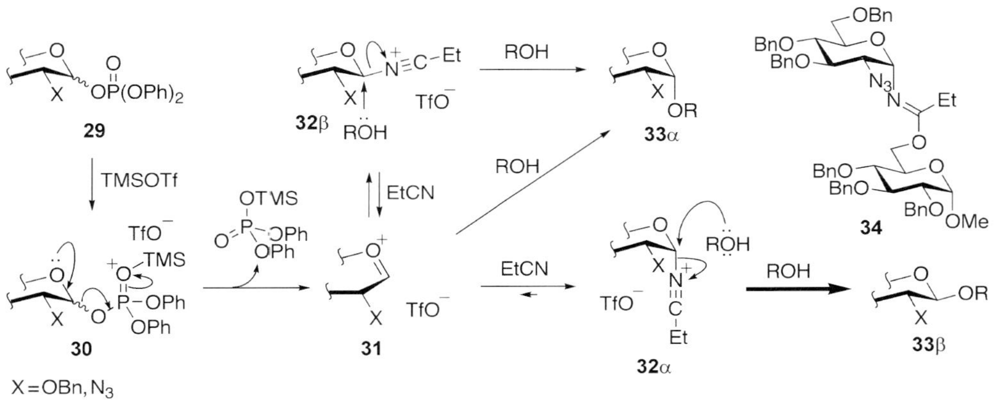

Scheme 3.131 A plausible mechanism for TMSOTf-promoted glycosidation of glycosyl phosphates in EtCN.

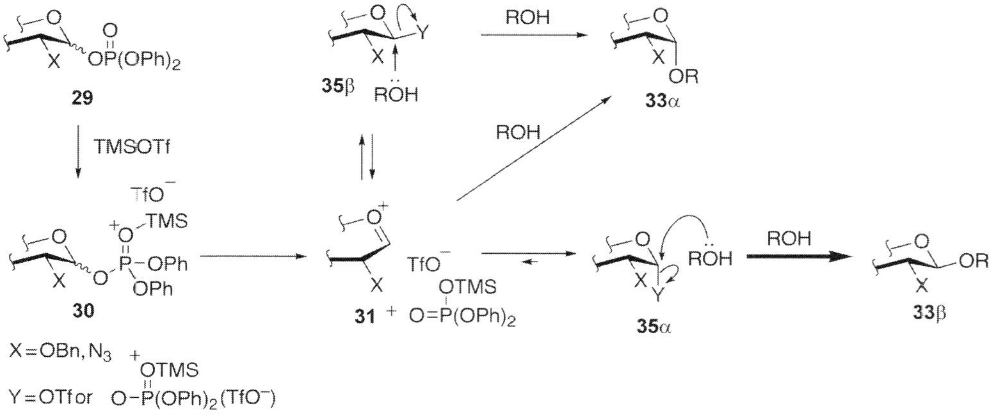

Scheme 3.132 A plausible mechanism for TMSOTf-promoted glycosidation of glycosyl phosphates in CH$_2$Cl$_2$.

excellent levels of β-selectivity observed with glycosylations of primary and reactive secondary alcohols can be explained by this mechanism. However, in the case of hindered secondary alcohols, the stereoelectronically favored axial attack of these alcohols from the α-face of the oxocarbenium ion intermediate **31** (known as the kinetic anomeric effect) competes with the S$_N$2-like mechanism to reduce β-selectivity [676].

3.4.3
Glycosyl Phosphites

3.4.3.1 Preparation of Glycosyl Phosphites
To date, dimethoxy-, diethoxy-, dibenzyloxy-, bis(2,2,2-trichloroethoxy)-, pinacolyl- and o-xylylenedioxyphosphino groups have been incorporated into the anomeric oxygen. Two methods are available for the efficient preparation of glycosyl phosphites, both of which include the phosphitylation of 1-O-unprotected sugars. For phosphitylation with phosphorochloridites, Et$_3$N or i-Pr$_2$NEt is employed as amine base (Scheme 3.133) [644]. On the contrary, the reaction with phosphoramidites proceeds in the presence of 1,2,4-triazole or 1H-tetrazole to give glycosyl phosphites [645]. In this reaction, Watanabe *et al.* posed the possibility of N-(glycosyl)tetrazole formation from the phosphite when using excess 1H-tetrazole [689]. The anomeric product ratio depends on many factors: the polarity of the solvent [645b], the nature of the starting sugars with various protecting groups [645b] and their anomeric

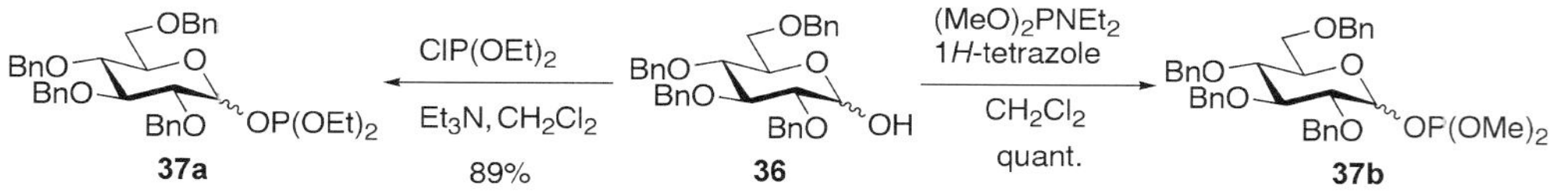

Scheme 3.133 Conversion of hemiacetals into glycosyl phosphites.

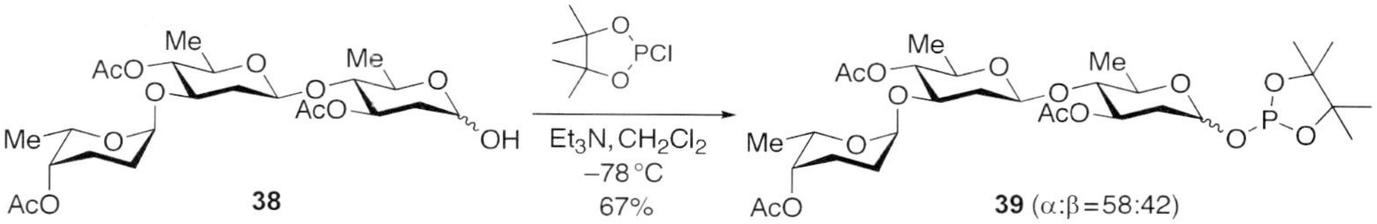

Scheme 3.134 Preparation of glycosyl pinacol phosphite.

composition [675,690]. Phosphitylation of N-acetyl-, N-phthaloyl- or N-2,2,2-trichloroethoxycarbonyl (Troc)-protected 2-amino-2-deoxyglycopyranoses led to the almost exclusive formation of α-phosphites [645b,691], whereas only β-anomers of sialyl diethyl and dibenzyl phosphites were obtained [644b,645b]. In general, it is difficult to isolate either anomer of glycosyl phosphites in pure form and in high yield. Thus, in most cases, an anomeric mixture of glycosyl phosphites has been employed for glycosidation. However, the anomeric configuration of the donor may influence the stereochemical course of the reaction [675].

Most glycosyl phosphites are moderately stable and easy to handle. Although diethyl phosphite derived from 2,6-dideoxysugar **38** could not be prepared owing to its hydrolytic instability, the corresponding pinacol phosphite **39** was found to possess sufficient stability to be isolated in acceptable yield (Scheme 3.134) [692]. The use of bis(2,2,2-trichloroethyl) phosphorochloridite was particularly effective for the phosphitylation of benzyl-protected fucose; diethyl phosphite, in contrast, was very unstable [693]. Attempts to prepare 2-azido-2-deoxyglycosyl diethyl and dibenzyl phosphites were unsuccessful, presumably owing to the propensity of the product to undergo Staudinger-type reactions [687,694], whereas bis(2,2,2-trichloroethyl) phosphite was uneventfully obtained [693]. It should be noted that the glycosidation of glycosyl phosphites with glycoside alcohols bearing an azido group proceeded without difficulty [695,696].

3.4.3.2 Glycosidation Using Glycosyl Phosphites

Construction of 1,2-*trans*-β-Glycosidic Linkages with a Participating Group at C-2 As already mentioned (see Section 3.4.2.2), neighboring-group participation by a C-2 substituent has been an obvious strategy to obtain 1,2-*trans*-β-glycosides. In this context, Wong and coworkers reported that glycosidation of acetyl-protected glycosyl dibenzyl phosphites in CH_2Cl_2 proceeded in the presence of 50 mol% of TMSOTf, providing 1,2-*trans*-β-glycosides in modest yields; one of the major problems was the formation of an orthoester-type phosphonate from the acyloxonium intermediate and the departing silyl phosphite [697]. Although the analogous by-product was also formed in the $BF_3 \cdot OEt_2$-mediated glycosidation of per-*O*-benzoylated glucosyl diethyl phosphite in CH_2Cl_2, this problem was effectively circumvented by using TMSOTf as a promoter (Scheme 3.135) [691]. Under these conditions, the phthalimido and 2,2,2-trichloroethyl carbamate groups also worked well as C-2 stereodirecting substituents, which are useful in the synthesis of oligosaccharides containing a variety of 2-amino-2-deoxysugars [691,698].

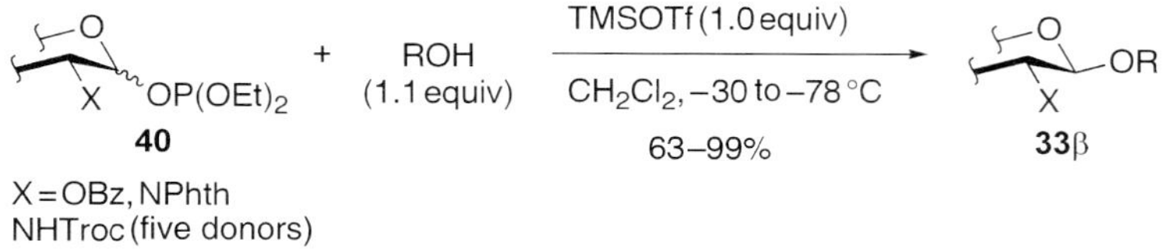

Scheme 3.135 Glycosidation of glycosyl phosphites containing a participating substituent at C-2.

Construction of 1,2-*trans*-β-Glycosidic Linkages Without a Participating Group at C-2 The so-called nitrile effect has also proven to be effective for the stereoselective construction of 1,2-*trans*-β-glycosidic linkages when employing benzyl-protected glycosyl phosphites as donors [675,689,693,699–702]. A particularly striking feature of the glycosyl phosphites is that a wide variety of promoters are available for their activation. TMSOTf [675], $BF_3 \cdot OEt_2$ [675,693], N-iodosuccinimide (NIS)–TfOH [689], $LiClO_4$ [699], I_2 [700], reusable heterogeneous solid acids such as montmorillonite K-10 [701] and triflimide (Tf_2NH) [702] activated glycosyl phosphites in MeCN or EtCN, affording glycosides with good to high β-selectivities. Glycosides with high β-selectivities were also formed in CH_2Cl_2 when performing glycosylation of reactive alcohols at low temperatures in the presence of $ZnCl_2$ [689,703], TMSOTf [697], TfOH [697], Tf_2O [697] and $BF_3 \cdot OEt_2$ [675]. In TMSOTf-promoted glycosidations, benzyl-protected glucosyl diethyl phosphite exhibited much higher reactivity than the corresponding diethyl phosphate [675]. Somewhat surprisingly, $BF_3 \cdot OEt_2$-promoted coupling of the diethyl phosphite (α:β = 90:10) with 6-OH or 4-OH glycosyl acceptors proceeded smoothly even at −78 °C. This example exhibits the highest 1,2-*trans*-β-selectivities known to date for glycosidations without neighboring-group participation, whereas no glycosidation with the diethyl phosphate was observed below 0 °C (Scheme 3.136) [675]. In this regard, it is particularly noteworthy that, with 6-OH glycosyl acceptor, virtually complete β-selectivities were independent of the solvent and the anomeric configuration of the donor. In contrast, with the less reactive 4-OH glycosyl acceptor in CH_2Cl_2, β-selectivities diminished with a decrease in the α:β anomeric ratio of the donor.

Preferential formation of β-glucosides from per-*O*-benzylated glucosyl diethyl phosphite was reported in the ionic liquid 1-hexyl-3-methylimidazolium trifluoromethanesulfonimidide, containing Tf_2NH at room temperature, albeit with a slight erosion in stereoselectivity [702].

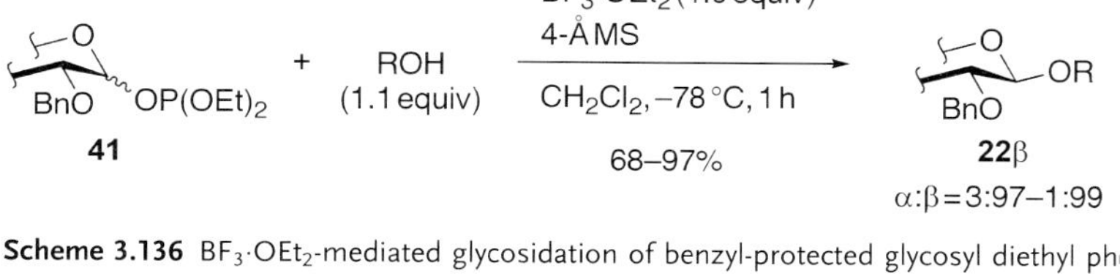

Scheme 3.136 $BF_3 \cdot OEt_2$-mediated glycosidation of benzyl-protected glycosyl diethyl phosphites.

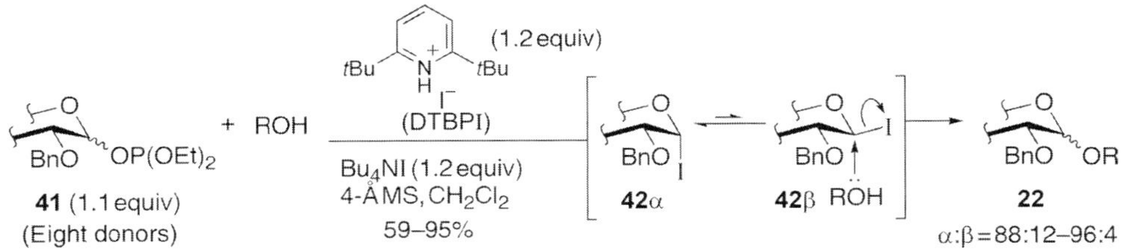

3.4 Phosphates, Phosphites and Other O–P Derivatives | 235

Scheme 3.137 α-Selective glycosidation of benzyl-protected glycosyl diethyl phosphites promoted by DTBPI in the presence of Bu$_4$NI.

Construction of 1,2-cis-α-Glycosidic Linkages Benzyl-protected glycosyl phosphites can be activated not only by TfOH and Tf$_2$NH but also by weaker Brønsted acids [697,704]. On the basis of salient donor characteristics, it was hypothesized that the activation of the phosphite group with promoters such as pyridinium iodides could lead to the formation of glycosyl iodides, the glycosidation of which, coupled with Lemieux's *in situ* anomerization method, might lead to the construction of 1,2-cis-α-glycosidic linkages. Indeed, the combined use of 2,6-di-*tert*-butylpyridinium iodide (DTBPI) as a promoter and Bu$_4$NI as an additive provided exceptionally high levels of 1,2-cis-α-selectivity and also permitted glycosylation of highly sensitive aglycones (Scheme 3.137) [704]. The glucosyl diethyl phosphite was converted into the corresponding α-glucosyl iodide **42α** under these conditions within 30 min, as judged by TLC and NMR, but extended reaction times (4–48 h) were required to obtain high yields of products, suggesting that glycosidation of the iodide **42** via *in situ* anomerization was the rate-determining step.

Although glycosyl dimethyl phosphites could be activated by ZnCl$_2$ alone, the reaction efficiency was improved by adding 2 equiv (relative to ZnCl$_2$) of AgClO$_4$, which aided not only in accelerating the reaction but also in suppressing the formation of relatively unreactive glycosyl chlorides [689,703]. The combined promoter permitted glycosylation of tertiary alcohols [705] and exhibited high α-selectivities when the reaction was performed in Et$_2$O [689]. This protocol was then successfully applied to the synthesis of adenophostin A and its analogs, wherein the exclusive formation of α-glucosides was realized using dioxane/toluene as the solvent (Scheme 3.138) [706].

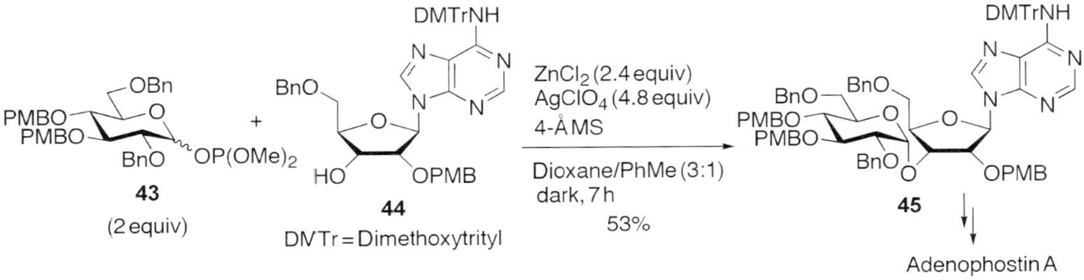

Scheme 3.138 Application of the ZnCl$_2$/AgClO$_4$-promoted glycosidation to the synthesis of adenophostin A.

Scheme 3.139 Acyl participation strategy for the stereoselective synthesis of α-galactos des.

Temp.	α:β
0 °C	6.2:1
−15 °C	6:1
−78 °C	1:2

Preferential formation of α-glucosides in Et$_2$O was also observed in Ba(ClO$_4$)$_2$-promoted glycosidation [699]. Interestingly, it has been suggested that per-O-benzylated glucosyl diethyl phosphite is a much better donor than the corresponding dimethyl and dibenzyl phosphites in terms of reactivity and product yield.

The effects of participating O-4 acyl groups on the anomeric ratio of glycosidic bond formation were first documented by van Boeckel et al. [707]. The remote group participation strategy has recently been extended to the phosphite method. The incorporation of acyl protecting groups at O-4 or O-6 enhanced the α-selectivity of the TMSOTf- or TfOH-catalyzed glycosidation of galactosyl dibenzyl phosphites (Scheme 3.139) [694,708]. Stereoselectivity was temperature dependent and higher α-selectivities were obtained at higher reaction temperatures.

As other fucosyl donors, phosphites displayed a striking preference for the formation of α-glycosides in either Et$_2$O or CH$_2$Cl$_2$ [693,694,709]. Schmidt and coworkers suggested that the glycosyl donor strength for the most frequently occurring sugars carrying identical leaving groups at C-1 under the same reaction conditions increases in the following order: sugar uronates (category I) < aldoses (II) < deoxysugars (III) ketoses (IV) < 3-deoxy-2-glyculosonates (V) [710]. Glycosyl phosphites complement glycosyl trichloroacetimidates in the higher reactivity range (categories III–V), as trichloroacetimidates tend to exhibit lower stabilities. Indeed, fucosylation proved this to be the case; the product yield obtained by the TMSOTf-catalyzed glycosidation of fucosyl bis(2,2,2-trichloroethyl) phosphite **49** with alcohol **48** in Et$_2$O surpassed that reported for the trichloroacetimidate method (Scheme 3.140).

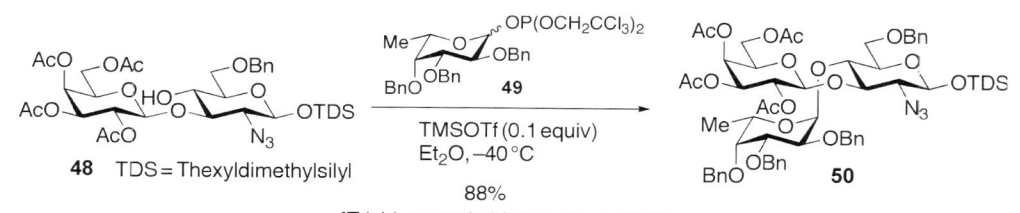

Scheme 3.140 α-Selective fucosylation: a comparative study.

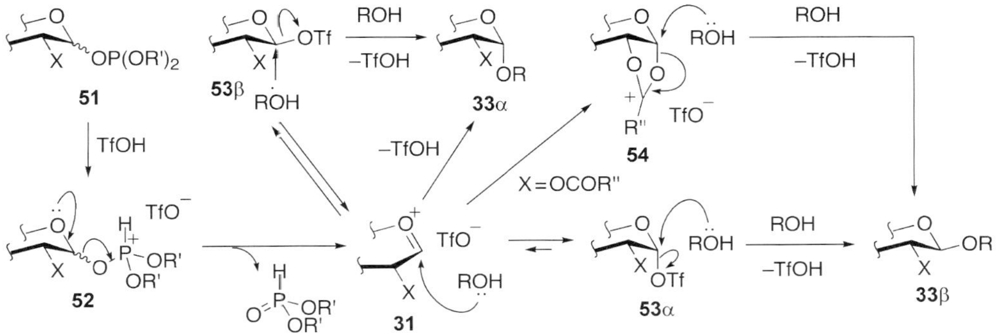

3.4 Phosphates, Phosphites and Other O–P Derivatives | 237

Scheme 3.141 A plausible mechanism for TfOH-catalyzed glycosidation of glycosyl phosphites.

3.4.3.3 Mechanism of Glycosidation Reaction with Glycosyl Phosphites

Two different mechanisms for the glycosidation reaction of glycosyl phosphites in CH_2Cl_2 have been originally proposed by the Wong's group [697]. When a Brønsted acid such as TfOH is employed as a catalyst, glycosyl phosphites **51** are activated by protonation of the phosphorus atom [711], resulting in the cleavage of the phosphite group to provide oxocarbenium ions **31** (dioxocarbenium ions **54** from 2-O-acylated donors) and hydrogen phosphites (Scheme 3.141) [697]. The oxocarbenium ions **31** are trapped by a counterion (e.g. TfO$^-$) and/or react with acceptor alcohols to lead to the formation of glycosides **33** and the regeneration of the catalyst (see Section 3.4.2.3 for the analogs' reaction mechanism).

For the $BF_3 \cdot OEt_2$-promoted reaction, the possibility of a similar mechanism is suggested by the following results: (1) per-O-benzylated glucosyl diethyl phosphite could not be directly activated by $BF_3 \cdot OEt_2$ at $-78\,^\circ$C, and (2) the 6-O-silylated glucoside did not act as a glycosyl acceptor [704]. These results reveal that the BF_3–acceptor alcohol complex functions as a proton donor to activate the phosphite group.

Wong and coworkers reported that the reaction of galactosyl and fucosyl dibenzyl phosphites with 1,3,5-trimethoxybenzene (TMB) proceeded in CH_2Cl_2 in the presence of 0.5 equiv of TMSOTf. This result implies that TMSOTf directly activates glycosyl phosphites, as no reaction between TMSOTf and TMB was observed [694]. This observation, together with the experimental evidence for the generation of trimethylsilyl dibenzyl phosphite in the coupling of acetyl-protected glycosyl dibenzyl phosphites with an oxygen nucleophile in the presence of TMSOTf [697], strongly suggests that TMSOTf activates glycosyl phosphites **51** by silylation of the oxygen atom to generate trimethylsilyl phosphites and oxocarbenium ion intermediates **31**, which react with the acceptor alcohol to give glycosides and TfOH (Scheme 3.142). Reaction of the released TfOH with the trimethylsilyl phosphites gives hydrogen phosphites and TMSOTf, which again functions as the catalyst. In this

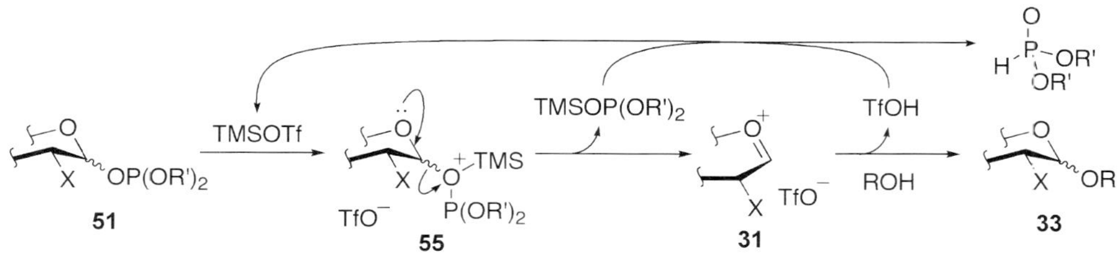

Scheme 3.142 A plausible mechanism for TMSOTf-catalyzed glycosidation of glycosyl phosphites.

pathway, the possibility that the released TfOH directly catalyzes glycosidation in the foregoing manner cannot be excluded.

3.4.4
Glycosyl Donors Carrying Other Phosphorus-Containing Leaving Groups

3.4.4.1 Glycosyl Dimethylphosphinothioates

The first recognized 1-O-phosphorus-substituted glycosyl donor was per-O-benzylated glucosyl dimethylphosphinothioate **56**, reported by Inazu and coworkers [712a] 4 years before the emergence of glycosyl phosphates. The lithiation of the glucose derivative **36** in THF at −30 °C, followed by the addition of Me$_2$P(S)Cl, exclusively gave the α-glucosyl donor **56**, which reacted with alcohols in benzene in the presence of AgClO$_4$ at room temperature to give glucosides in good yields with good to excellent α-selectivities (Scheme 3.143). The use of I$_2$ in the presence of 10 mol% of triphenylmethyl (Tr) perchlorate was also effective for this glycosidation reaction [712b]. Under the former conditions, phenols were effectively glycosylated, and in most cases the O → C glycoside rearrangement could not be detected [713]. The reactions are presumed to proceed through glycosyl perchlorate intermediates.

3.4.4.2 Glycosyl Phosphinimidates and Other N=P Derivatives

Glycosyl P,P-diphenyl-N-(p-toluenesulfonyl) phosphinimidates were prepared by the condensation of 1-O-lithium or thallium salts of hemiacetals with P,P-diphenyl-N-(p-toluenesulfonyl)phosphinimidic chloride [714]. Although the TMSOTf-promoted

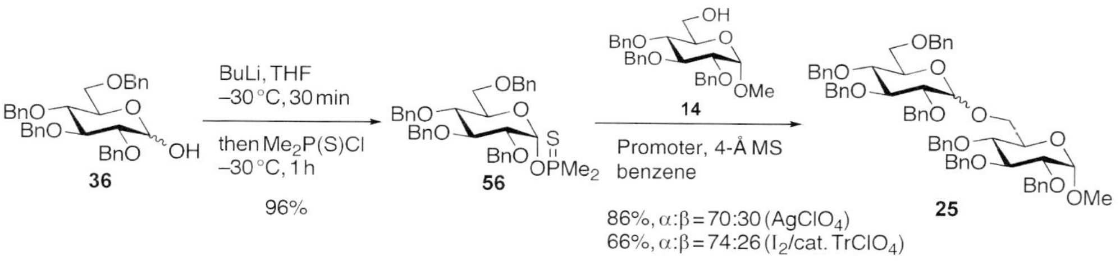

Scheme 3.143 Preparation and glycosidation of glycosyl dimethylphosphinothioate.

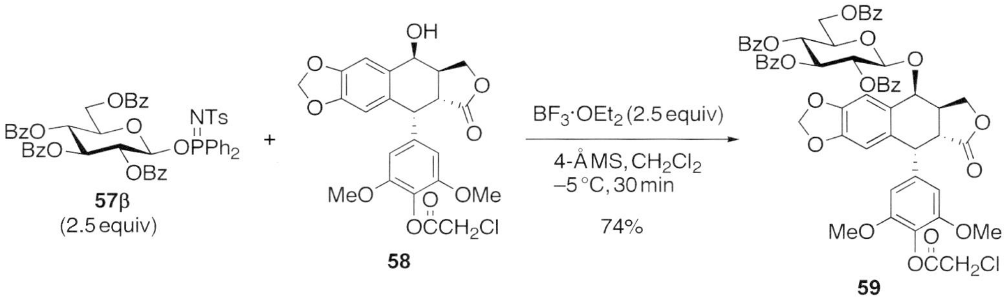

Scheme 3.144 Glycosidation of glycosyl *P,P*-diphenyl-*N*-(*p*-toluenesulfonyl)phosphinimidates with podophyllum lignan.

reaction of benzyl-protected glycosyl phosphinimidates in EtCN at −55 °C exhibited high β-selectivities [715], the synthetic utility of the donors could be extended by the combined use of $BF_3{\cdot}OEt_2$ as a promoter and CH_2Cl_2 as a solvent [714,715]. The $BF_3{\cdot}OEt_2$-promoted coupling of benzoyl-protected β-glycosyl phosphinimidates with a range of alcohols proceeded smoothly at −30 to −5 °C to exclusively afford 1,2-*trans*-linked glycosides [714]. The efficiency of this method has been well verified by the first successful glycosylations of highly acid-sensitive podophyllum lignan (Scheme 3.144) [716] and digitoxigenin [714].

Analogous glycosyl donors such as glycosyl phosphoramidimidates [717] and *N*-phenyl diethyl phosphorimidates [718] were prepared by Staudinger reaction from the corresponding glycosyl phosphoramidites and phosphites, respectively, with PhN_3 in benzene at 45–50 °C. These glycosyl donors were used in the TMSOTf- and $BF_3{\cdot}OEt_2$-promoted glycosidations in either EtCN or CH_2Cl_2. It is noteworthy that lutidinium *p*-toluenesulfonate (LPTS), originally used for α-selective glycosidation of *S*-glycosyl phosphorodiamidimidothioate **61** [646], also promoted the reaction of glycosyl phosphorimidate **60**, albeit with lower α-selectivity (Scheme 3.145) [718].

3.4.4.3 Glycosyl *N,N,N′,N′*-Tetramethylphosphorodiamidates

Owing to the poor reactivity of bis(dimethylamino)phosphorochloridate and the difficulty associated with removing excess phosphorylating agent, phosphorylation of 1-*O*-lithium salts of hemiacetals **62** was performed using equimolar amounts of the

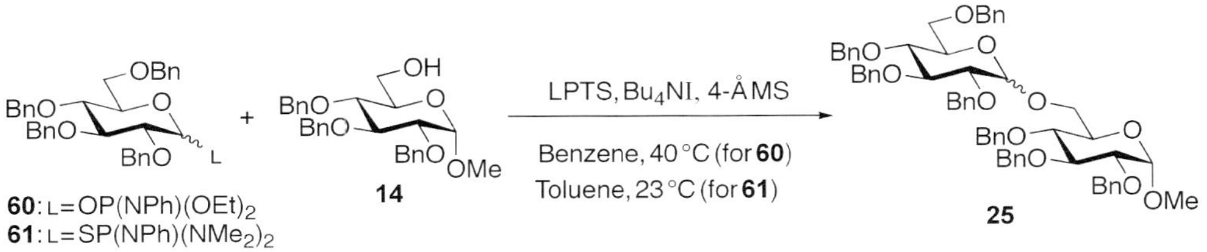

60: L = OP(NPh)(OEt)$_2$
61: L = SP(NPh)(NMe$_2$)$_2$

73%, α:β = 5:1 (from **60**)
83%, α:β = 10:1 (from **61**)

Scheme 3.145 LPTS-promoted glycosidations of per-*O*-benzylated glucosyl donors.

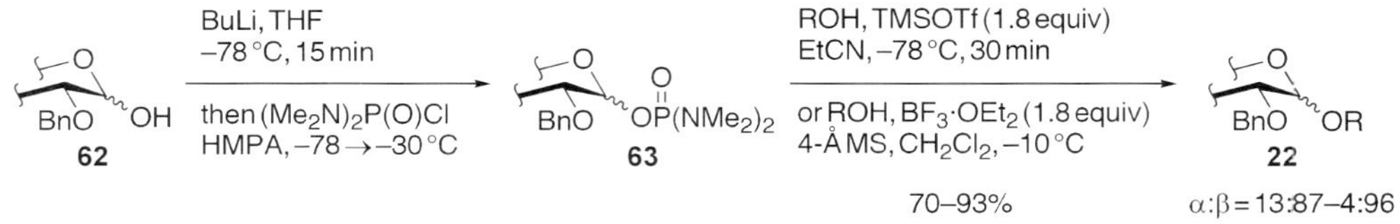

Figure 3.146 Preparation and glycosidation of glycosyl N,N,N',N'-tetramethylphosphorodiamidates.

reagent in THF in the presence of HMPA at $-30\,°C$ (Scheme 3.146) [719,720]. The excellent shelf life of per-O-benzylated α-glucosyl N,N,N',N'-tetramethylphosphorodiamidate was confirmed through methanolysis (23 °C), leading to the formation of 12% methyl β-glucoside after 30 days (cf. $t_{1/2} = 45$ min for the corresponding diphenyl phosphate). Somewhat surprisingly, in light of this result, glycosyl phosphorodiamidate **63** exhibited reactivity and stereoselectivity comparable to those observed with the corresponding diphenyl phosphate **21** in TMSOTf-promoted glycosidations [719–721]. An additional feature of phosphorodiamidates **63** is the fact that they can be activated by $BF_3 \cdot OEt_2$ in CH_2Cl_2 at $-10\,°C$, whereas the corresponding diphenyl phosphates are unreactive at below 10 °C. Steroidal alcohols such as estrone could be glycosylated using these conditions, leading to preferential formation of 1,2-trans-β-linked steroidal glycosides in good yields [719].

3.4.4.4 Miscellaneous O–P Derivatives

Singh and coworkers investigated TMSOTf-promoted glycosidation of per-O-benzylated glucosyl diphenylphosphinate, which had been prepared by the condensation of the corresponding glucose **36** with $Ph_2P(O)Cl$ in the presence of N-methylimidazole in CH_2Cl_2. However, the stereoselectivity was in the order of 1 : 4 favoring β-isomers [651].

Glucosyl diphenylphosphinite **64** was prepared from the corresponding hemiacetal **36** by the reaction with Ph_2PCl in the presence of i-Pr_2NEt in CH_2Cl_2. The reaction with 3β-cholestanol proceeded in the presence of $ZnCl_2$ in CH_2Cl_2 to provide glycoside with slightly higher β-selectivity than that obtained with the corresponding dimethyl phosphite, albeit with a prolonged reaction time (10 h versus 20 min) [689]. A new aspect of glycosidation with the donor **64** was revealed by Mukaiyama et al., wherein alkylating agents were used for donor activation (Scheme 3.147) [722]. With MeI as the promoter, the glycosidation of diphenylphosphinite **64** with a range of alcohols including glycosyl fluoride and thioglycoside in

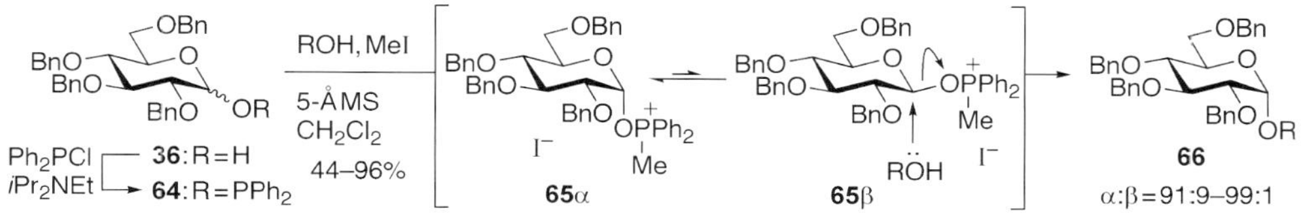

Scheme 3.147 Glycosidation with per-O-benzylated glucosyl diphenylphosphinite.

CH₂Cl₂ afforded disaccharides **66** in good to high yields with exceptionally high levels of α-selectivity. The reaction times of 1–4 days were necessary to complete the reaction assumed to proceed via glycosyl phosphonium iodide intermediates **65**.

The glycosyl donors, glycosyl phosphoramidites [723] and per-O-benzylated galactosyl difluorophosphate [724] were prepared and employed in glycosidation reactions.

3.4.5
Construction of Other Types of Glycosidic Linkages

Having discussed the methods for the stereoselective construction of 1,2-*cis*-α- and 1,2-*trans*-β-glycosidic linkages, the syntheses of β-mannosides, 2-acetamido-2-deoxy-β-glycosides, 2-deoxy-β-glucosides and α-sialosides are described in this section.

3.4.5.1 Construction of the β-Mannosidic Linkage

In view of the ubiquity of substructure Manβ (1–4) GlcNAc, in nearly all eukaryotic N-linked glycoproteins, β-mannosides have attracted considerable attention as the synthetic target. However, stereoselective construction of β-mannosidic linkages is recognized as one of the most challenging problems in carbohydrate chemistry [725]. This is because the anomeric effect as well as steric repulsion between an axially oriented nonparticipating group at C-2 and an incoming alcohol favor the formation of α-mannosidic linkages. Indeed, in contrast to the ready formation of α-mannosides [664b,689,703,726–728], attempts to favor β-mannoside synthesis from per-O-benzyl-protected mannosyl donors carrying phosphorus-containing leaving groups such as dimethylphosphinothioate [726,729], diphenyl phosphate [664b,728,730], diethyl phosphite [730] and phosphorodiamidate [730] met with only limited success (up to α:β = 19:81).

A major breakthrough in this field emerged in 1996 when Crich and Sun reported that the activation of 2,3-di-O-benzyl-4,6-O-benzylidene-protected mannosyl sulfoxide by Tf₂O in CH₂Cl₂ at −78 °C followed by the addition of acceptor alcohols gave β-mannosides in high yields with excellent levels of stereoselectivity [731]. The effectiveness of the 4,6-O-benzylidene acetal protection was not limited to the use of sulfoxide as the anomeric leaving group [732,733], and the protocol was successfully adapted to the phosphite method [730,734]. Mannosyl diethyl phosphite **67a** exhibited greater reactivity than the corresponding diphenyl phosphate **67b** and phosphorodiamidate **67c** in the presence of TMSOTf in CH₂Cl₂ and reacted with a wide range of acceptor alcohols at −45 °C to give β-mannosides in high yields with good to high stereoselectivities (up to α:β = 5:95) (Scheme 3.148). Similar β-selectivities were also observed by mannosylation with the donor **67a** in CH₂Cl₂ at −10 °C using montmorillonite K-10 as a promoter, suggesting that the counterion was not responsible for the stereoselectivity of the reaction [701b,735].

3.4.5.2 Construction of 2-Acetamido-2-deoxyglycosidic Linkages

As 2-acetamido-2-deoxyglycosides, mainly found in β-glycosidic linkages, are ubiquitous building blocks of glycolipids, glycoproteins, proteoglycans and

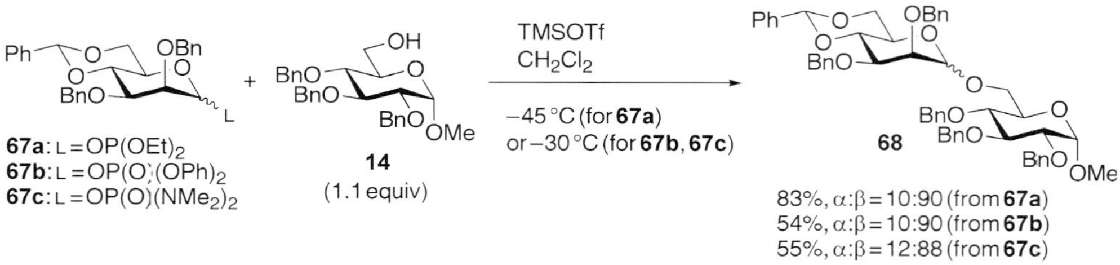

Scheme 3.148 Glycosidation with 4,6-O-benzylidene-protected mannosyl diethyl phosphite and its analogs.

peptidoglycans, numerous procedures for synthesizing 1,2-*trans*-β-linked 2-acetamido-2-deoxyglycosides have been reported [736]. A variety of different 2-amino protecting groups with an anchimeric assistance such as N-phthaloyl [691,698], N-2,2,2-trichloroethoxycarbonyl [691,698], N-benzyloxycarbonyl [737] or N-trichloroacetyl [682,738] were incorporated into donors with phosphorus-containing leaving groups and effectively utilized for this purpose.

An alternative approach to 2-acetamido-2-deoxy-β-glycosides involves the use of 2-azido-2-deoxyglycosyl donors [664b,687,728,739]. Per-O-benzylated glucosyl and galactosyl, and 4,6-O-benzylidene-protected galactosyl diphenyl phosphates **69** each reacted with a range of glycoside alcohols in the presence of TMSOTf in EtCN at −78 °C to afford β-disaccharides **70** in high yields with excellent stereoselectivities, regardless of the anomeric composition of the donor (Scheme 3.149) [687,739]. However, limitations of the phosphate method were recognized for per-O-acetylated glucosyl and galactosyl, and 4,6-O-benzylidene-protected glucosyl donors **71a–c**, indicating that the proper choice of protecting groups on 2-azido-2-deoxy-sugar components is crucial for the success of this method.

In terms of efficiency and practicality, a direct glycosidation method using donors with natural N-acetyl functionality would constitute an ideal procedure. In practice, however, the reaction of these donors generally leads to the preferential formation

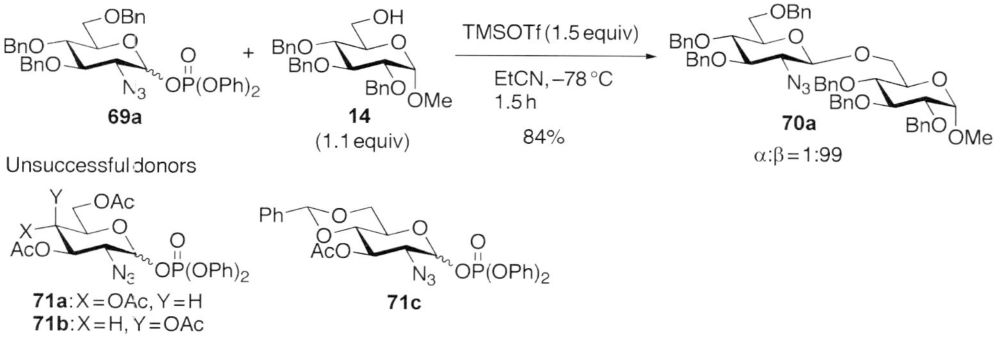

Scheme 3.149 Glycosidation with 2-azido-2-deoxyglycosyl diphenyl phosphates.

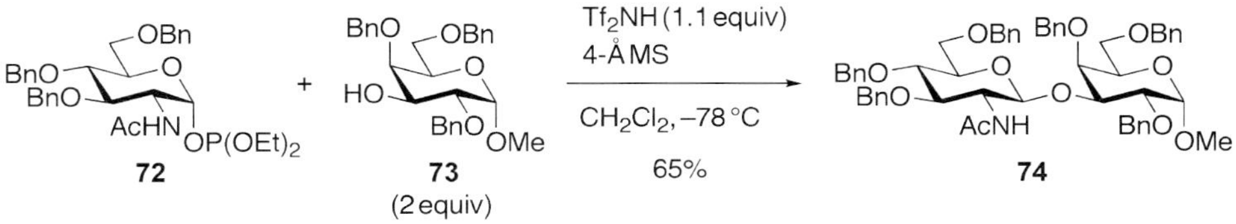

Scheme 3.150 Glycosidation with 2-acetamido-2-deoxyglycosyl diethyl phosphites.

of oxazoline derivatives via neighboring-group participation and subsequent abstraction of a proton from the formed oxazolinium ion intermediate. Hashimoto and coworkers recently found that the Tf$_2$NH-promoted glycosidation of 2-acetamido-2-deoxyglycosyl diethyl phosphites in CH$_2$Cl$_2$ at −78 °C proceeds without the intermediacy of an oxazolinium ion (Scheme 3.150) [740]. Using this method, 1,2-*trans*-β-linked glycosides were obtained in good to high yields with perfect stereoselectivity when reactive secondary and primary alcohols were employed as glycosyl acceptors.

3.4.5.3 Construction of 2-Deoxyglycosidic Linkages

The construction of 2-deoxy-β-glycosidic linkages has been a long-standing problem in carbohydrate chemistry, arising from the lack of both stereodirecting anchimeric assistance from the C-2 position and anomeric effect [741]. Attempts to solve this problem have focused on the introduction of a temporary participating substituent positioned equatorially at C-2 that requires removal after the glycosidation event. For example, the stereoselective construction of β-glycosidic linkages by TMSOTf-promoted glycosidations of 2-deoxy-2-(*p*-methoxyphenyl)thioglycosyl phosphorodiamidates **75** in CH$_2$Cl$_2$ at −78 °C was followed by the reduction with W-2 Raney nickel to provide 2-deoxy-β-glycosides **77** in good overall yields (Scheme 3.151) [742].

Direct glycosidations using 2-deoxyglycosyl donors have also been investigated to achieve this goal. However, benzyl-protected 2-deoxyglucosyl dimethylphosphinothioate [743] and 2-deoxyglycosyl *N,N*-diisopropyl phosphoramidite [744] exhibited excellent α-selectivities when the glycosidations were performed in the presence of TrClO$_4$ (benzene, room temperature) and TMSOTf (CH$_2$Cl$_2$, −78 °C), respectively. In contrast, TMSOTf-catalyzed reactions of 2-deoxyglycosyl diethyl phosphites in toluene at −94 °C afforded 2-deoxyglycosides in excellent yields with the highest β-selectivity known to date, albeit with a limited combination of coupling partners (Scheme 3.152) [745]. TMSOTf-catalyzed glycosidation was successfully applied to

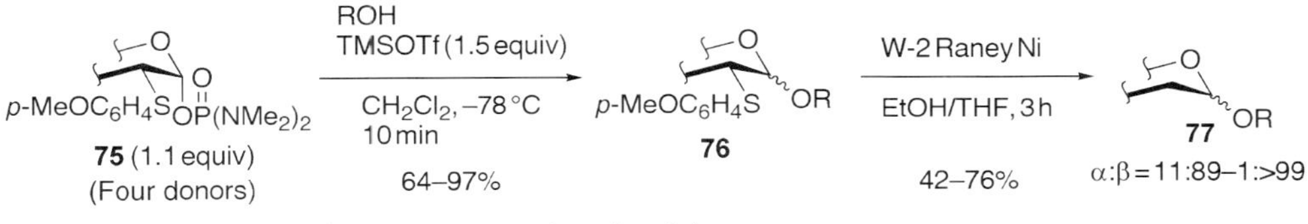

Scheme 3.151 Utilization of a directing (*p*-methoxyphenyl)thio group.

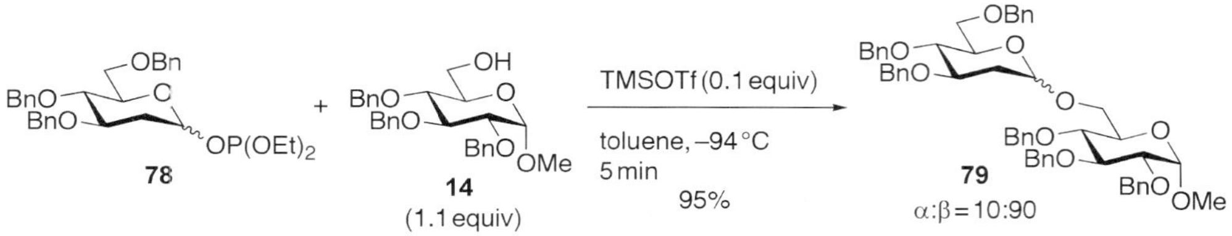

Scheme 3.152 TMSOTf-catalyzed glycosidation of 2-deoxyglycosyl diethyl phosphites.

the synthesis of hexasaccharide fragment of landomycin A by Guo and Sulikowski [692]. On the contrary, the use of $BF_3 \cdot OEt_2$ [690,710], $Sn(OTf)_2$ [710] and $LiClO_4$ [746] as promoters led to the preferential formation of α-glycosidic linkages.

The remote group participation strategy also proved effective in β-selective glycosidation of 2,6-dideoxyglucosyl diethyl phosphites mediated by montmorillonite K-10 in Et_2O at $-78\,°C$ [701b,747]; β-selectivity was enhanced (α : β = 20 : 80 → 10 : 90) by switching the protecting group at C-4 from a benzyl to a benzoyl group.

3.4.5.4 Construction of α-Sialosidic Linkages

Sialic acid residues are often linked to Gal or GalNAc via an α-glycosidic bond at the nonreducing terminus of oligosaccharides. Despite many advances in this area [748], α-selective and high-yielding coupling of sialyl donors with a wide range of glycoside alcohols remains a major problem owing to several factors: the steric congestion around the anomeric center, the presence of an electron-withdrawing carboxyl group that triggers 2,3-elimination and the lack of a participating substituent at C-3. Sialyl phosphites, which were introduced independently by Schmidt [644] and Wong [645] groups, are recognized as one of the most effective donors for this type of reaction. Sialyl phosphites are relatively stable and easy to handle, but they are reactive enough to be activated upon treatment with a catalytic amount of TMSOTf [644,645]. Although only β-anomers of sialyl diethyl and dibenzyl phosphites are obtained in most cases, α-phosphites produced in some cases are less reactive. The presence of unreacted α-phosphites leads to complications with isolation and purification [749]. NIS–TfOH [689], $ZnCl_2$–$AgClO_4$ [689], TfOH [750,751] and $Sn(OTf)_2$ [752] were also used to activate sialyl phosphites. High levels of α-selectivity have been achieved by the reaction in nitrile solvents at low temperatures by virtue of the nitrile effect, and the product yields are generally higher than (or comparable to) those obtained by using other donors such as thiosialosides and sialyl xanthates (Scheme 3.153). Although the use of CH_2Cl_2 led to the predominant formation of undesired β-isomers [689,695], THF was an effective alternative for α-sialylation in certain cases [695,753]. In general, reactive acceptors such as primary alcohols tend to give lower α : β ratios than less reactive acceptors [696,754]. In contrast to the phosphite method, 2 equiv of TMSOTf were required to activate the corresponding phosphates under identical conditions, affording sialosides in lower yields with diminished α-selectivities [644b,645].

Although β-sialyl diethyl and dibenzyl phosphites have many applications in the synthesis of sialooligosaccharides, direct comparison of their sialyl donor

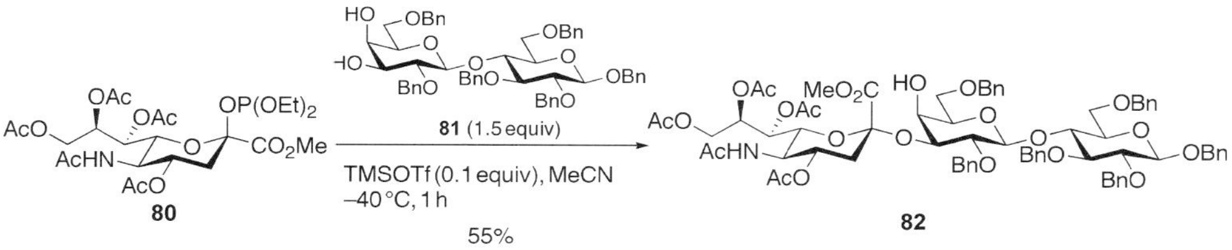

Scheme 3.153 TMSOTf-catalyzed glycosidation of sialyl phosphites. See ref. [644b].

properties in the synthesis of a tetrasaccharide sialyl Lewis X derivative revealed that the diethyl phosphite is more efficient than the dibenzyl phosphite, at least in this case (cat. TMSOTf, MeCN, −40 °C) [755]. The structure modifications of sialyl phosphites have been investigated to enhance α-selectivity. The superior result was obtained by simply changing the C-1 methyl ester to the benzyl ester (α : β = 6 : 1 → α only) [697]. Installation of *N,N*-dimethyl glycolamide ester auxiliary at C-1, which is capable of neighboring-group participation, dramatically improves α-sialylation in CH_2Cl_2 at −40 °C [756]. As in the case with other sialyl donors, participating auxiliaries at C-3 such as Br [757], thiobenzoyloxy [757], phenylseleno [758] and phenylthio groups [476] have been introduced to suppress glycal formation via 2,3-elimination and give good α-sialylation yields when unreactive acceptor alcohols are used. The effective use of thiobenzoyloxy and phenylthio groups was demonstrated by the construction of α(2–8)-linkages between two *N*-acetylneuraminic acid residues [476,757]. Over the last several years, remarkable progress in direct sialylations with a wide range of acceptors possessing different reactivities has been achieved by modifying the acetamido group at C-5 to reduce its nucleophilicity [759]. In accordance with the general trend, the replacement of the acetamido group at C-5 with azide [696], trifluoroacetamido [760] or (2,2,2-trichloroethoxy)carbonylamino group [760] enhanced the reactivity of sialyl phosphite donors, leading to the formation of sialosides in higher yields with excellent α-selectivities (Scheme 3.154).

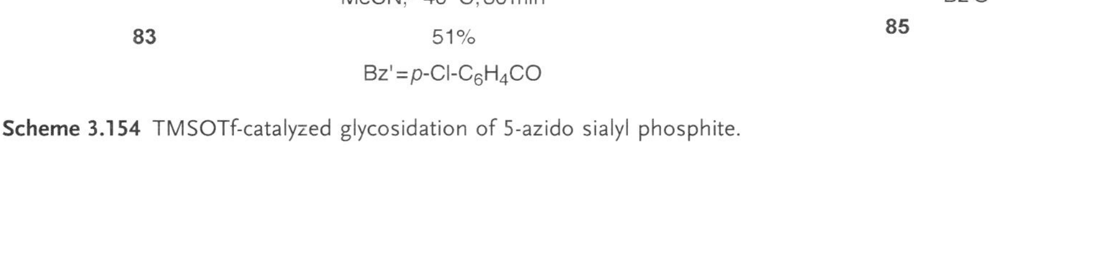

Scheme 3.154 TMSOTf-catalyzed glycosidation of 5-azido sialyl phosphite.

3.4.6
Chemoselective Glycosidation Strategies

With the advent of the armed–disarmed principle introduced by Fraser-Reid *et al.* [761], the development of innovative strategies for oligosaccharide synthesis has been the subject of intensive investigation in glycotechnology. The strategies reported to date involve chemoselective glycosidations based on anomeric leaving groups with different reactivities toward a given promoter [648b,762]. Because N,N,N',N'-tetramethylphosphorodiamidate group is sufficiently stable under reaction conditions used for routine functional and protecting group manipulations, it is a suitable leaving group for protection at O-1, thus allowing the preparation of disarmed donors that can be used as acceptors [763]. As in the case with other types of leaving groups, glycosyl phosphorodiamidates can be armed or disarmed, depending on the type of protecting group at C-2. Indeed, TMSOTf-promoted coupling of per-*O*-benzylated glucosyl phosphorodiamidate with 2,3,4-tri-*O*-benzoyl-α-glucosyl phosphorodiamidate proceeded smoothly in EtCN at −78 °C to afford the corresponding disaccharide in good yield with excellent β-selectivity, with no sign of self-condensation of the disarmed acceptor [763]. The armed–disarmed methodology was successfully applied to the stereocontrolled synthesis of Gb$_3$ (Scheme 3.155) [764].

Given the diversity of currently available phosphorus-containing leaving groups, the anomeric reactivity of glycosyl donors and acceptors can also be regulated by varying the type of leaving group [763]. This strategy allows for the coupling of donors and acceptors with identical protecting group patterns because per-*O*-benzoylated glucosyl phosphinimidate **57** and per-*O*-benzylated glucosyl diethyl phosphite **37a** are activated by BF$_3$·OEt$_2$ at lower temperatures than the corresponding

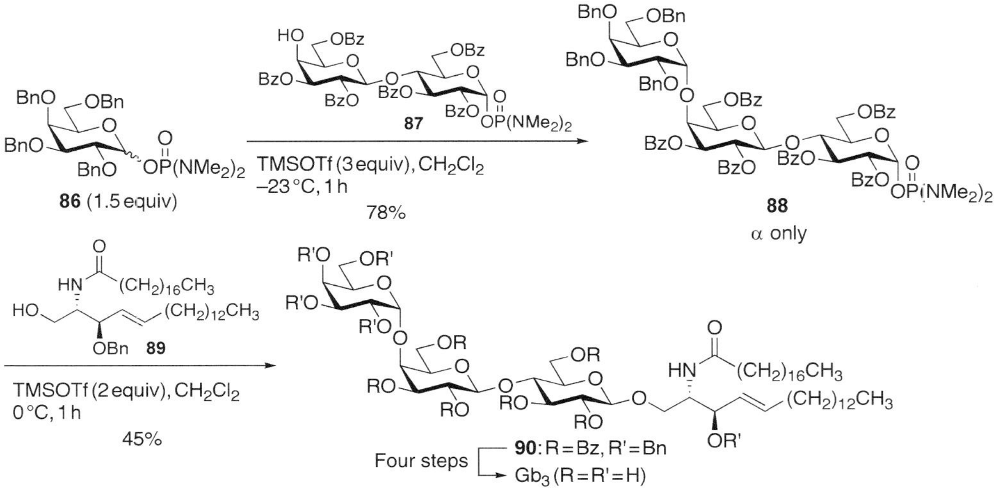

Scheme 3.155 Chemoselective glycosidation approach to the synthesis of Gb$_3$.

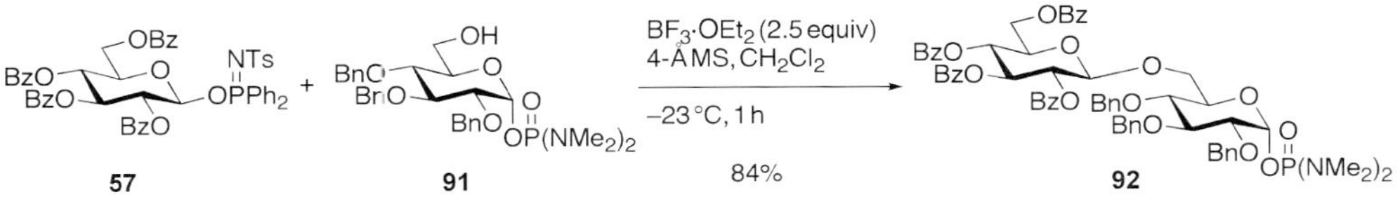

Scheme 3.156 Chemoselective glycosidation of disarmed donor with armed acceptor.

phosphorodiamidates. Of particular note is the realization of a reversal of the normal armed–disarmed pattern; the glycosidation of per-O-benzoylated glucosyl phosphinimidate **57** with 2,3,4-tri-O-benzylated α-glucosyl phosphorodiamidate **91** proceeded smoothly in the presence of BF$_3$·OEt$_2$ in CH$_2$Cl$_2$ at −23 °C, leading to the exclusive formation of the corresponding β-disaccharide **92** in excellent yield (Scheme 3.156).

The scope of this strategy could be extended to the glycosidation of sialyl phosphite **80**, resulting in a convergent synthesis of ganglioside GM$_3$ (Scheme 3.157) [751].

In addition, the chemoselective coupling of glycosyl phosphate and phosphite donors with thioglycoside acceptor could be achieved because of the compatibility of thioglycosides with TMSOTf (Scheme 3.154) [664b,696]. Further disaccharide elongation could be accomplished by thioglycoside activation. The one-pot glycosidation strategy demonstrated by Sulikowski and coworkers was based on the difference

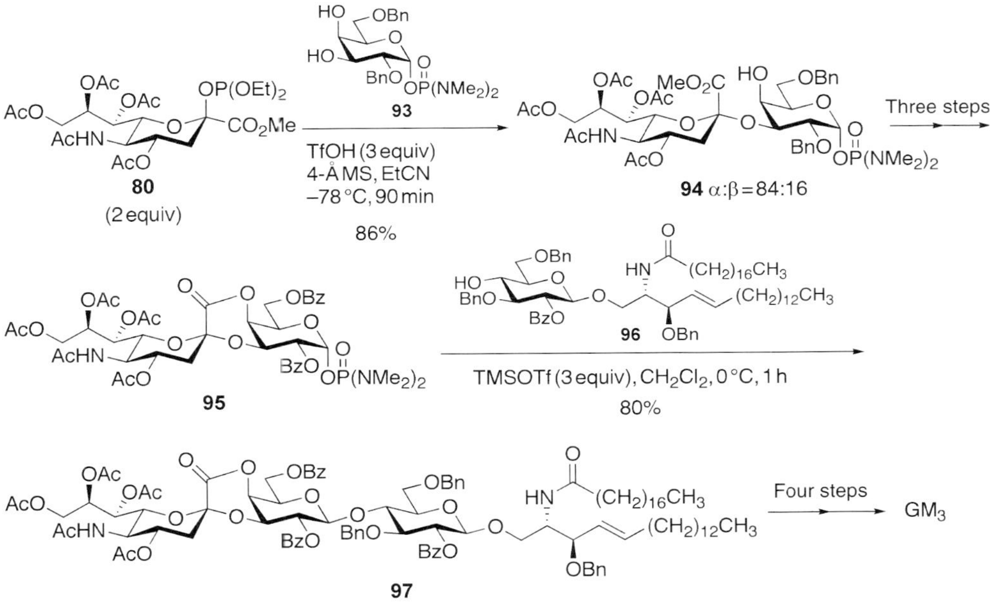

Scheme 3.157 Convergent synthesis of GM$_3$ by selective activation.

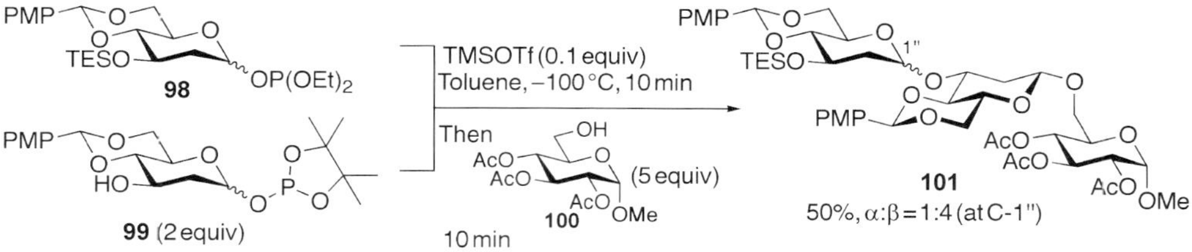

Scheme 3.158 One-pot glycosidation with glycosyl diethyl and pinacol phosphites.

in donor reactivity toward TMSOTf between diethyl phosphite **98** and pinacol phosphite **99** in the building blocks of the 2-deoxy series (Scheme 3.158) [765].

3.4.7
Application to the Synthesis of Natural Products

As mentioned above, glycosyl donors with phosphorus-containing leaving groups have been utilized for the synthesis of various oligosaccharides. In contrast, application of these methods to the synthesis of carbohydrate-containing natural products has been limited. Corey and Wu demonstrated the remarkable donor properties of per-O-benzylated glucosyl dimethyl phosphite **37b** in the total synthesis of paeoniflorin (Scheme 3.159) [705]. The glucosyl donor **37b** underwent coupling with racemic paeoniflorigenin derivative (±)-**102** (tertiary acceptor) in the presence of ZnCl$_2$/AgClO$_4$ in benzene to give an equimolar mixture of four possible isomers in 71% yield. It is noteworthy that a number of other processes for β-glycosylation (glucosyl sulfoxide, trichloroacetimidate, fluoride, thioglycoside and even phosphite when activated with TMSOTf) were not suitable for this coupling reaction.

The synthetic utility of the phosphate method was well documented by Boger and Honda in the total synthesis of bleomycin A$_2$ (Scheme 3.160) [766a]. The stereoselective construction of α-mannosidic linkage was achieved by TMSOTf-promoted glycosidation of 2-O-acetyl-protected α-mannosyl diphenyl phosphate **104** with gulose derivative **105**. After protective group manipulations and anomeric phosphorylation, the resultant β-diphenyl phosphate **107** was coupled with the suitably protected β-hydroxy-L-histidine derivative **108** in the presence of TMSOTf in Et$_2$O/CH$_2$Cl$_2$ (2:1) to afford glycoside **109** with an α:β ratio of >13:1, which was successfully incorporated into bleomycin A$_2$. Following this successful precedent, the phosphate method has been employed in the synthesis of bleomycin derivatives [766b,c].

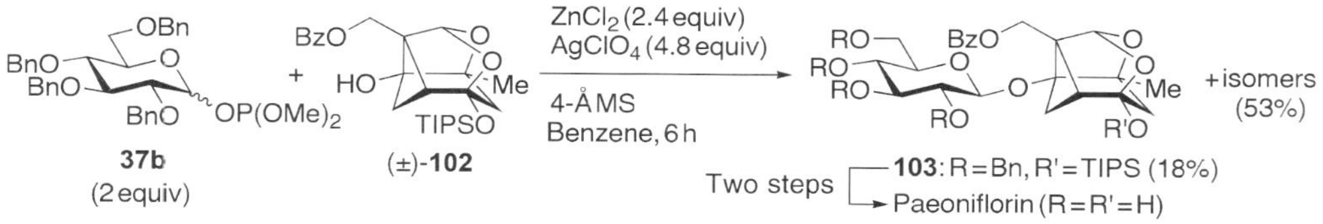

Scheme 3.159 Total synthesis of paeoniflorin.

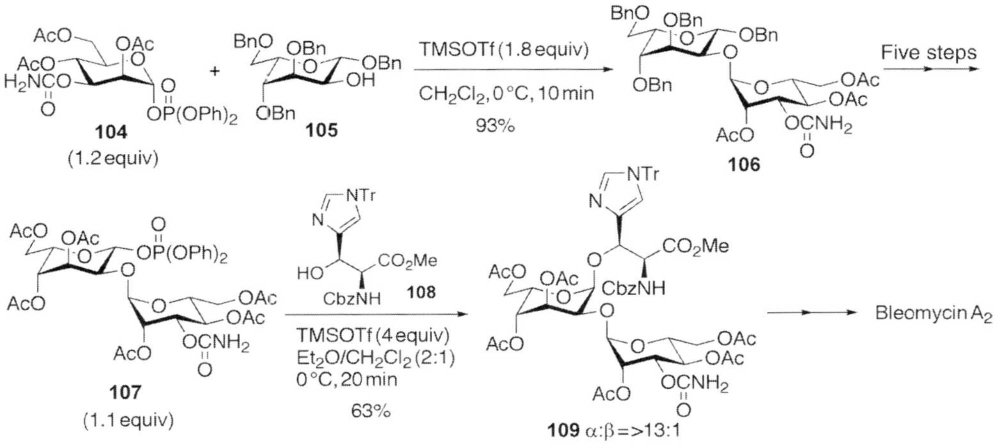

Scheme 3.160 Total synthesis of bleomycin A$_2$.

3.4.8
Conclusion

This review highlighted the recent progress in the development of glycosidation reactions capitalizing on phosphorus-containing leaving groups. These methods have distinguished themselves by varying the substituents on the phosphorus atom that are capable of tuning glycosyl donor properties. In addition, these methods have the advantages of operational simplicity and practical value as well as being high yielding and stereoselective. Thus, they are powerful alternatives to the existing glycosidation methods.

3.4.9
Experimental Procedures

3.4.9.1 Preparation of the Glycosyl Donors

Procedure for Preparation of 2,3,4,6-Tetra-O-Benzyl-D-Glucopyranosyl Diphenyl Phosphate (23a) from 1-O-Unprotected Sugar 36 [650]

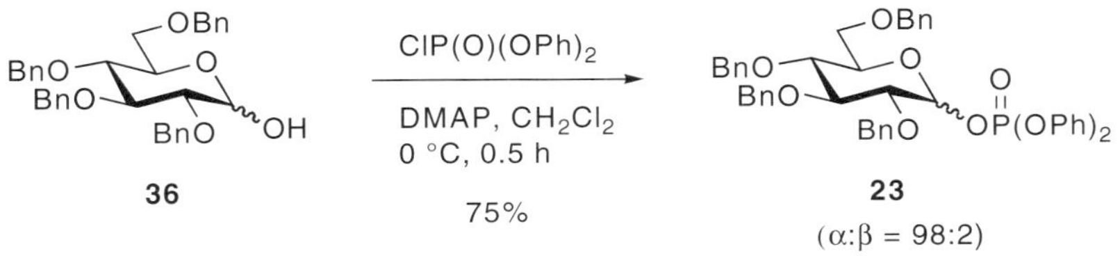

Diphenyl chlorophosphate (0.70 ml, 3.34 mmol) was added to a stirred solution of 2,3,4,6-tetra-O-benzyl-D-glucopyranose (36, 1.5 g, 2.78 mmol) and DMAP (1.02 g, 8.34 mmol) in CH$_2$Cl$_2$ (18 ml) under argon at 0 °C. After 0.5 h, the reaction was quenched with crushed ice, followed by stirring at 0 °C for 10 min. The mixture was poured into a two-layer mixture of ether and saturated aqueous NaHCO$_3$, and the formed product was extracted with EtOAc. The organic layer was successively washed with saturated aqueous NaHCO$_3$ and brine. The organic layer was separated, dried over Na$_2$SO$_4$ and concentrated *in vacuo*. The residue was purified by column chromatography on silica gel (1 : 3 EtOAc/hexane with 3% Et$_3$N) to afford the phosphate **23a** (1.54 g, 75%, α : β = 98 : 2).

Procedure for Preparation of 3,6-di-*O*-Benzyl-4-*O*-*tert*-Butyldimethylsilyl-2-*O*-Pivaloyl-β-D-Glucopyranosyl Dibutyl Phosphate (12) from Glycal 11 [664b]

To a solution of 1,5-anhydro-3,6-di-*O*-benzyl-4-*O*-*tert*-butyldimethylsilyl-2-deoxy-D-*arabino*-hex-1-enitol (**11**, 1.05 g, 2.39 mmol) in CH$_2$Cl$_2$ (24 ml) was added dimethyldioxirane (0.072 M in acetone, 40.0 ml, 2.87 mmol) under argon at 0 °C, and the reaction mixture was stirred for 15 min. After the solvent was removed *in vacuo*, the residue was dissolved in CH$_2$Cl$_2$ (48 ml). The solution was cooled to −78 °C for 15 min. A solution of dibutylphosphate (0.50 ml, 2.5 mmol) in CH$_2$Cl$_2$ (48 ml) was added dropwise over 5 min. After the addition was complete, the reaction mixture was warmed to 0 °C and DMAP (1.17 g, 9.56 mmol) and pivaloyl chloride (0.59 ml, 4.8 mmol) were added. The solution was warmed to room temperature over 1 h. The addition of 40% EtOAc–hexanes afforded a white precipitate that was filtered off through a pad of silica. The eluent was concentrated *in vacuo* and purified by column chromatography on silica gel (25% EtOAc–hexanes) to afford the phosphate **12** (1.48 g, 86%).

Procedure for Preparation of 2-*O*-Benzoyl-3,4,6-tri-*O*-Benzyl-α-D-Mannopyranosyl Dibutyl Phosphate (110) from 1,2-Orthoester 13 [666]

3,4,6-Tri-*O*-benzyl-1,2-*O*-[(*R*)-methoxyphenylmethylene]-β-D-mannopyranose (**13**, 200 mg, 0.35 mmol) and 4-Å MS (200 mg) were mixed under argon. CH$_2$Cl$_2$ (3.5 ml) was added, and the mixture was stirred at room temperature for 15 min. Dibutyl phosphate (0.21 ml, 1.05 mmol) was added dropwise. After stirring for 0.5 h at room temperature, the reaction was cooled to 0 °C and Et$_3$N (0.2 ml, 1.4 mmol) was added. The solution was warmed to room temperature and filtered through a pad of Et$_3$N-deactivated silica gel. The eluent was concentrated *in vacuo* and purified by column chromatography on silica gel (EtOAc–hexanes gradient elution) to afford phosphate **110** (253 mg, 97%).

Procedure for Preparation of 2,3,4,6-tetra-O-Benzyl-D-Glucopyranosyl Diethyl Phosphite (37a) Using Diethyl Phosphorochloridite [675]

To a solution of 2,3,4,6-tetra-O-benzyl-D-glucopyranose (**36**, 500 mg, 0.925 mmol, α : β = 9 : 1) in CH_2Cl_2 (5 ml) was added Et_3N (0.32 ml, 2.31 mmol) under argon at −78 °C. Diethyl phosphorochloridite (0.16 ml, 1.11 mmol) was then added dropwise and the reaction mixture was stirred for 0.5 h at −78 °C. The reaction was quenched with crushed ice, followed by stirring at 0 °C for 15 min. Saturated aqueous $NaHCO_3$ was added, and the whole was extracted with EtOAc. The organic layer was washed with brine, dried over Na_2SO_4 and concentrated *in vacuo*. The residue was purified by column chromatography on silica gel (1 : 7 EtOAc/hexane with 2% Et_3N) to afford phosphite **37a** (544 mg, 89%, α : β = 9 : 1).

Procedure for Preparation of 2,3,4,6-tetra-O-Acetyl-D-Glucopyranosyl Dibenzyl Phosphite (111) Using Dibenzyl N,N-Diethylphosphoramidite [645b]

Dibenzyl N,N-diethylphosphoramidite (0.86 g, 7.3 mmol) was added to a solution of 2,3,4,6-tetra-O-acetyl-D-glucopyranose (**1**, 1.0 g, 2.9 mmol) and 1,2,4-triazole (0.8 g, 11.5 mmol) in dry CH_2Cl_2 under nitrogen at room temperature. The mixture was allowed to stir at room temperature for 1–2 h before being diluted with ether. The mixture was successively washed with ice-cold saturated aqueous $NaHCO_3$, brine and water. The organic layer was separated, dried over Na_2SO_4 and concentrated *in vacuo*. The residue was purified by column chromatography on silica gel (1 : 4 EtOAc/hexane) to afford phosphite **111** (1.73 g, 97%, α : β = 1 : 4).

Procedure for Preparation of 2,3,4,6-tetra-O-Benzoyl-D-Glucopyranosyl P,P-Diphenyl-N-(p-Toluenesulfonyl)phosphinimidate (57) [714]

To a solution of 2,3,4,6-tetra-O-benzoyl-D-glucopyranose (**112**, 2.0 g, 3.36 mmol) in benzene (18 ml) was added thallous ethoxide (0.26 ml, 3.70 mmol) under argon, and the reaction mixture was stirred for 30 min. After the solvent was removed *in vacuo*, the residue was dissolved in benzene (5 ml) and *P,P*-diphenyl-*N*-(*p*-toluenesulfonyl)phosphinimidic chloride [767] (4.03 mmol in 10 ml of benzene, *in situ* prepared) was introduced, leading to the formation of a white precipitate. The suspension was stirred for 30 min and filtered through Celite. The filtrate was concentrated *in vacuo* and purified by column chromatography on silica gel (1:2 EtOAc/hexane with 1.5% Et$_3$N) to afford β-anomer **57β** (2.5 g, 80%) and α-anomer **57α** (0.27 g, 8.8%).

3.4.9.2 Glycosidation

General Procedure for TMSOTf-Promoted Glycosidation of Benzyl-Protected Glycosyl Diphenyl Phosphates 21 [643]

$$\underset{\underset{\text{(Three donors)}}{\textbf{21}}}{\text{BnO-OP(OPh)}_2} + \underset{(1.1 \text{ equiv})}{\text{ROH}} \xrightarrow[\text{EtCN, } -78\,°\text{C, <10 min}]{\text{TMSOTf (1.1 equiv)}} \underset{\underset{\alpha:\beta = 14:86-3:97}{\textbf{22}}}{\text{BnO-OR}} \quad 78-88\%$$

A solution of glycosyl diphenyl phosphate (**21**, 0.5 mmol) and glycosyl acceptor (0.55 mmol) in dry propionitrile (5 ml) was cooled to −78 °C under argon. TMSOTf (1 M in CH$_2$Cl$_2$, 0.55 mmol) was added to the mixture. After stirring for 5 min at −78 °C, the reaction was quenched by dropwise addition of Et$_3$N (0.3 ml). The reaction mixture was poured into a two-layer mixture of ether and saturated aqueous NaHCO$_3$, and the whole was extracted with EtOAc. The organic layer was washed with brine, dried over Na$_2$SO$_4$ and concentrated *in vacuo*. The residue was purified by column chromatography on silica gel (EtOAc–hexanes gradient elution) to provide the corresponding glycoside **22** (typical range 78–88% yield).

General Procedure for BF$_3$·OEt$_2$-Promoted Glycosidation of Benzyl-Protected Glycosyl Diethyl Phosphites 41 [675]

$$\underset{\underset{\text{(Three donors)}}{\textbf{41}}}{\text{BnO-OP(OEt)}_2} + \underset{(1.1 \text{ equiv})}{\text{ROH}} \xrightarrow[\text{CH}_2\text{Cl}_2,\, -78\,°\text{C, 1 h}]{\substack{\text{BF}_3\cdot\text{OEt}_2\ (1.0\ \text{equiv}) \\ 4\text{-Å MS}}} \underset{\underset{\alpha:\beta = 3:97-1:99}{\textbf{22β}}}{\text{BnO-OR}} \quad 68-97\%$$

A mixture of glycosyl diethyl phosphite (**41**, 0.15 mmol), glycosyl acceptor (0.165 mmol) and 4-Å MS (100 mg) in dry CH$_2$Cl$_2$ (2 ml) was cooled to −78 °C under argon. BF$_3$·OEt$_2$ (1 M in CH$_2$Cl$_2$, 0.15 mmol) was added to the mixture. After stirring for 1 h at −78 °C, the reaction was quenched by dropwise addition of Et$_3$N (0.21 ml), and the mixture was filtered through a Celite pad. Saturated aqueous NaHCO$_3$ was added to the filtrate and the whole was extracted with EtOAc. The organic layer was washed with brine, dried over Na$_2$SO$_4$, and concentrated *in vacuo*. The residue was purified by column chromatography on silica gel (EtOAc–hexane gradient elution) to provide the corresponding glycoside **22β** (typical range 68–97% yield).

General Procedure for DTBPI/Bu₄NI-Promoted Glycosidation of Benzyl-Protected Glycosyl Diethyl Phosphites 41 [704]

A solution of glycosyl diethyl phosphite (**41**, 0.11 mmol) in dry CH_2Cl_2 (1.0 ml) was added to a mixture of acceptor alcohol (0.10 mmol), DTBPI (0.12 mmol), n-Bu₄NI (0.12 mmol) and pulverized 4-Å MS (50 mg) under argon. The reaction vessel was sealed with stopper. The whole mixture was stirred at room temperature in dark. When TLC analysis indicated the completion of the reaction (typically 24–48 h), the reaction mixture was filtered through a Celite pad. The filtrate was poured into a two-layer mixture of EtOAc and saturated aqueous $NaHCO_3$. The separated organic layer was successively washed with 10% aqueous $Na_2S_2O_3$ and brine, dried over Na_2SO_4 and concentrated *in vacuo*. The residue was purified by column chromatography on silica gel (EtOAc–hexane gradient elution) to provide the corresponding glycoside **22α** (typical range 59–95% yield).

TMSOTf-Catalyzed Glycosidation of Sialyl Phosphites: A Typical Procedure [644b]

A solution of sialyl phosphite **80** (0.61 g, 1 mmol) and glycosyl acceptor **81** (1.32 g, 1.5 mmol) in dry acetonitrile (4 ml) was cooled to −40 °C. TMSOTf (0.018 ml, 0.1 mmol) dissolved in dry acetonitrile (0.5 ml) was added under nitrogen. After stirring for 1 h at −40 °C, the solution was neutralized with Et_3N and concentrated *in vacuo*. The residue was purified by column chromatography on silica gel (toluene–acetone gradient elution) to afford the corresponding oligosaccharide **82** (0.85 g, 55%).

Chemoselective Glycosidation of Benzoyl-Protected Glucosyl Phosphinimidate: A Typical Procedure [763]

To a mixture of 2,3,4,6-tetra-*O*-benzoyl-D-glucopyranosyl *P,P*-diphenyl-*N*-(*p*-toluenesulfonyl)phosphinimidate (**57**, 41.9 mg, 0.044 mmol), 2,3,4-tri-*O*-benzyl-α-D-glucopyranosyl *N,N,N′,N′*-tetramethylphosphorodiamidate (**91**, 23.4 mg, 0.04 mmol) and powdered 4-Å MS (40 mg) in CH_2Cl_2 (5 ml) was added $BF_3 \cdot OEt_2$ (1 M in CH_2Cl_2, 0.1 ml, 0.1 mmol) under argon at $-23\,°C$. After 1 h, the reaction was quenched by the addition of Et_3N (0.1 ml), and the reaction mixture was filtered through a Celite pad. The filtrate was poured into a two-layer mixture of EtOAc and saturated aqueous $NaHCO_3$. The separated organic layer was washed with brine, dried over Na_2SO_4 and concentrated *in vacuo*. The residue was purified by column chromatography on silica gel (2 : 1 to 4 : 1 EtOAc/hexane) to provide the corresponding disaccharide **92** (39.1 mg, 84%).

References

640 (a) Khan, S.H. and O'Neil, R.A. (eds) (1996) *Modern Methods in Carbohydrate Synthesis*, Harwood Academic, Amsterdam. (b) Hanesian, S. (ed.) (1997) *Preparative Carbohydrate Chemistry*, Marcel Dekker, New York. (c) Ernst, B., Hart, G.W. and Sinaÿ, P. (eds) (2000) *Carbohydrates in Chemistry and Biology*, vol. **1**, Wiley-VCH Verlag GmbH, Weinheim. (d) Wang, P.G. and Bertozzi, C.R. (eds) (2001) *Glycochemistry – Principles, Synthesis, and Applications*, Marcel Dekker, New York. (e) Levy, D.L. and Fügedi, P. (eds) (2006) *The Organic Chemistry of Sugars*, CRC Press, Boca Raton.

641 (a) Leloir, L.F. (1971) *Science*, **172**, 1299–1303. (b) Heidlas, J.E., Williams, K.W. and Whitesides, G.M. (1992) *Accounts of Chemical Research*, **25**, 307–314.

642 For a recent review, see Koeller, K.M. and Wong, C.-H. (2000) *Chemical Reviews*, **100**, 4465–4493.

643 Hashimoto, S., Honda, T. and Ikegami, S. (1989) *Journal of the Chemical Society, Chemical Communications*, 685–687.

644 (a) Martin, T.J. and Schmidt, R.R. (1992) *Tetrahedron Letters*, **33**, 6123–6126. (b) Martin, T.J., Brescello, R., Toepfer, A. and Schmidt, R.R. (1993) *Glycoconjugate Journal*, **10**, 16–25.

645 (a) Kondo, H., Ichikawa, Y. and Wong, C.-H. (1992) *Journal of the American Chemical Society*, **114**, 8748–8750. (b) Sim, M.M., Kondo, H. and Wong, C.-H. (1993) *Journal of the American Chemical Society*, **115**, 2260–2267.

646 Hashimoto, S., Honda, T. and Ikegami, S. (1990) *Tetrahedron Letters*, **31**, 4769–4772.

647 (a) Hashimoto, S., Honda, T., Yanagiya, Y., Nakajima, M. and Ikegami, S. (1995) *Journal of Synthetic Organic Chemistry Japan*, **53**, 620–632. (b) Zhang, Z. and Wong, C.-H. (2000) *Carbohydrates in Chemistry and Biology* (eds Ernst, B., Hart, G.W. and Sinaÿ, P.), vol. **1**, Wiley-VCH Verlag GmbH, Weinheim, pp. 117–134. (c) Vankayalapati, H., Jiang, S. and Singh, G. (2002) *Synlett*, 16–25. (d) Palmacci, E.R., Plante, O.J. and Seeberger, P.H. (2002) *European Journal of Organic Chemistry*, 595–606.

648 For general reviews covering this subject, see (a) Toshima, K. and Tatsuta, K. (1993) *Chemical Reviews*, **93**, 1503–1531. (b) Davis, B.G. (2000) *Journal of the Chemical Society, Perkin Transactions 1*, 2137–2160. (c) Demchenko, A.V. (2003) *Synlett*, 1225–1240. (d) Pellissier, H. (2005) *Tetrahedron*, **61**, 2947–2993. (e) Fügedi, P. (2006) *The Organic Chemistry of Sugars* (eds Levy, D.L. and Fügedi, P.), CRC Press, Boca Raton, pp. 90–179.

649 (a) Plante, O.J., Palmacci, E.R. and Seeberger, P.H. (2003) *Advances in Carbohydrate Chemistry and Biochemistry*,

650 Sabesan, S. and Neira, S. (1992) *Carbohydrate Research*, **223**, 169–185.
651 (a) Vankayalapati, H., Singh, G. and Tranoy, I. (1998) *Chemical Communications*, 2129–2130. (b) Vankayalapati, H., Singh, G. and Tranoy, I. (2001) *Tetrahedron: Asymmetry*, **12**, 1373–1381.
652 Inage, M., Chaki, H., Kusumoto, S. and Shiba, T. (1982) *Chemistry Letters*, 1281–1284.
653 Granata, A. and Perlin, A.S. (1981) *Carbohydrate Research*, **94**, 165–171.
654 For other procedures, see (a) Bogusiak, J. and Szeja, W. (1993) *Polish Journal of Chemistry*, **67**, 2181–2185. (b) Watanabe, Y., Hyodo, N. and Ozaki, S. (1988) *Tetrahedron Letters*, **29**, 5763–5764.
655 Garcia, B.A. and Gin, D.Y. (2000) *Organic Letters*, **2**, 2135–2138.
656 (a) Zervas, L. (1939) *Naturwissenschaften*, **27**, 317. (b) Wolfrom, M.L., Smith, C.S., Pletcher, D.E. and Brown, A.E. (1942) *Journal of the American Chemical Society*, **64**, 23–26.
657 MacDonald, D.L. (1962) *Journal of Organic Chemistry*, **27**, 1107–1109.
658 (a) Schmidt, R.R., Stumpp, M. and Michel, J. (1982) *Tetrahedron Letters*, **23**, 405–408. (b) Schmidt, R.R. and Stumpp, M. (1984) *Liebigs Annalen der Chemie*, 680–691. (c) Schmidt, R.R., Gaden, H. and Jatzke, H. (1990) *Tetrahedron Letters*, **31**, 327–330. (d) Schmidt, R.R. (1992) *Carbohydrates – Synthetic Methods and Applications in Medicinal Chemistry*, (eds Ogura, H., Hasegawa, A. and Suami, T.), Wiley-VCH Verlag GmbH, Weinheim, pp. 66–88.
659 (a) Illarionov, P.A., Torgov, V.I., Hancock, I.C. and Shibaev, V.N. (1999) *Tetrahedron Letters*, **40**, 4247–4250. (b) Illarionov, P.A., Torgov, V.I., Hancock, I. and Shibaev, V.N. (2000) *Russian Chemical Bulletin*, **49**, 1891–1894. (c) Illarionov, P.A., Torgov, V.I. and Shibaev, V.N. (2000) *Russian Chemical Bulletin*, **49**, 1895–1898.

58, 35–54. (b) Seeberger, P.H. and Werz, D.B. (2005) *Nature Reviews. Drug discovery*, **4**, 751–763.

660 Veeneman, G.H., Broxterman, H.J.G., van der Marel, G.A. and van Boom, J.H. (1991) *Tetrahedron Letters*, **32**, 6175–6178.
661 Pale, P. and Whitesides, G.M. (1991) *Journal of Organic Chemistry*, **56**, 4547–4549.
662 Boons, G.-J., Burton, A. and Wyatt, P. (1996) *Synlett*, 310–312.
663 (a) Timmers, C.M., van Straten, N.C.R., van der Marel, G.A. and van Boom, J.H. (1998) *Journal of Carbohydrate Chemistry*, **17**, 471–487. (b) Soldaini, G., Cardona, F. and Goti, A. (2003) *Tetrahedron Letters*, **44**, 5589–5592. (c) Soldaini, G., Cardona, F. and Goti, A. (2005) *Organic Letters*, **7**, 725–728.
664 (a) Plante, O.J., Andrade, R.B. and Seeberger, P.H. (1999) *Organic Letters*, **1**, 211–214. (b) Plante, O.J., Palmacci, E.R., Andrade, R.B. and Seeberger, P.H. (2001) *Journal of the American Chemical Society*, **123**, 9545–9554. (c) Love, K.R. and Seeberger, P.H. (2005) *Organic Syntheses*, **81**, 225–234.
665 Volkova, L.V., Danilov, L.L. and Evstigneeva, R.P. (1974) *Carbohydrate Research*, **32**, 165–166.
666 Ravidà, A., Liu, X., Kovacs, L. and Seeberger, P.H. (2006) *Organic Letters*, **8**, 1815–1818.
667 (a) Salo, W.L. and Fletcher, H.G., Jr (1970) *Biochemistry*, **9**, 878–881. (b) Khorlin, A.Y., Zurabyan, S.E. and Antonenko, T.S. (1970) *Tetrahedron Letters*, 4803–4804. (c) Busca, P. and Martin, O.R. (1998) *Tetrahedron Letters*, **39**, 8101–8104.
668 (a) Hoch, M., Heinz, E. and Schmidt, R.R. (1989) *Carbohydrate Research*, **191**, 21–28. (b) Schmidt, R.R., Wegmann, B. and Jung, K.-H. (1991) *Liebigs Annalen der Chemie*, 121–124.
669 Danishefsky, S.J. and Bilodeau, M.T. (1996) *Angewandte Chemie (International Edition in English)*, **35**, 1380–1419.
670 Westerduin, P., Veeneman, G.H., Marugg, J.E., van der Marel, G.A. and van Boom, J.H. (1986) *Tetrahedron Letters*, **27**, 1211–1214.

671 Vankayalapati, H. and Singh, G. (2000) *Journal of the Chemical Society, Perkin Transactions 1*, 2187–2193.

672 (a) Waldmann, H., Böhm, G., Schmid, U. and Röttele, H. (1994) *Angewandte Chemie (International Edition in English)*, **33**, 1944–1946. (b) Böhm, G. and Waldmann, H. (1996) *Liebigs Annalen der Chemie*, 621–625.

673 Hashimoto, S., Hayashi, M. and Noyori, R. (1984) *Tetrahedron Letters*, **25**, 1379–1382.

674 Zhang, Z., Ollmann, I.R., Ye, X.-S., Wischnat, R., Baasov, T. and Wong, C.-H. (1999) *Journal of the American Chemical Society*, **121**, 734–753.

675 Hashimoto, S., Umeo, K., Sano, A., Watanabe, N., Nakajima, M. and Ikegami, S. (1995) *Tetrahedron Letters*, **36**, 2251–2254.

676 Hashimoto, S., Yanagiya, Y. and Ikegami, S. unpublished result.

677 Honda, T. (1991) PhD thesis, the University of Tokyo.

678 Böhm, G. and Waldmann, H. (1996) *Liebigs Annalen der Chemie*, 613–619.

679 Bogusiak, J. and Szeja, W. (1994) *Polish Journal of Chemistry*, **68**, 2309–2314.

680 (a) Wulff, G. and Röhle, G. (1974) *Angewandte Chemie (International Edition in English)*, **13**, 157–170. (b) Igarashi, K. (1977) *Advances in Carbohydrate Chemistry and Biochemistry*, **34**, 243–283. (c) Demchenko, A., Stauch, T. and Boons, G.-J. (1997) *Synlett*, 818–820. (d) Demchenko, A.V., Rousson, E. and Boons, G.-J. (1999) *Tetrahedron Letters*, **40**, 6523–6526.

681 (a) Schmid, U. and Waldmann, H. (1996) *Tetrahedron Letters*, **37**, 3837–3840. (b) Schmid, U. and Waldmann, H. (1997) *Liebigs Annalen – Recueil*, 2573–2577.

682 Bosse, F., Marcaurelle, L.A. and Seeberger, P.H. (2002) *Journal of Organic Chemistry*, **67**, 6659–6670.

683 Duynstee, H.I., Wijsman, E.R., van der Marel, G.A. and van Boom, J.H. (1996) *Synlett*, 313–314.

684 Ratcliffe, A.J. and Fraser-Reid, B. (1990) *Journal of the Chemical Society, Perkin Transactions 1*, 747–750.

685 (a) Schmidt, R.R., Behrendt, M. and Toepfer, A. (1990) *Synlett*, 694–696. (b) Schmidt, R.R. (1996) *Modern Methods in Carbohydrate Synthesis* (eds Khan, S.H. and O'Neil, R.A.), Harwood Academic Amsterdam, pp. 20–54. (c) Schmidt, R.R. and Jung, K.-H. (1997) *Preparative Carbohydrate Chemistry* (ed. Hanessian, S.), Marcel Dekker, New York, pp. 283–312.

686 It has been concluded from the conformational analyses of glucosyl-anilinium and glucosyl-imidazolium derivatives that there is no firm evidence for the reverse anomeric effect. (a) Perrin, C.L., Fabian, M.A., Brunckova, J. and Ohta, B.K. (1999) *Journal of the American Chemical Society*, **121**, 6911–6918. (b) Perrin, C.L. and Kuperman, J. (2003) *Journal of the American Chemical Society*, **125**, 8846–8851.

687 Tsuda, T., Nakamura, S. and Hashimoto, S. (2004) *Tetrahedron*, **60**, 10711–10737.

688 Crich, D. and Chandrasekera, N.S. (2004) *Angewandte Chemie (International Edition)*, **43**, 5386–5389.

689 Watanabe, Y., Nakamoto, C., Yamamoto, T. and Ozaki, S. (1994) *Tetrahedron*, **50**, 6523–6536.

690 Paterson, I. and McLeod, M.D. (1995) *Tetrahedron Letters*, **36**, 9065–9068.

691 Hashimoto, S., Sano, A., Umeo, K., Nakajima, M. and Ikegami, S. (1995) *Chemical and Pharmaceutical Bulletin*, **43**, 2267–2269.

692 Guo, Y. and Sulikowski, G.A. (1998) *Journal of the American Chemical Society*, **120**, 1392–1397.

693 Müller, T., Hummel, G. and Schmidt, R.R. (1994) *Liebigs Annalen der Chemie*, 325–329.

694 Lin, C.-C., Shimazaki, M., Heck, M.-P., Aoki, S., Wang, R., Kimura, T., Ritzèn, H., Takayama, S., Wu, S.-H., Weitz-Schmidt, G. and Wong, C.-H. (1996) *Journal of the American Chemical Society*, **118**, 6826–6840.

695 Schwarz, J.B., Kuduk, S.D., Chen, X.-T., Sames, D., Glunz, P.W. and Danishefsky, S.J. (1999) *Journal of the American Chemical Society*, **121**, 2662–2673.

696 Yu, C.-S., Niikura, K., Lin, C.-C. and Wong, C.-H. (2001) *Angewandte Chemie (International Edition)*, **40**, 2900–2903.

697 Kondo, H., Aoki, S., Ichikawa, Y., Halcomb, R.L., Ritzen, H. and Wong, C.-H. (1994) *Journal of Organic Chemistry*, **59**, 864–877.

698 Sugai, T., Ritzén, H. and Wong, C.-H. (1993) *Tetrahedron: Asymmetry*, **4**, 1051–1058.

699 Schene, H. and Waldmann, H. (1998) *European Journal of Organic Chemistry*, 1227–1230.

700 Kartha, K.P.R., Kärkkäinen, T.S., Marsh, S.J. and Field, R.A. (2001) *Synlett*, 260–262.

701 (a) Nagai, H., Matsumura S. and Toshima, K. (2002) *Tetrahedron Letters*, **43**, 847–850. (b) Nagai, H., Sasaki, K., Matsumura, S. and Toshima, K. (2005) *Carbohydrate Research*, **340**, 337–353.

702 Sasaki, K., Nagai, H., Matsumura, S. and Toshima, K. (2003) *Tetrahedron Letters*, **44**, 5605–5608.

703 Watanabe, Y., Nakamoto, C. and Ozaki, S. (1993) *Synlett*, 115–116.

704 Tanaka, H., Sakamoto, H., Sano, A., Nakamura, S., Nakajima, M. and Hashimoto, S. (1999) *Chemical Communications*, 1259–1260.

705 Corey, E.J. and Wu, Y.-J. (1993) *Journal of the American Chemical Society*, **115**, 8871–8872.

706 (a) Marwood, R.D., Riley, A.M., Jenkins, D.J. and Potter, B.V.L. (2000) *Journal of the Chemical Society, Perkin Transactions 1*, 1935–1947. (b) Riley, A.M., Jenkins, D.J., Marwood, R.D. and Potter, B.V.L. (2002) *Carbohydrate Research*, **337**, 1067–1082.

707 van Boeckel, C.A.A., Beetz, T. and van Aelst, S.F. (1984) *Tetrahedron*, **40**, 4097–4107.

708 (a) Cheng, Y.-P., Chen, H.-T. and Lin, C.-C. (2002) *Tetrahedron Letters*, **43**, 7721–7723. (b) Fan, G.-T., Pan, Y., Lu, K.-C., Cheng, Y.-P., Lin, W.-C., Lin, S., Lin, C.-H., Wong, C.-H., Fang, J.-M. and Lin, C.-C. (2005) *Tetrahedron*, **61**, 1855–1862.

709 Wu, S.-H., Shimazaki, M., Lin, C.-C., Qiao, L., Moree, W.J., Weitz-Schmidt, G. and Wong, C.-H. (1996) *Angewandte Chemie (International Edition in English)*, **35**, 88–90.

710 Müller, T., Schneider, R. and Schmidt, R.R. (1994) *Tetrahedron Letters*, **35**, 4763–4766.

711 Olah, G.A. and McFarland, C.W. (1971) *Journal of Organic Chemistry*, **36**, 1374–1378.

712 (a) Inazu, T., Hosokawa, H. and Satoh, Y. (1985) *Chemistry Letters*, 297–300. (b) Yamanoi, T., Nakamura, K., Sada, S., Goto, M., Furusawa, Y., Takano, M., Fujioka, A., Yanagihara, K., Satoh, Y., Hosokawa, H. and Inazu, T. (1993) *Bulletin of the Chemical Society of Japan*, **66**, 2617–2622.

713 Yamanoi, T., Fujioka, A. and Inazu, T. (1994) *Bulletin of the Chemical Society of Japan*, **67**, 1488–1491.

714 Hashimoto, S., Honda, T. and Ikegami, S. (1990) *Heterocycles*, **30**, 775–778.

715 Hashimoto, S., Honda, T. and Ikegami, S. (1990) *Chemical and Pharmaceutical Bulletin*, **38**, 2323–2325.

716 Hashimoto, S., Honda, T. and Ikegami, S. (1991) *Tetrahedron Letters*, **32**, 1653–1654.

717 Chen, M.-J., Ravindran, K., Landry, D.W. and Zhao, K. (1997) *Heterocycles*, **45**, 1247–1250.

718 Pan, S., Li, H., Hong, F., Yu, B. and Zhao, K. (1997) *Tetrahedron Letters*, **38**, 6139–6142.

719 Hashimoto, S., Yanagiya, Y., Honda, T., Harada, H. and Ikegami, S. (1992) *Tetrahedron Letters*, **33**, 3523–3526.

720 Slotte, J.P., Östman, A.-L., Kumar, E.R. and Bittman, R. (1993) *Biochemistry*, **32**, 7886–7892.

721 Erukulla, R.K. and Bittman, R. (1994) *Synthetic Communications*, **24**, 2765–2770.

722 Mukaiyama, T., Kobashi, Y. and Shintou, T. (2003) *Chemistry Letters*, **32**, 900–901.

723 Niu, D., Chen, M., Li, H. and Zhao, K. (1998) *Heterocycles*, **48**, 21–24.

724 Neda, I., Sakhaii, P., Waßmann, A., Niemeyer, U., Günther, E. and Engel, J. (1999) *Synthesis*, 1625–1632.

725 (a) Gridley, J.J. and Osborn, H.M.I. (2000) *Journal of the Chemical Society, Perkin Transactions 1*, 1471–1491. (b) Pozsgay, V. (2000) *Carbohydrates in Chemistry and*

Biology (eds Ernst, B., Hart, G.W. and Sinaÿ, P.), **vol. 1**, Wiley-VCH Verlag GmbH, Weinheim, pp. 319–343. (c) ElAshry, E.S. H., Rashed, N. and Ibrahim, E.S.I. (2005) *Current Organic Synthesis*, **2**, 175–213.

726 Yamanoi, T., Nakamura, K., Takeyama, H., Yanagihara, K. and Inazu, T. (1994) *Bulletin of the Chemical Society of Japan*, **67**, 1359–1366.

727 (a) Watanabe, Y., Yamamoto, T. and Ozaki, S. (1996) *Journal of Organic Chemistry*, **61**, 14–15. (b) Watanabe, Y., Yamamoto, T. and Okazaki, T. (1997) *Tetrahedron*, **53**, 903–918. (c) Singh, G. and Vankayalapati, H. (2000) *Tetrahedron: Asymmetry*, **11**, 125–138.

728 Plante, O.J., Palmacci, E.R. and Seeberger, P.H. (2000) *Organic Letters*, **2**, 3841–3843.

729 Yamanoi, T., Nakamura, K., Takeyama, H., Yanagihara, K. and Inazu, T. (1993) *Chemistry Letters*, 343–346.

730 Tsuda, T., Arihara, R., Sato, S., Koshiba, M., Nakamura, S. and Hashimoto, S. (2005) *Tetrahedron*, **61**, 10719–10733.

731 (a) Crich, D. and Sun, S. (1996) *Journal of Organic Chemistry*, **61**, 4506–4507. (b) Crich, D. and Sun, S. (1997) *Journal of Organic Chemistry*, **62**, 1198–1199. (c) Crich, D. and Sun, S. (1997) *Journal of the American Chemical Society*, **119**, 11217–11223.

732 (a) Crich, D. and Sun, S. (1998) *Journal of the American Chemical Society*, **120**, 435–436. (b) Crich, D. and Sun, S. (1998) *Tetrahedron*, **54**, 8321–8348. (c) Crich, D. and Smith, M. (2001) *Journal of the American Chemical Society*, **123**, 9015–9020.

733 (a) Schmidt, R.R. and Weingart, R. (2000) *Tetrahedron Letters*, **41**, 8753–8758. (b) Kim, K.S., Kim, J.H., Lee, Y.J., Lee, Y.J. Park, J. (2001) *Journal of the American Chemical Society*, **123**, 8477–8481. (c) Tanaka, S., Takashina, M., Tokimoto, H., Fujimoto, Y., Tanaka, K. and Fukase, K. (2005) *Synlett*, 2325–2328. (d) Codée, J.D.C., Hossain, L.H. and Seeberger, P.H. (2005) *Organic Letters*, **7**, 3251–3254. (e) Baek, J.Y., Choi, T.J., Jeon, H.B. and Kim, K.S. (2006) *Angewandte Chemie (International Edition)*, **45**, 7436–7440.

734 Tsuda, T., Sato, S., Nakamura, S. and Hashimoto, S. (2003) *Heterocycles*, **59**, 509–515.

735 Nagai, H., Matsumura, S. and Toshima, K. (2003) *Carbohydrate Research*, **338**, 1531–1534.

736 (a) Banoub, J., Boullanger, P. and Lafont, D. (1992) *Chemical Reviews*, **92**, 1167–1195. (b) Debenham, J., Rodebaugh, R. and Fraser-Reid, B. (1997) *Liebigs Annalen – Recueil*, 791–802. (c) Bongat, A.F.G. and Demchenko, A.V. (2007) *Carbohydrate Research*, **342**, 374–406.

737 Inazu, T. and Yamanoi, T. (1989) *Chemistry Letters*, 69–72.

738 Kwon, Y.-U., Liu, X. and Seeberger, P.H. (2005) *Chemical Communications*, 2280–2282.

739 Tsuda, T., Nakamura, S. and Hashimoto, S. (2003) *Tetrahedron Letters*, **44**, 6453–6457.

740 Arihara, R., Nakamura, S. and Hashimoto, S. (2005) *Angewandte Chemie (International Edition)*, **44**, 2245–2249.

741 (a) Marzabadi, C.H. and Franck, R.W. (2000) *Tetrahedron*, **56**, 8385–8417. (b) Veyrières, A. (2000) *Carbohydrates in Chemistry and Biology* (eds Ernst, B., Hart, G.W. and Sinaÿ, P.), **vol. 1**, Wiley-VCH Verlag GmbH, Weinheim, pp. 367–405.

742 Hashimoto, S., Yanagiya, Y., Honda, T. and Ikegami, S. (1992) *Chemistry Letters*, 1511–1514.

743 Yamanoi, T. and Inazu, T. (1990) *Chemistry Letters*, 849–852.

744 Li, H., Chen, M. and Zhao, K. (1997) *Tetrahedron Letters*, **38**, 6143–6144.

745 Hashimoto, S., Sano, A., Sakamoto, H., Nakajima, M., Yanagiya, Y. and Ikegami, S. (1995) *Synlett*, 1271–1273.

746 Schene, H. and Waldmann, H. (1999) *Synthesis*, 1411–1422.

747 Nagai, H., Matsumura, S. and Toshima, K. (2002) *Chemistry Letters*, 1100–1101.

748 (a) Okamoto, K. and Goto, T. (1990) *Tetrahedron*, **46**, 5835–5857. (b) DeNinno, M.P. (1991) *Synthesis*, 583–593. (c) Schmidt, R.R., Castro-Palomino, J.C. and Retz, O. (1999) *Pure and Applied Chemistry*, **71**, 729–744. (d) Kiso,

M., Ishida, H. and Ito, H. (2000) *Carbohydrates in Chemistry and Biology* (eds Ernst, B., Hart, G.W. and Sinaÿ, P.), **vol 1**, Wilcy-VCH Verlag GmbH, Weinheim, pp. 345–365. (e) Boons, G.-J. and Demchenko, A.V. (2000) *Chemical Reviews*, **100**, 4539–4565. (f) Halcomb, R.L. and Chappell, M.D. (2002) *Journal of Carbohydrate Chemistry*, **21**, 723–768. (g) Ress, D.K. and Linhardt, R.J. (2004) *Current Organic Synthesis*, **1**, 31–46.

749 Bhattacharya, S.K. and Danishefsky, S.J. (2000) *Journal of Organic Chemistry*, **65**, 144–151.

750 Veeneman, G.H., van der Hulst, R.G.A., van Boeckel, C.A.A., Philipsen, R.L.A., Ruigt, G.S.F., Tonnaer, J.A.D.M., van Delft, T.M.L and Konings, P.N.M. (1995) *Bioorganic and Medicinal Chemistry Letters*, **5**, 9–14.

751 Sakamoto, H., Nakamura, S., Tsuda, T. and Hashimoto, S. (2000) *Tetrahedron Letters*, **41**, 7691–7695.

752 (a) Castro-Palomino, J.C., Ritter, G., Fortunate, S.R., Reinhardt, S., Old, L.J. and Schmidt, R.R. (1997) *Angewandte Chemie (International Edition in English)*, **36**, 1998–2001. (b) Castro-Palomino, J.C., Tsvetfkov, Y.E., Schneider, R. and Schmidt, R.R. (1997) *Tetrahedron Letters*, **38**, 6837–6840.

753 (a) Qiu, D. and Koganty, R.R. (1997) *Tetrahedron Letters*, **38**, 961–964. (b) Angus, D.I., Kiefel, M.J. and von Itzstein, M. (2000) *Bioorganic and Medicinal Chemistry*, **8**, 2709–2718. (c) Marcaurelle, L.A., Shin, Y., Goon, S. and Bertozzi, C.R. (2001) *Organic Letters*, **3**, 3591–3694.

754 Martichonok, V. and Whitesides, G.M. (1996) *Journal of Organic Chemistry*, **61**, 1702–1706.

755 (a) Singh, K., Fernández-Mayoralas, A. and Martín-Lomas, M. (1994) *Journal of the Chemical Society, Chemical Communications*, 775–776. (b) Coterón, J.M., Singh, K., Asensio, J.L., Domínguez-Dalda, M., Fernández-Mayoralas, A., Jiménez-Barbero, J., Martín-Lomas, M., Abad-Rodríguez, J. and Nieto-Sampedro, M. (1995) *Journal of Organic Chemistry*, **60**, 1502–1519.

756 (a) Haberman, J.M. and Gin, D.Y. (2001) *Organic Letters*, **3**, 1665–1668. (b) Galonić, D.P., van der Donk, W.A. and Gin, D.Y. (2003) *Chemistry – A European Journal*, **9**, 5997–6006.

757 Castro-Palomino, J.C., Tsvetkov, Y.E. and Schmidt, R.R. (1998) *Journal of the American Chemical Society*, **120**, 5434–5440.

758 Ercegovic, T., Nilsson, U.J. and Magnusson, G. (2001) *Carbohydrate Research*, **331**, 255–263.

759 De Meo, C., Demchenko, A.V. and Boons, G.-J. (2001) *Journal of Organic Chemistry*, **66**, 5490–5497.

760 Lin, C.-C., Huang, K.-T. and Lin, C.-C. (2005) *Organic Letters*, **7**, 4169–4172.

761 (a) Mootoo, D.R., Konradsson, P., Udodong, U. and Fraser-Reid, B. (1988) *Journal of the American Chemical Society*, **110**, 5583–5584. (b) Fraser-Reid, B., Udodong, U.E., Wu, Z., Ottosson, H., Merritt, J.R., Rao, C.S., Roberts, C. and Madsen, R. (1992) *Synlett*, 927–942.

762 (a) Boons, G.-J. (1996) *Tetrahedron*, **52**, 1095–1121. (b) Codée, J.D.C., Litjens, R.E.J.N., van den Bos, L.J., Overkleeft, H.S. and van der Marel, G.A. (2005) *Chemical Society Reviews*, **34**, 769–782.

763 Hashimoto, S., Sakamoto, H., Honda, T. and Ikegami, S. (1997) *Tetrahedron Letters*, **38**, 5181–5184.

764 Hashimoto, S., Sakamoto, H., Honda, T., Abe, H., Nakamura, S. and Ikegami, S. (1997) *Tetrahedron Letters*, **38**, 8969–8972.

765 Pongdee, R., Wu, B. and Sulikowski, G.A. (2001) *Organic Letters*, **3**, 3523–3525.

766 (a) Boger, D.L. and Honda, T. (1994) *Journal of the American Chemical Society*, **116**, 5647–5656. (b) Boger, D.L., Teramoto, S. and Zhou, J. (1995) *Journal of the American Chemical Society*, **117**, 7344–7356. (c) Thomas, C.J., Chizhov, A.O., Leitheiser, C.J., Rishel, M.J., Konishi, K., Tao, Z.-F. and Hecht, S.M. (2002) *Journal of the American Chemical Society*, **124**, 12926–12927.

767 Bock, H. and Wiegräbe, W. (1966) *Chemische Berichte*, **99**, 1068–1076.

4
Glycoside Synthesis from 1-Sulfur/Selenium-Substituted Derivatives

4.1
Thioglycosides in Oligosaccharide Synthesis
Wei Zhong, Geert-Jan Boons

4.1.1
Preparation and O-Glycosidation of Thioglycosides

Alkyl and aryl thioglycosides are versatile building blocks for oligosaccharide synthesis [1]. Owing to their excellent chemical stability, anomeric thio groups offer efficient protection of the anomeric center. However, in the presence of soft electrophiles, thioglycosides can be activated and used in direct glycosylations. Other attractive features of thioglycosides include their ability to be transformed into a range of other glycosyl donors and act as acceptors in glycosylation reactions, which make thioglycosides particularly suitable for use in chemoselective, orthogonal and iterative glycosylations [2]. This chapter reviews these properties of thioglycosides in detail.

4.1.2
Preparation of Thioglycosides

Many methods exist for the efficient preparation of thioalkyl and aryl glycosides (Table 4.1). Among these approaches, (Lewis) acid-mediated thiolysis of peracetylated sugars is the most commonly employed route. Lemieux and coworkers [3,4] were the first to demonstrate the efficiency of this reaction by preparing several 1,2-*trans* ethyl 1-thioglycopyranosides using ethanethiol as a solvent and zinc chloride as a nonprotic acid catalyst. A number of other catalysts have been reported, such as trimethylsilyl triflate (TMSOTf) [5], boron trifluoride diethyl etherate [5–10], tin(IV) chloride [11], titanium tetrachloride [12–14], iron(III) chloride [15], MoO_2Cl_2 [16] and *p*-toluenesulfonic acid [6]. The use of phosphorus oxychloride for the thioglycosidation of β-per-O-acetates has also been described [17], however, this procedure

Handbook of Chemical Glycosylation: Advances in Stereoselectivity and Therapeutic Relevance.
Edited by Alexei V. Demchenko.
Copyright © 2008 WILEY-VCH Verlag GmbH & Co. KGaA. All rights reserved.
ISBN: 978-3-527-31780-6

Table 4.1 Methods for the preparation of thioglycosides.

Starting material	Reagents	References
Peracetylated hexapyranoside	Thiol, (Lewis) acid	[3–20]
Acylated glycosyl halide	Thiolate anion	[24–34]
Acylated glycosyl halide	(i) (a) Thiourea (b) H_2O, K_2CO_3, (ii) alkylation	[36–40]
Acylated glycosyl halide	(i) Thioacetamide, (ii) RBr, phase transfer catalyst	[41,42]
Unprotected sugar	Ac_2O, acid, arene thiols	[21–23]
Dithioacetal	(i) Partial hydrolysis, (ii) alkylation	[37,45]
Acylated 1-thioaldose	Alkyl halides	[35]
Acylated 1-thioaldose	(i) Diazonium salt, (ii)	[46]
Acylated glycosyl xanthates	Sodium iodide	[48]
Acylated glycosyl thiocyanates	Grignard reagent	[49]
Acylated 1-thioaldoses	Alkene, AIBN	[47]
1-O-Alkyl glycosides	PhSSiMe$_3$, ZnI$_2$, Bu$_4$NI	[43,44]

gave poor selectivities and yields. Zirconium(IV) chloride [18,19] is an efficient catalyst in thioglycosylations leading mainly to the formation of peracetylated 1,2-*trans* 1-thioglycosides starting from the corresponding 1,2-*trans* acetylated saccharides. However, the preparation of peracetylated 1,2-*trans* 1-thiomannosides proceeded in a disappointing yield.

Treatment of *p*-methoxyphenyl (*pMP*) glycosides prepared from the corresponding 1-O-acetyl sugars using boron trifluoride etherate as promoter in combination with thiophenol gave the corresponding thioglycosides in high yield and high 1,2-*trans* selectivity [20]. The sequential per-O-acetylation and thioglycosidation of unprotected reducing sugars using a stoichiometric quantity of acetic anhydride and alkyl or aryl thiols have been reported. These reactions that are catalyzed by BF$_3$ etherate [21,22] or HClO$_4$ [23] constitute an efficient one-pot method for the synthesis of acetylated 1-thioglycosides.

1,2-*trans* Alkyl and aryl 1-thioglycosides have also been prepared by reaction of acylated glycosyl halides with thiols [24–33], disulfides [34] or, alternatively, by S-alkylation of tetra-O-acetyl-1-thiosugars [35]. A convenient and simple approach for the stereoselective synthesis of 1,2-*trans* 1-thioglycosides is based on the utilization of glycosyl isothiourea derivatives as precursors [36]. Conversion of 2,3,4,6-tetra-O-acetyl-1-thio-β-D-hexapyranoses into their pseudothiourea derivatives [37] followed by treatment with alkyl iodides (bromides) under basic conditions provides an efficient method for the synthesis of alkyl 1-thio-β-D-glucosides [38]. Recently, this procedure was successfully used for the synthesis of thio-linked oligosaccharides [39]. A simple and efficient procedure for the synthesis of thioglycosides has been achieved by the reaction of glycosylisothiouronium salts with alkyl or heteroaryl halides under microwave irradiation, which allows short reaction times. The yields of the products were comparable to conventional methods [40]. Mild and stereoselective aryl thioglycoside syntheses have also been accomplished by displacement of glycosyl halides under phase-transfer-catalyzed conditions [41,42]. Hanessian and Guindon reported a direct conversion of alkyl O-glycosides to their corresponding thioglycosides [43]. This reaction was applied recently by Liu *et al.* for the synthesis of various thioglycosyl building blocks [44].

Partial hydrolysis of dithioacetals has been found useful for the preparation of anomers, not obtained by the methods discussed above, and furanosidic thioglycosides [37,45].

Aryl thioglycosides can be obtained by the reaction of 1-thioglycopyranosides with diazonium salts, followed by thermal decomposition of the intermediate diazoproduct [46]. Acylated 1-thio-aldoses react with alkenes in the presence of azobis (isobutyronitrile) (*AIBN*) to give acylated alkyl 1-thio-glycopyranosides [47]. Thermal decomposition of glycosyl xanthates, which can be prepared by treating acylated glycopyranosyl halides with potassium alkyl or benzyl xanthate, gives the corresponding 1-thioglycosides [48]. Acylated glycopyranosyl thiocyanates can be prepared by reaction of acylated glycopyranosyl halides with potassium thiocyanate [49]. Treatment of the resulting product with Grignard reagents led to the formation of alkyl and acyl thioglycosides.

4.1.3
Indirect Use of Thioglycosides in Glycosidations

Thioglycosides can be transformed into a range of other glycosyl donors (Scheme 4.1). For example, treatment of a thioglycoside with bromine gives a glycosyl bromide, which after work up can be used in a Hg(II), Ag(I) [50] or phosphine oxide [51] promoted glycosylations. Iodine monobromide, an efficient reagent for the conversion of both activated and deactivated thioglycosides into glycosyl bromides, also permits the glycosylation via a bromide intermediate [52].

A glycosyl bromide can also be prepared *in situ* followed by glycosylation by reaction with $(Bu_4N)_2CuBr_4$ and AgOTf [8,53] or Et_4NBr and N,N,N',N'-tetramethylurea [54]. In an alternative approach, $AgOTf/Br_2$ was used as the activation reagent. A thioglycoside can also be converted into a glycosyl fluoride by treatment with *N*-bromosuccinimide/(diethylamino)sulfur trifluoride (NBS/DAST) [55–57], or hydrolyzed to give the corresponding aldose using a number of reagents such as NBS or *N*-iodosuccinimide (*NIS*) in wet acetone [58–60], $AgNO_3$ in wet acetone [61,62], $NBS/NaHCO_3$ (aq) or $CaCO_3$ (aq) in THF [63], NBS/HCl [64], nBu_4NIO_4/TrB $(C_6H_5)_4$, nBu_4NIO_4/trifluoromethanesulfonic acid (TfOH), $nBu_4NIO_4/HClO_4$

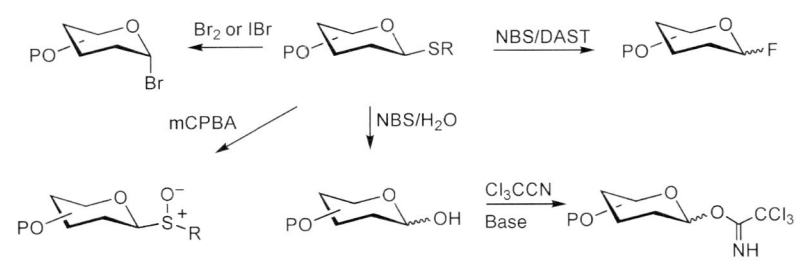

Scheme 4.1 Leaving-group interconversions of thioglycosides.

[65], $(NH_4)_6Mo_7O_{24} \cdot 4H_2O-H_2O_2$ with $HClO_4/NH_4Br$ [66], $V_2O_5-H_2O_2/NH_4Br$ [67], chloramine T [68], NIS/TfOH [69] and NIS/TFA [70]. The resulting hemiacetals are suitable substrates for the preparation of anomeric trichloroacetimidates [71–73]. Finally, another approach involves the oxidation of a thioglycoside to the corresponding sulfoxide using mCPBA [74–78], hydrogen peroxide–acetic anhydride–SiO_2 [76], oxone [79,80], selectfluor [81], magnesium monoperoxyphthalate (*MMPP*) [82] or tert-butyl hydroperoxide [83]. The resulting compound can then be activated with triflic anhydride at low temperature to give glycosides [74,76,84–88].

4.1.4
Direct Use of Thioglycosides in Glycosidations

Ferrier *et al.* reported, for the first time, the use of thioglycosides in direct glycosylations [89]. A number of phenyl 1-thioglucopyranosides were solvolyzed in methanol in the presence of $Hg(OAc)_2$ to give the corresponding methyl glycosides. These reactions proceeded with inversion of anomeric configuration and gave only acceptable yields when reactive sugar alcohols were employed. For example, the reaction of phenyl 1-thioglucopyranosides with 1,2:3,4-di-*O*-isopropylidene-α-D-galactopyranose gave an α-linked disaccharide in a yield of 54%. Several other heavy-metal-salt promoters (Table 4.2) have been proposed for the activation of thioglycosides: a notable example being $Pd(ClO_4)_2$, which was used by Woodward *et al.* [90] for the synthesis of Erythromycin A and by Wuts and Bigelow [91] for the preparation of Avermectin.

Despite these important achievements, heavy-metal-salt-mediated activation of thioglycosides did not give high yields consistently and consequently did not find wide application in glycosidic bond chemistry. Lönn demonstrated [12] that methyl triflate is an efficient thiophilic promoter and glycosylations mediated by this reagent usually gave good yields of glycosides. For example, thioglycosides activated with MeOTf were applied for the preparation of a saccharide component of a glycoprotein isolated from fucosidosis patients and for the preparation of phytoelicitor oligosaccharides involved in the recognition and defense of soybean plants against infections by *Phytophthora megasperma*.

Methyl triflate is highly toxic and can methylate hydroxyls when glycosyl acceptors of low reactivity are used. Intensive research has focused on finding alternative reagents with more favorable properties, and today the most commonly used reagents include dimethyl(methylthio) sulfonium triflate (*DMTST*) [92], *N*-iodosuccinimide-triflic acid (NIS-TfOH) or NIS/TMSOTf [93,94], iodonium dicollidine perchlorate (*IDCP*) [95,96] and phenylselenyl triflate (PhSeOTf) [97,98]. The activation of thioglycosides involves the reaction of an electrophilic species with the sulfur lone pair, resulting in the formation of a sulfonium intermediate. The latter intermediate is an excellent leaving group and can be displaced by a sugar hydroxyl.

Recently, a number of thiophilic activators that can activate thioglycosides of low reactivity at low temperature have been described. For example, thiophilic promoter systems, such as diphenylsulfoxide [87,99], *S*-(4-methoxyphenyl) benzenethiosulfinate

Table 4.2 Glycosidation of thioglycosides.

Activator	SR	References
$HgSO_4$	SPh	[89]
$HgCl_2$	SEt, SPh	[89,251]
PdHgOTf	SPh	[252]
$Hg(OBz)_2$	SPh	[253]
$Hg(NO_3)_2$	SPh	[254]
$Cu(OTf)_2$	S–(benzothiazol-2-yl)	[255]
$Pd(ClO_4)_2$	SPy	[90,91]
$CuBr_2/Bu_4NBr/AgOTf$	SMe, SEt	[8,53]
PhSeOTf	SMe	[97,98]
N-(Phenylseleno)phthalimide/TMSOTf	SMe, SPh	[256]
$AgOTf/Br_2$	SEt	[8]
NBS	SPh	[257]
NIS/TfOH	SMe, SEt, SPh	[93,94]
IDCP	SEt	[95,96]
IDCT	SEt	[258]
I_2	SMe	[259]
$PhIO/Tf_2O$	SMe	[260]
$NOBF_4$	SMe	[261,262]
MeI	SPy	[263]
MeOTf	SEt	[12]
PhSOTf	SMe, SEt, SPh	[264]
DMTST	SMe, SEt, SPh	[92]
$TrClO_4$	SCN Ph	[154–156]
AgOTf	S–(1-phenyl-1H-tetrazol-5-yl)	[265]
TBPA	SEt, SPh	[102,103]
e	SPh	[104–106]
$TrB(C_6F_5)_4/I_2/DDQ$	SEt	[107]
$TrB(C_6F_5)_4/NaIO_4$	SMe, SEt, SPh	[108,110]
$TrB(C_6F_5)_4/PhthNSEt$	SEt	[109]
NBS/TfOH	SPh	[266]
NBS/Me_3SiOTf	SEt, SPh	[267]
1-Fluoropyridinium triflates	SEt	[268]
$NIS/HClO_4$–Silica	STol	[269]
NBS/strong acid salts	SMe, SPh	[270]
IX/AgOTf (X = Cl, Br)	SMe, SEt, SPh	[271,272,111]
I_2/hexamethyldisilane (HMDS) or IX (X = Cl, Br)	SMe	[112,113]
NISorNBS/$TrB(C_6F_5)_4$	SEt	[273]
Ph_2SO/Tf_2O	SPh	[87,99]

(continued)

Table 4.2 (Continued)

Activator	SR	References
N-(Phenylthio)-|Å-caprolactam	STol	[274]
Benzenesulfinyl morpholine /triflic anhydride (BSM/Tf$_2$O)	STol	[101]
N-Phenylselenophthalimide–Mg(ClO$_4$)$_2$/PhIO–Mg(ClO$_4$)$_2$	SMe, SPh	[275]
S-(4-Methoxyphenyl) benzenethiosulfinate/triflic anhydride (MPBT/Tf$_2$O)	SPh	[100]
1-Benzenesulfinyl piperidine /2,4,6-tri-tertbutylpyrimidine /triflic anhydride (BSP/TTBP/Tf$_2$O)	SEt, SPh	[85]
AgOTf	(benzoxazolyl-S, thiazolinyl-S structures)	[114–118]

(MPBT) [100], benzenesulfinyl morpholine (BSM) [101] or 1-benzenesulfinyl piperidine/2,4,6-tri-*tert*-butylpyrimidine (BSP/TTBP) [85], in combination with triflic anhydride (Tf$_2$O) provide high yields of products for difficult glycosylations.

Thioglycosides can also be activated by a one-electron transfer reaction from sulfur to the activating reagent tris-(4-bromophenyl)ammoniumyl hexachloroantimonate (TBPA$^+$) [102,103]. The use of this promoter was inspired by an earlier report where activation was achieved under electrochemical conditions to give an intermediate S-glycosyl radical cation intermediate [104], and the reactivity and mechanism have also been explored [105,106].

A combined use of trityl tetrakis(pentafluorophenyl) borate [TrB(C$_6$F$_5$)$_4$], iodine (I$_2$) and 2,3-dichloro-5,6-dicyano-*p*-benzoquinone (DDQ) effectively activates thioglycosides of low reactivity, whereas a combined use of trityl tetrakis(pentafluorophenyl) borate and N-(ethylthio)phthalimide (PhthNSEt) activates highly reactive thioglycosides. The use of trityl tetrakis(pentafluorophenyl) borate and NaIO$_4$ as co-oxidant can activate thioglycosides as well. A selective use of trityl salts to activate thioglycosides has been applied in a one-pot glycosylation [107–110]. ICl/AgOTf works well for glycosylations with thioglycosyl donors having a participating group at C-2 of the glycosyl donor, whereas IBr/AgOTf is superior for glycosyl donors having a nonparticipating group at this position. The interhalogens in combination with silver triflate have been applied in the synthesis of bislactam analogs of Ganglioside GD3. IX promoter systems offer convenient handling of reagents and do not produce by-products such as N-succinimide, which is released in the popular NIS/TMSOTf-promoted glycosylations [111–113].

S-Benzoxazolyl (SBox) and, especially, S-thiazolinyl (STaz) moieties are sufficiently stable for use in anomeric protection. These derivatives can, however, be activated under mild conditions using silver triflate [114–118].

4.1.5
Anomeric Control in Glycosidations of Thioglycosides

The protecting group at C-2 of a glycosyl donor is an important determinant of the stereochemical outcome of a glycosylation [119–121]. In general, participating groups at C-2, such as O-acetyl, O-benzoyl and N-phthaloyl, lead to the formation of 1,2-*trans* glycosides, whereas nonparticipating groups, such as benzyl ethers, give mixtures of anomers (Scheme 4.2). The anomeric outcome of glycosylations with glycosyl donors having a nonparticipating group at C-2 is markedly influenced by the nature of the solvent [122]. In general, solvents of low polarity are thought to increase α-selectivity by suppressing the formation of oxacarbenium ions. Solvents of moderate polarity, such as mixtures of toluene and nitromethane, are highly beneficial when the glycosyl donors have participating C-2 substituents. It is likely that these solvents stabilize the positively charged intermediates.

Mechanistic studies [123] have shown that thioglycosides can undergo *in situ* anomerization in the presence of iodonium ion catalysts. It has been demonstrated that this anomerization proceeds by intermolecular exchange of alkyl thio groups. An increase in the steric bulk of the leaving group resulted in incomplete or no anomerization. It has been proposed that this anomerization process is important for the stereochemical outcome of glycosylations [123].

Some solvents form complexes with the oxacarbenium ion intermediates, thereby affecting the stereoselectivity of glycosylations. For example, diethyl ether is known to increase α-anomeric selectivity, presumably by formation of a diethyl oxonium-ion intermediate. The β-configuration of this intermediate is probably favored due to steric reasons. Nucleophilic displacement with inversion of configuration will then give an α-glycoside. Boons and coworkers showed that iodonium-ion-mediated glycosidations of thioglycosides in toluene/1,4-dioxane give much higher α-selectivities than when conventional glycosylation solvents are employed [124]. Furthermore, it was shown that the iodonium-ion source, glycosyl donor/acceptor ratio and presence of molecular sieves also have major impacts on the stereochemical outcome of a glycosylation.

Acetonitrile is another participating solvent, which in many cases leads to the formation of an equatorially linked glycoside [125–131]. It has been proposed that these reactions proceed via an α-nitrilium ion intermediate. It is not well understood why the nitrilium ion adopts an axial orientation; however, spectroscopic studies support the proposed anomeric configuration [130,131]. It is known that nucleophilic substitution of the α-nitrilium ion by an alcohol leads to β-glycosidic bonds and the best β-selectivities are obtained when reactive alcohols at low reaction temperatures are employed. Unfortunately, mannosides give poor anomeric selectivities under these conditions.

β-Mannosides are difficult to introduce because the axial C-2 substituent of a mannosyl donor sterically and electronically disfavors nucleophilic attack from the β-face. β-Mannosides have been obtained by the direct substitution of α-glycosyl triflates, which are conveniently prepared by the treatment of an anomeric sulfoxide with triflic anhydride (Tf_2O) or thioglycosides with NIS (Scheme 4.3a)

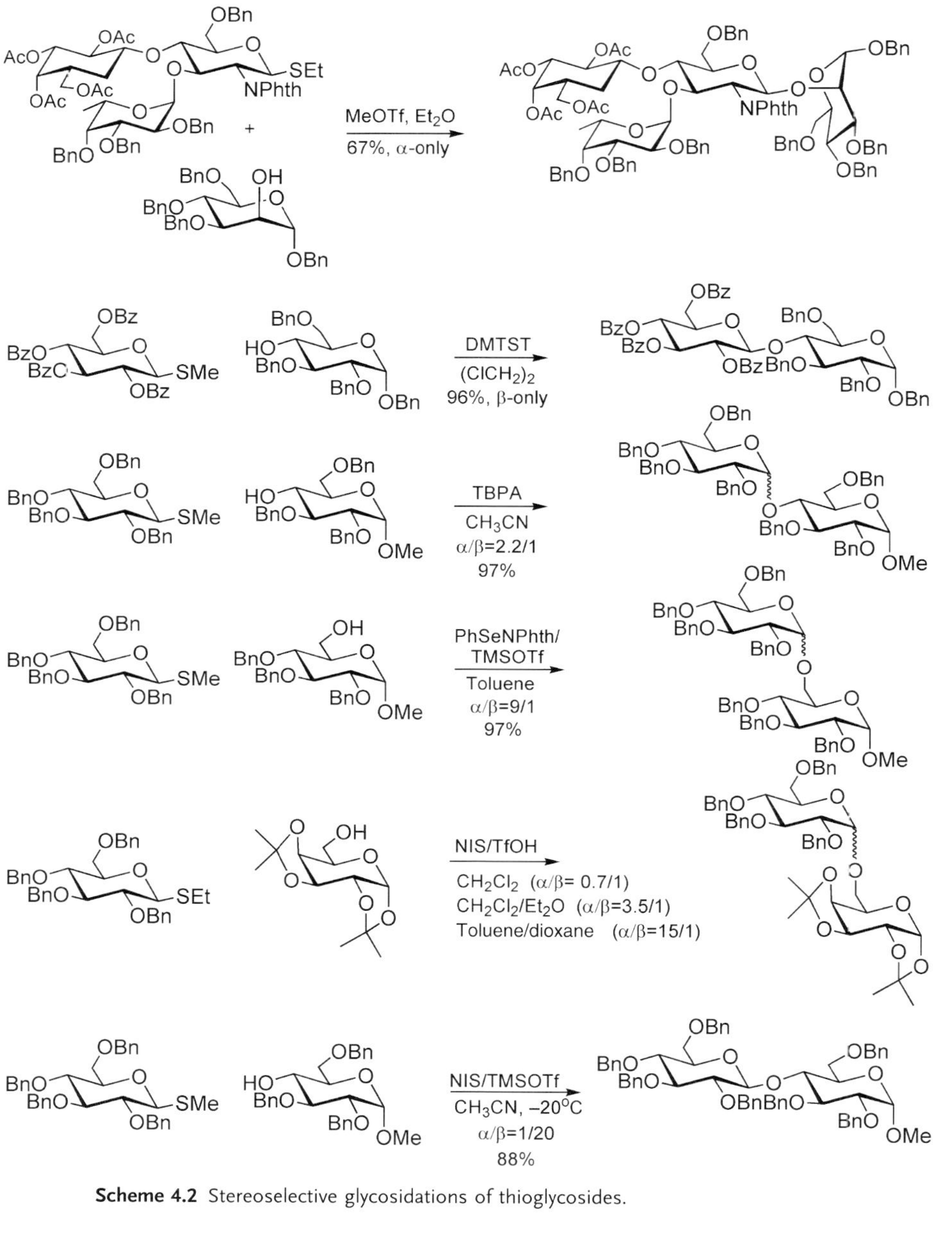

Scheme 4.2 Stereoselective glycosidations of thioglycosides.

[128,132–134]. An α-triflate is formed because this anomer is stabilized by a strong *endo*-anomeric effect. Upon addition of an alcohol, the triflate is displaced in an S$_N$2 fashion resulting in the formation of a α-mannoside. A mixture of anomers is obtained when triflic anhydride is added to a mixture of sulfoxide and alcohol. In

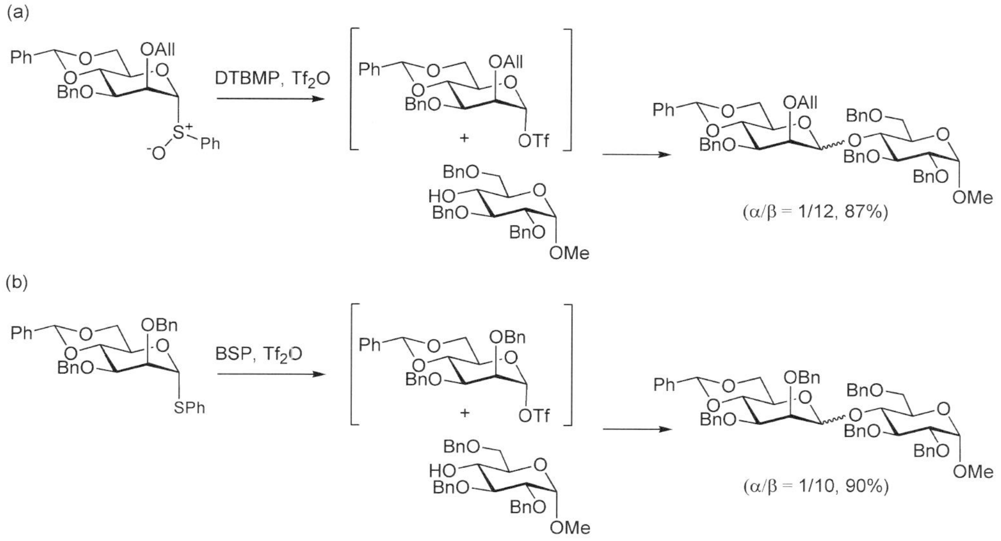

Scheme 4.3 Glycosidation of intermediate α-triflates.

this case, it is very likely that the glycosylation proceeds through an oxacarbenium ion because triflate formation is less likely owing to the greater nucleophilicity of an alcohol.

Another prerequisite of β-mannoside formation is the protection of the mannosyl donor as a 4,6-O-benzylidene acetal. Although this observation is difficult to rationalize, it has been suggested that oxacarbenium ion formation is disfavored because of the torsional strain engendered on going to the half-chair conformation of this intermediate. Crich and Chandrasekera employed α-deuterium kinetic isotope effects to unravel the mechanism of 4,6-O-benzylidene-directed β-mannosylation. It was found that a torsionally disarming benzylidene acetal opposed rehybridization at the anomeric carbon, thereby shifting the complete set of equilibria toward the covalent triflate and away from the solvent-separated ion pair (*SSIP*), resulting in minimization of α-glycoside formation [135].

Recently, powerful and metal-free thiophilic reagents have been shown to readily activate thioglycosides via glycosyl triflates leading to β-mannosides. For example, a combination of BSP and Tf$_2$O in the presence of TTBP [85] or MPBT and Tf$_2$O in the presence of DTBMP [100] at low temperature has been used to prepare β-mannosides in good yield and high β-anomeric selectivity. It was also found that 2-O-propargyl ethers were advantageous in the 4,6-O-benzylidene acetal-directed β-mannosylations (Scheme 4.3b) [136,137]. This approach has been applied to the synthesis of β-mannans from *Rhodotorula glutinis*, *Rhodotorula mucilaginosa* and *Leptospira biflexa* [138]. van Boom and coworkers developed the very potent thiophilic glycosylation promoter system, diphenylsulfoxide in combination with triflic anhydride, to activate thioglycosides for β-mannosylation [87,99,139]. Furthermore, Demchenko and coworkers found that the stereoselectivity of β-mannosylation could be improved when a participating moiety at C-4 (O-anisoyl, O-thiocarbamoyl) is employed. This improvement was achieved in glycosidations of S-ethyl and, especially, SBox mannosides [140].

Stork and coworkers [141,142] and Hindsgaul and coworkers [143–145] reported independently the preparation of β-mannosides in a highly stereoselective manner by intramolecular aglycon delivery (*IAD*). In this approach, a sugar alcohol (ROH) is first linked via an acetal or silicon tether to the C-2 position of a mannosyl donor and the subsequent activation of the anomeric center of this adduct forces aglycon delivery from the β-face of the glycosyl donor. The remnant of the tether hydrolyses during the work-up procedure (Scheme 4.4a). A silicon tether was easily introduced by the conversion of a glycosyl acceptor into a corresponding chlorodimethyl silyl ether and the subsequent reaction with the C-2 hydroxyl of a donor to give the silicon-tethered compound [141,142]. Oxidation of the phenylthio group yielded a phenylsulfoxide, which upon activation with Tf_2O resulted in the selective formation of a β-mannoside in a 61% overall yield. Alternatively, the direct activation of thioglycosides also resulted in the formation of β-mannosides. Acetal tethers could easily be prepared by the treatment of equimolar amounts of a 2-propenyl ether derivative of a saccharide with a sugar hydroxyl in the presence of a catalytic amount of acid (Scheme 4.4b) [143–145]. Activation of the anomeric thio moiety of the tethered compound with NIS in dichloromethane resulted in the formation of β-linked disaccharides. In this reaction, no α-linked disaccharide could be detected. It is of interest to note that when this reaction was performed in the presence of methanol, no methyl glycosides were obtained. This experiment indicates that the glycosylation proceeds through a concerted reaction and not by addition to an anomeric oxacarbenium ion.

Fairbanks modified the intramolecular aglycon delivery to achieve stereospecific 1,2-*cis* glycosylation via 2-O-vinyl thioglycosides, which were synthesized from the corresponding alcohols by Ir-catalyzed transvinylation with vinyl acetate, followed by iodine-mediated tethering of a range of primary and secondary carbohydrate acceptors and finally intramolecular aglycon delivery [146–150]. The use of such an intramolecular glycosylation strategy furnished the desired α-gluco and β-manno disaccharides in a stereoselective manner [146–149]. The methodology has been applied for the synthesis of a tetrasaccharide derived from *N*-linked glycans [150].

An intramolecular acetal has also been introduced by the treatment of a mixture of a 1-thio-mannoside, having a methoxybenzyl protecting group at C-2 and an alcohol with DDQ [71] (Scheme 4.4c). Activation of the thioglycoside with methyl triflate gave a β-mannoside as the only anomer. This approach was employed for the synthesis of the core pentasaccharide of *N*-linked glycoproteins.

Ziegler and coworkers prearranged a glycoside by employing a succinyl tether between C-6 of a mannosyl donor and C-3 of glucosyl acceptor [151,152]. They found that the nature of the glycosyl acceptor and the length of the tether affected the anomeric selectivity of the intramolecular mannosylation (Scheme 4.4d) [153].

Kochetkov and coworkers have reported [154–156] an efficient approach for the synthesis of 1,2-*cis* pyranosides employing 1,2-*trans*-glycosyl thiocyanates as glycosyl donors and tritylated sugar derivatives as glycosyl acceptors (Scheme 4.5). This

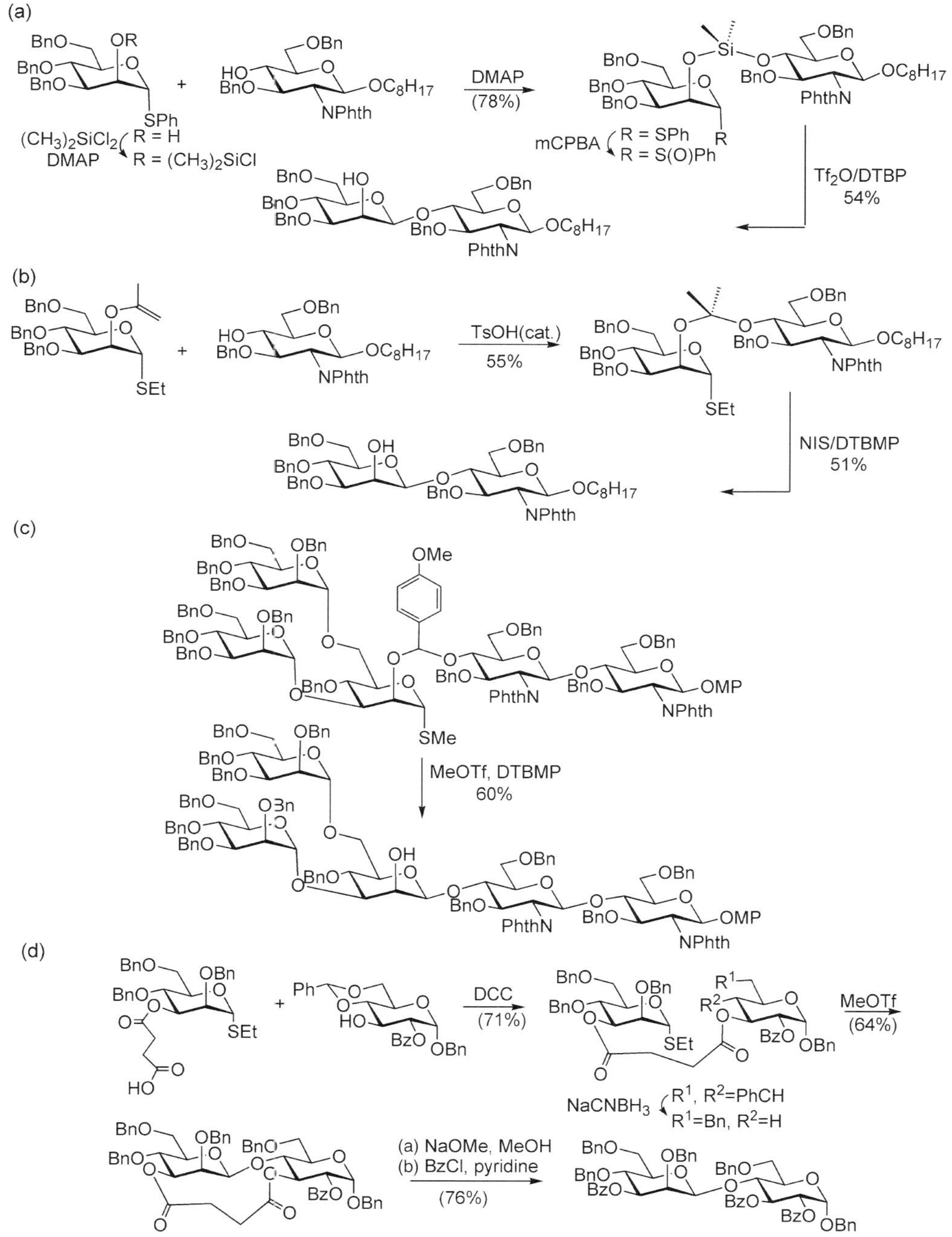

Scheme 4.4 Synthesis of β-mannosides by intramolecular aglycon delivery.

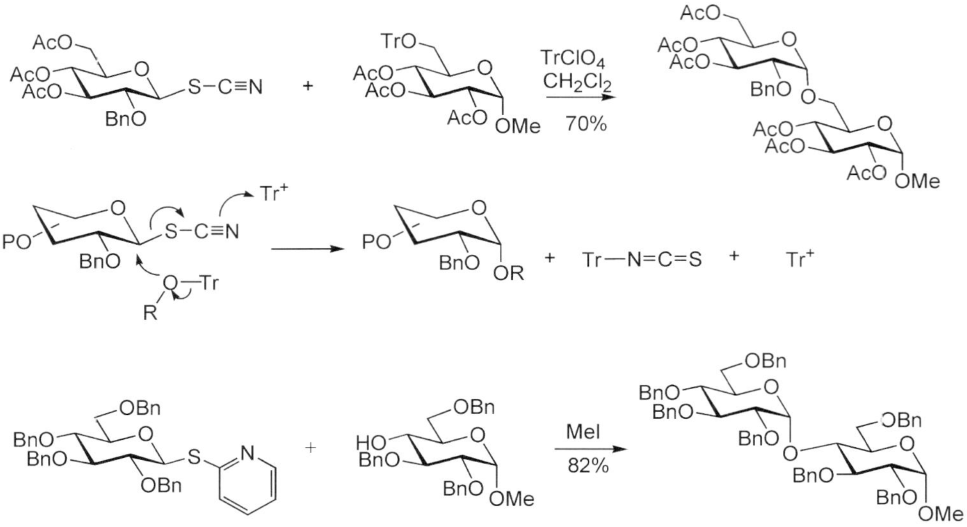

Scheme 4.5 Glycosidations of thioglycosides by inversion of configuration.

coupling is initiated by the reaction of the nitrogen of the thiocyanate with trityl cation from TrClO$_4$. This results in leaving-group departure accompanied by simultaneous nucleophilic attack by a trityl-protected sugar alcohol to give an α-glycoside. It appears that this reaction proceeds by clean S$_N$2 inversion of configuration at the anomeric center. Mereyala and coworkers [157–159] used 2-pyridyl 1-thioglycosides having a nonparticipating C2-substituent as glycosyl donors and methyl iodide as an activator to achieve stereoselective α-glycosylations in the D-gluco and D-galacto series. The reaction is proposed to proceed via the electrophilic activation of the glycosyl donor by methyl iodide, followed by the formation of sulfenium salt. An alcohol displaces the latter intermediate via an S$_N$2 mechanism.

2-Thio-sialyl glycosides (Scheme 4.6a) are commonly used for the preparation of α-sialyl glycosides [97,98,160–163]. The best yields and anomeric selectivities have been obtained when partially protected galactosyl acceptors are employed. Furthermore, it has been found that the reactivity of sialyl thioglycosyl donors can be significantly increased by acetylation of the acetamido group [164] (Scheme 4.6b). This modification enables the efficient synthesis of α-(2 → 8)-dimers of Neu5Ac [165].

Boons and coworkers [165,166] modified the C-5 amino group of 2-methyl and 2-thiophenyl sialosides into N-TFA derivatives, which provided a glycosyl donor that gives good yields and high α-anomeric selectivities in direct sialylations with a wide range of glycosyl acceptors of differing reactivities (Scheme 4.6c). Sialyl acceptor, protected as an N-TFA derivative, gave the best yields and it was postulated that lower nucleophilicity of the TFA-protected amino functionalities and enhanced reactivity affect the efficiency of the glycosylations.

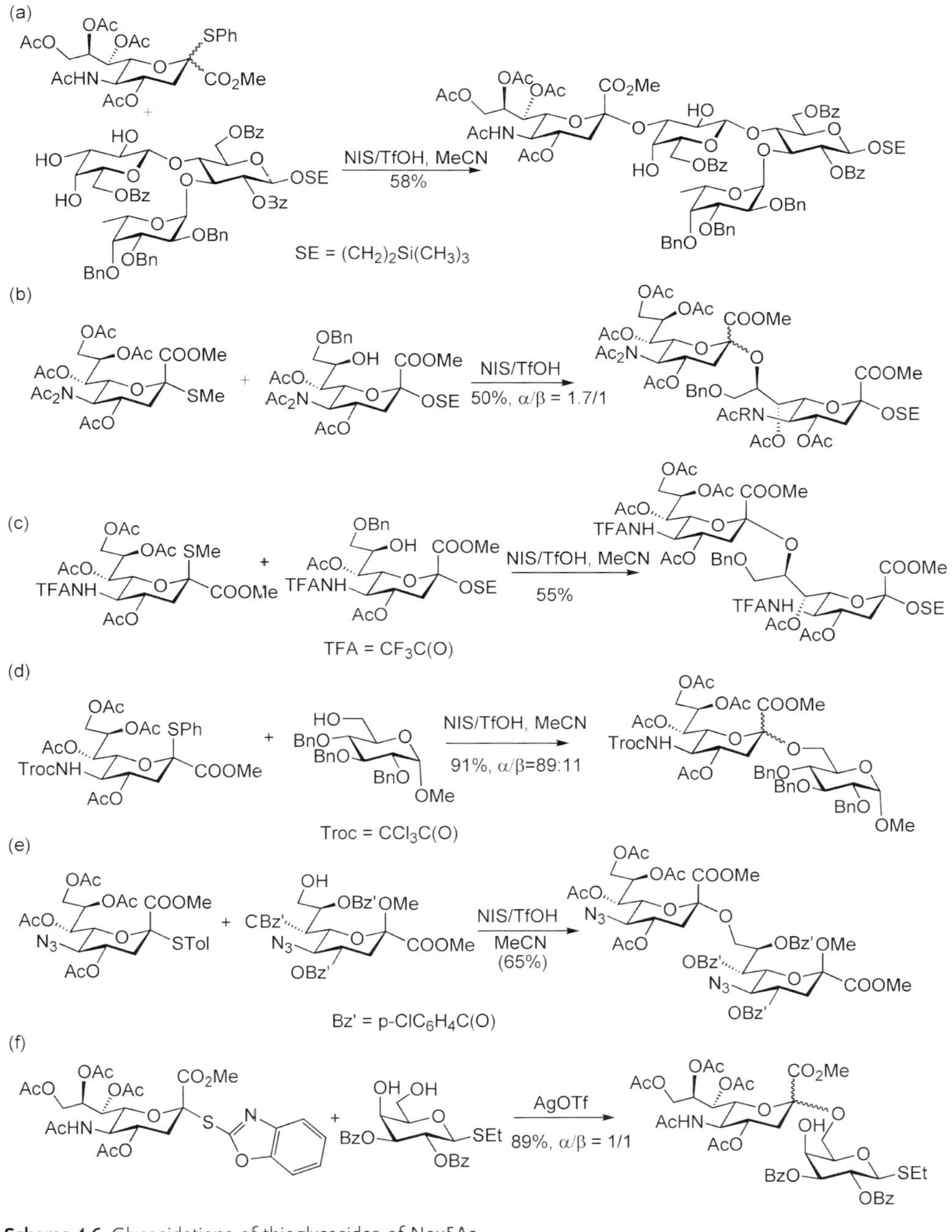

Scheme 4.6 Glycosidations of thioglycosides of Neu5Ac.

Takahashi and coworkers described an effective sialylation method utilizing the N-Fmoc, N-Troc and N-trichloroacetyl-β-thiophenyl sialosides (Scheme 4.6d) [167]. It was found that the N-Troc derivative of N-acetylneuraminic acid performed better than the corresponding N-Fmoc derivative. An N-Troc β-thiosialoside was applied for the synthesis of glycosyl amino acids by one-pot glycosylation [167]. Importantly, it was found that the N-Troc protecting group could be converted into an acetamido moiety without causing racemization of the peptide.

Another effective α-selective sialylation involves the use of a 5-N,4-O-carbonyl-protected sialyl donor, which could efficiently be used for the preparation of an α(2,8)-tetrasialoside. It was found that the 5-N,4-O-carbonyl protecting group improves the reactivity of the C-8 hydroxyl group of the sialyl acceptor [168].

Wong and coworkers showed that 5-azido sialyl donors protected with O-acetyl ester are useful for α-selective glycosylations of primary hydroxyls (Scheme 4.6e) [169]. It was proposed that the linear and electron-withdrawing nature of the C-5 azido moiety stabilizes the reactive axial acetonitrile adduct to allow the incoming nucleophile to approach the α-face in an S_N2-like fashion. In addition, a chemoselective glycosylation method has been developed for the synthesis of NeuAcα-(2 → 9) NeuAc as thioglycoside donor for use in the subsequent glycosylations [169].

Recently, De Meo and Parker described two novel sialyl donors bearing a thioimidoyl moiety as leaving group (Scheme 4.6f) [170]. The SBox and STaz sialosides proved to be excellent glycosyl donors when activated with MeOTf or AgOTf. In general, good yields and stereoselectivities were observed with a number of glycosyl acceptors ranging from highly reactive primary hydroxyls to less reactive secondary hydroxyls. The most attractive feature of thiomidoyl moieties is that they can be selectively activated in the presence of thioglycosides using AgOTf as promoter.

In brief, the use of acetonitrile as solvent and the selection of an appropriate C-5 amino protecting group and reactive promoter system are critical for achieving high α-selectivities and yields in the synthesis of sialosides.

4.1.6
Glycosylation Strategies Using Thioglycosides

4.1.6.1 Chemoselective Glycosylations

van Boom and coworkers showed that the reactivity of thioglycosides can be controlled by the selection of appropriate protecting groups. It was found that a C-2 ether protecting group activates and a C-2 ester deactivates the anomeric center [96]. This difference in reactivity was exploited for attractive chemoslective glycosylations. For example, iodonium-ion-mediated coupling of a fully benzylated thioglycoside with a partially benzoylated thioglycosyl acceptor gave a disaccharide mainly as the α-anomer in a yield of 84% (Scheme 4.7). It has been established that the resulting disarmed thioglycosyl disaccharide can be readily activated using the strong thiophilic promoter NIS/TfOH. The subsequent coupling with a glycosyl acceptor gives a trisaccharide [93,171–175]. The chemoselective glycosylation approach was rationalized as follows: the electron density of the anomeric sulfur atom in a 2-O-acyl ethylthio glycoside is decreased because of the inductive effects by the

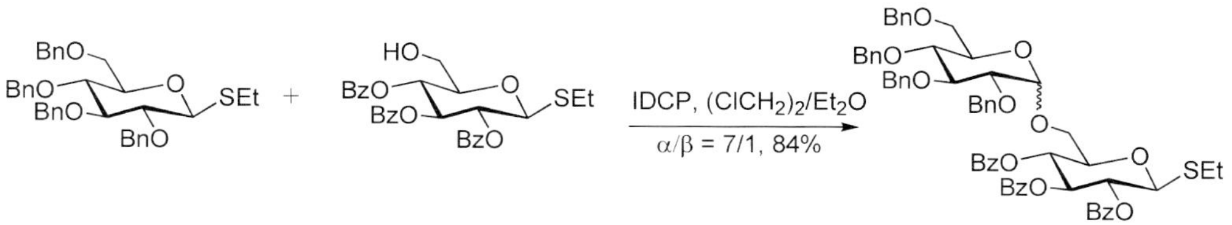

Scheme 4.7 Armed–disarmed glycosylations of thioglycosides.

electron-withdrawing ester functionality at C-2. As a result, nucleophilic complexation of the anomeric thio group with iodonium ions decreases and the thioglycoside can be regarded as disarmed with respect to an armed 2-O-alkyl thioglycoside. It is important to note that Fraser-Reid and coworkers introduced the armed–disarmed glycosylation protocol using n-pentenyl glycosides as glycosyl donors and acceptors [94,176,177].

Ley and coworkers proposed [178–186] that the armed–disarmed glycosylation strategy could gain versatility by further tuning of glycosyl donor leaving-group ability. In this respect, a dispiroketal or a butane-2,3-diacetal (BDA) protecting group has a marked effect on the reactivity of the anomeric center. It was found that thioglycosides protected with these functionalities have reactivities between an armed C-2-alkylated thioglycoside and a disarmed C-2-acylated thioglycoside (Scheme 4.8). For example, the three levels of anomeric reactivity were exploited for the preparation of a protected pseudopentasaccharide unit common to the variant surface glycoprotein of *Trypanosoma brucei* [178]. Thus, iodonium dicollidine perchlorate (IDCP)-mediated chemoselective glycosylation of benzylated-thioglycosyl donor with dispiroketal-protected acceptor gave a disaccharide in excellent yield (82%, α/β = 5/2). Further chemoselective glycosylation of the torsionally deactivated glycosyl donor with an electronically deactivated acceptor in the presence of the more powerful activator NIS/TfOH gave a 63% yield of trisaccharide as one isomer. The pseudopentasaccharide was obtained by NIS/TfOH-mediated condensation of the trisaccharide donor with a pseudodisaccharide acceptor.

In the armed–disarmed glycosylation approach, the leaving-group ability is controlled by protecting groups (ether/dispiroketal/ester). It may, however, be advantageous to control the anomeric reactivity by means of modifying the leaving group. Boons and coworkers [187,188] showed that the bulkiness of the anomeric thio group has a marked effect on the glycosyl donor reactivity and provides an opportunity to produce a new range of differentially reactive coupling substrates. For example, IDCP-mediated chemoselective glycosylation of a fully benzylated ethyl thioglycosyl donor with a partially benzylated dicyclohexylmethyl thioglycosyl acceptor gave a disaccharide in a yield of 45% as one anomer (Scheme 4.9). Further chemoselective coupling of the resulting sterically deactivated donor with an electronically deactivated glycosyl acceptor in the presence of the more powerful promoter system NIS/TfOH gave a trisaccharide in a yield of 70%. In both glycosylations, no self-condensed or polymeric products were detected (Scheme 4.9a). These experiments show that the reactivity of a C2-benzylated dicyclohexylmethyl thioglycoside is

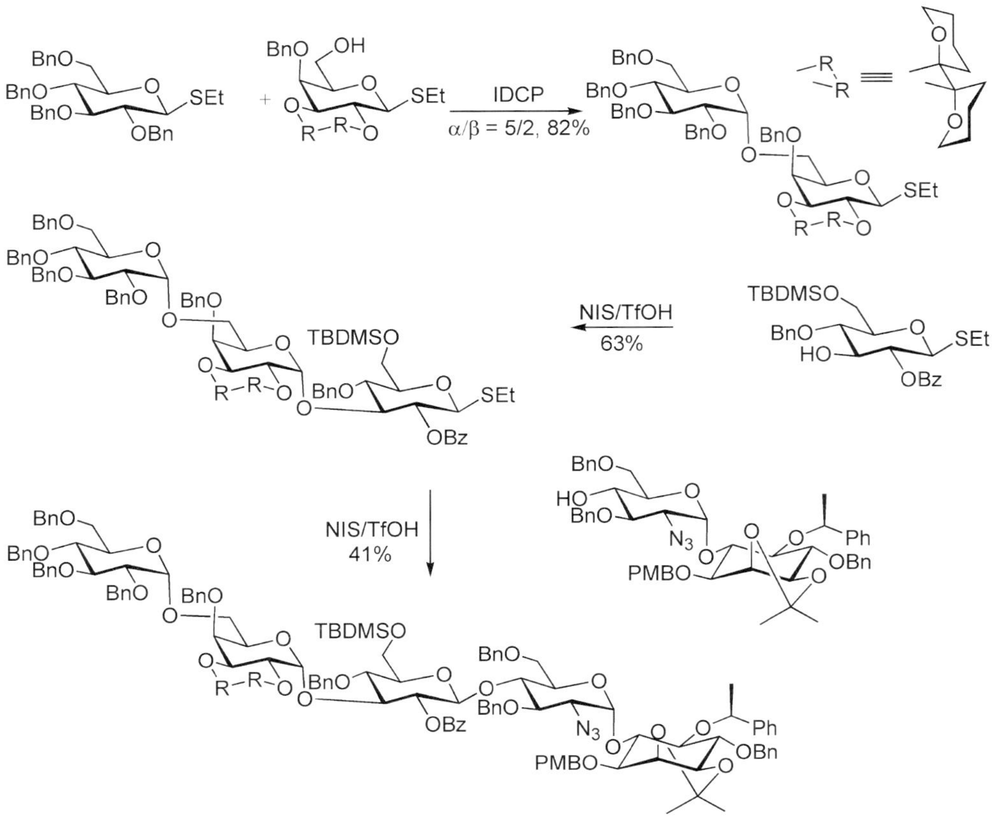

Scheme 4.8 Chemoselective glycosidations of thioglycosides.

between ethyl thioglycosides having a fully armed ether and disarmed ester protecting group on C-2. This new approach to tuning thioglycoside reactivity was employed for the preparation of a phytoalexin-elicitor active oligosaccharide and its photoreactive derivatives (Scheme 4.9b) [189].

The *trans*-2,3-cyclic carbonate function was introduced as a nonparticipating thioglycoside, which deactivates the anomeric center of thioglycosides by both electronic and conformational effects. These thioglycosides are significantly less reactive than corresponding ones having ester-protecting groups at C-2 [190]. Thioglycosides protected as a *trans*-2,3-cyclic carbonate remain intact upon treatment with thiophilic promoters such as NIS/TMSOTf, NIS/AgOTf and MeOTf. However, the activator PhSOTf, generated *in situ* by the reaction of PhSCl with AgOTf, can activate these thioglycosides. It was concluded that thioglycosides protected as *trans*-2,3-cyclic carbonates have significantly lower anomeric reactivities compared to the fully acylated and the *N*-acyl protected thioglycosides. As a result, these derivatives can be used as acceptors in chemoselective glycosylations

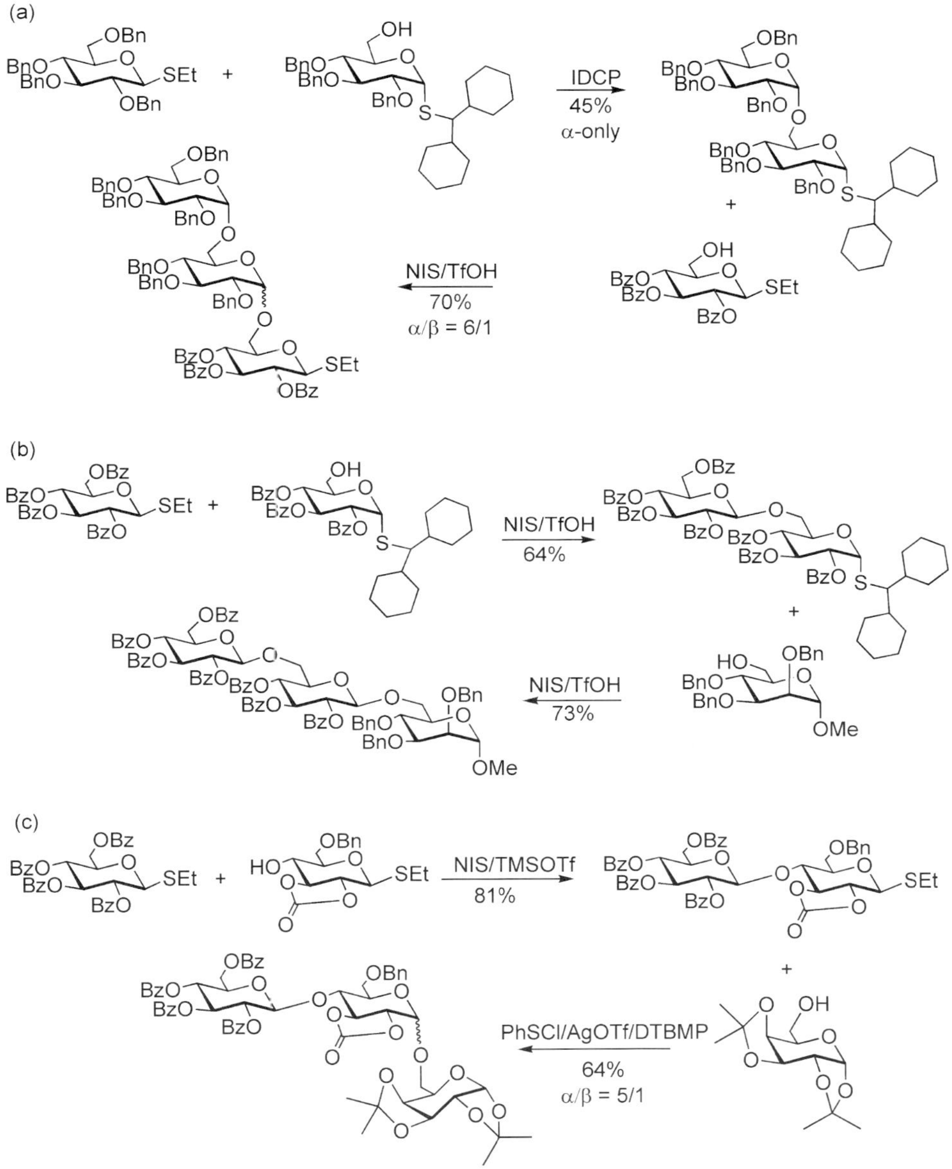

Scheme 4.9 Chemoselective glycosidations with sterically deactivated thioglycosides.

with a wide range of C2-alkylated or -acylated thioglycosyl donors (Scheme 4.9c). An interesting feature of these disarmed donors is that they permit the introduction of a 1,2-*cis* glycosides, whereas this is not possible with classical 2-acyl-disarming derivatives.

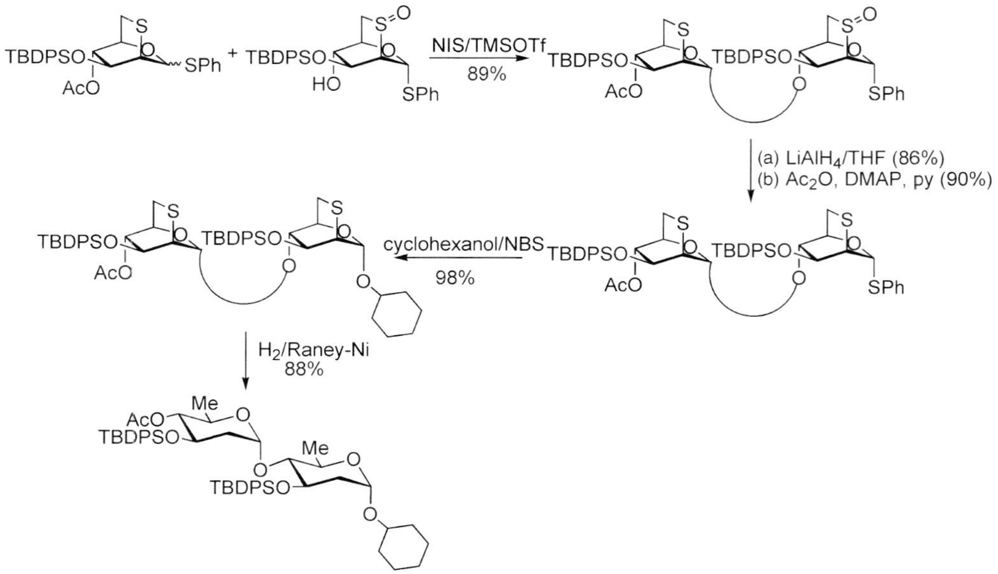

Scheme 4.10 Thioglycosides for the preparation of 2,6-di-deoxy-glycosides.

Toshima and coworkers developed a strategy for the chemoselective activation of thioglycosides for the preparation of 2,6-dideoxy glycosides [191–195]. Thus, the activated 2,6-anhydro-2-thioglycoside was coupled with the deactivated 2,6-anhydro-2-sulfinyl substrate to afford a disaccharide (Scheme 4.10). The resulting compound was converted into its active 2-thio-analog by reduction of the sulfinyl moiety and condensation with cyclohexanol. Reductive removal of the thio bridge afforded a 2,2′,6,6′-tetra-deoxy-disaccharide, which corresponds to the saccharide moiety of the biologically important Avermectin antibiotic.

Several methods for chemoselective glycosylations by one-pot procedures have been reported. For example, Kahne et al. [74] described a glycosylation method, which is based on selective activation of anomeric sulfoxides with triflic anhydride (Tf$_2$O) or triflic acid (TfOH). Mechanistic studies have revealed that the rate-limiting step in this reaction is triflation of the phenyl sulfoxide. Therefore, the reactivity of the glycosyl donor can be influenced by the nature of the substituent of the *para* position of the phenyl ring and the following reactivity order was established: OMe H > NO$_2$. Interestingly, the reactivity difference between a *p*-methoxyphenylsulfenyl glycoside and an unsubstituted phenylsulfenyl glycoside is sufficient to permit selective activation. In addition, silyl ethers are appropriate glycosyl acceptors when catalytic triflic acid is used as the activating reagent but these compounds react more slowly than the corresponding alcohols. These observations allowed for a one-pot synthesis of a trisaccharide from a mixture of three monosaccharides (Scheme 4.11) [84]. Thus, the treatment of the mixture with triflic acid resulted in the formation of

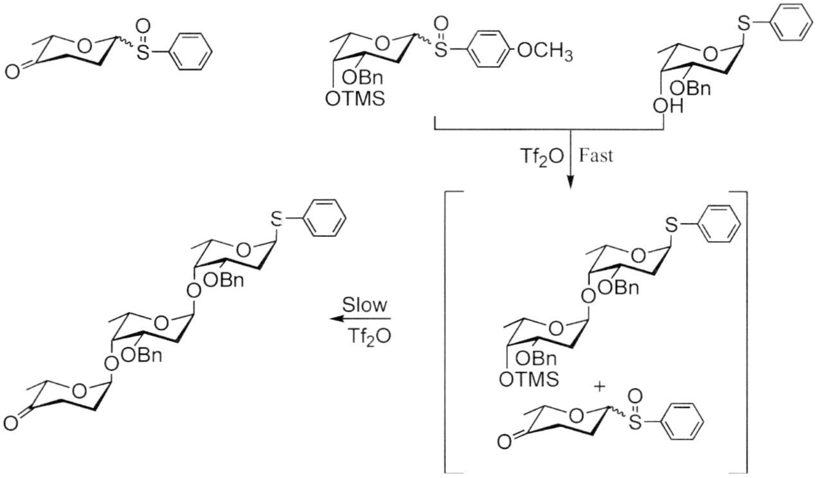

Scheme 4.11 Chemoselective glycosidations of anomeric sulfoxides by a one-pot procedure.

the expected trisaccharide in a 25% yield. No other trisaccharides were isolated and the only other coupling product was a disaccharide.

The products of the reaction indicate that the glycosylation takes place in a sequential manner. First, the most reactive *p*-methoxyphenylsulfenyl glycoside was activated and reacted with the sugar alcohol and not with the silyl ether. In the second stage of the reaction, the less reactive silyl ether of the disaccharide reacted with the less reactive sulfoxide to give the trisaccharide. The phenylthio group of the trisaccharide could be oxidized to a sulfoxide, which could be used in a subsequent glycosylation to give a part structure of the natural product Ciclumycin. Despite the relatively low yield of the coupling reactions, this methodology provides an efficient route to this compound.

Several variations of the one-pot multistep glycosylation concept have been reported. For example, Ley and coworkers [179,196] prepared a trisaccharide derived from the common polysaccharide antigen of group B *Streptococcus* by a facile one-pot two-step synthesis (Scheme 4.12). In this strategy, a benzylated activated thioglycosyl donor was chemoselectively coupled with the less reactive cyclohexane-1,2-diacetal (*CDA*)-protected thioglycosyl acceptor to give a disaccharide. Next, a second acceptor and additional activator were added to the reaction mixture, which resulted in the clean formation of a trisaccharide. The lower reactivity of the CDA-protected thioglycoside reflects the torsional strain inflicted upon the developing cyclic oxacarbenium ion, the planarity of which is opposed by the cyclic protecting group.

The one-pot two-step glycosylation strategy allows the construction of several glycosidic bonds without time-consuming work up and purification steps. It should, however, be realized that this type of reaction will only give satisfactory results when all the glycosylations are high yielding and highly stereoselective. For example, by exploiting neighboring-group participation, it is relatively easy to selectively install

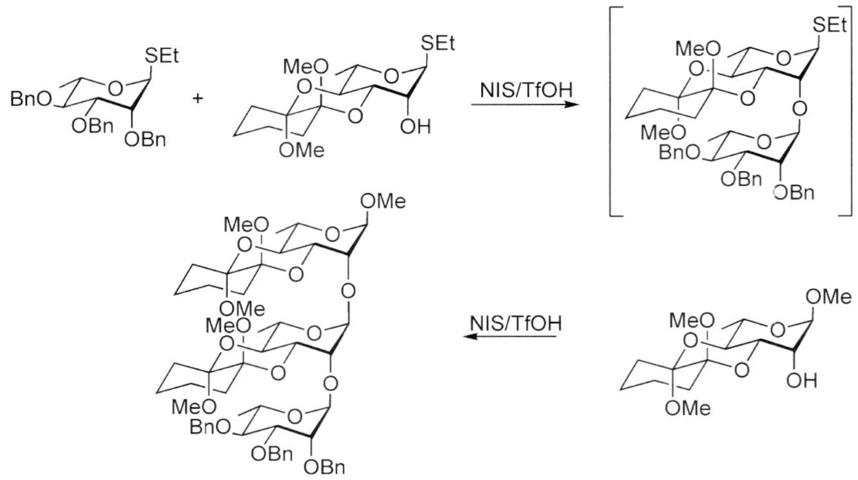

Scheme 4.12 One-pot multistep glycosidations of thioglycosides.

1,2-*trans* glycosides. Also, in general, mannosides give very high α-selectivities. Other types of glycosidic linkages may, however, pose problems.

Wong and coworkers have pursued an approach using HPLC for the rapid and precise measurement of relative reactivities of thioglycosyl donors. It was found that the nature of the saccharide, the position and the type of protecting groups contribute to anomeric reactivity. This information was employed to create a database of thioglycosyl reactivities, which can be used to select glycosyl donors and acceptors for easy and rapid one-pot assemblies of various linear and branched oligosaccharide structures [197]. The database has been successfully employed for one-pot multistep preparations of oligosaccharide libraries [198,199] and complex oligosaccharides such as Globo-H [200], fucosyl GM1 [201], sialyl Lewis X [202], oligolactosmine [203], α-Gal pentasaccharide [204], oligomannan [205] and Lewis Y [206]. The 'OptiMer' computer program was developed to guide the selection of appropriate thioglycosyl building blocks that have sufficiently different reactivities for one-pot multistep glycosylations. For example, the program aided in the selection of appropriate building blocks for the convenient synthesis of the tumor-associated hexasaccharide Globo-H. The reactivity of the building blocks was tuned by using electron-donating groups, such as benzyl ether and 2,2,2-trichloroethylcarbamate, and electron-withdrawing protecting groups, such as benzoyl, *p*-nitrobenzoyl (NBz) and *o*-chlorobenzyl ethers (ClBn) (Scheme 4.13).

A one-pot two-step glycosylation to give a trisaccharide was accomplished by simply changing the solvent system [207]. In this approach, the solvent controls the anomeric selectivity and also the rate of glycosylation. Thus, when a reactive ethyl thiorhamnoside and a less reactive thiophenyl mannoside were dissolved in diethyl ether, only the rhamnosyl donor was activated by promoter system NIS/AgOTf to give the corresponding thiophenyl disaccharides in an almost quantitative yield. After adding a

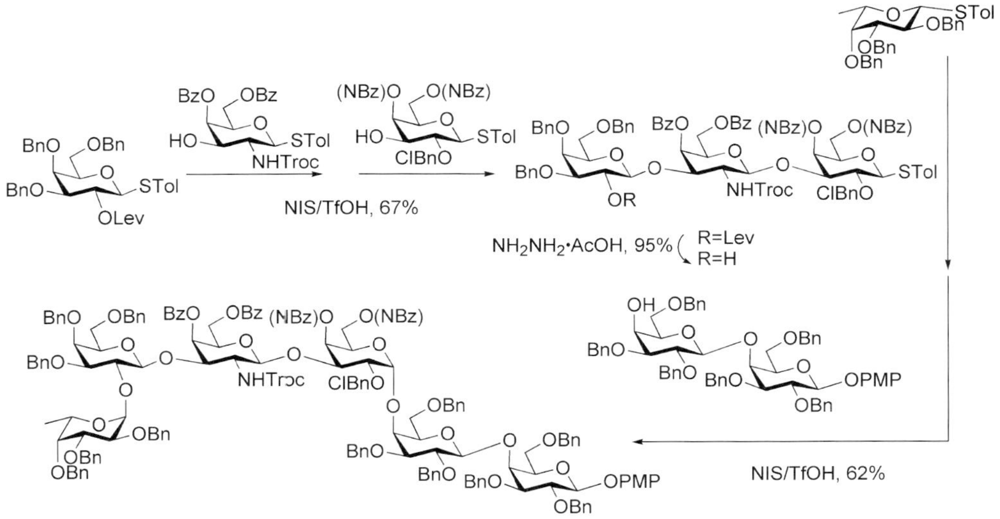

Scheme 4.13 One-pot synthesis of the Globo-H hexasaccharide.

glucosyl acceptor and an additional promoter dissolved in CH_2Cl_2, the intermediate thiophenyl disaccharide donor was activated by NIS/TMSOTf leading to the formation of a trisaccharide in high yield and stereoselectivity. Thus, by tuning the reactivity of acceptors and donors and performing the first glycosylation in diethyl ether (low glycosylation rate) and the second in CH_2Cl_2/Et_2O (higher glycosylation rate), a trisaccharide could be prepared by a one-pot two-step procedure (Scheme 4.14).

Baasov and coworkers achieved an efficient synthesis of an oligosaccharide using a one-pot procedure whereby the reactivity of glycosyl donors and acceptors were tuned by a combination of the nature of the C-2 amino protecting group (Troc, Phth) and anomeric leaving group (ethylthio and phenylthio) [208]. In addition, by exploiting solvent reactivity effects, an ethyl 1-thioglycoside could be activated in the presence of a phenyl 1-thioglucosyl acceptor. Thus, the synthesis exploited the observation that NTroc-protected thioglycosides are significantly more reactive than their NPhth-protected counterparts. Furthermore, successful synthesis of the target tetrasaccharide

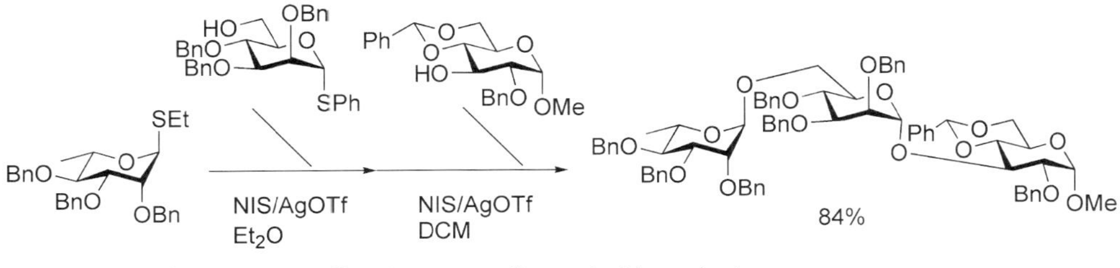

Scheme 4.14 Solvent reactivity effects in one-pot oligosaccharide synthesis.

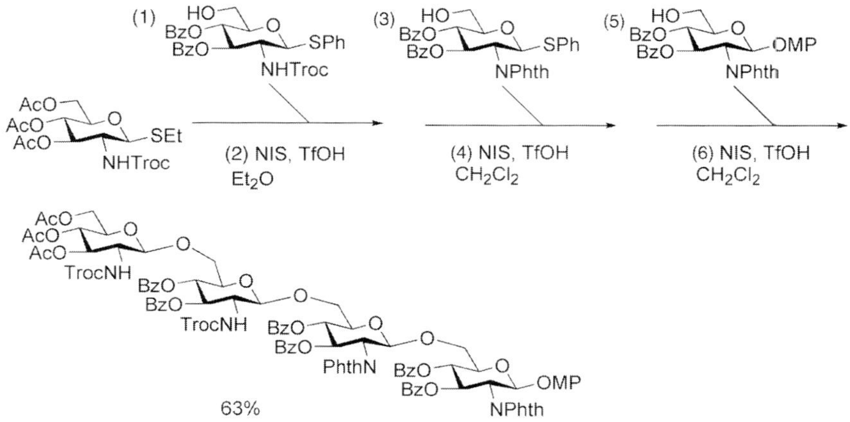

Scheme 4.15 One-pot synthesis of glucosamine oligosaccharide.

exploited the higher reactivity of thioethyl glucosides compared to similar thiophenyl glycosides. As a result, the desired tetraglucosamine could be prepared in an overall yield of 63% by a one-pot three-step glycosylation (Scheme 4.15).

4.1.6.2 Orthogonal and Semiorthogonal Glycosylations

Orthogonal glycosylations use glycosyl donors and acceptors that have different anomeric groups (e.g. X = F and Y = SR), which can be activated without affecting the other one. These synthetic approaches are attractive as no or very few protecting-group manipulations are involved during the assembly of a complex oligosaccharide.

Nicolaou and coworkers [209–214] have described a two-stage glycosylation strategy whereby a thioglycoside is converted into a glycosyl fluoride donor, which is then employed as a glycosyl donor for coupling with a thioglycosyl acceptor. The procedure can be repeated by the conversion of the anomeric thio group of the oligosaccharide into an anomeric fluoride that can be used in a further coupling reaction. This glycosylation strategy was exploited for the preparation of *Rhynchosporides* and key reactions are depicted in Scheme 4.16.

In a further improved orthogonal glycosylation strategy [215], thioglycosides and glycosyl fluorides act as glycosyl donors and acceptors and are coupled with each other in a chemoselective manner [215]. An example of this strategy is depicted in Scheme 4.17, in which the synthesis of β(1 → 4)-2-acetamido-2-deoxy-D-glucose-linked oligosaccharides is described. Thus, a thioglycosyl donor was coupled with a glycosyl fluoride acceptor using a thiophilic promoter. Next, the resulting glycosyl fluoride acted as a glycosyl donor and coupled with a thioglycosyl acceptor. Reiteration of the process leads to the rapid buildup of long-chain oligosaccharides.

Several other examples have been reported in which thioglycosides were used as glycosyl acceptors. Thus, thioglycosides containing free hydroxyls can be coupled chemoselectively with glycosyl bromides and chlorides in the presence of silver triflate or tin(II) chloride-silver perchlorate as the promoter system [12,53,56,162,218–221].

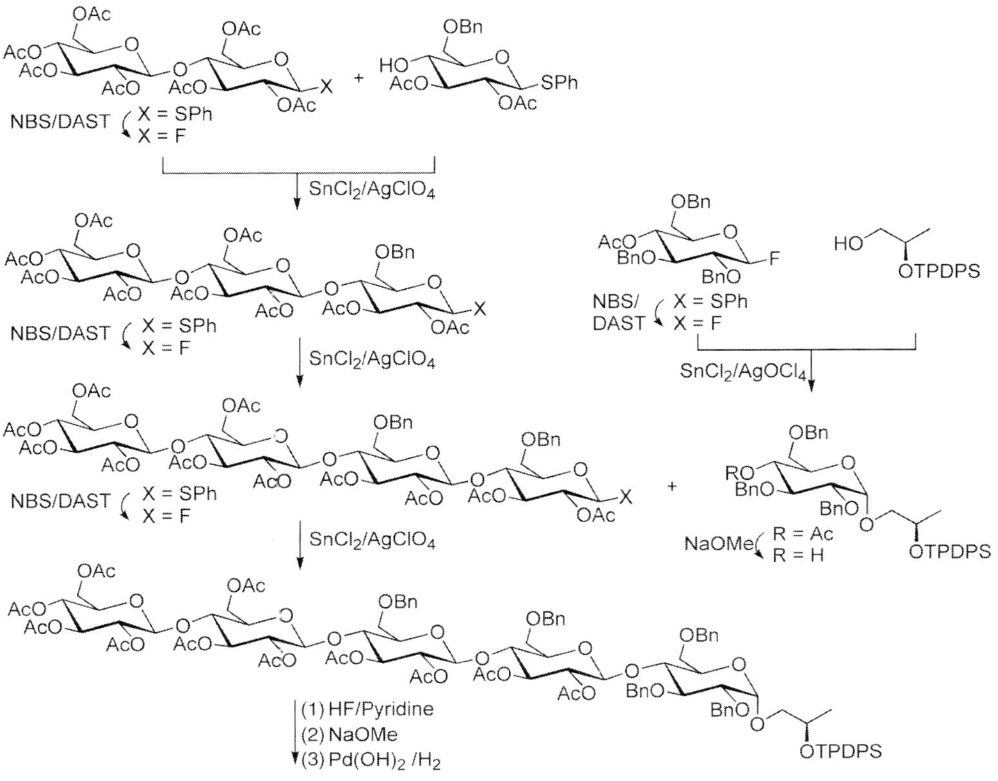

β-D-Glcp-(1-4)-β-D-Glcp-(1-4)-β-D-Glcp-(1-4)-β-D-Glcp-(1-4)-α-D-Glcp-(1-1)-1,2-propane diol

Scheme 4.16 Two-stage activation of thioglycosides.

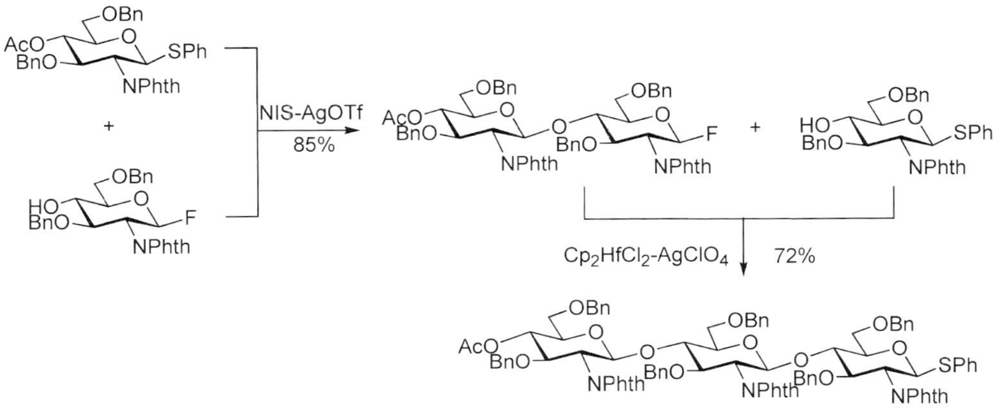

Scheme 4.17 Orthogonal glycosylations of thioglycosides and glycosyl fluorides.

Such a synthesis is shown in Scheme 4.18a, in which a glycosyl bromide is coupled with the thioglycosyl acceptor to afford a thioglycosyl disaccharide. The glycosyl bromide was obtained from the corresponding 1-thioglycoside by treatment with Br_2. It was shown [222–224] that phenyl selenoglycosides can be selectively activated in the

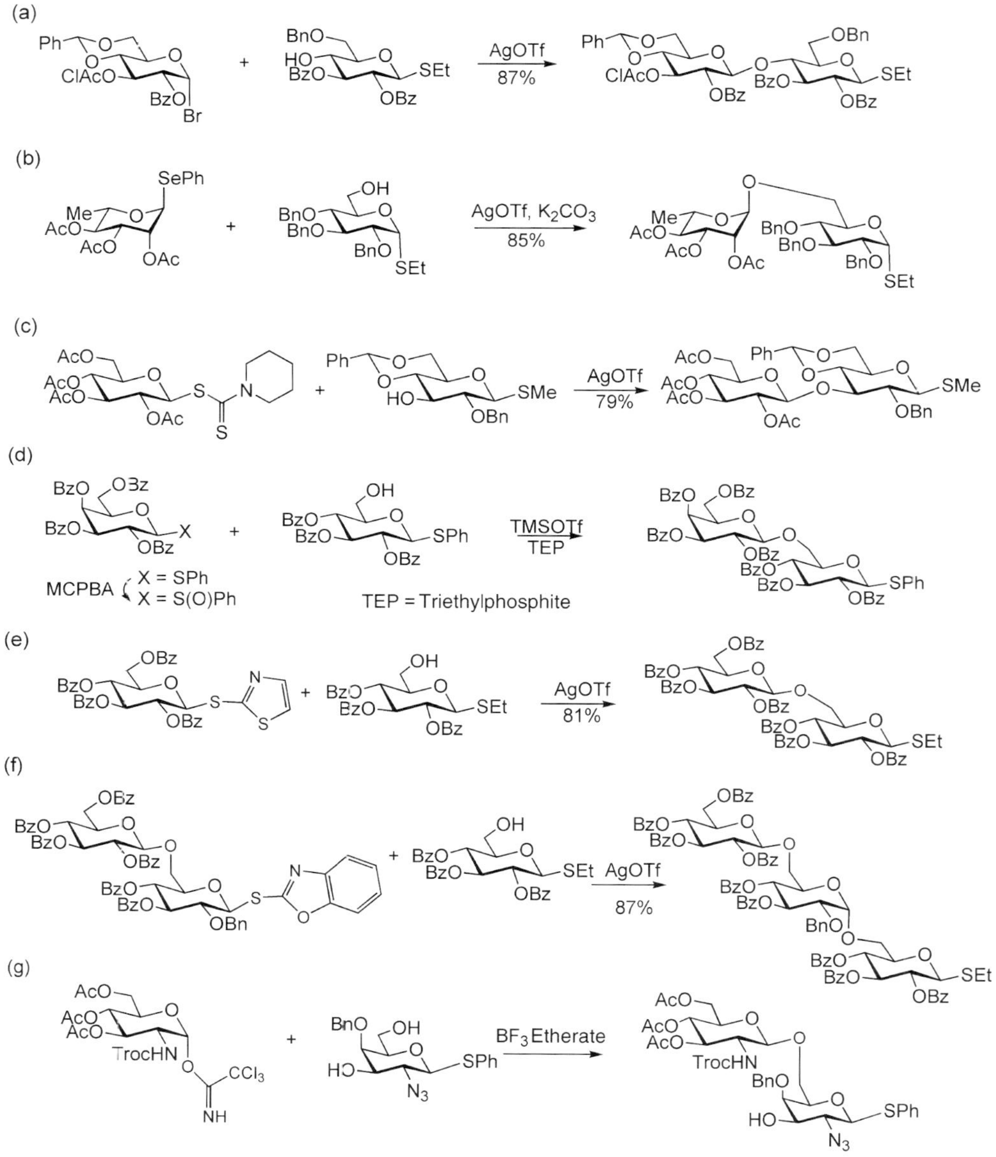

Scheme 4.18 Glycosylations with thioglycosyl acceptors.

presence of ethyl thioglycosides using silver triflate/potassium carbonate (or silver carbonate) as the promoter system (Scheme 4.18b). Garegg and coworkers reported the use of glycosyl 1-piperidinecarbodithioates in combination with thioglycosides (Scheme 4.18c) [225]. Peracetylated piperidinecarbodithioate donor could be selectively activated in the presence of the thioglycosyl acceptor by using silver triflate as the promoter to afford a thioglycosyl disaccharide. Kahne and coworkers reported a glycosylation approach in which glucosyl sulfoxides are activated in the presence of thioglycosides (Scheme 4.18d) [276–278]. Next, the resulting thioglycosyl disaccharides could be activated using a thiophilic promoter or converted into the corresponding sulfoxides. The reaction of thioglycosyl acceptors with trichloroacetimidates has also been described (Scheme 4.18g) [216,226]. Demchenko and coworkers have studied a series of thioimidate-based glycosyl donors, such as S-Benzoxazolyl and S-thiazolinyl (STaz) glycosides, which can be activated by AgOTf or Cu(OTf)$_2$ in the presence of ethyl 1-thioglycosides (Scheme 4.18e and f) [115,116]. In addition, a strategy was developed whereby anomeric reactivities were reduced by metal complexation with the anomeric group [227]. Furthermore, STaz glycosides can selectively be activated over conventional 1-thioglycosides and O-pentenyl glycosides, whereas bromides, trichloroacetimidates and 1-thioglycosides can be activated over the STaz moiety [228].

Orthogonol and semiorthogonal glycosylations have also been performed in one-pot multistep fashion. For example, Takahashi and coworkers [216] reported a one-pot two-step glycosylation in which the difference in reactivity between glycosyl donors and acceptors was accomplished through the use of two types of anomeric leaving groups with different reactivities (Scheme 4.19). Thus, a glycosyl bromide was coupled with a thioglycosyl acceptor in the presence of silver triflate to give a disaccharide. Although the anomeric thiophenyl groups are stable to silver triflate (AgOTf), an addition of both the second activator (NIS) and the glycosyl acceptor promoted selective activation of the glycosyl donor, resulting in the formation of a trisaccharide (84% overall yield). In this example, the stereochemical outcome of the glycosylations was controlled by the neighboring-group participation of the 2-O-

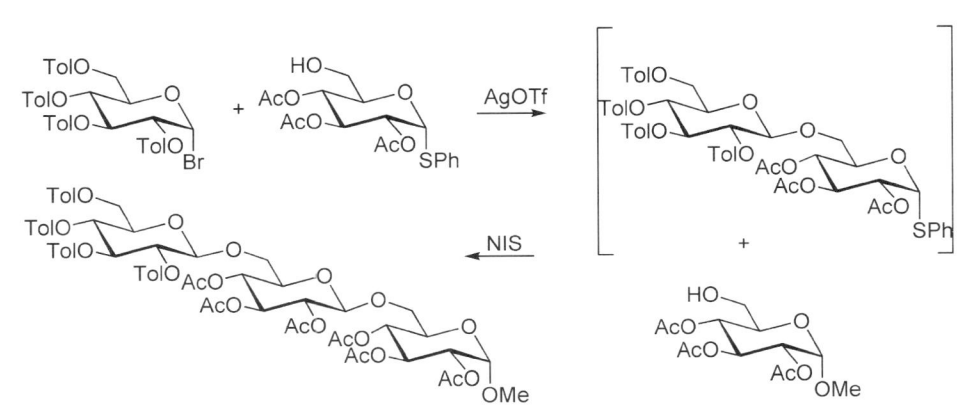

Scheme 4.19 One-pot multistep glycosylations with thioglycosides and glycosyl bromides.

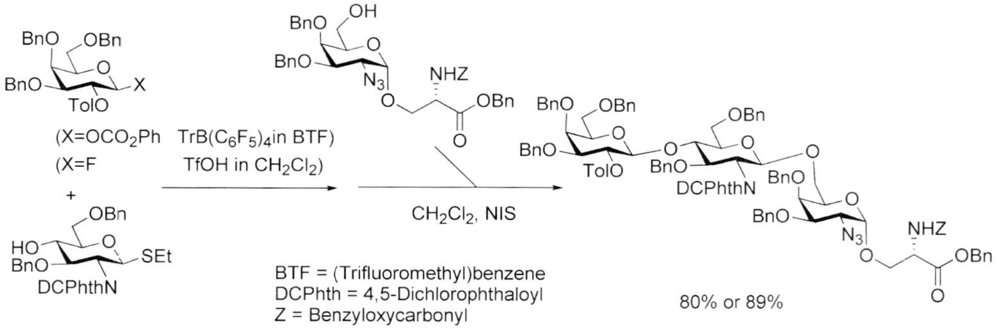

Scheme 4.20 One-pot synthesis of mucin-related F1α antigen.

toluoyl (Tol) and acetyl protecting groups. A similar one-pot two-step glycosylation procedure was employed for the preparation of an elicitor-active hexaglycoside.

Mukaiyama and Kobashi have reported a one-pot assembly of a mucin-related F1α antigen using anomeric fluorides and carbonates [51]. After a careful evaluation of solvent systems, promoters and reaction temperatures, a fully protected F1α antigen was synthesized by a one-pot sequential glycosylation using a galactosyl phenyl carbonate or fluoride, a thioglycoside and a glycosyl amino acid (Scheme 4.20). In the first step, the phenylcarbonate or fluoride donor was coupled with the ethyl thioglucoside in the presence of $TrB(C_6F_5)_4$ or TfOH, respectively. After TLC analysis indicated complete consumption of the glycosyl donor, consecutive addition of the terminal glycosyl amino acid and NIS provided the target trisaccharide in high yield (80 and 89%, respectively). In a similar manner, an anomeric fluoride donor and two different thioglycosides were employed for the preparation of a phytoalexin elicitor heptasaccharide (Scheme 4.21) [217]. Thus, TfOH-catalyzed double glycosylation of the fluoride with the ethyl thioglycosyl acceptor gave a trisaccharide, which was coupled with the highly deactivated p-(trifluoromethyl)benzoyl (CF_3Bz)-protected thioglycoside to afford a tetraglucoside intermediate as the major product. Next, the consecutive addition of a trisaccharide acceptor and NIS led to the formation of a heptasaccharide in an overall yield of 48%. Thus, four glycosidic linkages were stereoselectively introduced in a one-pot manner.

Huang et al. designed a general one-pot multistep glycosylation approach independent of differential glycosyl donor and acceptor reactivities. The new and elegant method is based on the preactivation of a thioglycosyl donor to give a reactive intermediate in the absence of the acceptor. Subsequently, a thioglycosyl acceptor can be added to the activated donor leading to the formation of a coupling product [229]. The resulting thioglycoside can be activated and employed in a subsequent glycosylation with a thioglycosyl acceptor. For example, a trisaccharide was assembled that contains the biologically relevant Fuc-α1,3-GlcNAc and GlcNAc-β1,3-Gal moieties. Thus, preactivation of the toluyl thiofucoside by p-TolSOTf at −60 °C was followed by the addition of a phthaloyl-protected thioglycosyl acceptor. The reaction

Scheme 4.21 Mukaiyama's one-pot synthesis of phytoalexin-elicitor active heptasaccharide.

mixture was allowed to warm to room temperature for over a period of 15 min, during which time a disaccharide was formed. After cooling the reaction mixture to −60 °C, the disaccharide was preactivated with *p*-TolSOTf, followed by the addition of the thiogalactosyl acceptor, producing a trisaccharide in 59% overall yield within a period of 1 h (Scheme 4.22). The trisaccharide carrying an anomeric *p*-thiotolyl moiety could be utilized as glycosyl donor for the synthesis of Le^x containing oligosaccharides. Excellent anomeric stereoselectivities were obtained in each glycosylation.

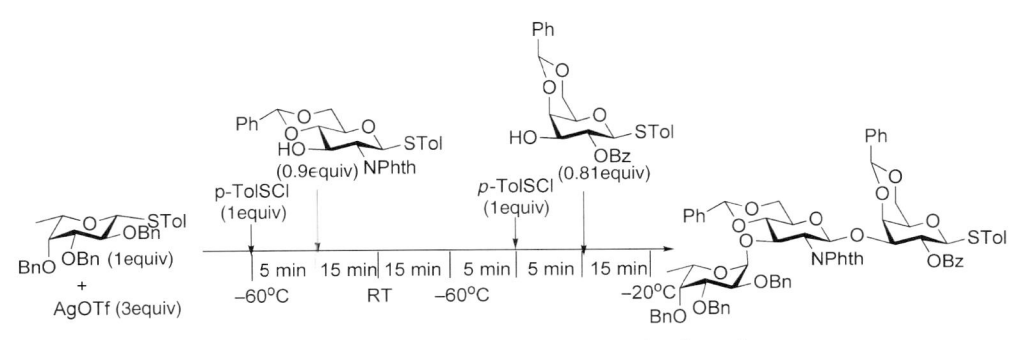

Scheme 4.22 Preactivation of *p*-tclyl thioglycoside in one-pot oligosaccharide synthesis.

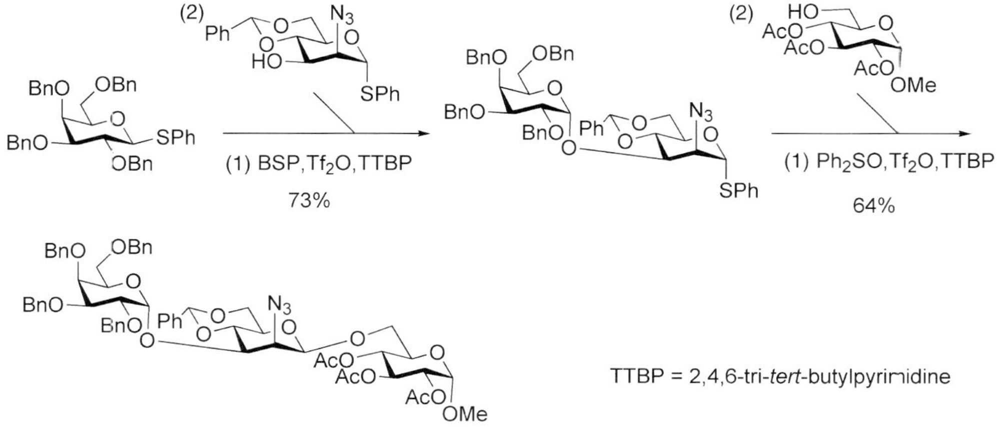

Scheme 4.23 Chemoselective glycosylation by preactivation strategy using Ph$_2$SO/Tf$_2$O promoter.

TTBP = 2,4,6-tri-*tert*-butylpyrimidine

van Boom and coworkers described a similar preactivation strategy for thioglycosides. It was found that diphenylsulfoxide in combination with triflic anhydride provides a very potent thiophilic reagent capable of activating deactivated thioglycosides [87,99]. A novel chemoselective condensation sequence was developed, in which a benzylated reactive thioglycosyl donor was selectively activated by a mild thiophilic promoter and chemoselectively condensed with a relatively unreactive thioglycosyl donor (Scheme 4.23). Addition of the acceptor and the more reactive promoter system Ph$_2$SO/Tf$_2$O led to the formation of a trisaccharide. The side products formed from the BSP/Tf$_2$O activation system were quenched by the addition of triethyl phosphite after each glycosylation to avoid activation of the acceptor and glycosylation product [87].

Finally, a novel sequential glycosylation procedure has been reported using 1-hydroxyl and thioglycosyl donors [230]. Yamago *et al.* reported a broad substrate scope utilizing the BSP/Tf$_2$O promoter system to preactivate thioglycosyl donors [231].

4.1.6.3 Two-Directional Glycosylation Strategies

The overall efficiency of chemoselective and orthogonal glycosylations is compromised by the linear nature of the glycosylation sequence and the fact that the growing oligosaccharide chain acts in each reaction as glycosyl donor. These problems can be addressed by two-directional glycosylation strategies. In such an approach, a thiosaccharide building block can act as glycosyl donor as well as glycosyl acceptor. These properties enable oligosaccharide assembly in a very flexible and highly convergent manner. For example, the coupling of a tritylated thioglycosyl donor with an acceptor having a 4-hydroxyl afforded a disaccharide in a yield of 62% ($\alpha/\beta = 6/1$) (Scheme 4.24) [232]. In this case, the 6-*O*-trityl group improved the α-selectivity of the glycosylation because of the steric effects. The trityl ether of the resulting product can act as an acceptor when it is glycosylated with a fully benzylated thioglycosyl donor using NIS

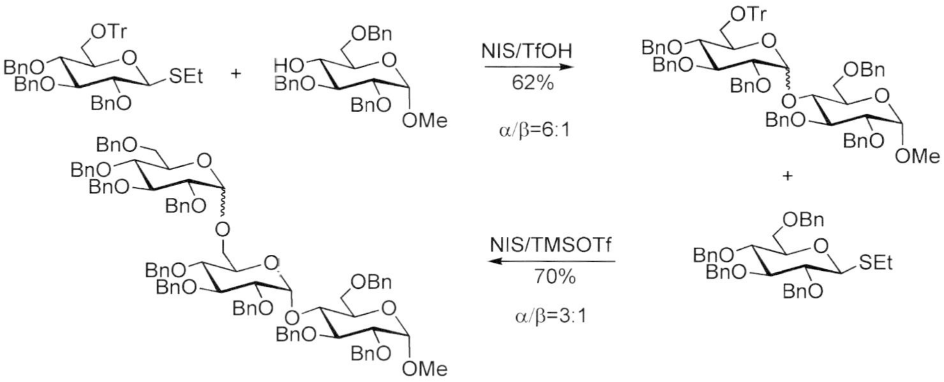

Scheme 4.24 Two-directional glycosylations with tritylated thioglycosides.

and a stochiometric amount of TMSOTf as the activator to give a trisaccharide in excellent yield as a mixture of anomers ($\alpha/\beta = 3/1$).

A two-direction glycosylation strategy can also be performed by regioselective glycosylation between glycosyl donors and acceptors, both of which contain a free hydroxyl group to give di- and trisaccharides [233]. The products of these glycosylations can immediately be employed as glycosyl acceptors in subsequent glycosylations without the need to perform protecting-group manipulations. In combination with the previously reported chemoselective glycosylations, this methodology provides a powerful method to assemble oligosaccharides in a highly convergent manner, avoiding protecting-group manipulations at the oligosaccharide stage. A prerequisite of regioselective glycosylation is that the acceptor's hydroxyl functionality must be substantially more reactive than the hydroxyl group of the glycosyl donor. Differences in reactivity may be achieved by primary versus secondary or equatorial versus axial disposition of hydroxyl groups. The new methodology was employed for the preparation of a pentasaccharide involved in the hyper-acute rejection response in xenotransplantation (Scheme 4.25) [234]. Thus, the α-linked Gal(1 → 3)Gal dimer was obtained by an armed–disarmed chemoselective glycosylation using NIS/TMSOTf as promoter and toluene/1,4-dioxane as reaction solvent. The right-hand trisaccharide was obtained in a good yield of 77% by a NIS/TMSOTf-mediated glycosylation between a lactoside acceptor and partially protected 2-deoxy-phthalimido-glucosyl acceptor. No self-condensation of the glycosyl donor was observed owing to the deactivation of the 4-OH by the neighboring benzoyl group. The pentasaccharide was obtained by coupling of the thiodisaccharide and trisaccharide acceptor in the presence of NIS/TMSOTf. Two-directional glycosylations using thioglycosides have also been performed on solid support [235].

Takahashi and coworkers described the one-pot synthesis of core 2 branched oligosaccharides [236]. It was found that boron trifluoride complexed with a trimethylsilyl ether would enhance the nucleophilicity of the silyl ether. As a result, glycosylations of the 6-O-TMS modified acceptor with a glycosyl fluoride provided

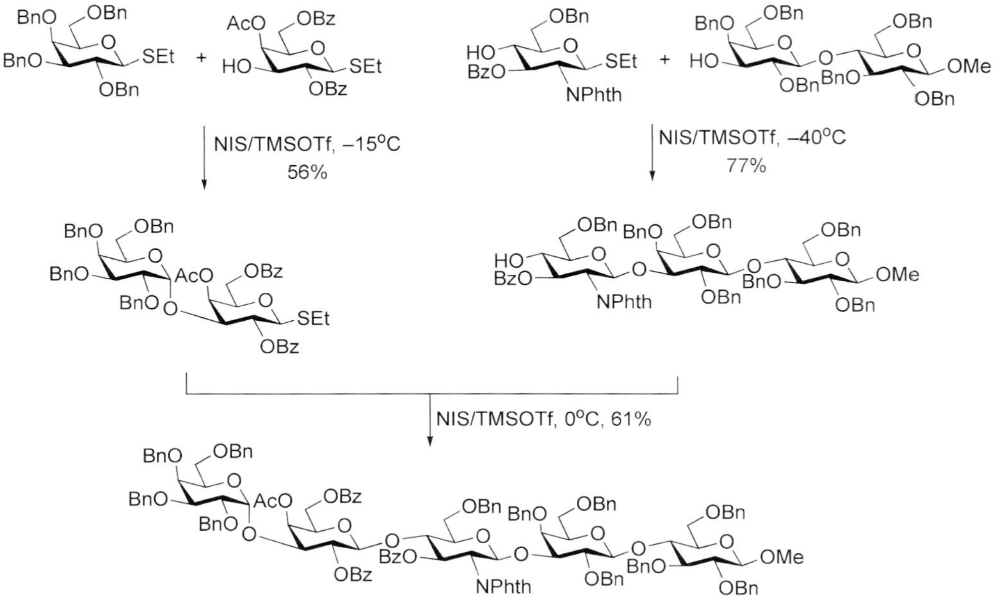

Scheme 4.25 The preparation of the Galili pentasaccharide using thioglycosyl donors and acceptors in a two-directional glycosylation strategy.

selectively glycosylation at C-6 of the thioglycosyl acceptor without the glycosylation of the C-3 hydroxyl. Therefore, a chemoselective glycosylation was performed between 6-O-silyl-4-benzyl-2-azido-thiogalactoside and a glycosyl fluoride in the presence of BF_3 etherate, followed by sequential coupling of the remaining secondary hydroxyl group with galactosyl fluoride in the presence of $ZrCp_2Cl_2/AgOTf$ to provide the desired trisaccharide. Subsequent NIS/TfOH-promoted glycosidation of the thioglycoside with amino acids provided products in good yield (Scheme 4.26).

The same group also prepared a phytoalexin elicitor heptamer by a one-pot six-step glycosylation protocol, providing the most impressive example of the potential of chemoselective glycosylation technology [216,237]. The sequential addition of seven reaction components with six appropriate activators resulted in the one-pot six-step glycosylation. First, a toluoyl-protected galactosyl bromide, in combination with AgOTf, ensured the regioselective glycosylation of the primary alcohol of thioglucoside diol. Second, the resulting 1-thioglycoside acted as a glycosyl donor in a coupling with a galactosyl fluoride acceptor using large excess of MeOTf to avoid self-condensation. Third, the glycosylation of the C3-hydroxyl was achieved using thioglucoside. Fourth, $HfCp_2Cl_2$–AgOTf-mediated coupling of the branched tetraglucoside to the third thioglycosidic building block was performed regioselectively. Fifth, the terminal glucose acceptor was condensed with the resulting pentasaccharide thiophenyl donor using a large excess of dimethyl(methylthio)

Scheme 4.26 One-pot synthesis of core 2 class amino acids.

sulfonium triflate. Finally, the heptasaccharide, the largest oligosaccharide amongst the reaction products, was formed via the second β-glucosidic linkage. The final compound was purified by size exclusion chromatography in a good yield of 24% (Scheme 4.27).

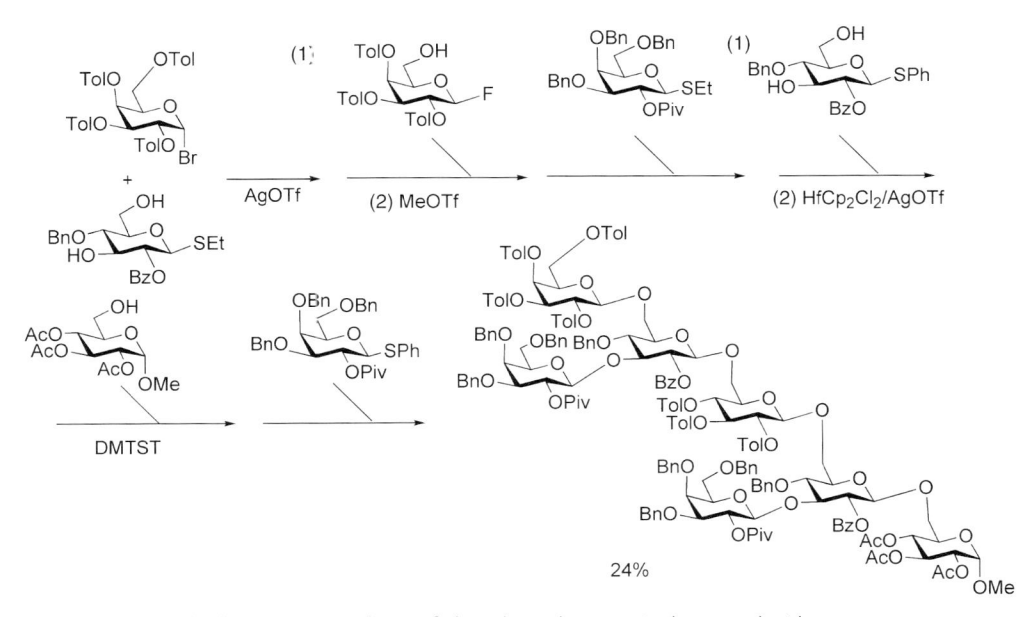

Scheme 4.27 Takahashi's one-pot synthesis of phytoalexin elicitor active heptasaccharide.

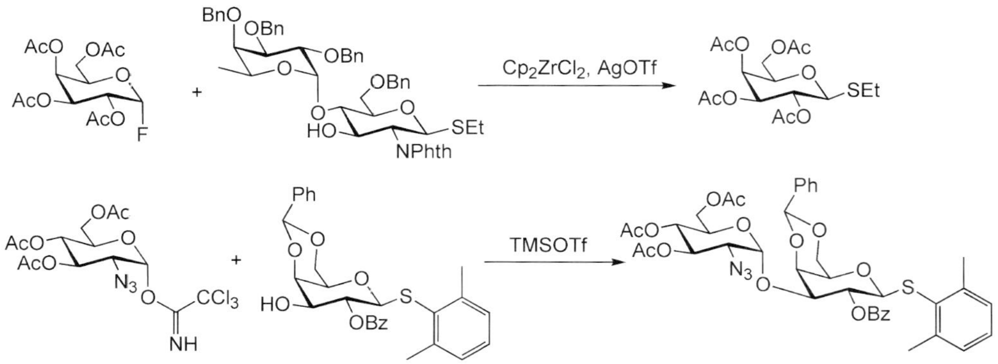

Scheme 4.28 Aglycon transfer of thioglycosides.

4.1.7
Aglycon Transfer

Although thioglycosides have been successfully employed as glycosyl acceptors, at times this type of glycosylation is plagued by aglycon transfer [2,236,238–249]. The aglycon transfer process is shown to affect both armed and disarmed thioglycosides, it causes anomerization of the carbon-sulfur bond of a thioglycoside and destroys the product of a glycosylation reaction. This side reaction is especially important to consider when carrying out complex reactions, such as solid-phase glycosylations, one-pot or orthogonal multicomponent glycosylations and construction of carbohydrate libraries. For example, an intermolecular aglycon transfer reaction was observed in a Cp_2ZrCl_2/AgOTf-mediated coupling of a glycosyl fluoride donor and a 1-thiodisaccharide acceptor (Scheme 4.28) [244]. It was rationalized as follows: the acyloxonium ion generated after activation of the glycosyl fluoride attacked the sulfur instead of the sterically hindered alcohol, leading to the formation of β-thioglycoside. Aglycon transfer could be avoided by employing a less reactive 1-thioglycosyl acceptor. Li and Gildersleeve examined a number of modified aglycons to prevent aglycon transfer [250]. It was found that the 2,6-dimethylphenyl 1-thio moiety was effectively blocking the transfer in a variety of model studies and glycosylation reactions. The DMP group can be installed in one step from a commercially available 2,6-dimethylthiophenol and is usable as a glycosyl donor.

4.1.8
General Procedure for Synthesis of Thioglycosides from Peracetylated Hexapyranosides Promoted by BF$_3$-Etherate [5–10]

To a solution of peracetylated hexapyranoside (1.0 equiv), ethanethiol (1.2 equiv) and freshly activated 4 Å powdered molecular sieves in dichloromethane (4.0 ml mmol^{-1}) was added BF_3–Et_2O (2.0 equiv) dropwise at 0 °C under an argon atmosphere.

The reaction mixture was stirred for 2 h at room temperature until TLC analysis indicated that the reaction was complete. The solution was filtered through Celite and washed with dichloromethane. The filtrate was washed with saturated aqueous NaHCO$_3$ and H$_2$O. The organic phase was dried (MgSO$_4$), filtered and the filtrate was concentrated to dryness. Purification of the crude product by column chromatography over silica gel afforded the target compound.

4.1.9
General Procedure for Synthesis of Thioglycosides by Displacement of Acylated Glycosyl Bromide with Thiolate Anion [34]

To a solution of diaryl or diaralkyldisulfide (1.0 equiv) in CH$_3$CN (5.0 ml mmol^{-1}) was added zinc dust (1.0 equiv) followed by fused ZnCl$_2$ (0.2 equiv). The reaction mixture was placed in a preheated oil bath at 70 °C for 45 min, during that time the reaction mixture became turbid indicating the formation of zinc thiolate. A solution of acylated glycosyl bromide (2.0 equiv) in CH$_3$CN (2.5 ml mmol^{-1}) was added to the turbid reaction mixture that was then stirred at 70 °C until TLC analysis indicated that the reaction was complete. The reaction mixture was concentrated *in vacuo* and the residue dissolved in dichloromethane. The organic layer was washed with saturated aqueous NaHCO$_3$ and H$_2$O, dried (MgSO$_4$), filtered and the filtrate was concentrated to dryness. Purification of the crude product by column chromatography over silica gel afforded the target compound.

4.1.10
General Procedure for Synthesis of Sialyl Thioglycosides Using TMSSMe and TMSOTf [165]

To a solution of methyl 2,4,7,8,9-penta-*O*-acetyl-5-(*N*-acetylacetamido)-3,5-dideoxy-D-glycero-α,β-D-*galacto*-non-2-ulopyranosonate (1.0 equiv) and freshly activated 4 Å powdered molecular sieves in 1,2-dichloroethane (2.0 ml mmol^{-1}) was added TMSSMe (1.4 equiv) and TMSOTf (0.75 equiv). The reaction mixture was stirred for 4.5 h at 50 °C and a further 16 h at room temperature. The solution was filtered through Celite and washed with dichloromethane. The filtrate was washed with saturated aqueous NaHCO$_3$ and H$_2$O, dried (MgSO$_4$), filtered and the filtrate was concentrated to dryness. Purification of the crude product by column chromatography over silica gel afforded methyl [methyl 4,7,8,9-tetra-*O*-acetyl-5-(*N*-acetylacetamido)-3,5-dideoxy-2-thiol-D-glycero-α,β-D-*galacto*-non-2-ulopyranosid]onate as α/β mixture (1 : 1).

4.1.11
General Procedure for Activation of Thioglycosides with Ph$_2$SO/Tf$_2$O [87,99]

To a solution of thioglycoside (1.0 equiv), Ph$_2$SO (2.8 equiv) and TTBP (3.0 equiv) in dichloromethane (4.0 ml mmol^{-1}) was added trifluoromethanesulfonic anhydride (1.4 equiv) at −60 °C under an argon atmosphere. The reaction mixture was stirred for 5 min, after which a solution of the glycosyl acceptor (1.5 equiv) in dichloromethane (2.0 ml mmol^{-1}) was added. The mixture was stirred at $-60\,°C$ for 1 h, after which it was slowly warmed to room temperature and quenched by the addition of saturated aqueous NaHCO$_3$. The organic layer was washed with brine, dried (MgSO$_4$), filtered and the filtrate was concentrated to dryness. Purification of the crude product by column chromatography over silica gel afforded the product.

4.1.12
General Procedure for Activation of Thioglycosides with BSP/TTBP/Tf$_2$O [85]

To a solution of thioglycoside (1.0 equiv), 1-benzenesulfinyl piperidine (1.0 equiv), TTBP (2.0 equiv), and freshly activated 3 Å powdered molecular sieves in dichloromethane (25.0 ml mmol^{-1}) was added trifluoromethanesulfonic anhydride (1.1 equiv) at $-60\,°C$ under an argon atmosphere. The reaction mixture was stirred for 5 min, after that a solution of the glycosyl acceptor (1.5 equiv) in dichloromethane (4.0 ml mmol^{-1}) was added. The reaction mixture was stirred at $-60\,°C$ for 2 min, after that it was slowly warmed to room temperature and quenched by the addition of saturated aqueous NaHCO$_3$. The organic layer was washed with brine, dried (MgSO$_4$), filtered and the filtrate was concentrated to dryness. Purification of the crude product by column chromatography over silica gel afforded the product.

4.1.13
General Procedure for Activation of Sialyl Thioglycosides with NIS/TfOH [165,166]

To a solution of sialyl thioglycoside (3.0 equiv), glycosyl acceptor (1.0 equiv) and freshly activated 3 Å powdered molecular sieves in MeCN (30.0 ml mmol^{-1}) was added NIS (6.0 equiv) and TfOH (0.6 equiv) at $-35\,°C$ under an argon atmosphere. The reaction mixture was stirred for 5 min until TLC analysis indicated that the reaction was complete. The solution was filtered through Celite and washed with dichloromethane. The filtrate was washed with aqueous Na$_2$S$_2$O$_3$ (20%) and H$_2$O. The organic phase was dried (MgSO$_4$), filtered and the filtrate was concentrated to dryness. Purification of the crude product by column chromatography over silica gel afforded the product.

References

1 Garegg, P.J. (1997) *Advances in Carbohydrate Chemistry and Biochemistry*, **52**, 179–205.
2 Codee, J.D.C., Litjens, R., van den Bos, L.J., Overkleeft, H.S. and van der Marel, G.A. (2005) *Chemical Society Reviews*, **34**, 769–782.
3 Lemieux, R.U. (1951) *Canadian Journal of Chemistry/Revue Canadiene de Chimie*, **29**, 1079–1091.
4 Lemieux, R.U. and Brice, C. (1955) *Canadian Journal of Chemistry/Revue Canadiene de Chimie*, **33**, 109–119.

5 Pozsgay, V. and Jennings, H.J. (1987) *Tetrahedron Letters*, **28**, 1375–1376.
6 Ferrier, R.J. and Furneaux, R.H. (1980) *Methods in Carbohydrate Chemistry*, vol. 8 (eds J.N. Bemiller and R.L. Whistler), Academic Press, New York, pp. 251–253.
7 Nilsson, M., Svahn, C.M. and Westman, J. (1993) *Carbohydrate Research*, **246**, 161–172.
8 Kihlberg, J.O., Leigh, D.A. and Bundle, D.R. (1990) *The Journal of Organic Chemistry*, **55**, 2860–2863.
9 Das, S.K. and Roy, N. (1996) *Carbohydrate Research*, **296**, 275–277.
10 Matsui, H., Furukawa, J., Awano, T., Nishi, N. and Sakairi, N. (2000) *Chemistry Letters*, **29**, 326–327.
11 Apparu, M., Blancmuesser, M., Defaye, J. and Driguez, H. (1981) *Canadian Journal of Chemistry/Revue Canadiene de Chimie*, **59**, 314–320.
12 Lönn, H. (1985) *Carbohydrate Research*, **139**, 105–113.
13 Olsson, L., Kelberlau, S., Jia, Z.Z.J. and Fraser-Reid, B. (1998) *Carbohydrate Research*, **314**, 273–276.
14 Das, S.K., Roy, J., Reddy, K.A. and Abbineni, C. (2003) *Carbohydrate Research*, **338**, 2237–2240.
15 Dasgupta, F. and Garegg, P.J. (1989) *Acta Chemica Scandinavica*, **43**, 471–475.
16 Weng, S.S., Lin, Y.D. and Chen, C.T. (2006) *Organic Letters*, **8**, 5633–5636.
17 Bhat, V.S., Sinha, B. and Bose, J.L. (1987) *Indian Journal of Chemistry Section B: Organic Chemistry Including Medicinal Chemistry*, **26**, 514–516.
18 Contour, M.O., Defaye, J., Little, M. and Wong, E. (1989) *Carbohydrate Research*, **193**, 283–287.
19 Defaye, J., Driguez, H., Ohleyer, E., Orgeret, C. and Viet, C. (1984) *Carbohydrate Research*, **130**, 317–321.
20 Zhang, Z.Y. and Magnusson, G. (1996) *Carbohydrate Research*, **295**, 41–55.
21 Agnihotri, G., Tiwari, P. and Misra, A.K. (2005) *Carbohydrate Research*, **340**, 1393–1396.
22 Tiwari, P., Agnihotri, G. and Misra, A.K. (2005) *Journal of Carbohydrate Chemistry*, **24**, 723–732.
23 Kumar, R., Tiwari, P., Maulik, P.R. and Misra, A.K. (2005) *European Journal of Organic Chemistry*, 74–79.
24 Schneider, W. and Sepp, J. (1918) *Chemische Berichte*, **51**, 220–234.
25 Purves, C. (1929) *Journal of the American Chemical Society*, **51**, 3619–3627.
26 Shah, R.H. and Bahl, O.P. (1974) *Carbohydrate Research*, **32**, 15–23.
27 Shah, R.H. and Bahl, O.P. (1978) *Carbohydrate Research*, **65**, 153–158.
28 Matta, K.L., Girotra, R.N. and Barlow, J.J. (1975) *Carbohydrate Research*, **43**, 101–109.
29 Chipowsky, S. and Yuan, Y.L. (1973) *Carbohydrate Research*, **31**, 339–346.
30 Durette, P.L. and Shen, T.Y. (1978) *Carbohydrate Research*, **67**, 484–490.
31 Montgomery, E.M., Richtmyer, N.K. and Hudson, C.S. (1946) *The Journal of Organic Chemistry*, **11**, 301–306.
32 Tsvetkov, Y.E., Byramova, N.E. and Backinowsky, L.V. (1983) *Carbohydrate Research*, **115**, 254–258.
33 Ogawa, T. and Matsui, M. (1977) *Carbohydrate Research*, **54**, C17–C21.
34 Mukherjee, C., Tiwari, P. and Misra, A.K. (2006) *Tetrahedron Letters*, **47**, 441–445.
35 Cerny, M. and Pacak, J. (1958) *Chemicke Listy*, **52**, 2090–2093.
36 Driguez, H. and Szeja, W. (1994) *Synthesis*, 1413–1414.
37 Horton, D. (1963) *Methods in Carbohydrate Chemistry*, vol. 2 (eds R.L. Whistler and M.L. Wolfrom), Academic Press, New York, pp. 433–437.
38 Erbing, B. and Lindberg, B. (1976) *Acta Chemica Scandinavica. Series B: Organic Chemistry and Biochemistry*, **30**, 611–612.
39 Ibatullin, F.M., Selivanov, S.I. and Shavva, A.G. (2001) *Synthesis*, 419–422.
40 El Ashry, E.S.H., Awad, L.F., Hamid, H.M.A. and Atta, A.I. (2006) *Synthetic Communications*, **36**, 2769–2785.
41 Cao, S.D., Meunier, S.J., Andersson, F.O., Letellier, M. and Roy, R. (1994) *Tetrahedron: Asymmetry*, **5**, 2303–2312.

42 Fujihira, T., Takido, T. and Seno, M. (1999) *Journal of Molecular Catalysis A – Chemical*, **137**, 65–75.
43 Hanessian, S. and Guindon, Y. (1980) *Carbohydrate Research*, **86**, C3–C6.
44 Liu, D.S., Chen, R., Hong, L.W. and Sofia, M.J. (1998) *Tetrahedron Letters*, **39**, 4951–4954.
45 Pacsu, E. (1963) *Methods in Carbohydrate Chemistry*, vol. 2 (eds R.L. Whistler and M.L. Wolfrom), Academic Press, New York, pp. 354–367.
46 Yde, M. and Debruyne, C.K. (1973) *Carbohydrate Research*, **26**, 227–229.
47 Lacombe, J.M., Rakotomanomana, N. and Pavia, A.A. (1988) *Tetrahedron Letters*, **29**, 4293–4296.
48 Sakata, M., Haga, M. and Tejima, S. (1970) *Carbohydrate Research*, **13**, 379–390.
49 Pakulski, Z., Pierozynski, D. and Zamojski, A. (1994) *Tetrahedron*, **50**, 2975–2992.
50 Wolfrom, M.L., Garg, H.G. and Horton, D. (1963) *The Journal of Organic Chemistry*, **28**, 2986–2988.
51 Mukaiyama, T. and Kobashi, Y. (2004) *Chemistry Letters*, **33**, 10–11.
52 Kartha, K.P.R. and Field, R.A. (1997) *Tetrahedron Letters*, **38**, 8233–8236.
53 Sato, S., Mori, M., Ito, Y. and Ogawa, T. (1986) *Carbohydrate Research*, **155**, C6–C10.
54 Shingu, Y., Nishida, Y., Dohi, H., Matsuda, K. and Kobayashi, K. (2002) *Journal of Carbohydrate Chemistry*, **21**, 605–611.
55 Mukaiyama, T., Murai, Y. and Shoda, S. (1981) *Chemistry Letters*, **10**, 431–432.
56 Nicolaou, K.C., Dolle, R.E., Papahatjis, D.P. and Randall, J.L. (1984) *Journal of the American Chemical Society*, **106**, 4189–4192.
57 Nicolaou, K.C. and Ueno, H. (1997) *Preparative Carbohydrate Chemistry* (ed. S. Hanessian), Dekker, New York, pp. 313–338.
58 Motawia, M.S., Marcussen, J. and Moller, B.L. (1995) *Journal of Carbohydrate Chemistry*, **14**, 1279–1294.
59 Damager, I., Olsen, C.E., Moller, B.L. and Motawia, M.S. (1999) *Carbohydrate Research*, **320**, 19–30.
60 Oshitari, T., Shibasaki, M., Yoshizawa, T., Tomita, M., Takao, K. and Kobayashi, S. (1997) *Tetrahedron*, **53**, 10993–11006.
61 Gomez, A.M., Company, M.D., Agocs, A., Uriel, C., Valverde, S. and Lopez, J.C. (2005) *Carbohydrate Research*, **340**, 1872–1875.
62 Garegg, P.J., Hultberg, H. and Lindberg, C. (1980) *Carbohydrate Research*, **83**, 157–162.
63 Davis, B.G. (2000) *Journal of the Chemical Society, Perkin Transactions*, **1**, 2137–2160.
64 Kasbeck, L. and Kessler, H. (1997) *Liebigs Annalen – Recueil*, **1997**, 169–173.
65 Uchiro, H., Wakiyama, Y. and Mukaiyama, T. (1998) *Chemistry Letters*, **27**, 567–568.
66 Mondal, E., Barua, P.M.B., Bose, G. and Khan, A.T. (2002) *Chemistry Letters*, **21**, 210–211.
67 Barua, P.M.B., Sahu, P.R., Mondal, E., Bose, G. and Khan, A.T. (2002) *Synlett*, 81–84.
68 Misra, A.K. and Agnihotri, G. (2004) *Carbohydrate Research*, **339**, 885–890.
69 Duynstee, H.I., de Koning, M.C., Ovaa, H., van der Marel, G.A. and van Boom, J.H. (1999) *European Journal of Organic Chemistry*, 2623–2632.
70 Dinkelaar, J., Witte, M.D., van den Bos, L.J., Overkleeft, H.S. and van der Marel, G.A. (2006) *Carbohydrate Research*, **341**, 1723–1729.
71 Schmidt, R.R. and Michel, J. (1980) *Angewandte Chemie (International Edition)*, **19**, 731–732.
72 Schmidt, R.R. and Kinzy, W. (1994) *Advances in Carbohydrate Chemistry and Biochemistry*, **50**, 21–123.
73 Schmidt, R.R. and Jung, K.H. (1997) *Preparative Carbohydrate Chemistry* (ed. S. Hanessian), Dekker, New York.
74 Kahne, D., Walker, S., Cheng, Y. and Vanengen, D. (1989) *Journal of the American Chemical Society*, **111**, 6881–6882.

75 Yan, L. and Kahne, D. (1996) *Journal of the American Chemical Society*, **118**, 9239–9248.
76 Kakarla, R., Dulina, R.G., Hatzenbuhler, N.T., Hui, Y.W. and Sofia, M.J. (1996) *The Journal of Organic Chemistry*, **61**, 8347–8349.
77 Kim, S.H., Augeri, D., Yang, D. and Kahne, D. (1994) *Journal of the American Chemical Society*, **116**, 1766–1775.
78 Agnihotri, G. and Misra, A.K. (2005) *Tetrahedron Letters*, **46**, 8113–8116.
79 Foti, C.J., Fields, J.D. and Kropp, P.J. (1999) *Organic Letters*, **1**, 903–904.
80 Hirano, M., Tomaru, J. and Morimoto, T. (1991) *Bulletin of the Chemical Society of Japan*, **64**, 3752–3754.
81 Vincent, S.P., Burkart, M.D., Tsai, C.Y., Zhang, Z.Y. and Wong, C.H. (1999) *The Journal of Organic Chemistry*, **64**, 5264–5279.
82 Chen, M.Y., Patkar, L.N. and Lin, C.C. (2004) *The Journal of Organic Chemistry*, **69**, 2884–2887.
83 Breton, G.W., Fields, J.D. and Kropp, P.J. (1995) *Tetrahedron Letters*, **36**, 3825–3828.
84 Raghavan, S. and Kahne, D. (1993) *Journal of the American Chemical Society*, **115**, 1580–1581.
85 Crich, D. and Smith, M. (2001) *Journal of the American Chemical Society*, **123**, 9015–9020.
86 Crich, D. and Lim, L.B.L. (2004) *Organic Reactions*, **64**, 115–251.
87 Codee, J.D.C., van den Bos, L.J., Litjens, R., Overkleeft, H.S., van Boeckel, C.A.A., van Boom, J.H. and van der Marel, G.A. (2004) *Tetrahedron*, **60**, 1057–1064.
88 Cow, D. and Li, W.J. (2006) *Organic Letters*, **8**, 959–962.
89 Ferrier, R.J., Hay, R.W. and Vethaviyasar, N. (1973) *Carbohydrate Research*, **27**, 55–61.
90 Woodward, R.B., Logusch, E., Nambiar, K.P., Sakan, K., Ward, D.E., Auyeung, B.W., Balaram, P., Browne, L.J., Card, P.J., Chen, C.H., Chenevert, R.B., Fliri, A., Frobel, K., Gais, H.J., Garratt, D.G., Hayakawa, K., Heggie, W., Hesson, D.P., Hoppe, D., Hoppe, I., Hyatt, J.A., Ikeda, D., Jacobi, P.A., Kim, K.S., Kobuke, Y., Kojima, K., Krowicki, K., Lee, V.J., Leutert, T., Malchenko, S., Martens, J., Matthews, R.S., Ong, B.S., Press, J.B., Babu, T.V.R., Rousseau, G., Sauter, H.M., Suzuki, M., Tatsuta, K., Tolbert, L.M., Truesdale, E.A., Uchida, I., Ueda, Y., Uyehara, T., Vasella, A.T., Vladuchick, W.C., Wade, P.A., Williams, R.M. and Wong, H.N.C. (1981) *Journal of the American Chemical Society*, **103**, 3215–3217.
91 Wuts, P.G.M. and Bigelow, S.S. (1983) *The Journal of Organic Chemistry*, **48**, 3489–3493.
92 Fugedi, P. and Garegg, P.J. (1986) *Carbohydrate Research*, **149**, C9–C12.
93 Veeneman, G.H., van Leeuwen, S.H. and van Boom, J.H. (1990) *Tetrahedron Letters*, **31**, 1331–1334.
94 Konradsson, P., Udodong, U.E. and Fraser-Reid, B. (1990) *Tetrahedron Letters*, **31**, 4313–4316.
95 Lemieux, R.U. and Morgan, A.R. (1965) *Canadian Journal of Chemistry*, **43**, 2205.
96 Veeneman, G.H. and Vanboom, J.H. (1990) *Tetrahedron Letters*, **31**, 275–278.
97 Ito, Y. and Ogawa, T. (1988) *Tetrahedron Letters*, **29**, 1061–1064.
98 Ito, Y., Ogawa, T., Numata, M. and Sugimoto, M. (1990) *Carbohydrate Research*, **202**, 165–175.
99 Codee, J.D.C., Litjens, R., den Heeten, R., Overkleeft, H.S., van Boom, J.H. and van der Marel, G.A. (2003) *Organic Letters*, **5**, 1519–1522.
100 Crich, D. and Smith, M. (2000) *Organic Letters*, **2**, 4067–4069.
101 Wang, C.N., Wang, H.S., Huang, X.F., Zhang, L.H. and Ye, X.S. (2006) *Synlett*, 2846–2850.
102 Amatore, C., Jutand, A., Mallet, J.M., Meyer, G. and Sinay, P. (1990) *Journal of the Chemical Society Chemical Communications*, 718–719.
103 Noyori, R. and Kurimoto, I. (1986) *The Journal of Organic Chemistry*, **51**, 4320–4322.

104 Marra, A., Mallet, J.M., Amatore, C. and Sinay, P. (1990) *Synlett*, 572–574.
105 van Well, R.M., Karkkainen, T.S., Kartha, K.P.R. and Field, R.A. (2006) *Carbohydrate Research*, **341**, 1391–1397.
106 Mehta, S. and Pinto, B.M. (1998) *Carbohydrate Research*, **310**, 43–51.
107 Takeuchi, K., Tamura, T., Jona, H. and Mukaiyama, T. (2000) *Chemistry Letters*, **29**, 692–693.
108 Uchiro, H. and Mukaiyama, T. (1997) *Chemistry Letters*, **26**, 121–122.
109 Jona, H., Takeuchi, K., Saitoh, T. and Mukaiyama, T. (2000) *Chemistry Letters*, **29**, 1178–1179.
110 Mukaiyama, T., Wakiyama, Y., Miyazaki, K. and Takeuchi, K. (1999) *Chemistry Letters*, **28**, 933–934.
111 Meijer, A. and Ellervik, U. (2004) *The Journal of Organic Chemistry*, **69**, 6249–6256.
112 Cura, P., Aloui, M., Kartha, K.P.R. and Field, R.A. (2000) *Synlett*, 1279–1280.
113 Kartha, K.P.R., Cura, P., Aloui, M., Readman, S.K., Rutherford, T.J. and Field, R.A. (2000) *Tetrahedron: Asymmetry*, **11**, 581–593.
114 Demchenko, A.V., Malysheva, N.N. and De Meo, C. (2003) *Organic Letters*, **5**, 455–458.
115 Demchenko, A.V., Kamat, M.N. and De Meo, C. (2003) *Synlett*, 1287–1290.
116 Demchenko, A.V., Pornsuriyasak, P., De Meo, C. and Malysheva, N.N. (2004) *Angewandte Chemie (International Edition)*, **43**, 3069–3072.
117 Pornsuriyasak, P. and Demchenko, A.V. (2005) *Tetrahedron: Asymmetry*, **16**, 433–439.
118 Ramakrishnan, A., Pornsuriyasak, P. and Demchenko, A.V. (2005) *Journal of Carbohydrate Chemistry*, **24**, 649–663.
119 Paulsen, H. (1982) *Angewandte Chemie (International Edition)*, **21**, 155–173.
120 Schmidt, R.R. (1986) *Angewandte Chemie (International Edition)*, **25**, 212–235.
121 Boons, G.J. (1996) *Contemporary Organic Synthesis*, **3**, 173–200.
122 Wulff, G. and Rohle, G. (1974) *Angewandte Chemie (International Edition)*, **13**, 157–170.
123 Boons, G.J. and Stauch, T. (1996) *Synlett*, 906–908.
124 Demchenko, A.V., Stauch, T. and Boons, G.J. (1997) *Synlett*, 818–820.
125 Schmidt, R.R. and Michel, J. (1985) *Journal of Carbohydrate Chemistry*, **4**, 141–169.
126 Schmidt, R.R., Behrendt, M. and Toepfer, A. (1990) *Synlett*, 694–696.
127 Vankar, Y.D., Vankar, P.S., Behrendt, M. and Schmidt, R.R. (1991) *Tetrahedron*, **47**, 9985–9992.
128 Schmidt, R.R., Gaden, H. and Jatzke, H. (1990) *Tetrahedron Letters*, **31**, 327–330.
129 Pougny, J.R. and Sinay, P. (1976) *Tetrahedron Letters*, **17**, 4073–4076.
130 Lemieux, R.U. and Ratcliffe, R.M. (1979) *Canadian Journal of Chemistry/Revue Canadiene de Chimie*, **57**, 1244–1251.
131 Ratcliffe, A.J. and Fraser-Reid, B. (1990) *Journal of the Chemical Society Perkin Transactions*, **1**, 747–750.
132 Crich, D. and Sun, S.X. (1996) *The Journal of Organic Chemistry*, **61**, 4506–4507.
133 Crich, D. and Sun, S.X. (1997) *The Journal of Organic Chemistry*, **62**, 1198–1199.
134 Crich, D. and Sun, S.X. (1997) *Journal of the American Chemical Society*, **119**, 11217–11223.
135 Crich, D. and Chandrasekera, N.S. (2004) *Angewandte Chemie (International Edition)*, **43**, 5386–5389.
136 Crich, D., Jayalath, P. and Hutton, T.K. (2006) *The Journal of Organic Chemistry*, **71**, 3064–3070.
137 Crich, D. and Jayalath, P. (2005) *Organic Letters*, **7**, 2277–2280.
138 Crich, D., Li, W.J. and Li, H.M. (2004) *Journal of the American Chemical Society*, **126**, 15081–15086.
139 Litjens, R., Leeuwenburgh, M.A., van der Marel, G.A. and van Boom, J.H. (2001) *Tetrahedron Letters*, **42**, 8693–8696.
140 De Meo, C., Kamat, M.N. and Demchenko, A.V. (2005) *European Journal of Organic Chemistry*, 706–711.
141 Stork, G. and Kim, G. (1992) *Journal of the American Chemical Society*, **114**, 1087–1088.

142 Stork, G. and LaClair, J.J. (1996) *Journal of the American Chemical Society*, **118**, 247–248.

143 Barresi, F. and Hindsgaul, O. (1991) *Journal of the American Chemical Society*, **113**, 9376–9377.

144 Barresi, F. and Hindsgaul, O. (1992) *Synlett*, 759–761.

145 Barresi, F. and Hindsgaul, O. (1994) *Canadian Journal of Chemistry/Revue Canadiene de Chimie*, **72**, 1447–1465.

146 Chayajarus, K., Chambers, D.J., Chughtai, M.J. and Fairbanks, A.J. (2004) *Organic Letters*, **6**, 3797–3800.

147 Aloui, M., Chambers, D.J., Cumpstey, I., Fairbanks, A.J., Redgrave, A.J. and Seward, C.M.P. (2002) *Chemistry – A European Journal*, **8**, 2608–2621.

148 Ennis, S.C., Fairbanks, A.J., Slinn, C.A., Tennant-Eyles, R.J. and Yeates, H.S. (2001) *Tetrahedron*, **57**, 4221–4230.

149 Seward, C.M.P., Cumpstey, I., Aloui, M., Ennis, S.C., Redgrave, A.J. and Fairbanks, A.J. (2000) *Chemical Communications*, 1409–1410.

150 Attolino, E., Cumpstey, I. and Fairbanks, A.J. (2006) *Carbohydrate Research*, **341**, 1609–1618.

151 Ziegler, T., Lemanski, G. and Rakoczy, A. (1995) *Tetrahedron Letters*, **36**, 8973–8976.

152 Ziegler, T. and Lemanski, G. (1998) *European Journal of Organic Chemistry*, 163–170.

153 Ziegler, T. and Lemanski, G. (1998) *Angewandte Chemie (International Edition)*, **37**, 3129–3132.

154 Kochetkov, N.K., Klimov, E.M. and Malysheva, N.N. (1989) *Tetrahedron Letters*, **30**, 5459–5462.

155 Kochetkov, N.K., Klimov, E.M., Malysheva, N.N. and Demchenko, A.V. (1991) *Carbohydrate Research*, **212**, 77–91.

156 Kochetkov, N.K., Klimov, E.M., Malysheva, N.N. and Demchenko, A.V. (1992) *Carbohydrate Research*, **232**, C1–C5.

157 Reddy, G.V., Kulkarni, V.R. and Mereyala, H.B. (1989) *Tetrahedron Letters*, **30**, 4283–4286.

158 Mereyala, H.B. and Reddy, G.V. (1991) *Tetrahedron*, **47**, 9721–9726.

159 Mereyala, H.B. and Reddy, G.V. (1991) *Tetrahedron*, **47**, 6435–6448.

160 Hasegawa, A., Nagahama, T., Ohki, H., Hotta, K., Ishida, H. and Kiso, M. (1991) *Journal of Carbohydrate Chemistry*, **10**, 493–498.

161 Hasegawa, A., Ohki, H., Nagahama, T., Ishida, H. and Kiso, M. (1991) *Carbohydrate Research*, **212**, 277–281.

162 Birberg, W., Fügedi, P., Garegg, P.J. and Pilotti, Å. (1989) *Journal of Carbohydrate Chemistry*, **8**, 47–57.

163 Boons, G.J. and Demchenko, A.V. (2000) *Chemical Reviews*, **100**, 4539–4566.

164 Demchenko, A.V. and Boons, G.J. (1998) *Tetrahedron Letters*, **39**, 3065–3068.

165 Demchenko, A.V. and Boons, G.J. (1999) *Chemistry – A European Journal*, **5**, 1278–1283.

166 De Meo, C., Demchenko, A.V. and Boons, G.J. (2001) *The Journal of Organic Chemistry*, **66**, 5490–5497.

167 Tanaka, H., Adachi, M. and Takahashi, T. (2005) *Chemistry – A European Journal*, **11**, 849–862.

168 Tanaka, H., Nishiura, Y. and Takahashi, T. (2006) *Journal of the American Chemical Society*, **128**, 7124–7125.

169 Yu, C.S., Niikura, K., Lin, C.C. and Wong, C.H. (2001) *Angewandte Chemie (International Edition)*, **40**, 2900–2903.

170 De Meo, C. and Parker, O. (2005) *Tetrahedron: Asymmetry*, **16**, 303–307.

171 Zuurmond, H.M., Vanderlaan, S.C., Vandermarel, G.A. and Vanboom, J.H. (1991) *Carbohydrate Research*, **215**, C1–C3.

172 Zuurmond, H.M., Vandermarel, G.A. and Vanboom, J.H. (1991) *Recueil Des Travaux Chimiques Des Pays-Bas (Journal of the Royal Netherlands Chemical Society)*, **110**, 301–302.

173 Zegelaarjaarsveld, K., Vandermarel, G. and Vanboom, J.H. (1992) *Tetrahedron*, **48**, 10133–10148.

174 Zuurmond, H.M., Vanderklein, P.A.M., Vandermeer, P.H., Vandermarel, G.A. and Vanboom, J.H. (1992) *Recueil Des*

Travaux Chimiques Des Pays-Bas (Journal of the Royal Netherlands Chemical Society), **111**, 365–366.
175 Zuurmond, H.M., Veeneman, G.H., Vandermarel, G.A. and Vanboom, J.H. (1993) Carbohydrate Research, **241**, 153–164.
176 Mootoo, D.R., Konradson, P., Udodong, U. and Fraser-Reid, B. (1988) Journal of the American Chemical Society, **110**, 5583–5584.
177 Konradsson, P., Mootoo, D.R., McDevitt, R.E. and Fraser-Reid, B. (1990) Journal of the Chemical Society: Chemical Communications, 270–272.
178 Boons, G.J., Grice, P., Leslie, R., Ley, S.V. and Yeung, L.L. (1993) Tetrahedron Letters, **34**, 8523–8526.
179 Ley, S.V. and Priepke, H.W.M. (1994) Angewandte Chemie (International Edition), **33**, 2292–2294.
180 Ley, S.V., Downham, R., Edwards, P.J., Innes, J.E. and Woods, M. (1995) Contemporary Organic Synthesis, **2**, 365–392.
181 Grice, P., Ley, S.V., Pietruszka, J. and Priepke, H.W.M. (1996) Angewandte Chemie (International Edition), **35**, 197–200.
182 Grice, P., Ley, S.V., Pietruszka, J., Osborn, H.M.I., Priepke, H.W.M. and Warriner, S.L. (1997) Chemistry – A European Journal, **3**, 431–440.
183 Grice, P., Ley, S.V., Pietruszka, J., Priepke, H.W.M. and Warriner, S.L. (1997) Journal of the Chemical Society Perkin Transactions, **1**, 351–363.
184 Green, L., Hinzen, B., Ince, S.J., Langer, P., Ley, S.V. and Warriner, S.L. (1998) Synlett, 440–442.
185 Cheung, M.K., Douglas, N.L., Hinzen, B., Ley, S.V. and Pannecoucke, X. (1997) Synlett, 257–260.
186 Douglas, N.L., Ley, S.V., Lucking, U. and Warriner, S.L. (1998) Journal of the Chemical Society Perkin Transactions, **1**, 51–65.
187 Boons, G.J., Geurtsen, R. and Holmes, D. (1995) Tetrahedron Letters, **36**, 6325–6328.
188 Geurtsen, R., Holmes, D.S. and Boons, G.J. (1997) The Journal of Organic Chemistry, **62**, 8145–8154.
189 Geurtsen, R., Côté, F., Hahn, M.G. and Boons, G.J. (1999) The Journal of Organic Chemistry, **64**, 7828–7835.
190 Zhu, T. and Boons, G.J. (2001) Organic Letters, **3**, 4201–4203.
191 Toshima, K., Mukaiyama, S., Ishiyama, T. and Tatsuta, K. (1990) Tetrahedron Letters, **31**, 6361–6362.
192 Toshima, K., Mukaiyama, S., Ishiyama, T. and Tatsuta, K. (1990) Tetrahedron Letters, **31**, 3339–3342.
193 Toshima, K., Mukaiyama, S., Yoshida, T., Tamai, T. and Tatsuta, K. (1991) Tetrahedron Letters, **32**, 6155–6158.
194 Toshima, K., Nozaki, Y., Mukaiyama, S. and Tatsuta, K. (1992) Tetrahedron Letters, **33**, 1491–1494.
195 Toshima, K., Nozaki, Y., Inokuchi, H., Nakata, M., Tatsuta, K. and Kinoshita, M. (1993) Tetrahedron Letters, **34**, 1611–1614.
196 Grice, P., Ley, S.V., Pietruszka, J., Priepke, H.W.M. and Walther, E.P.E. (1995) Synlett, 781–784.
197 Zhang, Z., Ollmann, I.R., Ye, X.S., Wischnat, R., Baasov, T. and Wong, C.H. (1999) Journal of the American Chemical Society, **121**, 734–753.
198 Ritter, T.K., Mong, K.K.T., Liu, H.T., Nakatani, T. and Wong, C.H. (2003) Angewandte Chemie (International Edition), **42**, 4657–4660.
199 Ye, X.S. and Wong, C.H. (2000) The Journal of Organic Chemistry, **65**, 2410–2431.
200 Burkhart, F., Zhang, Z.Y., Wacowich-Sgarbi, S. and Wong, C.H. (2001) Angewandte Chemie (International Edition), **40**, 1274–1277.
201 Mong, T.K.K., Lee, H.K., Duron, S.G. and Wong, C.H. (2003) Proceedings of the National Academy of Sciences of the United States of America, **100**, 797–802.
202 Zhang, Z.Y., Niikura, K., Huang, X.F. and Wong, C.H. (2002) Canadian Journal of Chemistry/Revue Canadiene de Chimie, **80**, 1051–1054.

203 Mong, T.K.K., Huang, C.Y. and Wong, C.H. (2003) *The Journal of Organic Chemistry*, **68**, 2135–2142.

204 Wang, Y.H., Huang, X.F., Zhang, L.H. and Ye, X.S. (2004) *Organic Letters*, **6**, 4415–4417.

205 Lee, H.K., Scanlan, C.N., Huang, C.Y., Chang, A.Y., Calarese, D.A., Dwek, R.A., Rudd, P.M., Burton, D.R., Wilson, I.A. and Wong, C.H. (2004) *Angewandte Chemie (International Edition)*, **43**, 1000–1003.

206 Mong, K.K.T. and Wong, C.H. (2002) *Angewandte Chemie (International Edition)*, **41**, 4087–4090.

207 Lahmann, M. and Oscarson, S. (2000) *Organic Letters*, **2**, 3881–3882.

208 Fridman, M., Solomon, D., Yogev, S. and Baasov, T. (2002) *Organic Letters*, **4**, 281–283.

209 Nicolaou, K.C., Randall, J.L. and Furst, G.T. (1985) *Journal of the American Chemical Society*, **107**, 5556–5558.

210 Nicolaou, K.C., Caulfield, T.J., Kataoka, H. and Stylianides, N.A. (1990) *Journal of the American Chemical Society*, **112**, 3693–3695.

211 Nicolaou, K.C., Hummel, C.W. and Iwabuchi, Y. (1992) *Journal of the American Chemical Society*, **114**, 3126–3128.

212 Nicolaou, K.C., Bockovich, N.J. and Carcanague, D.R. (1993) *Journal of the American Chemical Society*, **115**, 8843–8844.

213 Nicolaou, K.C., Caulfield, T., Kataoka, H. and Kumazawa, T. (1988) *Journal of the American Chemical Society*, **110**, 7910–7912.

214 Nicolaou, K.C., Caulfield, T.J. and Katoaka, H. (1990) *Carbohydrate Research*, **202**, 177–191.

215 Kanie, O., Ito, Y. and Ogawa, T. (1994) *Journal of the American Chemical Society*, **116**, 12073–12074.

216 Yamada, H., Harada, T. and Takahashi, T. (1994) *Journal of the American Chemical Society*, **116**, 7919–7920.

217 Hashihayata, T., Ikegai, K., Takeuchi, K., Jona, H. and Mukaiyama, T. (2003) *Bulletin of the Chemical Society of Japan*, **76**, 1829–1848.

218 Lonn, H. (1984) *Chemical Communications (Stockholm University)*, **2**, 1–30.

219 Lonn, H. (1985) *Carbohydrate Research*, **139**, 115–121.

220 Lonn, H. (1987) *Journal of Carbohydrate Chemistry*, **6**, 301–306.

221 Hallgren, C. and Widmalm, G. (1993) *Journal of Carbohydrate Chemistry*, **12**, 309–333.

222 Mehta, S. and Pinto, B.M. (1991) *Tetrahedron Letters*, **32**, 4435–4438.

223 Mehta, S. and Pinto, B.M. (1993) *The Journal of Organic Chemistry*, **58**, 3269–3276.

224 Sliedregt, L., Vandermarel, G.A. and Vanboom, J.H. (1994) *Tetrahedron Letters*, **35**, 4015–4018.

225 Fugedi, P., Garegg, P.J., Oscarson, S., Rosen, G. and Silwanis, B.A. (1991) *Carbohydrate Research*, **211**, 157–162.

226 Yamada, H., Harada, T., Miyazaki, H. and Takahashi, T. (1994) *Tetrahedron Letters*, **35**, 3979–3982.

227 Pornsuriyasak, P., Gangadharmath, U.B., Rath, N.P. and Demchenko, A.V. (2004) *Organic Letters*, **6**, 4515–4518.

228 Pornsuriyasak, P. and Demchenko, A.V. (2006) *Chemistry – A European Journal*, **12**, 6630–6646.

229 Huang, X.F., Huang, L.J., Wang, H.S. and Ye, X.S. (2004) *Angewandte Chemie (International Edition)*, **43**, 5221–5224.

230 Codee, J.D.C., van den Bos, L.J., Litjens, R., Overkleeft, H.S., van Boom, J.H. and van der Marel, G.A. (2003) *Organic Letters*, **5**, 1947–1950.

231 Yamago, S., Yamada, T., Maruyama, T. and Yoshida, J. (2004) *Angewandte Chemie (International Edition)*, **43**, 2145–2148.

232 Boons, G.J., Bowers, S. and Coe, D.M. (1997) *Tetrahedron Letters*, **38**, 3773–3776.

233 Boons, G.J. and Zhu, T. (1997) *Synlett*, 809–811.

234 Zhu, T. and Boons, G.J. (1998) *Journal of the Chemical Society Perkin Transactions*, **1**, 857–861.

235 Zhu, T. and Boons, G.J. (1998) *Angewandte Chemie (International Edition)*, **37**, 1898–1900.
236 Tanaka, H., Adachi, M. and Takahashi, T. (2004) *Tetrahedron Letters*, **45**, 1433–1436.
237 Tanaka, H., Adachi, M., Tsukamoto, H., Ikeda, T., Yamada, H. and Takahashi, T. (2002) *Organic Letters*, **4**, 4213–4216.
238 Kihlberg, J., Eichler, E. and Bundle, D.R. (1991) *Carbohydrate Research*, **211**, 59–75.
239 Knapp, S. and Nandan, S.R. (1994) *The Journal of Organic Chemistry*, **59**, 281–283.
240 Leigh, D.A., Smart, J.P. and Truscello, A.M. (1995) *Carbohydrate Research*, **276**, 417–424.
241 Belot, F. and Jacquinet, J.C. (1996) *Carbohydrate Research*, **290**, 79–86.
242 Du, Y.G., Lin, J.H. and Linhardt, R.J. (1997) *Journal of Carbohydrate Chemistry*, **16**, 1327–1344.
243 Yu, H., Yu, B., Wu, X.Y., Hui, Y.Z. and Han, X.W. (2000) *Journal of the Chemical Society Perkin Transactions*, **9**, 1445–1453.
244 Zhu, T. and Boons, G.J. (2000) *Carbohydrate Research*, **329**, 709–715.
245 Sherman, A.A., Yudina, O.N., Mironov, Y.V., Sukhova, E.V., Shashkov, A.S., Menshov, V.M. and Nifantiev, N.E. (2001) *Carbohydrate Research*, **336**, 13–46.
246 Cheshev, P.E., Kononov, L.O., Tsvetkov, Y.E., Shashkov, A.S. and Nifantiev, N.E. (2002) *Russian Journal of Bioorganic Chemistry*, **28**, 419–429.
247 Geurtsen, R. and Boons, G.J. (2002) *Tetrahedron Letters*, **43**, 9429–9431.
248 Xue, J., Khaja, S.D., Locke, R.D. and Matta, K.L. (2004) *Synlett*, 861–865.
249 Sun, J.S., Han, X.W. and Yu, B. (2005) *Organic Letters*, **7**, 1935–1938.
250 Li, Z.T. and Gildersleeve, J.C. (2006) *Journal of the American Chemical Society*, **128**, 11612–11619.
251 Tsai, T.Y.R., Jin, H. and Wiesner, K. (1984) *Canadian Journal of Chemistry*, **62**, 1403–1405.
252 Garegg, P.J., Henrichson, C. and Norberg, T. (1983) *Carbohydrate Research*, **116**, 162–165.
253 Van Cleve, J.W. (1979) *Carbohydrate Research*, **70**, 161–164.
254 Hanessian, S., Bacquet, C. and Lehong, N. (1980) *Carbohydrate Research*, **80**, C17–C22.
255 Mukaiyama, T., Nakatsuki, T. and Shoda, S. (1979) *Chemistry Letters*, **8**, 487–490.
256 Shimizu, H., Ito, Y. and Ogawa, T. (1994) *Synlett*, 535–536.
257 Nicolaou, K.C., Seitz, S.P. and Papahatjis, D.P. (1983) *Journal of the American Chemical Society*, **105**, 2430–2434.
258 Veeneman, G.H., van Leeuwen, S.H., Zuurmond, H. and van Boom, J.H. (1990) *Journal of Carbohydrate Chemistry*, **9**, 783–796.
259 Kartha, K.P.R., Aloui, M. and Field, R.A. (1996) *Tetrahedron Letters*, **37**, 5175–5178.
260 Fukase, K., Hasuoka, A., Kinoshita, I. and Kusumoto, S. (1992) *Tetrahedron Letters*, **33**, 7165–7168.
261 Pozsgay, V. and Jennings, H.J. (1987) *The Journal of Organic Chemistry*, **52**, 4635–4637.
262 Pozsgay, V. and Jennings, H.J. (1988) *The Journal of Organic Chemistry*, **53**, 4042–4052.
263 Reddy, G.V., Kulkari, V.R. and Mereyala, H.B. (1989) *Tetrahedron Letters*, **30**, 4283–4286.
264 Dasgupta, F. and Garegg, P.J. (1988) *Carbohydrate Research*, **177**, C13–C17.
265 Tsuboyama, K., Takeda, K., Torri, K., Ebihara, M., Shimizu, J., Suzuki, A., Sato, N., Furuhata, K. and Ohura, H. (1990) *Chemical & Pharmaceutical Bulletin*, **38**, 636.
266 Sasaki, M. and Tachibana, K. (1991) *Tetrahedron Letters*, **32**, 6873–6876.
267 Qin, Z.H., Li, H., Cai, M.S. and Li, Z.J. (2002) *Carbohydrate Research*, **337**, 31–36.
268 Tsukamoto, H. and Kondo, Y. (2003) *Tetrahedron Letters*, **44**, 5247–5249.
269 Mukhopadhyay, B., Collet, B. and Field, R.A. (2005) *Tetrahedron Letters*, **46**, 5923–5925.
270 Fukase, K., Hasuoka, A., Kinoshita, L., Aoki, Y. and Kusumoto, S. (1995) *Tetrahedron*, **51**, 4923–4932.

271 Ercegovic, T., Meijer, A., Ellervik, U. and Magnusson, G. (2001) *Organic Letters*, **3**, 913–915.
272 Meijer, A. and Ellervik, U. (2002) *The Journal of Organic Chemistry*, **67**, 7407–7412.
273 Takeuchi, K., Tamura, T. and Mukaiyama, T. (2000) *Chemistry Letters*, **29s**, 124–125.
274 Duron, S.G., Polat, T. and Wong, C.H. (2004) *Organic Letters*, **6**, 839–841.
275 Fukase, K., Nakai, Y., Kanoh, T. and Kusumoto, S. (1998) *Synlett*, **1**, 84–86.
276 Yang, D., Kim, S.H. and Kahne, D. (1991) *Journal of the American Chemical Society*, **113**, 4715–4716.
277 Yan, L., Taylor, C.M., Goodnow, J.R. and Kahne, D. (1994) *Journal of the American Chemical Society*, **116**, 6953–6954.
278 Soong-Hoon, K., Augeri, D., Yang, D. and Kahne, D. (1994) *Journal of the American Chemical Society*, **116**, 1766–1775.

4.2
Sulfoxides, Sulfimides and Sulfones
David Crich, Albert A. Bowers

4.2.1
Introduction

The combination of high reactivity and mildness of the reaction conditions renders the sulfoxide method one of the most powerful glycosylation procedures, and it is surprising that it has not been more widely adopted since its introduction in 1989 [279]. However, despite the fact that this method is a relative newcomer in the glycosylation arena, it has been more closely scrutinized than any other protocol in terms of its mechanism, and so is one of the better understood reactions [135,280,281]. All major classes of glycosidic bond have been successfully prepared by the sulfoxide method with the single exception of sialic acid glycosides. The related glycosyl sulfimides, sulfones and cyclic sulfites have been much less extensively studied compared to glycosyl sulfoxides.

The field has been reviewed [282,283] most recently and comprehensively in 2004 when extensive tables of examples and numerous experimental parts were provided [86].

4.2.2
Donor Preparation

4.2.2.1 **Sulfoxides**
Glycosyl sulfoxides are generally shelf-stable entities that are readily prepared by the oxidation of the parent thioglycosides (discussed in Section 4.1) [284]. It may be that the additional step required to convert a thioglycoside into a sulfoxide deters some in the field, but in reality this is no more than is needed to convert typical anomeric esters or silyl ethers into the corresponding hemiacetals and then to imidate esters. Historically, Micheel and Schmitz were the first to prepare a glycosyl sulfoxide by the oxidation of ethyl α-D-thioglucopyranoside with wet hydrogen peroxide. This reaction resulted in high yields of a single diastereomer at the sulfur center, the

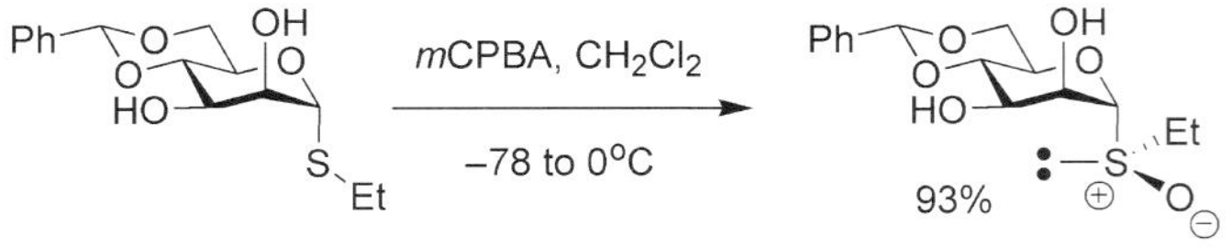

Scheme 4.29 Diastereoselective sulfoxide formation in the axial series.

stereochemical sense and rationale for which were not determined at that time. Subsequently, it was found that high selectivity is the norm in the oxidation of axial thioglycosides to the corresponding sulfoxides, and it has been suggested that this is a consequence of the differential shielding of the two lone pairs imposed by the *exo*-anomeric effect, compounded by the steric contributions of the C-2 substituent (Scheme 4.29) [285–289]. In the equatorial series, where both lone pairs are exposed, selectivities are much lower, not withstanding the *exo*-anomeric effect, and depend to a greater extent on the configuration and the nature of the substituent at C-2 (Scheme 4.30) [75,288,290,291].

In contrast to the kinetic selectivities obtained by the oxidation of thioglycosides, thermodynamic selectivities were obtained in a series of *S*-allyl glycosyl sulfoxides by taking advantage of the allyl sulfoxide–allyl sulfenate equilibrium. Occasionally, the pyranose ring conformation is seen to be dependent on the configuration at sulfur. For example, the oxidation of allyl tri-*O*-benzoyl-thio-α-D-xylopyranoside with *m*CPBA in dichloromethane selectively gave (*R*)-sulfoxide, which was crystallographically and spectroscopically shown to adopt the inverted $^1C_4$ conformation. On heating in benzene, the inversion of stereochemistry at sulfur was observed along with a return to the $^4C_1$ conformation. In methanolic solution, the (*R*)-configuration and the $^1C_4$ conformation were preferred thermodynamically, reflecting the increased steric bulk of the solvated sulfoxide (Scheme 4.31). In the more commonly studied hexose sugars, the $^4C_1$ conformation is typically retained regardless of the configuration at sulfur [288].

In addition to the standard *m*CPBA, many oxidants have been applied in the conversion of thioglycosides into glycosyl sulfoxides, including sodium metaperio-

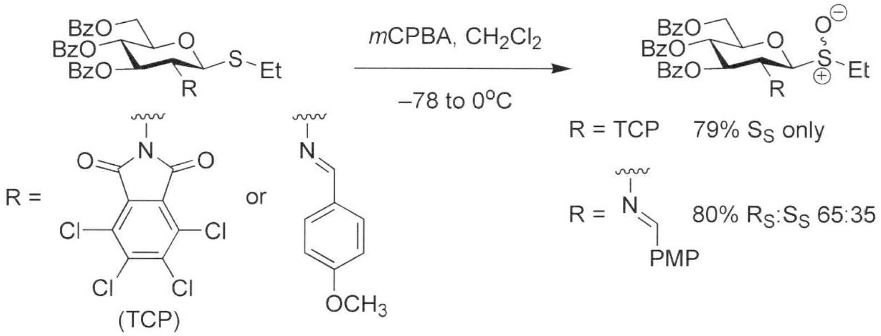

Scheme 4.30 Protecting-group-dependent stereoselectivity in the equatorial series.

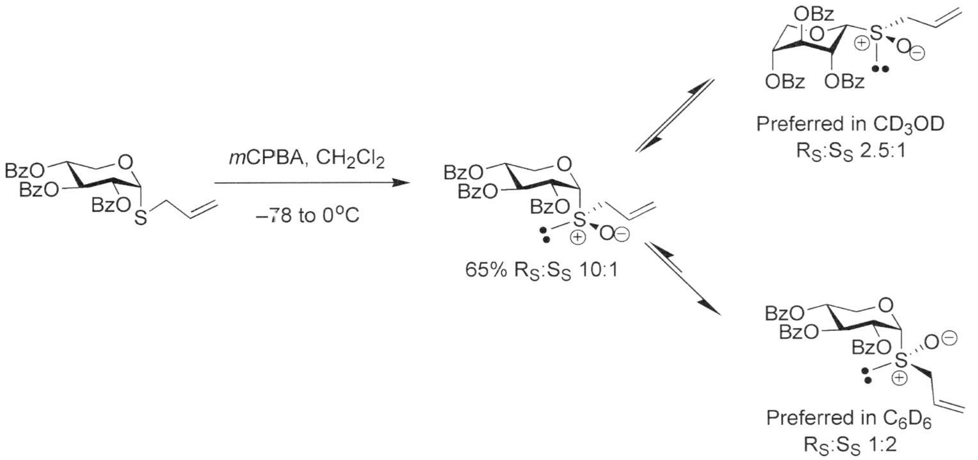

Scheme 4.31 Sulfoxide equilibration and pyranose ring inversion.

date [287], perbenzoic acid [292], OXONE® [287,293], urea/H_2O_2 complex [76], H_2O_2/acetic acid [294] and magnesium monoperoxyphthalate (MMPP)/wet THF [285,295]. All of these provide varying degrees of improvement over the common problems of the mCPBA oxidation: (1) need for low temperatures (−78 to −30 °C), (2) removal of the by-product m-chlorobenzoic acid and (3) overoxidation to the sulfone [75,279,296]. In this regard, it is noteworthy that t-butyl hydrogen peroxide, OXONE® and H_2O_2/HOAc achieve rapid oxidations at room temperature with little observance of sulfone by-products when employed with SiO_2 as a solid support [76,297]. The H_2O_2/HOAc/SiO_2 protocol has been applied effectively on the kilogram scale [298]. Protic, perfluorinated solvents, such as 1,1,1,3,3,3-hexafluoro-2-propanol (HFIP), have also been shown to suppress the formation of sulfone by-products when used in conjunction with 30% H_2O_2 [299,300]. Quantum mechanical calculations have been used to rationalize this effect in terms of the acidity of HFIP, which appears appropriately tuned to activate the oxidant while simultaneously deactivating the sulfoxide products, both through hydrogen bonding [299]. Recent work has demonstrated that MMPP is an effective reagent for the rapid conversion of thioglycosides into sulfoxides in dichloromethane when used in conjunction with microwave irradiation [301].

Methods for indirect oxidation have also been developed. The combination of KF/mCPBA in acetonitrile and water has been used to generate KOF·CH_3CN reagent, a mild and selective oxidant that reacts at 0 °C with no overoxidation [78]. This reagent functions by providing a fluorosulfonium ion intermediate, which is hydrolyzed in the presence of water to the desired sulfoxides. As a result of the indirect oxidation method, the typical stereoselectivity of mCPBA-type oxidations is not observed here. The KOF·CH_3CN oxidant is similar in scope and mechanism to 1-fluoropyridinium triflates, Selectfluor® [302] and the more classical t-butyl hypochlorite [288].

Typically, armed thioglycosides are more rapidly converted into sulfoxides than their disarmed congeners, irrespective of the oxidant used.

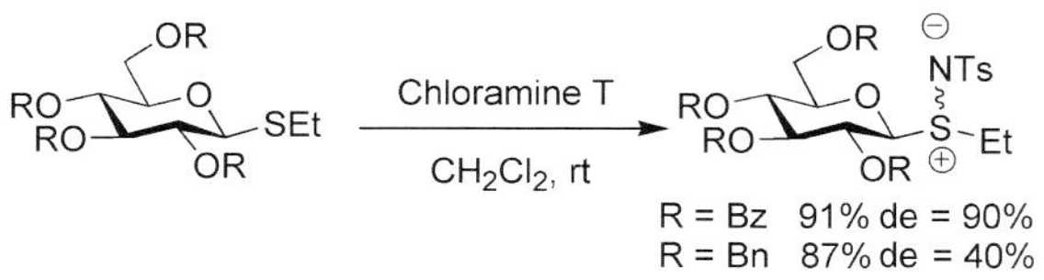

Scheme 4.32 Preparation of sulfimides with chloramine T.

R = Bz 91% de = 90%
R = Bn 87% de = 40%

4.2.2.2 Sulfimides

The glycosyl sulfimides are prepared by the oxidation of thioglycosides with chloramine T (Scheme 4.32) [303,304]. This method can be plagued by side formation of sulfoxides, possibly resulting from the hydrolysis of intermediates by adventitious water. This tends to be more problematic with armed thioglycosides than with disarmed ones, and the S-ethyl moiety is more readily oxidized than the S-phenyl moiety. Similar to the preparation of sulfoxides, the oxidation to sulfimides results in high diastereoselectivities at the stereogenic sulfur, but the configuration of the major isomer is yet to be determined [304]. As with the sulfoxides [291], diastereoselectivity in the equatorial series is strongly dependent on the protecting group at O-2 [304]. Because of their intrinsic instability, sulfimides are generally prepared and glycosidated without intervening purification.

4.2.2.3 Sulfones

Although the exhaustive oxidation of thioglycosides with peracids is the standard entry into the glycosyl sulfones [305–308], they have also been produced from glycals and lactols [307]. From the glycal, the Ferrier reaction with benzenesulfinic acid proceeds readily at room temperature for standard armed protecting-group patterns. The disarmed glycals require activation with a Lewis acid such as BF_3OEt_2 (Scheme 4.33) [307,309].

Alternatively, lactols react with benzenesulfinic acid in the presence of $CaCl_2$ to yield the sulfones, again at room temperature [307,309]. In the axial series, the bulk of the sulfone group is such that the $^4C_1$ chair is not always the preferred conformation, and it has been shown that a twist-boat conformer is adopted in at least one instance [305]. Nevertheless, equilibration studies have shown that the sulfonyl group has a small anomeric effect and that the 'axial' anomer is preferred [310].

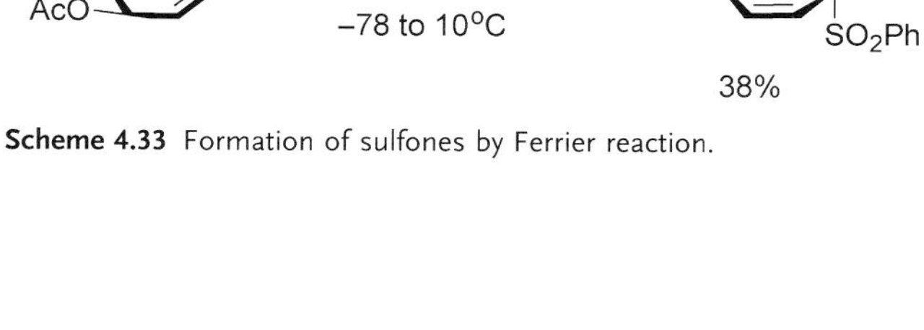

Scheme 4.33 Formation of sulfones by Ferrier reaction.

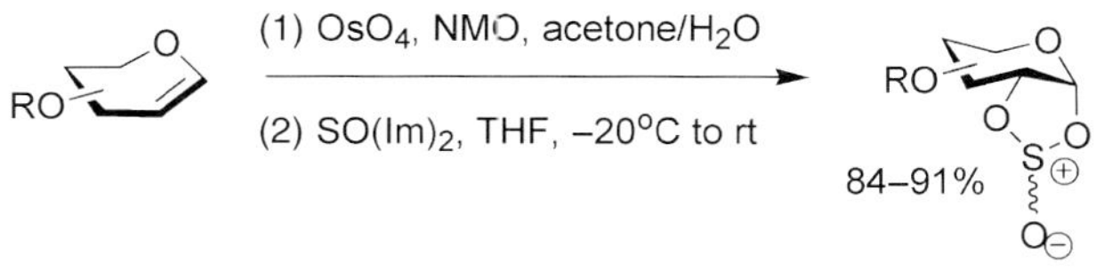

Scheme 4.34 Preparation of cyclic sulfites from glycols.

4.2.2.4 Other Oxidized Derivatives of Thioglycosides

A range of other sulfur(IV) and (VI) derivatives formally obtained by the oxidation of thioglycosides have been prepared, but they are apparently not employed to date in *O*-glycosylation reactions. These include glycosyl sulfenamides and sulfonamides, as well as sulfinates and sulfonates [289]. *S*-Glycosyl sulfenic acids have been prepared as transients by the *syn*-elimination of *S*-(2-cyanoethyl) glycosyl sulfoxides [311].

4.2.2.5 1,2-Cyclic Sulfites

The 1,2-cyclic sulfites are readily obtained from anomeric mixtures of 1,2-diols by reaction with sulfinyl diimidazolide as mixtures of *exo*- and *endo*-isomers at sulfur (Scheme 4.34) [312–314].

4.2.3
Glycosylation

4.2.3.1 Sulfoxides

The most common activator for the glycosyl sulfoxides is trifluoromethanesulfonic anhydride (triflic anhydride), which, in the absence of nucleophiles, rapidly and cleanly converts most sulfoxides into the corresponding glycosyl triflates in a matter of minutes at −78 °C in dichloromethane solution [86,280,315,316]. In the more extensively studied mannopyranose series, only the α-mannosyl triflate is observed by low-temperature NMR spectroscopy (Scheme 4.35) [280]. In the glucopyranose series, mixtures of α- and β-triflates are observed, in which the α-anomer nevertheless predominates (Scheme 4.36) [280].

After the formation of glycosyl triflate, addition of an acceptor alcohol, still at low temperature, results in the rapid formation of the desired glycosidic bond.

In the critical area of β-mannoside synthesis [317–321], the evidence strongly suggests that α-mannosyl triflate serves as a reservoir for a transient contact ion pair (CIP), which is the glycosylating species (Scheme 4.37), although the possibility of an S_N2-like mechanism with an exploded transition state cannot be completely excluded [135]. In view of the probable operation of the contact ion-pair mechanism

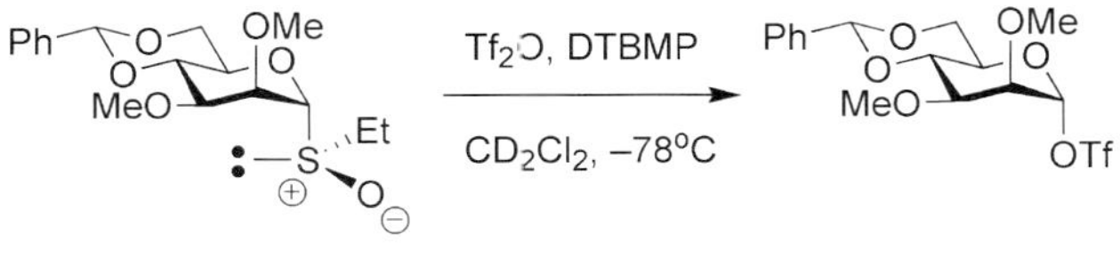

Scheme 4.35 Glycosyl triflate formation in the mannose series.

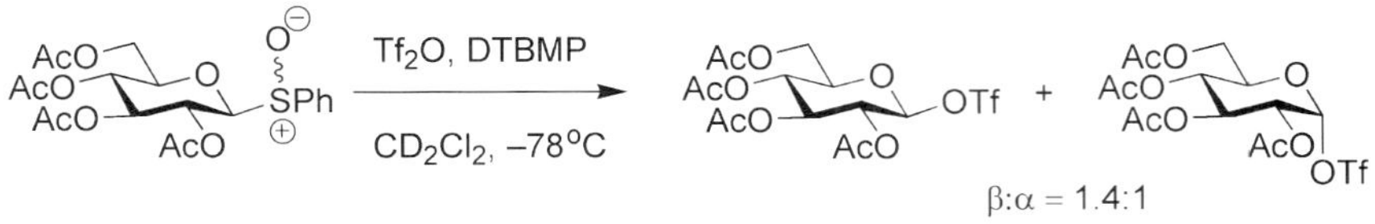

Scheme 4.36 Formation of an anomeric mixture of glucosyl triflates.

for the highly β-selective 4,6-O-benzylidene-protected mannose series, it is highly likely that closely related mechanisms are the rule in this type of glycosylation reaction. Differences in the selectivity likely arise from variations in the tightness of the contact ion pair and in the extent of equilibration with the solvent system.

When substoichiometric triflic anhydride is employed, an alternative mechanism comes into play, at least for the highly armed 2,3,4-tri-O-benzyl-fucose system [281]. Under these conditions, the oxacarbenium formed in the initial activation step of the sulfoxide is trapped not by the triflate anion but by another molecule of sulfoxide. Overall, this process results in the isomerization of the glycosyl sulfoxide to the corresponding O-glycosyl sulfenate, which is much less reactive as a donor and can be isolated (Scheme 4.38) [281]. O-Glycosyl sulfoxonium ions have subsequently been detected spectroscopically in the activation of hemiacetals with triflic anhydride and excess diphenyl sulfoxide, and are likely the intermediates in the activation of thioglycosides by the same combination [322]. The triflic-anhydride-catalyzed isomerization of glycosyl sulfoxides to O-glycosyl sulfenates can be suppressed by the inverse addition of the glycosyl sulfoxide to triflic anhydride [281].

When triflic anhydride is added to a preformed mixture of glycosyl sulfoxide and acceptor alcohol, it seems apparent that the first formed oxacarbenium ion is directly trapped by the alcohol, without the need for the implication of glycosyl triflates [75,280,323].

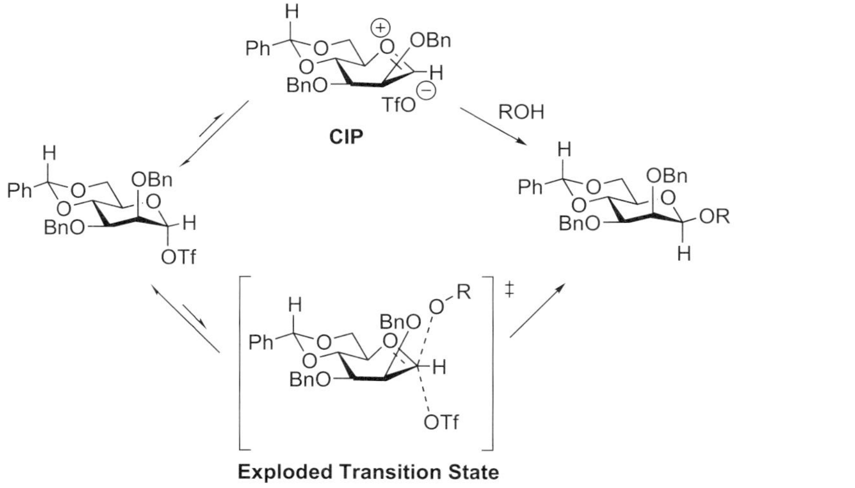

Scheme 4.37 Mechanism of β-mannosylation.

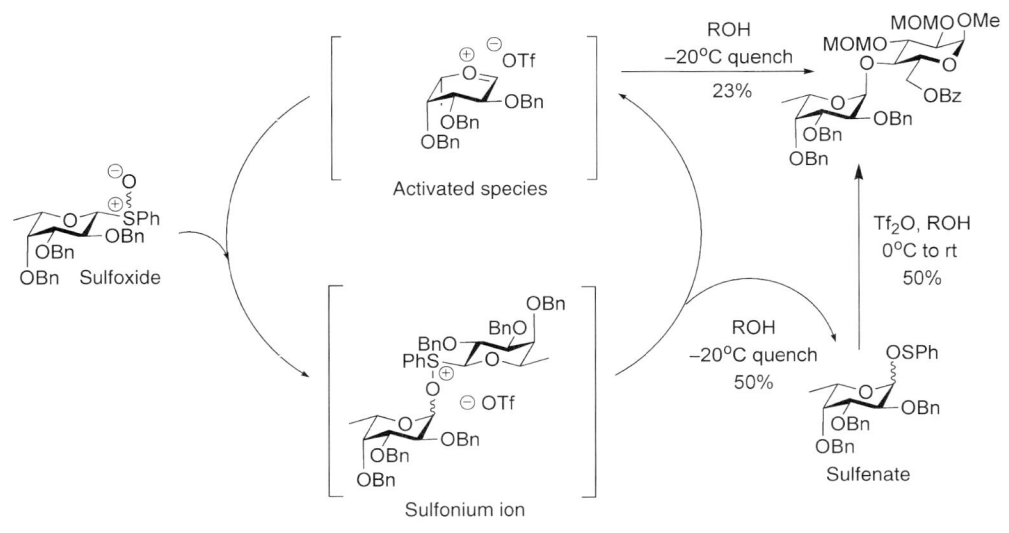

Scheme 4.38 Alternative mechanism for sulfoxide glycosylation.

Although most glycosyl sulfoxides are cleanly and rapidly activated by triflic anhydride at −78 °C, occasional exceptions are found. Examples include the 2,3-anhydro lyxo- and ribo-pentofuranosyl sulfoxides **1** and **2** that were demonstrated by NMR spectroscopy to not proceed to completion below −40 °C [324].

Although glycosyl triflates have been demonstrated to be intermediates with a number of armed donors, and even with disarmed donors not capable of neighboring-group participation, such as the sulfonate esters, typical disarmed donors with esters in the 2-position function in the anticipated manner through anchimeric

Scheme 4.39 Glycosylation with neighboring-group participation.

Scheme 4.40 Orthoester formation from a 2-O-benzoate.

assistance (Scheme 4.39) [75,86,279,325]. Moreover, low-temperature ^{13}C NMR experiments enabled the clear identification of an intermediate dioxalenium ion and a subsequent orthoester in the activation of a benzoylated sulfoxide with triflic anhydride (Scheme 4.40) [326].

The stereodirecting nitrile effect [131,327–330] is also applicable to sulfoxide glycosylations with activation by triflic anhydride (Scheme 4.41) [279], even if selectivities remain modest for the formation of the 1,2-*cis*-equatorial class of glycosidic bond [329,331].

A variety of other activating systems have been employed for the promotion of sulfoxide-based glycosylation reactions, but none have been studied to the same extent as the triflic-anhydride-mediated reaction [86]. One of the most potent activators, benzenesulfenyl triflate, a by-product of the activation with triflic anhydride, has been shown to bring about rapid conversion of sulfoxides into glycosyl triflates [280]. Unfortunately, this reagent is unstable and has to be prepared *in situ* from silver triflate and benzenesulfenyl chloride.

Although the existence of glycosyl triflates has only been demonstrated for the triflic anhydride and benzenesulfenyl triflate promoter systems, presumably the same intermediates may be invoked on preactivation with other triflate-incorporating systems such as TMSOTf (Scheme 4.42) [332] and triflic acid (Scheme 4.43) [84,333].

Molecular iodine has also been demonstrated to activate certain mannosyl sulfoxides albeit over extended periods of time and at higher temperatures. Mannosyl iodides may be involved as intermediates here, but O-iodyl mannosyl sulfonium salts have also been discussed [334]. The yield and selectivity of these reactions vary with respect to the potency of the iodonium ion source, IBr yielding much less

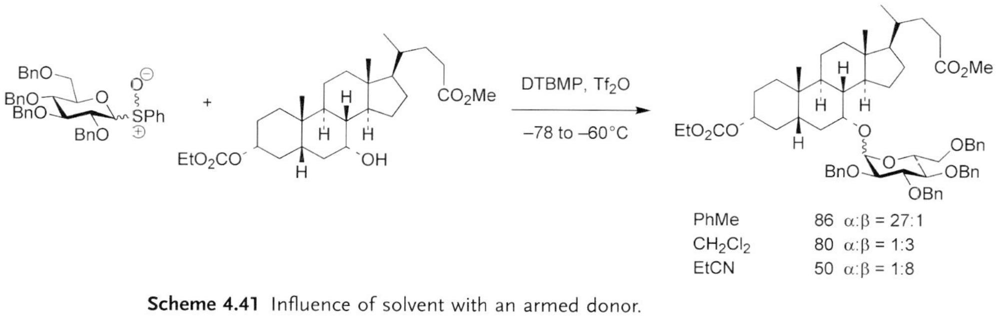

PhMe	86	α:β = 27:1
CH$_2$Cl$_2$	80	α:β = 1:3
EtCN	50	α:β = 1:8

Scheme 4.41 Influence of solvent with an armed donor.

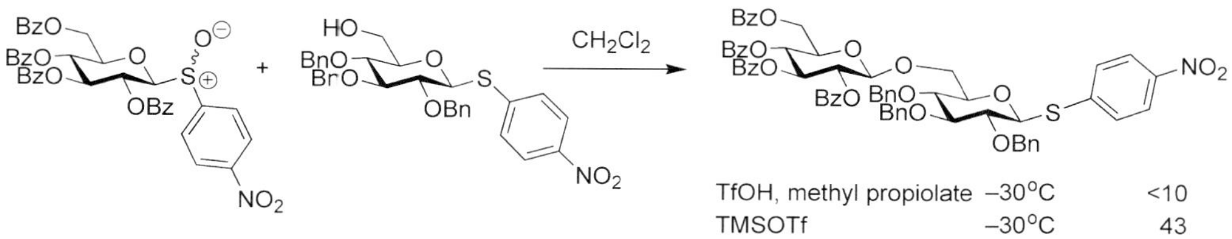

Scheme 4.42 Activation with triflic acid, TMS triflate.

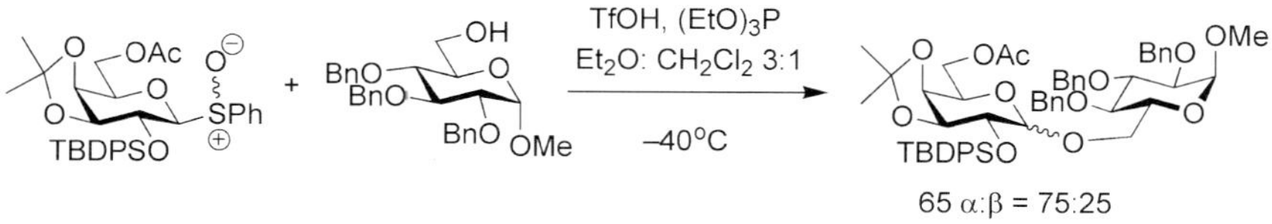

TfOH, methyl propiolate −30 °C	<10
TMSOTf −30 °C	43
TMSOTf, (MeO)$_3$P −30 to 20 °C	51

Scheme 4.43 Activation by means of triflic acid.

pronounced β-mannoside selectivity relative to molecular iodine and ICl affording somewhat unreactive mannosyl chloride (Scheme 4.44).

The combination of dicyclopentadienylzirconium dichloride and silver perchlorate activates armed glycosyl sulfoxides in dichloromethane between −20 °C and room temperature, but only very simple acceptors were studied [335]. Other Lewis and Brønsted acids studied include the environmentally benign europium, lanthanum and ytterbium triflates [336], certain polyoxometallates [337], sulfated zirconia [338] and Nafion H [338].

It has been demonstrated that the warming of glycosyl triflates above the decomposition temperature (which varies according to structure) results in the formation of two types of products, both of which appear to arise from the oxacarbenium ion. One pathway involves cyclization onto the O-2 protecting group, as in the example shown in Scheme 4.45 [339], whereas the other is deprotonation resulting in the formation of a glycal-type derivative, as shown in Scheme 4.46 [280]. As the temperature at which such by-products form for any given substrate can be readily

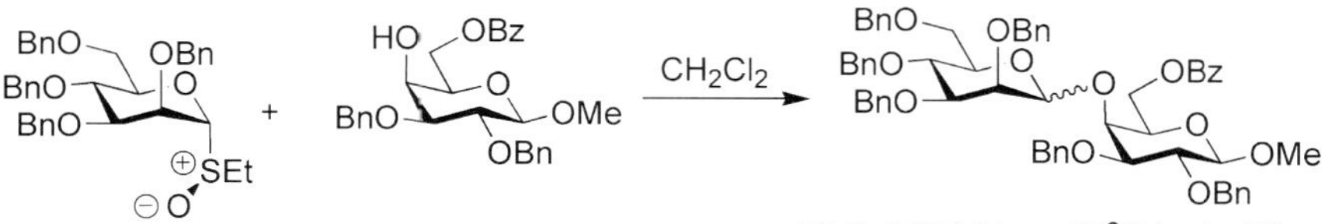

Tf$_2$O, DTBMP	−78 °C to rt	79	α:β = 3:2:1
I$_2$, K$_2$CO$_3$	−5 °C to rt	55	α:β = 1:6.6
IBr, no base	−5 °C to rt	62	α:β = 1:2.7
ICl, no base	−5 °C to rt	0	

Scheme 4.44 Activation with iodine and interhalogen compounds.

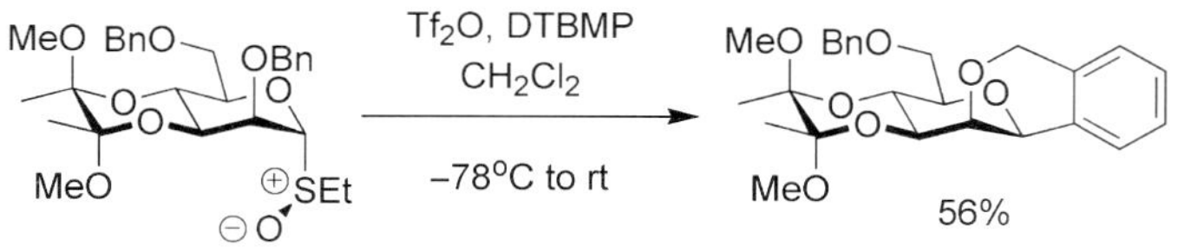

Scheme 4.45 Trapping of the oxacarbenium ion by a 2-O-benzyl ether.

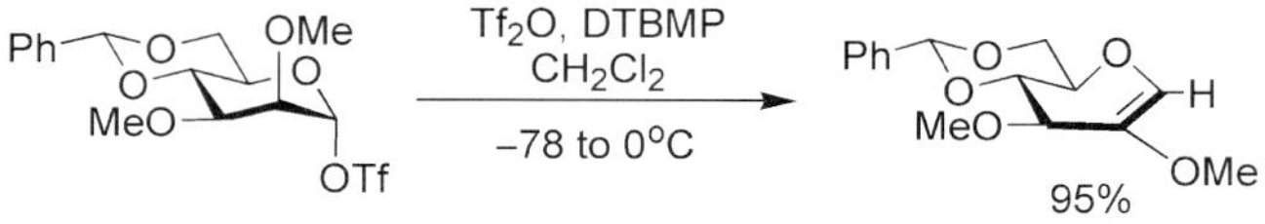

Scheme 4.46 Decomposition with glycal formation.

ascertained by a simple variable temperature NMR experiment, these side reactions are easily avoided in practice.

The presence of strongly nucleophilic groups in either the donor or the acceptor can be problematic in sulfoxide-type glycosylations when activation is conducted with triflic anhydride. The most common culprit is the amide group [340,341], which is illustrated by the formation of a dihydrooxazine when a 3-acetamido alcohol was employed as acceptor (Scheme 4.47) [342]. Self-evidently, acceptor-based problems of this type could be avoided by the preactivation of the sulfoxide with triflic anhydride.

In comparative studies, it has been shown that the azide group is a better surrogate for the acetamido moiety than the phthalimido system (Scheme 4.48) in sulfoxide-type couplings [340]. Subsequent work with trichloroacetimidates, however, suggests that the *N*-trichloroethoxycarbamate group should also function well [343].

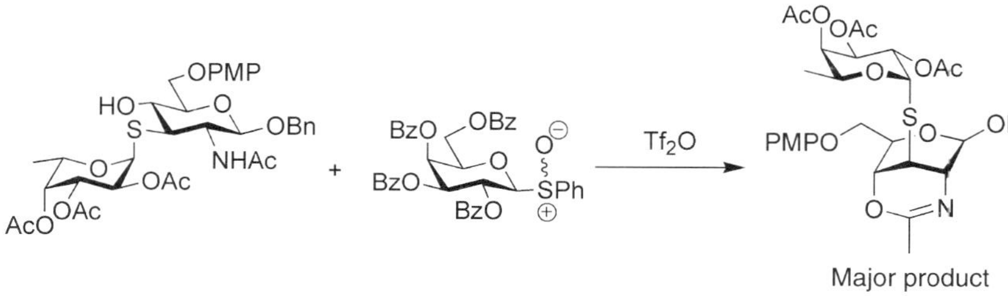

Scheme 4.47 Activation of an amide with triflic anhydride.

X = NHAc, α-OMe   9%
X = NPhth, β-OMe 53%
X = N$_3$, β-OMe   70%

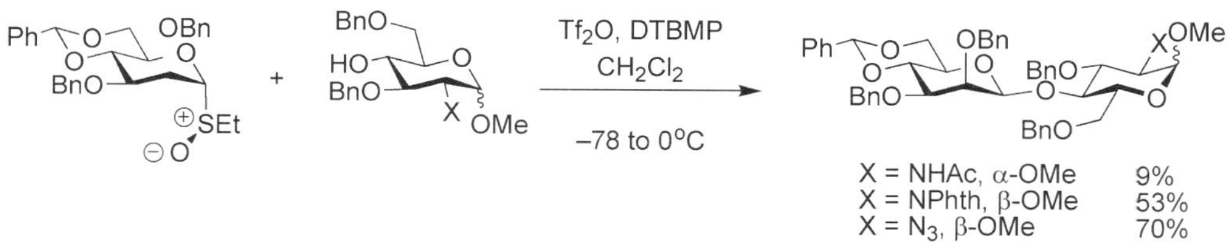

Scheme 4.48 Comparative study of nitrogen protecting groups.

The problem of the nucleophilicity of amides in glycosylation reactions is not limited to the sulfoxide method and has been shown to result in the formation of glycosyl imidates from intermolecular reaction with activated donors. It appears that this problem may be suppressed by the prior silylation of the amide [348,349]. Accordingly, it may be sufficient to operate the sulfoxide method with an excess of triflic anhydride when amides are present so as to convert all amides into O-triflyl imidates, which are then hydrolyzed on work-up. Despite these problems, several examples have been published of successful sulfoxide glycosylation reactions with acceptors carrying remote peptide bonds [344,345] and with donors coupled to resins via amide-based linkages [346,347], with no apparent problems reported. Sulfonamides and tertiary amides appear to be well tolerated by the sulfoxide method [340,350].

Further sources of potential problems in sulfoxide glycosylations are the electrophilic by-products generated on activation. These include the sulfenyl triflates and the products of their own reaction with glycosyl sulfoxides [280]. Unavoidable electrophiles of this type may result in the prior activation of acceptor-based sulfoxides, thioglycosides [351] and some [352], but not all [353], pentenyl glycosides. A number of scavengers have been developed to circumvent problems of this kind. Thus, methyl propiolate was first employed to this end in a one-pot synthesis of ciclamycin 0 trisaccharide [84], with the alkyne acting as an effective trap for sulfenic acid. Alkynes can also serve to capture electrophiles generated on activation of sulfoxides with triflic acid, and so prevent them from reacting prematurely with acceptor-based sulfoxides. However, it was shown that the dehydration of two molecules of sulfenic acid to the anhydride competes effectively with trapping by the alkyne and that the molecule of water generated in this process reacts effectively with the activated glycosylating species and lowers the overall yield (Scheme 4.49) [333]. The use of *tert*-butyl mercaptan as a nucleophile for the removal of sulfenic acid by-products presumably also suffers from the formation of a molecule of water [337]. More recent applications of the sulfoxide method have seen alkynes replaced by double bonds, particularly 4-allyl-1,2-dimethoxybenzene (ADMB), as effective scavengers of

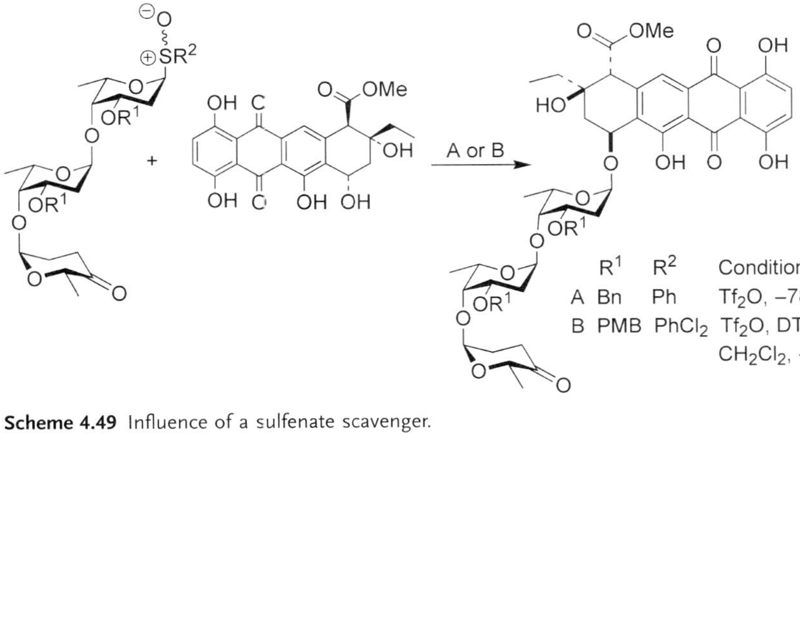

	R^1	R^2	Conditions	Yield
A	Bn	Ph	Tf$_2$O, –78 to –50°C	16%
B	PMB	PhCl$_2$	Tf$_2$O, DTBMP, ADMB CH$_2$Cl$_2$, –78°C	75%

Scheme 4.49 Influence of a sulfenate scavenger.

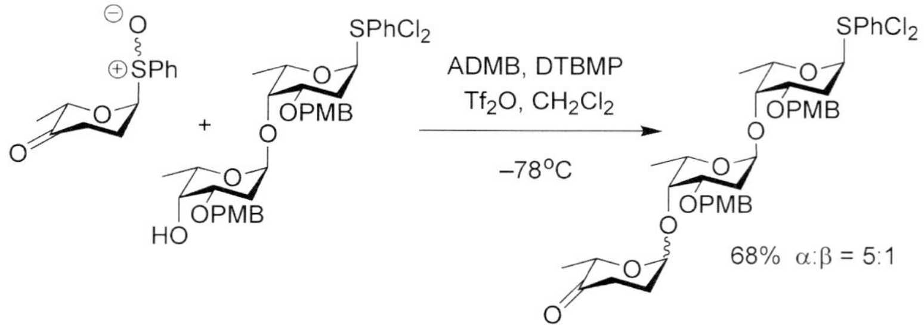

Scheme 4.50 Synthesis of the ciclamycin 0 trisaccharide assisted by a sulfenate scavenger.

electrophilic by-products (Scheme 4.49) [354]. Triethyl phosphite and trimethyl phosphite have also been employed as effective scavengers of electrophilic by-products generated in sulfoxide glycosylation reactions (Schemes 4.42 and 4.43) [332,333].

Problems with the reactivity of highly armed thioglycosides in the acceptor toward by-products from the sulfoxide activation may also be suppressed by the introduction of steric bulk in the *ortho*-positions of the thioglycoside. For example, a synthesis of the ciclamycin 0 trisaccharide was carried out with the assistance of ADMB as scavenger, leaving the 2,6-dichlorophenyl thioglycoside intact (Scheme 4.50) [354]. However, it should be noted that with less armed thioglycosides such precautions are not always necessary, and there are numerous sulfoxide glycosylations in which thioglycosides have been successfully carried [86], excellent examples of which are provided in Schemes 4.42 and 4.51 [355].

Hindered nonnucleophilic bases are typically added to sulfoxide glycosylations to buffer the acidic by-products. Classically, the 2,6-di-*tert*-butylpyridines have been employed for this purpose [86], but the more highly crystalline and easily handled 2,4,6-tri-*tert*-butylpyrimidine is finding increasing favor in this regard [356].

When neighboring-group participation is a feature of the glycosylation reaction, the use of a base in this manner frequently results in the isolation of orthoesters rather than the desired glycosides. In the case of activation by triflic anhydride, it is possible to avoid this problem by simply omitting the base. Alternatively, with more sensitive substrates, the hindered base may be retained and boron trifluoride etherate be added to promote the rearrangement of the orthoester to the glycoside, as in

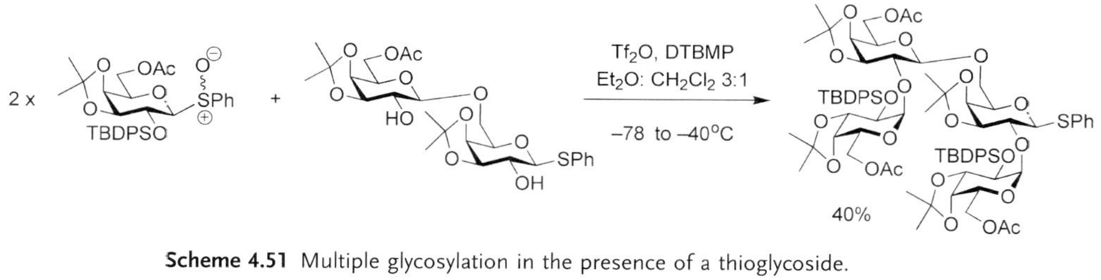

Scheme 4.51 Multiple glycosylation in the presence of a thioglycoside.

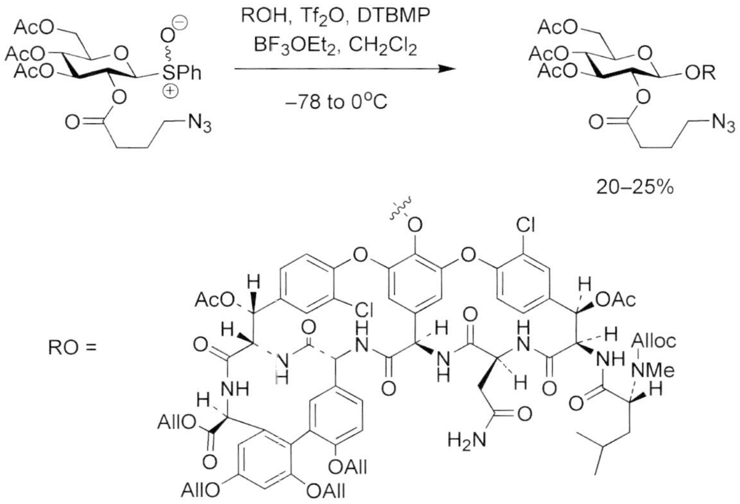

Scheme 4.52 Overcoming orthoester formation in the presence of a base.

the example of Scheme 4.52. This example also illustrates the use of a hindered phenol as glycosyl acceptor and the ability to function in the presence of remote amide bonds [344]. The success of this protocol derives from the inability of the highly hindered bases to form amine–borane adducts. The 4-azido butanoyl protecting group could be selectively removed with triphenylphosphine.

4.2.3.2 Sulfimides

This class of donor is activated by soft Lewis acids, such as copper triflate at room temperature, and despite their hydrolytic instability, they appear inert to conditions of sulfoxide activation, TMSOTf or Tf$_2$O (Scheme 4.53). Activation is achieved with stoichiometric promoter in the presence of the acceptor alcohol, and although the mechanism has not been investigated, presumably it proceeds via coordination followed by collapse to a stabilized oxacarbenium ion. The method is compatible with standard glycosidation solvents such as dichloromethane, acetonitrile and diethyl ether, and ester-directed couplings do not lead to orthoesters, perhaps as a result of the presence of the Lewis acid promoter [303,304].

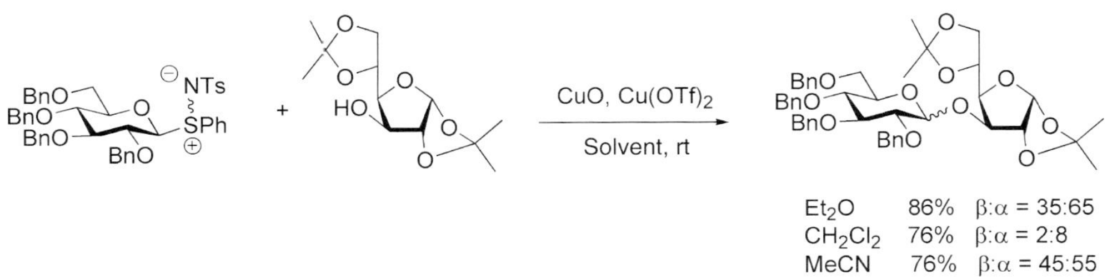

Scheme 4.53 Glycosylation with a sulfimide.

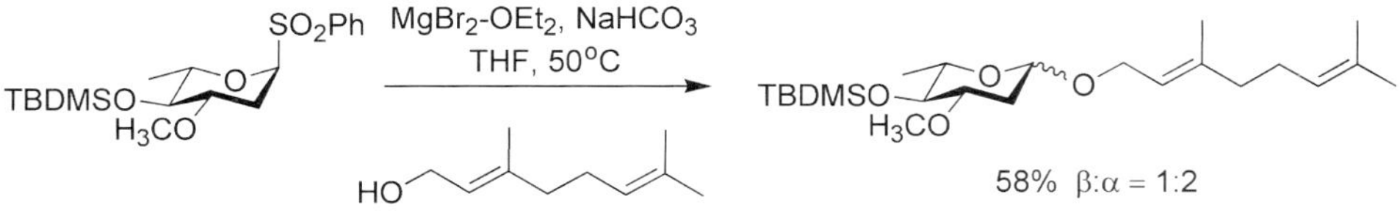

Scheme 4.54 Glycosylation with a phenyl sulfone.

4.2.3.3 Sulfones

Both glycosyl phenyl sulfones and glycosyl 2-pyridyl sulfones have been employed as donors in glycosylation reactions. The phenyl sulfones are activated with MgBr$_2$ etherate in THF at room temperature (Schemes 4.54 and 4.55) [307,309]. Considerable rate enhancement has been reported either by heating at reflux or by the use of ultrasonication.

The 2-pyridyl sulfones have been activated with Sm(OTf)$_3$ in toluene at 70 °C. The reaction also proceeds in refluxing methylene chloride, albeit with slightly diminished yields (Scheme 4.56) [308]. The mechanism has not been studied in either case, but activation has been suggested to involve the complexation of the metal ion with the pyridyl nitrogen and one of the sulfur oxygens, followed by the cleavage of the C1—S bond leading to an oxacarbenium ion, for the pyridyl sulfones.

4.2.3.4 Cyclic Sulfites

The cyclic sulfites were first found to react with lithium phenoxides as nucleophiles in DMF in a one-pot procedure commencing from the unprotected diol [357]. Subsequent work opened up this class of donor to alcohol nucleophiles in conjunction with the use of a Lewis acid, such as Yb(OTf)$_3$ or Ho(OTf)$_3$, to activate the donor in refluxing toluene (Scheme 4.57) [314,358,359]. The very high degree of β-selectivity observed in these reactions is consistent with an S$_N$2-like displacement of the sulfite oxygen.

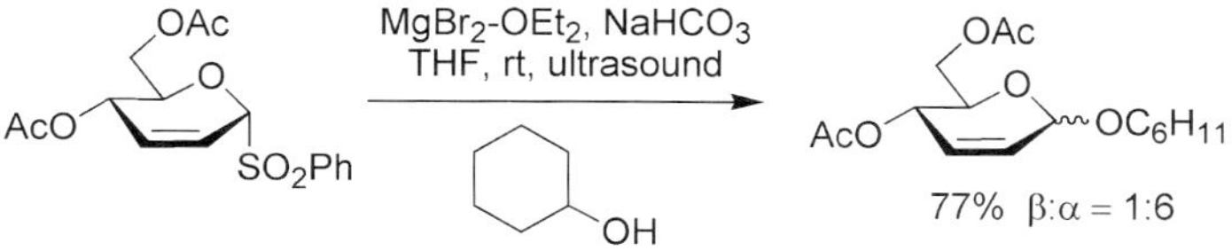

Scheme 4.55 Glycosylation with a 2,3-unsaturated glycosyl sulfone.

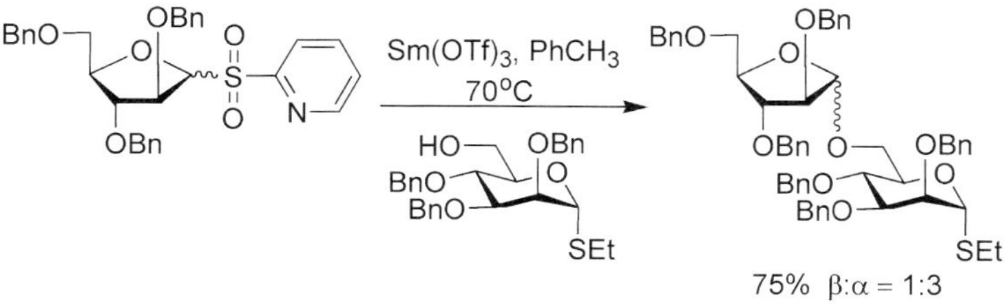

Scheme 4.56 Stability of a thioglycoside toward sulfone activation.

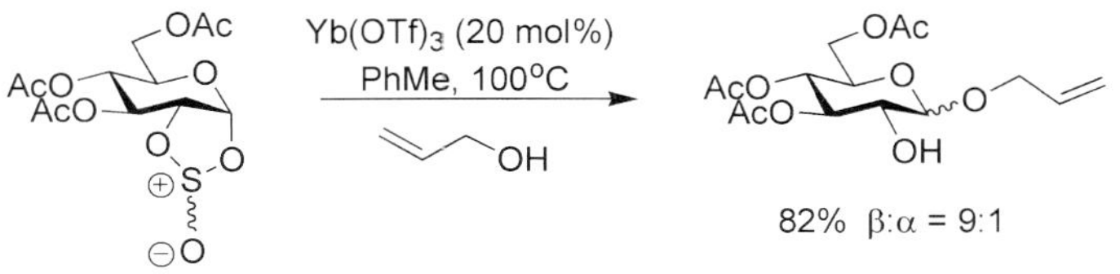

Scheme 4.57 Glycosylation with a cyclic sulfite.

4.2.4
Applications in Total Synthesis

In addition to the examples laid out in the above schemes, the sulfoxide method has been employed in the synthesis of numerous natural products. The examples presented below are chosen to illustrate the power of the method and the broad functional group compatibility.

The glycosylation of an unreactive alcohol, for which several other methods were reported to have failed, constituted a key step in the synthesis of hikizimycin (Scheme 4.58) [360].

A remarkable example of the compatibility of the sulfoxide method with polyene functionality is taken from a key step in the synthesis of apoptolidin (Scheme 4.59) [361].

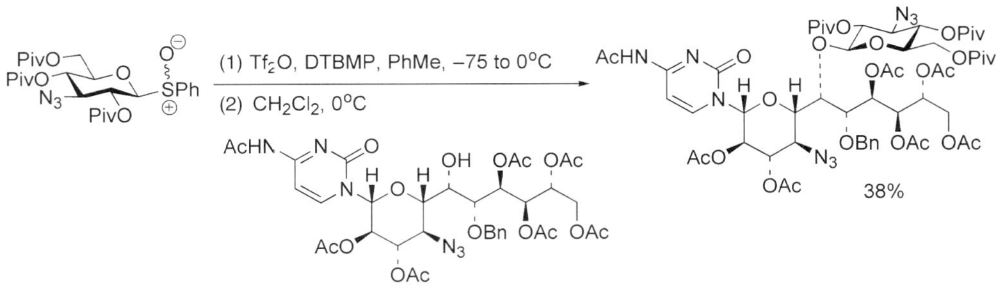

Scheme 4.58 Glycosylation of hindered substrate in a hikizimycin synthesis.

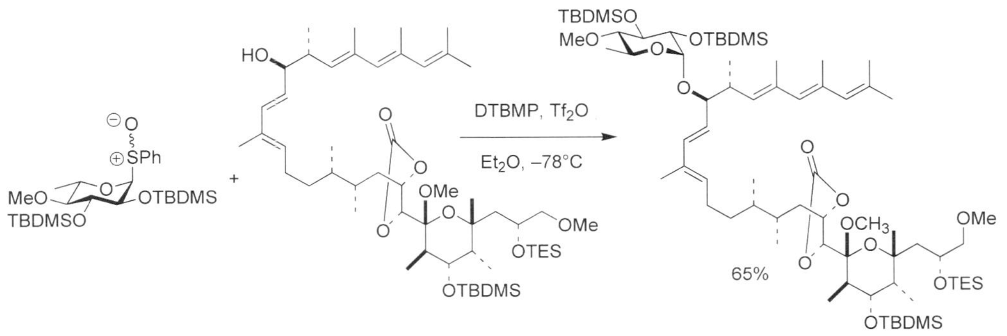

Scheme 4.59 Tolerance of polyene functionality *en route* to apoptolidin.

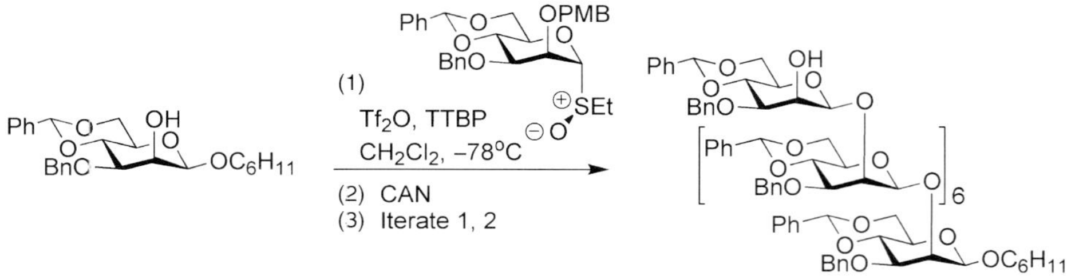

Scheme 4.60 Synthesis of a β-mannan.

The sulfoxide method was employed in the direct synthesis of a β-1,2-mannooctaose (Scheme 4.60) [362–364]. The synthesis of a β-mannosyl phosphoisoprenoid illustrates the possibility of employing even such weak nucleophiles as phosphates (Scheme 4.61) [365]. Both syntheses rely on the presence of 4,6-O-benzylidene acetal, and its effect on the covalent triflate–contact ion-pair equilibrium [366,367], to influence the stereochemistry of the glycosylation process [295,323].

In the furanoside field, the introduction of the last two β-arabino units of a hexasaccharide motif from a bacterial cell wall arabinogalactan was achieved by the sulfoxide method with stereocontrol achieved because of the presence of the 2,3-anhydro group (Scheme 4.62) [368]. More recently, the direct stereocontrolled synthesis of arabinofuranosides has been achieved by the sulfoxide method with the aid of a 3,5-O-(di-*tert*-butylsilylene)-protected donor [369].

Related 2,3-anhydrogulofuranosyl sulfoxides have been employed in the stereocontrolled synthesis of β-arabinofuranosides [370]. The epimeric α-arabinofuranosides have also been synthesized by the sulfoxide method, with the aid of the neighboring-group participation [371,372].

The crown jewel in the application of the sulfoxide method to the assembly of natural products is the synthesis of a pentasaccharide from the antibiotic moenomycin A, wherein each glycosidic bond was formed in a stereocontrolled manner by one variant or another by the sulfoxide method (Scheme 4.63) [373].

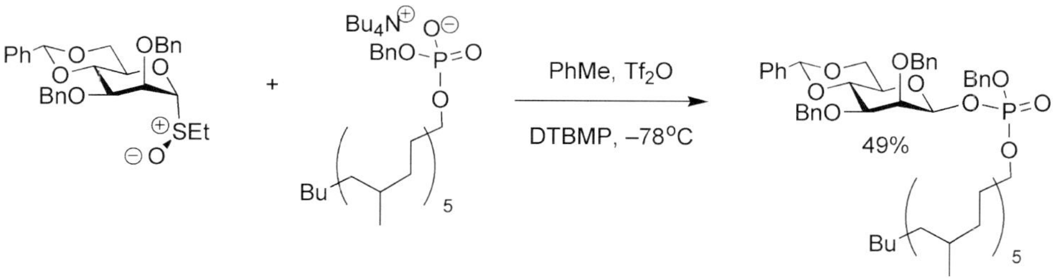

Scheme 4.61 Synthesis of a β-mannosyl phosphoisoprenoid.

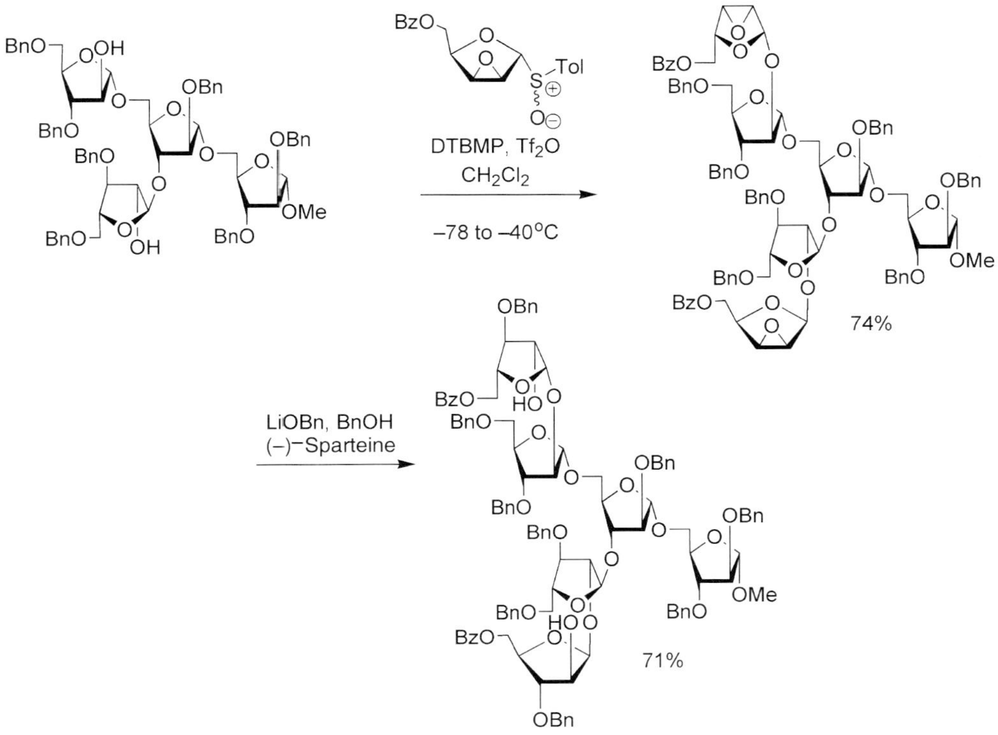

Scheme 4.62 Stereocontrolled β-arabinofuranoside formation in an arabinogalactan synthesis.

4.2.5
Special Topics

4.2.5.1 Intramolecular Aglycone Delivery (IAD)

The sulfoxide method has been applied to the concept [319,374] of intramolecular aglycone delivery for the formation of β-mannosides by means of a silylene linker. In the original work, the acceptor and a thioglycoside donor were joined by means of a silylene group before the oxidation to the sulfoxide [141]. However, it was later found that the preformed sulfoxide was tolerated by the chemistry for the introduction of the linker [286,375]. The intramolecular aglycone delivery step was shown to function effectively for the transfer of the donor to the 2-, 3- and 6-position of glucopyranosides, as exemplified in Scheme 4.64.

Problems were encountered, however, when the transfer to the 4-position of glucopyranosides was investigated. In these cases, the major product was that of trapping of the activated donor by the benzyloxy group at the 6-position, via a nine-membered cyclic transition state, with concomitant loss of the benzyl group (Scheme 4.65). Interestingly, no such problems were described for closely related intramolecular delivery of the glucose and glucosamine 4-OH groups with other linkers and methods of donor activation [319,374,376].

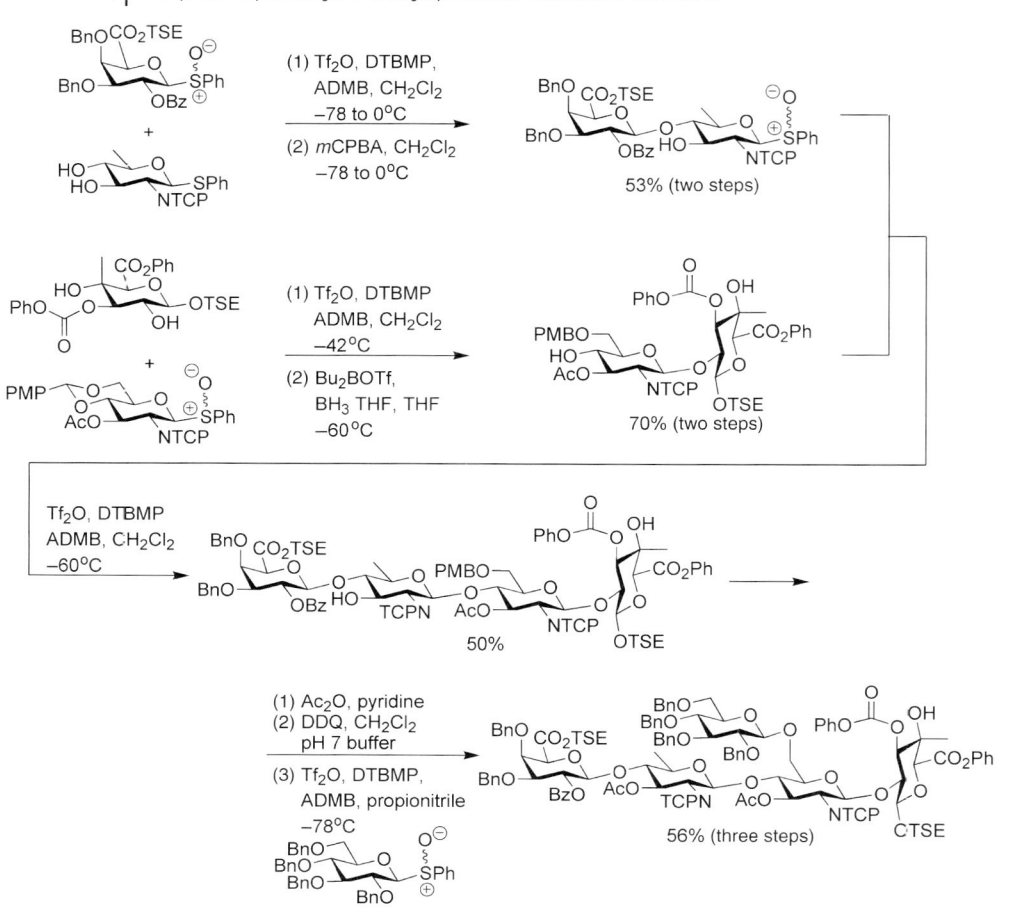

Scheme 4.63 Multiple use of the sulfoxide glycosylation in the synthesis of the moenomycin A pentasaccharide.

Sulfoxide-mediated intramolecular aglycone delivery has been conducted with a temporary linker formed *in situ* by the reaction of lanthanide triflates with the donor and acceptor-based alcohols (Scheme 4.66) [336]. However, as the selectivities recorded were modest, it has to be assumed that intermolecular glycosylation was an important side reaction in this chemistry.

Scheme 4.64 Sulfoxide-mediated intramolecular aglycone delivery.

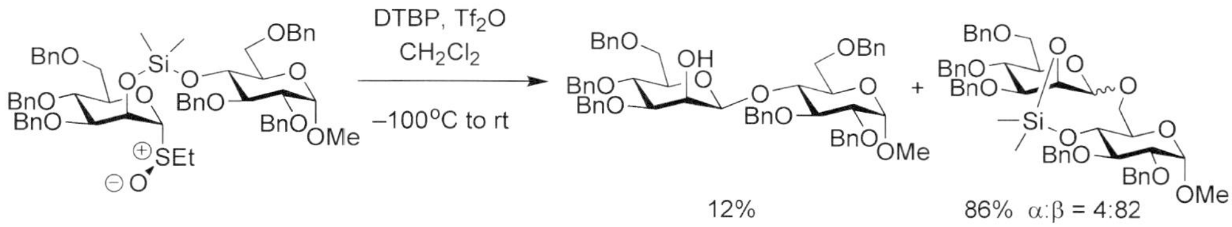

Scheme 4.65 Diverted intramolecular aglycone delivery.

Additional aspects of intramolecular aglycone delivery are discussed in Section 5.4.

4.2.5.2 Polymer-Supported Synthesis

Sulfoxide donors have been employed in the glycosylation of soluble and insoluble polymer-supported glycosyl acceptors, and the area has been reviewed [86,283].

In the original report, which employed a thioglycoside linker to an insoluble cross-linked polystyrene resin, the donor carried a labile 6-*O*-triphenylmethyl ether protecting group such that after the treatment with acid the coupling could be iteratively operated (Scheme 4.67) [377]. In addition to the example given, a range of axial and equatorial glycosidic linkages to typical carbohydrate acceptors were formed in this manner [377]. This general method was also applied to the parallel synthesis of an approximately 1300-member combinatorial library of disaccharides using a split-and-pool technique, a tentagel resin and the thioglycoside-type linker [346].

In a later study, an insoluble Rink amide resin was employed with the linkage of the glucuronic-acid-based acceptor through an amide bond (Scheme 4.68) [347,378].

A soluble aminomethylated polyethylene glycol and a succinoyl linker were used to support a 9-fluorenylmethyl group for solution-phase glycosylation by the sulfoxide method. With the help of temporary protection by a 6-*O*-triphenylmethyl ether, the method could be carried out iteratively to form disaccharides (Scheme 4.69) [379].

4.2.5.3 Ring Closing and Glycosylation

An elegant method for the formation of glycosidic bonds from acyclic dithioacetal monosulfoxides and glycosyl acceptors with triflic anhydride has been developed. This method takes advantage of the sulfenyl triflate generated from the reaction of

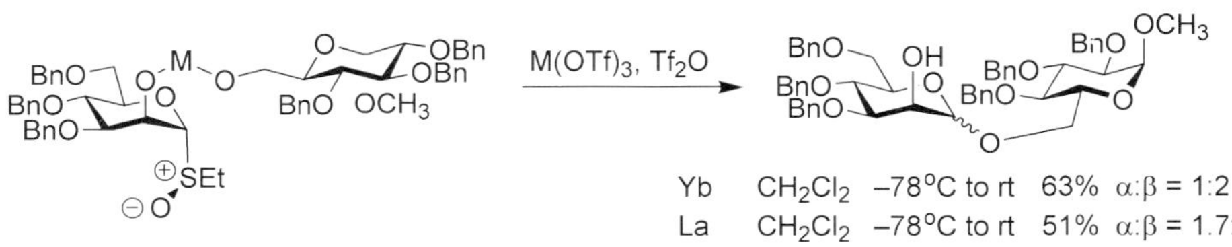

Yb	CH_2Cl_2	−78°C to rt	63%	α:β = 1:2
La	CH_2Cl_2	−78°C to rt	51%	α:β = 1.7:1
Eu	CH_2Cl_2	−78°C to rt	76%	α:β = 1:1.5
Eu	MeCN	−40°C to rt	82%	α:β = 1:4.3
Yb	MeCN	−40°C to rt	63%	α:β = 1:2.8

Scheme 4.66 Intramolecular aglycone delivery via metal complexes.

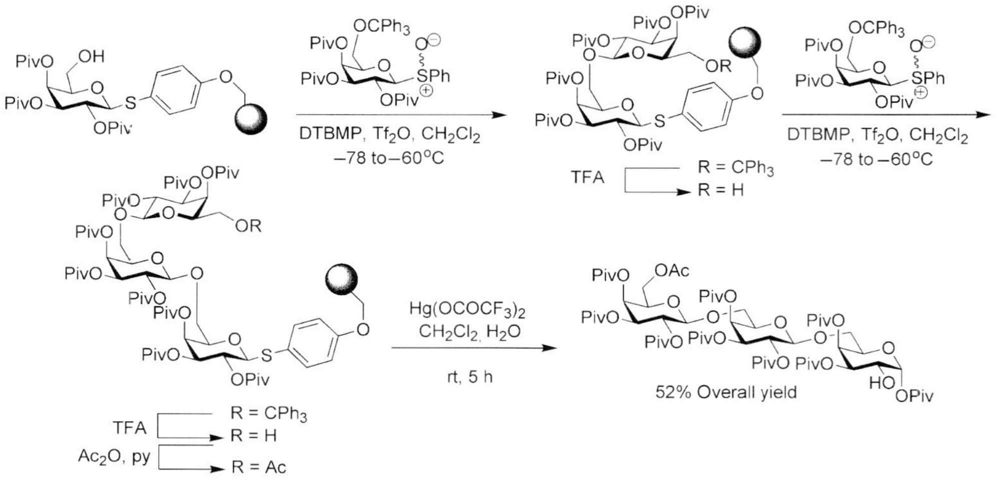

Scheme 4.67 Solid-phase oligosaccharide synthesis on a Merrifield resin.

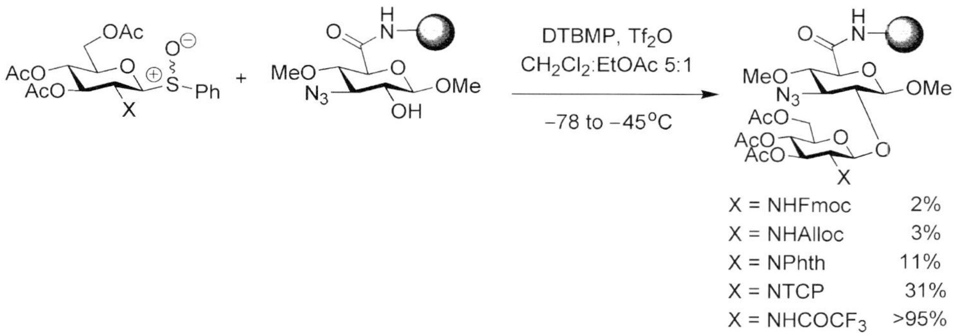

Scheme 4.68 Solid-phase synthesis on a Rink resin.

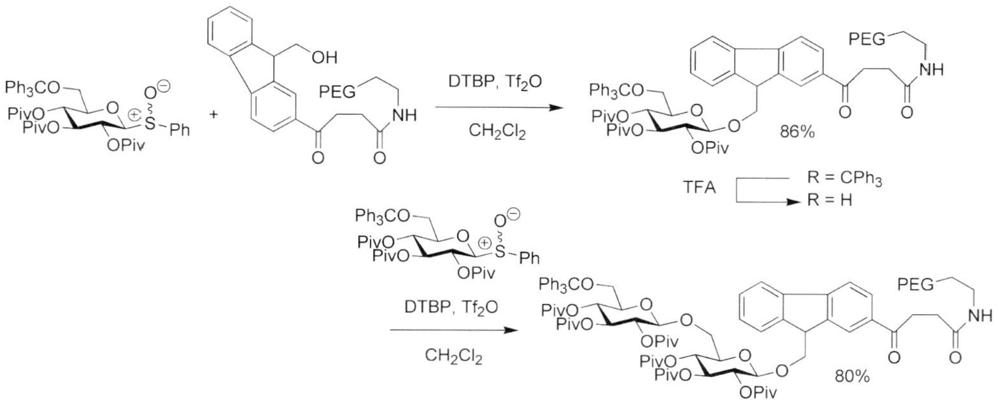

X = NHFmoc	2%
X = NHAlloc	3%
X = NPhth	11%
X = NTCP	31%
X = NHCOCF$_3$	>95%

Scheme 4.69 Disaccharide synthesis on a soluble polyethylene glycol support.

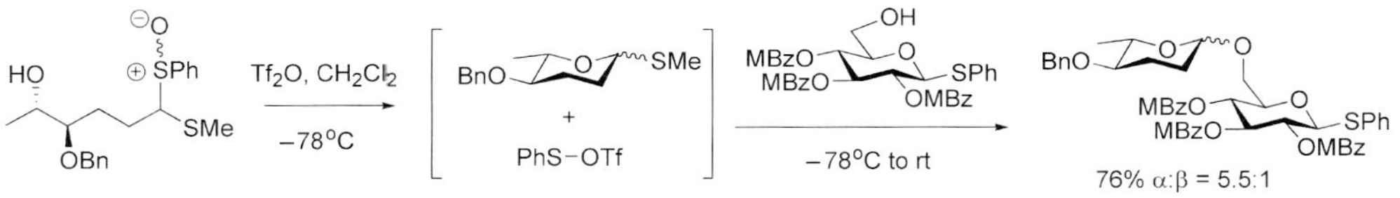

Scheme 4.70 Disaccharide synthesis by ring-closing glycosylation.

sulfoxide and triflic anhydride in the initial ring-closing step to bring about the activation of the intermediate thioglycoside for the coupling reaction (Scheme 4.70) [380].

A variation on this theme was applied to the synthesis of oligo-α-(2,8)-3-deoxy-D-manno-2-octulosonic acid derivatives. In this sequence, the monosulfoxide of α-keto ester-derived dithianes is activated with triflic anhydride resulting in the ring closure onto an adjacent alcohol and the elimination of the final sulfur residue. The glycal-like 2,3-unsaturated octulosonate ester generated in this manner subsequently serves as electrophile in an iodoalkoxylation reaction with a second molecule of the dithiane sulfoxide, thereby setting the stage for iteration of the entire sequence and synthesis of oligomeric derivatives (Scheme 4.71) [381].

4.2.5.4 Activation of Thioglycosides by Sulfoxides and Related Reagents

The power of the sulfoxide method and the mildness of the conditions have stimulated the development of a number of reagents capable of activating simple thioglycosides under comparable circumstances. Benzenesulfenyl triflate **3** and the analogous *p*-toluenesulfenyl triflate **4** perform admirably in this respect [295,351,382,383], but the need to generate these unstable reagents *in situ* from the corresponding sulfenyl chlorides and silver triflate drove the search for more

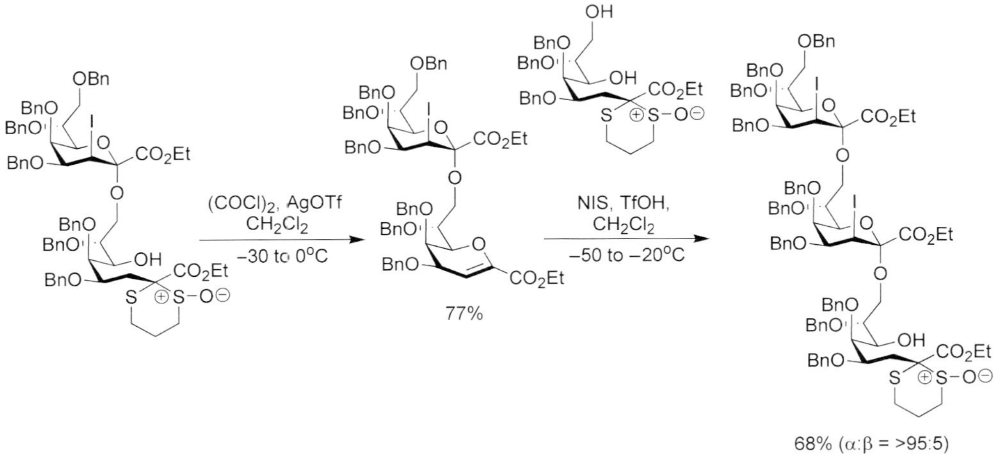

Scheme 4.71 Iterative ring-closing glycosylation approach to mannooctulsonic acid derivatives.

convenient reagents. The first among these was S-(methoxyphenyl)benzenethiosulfenate **5**, which in conjunction with triflic anhydride was capable of rapidly activating armed thioglycosides at −78 °C [384]. 1-Benzenesulfinylpiperidine **6** was subsequently developed [85] as a readily prepared [385] crystalline substance capable, together with triflic anhydride, of activating all but the most highly disarmed thioglycosides at −60 °C in dichloromethane. This reagent has been applied successfully in the stereocontrolled synthesis of a number of complex oligosaccharides [386,387] and in the polymer-supported synthesis of β-mannosides with the aid of a 4,6-*O*-polystyrylboronate-supported thiomannoside donor [388]. The success of the BSP method has spawned the development of a number of alternatives, all of which work in conjunction with triflic anhydride, including the liquid 1-benzenesulfinyl pyrrolidine **7**, which has a higher solubility at lower temperatures [385], a sulfenamide version **8** [389] and the apparently still more potent 1-benzenesulfinyl morpholine **9** [390]. The combination of diphenyl sulfoxide **10** with triflic anhydride also activates both thioglycosides [391,392] and 1-hydroxy sugars (hemiacetals) [393] under conditions comparable to the sulfoxide glycosylation method and is somewhat more potent than **6** and its variants toward strongly disarmed donors. Most recently, the combination of dimethyl disulfide **11** and triflic anhydride has been demonstrated to rapidly activate both armed and disarmed thioglycosides toward glycosylation at low temperatures [394].

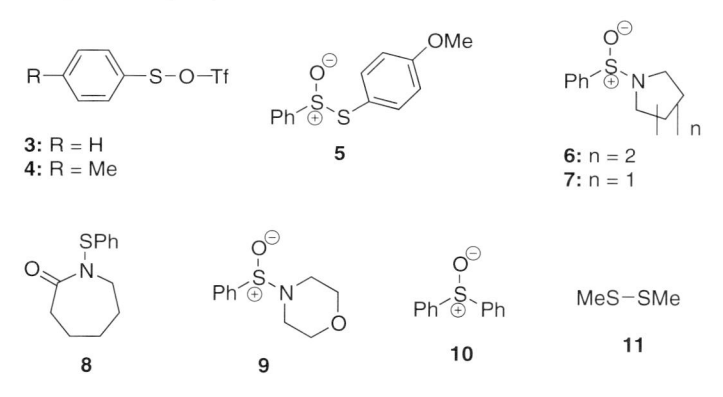

3: R = H
4: R = Me

5

6: n = 2
7: n = 1

8

9

10

11

4.2.6
Experimental Procedures

4.2.6.1 General Procedure for the Preparation of Glycosyl Sulfoxides
The thioglycoside donor is dissolved in CH$_2$Cl$_2$ (∼0.1 M) and cooled to −78 °C under an inert atmosphere. *m*CPBA (70 wt%, 1.2 equiv) is then added portionwise with minimal exposure to the atmosphere. The reaction mixture is warmed to room temperature over 1 h, at which time TLC shows the dissappearance of starting material and the formation of more polar compounds. The sulfoxides are purified by column chromatography over silica.

4.2.6.2 General Procedure for Sulfoxide Glycosidation

The sulfoxide donor (1.0 equiv) and 2,4,6-tri-*tert*-butylpyrimidine (2.0 equiv) are dissolved in dry CH_2Cl_2 (0.04 M) together with crushed 4-Å molecular sieves under argon and cooled to $-78\,°C$. Tf_2O (1.2 equiv) is added and the solution is allowed to stir for ~30 min. The acceptor (1.0 equiv) in CH_2Cl_2 (0.1 M) is then added dropwise. The reaction mixture is allowed to warm to $-30\,°C$, when it is quenched by the addition of Et_3N and filtered through Celite. The solvent is removed and the crude mixture is directly purified by column chromatography on silica gel.

4.2.7 Conclusion

The sulfoxide method has been successfully applied to the formation of most common classes of glycosidic bond with the exception of the sialosides. The high yields and relative lack of sensitivity to steric hindrance in the acceptor evident from the examples presented more than outweigh the minor inconvenience of the extra step required to convert thioglycoside into sulfoxide. Mechanistically, the sulfoxide method is one of the easiest glycosylation reactions to study, making it one of the better understood methods and, therefore, one of the better candidates for rationale improvement. Taking all of these factors into consideration, it is to be expected that this method will continue to develop into one of the most widely applied and reliable glycosylation protocols.

References

279 Kahne, D., Walker, S., Cheng, Y. and Engen, D.V. (1989) *Journal of the American Chemical Society*, **111**, 6881–6882.

280 Crich, D. and Sun, S. (1997) *Journal of the American Chemical Society*, **119**, 11217–11223.

281 Gildersleeve, J., Pascal, R.A. and Kahne, D. (1998) *Journal of the American Chemical Society*, **120**, 5961–5969.

282 Norberg, T. (1996) *Modern Methods in Carbohydrate Synthesis* (eds S.H. Khan and R.A. O'Neill), Harwood Academic Publishers, Amsterdam, pp. 82–106.

283 Taylor, C.M. (2001) *Solid Support Oligosaccharide Synthesis and Combinatorial Libraries* (ed. P.H. Seeberger), Wiley Interscience, New York, pp. 41–65.

284 Micheel, F. and Schmitz, H. (1939) *Chemische Berichte*, **72**, 992–994.

285 Crich, D., Sun, S. and Brunckova, J. (1996) *The Journal of Organic Chemistry*, **61**, 605–615.

286 Stork, G. and La Clair, J.J. (1996) *Journal of the American Chemical Society*, **118**, 247–248.

287 Crich, D., Mataka, J., Sun, S., Lam, K.-C., Rheingold, A.R. and Wink, D.J. (1998) *Journal of the Chemical Society: Chemical Communications*, 2763–2764.

288 Crich, D., Mataka, J., Zakharov, L.N., Rheingold, A.L. and Wink, D.J. (2002) *Journal of the American Chemical Society*, **124**, 6028–6036.

289 Knapp, S., Darout, E. and Amorelli, B. (2006) *The Journal of Organic Chemistry*, **71**, 1380–1389.

290 Khiar, N. (2000) *Tetrahedron Letters*, **41**, 9059–9063.

291 Khiar, N., Fernandez, I., Araujo, C.S., Rodriguez, J.-A., Suarez, B. and Alvarez, E. (2003) *The Journal of Organic Chemistry*, **68**, 1433–1442.

292 Wagner, G. and Wagler, M. (1964) *Archives of Pharmacie*, **297**, 206–218.

293 Crich, D. and Dai, Z. (1999) *Tetrahedron*, **55**, 1569–1580.
294 Wagner, G. and Kuhmstedt, H. (1959) *Naturwissenschaften*, **46**, 425–426.
295 Crich, D. and Sun, S. (1998) *Tetrahedron*, **54**, 8321–8348.
296 Schmidt, R.R. and Kast, J. (1986) *Tetrahedron Letters*, **27**, 4007–4010.
297 Chen, M.-Y., Patkar, L.N., Chen, H.-T. and Lin, C.-C. (2003) *Carbohydrate Research*, **338**, 1327–1332.
298 Sofia, M.J., Kakarla, R., Kogan, N., Dulina, R., Hui, Y.W., Hatzenbuhler, N.T., Liu, D., Chen, A. and Wagler, T. (1997) *Bioorganic & Medicinal Chemistry Letters*, **7**, 2251–2254.
299 Ravikumar, K.S., Zhang, Y.M., Begue, J.-P. and Bonnet-Delpon, D. (1998) *European Journal of Organic Chemistry*, 2937–2940.
300 Misbahi, K., Lardic, M., Ferrieres, V., Noiret, N., Kerbal, A. and Plusquellec, D. (2001) *Tetrahedron: Asymmetry*, **12**, 2389–2393.
301 Chen, M.-Y., Patkar, L.N. and Lin, C.-C. (2004) *The Journal of Organic Chemistry*, **69**, 2884–2887.
302 Vincent, S.P., Burkart, M.D., Tsai, C.-Y., Zhang, Z. and Wong, C.-H. (1999) *The Journal of Organic Chemistry*, **64**, 5264–5279.
303 Cassel, S., Plessis, I., Wessel, H.P. and Rollin, P. (1998) *Tetrahedron Letters*, **39**, 8097–8100.
304 Chery, F., Cassel, S., Wessel, H.P. and Rollin, P. (2002) *European Journal of Organic Chemistry*, 171–180.
305 Beau, J.-M. and Sinaÿ, P. (1985) *Tetrahedron Letters*, **26**, 6193–6196.
306 Skrydstrup, T., Urban, D. and Beau, J.-M. (1998) *The Journal of Organic Chemistry*, **63**, 2507.
307 Brown, D.S., Bruno, M., Davenport, R.J. and Ley, S.V. (1989) *Tetrahedron*, **45**, 4293–4308.
308 Chang, G.X. and Lowary, T.L. (2000) *Organic Letters*, **2**, 1505–1508.
309 Brown, D.S., Ley, S.V., Vile, S. and Thompson, M. (1991) *Tetrahedron*, **47**, 1329–1342.
310 Chen, G., Franck, R.W., Yang, G. and Blumenstein, M. (2002) *Canadian Journal of Chemistry*, **80**, 894–899.
311 Aversa, M.C., Barattucci, A., Bilardo, M.C., Bonaccorsi, P., Giannetto, P., Rollin, P. and Tatibouët, A. (2005) *The Journal of Organic Chemistry*, **70**, 7389–7396.
312 Guiller, A., Gagnieu, C.H. and Pacheco, H. (1986) *Journal of Carbohydrate Chemistry*, **2**, 153–160.
313 Gagnieu, C.H., Guiller, A. and Pacheco, H. (1986) *Carbohydrate Research*, **180**, 223–231.
314 Sanders, W.J. and Kiessling, L.L. (1994) *Tetrahedron Letters*, **35**, 7335–7338.
315 Crich, D. (2001) *Glycochemistry: Principles, Synthesis, and Applications* (eds P.G. Wang and C.R. Bertozzi), Dekker, New York, pp. 53–75.
316 Crich, D. (2002) *Journal of Carbohydrate Chemistry*, **21**, 663–686.
317 Pozsgay, V. (2000) *Carbohydrates in Chemistry and Biology*, vol. **1** (eds B. Ernst, G.W. Hart and P. Sinaÿ), Wiley-VCH Verlag GmbH, Weinheim, pp. 319–343.
318 Ito, Y. and Ohnishi, Y. (2001) *Glycoscience: Chemistry and Chemical Biology*, vol. **2** (eds B. Fraser-Reid, T. Kuniaki and J. Thiem), Springer-Verlag, Berlin, pp. 1589–1619.
319 Barresi, F. and Hindsgaul, O. (1996) *Modern Methods in Carbohydrate Synthesis* (eds S.H. Khan and R.A. O'Neill), Harwood Academic Publishers, Amsterdam, pp. 251–276.
320 Demchenko, A.V. (2003) *Synlett*, 1225–1240.
321 Gridley, J.J. and Osborn, H.M.I. (2000) *Journal of the Chemical Society Perkin Transactions*, **1**, 1471–1491.
322 Garcia, B.A. and Gin, D.Y. (2000) *Journal of the American Chemical Society*, **122**, 4269–4279.
323 Crich, D. and Sun, S. (1997) *The Journal of Organic Chemistry*, **62**, 1198–1199.

324 Callam, C.S., Gadikota, R.R., Krein, D.M. and Lowary, T.L. (2003) *Journal of the American Chemical Society*, **125**, 13112–13119.

325 Bousquet, E., Khitri, M., Lay, L., Nicotra, F., Panza, L. and Russo, G. (1998) *Carbohydrate Research*, **311**, 171–181.

326 Crich, D., Dai, Z. and Gastaldi, S. (1999) *The Journal of Organic Chemistry*, **64**, 5224–5229.

327 Pougny, J.-R. and Sinaÿ, P. (1976) *Tetrahedron Letters*, **17**, 4073–4076.

328 Braccini, J., Derouet, C., Esnault, J., de Penhoat, C.H., Mallet, J.-M., Michon, V. and Sinaÿ, P. (1993) *Carbohydrate Research*, **246**, 23–41.

329 Schmidt, R.R., Behrendt, M. and Toepfer, A. (1990) *Synlett*, 694–696.

330 Majumdar, D. and Boons, G.-J. (2005) *Handbook of Reagents for Organic Synthesis: Reagents for Glycoside Nucleotide, and Peptide Synthesis* (ed. D. Crich), Wiley, Chichester, pp. 11–15.

331 Crich, D. and Patel, M. (2006) *Carbohydrate Research*, **341**, 1467–1475.

332 Sliedregt, L.A.J.M., van der Marel, G.A. and van Boom, J.H. (1994) *Tetrahedron Letters*, **35**, 4015–4018.

333 Alonso, I., Khiar, N. and Martin-Lomas, M. (1996) *Tetrahedron Letters*, **37**, 1477–1480.

334 Marsh, S.J., Kartha, K.P.R. and Field, R.A. (2003) *Synlett*, 1370–1372.

335 Wipf, P. and Reeves, J.T. (2001) *The Journal of Organic Chemistry*, **66**, 7910–7914.

336 Chung, S.-K. and Park, K.-H. (2001) *Tetrahedron Letters*, **42**, 4005–4007.

337 Nagai, H., Matsmura, S. and Toshima, K. (2000) *Tetrahedron Letters*, **41**, 10233–10237.

338 Nagai, H., Kawahara, K., Matsumura, S. and Toshima, K. (2001) *Tetrahedron Letters*, **42**, 4159–4162.

339 Crich, D., Cai, W. and Dai, Z. (2000) *The Journal of Organic Chemistry*, **65**, 1291–1297.

340 Crich, D. and Dudkin, V. (2001) *Journal of the American Chemical Society*, **121**, 6819–6825.

341 Broddefalk, J., Forsgren, M., Sethson, I. and Kihlberg, J. (1999) *The Journal of Organic Chemistry*, **64**, 8948–8953.

342 Aguilera, B. and Fernando-Mayoralas, A. (1998) *The Journal of Organic Chemistry*, **63**, 2719–2723.

343 Xue, J., Khaja, S.D., Locke, R.D. and Matta, K.L. (2004) *Synlett*, 861–865.

344 Thompson, C., Ge, M. and Kahne, D. (1999) *Journal of the American Chemical Society*, **121**, 1237–1244.

345 Ge, M., Thompson, C. and Kahne, D. (1998) *Journal of the American Chemical Society*, **120**, 11014–11015.

346 Liang, R., Yan, L., Loebach, J., Ge, M., Uozumi, Y., Sekanina, K., Horan, N., Gildersleeve, J., Thompson, C., Smith, A., Biswas, K., Still, W.C. and Kahne, D. (1996) *Science*, **274**, 1520–1522.

347 Silva, D.J., Wang, H., Allanson, N.M., Jain, R.K. and Sofia, M.J. (1999) *The Journal of Organic Chemistry*, **64**, 5926–5929.

348 Liao, L. and Auzanneau, F.-I. (2003) *Organic Letters*, **5**, 2607–2610.

349 Liao, L. and Auzanneau, F.-I. (2005) *The Journal of Organic Chemistry*, **70**, 6265–6273.

350 Dudkin, V.Y., Miller, J.S. and Danishefsky, S.J. (2003) *Tetrahedron Letters*, **44**, 1791–1793.

351 Crich, D. and Sun, S. (1998) *Journal of the American Chemical Society*, **120**, 435–436.

352 Baek, J.Y., Choi, T.J., Jeon, H.B. and Kim, K.S. (2006) *Angewandte Chemie (International Edition)*, **45**, 7436–7440.

353 Crich, D. and Dudkin, V. (2000) *Tetrahedron Letters*, **41**, 5643–5646.

354 Gildersleeve, J., Smith, A., Sakurai, D., Raghavan, S. and Kahne, D. (1999) *Journal of the American Chemical Society*, **121**, 6176–6182.

355 Khiar, N. and Martin-Lomas, M. (1995) *The Journal of Organic Chemistry*, **60**, 7017–7021.

356 Crich, D., Smith, M., Yao, Q. and Picione, J. (2001) *Synthesis*, 323–326.

357 Aouard, M.E.A., Meslouti, A.E., Uzan, R. and Beaupère, D. (1994) *Tetrahedron Letters*, **35**, 6279–6282.

358 Manning, D.D., Bertozzi, C.R., Pohl, N. L., Rosen, S.D. and Kiessling, L.L. (1995) *The Journal of Organic Chemistry*, **60**, 6254–6255.

359 Sanders, W.J., Manning, D.D., Koeller, K. M. and Kiessling, L.L. (1997) *Tetrahedron*, **53**, 16391–16422.

360 Ikemoto, N. and Schreiber, S.L. (1992) *Journal of the American Chemical Society*, **114**, 2524–2536.

361 Nicolaou, K.C., Li, Y., Fylaktakidou, K.C., Mitchell, H.J. and Sugita, K. (2001) *Angewandte Chemie (International Edition)*, **40**, 3854–3857.

362 Crich, D., Li, H., Yao, Q., Wink, D.J., Sommer, R.D. and Rheingold, A.L. (2001) *Journal of the American Chemical Society*, **121**, 5826–5828.

363 Crich, D., Banerjee, A. and Yao, Q. (2004) *Journal of the American Chemical Society*, **126**, 14930–14934.

364 Dromer, F., Chevalier, R., Sendid, B., Improvisi, L., Jouault, T., Robert, R., Mallet, J.M. and Poulain, D. (2002) *Antimicrobial Agents and Chemotherapy*, **46**, 3869–3876.

365 Crich, D. and Dudkin, V. (2002) *Journal of the American Chemical Society*, **124**, 2263–2266.

366 Jensen, H.H., Nordstrom, M. and Bols, M. (2004) *Journal of the American Chemical Society*, **126**, 9205–9213.

367 Andrews, C.W., Rodebaugh, R. and Fraser-Reid, B. (1996) *The Journal of Organic Chemistry*, **61**, 5280–5289.

368 Gadikota, R.R., Callam, C.S., Wagner, T., Del Fraino, B. and Lowary, T.L. (2003) *Journal of the American Chemical Society*, **125**, 4155–4165.

369 Crich, D., Pedersen, C.M., Bowers, A.A. and Wink, D.J. (2007) *The Journal of Organic Chemistry*, **72**, 1553–1565.

370 Bai, Y. and Lowary, T.L. (2006) *The Journal of Organic Chemistry*, **71**, 9658–9671.

371 Ferrieres, V., Joutel, J., Boulch, R., Roussel, M., Toupet, L. and Plusquellec, D. (2000) *Tetrahedron Letters*, **41**, 5515–5519.

372 Codée, J.D.C., van den Bos, J., Litjens, R. E.J.N., Overkleeft, H.S., van Boom, J.H. and van der Marel, G.A. (2003) *Organic Letters*, **5**, 1947–1950.

373 Taylor, J.G., Li, X., Oberthür, M., Zhu, W. and Kahne, D.E. (2006) *Journal of the American Chemical Society*, **128**, 15084–15085.

374 Jung, K.H., Muller, M. and Schmidt, R.R. (2000) *Chemical Reviews*, **100**, 4423–4442.

375 Packard, G.K. and Rychnovsky, S.D. (2001) *Organic Letters*, **3**, 3393–3396.

376 Lergenmuller, M., Nukada, T., Kuramochi, K., Dan, A., Ogawa, T. and Ito, Y. (1999) *European Journal of Organic Chemistry*, 1367–1376.

377 Yan, L., Taylor, C.M., Goodnow, R. and Kahne, D. (1994) *Journal of the American Chemical Society*, **116**, 6953–6954.

378 Sofia, M.J., Allanson, N., Hatzenbuhler, N. T., Jain, R., Kakarla, R., Kogan, N., Liang, R., Liu, D., Silva, D.J., Wang, H., Gange, D., Anderson, J., Chen, A., Chi, F., Dulina, R., Huang, B., Kamau, M., Wang, C., Baizman, E., Branstrom, A., Bristol, N., Goldman, R., Han, K., Longley, C., Midha, S. and Axelrod, H.R. (1999) *Journal of Medicinal Chemistry*, **42**, 3193–3198.

379 Wang, Y., Zhang, H. and Voelter, W. (1995) *Chemistry Letters*, 273–274.

380 Amaya, T., Takahashi, D., Tanaka, H. and Takahashi, T. (2003) *Angewandte Chemie (International Edition)*, **42**, 1833–1836.

381 Tanaka, H., Takahashi, D. and Takahashi, T. (2006) *Angewandte Chemie (International Edition)*, **45**, 770–773.

382 Huang, X., Huang, L., Wang, H. and Ye, X.-S. (2004) *Angewandte Chemie (International Edition)*, **43**, 5221–5224.

383 Huan, L. and Huang, X. (2007) *Chemistry – A European Journal*, **13**, 529–540.

384 Crich, D. and Smith, M. (2000) *Organic Letters*, **2**, 4067–4069.

385 Crich, D., Banerjee, A., Li, W. and Yao, Q. (2005) *Journal of Carbohyarate Chemistry*, **24**, 415–424.

386 Crich, D. (2007) ACS Symposium Series, American Chemical Society, Washington vol. 960,, pp. 60–72.

387 Crich, D. and Banerjee, A. (2006) *Journal of the American Chemical Society*, **128**, 8078–8086.

388 Crich, D. and Smith, M. (2002) *Journal of the American Chemical Society*, **124**, 8867–8869.

389 Durón, S.G., Polat, T. and Wong, C.-H. (2004) *Organic Letters*, **6**, 839–841.

390 Wang, C., Wang, H., Huang, X., Zhang, L.-H. and Ye, X.-S. (2006) *Synlett*, 2846–2850.

391 Codée, J.D.C., van den Bos, L.J., Litjens, R.E.J.N., Overkleeft, H.S., van Boeckel, C.A.A., van Boom, J.H. and van der Marel, G.A. (2004) *Tetrahedron*, **60**, 1057–1064.

392 Codée, J.D.C., Litjens, R.E.J.N., Van den Bos, L.J., Overkleeft, H.S. and Van der Marel, G. (2005) *Chemical Society Reviews*, **34**, 769–782.

393 Garcia, B.A., Poole, J.L. and Gin, D.Y. (1997) *Journal of the American Chemical Society*, **119**, 7597–7598.

394 Fügedi, P. (2006) 23rd International Symposium of Carbohydrate Chemistry, Whistler, p. 122.

4.3
Xanthates, Thioimidates and Other Thio Derivatives

Wiesław Szeja, Grzegorz Grynkiewicz

4.3.1
Introduction

Carbohydrate derivatives, in which one or more of the oxygen atoms bonded directly to the carbon skeleton have been replaced by sulfur, are termed *thiosugars*. The placement of the sulfur atom at the anomeric position constitutes a special case, because thioglycosides, alkyl, aryl and heterocyclic, occupy a very important place as versatile glycosyl donors in glycosidation methodology. Anomeric thiocarbonyl compounds, on the contrary, have been less explored, although their potential and scope is likely to be similar.

The versatility of thiosugars in carbohydrate chemistry derives from the fact that the sulfur atom is a 'soft base' and is therefore able to react selectively with soft acids such as heavy-metal cations, halogens, alkylating reagents and carbonium ions. The oxygen-bearing groups, in relation to the former, are 'hard bases', which can be functionalized with 'hard acids', usually without affecting the thiofunction. The sulfur-bearing substituents at the anomeric center can be selectively activated with soft electrophilic promoters to form reactive glycosylating species. Thus, the thiosugars are useful building blocks for differentiating selected steps in the glycoside synthesis. The compounds, which constitute the scope of this chapter (Figure 4.1), add to this versatility because their activation conditions and reactivity differ from conventional alkyl and aryl glycosides, discussed in Section 4.1. Methods for their

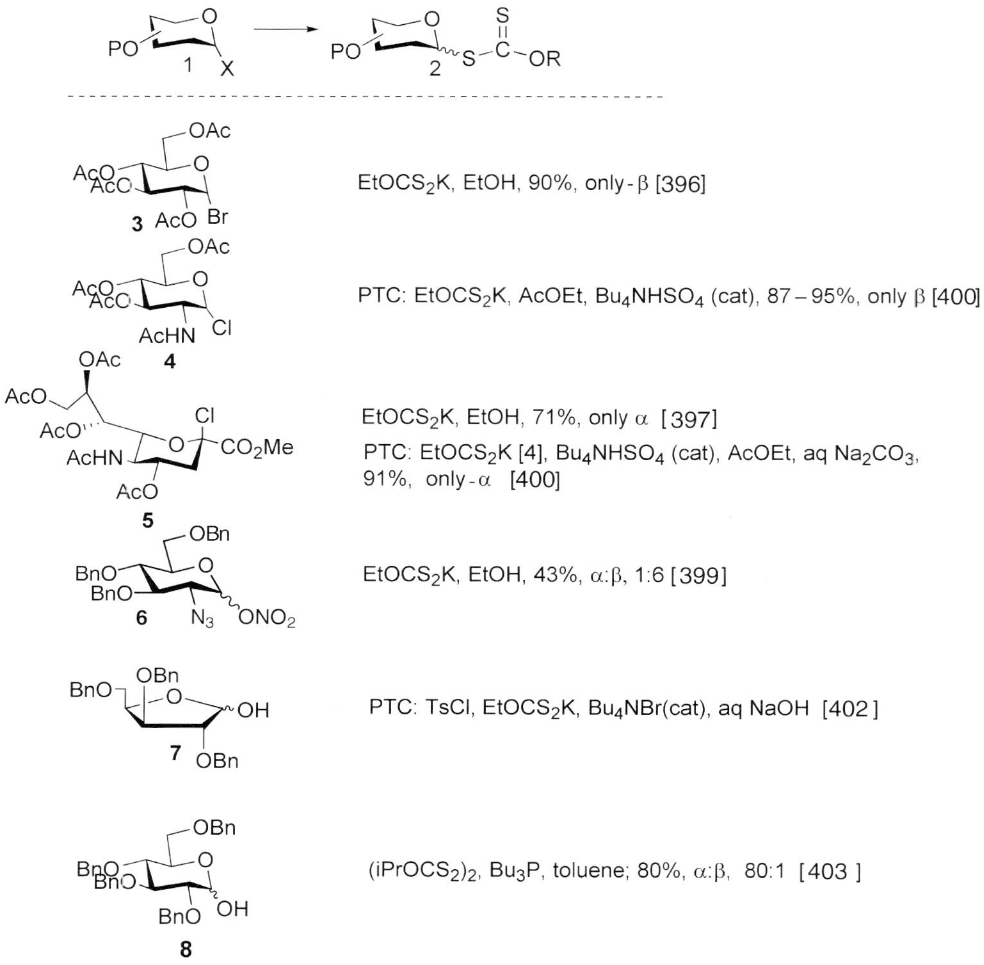

Scheme 4.72 Synthesis of S-glycosyl dithiocarbonates (glycosyl xanthates).

preparation and reactivity characteristics of particular thiocarbonyl systems that follow general rules have been previously reviewed [395].

4.3.2
Dithiocarbonates – Preparation and Application as Glycosyl Donors

Glycosyl dithiocarbonates (xanthates) are attractive glycosyl donors because of their high reactivity in glycosylation reactions. Most procedures for the synthesis of dithiocarbonates involve reaction of per-O-acetylglycosyl halide **3** or **4** with potassium ethoxydithiocarbonate or related thio reagent [396] as illustrated on Scheme 4.72. The synthesis of glycosyl xanthates and their application as glycosyl donors were done by Sinay and coworkers [397–399]. 2-(Ethoxy)dithiocarbonate (2-xantho) derivative of Neu5Ac had been synthesized from the 2-chloride **5** by the reaction with potassium ethoxydithiocarbonate in EtOH [397]. Sialyl xanthate is a stable crystalline material that has a long shelf life. Roy and coworkers demonstrated [400] that phase-transfer-catalyzed (*PTC*) nucleophilic displacement of a wide range of glycosyl halides could secure the entry into *S*-glycosyl xanthates. Per-*O*-acetylated glycosyl bromides or chlorides were selectively transformed into their corresponding *S*-glycosyl xanthates in high yield (91–98%) using tetrabutyl ammonium hydrogen sulfate (*TBAHS*) as the catalyst and ethyl acetate as the solvent with aqueous sodium carbonate as the counter phase. The substitution proceeded with complete inversion of the configuration. Substituents other than anomeric halides have also been employed for such syntheses. The azidonitration of tri-*O*-benzyl-D-galactal gave a mixture of anomeric nitrates of 2-azido-2-deoxy-D-galactopyranose **6** (Scheme 4.72). Treatment of the reaction mixture with *O*-ethyl-*S*-potassium dithiocarbonate led to the mixture of *O*-ethyl-*S*-(2-azido-3,4,6-tri-*O*-benzyl-2-deoxy-β-D-galactopyranosyl) dithiocarbonates [399]. The common method for the synthesis of dithiocarbonates by displacement of an anomeric substituent is ineffective when applied to reactive glycosyl bromides, derivatives of 2-deoxysugars, because of the competing elimination reaction.

Transformation of 2-deoxysugar derivatives into glycosyl xanthates can be performed by the treatment of *O*-benzyl-protected hemiacetal derivative with diphenylphosphoryl chloride, followed by the reaction with *O*-ethyl potassium xanthate in the presence of a base (NaOH, PTC reaction or NaH in appropriate organic solvent). High yields and selectivities in such reactions were observed when using sodium hydride in anhydrous THF [401].

S-(Glycofuranosyl)-*O*-ethyl dithiocarbonates derivatives of L-arabinose, D-xylose **7** and D-ribose were conveniently prepared by treatment of *O*-benzyl-protected 1-OH pentoses with diphenylphosphoryl chloride and potassium *O*-alkyl dithiocarbonate under PTC conditions [402]. In these reactions, the initially formed glycosyl 1-*O*-diphenylphosphate reacts with sulfur nucleophile present in the organic phase in the form of ion pair with tetrabutylammonium ion. An efficient, simple 'one-pot' procedure to obtain glycosyl xanthate was reported by Rollin and coworkers [403]. Treatment of a solution of tetra-*O*-benzyl-D-glucose **8** in toluene with tri-*n*-butyl phosphine and diisopropyl dithiocarbonate disulfide gave a mixture of glycosyl xanthates in good yields (82–87%).

The use of glycosyl xanthate as a glycosyl donor in the presence of $BF_3 \cdot Et_2O$ was first reported by Pougny [404]. Sinay and coworkers used *S*-glycosyl xanthates for the stereoselective synthesis of biologically important galactosamine-containing oligosaccharides [399]. The xanthate prepared was reacted with methyl 2,3,4-tri-*O*-benzyl-α-D-glucopyranoside **11** in acetonitrile in the presence of copper(II) triflate to give a mixture of disaccharides **12** (α/β = 1/6). The β-selectivity observed is because of the formation of the α-nitrilium intermediate in the rate-determining step that upon substitution leads to the β-anomer. In contrast, excellent α-stereoselectivity was

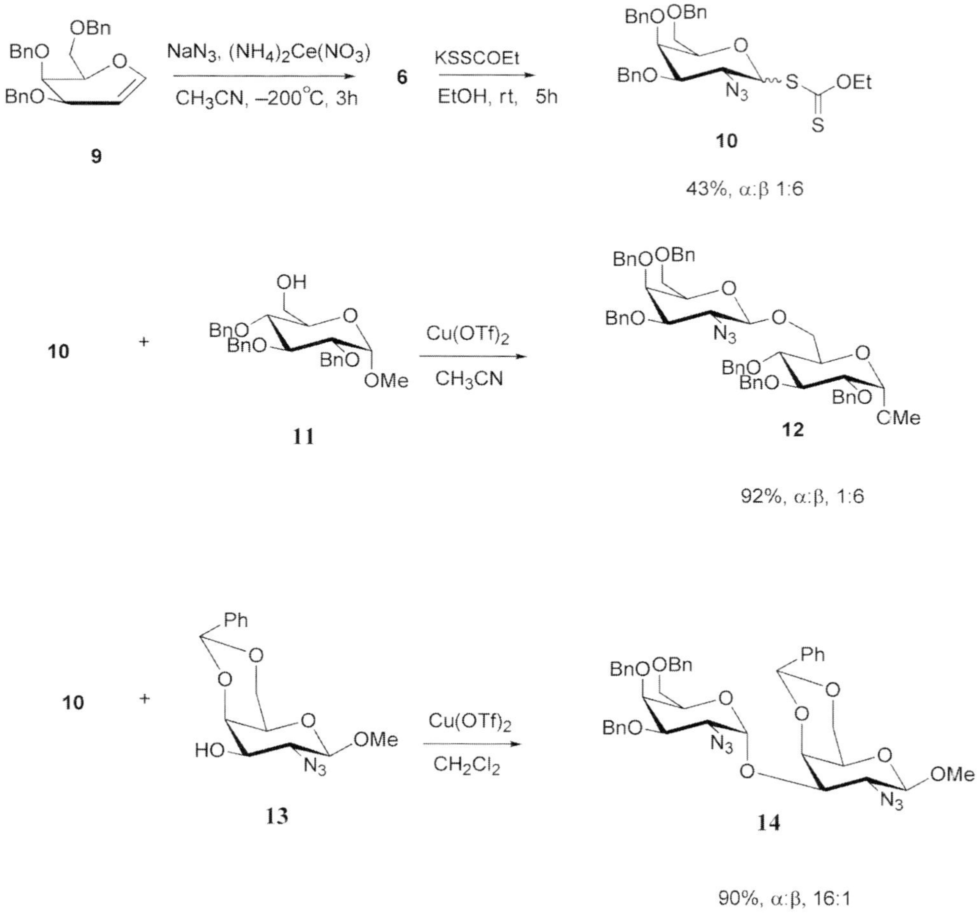

Scheme 4.73 Glycosyl xanthates in stereoselective synthesis of galactosamine-containing oligosaccharides [399].

observed ($\alpha/\beta = 16/1$) for the reaction of **10** with acceptor **13** carried out in dichloromethane (Scheme 4.73).

The application of sialyl xanthate **15** as glycosyl donor was done by Marra and Sinay [397,398]. Dimethyl(methylthio)sulfonium triflate (*DMTST*)-promoted reaction of **15** with 6-hydroxyl of a galactosyl acceptor **16** (Scheme 4.74) afforded α-(2,6)-linked disaccharide **17** in 48% yield, contaminated with β-isomer (4%). A 2-thioalkyl glycosyl donor gave a lower yield (32%) of the disaccharide **17** ($\alpha:\beta = 3:1$) when reacted under similar conditions, illustrating the advantageous properties of 2-xanthates. When glycosylation was performed in dichloromethane, β-isomer was the only product that was isolated (25% yield). Regioselective glycosylation of **18** led to (2,3)-linked disaccharide **19**. In dichloromethane, the glycosylation probably

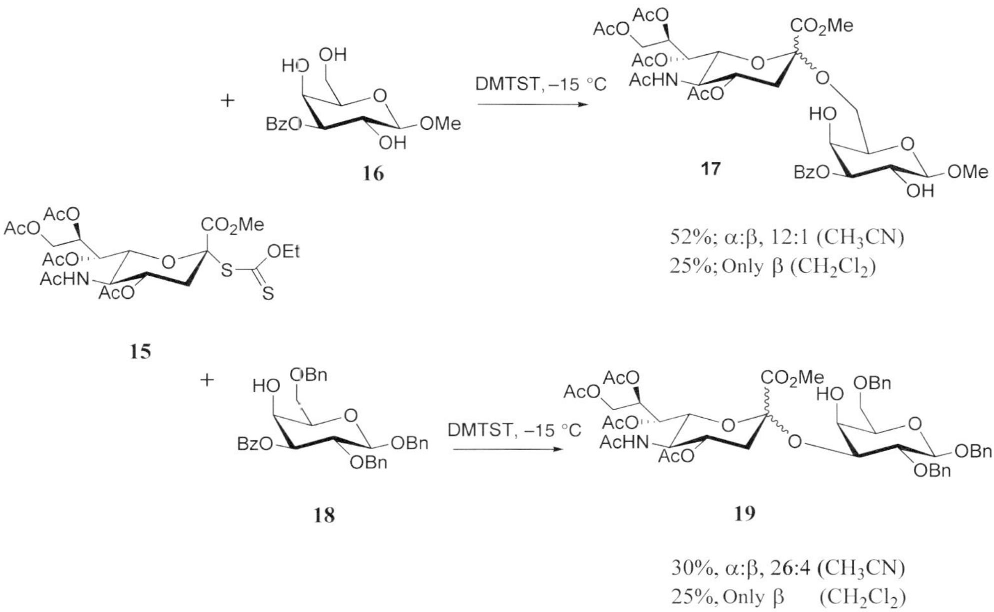

Scheme 4.74 Sialyl xanthate in regioselective and stereoselective glycosylations [398].

involves a reactive anomeric oxocarbenium ion, which is subject to axial attack, to give β-isomer. In acetonitrile, the reaction may involve the β-nitrilium ion, which is more reactive than the α-isomer as a consequence of the reverse anomeric effect displayed by a positively charged leaving group [405,406,410].

The application of highly reactive thiophilic reagents, such as methylsulfenyl triflate (MeSOTf) [407–409], as the promoter greatly improved the synthesis of sialyl glycosides. The reactive thiophilic reagent can be generated *in situ* by the reaction of methylsulfenyl bromide (MeSBr) with AgOTf. MeSOTf activates 2-xanthate **15** at low temperature (−70 °C), and the best results of glycosylation were obtained when a mixture of MeCN/CH$_2$Cl$_2$ (3/2 vol/vol) was used as the reaction solvent [407]. The reaction mechanism of xanthate activation is similar to that of thioglycosides [410]. The oxonium cation formed in the first step of reaction of xanthate with PhSOTf is stabilized by interaction with acetonitrile. In the next step, nucleophile reacts with more reactive, less hindered cation to mainly give α-product. This approach was used in the synthesis of ganglioside GM$_3$ analog **21** as shown in Scheme 4.75 [409]. An attractive feature of this glycosylation protocol is that sialyl xanthate **15** can be selectively activated in the presence of thioglycosides [407].

It was found that PhSOTf is superior to MeSOTf in terms of both yield and stereoselectivity of sialylation, especially when applied in combination with the hindered base 2,6-di(*tert*-butyl)-pyridine (*DTBP*) at low temperatures (−70 °C) [410]. Promoter was generated by the reaction of phenylsulfenyl chloride (PhSCl)

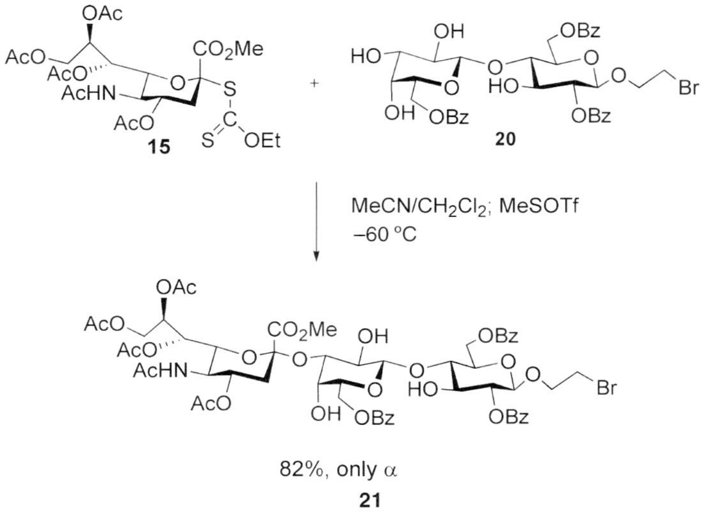

Scheme 4.75 Sialyl xanthate in the synthesis of GM$_3$ ganglioside [409].

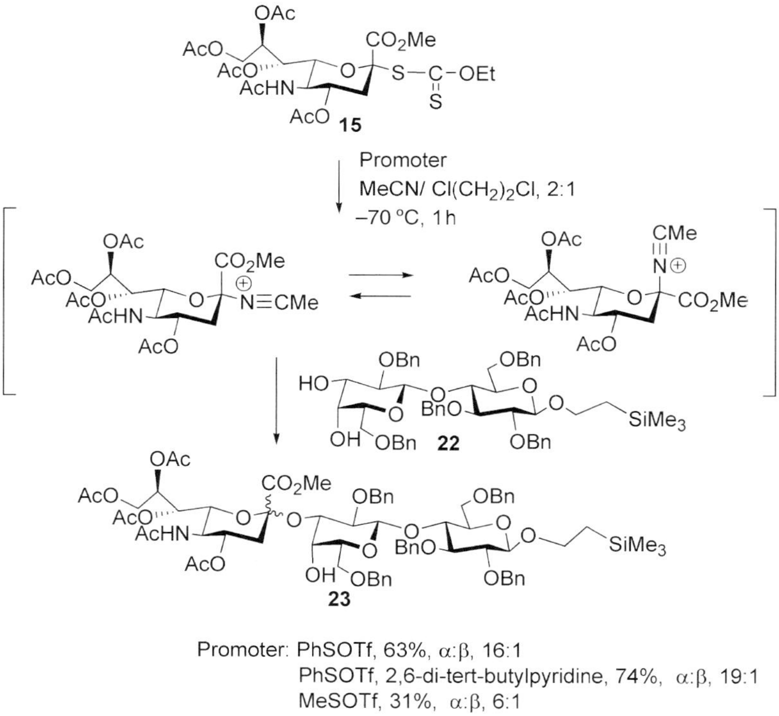

Scheme 4.76 Sialyl xanthate in stereoselective glycosylation [410].

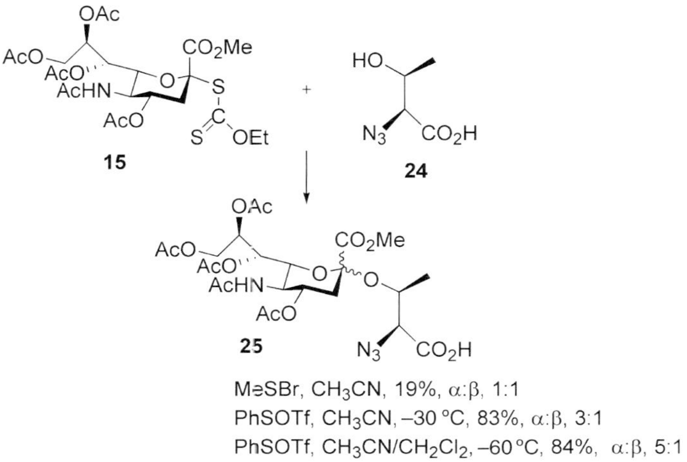

MeSBr, CH₃CN, 19%, α:β, 1:1
PhSOTf, CH₃CN, −30 °C, 83%, α:β, 3:1
PhSOTf, CH₃CN/CH₂Cl₂, −60 °C, 84%, α:β, 5:1

Scheme 4.77 Preparation of solid-phase glycopeptide library [411,412].

with AgOTf. A relevant example is shown in Scheme 4.76: thus, protected GM$_3$ trisaccharide **23** was obtained on a gram scale in an excellent yield of 74%, mainly as the α-anomer.

One of the disadvantages of this method is a reduced yield because of the formation of sialyl glycal during both the preparation of sialyl xanthates and their glycosidation. Also, some amount of the β-sialoside is often formed, especially when reaction is performed with more reactive alcohols. Separation of the anomeric mixture is usually difficult and requires careful chromatography [410]. Salic-acid-containing amino acid building blocks were used for the preparation of glycopeptide libraries on the solid phase [411,412]. It was concluded that PhSOTf is the promoter of choice for the sialylation of amino acid acceptors. Thus, the coupling of the sialyl donor **15** with an easily available acceptor **24** in acetonitrile/methylene chloride gave compound **25** in a high yield (Scheme 4.77).

Several other examples of successful application of sialyl xanthates for oligosaccharide synthesis have been reported [411,413–427].

4.3.3
Glycosyl Thioimidates – Preparation and Application as Glycosyl Donors

Glycosyl thioimidates (heteroaryl thioglycosides) were proposed as a new class of glycosyl donors [428,429]. Similar to the thioglycosides, glycosyl thioimidates are easily accessible. General approach to the synthesis of these compounds may proceed through the nucleophilic displacement of a leaving group of the glycosyl donor by a thioimidate anion. The convenient method of synthesis consists of the reaction between a per-O-acetyl **3**, per-O-benzoyl **26**, per-O-benzyl glycosyl halide **27** and appropriate thiolate anion. Owing to the high nucleophilicity of thioimidate anion, it is possible to perform the reaction in polar solvents such as acetone, acetonitrile, alcohol or even acetone–water. The reaction is performed in the presence of inorganic bases such as sodium or potassium hydroxide, sodium hydride or potassium carbonate, and usually 1,2-*trans* glycoside is obtained [430]. By this procedure, Bertram and coworkers prepared peracetylated *N,N*-diethyl and *N,N*-diallyl *S*-glycosyl dithiocarbamate derivatives of D-glucose, lactose and cellobiose [431,432].

Zinner used sodium as a base [433,434] for the exchange reaction. Demchenko performed reaction of glycosyl bromide **26** with potassium *S*-benzoxazolyl (SBox) in the presence of 18-crown-6 [435]. Similarly, benzoylated *S*-thiazolinyl (STaz) glycosides were obtained when NaSTaz or KSTaz reacted in the presence of a crown ether with per-*O*-benzoyl glycosyl bromide **26** or chloride **27**. The latter reaction mixture containing the desired *S*-glycosyl derivative was contaminated with the products of *N*-glycosylation and β-elimination. Alternatively, glycosyl bromides were reacted directly with HSBox in the presence of K_2CO_3 in acetone [435]. Similarly, benzoylated *S*-thiazolinyl glycosides were obtained when NaSTaz or KSTaz reacted in the presence of a crown ether with per-*O*-benzoyl glycosyl bromide or chloride. Again, formation of some *N*-glycosylation and β-elimination products [437] was observed. Per-*O*-benzylated *S*-benzothiazolyl glycoside (SBtaz) was prepared from the anomeric chloride **27** and mercaptobenzothiazole **32** in the presence of 1,8-bis(dimethylamino)naphthalene [438], indicating significant influence of the base on the reaction outcome. A new approach to the preparation of these types of compounds was developed by Szeja and Bogusiak [439]. It is especially suited for acetal- and benzyl-protected sugars and involves generating glycosyl tosylates under phase-transfer conditions *in situ*, followed by substitution with thioimidate ion. Under the PTC conditions, a mixture of thioimidates was obtained when 2-mercaptobenzothiazole was reacted with 2,3,4,6-tetra-*O*-benzyl-D-glucopyranose **8** with tosyl chloride in a two-phase system (50% aqueous sodium hydroxide/benzene) in the presence of catalytic amounts of tetrabutyl ammonium chloride. Another method for the synthesis of glycosyl thioimidates involves the treatment of protected hemiacetal precursors with disulfide in the presence of trialkyl phosphines to give a mixture of heteroaryl thioglycosides [157,440,441]. Thus, the reaction of **8** with piridinium disulfide in the presence of tributyl phosphine afforded the pyridyl 1-thio-D-glucopyranosides [440]. Benzylated β-D-mannopyranose orthoacetate **29** was coupled with 2-mercaptopyridine **36**, 2-mercaptopyrimidine **35**, 2-mercaptobenzoxazole **33** and 2-mercaptobenzothiazole **32** to give the 1-thio-α-D-mannopyranosides in excellent yields (89–93%) [442]. Easily available per-*O*-acetylated sugars are convenient substrates for the synthesis of glycosyl thioimidates [435,437,443]. The standard procedure involves the reaction of such sugar derivative with a slight excess of mercapto imidate derivative using Lewis acid as the promoter, and usually, 1,2-*trans* product predominates [435,437,443]. For example, D-glucose pentaacetate **28** reacted with 2-mercaptobenzoxazole (HSBox) **33** in the presence of BF_3–Et_2O to afford the corresponding thioglycopyranoside as an anomeric mixture [435].

Similarly, the reaction of per-*O*-acetyl D-ribofuranose **38** with 2-mercaptopyridine **36** afforded the corresponding thioglycoside in a good yield (Scheme 4.79) [446]. Ferrieres

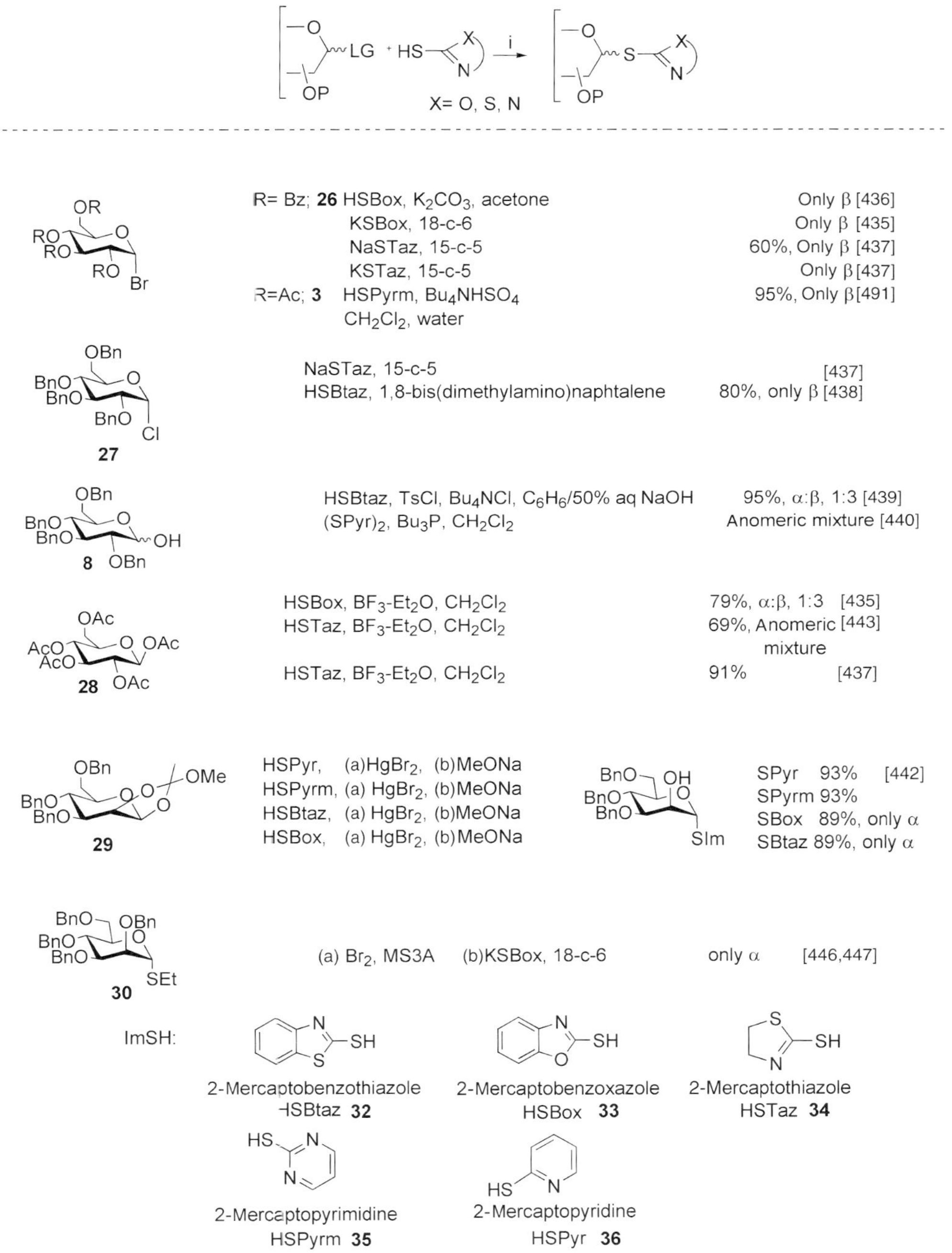

4.3 Xanthates, Thioimidates and Other Thio Derivatives | 337

Substrate	Conditions	Result
26 (R=Bz)	HSBox, K_2CO_3, acetone	Only β [436]
	KSBox, 18-c-6	Only β [435]
	NaSTaz, 15-c-5	60%, Only β [437]
	KSTaz, 15-c-5	Only β [437]
3 (R=Ac)	HSPyrm, Bu_4NHSO_4, CH_2Cl_2, water	95%, Only β [491]
27	NaSTaz, 15-c-5	[437]
	HSBtaz, 1,8-bis(dimethylamino)naphtalene	80%, only β [438]
8	HSBtaz, TsCl, Bu_4NCl, C_6H_6/50% aq NaOH	95%, α:β, 1:3 [439]
	(SPyr)$_2$, Bu_3P, CH_2Cl_2	Anomeric mixture [440]
28	HSBox, BF_3-Et_2O, CH_2Cl_2	79%, α:β, 1:3 [435]
	HSTaz, BF_3-Et_2O, CH_2Cl_2	69%, Anomeric mixture [443]
	HSTaz, BF_3-Et_2O, CH_2Cl_2	91% [437]
29	HSPyr, (a)$HgBr_2$, (b)MeONa	SPyr 93% [442]
	HSPyrm, (a) $HgBr_2$, (b)MeONa	SPyrm 93%
	HSBtaz, (a) $HgBr_2$, (b)MeONa	SBox 89%, only α
	HSBox, (a) $HgBr_2$, (b)MeONa	SBtaz 89%, only α
30	(a) Br_2, MS3A (b)KSBox, 18-c-6	only α [446,447]

ImSH:

2-Mercaptobenzothiazole HSBtaz **32**
2-Mercaptobenzoxazole HSBox **33**
2-Mercaptothiazole HSTaz **34**
2-Mercaptopyrimidine HSPyrm **35**
2-Mercaptopyridine HSPyr **36**

Scheme 4.78 Synthesis of S-glycopyranosyl imidates.

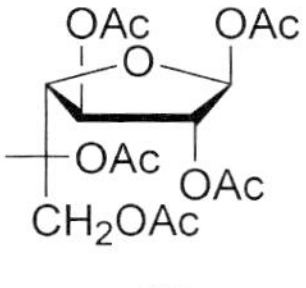

Scheme 4.79 reactions

Compound 37 (OAc, OAc, OAc, OAc, CH₂OAc furanose):
- HSPyrm, BF$_3$-Et$_2$O, CH$_2$Cl$_2$ → 99%, Only α [444]
- HSBtaz, BF$_3$-Et$_2$O, CH$_2$Cl$_2$ → 83%, α:β, 1:6.7 [444]

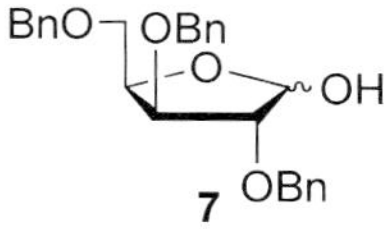

Compound 7 (BnO, OBn, OBn furanose with anomeric OH):
- HSBtaz, (PhO)$_2$P(=O)Cl, Bu$_4$NBr, CH$_2$Cl$_2$/aq NaOH → Anomeric mixture [445]

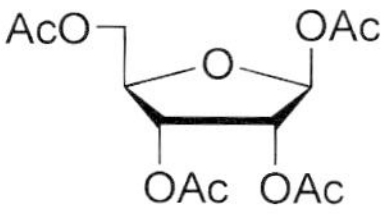

Compound 38 (AcO, OAc, OAc, OAc furanose):
- HS Pyr, BF$_3$-Et$_2$O, ClCH$_2$CH$_2$Cl → 85%, β-only [446]

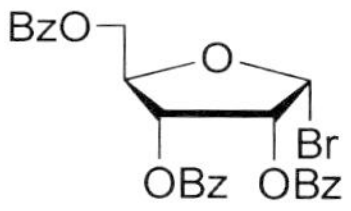

Compound 39 (BzO, OBz, OBz furanosyl bromide):
- HSPyr, K$_2$CO$_3$, Toluene–acetone → 88%, β-only [440]

Scheme 4.79 Synthesis of S-glycofuranosyl imidates.

and coworkers described the synthesis of per-O-acetylated S-benzothiazolyl galactofuranosides by the treatment of acetylated galactofuranose **37** with 2-mercaptobenzothiazole **32** in the presence of BF$_3$–Et$_2$O [444]. Bogusiak reported the synthesis of S-benzothiazolyl furanosides of the L-arabino, D-ribo and D-xylo series from the reducing sugars **7** and **32** in the presence of diphenyl phosphoryl chloride [(PhO)$_2$P(=O)Cl] under phase-transfer conditions [445]. A number of acetylated S-benzothiazolyl D-gluco and D-galactofuranosides have been synthesized from per-O-acetylated glycofuranosyl bromides and sodium thiolates in MeCN [443]. 2,3,5-Tri-O-benzoyl-α-D-ribofuranosyl bromide **39** was reacted with **36** in the presence of K$_2$CO$_3$ in hot toluene–acetone as the solvent to give the expected thioglycoside in a high yield [440].

The main efforts in the field of synthetic carbohydrate chemistry have been focusing on the development of new glycosylation methodologies and convergent strategies for oligosaccharide synthesis [1,63,447–453]. The development of new and efficient strategies for the assembly of oligosaccharides and glycoconjugates is an intensive field of research. The synthesis of oligosaccharides is traditionally a time-consuming process, mainly because of the extensive need for protective-group manipulations. Most of the contemporary researches in this field are therefore focused on the development of glycosylation approaches in which the number of synthetic and purification steps is reduced. Recent solution-phase methodologies that omit the need for the intermediate installation of suitable anomeric leaving-group and/or protecting-group manipulations include chemoselective [454–208,459,460], orthogonal [237,436,450], iterative [461–463] and one-pot glycosylations [464–466]. The majority of these approaches are based on selective activation of one leaving group over another. Among these, one-pot strategies perhaps offer the shortest pathway to oligosaccharides, as the sequential glycosylation reactions are performed in a single flask and do not require isolation and purification of the intermediates.

Although it is difficult to imagine realization of such strategies with the use of classic glycosyl donors such as acetylated glycosyl halides, glycosyl imidates, efforts are underway to design appropriate procedures on the basis of diversified thioglycosyl synthons. Garegg and coworkers have examined the potential of glycosyl 1-piperidinecarbothioates as glycosyl donors in oligosaccharide synthesis. Thiophilic promoters such as methyl triflate (MeOTf) and silver triflate, as well as metal salts such as tin(IV) chloride and iron(III) chloride gave good yields of the desired disaccharides. In their approach, activation of carbothioate over ethyl thioglycosides exemplifies the principle of selective activation of anomeric thio substituents [467].

Demchenko and coworkers have demonstrated that glycosyl thioimidates, a class of glycosyl donors, with the generic leaving group $SCR^1 = NR^2$ fulfill the requirements for modern building blocks and are suitable for a variety of convergent synthetic strategies for oligosaccharide synthesis [118,227,428,435,436,446,447,468,469]. Considering the multifunctional character of the thioimidoyl moiety, three major activation pathways (a–c, Scheme 4.80) for their glycosidation have been postulated.

In the first pathway, thiophilic reagents (NIS/TMSOTf) activate the anomeric leaving group via complexation to the sulfur atom (pathway a). In the second approach, electrophilic promoters such as MeOTf target the thioimidoyl nitrogen (pathway b). Finally, metal-salt-based promoters (AgOTf or $Cu(OTf)_2$) can complex to both the sulfur and nitrogen atoms intra- or intermolecularly (pathway c) and stimulate the anomeric activation. Some of these promoters are already commonly used for thioglycoside activation [1]. However, there is a possibility of using metal-salt-based activation that distinguishes the thioimidates from their S-alkyl/aryl counterpart. Examples of practical and highly stereoselective glycosylations [435,436] are presented in Scheme 4.81. Thus, when per-O-benzoylated SBox derivatives of the D-gluco **39**, D-galacto and D-manno series were reacted with glycosyl acceptor **40** in the presence of AgOTf, the corresponding disaccharides such as **41** were obtained [435]. Glucosyl donor **42** and its galactosyl counterpart bearing a nonparticipating group at C-2 were used in stereoselective synthesis of 1,2-cis disaccharides, for example **44** [436].

It was reported that selective activation of the SBox glycosyl donors over SEt and O-pentenyl glycoside acceptors can also be achieved in the presence of AgOTf

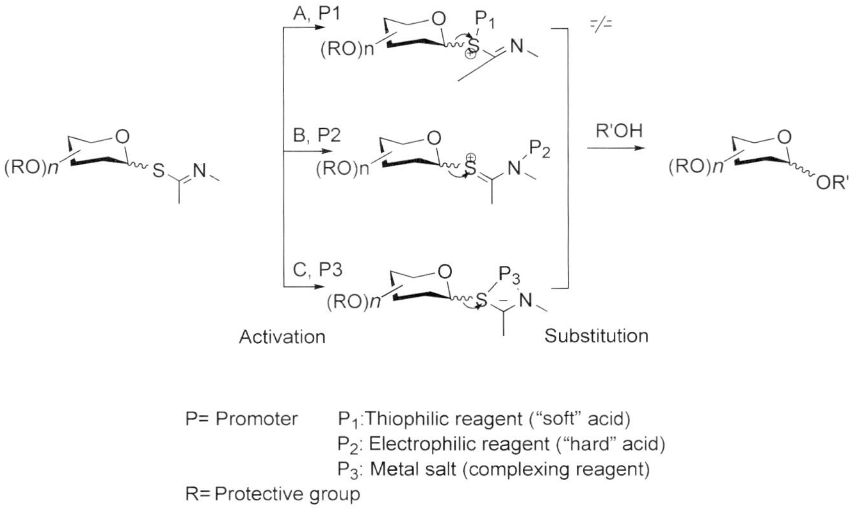

Scheme 4.80 Activation pathways for thioimidate glycosidation [227].

[435,436]. Thus, the reaction of glycosyl acceptors **45** and **47** with SBox glycosyl donor **42** provided a complete stereoselectivity in 1,2-*cis* glycosidations of **45** and high stereoselectivity was achieved in the reaction with **47** [435,436] (Scheme 4.82).

It has been found that 2-O-benzyl-3,4,6-tri-O-acyl SBox glycosides are significantly less reactive than 'disarmed' peracylated derivatives. Taking into account this observation, a convergent synthesis of oligosaccharides was developed. Activation of the *armed* **49** SBox glycoside over *moderately disarmed* **50** provided disaccharide **52** in a good yield (Scheme 4.83). Disaccharide **52** was then activated over the disarmed glycosyl acceptor **53** to afford the trisaccharide **54**.

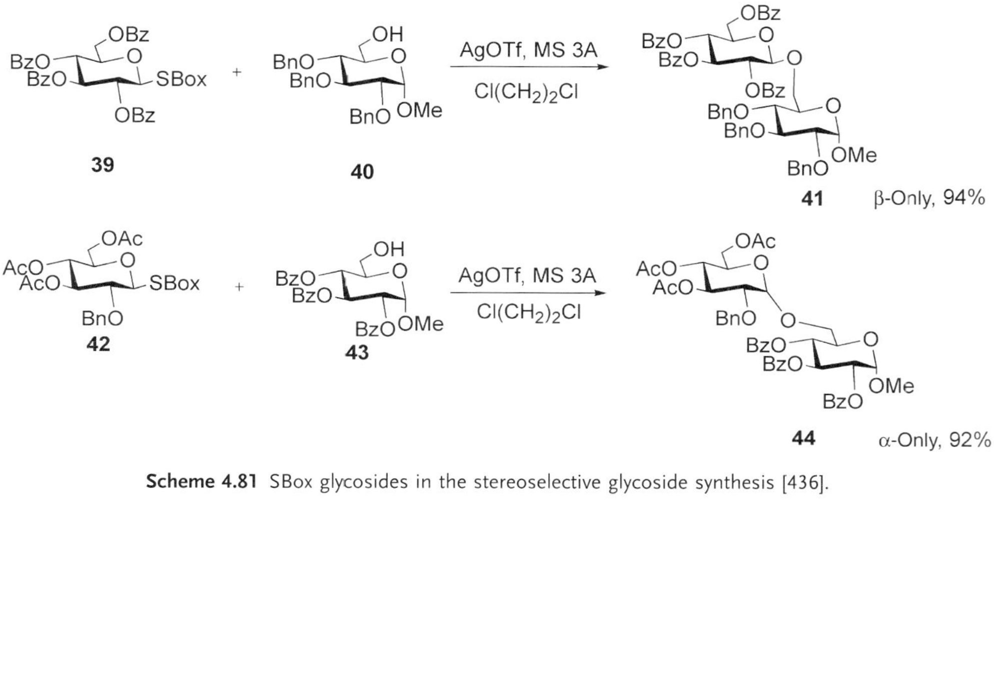

Scheme 4.81 SBox glycosides in the stereoselective glycoside synthesis [436].

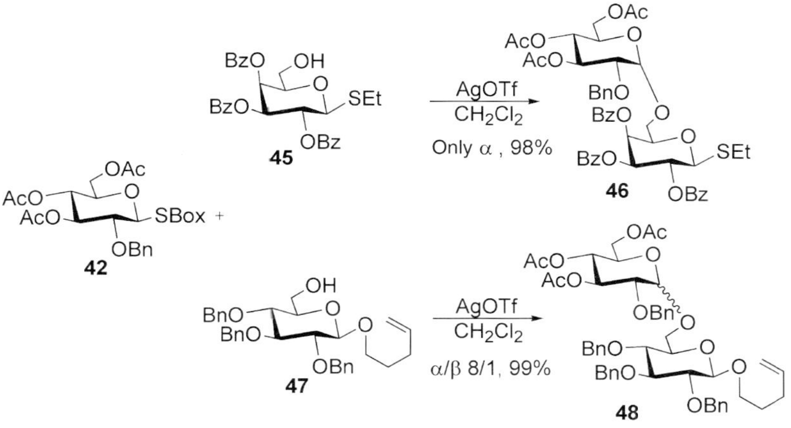

Scheme 4.82 Orthogonality of the SBox glycosides [435,436].

Continuing this work, Demchenko et al. reported the application of the SBox glycosides to the high-yielding synthesis of disaccharides of the 2-amino-2-deoxy series [469]. The N-substituted SBox glycosides, 2-NPth, 2-NHTFA, 2-N-trichloroethoxy carbonyl (NHTroc) and NHAc, were activated with AgOTf or MeOTf, affording the disaccharide derivatives in high yields and with complete stereoselectivity. The reactivity of glycosyl donors strongly depends on the N-protecting groups and promoters. It was found that in MeOTf-promoted glycosylation NPth derivative is less reactive than NHTroc glycosyl donor. This observation gave rise to a complementary glycosylation approach for chemoselective glycosidation of 2-aminosugars. Glycosyl acceptor **56** was glycosylated with glycosyl donor **55**, as illustrated in Scheme 4.84. This coupling is best accomplished at a reduced temperature (5 °C); under these reaction conditions, the disaccharide **57** was obtained with complete

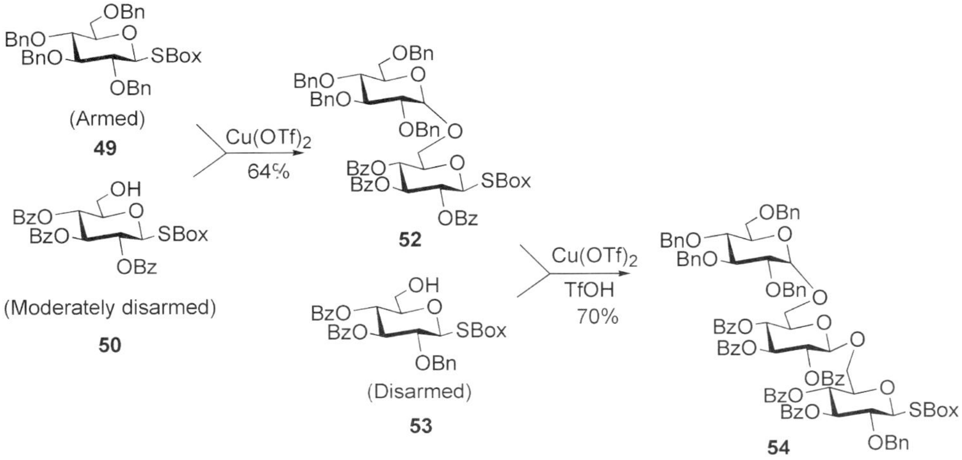

Scheme 4.83 Chemoselective activation of the SBox glycoside [428].

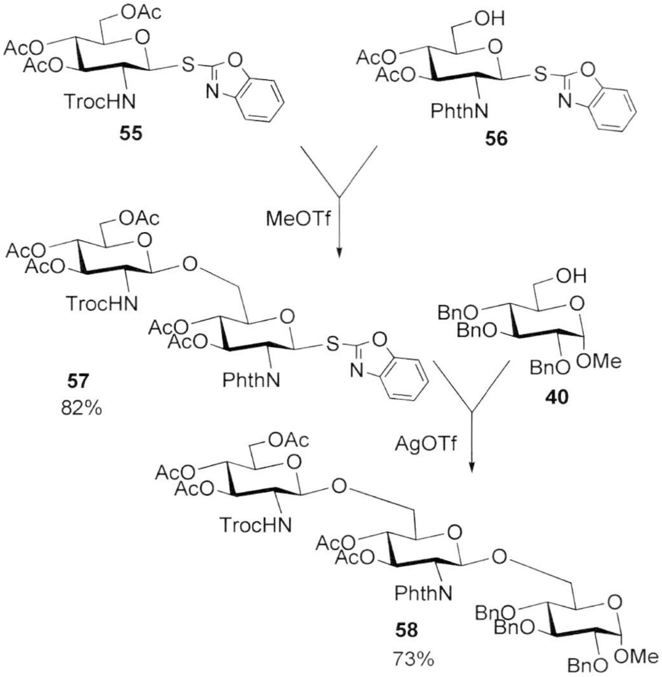

Scheme 4.84 Synthesis of oligosaccharides of 2-deoxy-2-aminosugars [469].

1,2-*trans* stereoselectivity and a high yield of 82%. Subsequent AgOTf-promoted glycosidation of disaccharide **57** with glycosyl acceptor **40** at room temperature gave trisaccharide **58** in 73% yield and with complete β-selectivity. This two-step sequential activation leads to the *trans–trans*-linked oligosaccharides.

Investigation of the glycosyl donor properties of STaz glycosides resulted in the development of another general approach to 1,2-*cis* and 1,2-*trans* glycosylation [437]. Thiophilic reagents, AgOTf, MeOTf, NIS/TfOH and Cu(OTf)$_2$, were found to be effective as the promoters in glycosylation reactions. Perbenzoylated STaz derivatives of the D-glucose and D-galactose were selected to probe 1,2-*trans*-glycosidation experiments with *S*-ethyl glycosyl acceptors. These glycosylations proceeded smoothly and afforded good results consistently. Perbenzylated STaz glycosides were employed in 1,2-*cis* glycosylations of the SEt-moiety-containing glycosyl acceptors. In all the cases high yields were achieved, and desired disaccharides were obtained as anomeric mixtures in 86–99% yield.

As the activation of STaz glycoside is observed in the presence of NIS and stoichiometric amount of TfOH, it has been reasoned that it might be also possible to activate SEt or SPh glycosyl donors over STaz glycosides [437]. Indeed, glycosidation of the *S*-ethyl glycosyl donor **63** over STaz glycosyl acceptor **64** was activated with NIS in the presence of catalytic amounts of TfOH. This allowed the use of STaz glycosides **59** and **64** in orthogonal glycosylations in combination with *S*-ethyl thioglycosides **60** and **63** (Scheme 4.85). The activation of **59** over SEt glycosyl acceptor **60** with

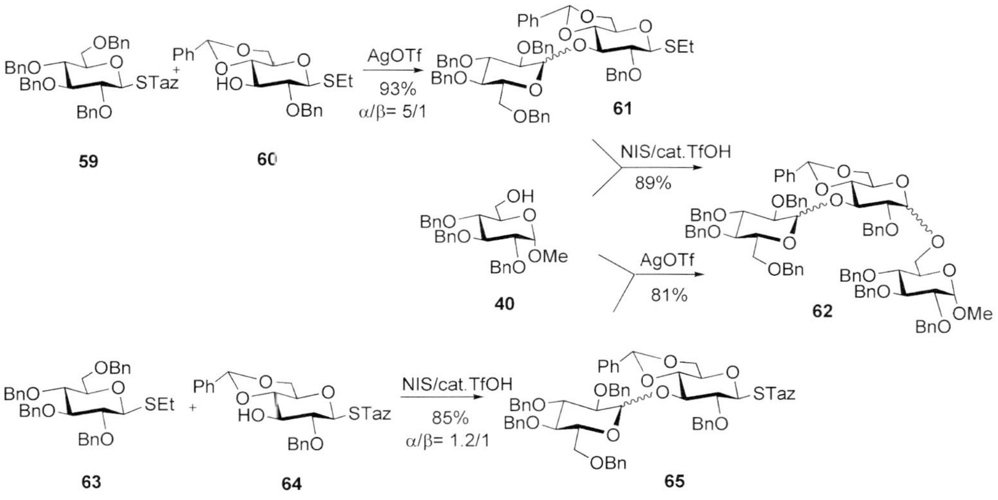

Scheme 4.85 Orthogonality of the STaz and SEt glycosides [437].

AgOTf gave disaccharide **61**. Under these reaction conditions, ethylthio glucoside was stable. In the next step, the activation of the SEt glycosyl donor **61** with NIS and catalytic amount of TfOH gave trisaccharide **62** in a good yield. The second pathway involved glycosidation of the SEt glycosyl donor **63** with STaz glycosyl acceptor **64** in the presence of NIS and a catalytic amount of TfOH. The disaccharide **65** formed was then activated with AgOTf to afford trisaccharide **62**. These results imply a fully orthogonal character of the two classes of leaving groups, STaz and SEt.

On the basis of the results of selective activations, Demchenko and coworkers performed a one-pot synthesis of tetrasaccharides [468]. This was achieved by the stepwise activation of SBox over S-ethyl, S-ethyl over STaz and, finally, STaz over stable glycosyl acceptor (OMe). To execute this one-pot sequence, authors chose the following building blocks: SBox glycosyl donor **39** was glycosidated at the first step with S-ethyl glycosyl acceptor **66**. This resulted in the formation of a disaccharide derivative **67**, the SEt moiety of which was further activated over the STaz moiety of the second step (Scheme 4.86). The resulting trisaccharide **71** could then be glycosidated with 6-OH glycosyl acceptor **40**, bearing a stable O-methyl moiety at the anomeric center to afford a linear tetrasaccharide **72**. The promoters of choice were AgOTf for the activation of the SBox and NIS/catalytic TfOH for the activation of S-ethyl and, finally, more AgOTf should be added for the activation of STaz moiety of the trisaccharide **71**.

Demchenko and coworkers demonstrated that some glycosyl imidates (STaz) can serve as both the glycosyl donor and glycosyl acceptor in accordance with the so-called temporary deactivation concept [227]. It was found that the STaz moiety forms a stable, nonionizing metal complex with palladium bromide. This allows chemoselective activation of a 'free' STaz leaving group (glycosyl donor) over a deactivated (complexed) STaz moiety (glycosyl acceptor). It was demonstrated that either benzoylated or benzylated STaz glycosyl donors could be activated over temporarily deactivated benzoy-

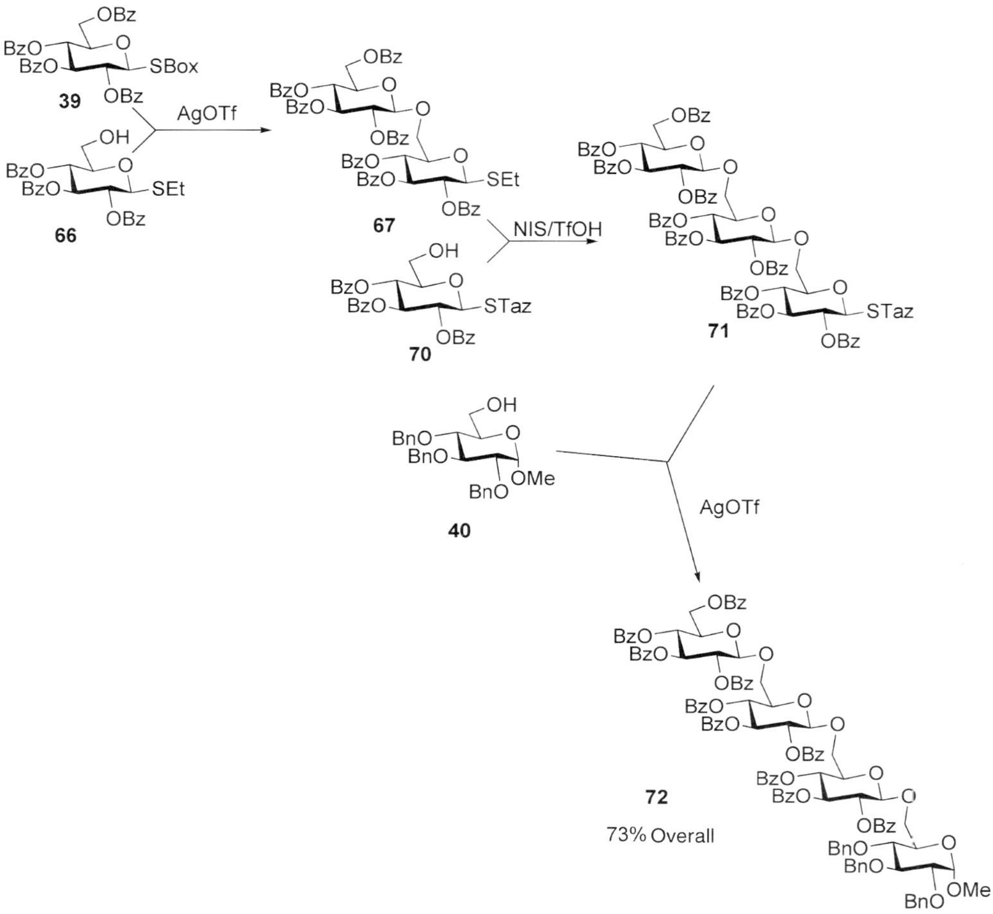

Scheme 4.86 Orthogonal thioimidate-based, one-pot oligosaccharide synthesis [468].

lated or benzylated glycosyl acceptors. The concept is illustrated in Scheme 4.87. An attractive feature of this strategy is that it does not rely on protecting groups to control the leaving-group ability and, thus, glycosyl donor reactivity.

Upon glycosylation of the complex **75** with STaz glycosyl donor **73**, the obtained disaccharide **76** was decomplexed by the treatment with NaCN in acetone. As a result, the β-linked disaccharide **77** was isolated in good yield. These results provide encouraging support for the idea of the temporary deactivation that allows for the preferential activation of one leaving group over another without the requirement for altering the protecting-group pattern. As protecting groups also control the anomeric stereoselectivity of glycosylation, the new approach offers more flexibility in this respect.

Novel sialosyl donors, *S*-benzoxazolyl and *S*-thiazolyl sialosides, have been synthesized [170]. Both SBox and STaz sialosides proved to be excellent glycosyl donors

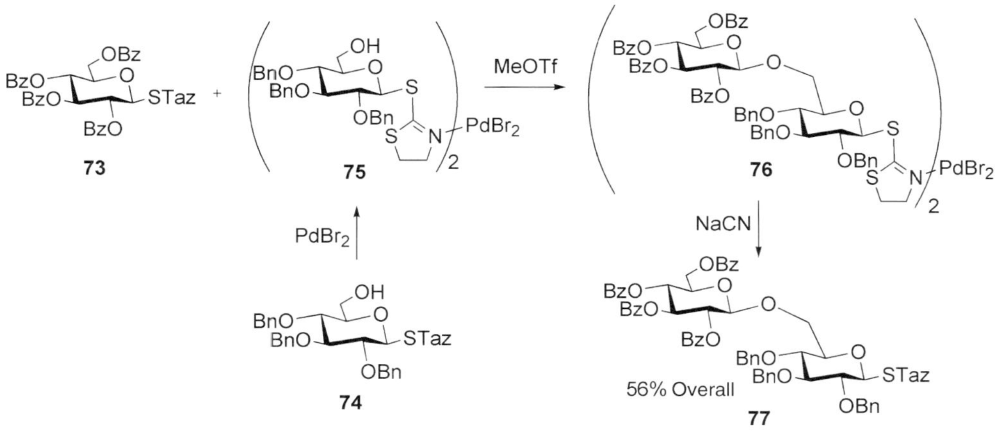

Scheme 4.87 Temporary deactivation method [227].

when activated with MeOTf and AgOTf. In general, good yields and stereoselectivities were observed with a number of glycosyl acceptors, ranging from highly reactive primary to less reactive secondary alcohols. The most attractive feature of the thioimidoyl moieties is that they can be selectively activated over thioglycosides in the presence of AgOTf as the promoter. It was demonstrated that the selective activation of the SBox sialyl donor over ethyl thioglycoside allows the synthesis of disaccharides that can be used in subsequent glycosylations without further manipulations [170].

Recently, various studies of anomeric stereocontrol have been reported using the intramolecular reaction between a glycosyl donor and a glycosyl acceptor, which are connected via a suitable linker (Scheme 4.88) [470]. The intramolecular glycosylation approach by 'linking the accepting atom to the donor via a bifunctional group' (intramolecular aglycon delivery, *IAD*) was originally developed for the synthesis of β-mannopyranosides and later extended to the synthesis of other glycosides. The synthesis of the β-mannosidic linkage is a difficult task [472–475] because both anomeric effect and participating neighboring groups at 2-O favors the formation of α-mannosides (1,2-*trans* configuration). To obtain the desired proximity between the glycosyl donor and acceptor moiety, the rigid spacer concept was designed [476–478]. As a powerful example for a rigid spacer, the *m*-xylylene moiety and derivatives were chosen. Taking this into account, Schmidt and coworkers have developed an efficient and highly regio- and stereoselective protocol for the intramolecular β-mannopyranoside synthesis by using glycosyl thioimidates as glycosyl donors and *m*-xylylene moieties as rigid spacers linked to the 2-hydroxy group of the mannose residue [442].

Reaction of glycosyl thioimidate **79** with αα'-dibromo-*m*-xylene in the presence of NaH as a base and 15-crown-5 as a supporting reagent allowed the intermediate **80** (Scheme 4.88). Treatment of the diol **81** with dibutyltin oxide in dry toluene and then reaction with **80** in the presence of tetrabutylammonium iodide afforded the desired O-linked intermediate **82**. Activation of this compound with NIS–TMSOTf afforded **83** in a good yield. Hydrogenolysis followed by acetylation gave the desired disac-

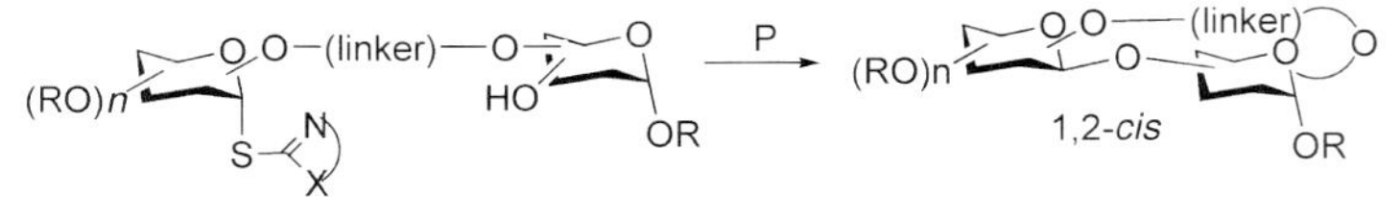

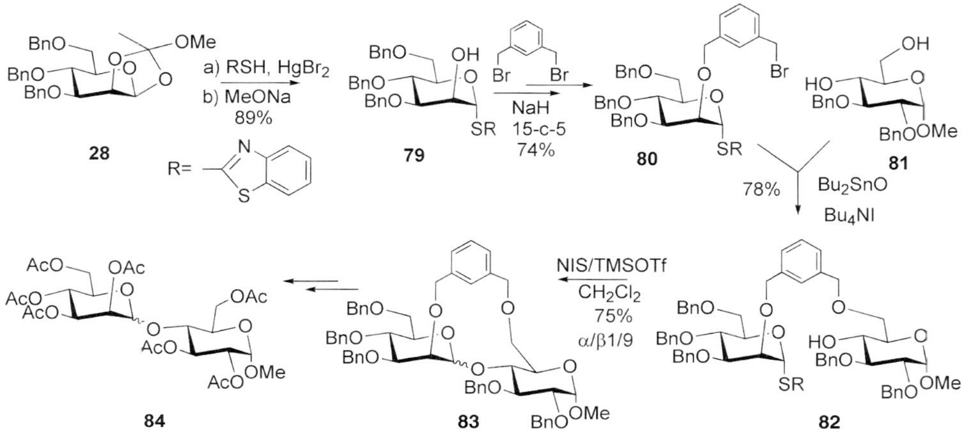

Scheme 4.88 Intramolecular β-mannosylation [158].

charide **84** with good stereoselectivity. The method developed permits the synthesis of a variety of differently functionalized β-mannopyranoside derivatives.

Application of the SBox glycosides to β-mannosylations was also investigated [140]. It was determined that the use of the SBox glycosyl donors protected with either *p*-methoxybenzoyl or *N,N*-diethylthiocarbamoyl moieties at C-4 improve the stereoselectivity. Thus, coupling of the SBox glycoside **86** with glycosyl acceptor **87** afforded disaccharide **89** in good yield and stereoselectivity. It has been postulated that the improved stereoselectivity is because of the long-range participation of a substituent at C-4 and the formation of cyclic carboxonium ion followed by the nucleophilic attack from the β-face of the sugar ring. In contrast, when perbenzylated glycosyl donor **85** was used, disaccharide **88** was obtained with low stereoselectivity [140] (Scheme 4.89).

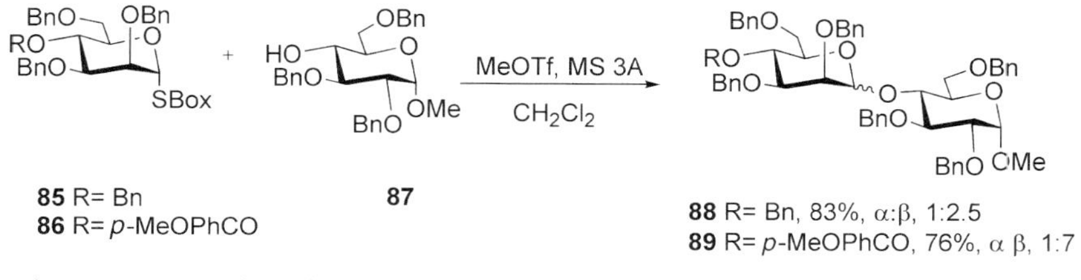

85 R= Bn
86 R= *p*-MeOPhCO

87

88 R= Bn, 83%, α:β, 1:2.5
89 R= *p*-MeOPhCO, 76%, α β, 1:7

Scheme 4.89 SBox glycoside in the remote-assisted β-mannosylation [146].

Hanessian explored the scope of 2-thiopyridyl (SPyr) and 2-thiopyrimidinyl (SPyrm) compounds as glycosyl donors by developing a 'remote activation' concept [254,479]. It was assumed that a suitable substituent at the anomeric center such as sulfur (soft base) in combination with a nitrogen (hard base) could satisfy the requirements for the remote activation. Bidendate activation through a chelation of nitrogen and sulfur with the metal cation and formation of an oxocarbenium intermediate were postulated. Subsequent nucleophilic attack by an alcohol would form the glycoside bond. Treatment of O-unprotected glycosyl donors, SPyrm and SPyr-β-D-glucopyranoside, with a variety of alcohols in the presence of mercuric nitrate in acetonitrile solution, led to the anomeric mixture of glycosides within a few minutes [254]. However, the necessity to use an excess of acceptor, the formation of anomeric mixture of glycosides, and the use of mercuric salts limited the generality of this method of glycoside synthesis. The glycosyl-donating properties of protected thiopirydyl and thiopyrimidinyl glycosides activated with other thiophilic salts such as silver triflate [480] and lead perchlorate [481] were also studied.

The most impressive application of 2-thiopyridyl and 2-thiopyrimidinyl donors is in the area of antibiotics. Thus, Woodward *et al.* [481] successfully completed the total synthesis of erythromycin by using SPyrm glycoside of D-desosamine and SPyr-glycoside of L-cladinose as glycosyl donors to the subsequent glycosylation with erythronalide A. This methodology was also successfully used in the synthesis of oleandomycin [482,483], erythromycin A [484] and erythromycin B [485].

A thorough investigation of the glycosyl donor properties of the 2-thiopyridyl glycosides was performed by Mereyala and coworkers using MeI as the activator [159,486,158,487–490]. Although in some cases the reaction was rather sluggish and required prolonged reaction times at elevated temperatures, the benefit of high yield and excellent to complete stereoselectivities was apparent. One of the most valuable applications of the SPyr and SPyrm glycosides is for chemoselective oligosaccharide synthesis in accordance with 'armed–disarmed' strategy [457,458]. Thus, the benzylated S-pyridyl glycoside **90** (armed donor) can be chemoselectively activated with MeI over the partially acetylated **91** (disarmed) acceptor to afford the α-linked oligosaccharide **92** in complete stereoselectivity [158] (Scheme 4.90).

The application of per-O-benzylated SPyrm-D-gluco, D-galacto, D-xylo and D-arabinopyranosides to stereoselective glycosylation was reported by Kong and coworkers [491,492]. The results are qualitatively similar to those obtained with thiopyridyl derivatives. Thus, glycosylation of methyl 2,4,6-tri-O-benzyl-α-D-mannopyranoside with benzylated SPyr–β-D-glucopyranoside in the presence of TMSOTf afforded disaccharide in good yields with complete 1,2-*cis* stereoselectivity. The SPyr glycosides were successfully applied for the synthesis of 2-deoxyglycosides [486,158,487], aminosugars [490], furanosides [440] and α-fucosides [490].

It was reported that 5-nitro-2-thiopyridyl group can be used for the anomeric protection of the glycosyl acceptor unit [493]. As this anomeric group was found to be stable in the presence of AgOTf, NIS/TfOH or TMSOTf, a variety of glycosyl donors such as bromides, dithiocarbamates and even S-alkyl glycosides could be selectively activated in the presence of 5-nitropyridyl moiety. Recently, Lowary reported the use of this class of glycosyl donors for the O-glycosidation of furanoses and pyranoses.

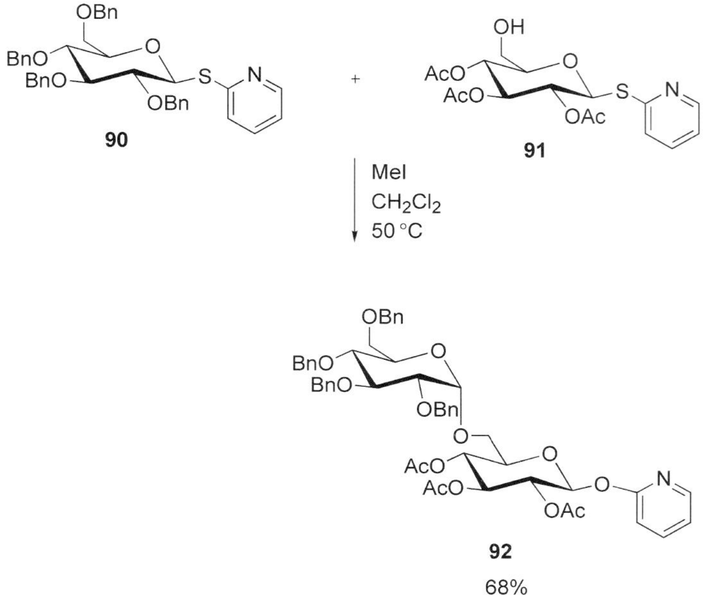

Scheme 4.90 Arming–disarming properties of the SPyr glycosides [158].

Thus, samarium(III)-triflate-promoted reactions provided O-linked disaccharides in moderate to excellent yields [308]. Szeja and Niemiec reported the use of 5-nitro-S-pirydyl glucoside as glycosyl donors in β-glucosidase-catalyzed glycosylation of alcohols [494].

Glycofuranosyl thioimidates have attracted attention as versatile and flexible building blocks for the synthesis of simple as well as complex oligosaccharides. These compounds can be activated with a number of thiophilic reagents to act as glycosyl donors [1]. The use of S-pyridyl thioglycosides in the synthesis of oligofuranosides was introduced by Mereyala et al. [488]. Direct coupling of S-pyridiyl ribofuranoside and S-pyridyl 2-deoxy-ribofuranoside derivatives with a variety of protected sugars promoted by methyl iodide in methylene chloride exclusively gives 1,2-*cis* glycosidic linkages. However, small amounts of 1,4-anhydro-2-deoxy-D-*erythro*-pent-1-enitol are formed along with 2-deoxy-ribofuranosido disaccharides. The properties of glycofuranosyl S-xanthates, N,N-diethyldithiocarbamate and O,O-diethyldithiophosphate derivatives of benzylated D-ribo, D-xylo and L-arabinofuranoses as glycosyl donors have been studied by Bogusiak and Szeja [495–497]. These functional groups can be effectively activated by silver triflate. Independent of the anomeric ratio of the glycosyl donors, the 1,2-*trans* glycofuranosides were formed with moderate to high stereoselectivity. These donors can also be converted into 1,2-*cis* glycofuranosides using a combination of silver triflate and stoichiometric amount of polar organic reactants such as DMSO, tetramethylurea (*TMU*) or hexamethylphosphoric triamide (*HMPA*). Glycosylation has been found to proceed most stereoselectively with HMPA [497]. It was found that NIS/TfOH-mediated coupling of

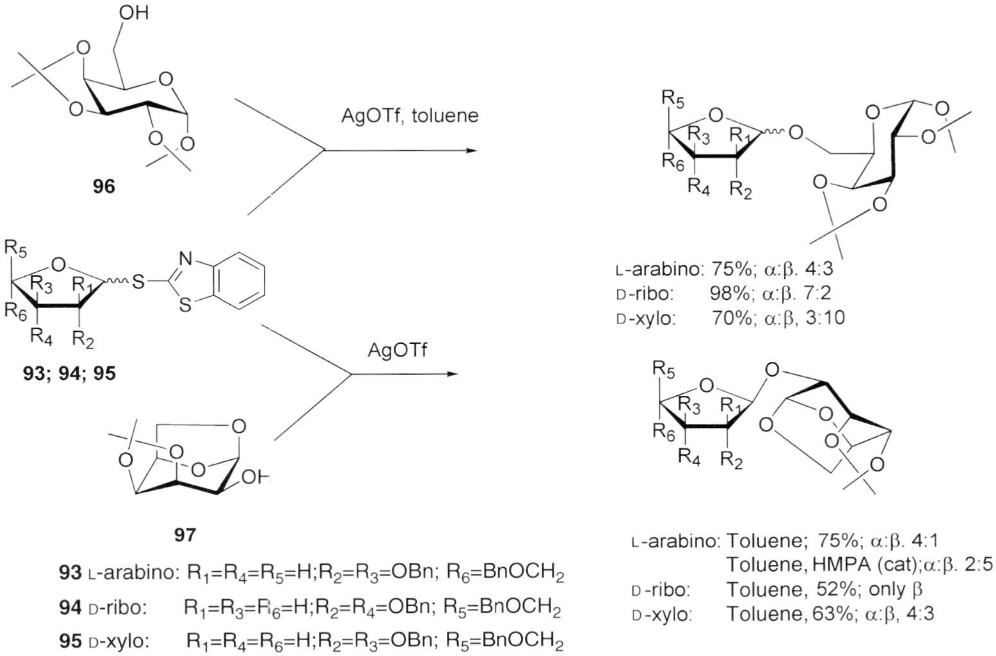

Scheme 4.91 Glycosidation of S-benzothiazolyl glycofuranosides [445].

S-glycofuranosyl dithiocarbamates with 5-nitro-2-pyridyl 2,3,4-tri-O-benzoyl-1-thio-β-D-glucopyranoside as an acceptor gives access to valuable 1,2-cis-linked furanosyl-1-thiopyranosides [498]. Recently, Bogusiak reported that perbenzylated S-benzothiazolyl glycosides of L-arabino, D-ribo and D-xylofuranose can be used as glycosyl donors (Scheme 4.91).

A satisfactory result was obtained in AgOTf- or NIS/TfOH-activated condensation of S-benzothiazolyl pentofuranosides **93**, **94** and **95** with 1,2:3,4-di-isopropylidene-α-D-galactopyranose **96** and 1,6-anhydro-3,4-isopropylidene-β-D-galactopyranose **97** as the acceptor to afford the corresponding disaccharides in moderate to good yields and stereoselectivity [445].

4.3.4
Glycosyl Thiocyanates as Glycosyl Donors

Kochetkov and coworkers have reported the use of 1,2-trans-glycosyl thiocyanates, having a nonparticipating group (methyl, benzyl) at C-2, for highly stereoselective 1,2-cis glycosylation [499–501]. The synthesis of 1,2-trans thiocyanates, accomplished by the reaction of α-D-glucopyranosyl, α-D-galactopyranosyl (**98**), α-D-mannopyranosyl and β-L-arabinopyranosyl bromides with KSCN in the presence of crown ether, was typically accompanied by the formation of the corresponding isothiocyanates (8–18%) [499]. Acetylated glycosyl thiocyanates kept at low temperature remain stable for at least 1 month.

The glycosylations were carried out in dry dichloromethane in the presence of 0.1 equiv of triphenylmethylium perchlorate. It was assumed that the triphenylmethylium cation attacks the nitrogen of the thiocyanate group whereas the oxygen of the trityl ether attacks the anomeric carbon in a 'push–pull' fashion. The trityl isothiocyanate formed was found to be unreactive under the reaction conditions. The disaccharide derivatives with (1–6), (1–4), (1–3) and (1–2) linkages have been obtained in good yields.

The method was also applied in the stereoselective synthesis of 1,2-*cis* glycosidic linkage of 2-aminosugars. The 1,2-*trans*-2-azido-2-deoxyglycosyl thiocyanate was used as the glycosyl donor with subsequent reduction of the azido group. The α-D-glucosaminyl-D-glucoses with (1–6), (1–3) and (1–2) glycosidic linkages were obtained [499].

4.3.5
Glycosyl Dithiophosphates as Glycosyl Donors

Glycosyl dithiophosphates are useful intermediates in the synthesis of various classes of sugar derivatives. The breaking of the C–S bond in the presence of suitable nucleophiles results in glycosides. The methods for the synthesis of glycosyl dithiophosphates are shown in Scheme 4.92. The phosphorothioate anion is a very active nucleophile, and the displacement of an anomeric leaving group is a convenient method for the synthesis of *S*-glycosyl phosphorothioates. Reaction of per-*O*-acetylglycosyl halides with *O,O*-dialkyl phosphorothioates led to a mixture of anomeric phosphate esters [502].

This method is very efficient when applied to hexopyranose series. Some difficulties have been experienced in attempted syntheses of glycosyl dithiophosphates of pentofuranoses. Michalska *et al.* [503,504] have found that the reaction of per-acetylated mono- and disaccharides with *O,O*-dialkylphosphorothioates, its trimethylsilil esters or trialkylammonium salts in the presence of boron trifluoride etherate allows the synthesis of *S*-glycosyl phosphorothioates with high yield. Stereochemical course of nucleophilic substitution was influenced by the anomeric configuration, the kind of substituent used at C-2 and the molar ratio of reagents. In the case of D-glucose pentaacetate, the β-anomer was significantly more reactive than the α-anomer. In the presence of a participating acetyl group, the 1,2-*trans* products were formed stereoselectivity.

The synthetic route proposed by Szeja and Bogusiak [505,506] involves a one-pot conversion of reducing sugar derivatives into 1-thiosugar by intermolecular nucleophilic substitution of the intermediate glycosyl 1-*O*-sulfonates. Generation of a good leaving group from the free anomeric hydroxyl was achieved by the treatment with tosyl chloride under phase-transfer conditions (aqueous sodium hydroxide/organic solvent and tetrabutylammonium bromide as a catalyst). The treatment of the glycal derivatives with an equimolar amount of *O,O*-dialkyl *S*-hydrogen phosphorothioates in benzene for 24 h quantitatively afforded an anomeric mixture of *S*-(2-deoxyglycosyl)phosphorothioates [507–509]. Glycosyl dithiophosphates can also be synthesized from glycals in a two-step procedure. Epoxidation of the glycals with

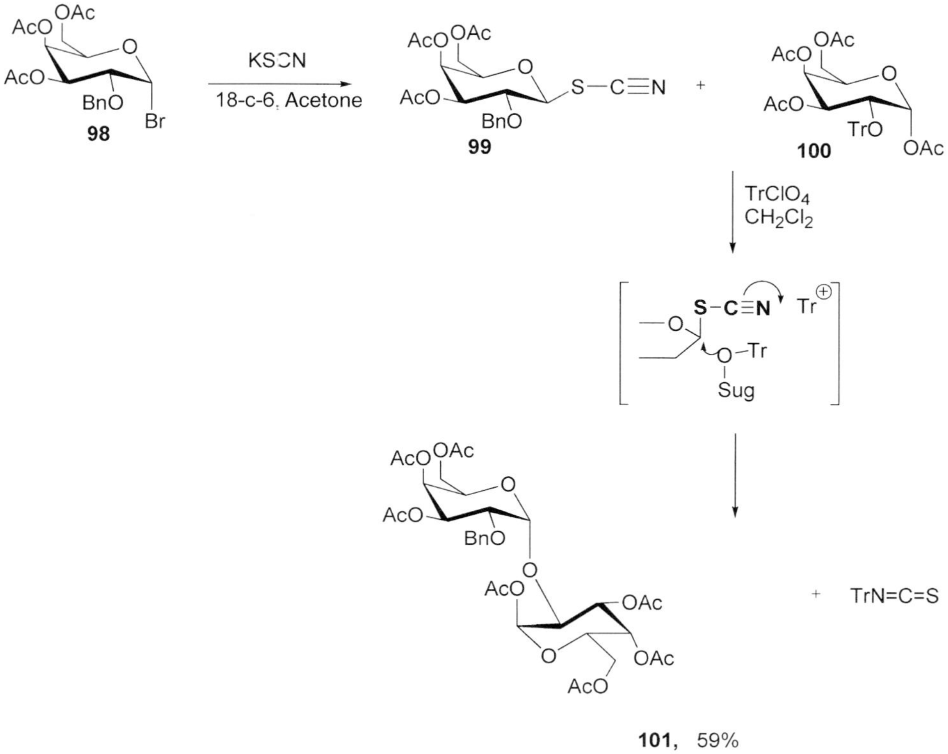

Scheme 4.92 Stereoselective glycosylation via glycosyl thiocynate [155].

dimethoxydioxirane (*DMDO*) afforded the 1,2-anhydro sugar that underwent oxirane ring opening with commercially available *O,O*-diethyldithiophosphate to afford glycosyl dithiophosphate in good yields (82–88%) [510] (Scheme 4.93) .

Bielawska and Michalska [511] have introduced *S*-(2-deoxyglycosyl)phosphorothioates as glycosyl donors. Activation of these compounds for glycosidation with silver salts (AgF, AgClO$_4$ or AgOTf) resulted in the formation of 2′-deoxydisaccharides as anomeric mixtures. Thiem and coworkers reported [507] an efficient activation method for *S*-(2-deoxyglycosyl)phosphorothioate donors **102** by using *N*-iodosuccinimide (*NIS*) or iodonium bis(2,4,6-trimethylpyridine) perchlorate (*IDCP*) (Scheme 4.94).

This improvement allowed achieving the synthesis of 2-deoxy disaccharide **103** with high stereoselectivity and good yield in short reaction times. The composition of the reaction mixture was found to be practically independent of the configuration of the *S*-(2-deoxy-D-glucopyranosyl)phosphorothioate. These results seem to provide an evidence that this glycosylation procedure proceeds via the S$_N$1 displacement reaction mechanism.

In the search for more efficient promoters for β-selective glycosylation, Seeberger and coworkers [510] explored coupling conditions used for thioglycoside donors.

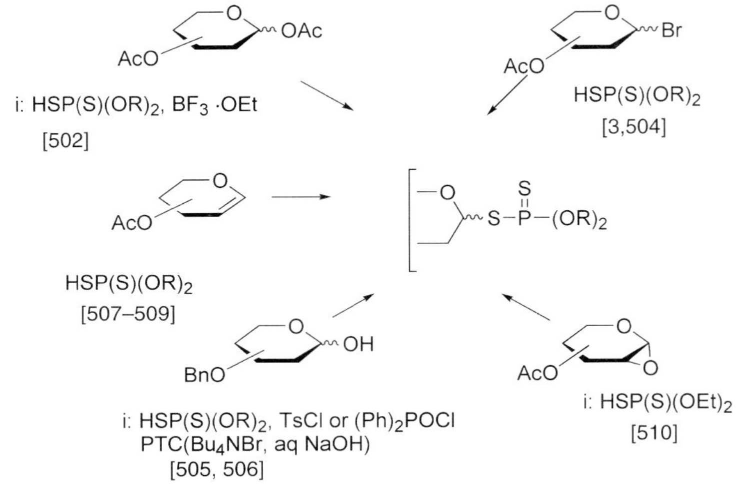

Scheme 4.93 Synthesis of glycosyl dithiophosphates.

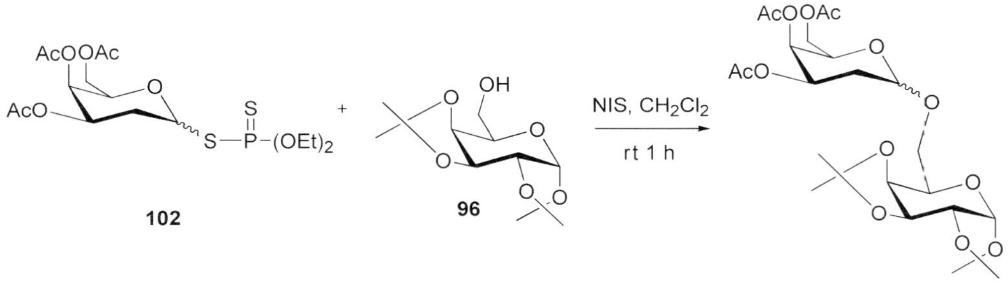

Scheme 4.94 Stereoselective synthesis of 2-deoxy-α-glycosides [507].

Although the coupling of glycosyl dithiophosphates using an excess methyl triflate as an activator in the presence of 2,6-di-tertbutylpyridine proceeded in a satisfactory yield (70%), the application of dimethylsulfonium triflate resulted in the formation of disaccharides in a considerably more effective way (94% yield).

4.3.6
Conclusions

Chemical entities discussed in this chapter as glycosyl donors share the principal structural feature: C(anomeric)−sulfur atom bond with thioglycosides, discussed earlier. However, the electron density on the sulfur atom is diminished, and consequently its chemical reactivity differs considerably, because of substitution with electron-withdrawing groups such as carboxylic or phosphoric acid residues. This marked difference in the reactivity of *m*-alkyl (or some aryl) glycosides can be taken as an advantage in the design of procedures for one-pot, armed–disarmed or orthogonal glycosylations. Moreover, apart from chemoselective activation, some of the *S*-glycosidic compounds described herein are also prone to selective deactivation through metal complexation (e.g. heteroaryl thioglycosides).

Xanthates, thioimidates and other anomeric thio derivatives discussed in this chapter constitute convenient glycosyl donors because they are easy to prepare, reasonably stable, as well as afford excellent yields and stereoselectivities of *O*-glycosides upon activation with a variety of mildly acidic or thiophilic activators, such as Lewis acids and heavy-metal salts (particularly triflates). These glycosyl donors, in combination with more traditional glycosylating reagents, allow for the fine tune-up of an anomeric leaving-group reactivity, which in turn makes possible to design the synthesis of challenging glycosidic linkages and complex oligosaccharides needed as molecular probes for biological studies and new chemical entities for drug discovery.

4.3.7
Typical Experimental Procedures

4.3.7.1 Preparation of Xanthates

O-Ethyl S-(2-Azido-3,4,6-Tri-O-Benzyl-2-Deoxy-β-D-Galactopyranosyl) Dithiocarbonate
A solution of 2-azido glycosyl nitrate 6 (2.5 mmol) and *O*-ethyl *S*-potassium dithiocarbonate 5 mmol) in acetonitrile (20 ml) was kept for 5 h at room temperature, diluted with dichloromethane (200 ml), washed with water (30 ml), dried (MgSO$_4$) and concentrated. Column chromatography of the residue (4:1 hexane–ethyl acetate, 0.3% triethylamine) gave the title compound in 97% yield.

O-Ethyl S-[Methyl (5-Acetamido-4,7,8,9-Tetra-O-Acetyl-3,5-Dideoxy-α-D-Glycero-D-Galacto-2-Nonulopyranosyl)onate] Dithiocarbonate (15) To a solution of sialyl chloride (1 mmol), tetrabutylammonium hydrogen sulfate (1 mmol) and *S*-potassium *O*-ethyldithiocarbonate in ethyl acetate (5 ml) was added 2 M sodium carbonate (5 ml). The two-phase mixture was vigorously stirred at room temperature until TLC showed disappearance of the halide. Ethyl acetate (50 ml) was then added, the organic phase was separated and successively washed with saturated sodium hydrogen carbonate, water and brine. The combined organic extracts were dried (sodium sulfate), filtered and concentrated to afford *S*-glycosyl dithiocarbonate. The oily residue obtained was purified by column chromatography over silica gel (toluene–ethyl acetate, 1:2) as eluant to provide pure product **15** as foam in 91% yield. Crystallization from benzene–pentane afforded **15** (72%).

4.3.7.2 Glycosidation of Xanthates (Scheme 4.95)

Methyl 6-O-(3-O-Acetyl-2-Azido-4,6-O-Benzylidene-2-Deoxy-α- and β-D-Galactopyranosyl)-2,3,4-Tri-O-Benzyl-α-D-Galactopyranoside (110) A solution of xanthate **108** (1.2 mmol), glycosyl acceptor **11** (372 mg, 0.8 mmol) and activated 4-Å powdered

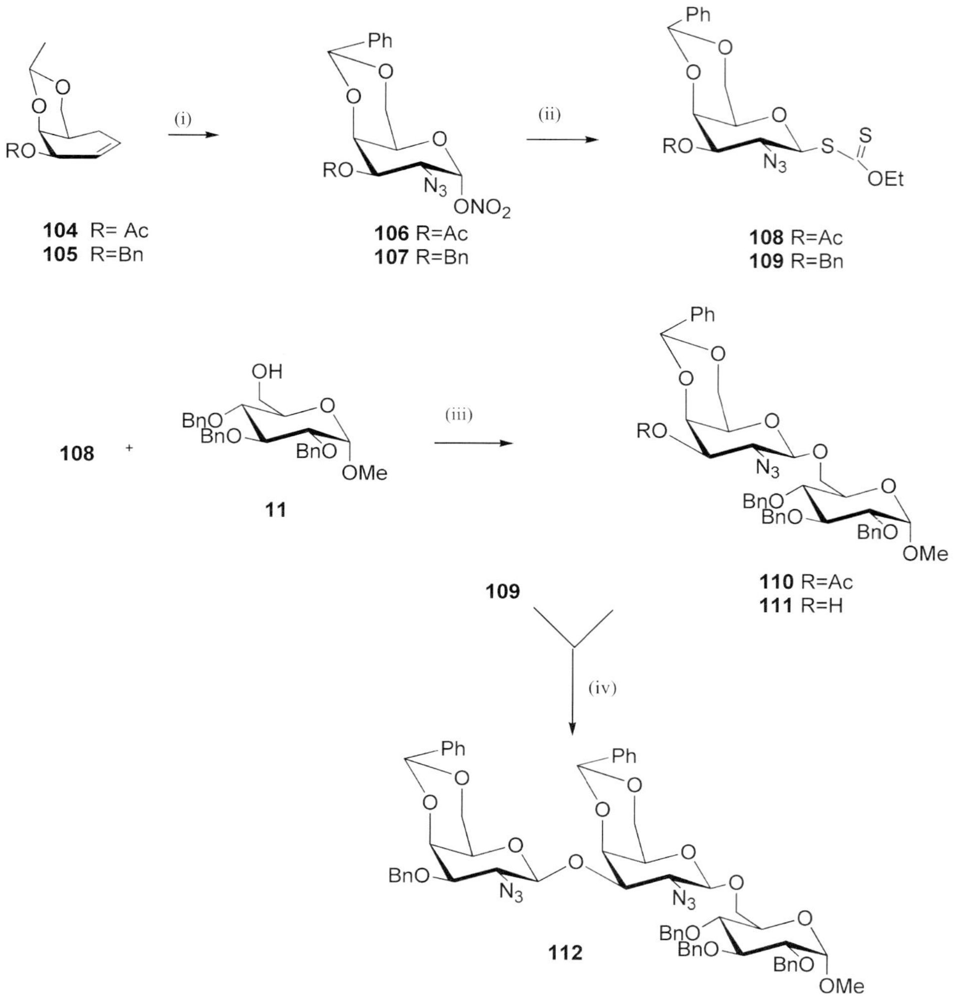

Scheme 4.95 Stereoselective synthesis of oligosaccharides via glycosyl xanthate [399].

molecular sieves (0.60 g) in anhydrous acetonitrile (10 ml) was stirred for 15 min at room temperature. Dimethyl(methylthio)sulfonium triflate was added and the stirring was continued for 1 h at room temperature. The suspension was treated with an excess of diisopropylamine, diluted with dichloromethane, filtered (Celite) and concentrated. Column chromatography of the residue (10 : 1 → 5 : 1 toluene–ethyl acetate) gave the title oligosaccharides **110** as separate anomers in 13 (α) and 73% (β) yield.

Methyl O-(2-Azido-3-O-Benzyl-4,6-O-Benzylidene-2-Deoxy-α-D-Galactopyranosyl)-(1–3)-O-(2-Azido-4,6-O-Benzylidene-2-Deoxy-β-D-Galactopyranosyl)-(1–6)-2,3,4-Tri-O-Benzyl-α-D-Glucopyranoside (112) A mixture of the disaccharide acceptor **111** (0.1 mmol),

4.3 Xanthates, Thioimidates and Other Thio Derivatives | 355

xanthate donor **109** (0.15 mmol), activated 4-A powdered molecular sieves (0.20 g) and anhydrous dichloromethane (2 ml) was stirred for 15 min at room temperature. Copper(II) triflate (0.43 g, 1.2 mmol) was then added and stirring was continued for 6 h at room temperature. The mixture was diluted with dichloromethane (200 ml), washed with water (30 ml), dried (MgSO$_4$) and concentrated. Column chromatography of the residue (2 : 1 toluene–ethyl acetate, containing 0.3% triethylamine) gave the title α-trisaccharide **112** in 80% yield.

2-(Trimethylsilyl)-Ethyl-[Methyl (5-Acetamido-4,7,8,9-Tetra-O-Acetyl-3,5-Dideoxy-α-D-Glycero-D-Galacto-2-Nonulopyranosyl)onate] (2–3)-(2,6-Di-O-benzyl-β-D-Galactopyranosyl)-(1–4)-2,3,6-Tri-O-Benzyl-β-D-Glucopyranoside (114) (Scheme 4.96)

A solution of dithiocarbonate **15** (1.44 mmol) and disaccharide acceptor **113** (0.96 mmol) in a mixture of dry acetonitrile (20 ml)/methylene chloride (10 ml) and powdered molecular sieves (3 g, 4A) was stirred under nitrogen for 1 h. AgOTf (1.58 mmol) and 2,6-di-*tert*-butylpyridine (1.70 mmol) were added and the mixture was cooled to −70 °C and kept protected from light. Benzenesulfenyl chloride (1.55 mmol) was then added, and stirring was continued for 2 h at −70 °C. The mixture was diluted with a suspension of silica gel (5 g) in ethyl acetate (30 ml), filtered (Celite), washed with saturated aqueous sodium hydrogen carbonate and water, dried (sodium sulfate) and concentrated. The residue was chromatographed (chloroform–acetone, 9 : 1, 4 : 1) to give trisaccharide **114** as foam (74%).

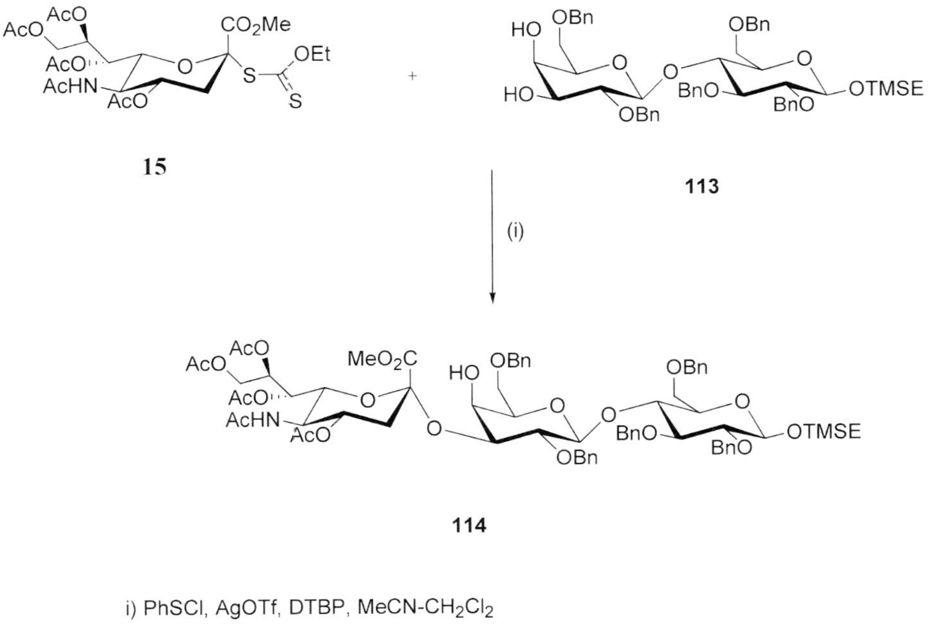

i) PhSCl, AgOTf, DTBP, MeCN-CH$_2$Cl$_2$

Scheme 4.96 Stereoselective sialylation with sialyl xanthate [410].

4.3.7.3 Preparation of Thioimidates

2-Thiazolinyl 2,3,4,6-Tetra-O-Benzoyl-1-thio-β-D-Glucopyranoside (3) Crown ether [15]-crown-5 and salt NaSTaz (6 mmol), prepared from NaOMe and 2-mercaptothiazoline in methanol, were added to a stirred solution of a tetra-O-benzoyl-α-D-glucopyranosyl bromide (3.0 mmol) in a dry acetonitrile (24 ml) under argon. The reaction mixture was stirred for 1 h at room temperature. Upon completion, the mixture was diluted with toluene (30 ml) and washed with 1% aq NaOH (15 ml) and water (3 × 10 ml), and the organic phase was separated, dried over MgSO$_4$ and concentrated *in vacuo*. The residue was purified by column chromatography on silica gel (ethyl acetate/hexane gradient elution) to afford the STaz glycoside (53%).

2-Benzoxazolinyl 2,3,4,6-Tetra-O-Benzoyl-1-thio-β-D-Glucopyranoside (39) 18-Crown-6 (0.6 mmol) and KSBox (3.45 mmol, prepared from HSBox and K$_2$CO$_3$) were added to a stirred solution of a glycosyl bromide (3.0 mmol) in dry acetone (4 ml) under an atmosphere of argon. The reaction mixture was stirred for 1 h at 55 °C. Upon completion, the mixture was diluted with CH$_2$Cl$_2$ (30 ml) and washed with 1% aq NaOH (15 ml) and water (3 × 10 ml). The organic phase was separated, dried over MgSO$_4$ and concentrated *in vacuo*. The residue was purified by column chromatography on silica gel (ethyl acetate–hexane gradient elution) to afford the title SBox glycoside (87%).

One-Pot Synthesis of Methyl O-(2,3,4,6-Tetra-O-Benzoyl-β-D-Glucopyranosyl)-(1–6)-O-(2,3,4-Tri-O-Benzoyl-β-D-Galactopyranosyl)-(1–6)-O-(2,3,4-Tri-O-Benzoyl-β-D-Glucopyranosyl)-(1–6)-2,3,4-Tri-O-Benzyl-β-D-Glucopyranoside (72) (Scheme 4.86) A mixture of SBox donor **39** (0.0274 mmol), S-ethyl acceptor **66** (0.0249 mmol) and freshly activated molecular sieves (3 A, 60 mg) in (ClCH$_2$)$_2$ (0.5 ml) was stirred under argon for 1 h. AgOTf (0.0603 mmol) was added. The reaction mixture was then stirred for 10 min at room temperature. Upon completion, the reaction mixture was cooled to −20 °C, and STaz acceptor **70** (0.0224 mmol), NIS (0.05 mmol) and TfOH (0.005 mmol) were added to it. The reaction mixture was stirred for 30 min. Upon completion, the reaction mixture was warmed up to room temperature, and then glycosyl acceptor **40** (0.0249 mmol) and AgOTf (0.05 mmol) were added. Upon completion (2 h), the solid was filtered and washed with CH$_2$Cl$_2$. The combined filtrate (30 ml) was washed with 20% aq Na$_2$S$_2$O$_3$ (15 ml) and water (3 × 10 ml). The organic phase was then separated, dried with MgSO$_4$ and concentrated *in vacuo*. The residue was purified by column chromatography on silica gel (acetone/toluene gradient elution) to allow the title tetrasaccharide **72** in 73% yield.

4.3.7.4 Synthesis of Glycosyl Thiocyanates (Scheme 4.92)

3,4,6-Tri-O-Acetyl-2-O-Benzyl-β-D-Galactopyranosyl Thiocyanate (99) 2-O-Benzyl-3,4,6-tri-O-acetyl-α-D-galactopyranosyl bromide (**98**) (4.0 mmol), potassium thiocyanate (12.0 mmol), dried *in vacuo* at 110 °C, and 18-crown-6 (0.4 mmol), dried *in vacuo* for 24 h at 20 °C, were dissolved in dry acetone (10 ml). The reaction was monitored by TLC and when complete, the reaction mixture was concentrated, and traces of acetone was removed by coevaporation with benzene, filtered and concentrated. Residue was subjected to column chromatography (elution with benzene–ether) to allow the title compound in 54% yield.

4.3.7.5 Glycosidation of Thiocyanates

1,3,4,6-Tetra-O-Acetyl-2-O-(3,4,6-Tri-O-Acetyl-2-O-Benzyl-α-D-Galactopyranosyl)-α-D-Galactopyranose (101) (Scheme 4.92) Thiocyanate **99** (0.20 mmol), tritylated acceptor derivative **100** (0.20 mmol) and triphenylmethylium perchlorate (0.02 mmol) were dissolved in dry methylene chloride (2.5 ml). The reaction was monitored by TLC and when completed, a few drops of pyridine were added to quench the reaction. Reaction mixture was diluted with chloroform (30 ml), washed with water and concentrated to dryness. The residue was then dissolved in dry pyridine (2 ml), Ac_2O (1 ml) was added and the mixture was stored overnight at room temperature. Thereafter, methanol was added, the mixture was coconcentrated several times with toluene and the residue was subjected to column chromatography. Elution with benzene–ether gave disaccharide **101** as syrup in 59% yield.

4.3.7.6 Synthesis of S-(2-Deoxyglycosyl) Phosphorodithioates (Scheme 4.93)

Treatment of a 1,2-dehydro derivative (1 mmol) with an equimolar amount of O,O-diethyl S-hydrogen phosphorodithioate (1 mmol) in benzene (3 ml) for 24 h quantitatively yielded a crude anomeric mixture of the title compounds after evaporation *in vacuo*.

4.3.7.7 Glycosidation of Glycosyl Phosphorodithioates

1,2:3,4-Di-O-Isopropylidene-6-O-(2-Deoxy-3,4,6-Tri-O-Acetyl-α,β-D-Galactopyranosyl)-α-D-Galactopyranose 103 (Scheme 4.94) Glycosyl donor **102** (1 mmol) and the acceptor **96** (1.2 mmol) in CH_2Cl_2 (2 ml) were allowed to react in the presence of 4A molecular sieves and NIS (1.1 mmol) at room temperature in a dry N_2 atmosphere. After 1 h the mixture was diluted with $CHCl_3$, washed with 10% $Na_2S_2O_3$ solution and water, dried ($MgSO_4$), filtered, concentrated *in vacuo* and chromatographed (silica gel, flash technique). This afforded the disaccharide **103** in 80% yield as anomeric mixture.

References

395 (a) Duus, F. (1979) Thiocarbonyl compounds, in *Comprehensive Organic Chemistry* vol. 3, (eds D. Barton and W.D. Ollis), Pergamon, pp. 373–487. (b) Crich, D. and Quintero, L. (1989) *Chemical Reviews*, **89**, 1413–1432. (c) Walter, W. and Voss, J. (1970) The chemistry of thioamides in *The Chemistry of Amides* (ed. J. Zabicky), Wiley Interscience, London, pp. 383–475. (d) Griffin, T.S., Woods, T.S. and Klayman, D.J. (1975) *Advances in Heterocyclic Chemistry*, **18**, 99–158.

(e) Witczak, Z.J. (1986) *Advances in Carbohydrate Chemistry and Biochemistry*, **44**, 91–145. (f) Manuel, J., Fernandez, G. and Mellet, C.O. (2000) *Advances in Carbohydrate Chemistry and Biochemistry*, **55**, 35–135.

396 Horton, D. (1963) *Methods in Carbohydrate Chemistry*, **2**, 433–437.

397 Marra, A. and Sinay, P. (1989) *Carbohydrate Research*, **187**, 35–42.

398 Marra, A. and Sinay, P. (1990) *Carbohydrate Research*, **195**, 303–308.

399 Marra, A., Gauffeny, F. and Sinay, P. (1991) *Tetrahedron*, **47**, 5149–5160.

400 Trapper, F.D., Anderson, F.O. and Roy, R. (1992) *Journal of Carbohydrate Chemistry*, **11**, 741–750.

401 Wandzik, I. and Szeja, W. (1998) *Polish Journal of Chemistry*, **72**, 703–709.

402 Bogusiak, J., Wandzik, I. and Szeja, W. (1996) *Carbohydrate Letters*, **1**, 411–416.

403 Guerard, D., Tatibouet, A., Gareau, Y. and Rollin, P. (1999) *Organic Letters*, **1**, 521–522.

404 Pougny, J.-R. (1986) *Journal of Carbohydrate Chemistry*, **5**, 529–535.

405 Nifantev, N.E., Tsvetkov, Y.E., Shashkov, A.S., Kononov, L.O., Menshov, V.M., Tuzikov, A.B. and Bovin, N. (1996) *Journal of Carbohydrate Chemistry*, **15**, 939–953.

406 Greilich, U., Brescello, R., Jung, K.H. and Schmidt, R.R. (1996) *Liebigs Annalen der Chemie*, 663–672.

407 Bilberg, W. and Lonn, H. (1991) *Tetrahedron Letters*, **32**, 7453–7456.

408 Bilberg, W. and Lonn, H. (1991) *Tetrahedron Letters*, **32**, 7457–7458.

409 Lonn, H. and Stenvall, K. (1992) *Tetrahedron Letters*, **33**, 115–116.

410 Martichonok, V. and Whitesides, G.M. (1996) *The Journal of Organic Chemistry*, **61**, 1702–1706.

411 Boons, G.-J. and Demchenko, A.V. (2000) *Chemical Reviews*, **100**, 4539–4565, and references therein.

412 Halkes, K.M., Hilaire, P.H.S., Jansson, A.H., Gotfreidsen, C.H. and Meldal, M. (2000) *Journal of the Chemical Society Perkin Transactions*, 2127–2133.

413 Sabesan, S., Neira, S., Davidson, F., Duus, J.O. and Bock, K. (1994) *Journal of the American Chemical Society*, **116**, 1616–1634.

414 Ray, A.K., Nilsson, U. and Magnusson, G. (1992) *Journal of the American Chemical Society*, **114**, 2256–2257.

415 Elofsson, M. and Kihlberg, J. (1995) *Tetrahedron Letters*, **36**, 7499–7502.

416 Ercegovic, T. and Magnusson, G. (1995) *The Journal of Organic Chemistry*, **60**, 3378–3384.

417 Wilstermann, M., Kononov, L.O., Nilsson, U., Ray, A.K. and Magnusson, G. (1995) *Journal of the American Chemical Society*, **117**, 4742–4754.

418 Elofsson, M., Salvador, L.A. and Kihlberg, J. (1997) *Tetrahedron*, **53**, 369–390.

419 Liebe, B. and Kunz, H. (1997) *Angewandte Chemie (International Edition in English)*, **36**, 618–621.

420 van Seeventer, P.B., Kerekgyarto, J., van Dorst, J.A.L.M., Halkes, K.M., Kamerling, J.P. and Vliegenthart, J.F.G. (1997) *Carbohydrate Research*, **300**, 127–138.

421 Wilstermann, M. and Magnusson, G. (1997) *The Journal of Organic Chemistry*, **62**, 7961–7971.

422 Ellervick, U. and Magnusson, G. (1998) *The Journal of Organic Chemistry*, **63**, 9314–9322.

423 Ellervick, U., Grundberg, H. and Magnusson, G. (1998) *The Journal of Organic Chemistry*, **63**, 9323–9338.

424 Salvatore, B.A. and Prestegard, J.H. (1998) *Tetrahedron Letters*, **39**, 9319–9322.

425 Tietze, L.F., Janssen, C.O. and Gewert, J.A. (1998) *European Journal of Organic Chemistry*, 1887–1894.

426 Tietze, L.F. and Gretzke, D. (1998) *European Journal of Organic Chemistry*, 1895–1899.

427 Liebe, B. and Kunz, H. (1994) *Tetrahedron Letters*, **35**, 8777–8778.

428 Pornsuriyasak, P., Kamat, M.N. and Demchenko, A.V. (2007) *ACS Series*, **960**, 165–189.

429 ElAshry, H.S.H., Awad, L.F. and Atta, A.I. (2006) *Tetrahedron*, **62**, 2943–2998.
430 Tejima, S. and Ishiguro, S. (1967) *Chemical & Pharmaceutical Bulletin*, **15**, 255–263.
431 Lee, B.-H., Bertram, B., Schmezer, P., Frank, N. and Wiessler, M (1994) *Journal of Medicinal Chemistry*, **123**, 3154–3162.
432 Lee, B.-H., Bertram, B., Schmezer, P., Frank, N. and Wiessler, M. (1993) *Royal Society of Chemistry*, **123**, 53–57.
433 Zinner, H. and Pfeifer, M. (1963) *Chemische Berichte*, **96**, 432–437.
434 Zinner, H. and Peseke, K. (1965) *Chemische Berichte*, **98**, 3515–3519.
435 Demchenko, A.V., Kamat, M.N. and De Meo, C. (2003) *Synlett*, 1237–1290.
436 Demchenko, A.V., Malysheva, N.N. and and De Meo, C. (2003) *Organic Letters*, **5**, 455–458.
437 Demchenko, A.V., Pornsuriyasak, P., De Meo, C. and Malysheva, N.N. (2004) *Angewandte Chemie (International Edition)*, **43**, 3069–3072.
438 Mukaiyama, T., Nakatsuka. T. and Shoda, S.I. (1979) *Chemistry Letters*, 487–490.
439 Bogusiak, J. and Szeja, W. (1988) *Chemistry Letters*, 1975–1976.
440 Mereyala, H.B., Kulkarni, V.R., Ravi, D., Sharma, G.V.M., Rao, B.V. and Reddy, G.B. (1992) *Tetrahedron*, **48**, 545–562.
441 Stewart, A.O. and Williams, R.M. (1985) *Journal of the American Chemical Society*, **107**, 4289–4296.
442 Abdel-Rahman, A.A.H., El Ashry, E.S.H. and Schmidt, R.R. (2002) *Carbohydrate Research* **337**, 195–206.
443 Khodair, A.I., Al-Masoudi, N.A. and Gesson, J.P. (2003) *Nucleosides Nucleotides & Nucleic Acids*, **22**, 2061–2076.
444 Euzen, R., Ferrieres, V. and Plusquellec, D. (2005) *The Journal of Organic Chemistry*, **70**, 847–855.
445 Bogusiak, J. (2005) *Letters in Organic Chemistry*, **2**, 271–273.
446 Fourrey, J.-L. and Jouin, P. (1979) *The Journal of Organic Chemistry*, **44**, 1892–1894.
447 Ernst B.,Hart G.W. andSinay P. (eds) (2000) *Carbohydrates in Chemistry and Biology*, vol. **1**,Wiley-VCH Verlag GmbH, Weinheim, Germany. Chapters 2–9.
448 Jung, K.-H. and Schmidt, R.R. (2000) *Chemical Reviews*, **100**, 4423–4442.
449 Davis, B.G. (2000) *Journal of the Chemical Society, Perkin Transactions*, **1**, 2137–2160.
450 Boons, G.-J. (1996) *Tetrahedron*, **52**, 1095–1121.
451 Schmidt, R.R. and Kinzy, W. (1994) *Advances in Carbohydrate Chemistry and Biochemistry*, **50**, 21–123.
452 Danishefsky, S.J. and Bilodeau, M.T. (1996) *Angewandte Chemie (International Edition in English)*, **35**, 1380–1419.
453 Toshima, K. and Tatsuta, K. (1993) *Chemical Reviews*, **93**, 1503–1531.
454 Douglas, N.L., Ley, S.V., Lucking, U. and Warriner, S.L. (1998) *Journal of the Chemical Society Perkin Transactions*, **1**, 51–65.
455 Zhang, Z., Ollmann, I.R., Ye, X.-S., Wischnat, R., Baasov, T. and Wong, C.-H. (1999) *Journal of the American Chemical Society*, **121**, 734–753.
456 Zhu, T. and Boons, G.-J. (2001) *Organic Letters*, **3**, 4201–4203.
457 Mootoo, D.R., Konradsson, P., Udodong, U. and Fraser-Reid, B. (1988) *Journal of the American Chemical Society*, **110**, 5583–5584.
458 Fraser-Reid, B., Wu, Z., Udodong, U.E. and Ottosson, H. (1990) *The Journal of Organic Chemistry*, **55**, 6068–6070.
459 Baeschlin, D.K., Green, L.G., Hahn, M.G., Hinzen, B., Ince, S.J. and Ley, S.V. (2000) *Tetrahedron: Asymmetry*, **11**, 173–197.
460 Veeneman, G.H. and van Boom, J.H. (1990) *Tetrahedron Letters*, **31**, 275–278.
461 Friesen, R.W. and Danishefsky, S.J. (1989) *Journal of the American Chemical Society*, **111**, 6656–6660.
462 Nguyen, H.M., Poole, J.L. and Gin, D.Y. (2001) *Angewandte Chemie (International Edition)*, **40**, 414–417.
463 Yamago, S., Yamada, T., Hara, O., Ito, H., Mino, Y. and Yoshida, J.-I. (2001) *Organic Letters*, **3**, 3867–3870.

464 Burkhart, F., Zhang, Z., Wacowich-Sgarbi, S. and Wong, C.-H. (2001) *Angewandte Chemie (International Edition)*, **40**, 1274–1277.

465 Mong, K.-K.T. and Wong, C.-H. (2002) *Angewandte Chemie (International Edition)*, **41**, 4087–4090.

466 Mong, K.-K.T., Lee, H.-K., Duron, S.G. and Wong, C.-H. (2003) *Proceedings of the National Academy of Sciences of the United States of America*, **100**, 797–802.

467 Fugedi, P., Garegg, P.J., Oscarson, S., Rosen, G. and Silvanis, B.A. (1991) *Carbohydrate Research*, **211**, 157–162.

468 Pornsuriyasak, P. and Demchenko, A.V. (2005) *Tetrahedron: Asymmetry*, **16**, 433–439.

469 Bongat, A.F.G., Kamat, M.N. and Demchenko, A.V. (2007) *The Journal of Organic Chemistry*, **72**, 1480–1483.

470 Jung, K.-H., Muller, M. and Schmidt, R.R. (2000) *Chemical Reviews*, **100**, 4423–4442.

471 Weingart, R. and Schmidt, R.R. (2000) *Tetrahedron Letters*, **41**, 8753–8758.

472 Khan S.H. and O'Neill R.A. (eds) (1996) *Modern Methods in Carbohydrate Chemistry*, Harwood Academic, Amsterdam.

473 Pozsgay, V. (2000) *Carbohydrates in Chemistry and Biology*, vol **1**, (eds B. Ernst, G.W. Hart and P. Sinay), Wiley-VCH Verlag GmbH, Weinheim, pp. 319–343.

474 Crich, D. (2002) *Journal of Carbohydrate Chemistry*, **21**, 667–690.

475 El Ashry, E.S.H., Rashed, N. and Ibrahim, E.S.I. (2005) *Current Organic Synthesis*, **2**, 175–213.

476 Huchel, U. and Schmidt, R.R. (1998) *Tetrahedron Letters*, **39**, 7693–7694.

477 Muller, M., Huchel, U., Geyer, A. and Schmidt, R.R. (1999) *The Journal of Organic Chemistry*, **64**, 6190–6201.

478 Muller, M. and Schmidt, R.R. (2001) *European Journal of Organic Chemistry*, 2055–2066.

479 Hanessian, S. and Lou, B. (2000) *Chemical Reviews*, **100**, 4443–4463.

480 Hanessian, S., Ugolini, A., Dube, D., Hodges, P.J. and Andre, C. (1986) *Journal of the American Chemical Society*, **108**, 2776–2778.

481 Woodward, R.B., Logusch, E., Nambiar, K.P., Sakan, K., Ward, D.E., Au-Yeung, B.W., Balaram, P., Browne, L.J., Card, P.J. and Chen, C.H. (1981) *Journal of the American Chemical Society*, **103**, 3215–3217.

482 Tatsuta, K., Kobayashi, Y. and Gunji, H. (1988) *The Journal of Antibiotics*, **41**, 1520–1523.

483 Tatsuta, K., Kobayashi, Y., Gunji, H., Masuda, H. and Kohoku, H. (1988) *Tetrahedron Letters*, **29**, 3975–3978.

484 Toshima, K., Mukaiyama, S., Yoshida, T., Tamai, T. and Tatsuta, K. (1991) *Tetrahedron Letters*, **32**, 6155–6158.

485 Hergenrother, P.J., Hodgson, A., Judd, A.S., Lee, W.-C. and Martin, S.F. (2003) *Angewandte Chemie (International Edition)*, **42**, 3278–3281.

486 Ravi, D., Kulkarni, V.R. and Mereyala, H.B. (1989) *Tetrahedron Letters*, **30**, 4287–4290.

487 Mereyala, H.B. and Ravi, D. (1991) *Tetrahedron Letters*, **32**, 7317–7320.

488 Mereyala, H.B., Hotha, S. and Gurjar, M.K. (1998) *Chemical Communications*, 685–686.

489 Mereyala, H.B. and Gurijala, V.R. (1993) *Carbohydrate Research*, **242**, 277–280.

490 Mereyala, H.B., Gurrala, S.R. and Gadikota, R.R. (1997) *Journal of the Chemical Society Perkin Transactions*, **1**, 3179–3182.

491 Chen, Q. and Kong, F. (1995) *Carbohydrate Research*, **272**, 149–157.

492 Ning, J., Xing, Y. and Kong, F. (2003) *Carbohydrate Research*, **338**, 55–60.

493 Pastuch, G., and Wandzik, I. and Szeja, W. (2000) *Tetrahedron Letters*, **41**, 9923–9926.

494 Niemiec-Cyganek, A. and Szeja, W. (2003) *Polish Journal of Chemistry*, **77**, 969–973.

495 Bogusiak, J. and Szeja, W. (1996) *Carbohydrate Research*, **295**, 235–243.

496 Bogusiak, J. and Szeja, W. (1997) *Synlett*, 661–662.

497 Bogusiak, J. and Szeja, W. (2001) *Carbohydrate Research*, **330**, 141–144.
498 Bogusiak, J. and Szeja, W. (2001) *Tetrahedron Letters*, **42**, 141–144.
499 Kochetkov, N.K., Klimov, E.M., Malysheva, N.N. and Demchenko, A.V. (1993) *Carbohydrate Research* **242** C7–C10.
500 Kochetkov, N.K., Klimov, E.M., Malysheva, N.N. and Demchenko, A.V. (1989) *Tetrahedron Letters*, **30**, 5499–5462.
501 Kochetkov, N.K., Klimov, E.M., Malysheva, N.N. and Demchenko, A.V. (1990) *Bioorganicheskaia Khimiia*, **16**, 701–710.
502 Michalska, M., Orlich-Keżel, I. and Michalski, J. (1978) *Tetrahedron*, **34**, 617–622.
503 Kudelska, W. and Michalska, M. (1994) *Tetrahedron Letters*, **40**, 7459–7462.
504 Kudelska, W. and Michalska, M. (1995) *Synthesis*, 1539–1544.
505 Szeja, W. and Bogusiak, J. (1987) *Carbohydrate Research*, **170**, 235–239.
506 Szeja, W. and Bogusiak, J. (1988) *Synthesis*, 224–226.
507 Laupichler, L., Sajus, H. and Thiem, J. (1992) *Synthesis*, 1133–1136.
508 Borowiecka, J. and Michalska, M. (1979) *Carbohydrate Research*, **86**, C8–C10.
509 Borowiecka, J., Lipka, P. and Michalska, M. (1988) *Tetrahedron*, **44**, 2067–2076.
510 Plante, O.J., Palmacci, E.R., Andrade, R.B. and Seeberger, P.H. (2001) *Journal of the American Chemical Society*, **123**, 9545–9554.
511 Bielawska, H. and Michalska, M. (1991) *Journal of Carbohydrate Chemistry*, **10**, 107–112.

4.4
Selenoglycosides

Robert A. Field

4.4.1
Background

Interest in the use of selenoglycosides as donors for *O*-glycosylation mainly emanates from Pinto's work in the early 1990s, which introduced this versatile class of compound to the armory of donor molecules for glycosylation chemistry [222,223,512]. Earlier studies on selenium-containing sugars have been comprehensively reviewed by Witczak and Czernecki [513]. As with the difference in chemistry and reactivity of oxygen versus sulfur, migration down the periodic table to selenium, and indeed beyond to tellurium, results in increased reactivity per se and a greater propensity for homolysis of the anomeric carbon–chalcogen bond. The latter provides ready access to anomeric radical chemistry *en route* to *C*-glycosides [513,514]. The former has attracted attention in relation to expanding the repertoire of tunable sugar donor functionalities (protecting groups; nature and substitution of the anomeric leaving group) in relation to *O*-glycosylation. Specifically, O versus S versus Se reactivity differences have received attention for the exploitation of armed–disarmed [515,474] approaches to one-pot, multiple-glycosylation reaction sequences. This chapter outlines common methods for the preparation of selenoglycosides and their activation for *O*-glycoside synthesis using either the conventional or single-electron transfer (SET) process. The application and integration of selenoglycosides in multistep oligosaccharide syntheses is also highlighted.

4.4.2
Selenoglycoside Preparation

Selenoglycosides are typically prepared by treating a glycosyl acetate with phenylselenol, derived from the hypophosphorus acid reduction of PhSeSePh in the presence of $BF_3 \cdot OEt_2$ [222] or by treating glycosyl halides with $NaBH_4$-reduced PhSeSePh (Scheme 4.97) [516]. The latter chemistry is also effective for the preparation of the related telluroglycosides [517]. Selenoglycosides can also be prepared from orthoesters by treatment with phenylselenol in the presence of $HgBr_2$ [147,518] or $SnCl_4$ (Scheme 4.97) [518].

Selenoglycosides of sialic acid have been successfully prepared in excellent yields from the corresponding peracetylated glycosyl chloride with phenylselenol in the presence of N,N-di-isopropylethylamine [519]. This reagent combination succeeded where others were less effective or failed (Scheme 4.98; [520].

The exposure of selenostannane to glycosyl acetates in the presence of catalytic $Bu_2Sn(OTf)_2$ also provides selenoglycosides in good yield and with reasonable stereocontrol (Scheme 4.99) [520]. Thioglycosidation can be achieved in a similar manner by the use of thiostannane; this reaction is also reported to be effective for the conversion of methyl glycosides into thioglycosides. A recent alternative to this approach employs indium(I)-iodide-mediated cleavage of diselenides for reaction with glycosyl bromides. This convenient and odorless methodology gives selenoglyco-

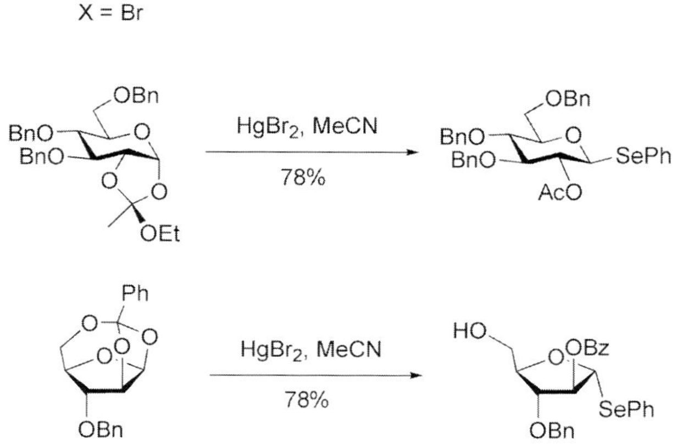

Scheme 4.97 Selenoglycoside synthesis.

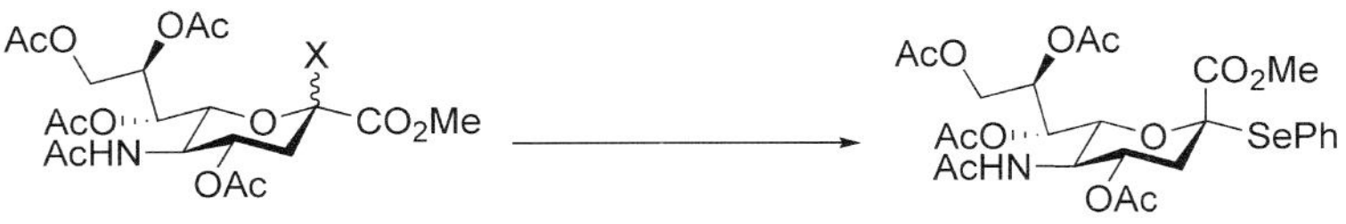

Scheme 4.98 α-Selenosialoside synthesis.

X	Conditions	Yield
OAc	PhSeH, BF$_3$·OEt$_2$	0 %
Cl	PhSeSePh, NaH	20 %
Cl	PhSeH, BnEt$_3$N$^+$Cl$^-$	60 %
Cl	PhSeH, iPr$_2$NEt	97 %

sides in excellent yield. The method is successful across a range of sugars, including sialic acid (Scheme 4.99) [521].

The stereoselective synthesis of α-selenoglycosides can also be achieved through recently described reactions of selenocarboxylates [522]. Reaction of β-glycosyl chlorides with potassium p-methylselenobenzoate gives the selenoglycosyl p-methylbenzoate, which upon reaction with an amine nucleophile gives the α-anomeric selenolate anion. Subsequent *in situ* reaction with various electrophiles in the presence of Cs$_2$CO$_3$ gives α-selenoglycosides, including alkyl and aryl selenoglycosides, selenoglycosyl amino acid and selenodisaccharides (Scheme 4.100) [523].

For the synthesis of glycans containing 2-amino-2-deoxysugars, the regioselective azidophenylselenation of glycals is popular [524–528]. This methodology follows on from the classic azidonitration work of Lemieux and Ratcliffe, first reported in the

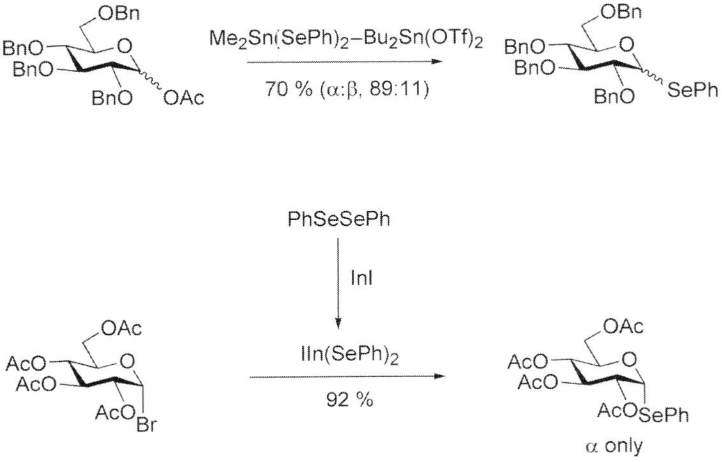

Scheme 4.99 Tin and indium in selenoglycosides synthesis.

Scheme 4.100 Selenoglycosides from selenobenzoates.

late 1970s [529]. Radical-mediated, anti-Markovnikov azidophenylselenylation of variously protected glycals affords phenyl 2-azido-2-deoxy-α-selenoglycopyranosides (Scheme 4.101).

A mixture of the *manno* and *gluco* isomers was obtained from D-glucal, albeit in very high yield (>80%). In contrast, azidophenylselenylation of D-galactal gave complete stereocontrol, affording exclusively the α-*galacto* isomer in 70–80% yield. More recent work by Nifantiev and colleagues [530] reports an improved preparative method for homogeneous azidophenylselenylation of glycals, consisting of the reaction with TMSN$_3$ and (PhSe)$_2$ in the presence of PhI(OAc)$_2$. The use of TMSN$_3$ instead of NaN$_3$, as in the earlier heterogeneous procedure [524–528], allowed both a reduced reaction time and scale-up that was not previously achievable on a reliable basis.

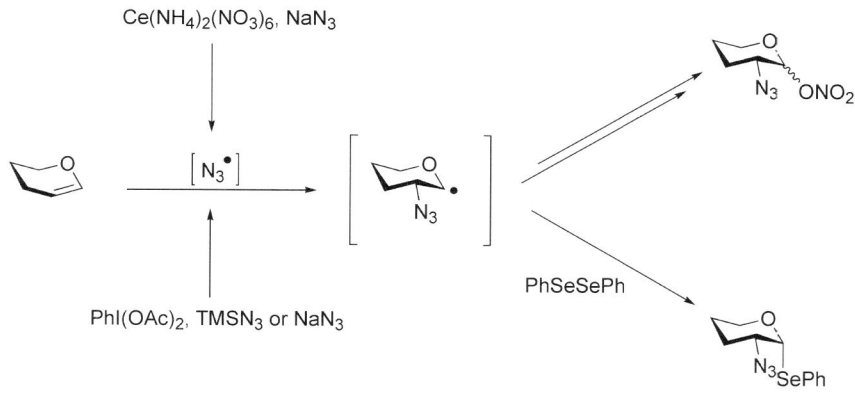

Scheme 4.101 Azidonitration and azidophenylselenation.

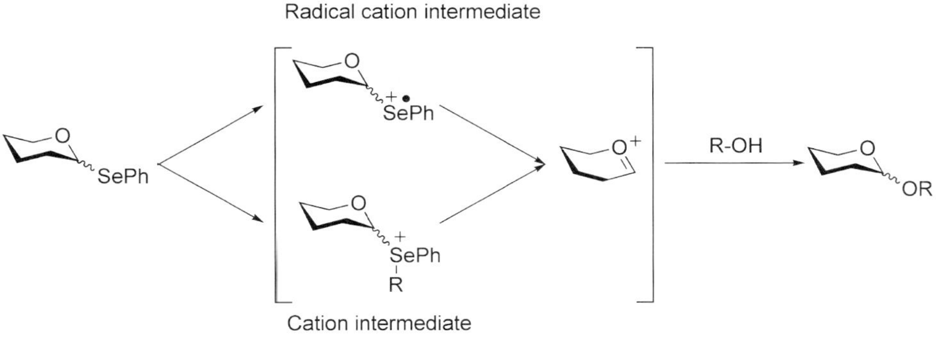

Scheme 4.102 Cation and radical-cation pathways for selenoglycoside glycosylation.

4.4.3
Selenides as Donors

Selenoglycosides are attractive donor species owing to the wide range of reagents that can promote cation- and radical-cation-based processes for their activation and subsequent O-glycosylation (Scheme 4.102) [223,512].

In addition to their direct activation for glycosylation, selenoglycosides are versatile building blocks that can easily be converted into other glycosyl donor species (Scheme 4.103). For instance, hydrolysis to hemiacetal [223] provides access to glycosyl N-phenyltrifluoroacetimidates [531]. Treatment with molecular iodine gives rise to the thermodynamically favored glycosyl iodides [532], whereas treatment with bromine kinetically favored β-bromides [533]. Reaction of 2-hydroxy phenyl selenoglycosides with diethylaminosulfur trifluoride (DAST) leads to the corresponding 2-phenylselenoglycosyl fluorides, which react stereoselectively with carbohydrate acceptors to afford 2-phenylselenoglycosides *en route* to 2-deoxy glycosides (by reduction) or orthoesters (by oxidative elimination and rearrangement) [534].

4.4.3.1 Promoters for Selenoglycoside Activation

Cation-Based Activation Pinto's works not only described AgOTf/K_2CO_3 as the promoter of choice for the activation of phenyl selenoglycosides but also indicated

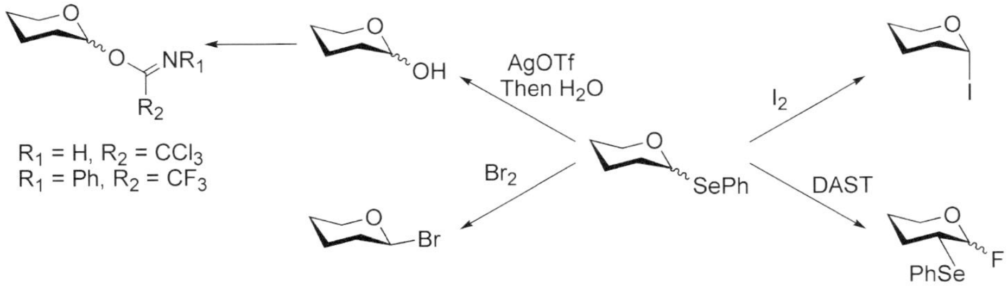

Scheme 4.103 Transformation of selenoglycosides to other potential donor species.

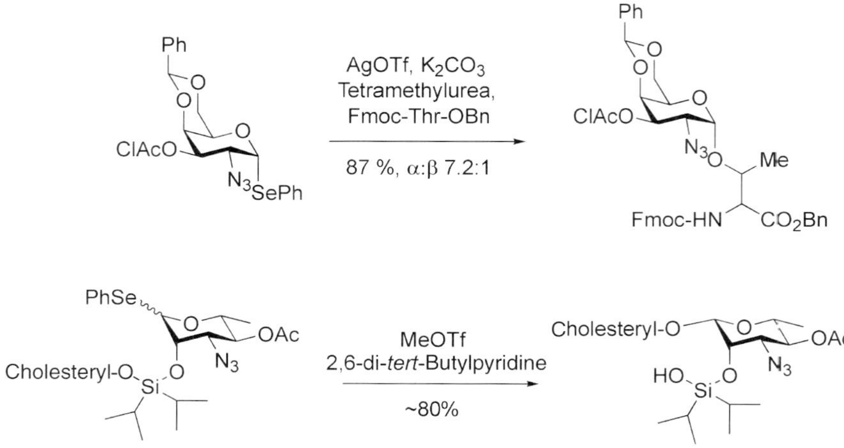

Scheme 4.104 AgOTf/K$_2$CO$_3$ and MeOTf promoters for selenoglycoside glycosylation.

that donors of this type can be activated by other promoters such as MeOTf, PhSeOTf or CuBr$_2$–Bu$_4$NBr–AgOTf [223,512]. It is also to be noted that AgOTf in the presence of an organic base, such as collidine or 1,1,3,3,-tetramethylurea, is *not* capable of activating phenyl selenoglycosides. This offers an additional opportunity for glycosylation of acceptors based on partially protected phenyl selenoglycosides (see Sections 4.4.4 and 4.4.5).

Although AgOTf/K$_2$CO$_3$ had proved to be a suitable promoter in connection with the synthesis of the *O*-glycosylated amino acid building blocks related to TF antigen

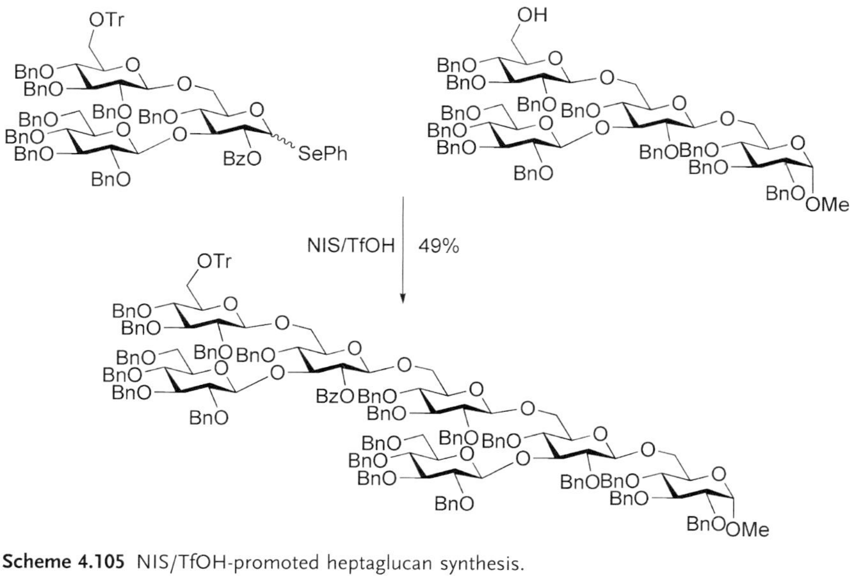

Scheme 4.105 NIS/TfOH-promoted heptaglucan synthesis.

[535], it proved ineffective in selenoglycoside-based glycosylation using an intramolecular aglycone delivery approach (Scheme 4.104). However, MeOTf was successfully employed in this latter study [386], which aimed to develop generic methods for the stereoselective coupling of D-mycosamine, a common sugar often found attached to macrolide antibiotics.

As noted by Pinto, electrophilic selenylating agents, and in particular PhSeOTf, are useful promoters for the activation of the phenylseleno group in glycosylation reactions, including those of 2-azido-2-deoxysugar donors [536]. However, NO·BF$_4$ proved only moderately effective when used in conjunction with the phenyl selenoglycoside of a 5-deoxy-5-thioglucose-based donor. An extensive transfer of acetate from the donor to the acceptor alcohol was noted, in addition to modest yields of disaccharides [537].

Extending the application of glycosylation chemistry initially developed for thioglycosides [541,538,539], the van Boom group have shown the potential of iodonium-ion-mediated glycosidations of phenyl selenoglycosides in the chemoselective synthesis of 1,2-*cis*- or 1,2-*trans*-linked disaccharides. Specifically, fully benzylated or benzoylated phenyl selenoglycosides can be activated by the promoters NIS/TfOH and IDCP [540,541]. The former reagent has been employed in several selenoglycoside-based syntheses, including the bis-*O*-glycosylation of 4,4′-dihydroxybiphenyl that was achieved in moderate yield [542]. NIS/TfOH has also been used in the sialylation of phenolic and sugar alcohols, in moderate yields; dimethyl methylthiosulfonium triflate (DMTST) gave similar modest results with phenols [519]. The NIS/TfOH promoter combination is also used for the preparation of a tetrasaccharide fragment of the cell wall from a *Proteus vulgaris* strain [543] and the blockwise coupling of a trisaccharide phenyl selenoglucosyl donor with a tetrasaccharide to furnish a heptaglucan phytoalexin elicitor (Scheme 4.105; see also Section 4.4.5) [544].

Molecular iodine has also been investigated as a promoter iodonium ion equivalent. It has been shown to be an effective general promoter for the activation of armed glycosyl donors based on thio-orthoesters, glycosyl-sulfoxides, -selenides and -phosphites, trichloroacetimidates and pentenyl glycosides [545]. *N*-iodosuccinimide (NIS) alone is also capable of activating armed thioglycosides and selenoglycosides, but glycosylation yields are somewhat erratic [546]. The stereochemical outcome of glycosylation reactions with model thioglycosides and selenoglycosides has been shown to be dependent on the solvent and the source of promoter iodonium ion, with iodine giving different results to NIS alone and to NIS/TMSOTf (Scheme 4.106).

Further investigation of donor activation with iodine in the absence of alcohol nucleophile showed that, in contrast to armed thioglycosides that anomerize and disarmed thioglycosides that do not react [547], both armed and disarmed selenoglycosides give rise to the corresponding glycosyl iodides [546]. This may point to the involvement of glycosyl iodides as intermediates in iodine-promoted glycosylation with armed thio- and seleno-glycosides, which potentially has an impact on the stereochemical outcome of the glycosylation process. Although this may be an issue when solvent-assisted stereocontrol with acetonitrile is being used [548,549], it can be overcome by *in situ* oxidation of iodide to iodine by the use of iodine in

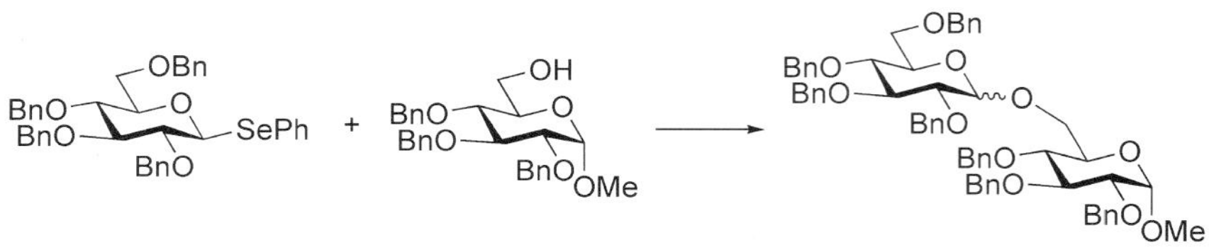

368 | *4 Glycoside Synthesis from 1-Sulfur/Selenium-Substituted Derivatives*

Conditions	Yield	α:β
NIS, MeCN	64%	1/5
NIS, CH$_2$Cl$_2$	32%	1/1.2
NIS, TMSOTf, MeCN	83%	1/5
NIS, TMSOTf, CH$_2$Cl$_2$	92%	1/1.4
I$_2$ (2 equiv), MeCN	91%	1/1.5
I$_2$ (2 equiv), CH$_2$Cl$_2$	52%	2/1

Scheme 4.106 Promoter- and solvent-dependent selenoglycosylation.

combination with DDQ as the glycosylation promoter (DDQ itself is not an effective activator of thio- and seleno-glycosides, Scheme 4.107).

Synthesis of the trisaccharide repeating unit of the acidic polysaccharide of the bacteriolytic complex of lysoamidase provided a vehicle for the identification of a new single set of activation conditions for use with both thio- and selenoglycosides [550]. The powerful 1-benzenesulfinyl piperidine (BSP)/Tf$_2$O sulfonium-activator system developed by Crich and Smith [85] enabled the construction process to be based on a linear glycosylation strategy starting from the reducing end (Scheme 4.108).

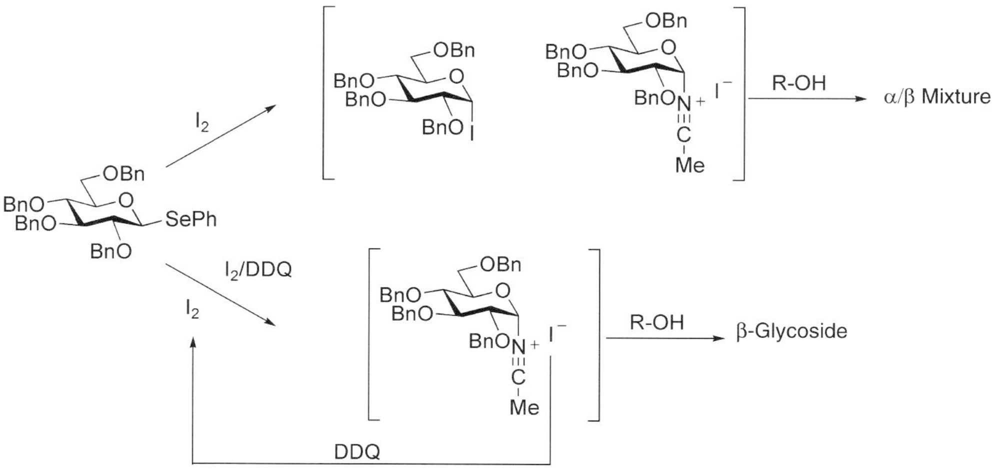

Scheme 4.107 Iodine/DDQ-enhanced β-glycosylation with selenoglycoside donor.

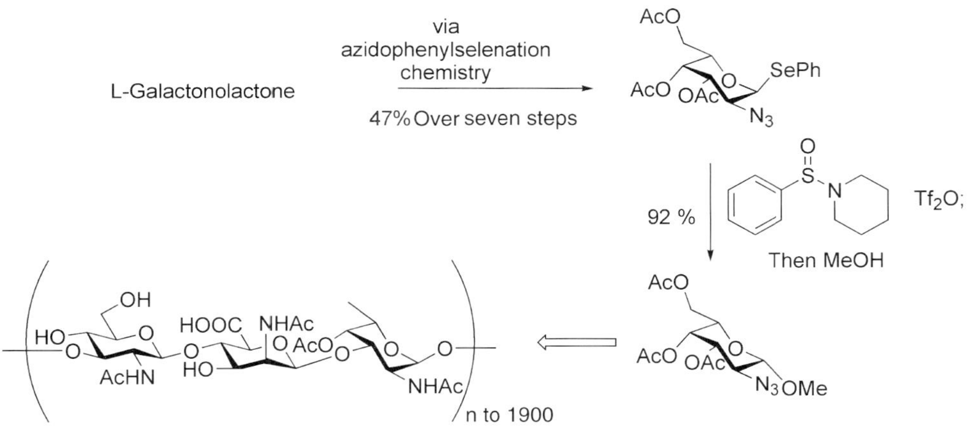

Scheme 4.108 *En route* to the lysoamidase trisaccharide repeat unit.

The selective activation of phenyl selenoglycosides over ethyl thioglycosides with AgOTf and anhydrous K_2CO_3 gives an efficient synthesis of disaccharides from selenoglycoside donors and thioglycoside acceptors [222,223]. In addition, the selective activation of telluro- over selenoglucoside donors identifies a clear reactivity scale for various telluro-, seleno- and thiosugars [551], although this is yet to be fully exploited in oligosaccharide synthesis.

Radical-Cation-Based Activation As for thioglycosides and also telluroglycosides, phenyl selenoglycosides are amenable to single-electron transfer activation. This can lead to photochemical oxidation, electrochemical activation or the use of organic single-electron transfer agents, such as tris(4-bromophenyl)ammoniumyl hexachloroantimonate (*BAHA*, Scheme 4.109).

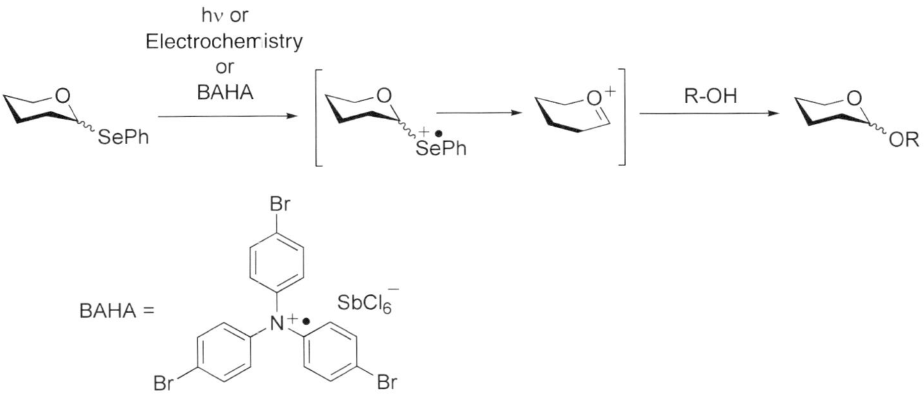

Scheme 4.109 Single-electron transfer activation of selenoglycosides.

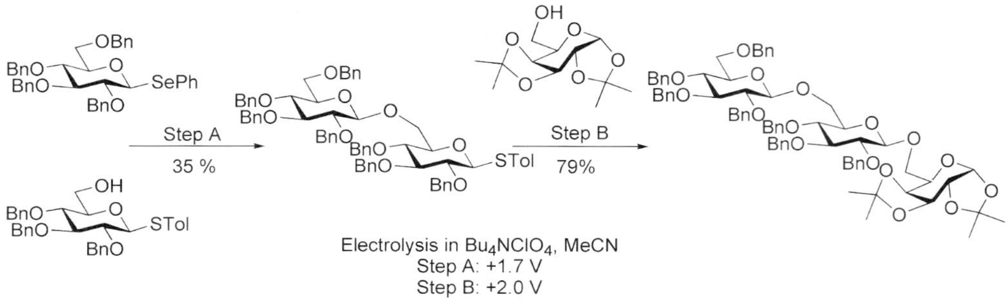

Scheme 4.110 Electrochemical synthesis employing O-, S- and Se-glycosides.

Photochemical oxidation has been employed for the activation of O-glycosides [552], thioglycosides [553] and telluroglycosides [554]. It is also effective for O-glycosylation with the phenyl selenoglycoside donors [555], although this approach has not been widely taken up for preparative use.

Similarly, the electrochemical glycosylation methods developed for use with aryl O-glycosides [103] and thioglycosides [556–558] can also be applied to the phenyl selenoglycosides and telluroglycosides [559]. The lower oxidation potentials of the phenyl Se-glycosides compared to phenyl S-glycosides, which in turn are lower than that for phenyl O-glycosides, offers compatibility with a wider range of protecting groups [106,560]. Ionization potentials (kJ mol^{-1}) for a series of typical glycosides are O (314) > S (239) > Se (225) > Te (208) [561]. The selection of selenoglycosides with ionization potentials that are dependent on aglycone and/or sugar protecting groups is attractive in the context of iterative electrochemical glycosylation. However, application of this information is not entirely straightforward. Although disaccharide syntheses have been achieved in good yield, the relative reactivity of selenoglycosides in preparative glycosylation has proved rather insensitive to oxidation potential values [561]. Furthermore, despite reporting the first electrochemical trisaccharide synthesis (Scheme 4.110) and that a variety of disaccharides are readily synthesized in high yield, Fairbanks and coworkers have also noted limitations in the use of selenoglycosides for the selective electrochemical glycosylation of thioglycoside acceptors [562].

Chemically induced glycosylation reactions with phenyl selenoglycosides promoted by BAHA [548], as previously described for thioglycosides [563–565], are thought to proceed via a single-electron transfer mechanism. Studies have provided support for the SET mechanism, but an alternative involving electrophilic activation cannot be discounted [106]. BAHA-promoted glycosylation with per-O-benzylated phenyl β-selenoglucoside gives α-1,6-linked glucosides in moderate to good yield but with poor stereocontrol in either dichloromethane or acetonitrile. In contrast, the same reaction promoted by iodine in combination with DDQ gave better acetonitrile-assisted β-stereoselectivity and higher yields with both thioglycoside and selenoglycoside donors than by reactions promoted by BAHA in acetonitrile [546].

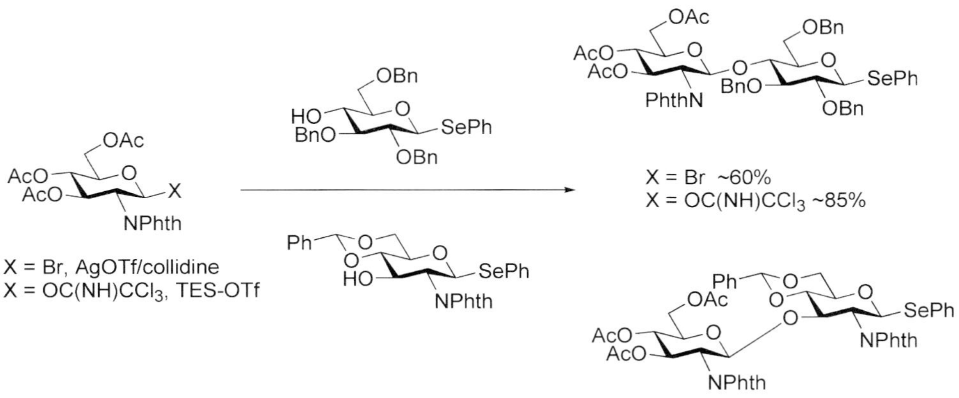

Scheme 4.111 Selenide acceptors with bromide and trichloroacetimidate donors.

4.4.4
Selenoglycosides as Acceptors

The fact that phenyl selenoglycosides are rendered unreactive toward AgOTf in the presence of an organic base, such as collidine, suggested that the preferential coordination of silver cation to the base left it unavailable for coordination to and activation of the selenoglycoside. This, in turn, highlighted the possibility of selective activation of another donor over a selenoglycoside, leaving the latter to serve as a glycosyl acceptor [223]. Indeed, reaction of selenoglycoside acceptors with a glycosyl bromide donor gave disaccharides in good yield (Scheme 4.111). The selective activation of glycosyl trichloroacetimidate over selenoglycoside has also been demonstrated: in the presence of a catalytic amount of TESOTf, disaccharides were obtained in excellent yield (Scheme 4.111) [223].

Dehydrative glycosylation of 1-hydroxy donors with Ph_2SO/Tf_2O in conjunction with thioglycoside acceptors opens the way for sequential double glycosylation, one-pot procedures for trisaccharide synthesis, as exemplified by the efficient one-pot synthesis of the α-Gal epitope and a hyaluronan trisaccharide [566]. This study also shows the potential of selenoglycoside as acceptors in dehydrative glycosylation (Scheme 4.112).

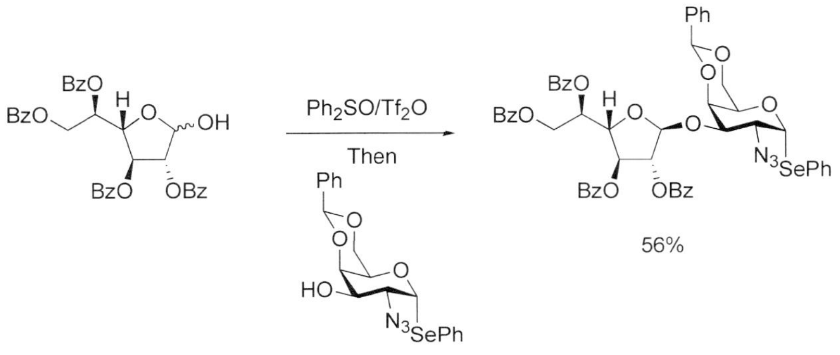

Scheme 4.112 Selenide acceptors for dehydrative glycosylation.

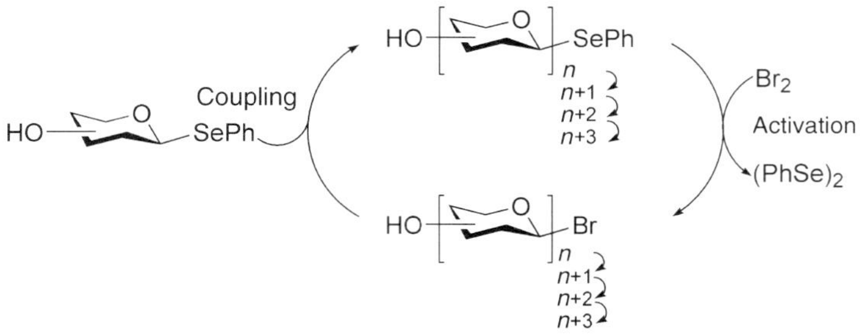

Scheme 4.113 Iterative glycosylation based on a glycosylselenide acceptor and its facile conversion into reactive β-glycosyl bromide donor.

The use of selenoglycosides as acceptors in iterative glycosylation with reactive β-glycosyl bromides, themselves derived from selenoglycosides, has also been investigated recently (Scheme 4.113; see also Section 4.4.5) [533,567].

4.4.5
Exploiting Selenoglycoside Relative Reactivity in Oligosaccharide Synthesis

The versatility of selenoglycosides in oligosaccharide synthesis is illustrated by the synthesis of galactofuranosyl-containing oligosaccharides corresponding to the glycosylinositolphospholipid of the protozoan parasite *Trypanosoma cruzi*, the causative agent of Chagas' disease [568]. The synthesis employs the NIS/TfOH-mediated selective activation of a phenyl selenogalactofuranoside or a phenyl selenomannopyranoside donor over ethyl thioglycoside acceptors. The need for careful control of reaction conditions and reagent stoichiometry to achieve such selectivity is crucial; NIS/TfOH is also capable of thioglycoside activation, as exploited later in the same synthesis (Scheme 4.114).

Although selective activation of selenoglycosides in the presence of thioglycosides is well precedented, it cannot always be relied on. In studies on the preparation of 2-*O*-ribofuranosyl-ribofuranosides from selenoglycoside donors and various thioglycoside acceptors, the formation of the desired disaccharide thioglycoside was accompanied by side products arising from *trans*-glycosylation and the formation of 1,1'-linked donor-derived disaccharides ([513] [569]. Furthermore, in the synthesis of a tetrasaccharide portion of the glucose-terminated arm of the *N*-glycan tetradecasaccharide, several attempted glycosylation reactions of prospective selenoglyside donor and thioglycoside acceptor met with failure [570]. Despite assessing a range of promoters (IDCP, AgOTf/K_2CO_3 and NIS/TfOH), TLC analysis indicated the formation of many reaction products, presumably because of the competitive activation of the thioglycoside and subsequent self-condensation (Scheme 4.115).

On a positive note, tuning the reactivity of glycosyl donors by selective introduction of different protecting and leaving groups has enabled highly efficient

Scheme 4.114 Selenide donors and acceptors for multistep oligosaccharide synthesis.

oligosaccharide synthesis. Utilizing both phenylseleno and ethylthio glycosides in combination with the cyclohexane-1,2-diacetal (*CDA*) protecting group provided four different levels of reactivity. One-pot sequential glycosidation of three components gave trisaccharides and tetrasaccharides, whereas the further extension

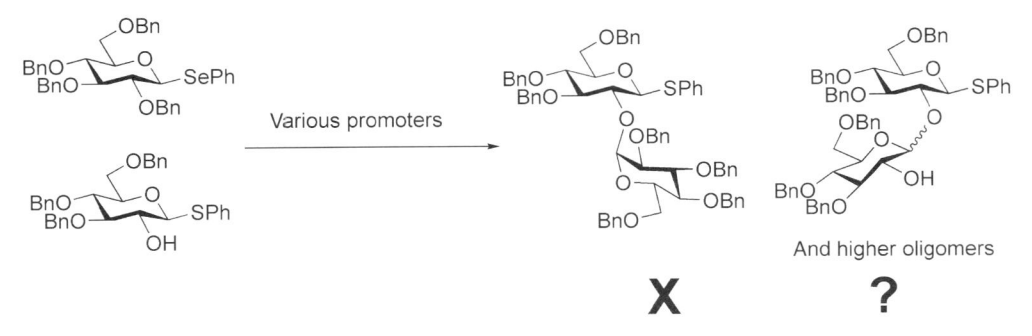

Scheme 4.115 Selenoglycoside donors and thioglycoside acceptors do not always behave similarly.

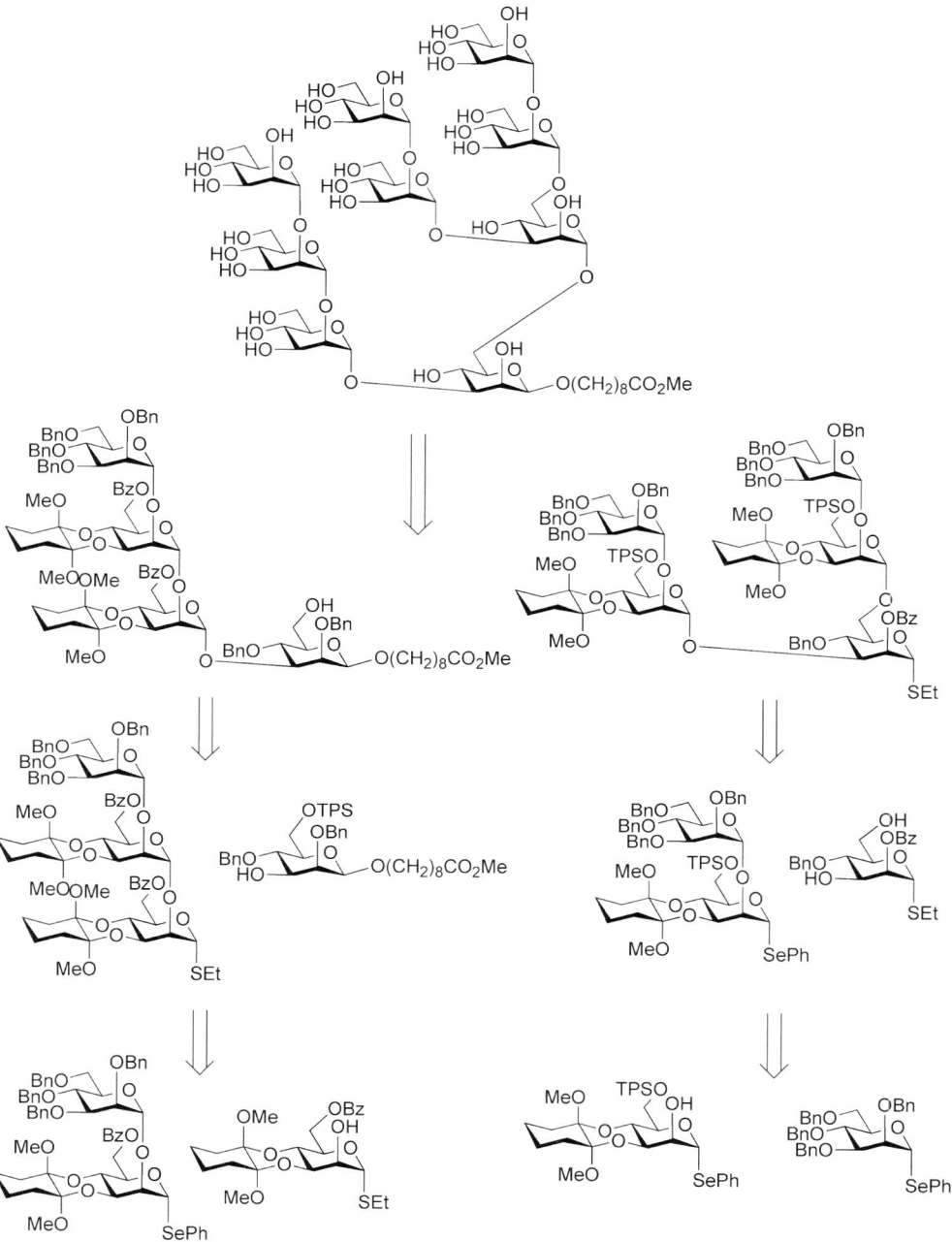

Scheme 4.116 Tuning building block reactivity for stream-lined high-mannose nonasaccharide synthesis.

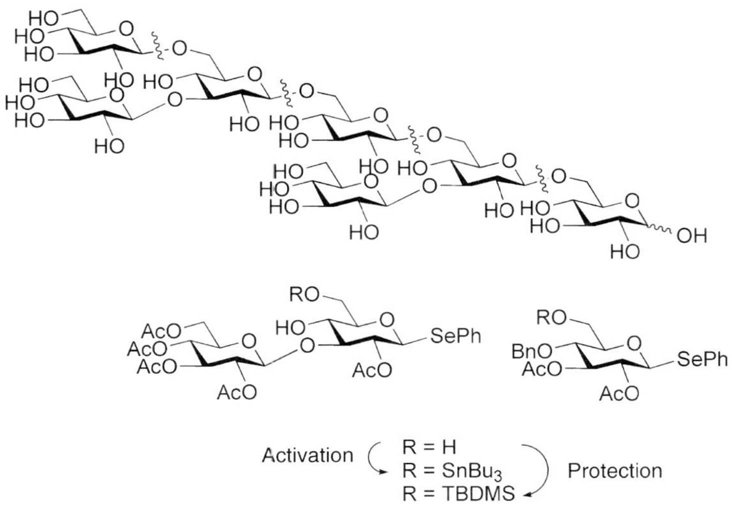

Scheme 4.117 Retrosynthetic analysis of heptaglucan phytoalexin synthesis achieved using an iterative approach.

of the approach reduced the number of steps from the monosaccharide building blocks to a nonasaccharide, triantennary mannan to only five (Scheme 4.116) [571,182,572].

On a similar grand scale, the ease of conversion of selenoglycosides into β-glycosyl bromides enables the iterative glycosylation of selenoglycosides [533,567]. Treatment of 2-O-acyl-protected selenoglycosides with bromine selectively generates β-glycosyl bromides for use as glycosyl donors. Coupling the β-glycosyl bromide with another selenoglycoside then affords the corresponding glycosylated selenoglycoside, which can be directly used in the next round of glycosylation. The iteration of this sequence allows the synthesis of a variety of oligosaccharides, including a set of nested oligoglucoside fragments of a heptasaccharide phytoalexin (Scheme 4.117). A feature of the iterative approach is that glycosyl donors and acceptors with the same anomeric reactivity can be selectively coupled by the activation of the glycosyl donor prior to the introduction of the glycosyl acceptor. The same selenoglycosides can, therefore, in practice be used as both glycosyl donors and acceptors. This approach exemplifies the structural diversity that can be constructed solely from selenoglycoside building blocks.

4.4.6
Summary

In conclusion, the phenyl selenoglycosides are versatile building blocks for oligosaccharide synthesis. Through judicious choice of protecting groups, and particularly promoters, they can be exploited as either donors or acceptors in glycosylation reactions. Although a number of syntheses have exploited the differential reactivity of selenoglycosides over thioglycosides, it is evident that, using either cation-based or electrochemical activation, the expected difference in reactivity of such building blocks cannot always be relied upon. Nonetheless, examples of selenoglycoside-based oligosaccharide syntheses continue to appear in the literature. Thus, further investigation of selenoglycosides is justified and is necessary, to realize their full potential.

4.4.7
Examples of Experimental Procedures

4.4.7.1 Typical Procedure for the Preparation of Selenoglycosides from Glycosyl Bromides

A solution of diphenyldiselenide (20 mmol) in dry ethanol (100 ml) was cooled by an ice bath. To this solution was 'carefully' added a preformed, cold solution of sodium borohydride (40 mmol) in dry ethanol (20 ml). As the mixture was stirred over 30 min, the yellow diselenide solution became colorless, reflecting selenol formation. To this mixture was added a solution of glycosyl bromide (30 mmol) in CH_2Cl_2 (20 ml) and the stirring was continued for 2 h at room temperature. After aspirating the solution to oxidize the remaining selenol, the solvent was evaporated *in vacuo* and the resulting material was dissolved in CH_2Cl_2 (200 ml) and washed with water (2×100 ml), and the organic extract was dried over $MgSO_4$ and concentrated *in vacuo*. The desired selenoglycoside product was obtained by column chromatography (EtOAc/hexane, 1/10 to 1/2, vol/vol) in high yield (typically more than 90%)

4.4.7.2 Typical Procedure for the Preparation of Selenoglycosides from Glycals

Sodium azide (10.0 mmol), diphenyldiselenide (3.0 mmol) and (diacetoxyiodo)benzene (5.6 mmol) were added to a solution of acetylated glycal (4.0 mmol) in CH_2Cl_2 (60 ml) under nitrogen. The reaction mixture was stirred at room temperature for 48 h and washed with water (50 ml), and the aqueous layer was back-extracted with CH_2Cl_2 (3×50 ml). The combined organic extracts were then dried over $MgSO_4$ and concentrated to dryness. The resulting oil was purified by column chromatography (EtOAc/hexane, 1/10 to 1/2, vol/vol) to give the phenyl 2-azido-2-deoxy-selenoglycoside in good yield (typically ~80%).

4.4.7.3 Typical Procedure for NIS/TfOH-Promoted Glycosylation with Selenoglycosides

Powdered 4-Å molecular sieves (100 mg) were added to a solution of glycosyl donor (0.2 mmol) and glycosyl acceptor (0.16 mmol, 0.8 equiv) in dry CH_2Cl_2 (5.0 ml). The resulting mixture was stirred at room temperature under nitrogen for 30 min and then NIS (1.1 mmol, 5 equiv) and TMSOTf (0.02 mmol) were added to it. When TLC analysis indicated the completion of the reaction (typically ~30 min), the mixture was diluted with EtOAc (10 ml) and washed with 10% aq $Na_2S_2O_3$ (2×5 ml) and brine (5 ml). The organic phase was then separated, dried over $MgSO_4$ and concentrated to dryness. The resulting residue was purified by column chromatography on silica gel (EtOAc/hexane, 0/1 to 1/4, vol/vol) to provide the desired disaccharide (typical range 60–85%).

4.4.7.4 Typical Procedure for BAHA-Promoted Glycosylation with Selenoglycosides

Powdered 4-Å molecular sieves (100 mg) were added to a solution of glycosyl donor (0.2 mmol) and glycosyl acceptor (0.16 mmol, 0.8 equiv) in dry CH_2Cl_2 (5.0 ml). The resulting mixture was stirred at room temperature under nitrogen for 30 min and then BAHA (0.6 mmol, 3 equiv) was added. When TLC analysis indicated the completion of the reaction (typically ∼30–60 min), the reaction mixture was cooled by an ice bath and neutralized with Et_3N, filtered through Celite with the help of CH_2Cl_2 and concentrated to dryness. The resulting residue was purified by column chromatography on silica gel (EtOAc/hexane, 0/1 to 1/4, vol/vol) to provide the desired disaccharide (typical range 60–85%).

References

512 Mehta, S. and Pinto, B.M. (1996) *Modern Methods in Carbohydrate Synthesis* (eds S.H. Khan and R.A. O'Neill), Harwood Academic Publishers, Amsterdam, pp. 107–129.

513 Witczak, Z.J. and Czernecki, S. (1998) *Advances in Carbohydrate Chemistry and Biochemistry*, **53**, 143–199.

514 Praly, J.-P. (2001) *Advances in Carbohydrate Chemistry and Biochemistry*, **56**, 65–151.

515 Fraser-Reid, B., Udodong, U.E., Wu, Z.F., Ottosson, H., Merritt, J.R., Rao, C.S., Roberts, C. and Madsen, R. (1992) *Synlett*, 927–942.

516 Crich, D., Suk, D.-H. and Sun, S. (2003) *Tetrahedron: Asymmetry*, **14**, 2861–2864.

517 Yamago, S., Kokubo, K., Masuda, S. and Yoshida, J. (1996) *Synlett*, 929–930.

518 Sanchez, S., Bamhaoud, T. and Prandi, J. (2002) *European Journal of Organic Chemistry*, 3864–3873.

519 Ikeda, K., Sugiyama, Y. Tanaka, K. and Sato, M. (2002) *Bioorganic & Medicinal Chemistry Letters*, **12**, 2309–2311.

520 Sato, T., Fujita, Y., Otera J. and Nozaki, H. (1992) *Tetrahedron Letters*, **33**, 239–242.

521 Tiwari, P. and Misra, A.K. (2006) *Tetrahedron Letters*, **47**, 2345–2348.

522 Knapp, S. and Darout, E. (2005) *Organic Letters*, **7**, 203–206.

523 Nanami, M., Ando, H., Kawai, Y., Koketsu, M. and Ishihara, H. (2007) *Tetrahedron Letters*, **48**, 1113–1116.

524 Czernecki, S. and Randriamandimby, D. (1993) *Tetrahedron Letters*, **34**, 7915–7916.

525 Santoyo-Gonzales, F., Calvo-Flores, F.G., Garcia-Mendoza, P., Hernandez-Mateo, F., Isac-Garcia, J. and Robles-Diaz, R. (1993) *The Journal of Organic Chemistry*, **58**, 6122–6125.

526 Czernecki, S., Ayadi, E. and Randriamandimby, D. (1994) *The Journal of Organic Chemistry*, **59**, 8256–8260.

527 Czernecki, S., Ayadi, E. and Randriamandimby, D. (1994) *Journal of the Chemical Society: Chemical Communications*, 35–36.

528 Czernecki, S. and Ayadi, E. (1995) *Canadian Journal of Chemistry*, **73**, 343–350.

529 Lemieux, R.U. and Ratcliffe, R.M. (1979) *Canadian Journal of Chemistry*, **57**, 1244–1251.

530 Mironov, Y.V., Sherman, A.A. and Nifantiev, N.E. (2004) *Tetrahedron Letters*, **45**, 9107–9110.

531 Bedini, E., Esposito, D. and Parrilli, M. (2006) *Synlett*, 825–830.

532 van Well, R.M., Kartha, K.P.R. and Field, R.A. (2005) *Journal of Carbohydrate Chemistry*, **24**, 463–474.

533 Yamago, S., Yamada, T., Hara, O., Ito, H., Mino, Y. and Yoshida, J. (2001) *Organic Letters*, **3**, 3867–3870.

534 Nicolaou, K.C., Fylaktakidou, K.C., Mitchell, H.J., van Delft, F.L., Rodriguez, R.M., Conley, S.R. and Jin, Z.D. (2000) *Chemistry – A European Journal*, **6**, 3166–3185.

535 Jiaang, W.T., Chang, M.Y., Tseng, P.H. and Chen, S.T. (2000) *Tetrahedron Letters*, **41**, 3127–3130.

536 Tingoli, M., Tiecco, M., Testaferri, L. and Temperini, A. (1994) *Journal of the Chemical Society: Chemical Communications*, 1883–1884.

537 Mehta, S., Jordan, K.L., Weimar, T., Kreis, U.C., Batchelor, R.J., Einstein, F.W.B. and Pinto, B.M. (1994) *Tetrahedron: Asymmetry*, **5**, 2367–2396.

538 Veeneman, G.H., van Leeuwen, S.H. and van Boom, J.H. (1990) *Tetrahedron Letters*, **31**, 1331–13334.

539 Konradsson, P., Udodong, U.E. and FraserReid, B. (1990) *Tetrahedron Letters*, **31**, 4313–4316.

540 Zuurmond, H.M., van der Klein, P.A.M., van der Meer, P.H., van der Marel, G.A. and van Boom, J.H. (1992) *Recueil Des Travaux Chimiques Des Pays-Bas (Journal of the Royal Netherlands Chemical Society)*, **111**, 365–366.

541 Zuurmond, H.M., van der Meer, P.H., van der Klein, P.A.M., van der Marel, G.A. and van Boom, J.H. (1993) *Journal of Carbohydrate Chemistry*, **12**, 1091–1103.

542 Hayes, W., Osborn, H.M.I., Osborne, S.D., Rastall, R.A. and Romagnoli, B. (2003) *Tetrahedron*, **59**, 7983–7996.

543 Zuurmond, H.M., van der Klein, P.A.M., de Wildt, J., van der Marel, G.A. and van Boom, J.H. (1994) *Journal of Carbohydrate Chemistry*, **13**, 323–339.

544 Timmers, C.M., van der Marel, G.A. and van Boom, J.H. (1995) *Chemistry – A European Journal*, **1**, 161–164.

545 Kartha, K.P.R., Kärkkäinen, T.S., Marsh, S.J. and Field, R.A. (2001) *Synlett*, 260–262.

546 van Well, R.M., Kärkkäinen, T.S., Kartha, K.P.R. and Field, R.A. (2006) *Carbohydrate Research*, **341**, 1391–1397.

547 Kartha, K.P.R., Cura, P., Aloui, M., Readman, S.K., Rutherford, T.J. and Field, R.A. (2000) *Tetrahedron: Asymmetry*, **11**, 581–593.

548 Ratcliffe, A.J. and Fraser-Reid, B. (1990) *Journal of the Chemical Society Perkin Transactions*, **1**, 747–750.

549 Braccini, I., Derouet, C., Esnault, J., Herve du Penhoat, C., Mallet, J.-M., Michon, V. and Sinay, P. (1993) *Carbohydrate Research*, **246**, 23–41.

550 Litjens, R.E.J.N., den Heeten, R., Timmer, M.S.M., Overkleeft, H.S. and van der Marel, G.A. (2005) *Chemistry – A European Journal*, **11**, 1010–1016.

551 Stick, R.V., Tilbrook, D.M.G. and Williams, S.J. (1997) *Australian Journal of Chemistry*, **50**, 237–240.

552 Timpa, J.D. and Griffin, G.W. (1984) *Carbohydrate Research*, **131**, 185–196.

553 Griffin, G.W., Bandara, N.C., Clarke, M.A., Tsang, W.-S., Garegg, P.J., Oscarson, S. and Silwanis, B.A. (1990) *Heterocycles*, **30**, 939–947.

554 Yamago, S., Hashidume, M. and Yoshida, J.-I. (2002) *Tetrahedron*, **58**, 6805–6813.

555 Furuta, T., Takeuchi, K. and Iwamura, M. (1996) *Chemical Communications*, 157–158.

556 Balavoine, G., Greg, A., Fischer, J.-C. and Lubineau, A. (1990) *Tetrahedron Letters*, **31**, 5761–5764.

557 Amatore, C., Jutand, A., Mallet, J.-M., Meyer, G. and Sinay, P. (1990) *Journal of the Chemical Society: Chemical Communications*, 718–719.

558 Mallet, J.-M., Meyer, G., Yvelin, F., Jutand, A., Amatore, C. and Sinay, P. (1993) *Carbohydrate Research*, **244**, 237–246.

559 Yamago, S., Kokubo, K. and Yoshida, J. (1997) *Chemistry Letters*, 111–112.

560 Bard, A.J. and Faulkner, L.R. (1980) *Electrochemical Methods Fundamentals and Applications*, John Wiley & Sons, New York.

561 Yamago, S., Kokubo, K., Hara, O., Masuda, S. and Yoshida, J. (2002) *The Journal of Organic Chemistry*, **67**, 8584–8592.

562 France, R.R., Compton, R.G., Davis, B.G., Fairbanks, A.J., Rees, N.V. and Wadhawan, J.D. (2004) *Organic & Biomolecular Chemistry*, **2**, 2195–2202.

563 Marra, A., Mallet, J.-M., Amatore, C. and Sinay, P. (1990) *Synlett*, 572–574.

564 Sinay, P. (1991) *Pure and Applied Chemistry (Chimie pure et Appliquee)*, **63**, 519–529.

565 Zhang, Y.-M., Mallet, J.-M. and Sinay, P. (1992) *Carbohydrate Research*, **236**, 73–88.

566 Codee, J.D.C., van den Bos, L.J., Litjens, R.E.J.N., Overkleeft, H.S. van Boom, J.H. and van der Marel, G.A. (2003) *Organic Letters*, **5**, 1947–1950.

567 Yamago, S., Yamada, T., Ito, H., Hara, O., Mino, Y. and Yoshida, J. (2005) *Chemistry – A European Journal*, **11**, 6159–6174.

568 Randell, K.D., Johnston, B.D., Brown, P.N. and Pinto, B.M. (2000) *Carbohydrate Research*, **325**, 253–264.

569 Sliedregt, L.A.J.M., Broxterman, H.J.G., van der Marel, G.A. and van Boom, J.H. (1994) *Carbohydrate Letters*, **1**, 61–68.

570 Ennis, S.C., Cumpstey, I., Fairbanks, A.J., Butters, T.D., Mackeen, M. and Wormald, M.R. (2002) *Tetrahedron*, **58**, 9403–9411.

571 Grice, P., Ley, S.V., Pietruszka, J., Priepke, H.W.M. and Walther, E.P.E. (1995) *Synlett*, 781–784.

572 Grice, P., Ley, S.V., Pietruszka, J. and Priepke, H.W.M. (1996) *Angewandte Chemie (International Edition in English)*, **35**, 197–200.

5
Other Methods for Glycoside Synthesis

5.1
Orthoesters and Related Derivatives
Bert Fraser-Reid, J. Cristóbal López

5.1.1
Introduction

The difficulties in the preparation of complex oligosaccharides, compared to other biopolymers such as peptides or nucleic acids, are the result of a greater number of possibilities for the combination of monomeric units. For instance, the hypothetical 'simple' glycosidation shown in Scheme 5.1 entails three of the four modes of selectivity: stereo, chemo and regio. These modes, according to Trost [1], confront organic synthesis in general and any reliable glycosylation is expected to comply with them. The fourth one, enantioselectivity, is usually not encountered in oligosaccharide synthesis because the chiralities of donor and acceptor are usually specified by nature.

Contributions in twentieth-century chemical syntheses of oligosaccharides have been aimed at stereoselective, chemoselective and regioselective saccharide couplings. In fact, the search for stereoselective couplings occupied most of the last century, and it was only in the early 1990s that the relevance of chemoselective couplings was addressed. The advent of regioselective coupling processes can be considered an even more recent development.

As a result of the explosion of interest in oligosaccharide synthesis, a plethora of different glycosylation methods is now available, and yet chemical oligosaccharide synthesis still remains a formidable problem at the beginning of the twenty-first century. Clearly, the preparation of large oligosaccharides demands the use of more than one reliable method for the construction of glycosidic linkages.

From a historic perspective glycosyl chlorides and bromides introduced, respectively, by Michael [2] and Koenigs and Knorr [3] were the most widely used donors in the saccharide synthesis for a very long time. The introduction of 1,2-orthoesters in 1964 [4] was the first important attempt to find an alternative to the Koenigs–Knorr method. However, Paulsen in his 1990 review of reliable donors for glycosyl

Handbook of Chemical Glycosylation: Advances in Stereoselectivity and Therapeutic Relevance.
Edited by Alexei V. Demchenko.
Copyright © 2008 WILEY-VCH Verlag GmbH & Co. KGaA. All rights reserved.
ISBN: 978-3-527-31780-6

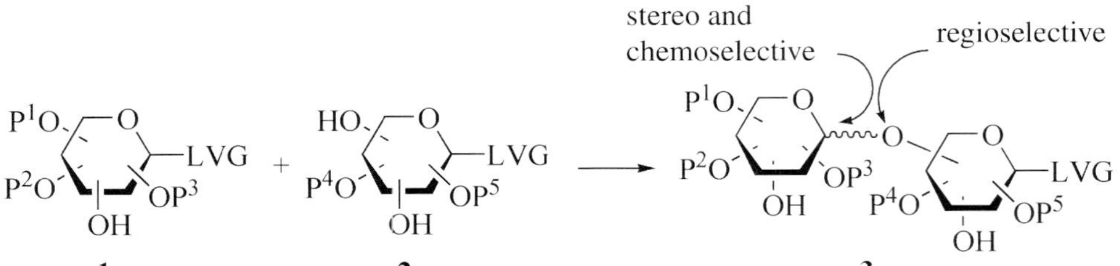

Scheme 5.1 Glycosyl coupling and selectivity.

couplings cited only halogenides, imidates and thioglycosides. [5] The aim of this chapter is to provide some insight into recent developments in 1,2-orthoester chemistry that recognize the trustworthiness of these donors for stereoselective, chemoselective and regioselective glycosyl couplings.

The concept of neighboring-group participation and the formation and isolation of 1,2-orthoesters are intimately linked, as evident from the title of Isbell's monumental 1941 paper 'Sugar acetates, acetylglycosyl halides and orthoacetates in relation to the Walden inversion' [6]. Indeed, in today's terminology it could be said that 1,2-orthoesters benefit from the 1,2-*trans* stereodirecting properties of 2-*O*-acyl donors without being disarmed. Thus, they can be regarded as 'glycosyl donors with a nondisarming participating group at C-2'. Their high reactivity makes them amenable to *chemoselective* activation in the presence of different glycosyl donors, and they have recently proved themselves as privileged donors for *regioselective* couplings [7].

5.1.2
Sugar 1,2-Orthoesters

Carbohydrate orthoesters, first reviewed by Pacsu more than 60 years ago [8], were reported by Fischer *et al.* [9] as by-products of the Koenigs–Knorr reaction [3] of acetobromo-L-rhamnose (**1**) with methanol. Orthoacetate **3** was isolated along with the expected α- and β-methyl rhamnosides **2** (Scheme 5.2). However, its 'true structure' was assigned only 10 years later by several research groups [10–12].

In this connection, early syntheses of 1,2-orthoesters employed 1,2-*trans* acylhalogenoses as the starting materials in the presence of alcohols and a base (Scheme 5.3) [13].

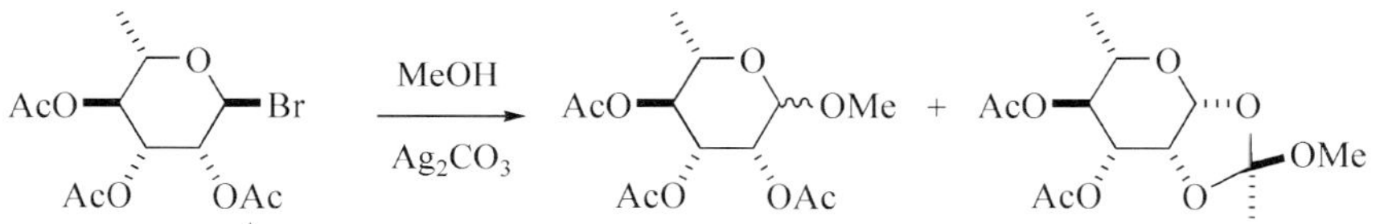

Scheme 5.2 First preparation of a sugar 1,2-orthoester.

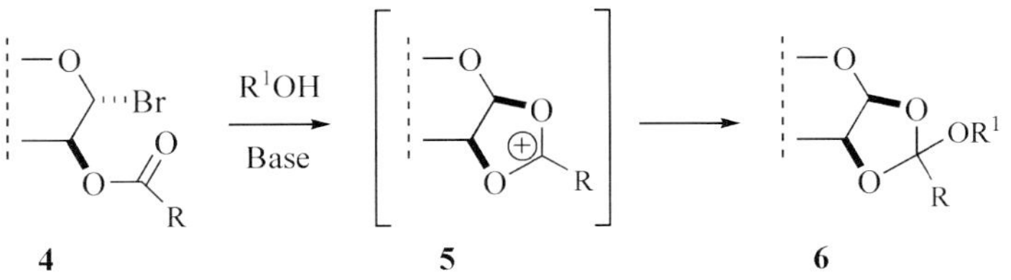

Scheme 5.3 Othoesters format on from 1,2-*trans* acylhalogenoses.

It is also possible to prepare 1,2-orthoesters from 1,2-*cis*-glycosyl halides, as reported by Lemieux and Morgan, using alcohols in the presence of tetraalkylammonium halides and *sym*-collidine [14]. This reaction is believed to occur with double inversion at C-1, the initial step being the anomerization of α-bromide (Lemieux's halide ion-catalyzed method) [15], and the orthoester is then produced by the attack at C-1 of the participating *trans*-ester group at C-2 with subsequent reaction of the alcohol at the acyloxonium center (as in Scheme 5.2). The formation of the substituted ethylidene derivative is acccmpanied by the appearance of a new chiral center at the dioxolane ring with the possible formation of *exo*- and *endo*-orthoesters. However, normally the *exo*- isomer is preferred owing to the greater accessibility of the *exo*- side in the dioxolenium cation to attack by nucleophile (alcohol) (Scheme 5.4).

Additional methods for the synthesis of 1,2-orthoesters from acetohalogeno sugars include the treatment with lead carbonate and calcium sulfate in ethyl acetate [16], silver nitrate and 2,4,6-trimethylpyridine [17], *N,N*-dimethylformamide dialkylacetals–AgOTf [18], silver triflate–2,4,6-collidine [19], trialkylstannyl methoxide–Et$_4$NBr [20], silver salicilate [21], silver nitrate–2,4,6-collidine [22], mercury(II) bromide–2,4,6-collidine [23], *N,N*-dimethylformamide dialkylacetals–Et$_4$NBr [24], potassium fluoride [25] and 2,6-lutidine in ionic liquid [bmin]PF$_6$ [26]. Other starting materials have also been used to prepare orthoesters: (a) pyranose hemiacetals have been converted into 1,2-orthoesters by the treatment with 1-chloro-2,*N,N*-trimethylpropenylamine followed by alcohol in the presence of triethylamine [27], (b) peracetylated sugars can be changed into 1,2-orthoesters via the *in situ* generation of glycosyl iodides promoted by I$_2$/Et$_3$SiH followed by the treatment with lutidine and Bu$_4$NBr [28] and (c) 1,2-*O*-vinylidene acetals can yield 1,2-orthoesters on treatment with alcohols in the presence of *p*-toluenesulfonic or camphorsulfonic acids [29].

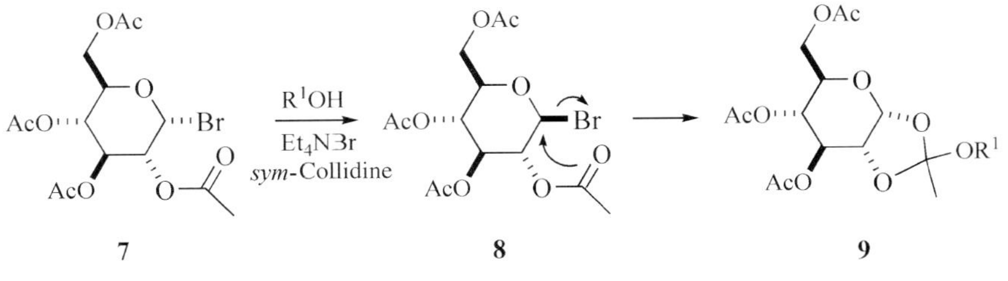

Scheme 5.4 Orthoester formation from 1,2-*cis* acylhalogenoses.

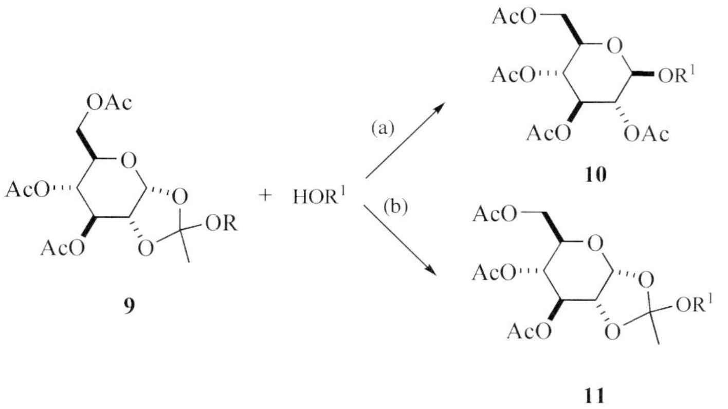

Scheme 5.5 1,2-O-Alkyl orthoesters in glycosylation: (a) polar or nonpolar solvents, high amount of acid; (b) low polarity solvents, minor amounts of acid catalyst.

5.1.2.1 1,2-O-Alkyl Orthoesters as Glycosyl Donors – Early Developments

In 1964, Kochetkov, Khorlin and Bochkov reported that the reaction of 1,2-alkylorthoacetates with alcohols in the presence of catalytic amounts of $HgBr_2$ and pTsOH furnished acetylated 1,2-trans glycosides or isomeric orthoesters depending on the reaction conditions [4]. Polar solvents (nitromethane, acetonitrile) and large amounts of catalyst promoted glycosylation (a, Scheme 5.5), whereas solvents of low polarity (dichloroethane) and the use of small amounts of catalyst favored trans-orthoesterification (b, Scheme 5.5) [16].

The application of these optimized conditions permitted Kochetkov et al. to prepare, among others, β-cholesteryl glucoside in 45% yield. However, with low reactive aglycons (R^1OH) the yield was poor (10–20%) owing to the competing glycosylation of the extruded alcohol (ROH) from the initial orthoesters (Scheme 5.6).

To address this problem, the authors devised two modifications. [30] The first one, two-stage glycosylation (Scheme 5.7), employed an initial, reversible trans-orthoesterification step (**12** → **14**, Scheme 5.7) in which the departing alcohol was removed either azeotropically or by molecular sieves [31]. The new orthoester **14** was then processed to give the glycoside (**6**) under the conditions developed in their previous work. The second variation consisted of the use of orthoacetates of hindered alcohols (isopropyl and tert-butyl) that minimize the return of the alcohol that is split off.

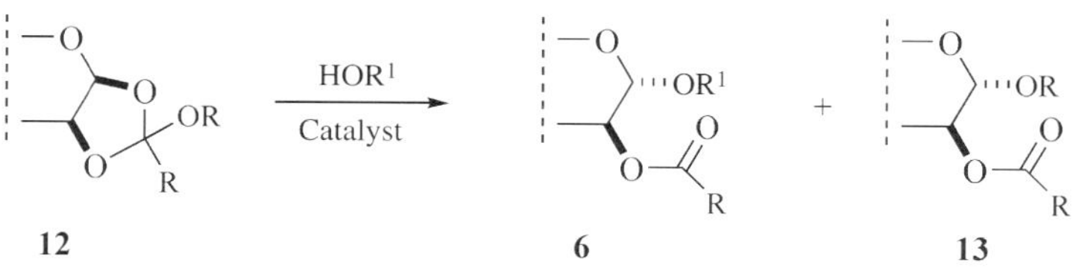

Scheme 5.6 The competing glycosylation issue with orthoesters.

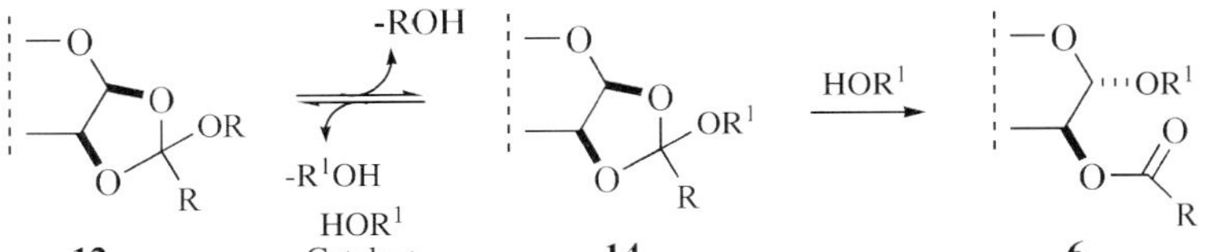

Scheme 5.7 The two-stage modification for glycosylation with orthoesters.

These modifications met with only limited success; however, they set the basis for the future applications of orthoesters as glycosyl donors.

Another drawback, soon noted by Kochetkov's school, was the presence of a variety of other side products [32]. These included 1,2-*cis* counterparts of the expected 1,2-*trans* glycosides, appreciable amounts of acylated aglycons, glycosides (*cis* and *trans*) with a free hydroxyl group at C-2 and compounds arising from the latter after further glycosylation of the 2-OH group. All of these side products can be assembled under a unified mechanistic interpretation, resulting from the efforts of several groups that cataloged some of the perils of working with 1,2-orthoester glycosyl donors [19,33–35]. The nature of the side products was rationalized by reference to the different possible pathways depicted in Scheme 5.8. The majority of the oxonium and acyloxonium ion intermediates are connected with starting orthoester by fast reversible equilibria. The main reaction pathway is the one arising from the equilibrating intermediates **14** and **15**, and the rate-determining steps are the ones shown by bold arrows. The product composition has been shown to be dependant on parameters such as the promoter (E), the solvent and the structure of the alcohol.

5.1.2.2 1,2-O-Cyanoethylidene Derivatives

Further modifications of the original orthoester glycosylation method were developed to eliminate these disadvantages. Among them the use of 1,2-O-(1-cyanoethylidene) derivatives **24** [36,37] was particularly noteworthy. Cyanoethylidene derivatives could be chemoselectively activated by trityl ion, thereby allowing chemoselective entry to bridging cation **15**, and thus minimizing the side products arising from intermediates **19** and **21** (see Scheme 5.8). The implementation of this method, however, requires the use of trityl ethers of alcohols as the glycosyl acceptors.

Cyanoethylidene derivatives can be prepared from the corresponding glycosyl halides by the treatment with KCN and $n\text{-Bu}_4\text{NBr}$ in CH_3CN or from the corresponding peracetates and TMSCN in the presence of stannous chloride (Scheme 5.9).

A recent synthesis of mannodendrimers by Backinowsky *et al.* illustrates the state of the art of this methodology (Scheme 5.10) [38]. Acetobromomannose (**25**) could be chemoselectively activated in the presence of cyanoethylidene-derived triol **26** to afford trimannan **27a** in a regioselective manner. The acetylation of the latter led to **27b**, which could regioselectively glycosylate di-trityl derivative **28** at the secondary trityl group to yield tetramannan **29**, in keeping with the greater reactivity of secondary versus primary trityl groups with various glycosyl donors [39,40]. Alternatively, **28** could be bis-glycosylated with 2 equiv of **26b** to furnish heptamannan **30**.

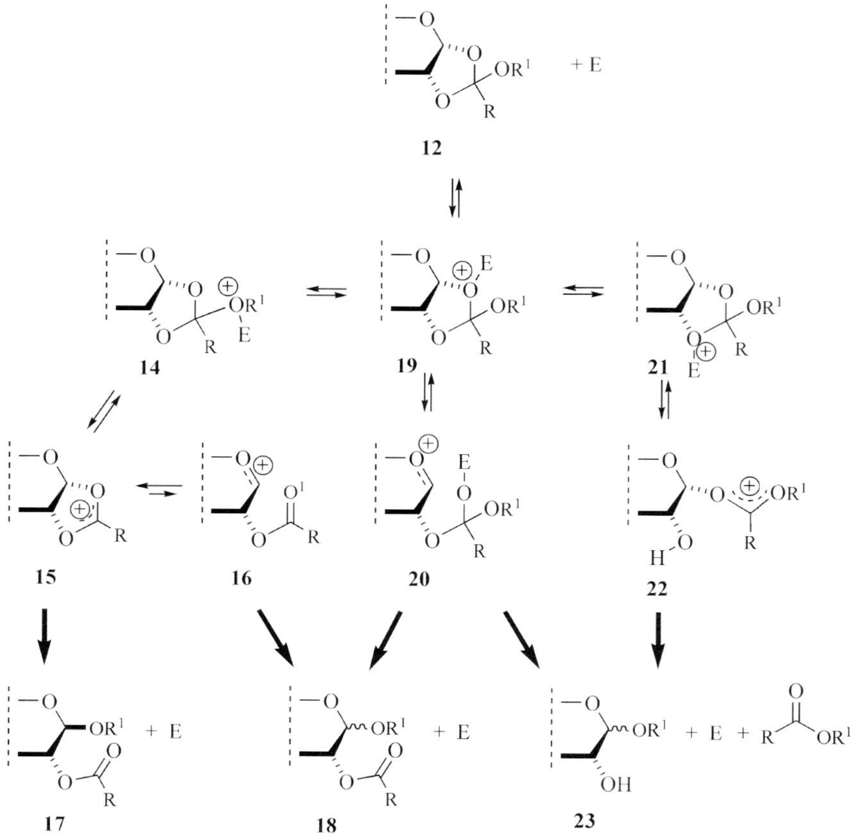

Scheme 5.8 Proposed reaction pathways for the reaction of orthoester **12**.

The 1,2-O-cyanoethylidene method also became the basis for a polycondensation reaction, which allowed the synthesis of many polysaccharides [37]. When both the O-trityl and cyanoethylidene groups are present in the same molecule as in **31**, polycondensation takes place under glycosidation conditions giving a polysaccharide chain, for example **32** (Scheme 5.11).

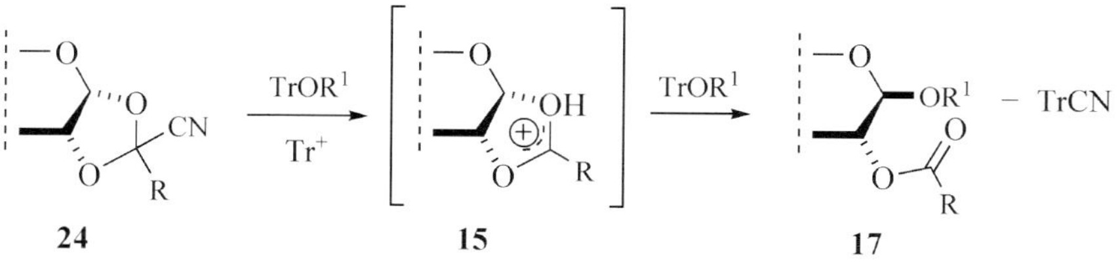

Scheme 5.9 1,2-O-Cyanoethylidene derivatives in glycosyl couplings.

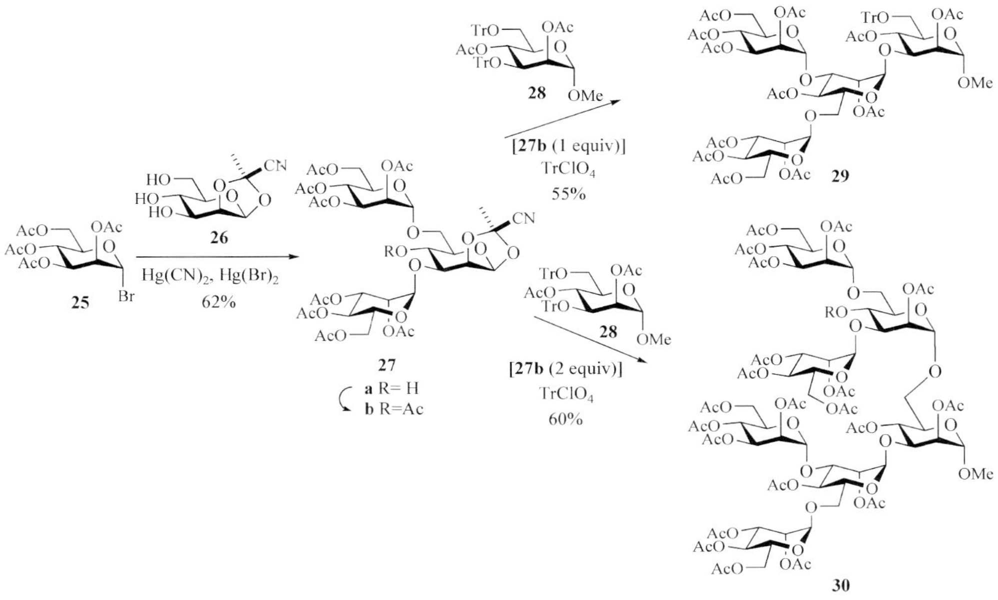

Scheme 5.10 Synthesis of tetra- and heptamannan derivatives.

5.1.2.3 1,2-Thioorthoester Derivatives

Related to the above cyanoethylidene derivatives, 1,2-thioorthoesters, for example **33**, obtained by the reaction of peracetylglycosyl bromides with thiols in the presence of 2,6-lutidine or 2,4,6-collidine, were also introduced by Kochetkov et al. as glycosyl donors [41]. Primary and secondary trityl ethers of monosaccharides could be coupled under the agency of triphenylmethylium perchlorate (Scheme 5.12a) [42]. However, some 4-O-trityl ether derivatives failed to undergo glycosyl coupling, and anomeric mixtures of acetylated thioglycosides were isolated instead. More recently, iodonium-based promoters iodonium dicollidine perchlorate (IDCP) [43], NIS/TfOH [44–46] and I$_2$ [47] have been assayed for the activation of 1,2-thioorthoesters. Although these promoters allowed the use of hydroxyl groups (rather than trityl ethers) as acceptors, the glycosyl coupling proved successful only with primary OH (Scheme 5.12a) [46,47]. Thioorthoesters have recently been used as starting materials for the preparation of S-thiazolinyl (STaz) glycosides, for example **34** (Scheme 5.12b) [48].

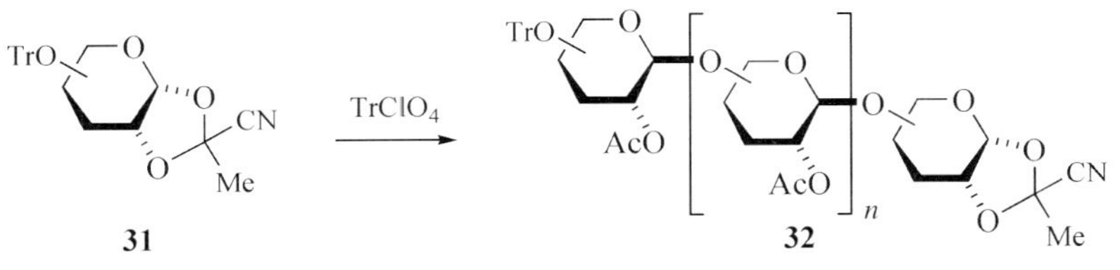

Scheme 5.11 Polycondensation of a monosaccharide unit leading to a homopolysaccharide.

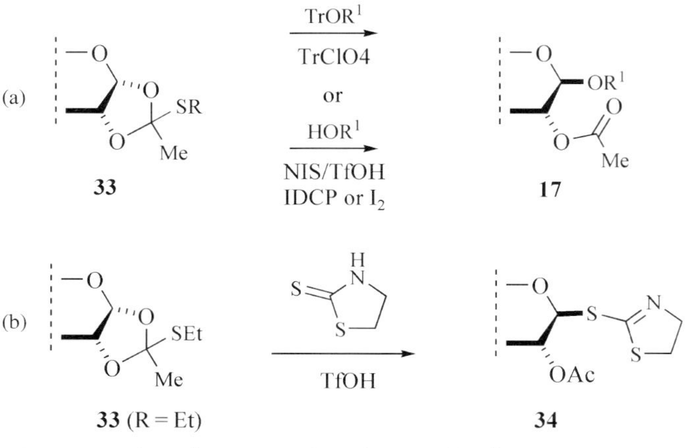

Scheme 5.12 Thioorthoesters in glycosylation and in the synthesis of (STaz)-glycosyl donors.

5.1.2.4 Internal Orthoesters

Arabinofuranose 1,2,5-orthoesters and mannopyranose 1,2,6-orthoesters have been evaluated as glycosyl donors with limited success. Kochetkov *et al.* used β-L-arabinofuranose 1,2,5-orthobenzoate **35** as a monomer unit in a polymerization strategy (Scheme 5.13a), the implementation of which demanded the use of an alcohol initiator, **36** [49]. Wong and coworkers described the preparation of mannose 1,2,6-orthoester **38** and its glycosylation with primary hydroxyl acceptors. Glycosylation of **38** with BF$_3$·Et$_2$O yielded disaccharide **40a** (Scheme 5.13b) [50], whereas other catalysts gave variable amounts (4–14%) of trisaccharide **40b** arising from further glycosyl coupling of the liberated 6-OH group.

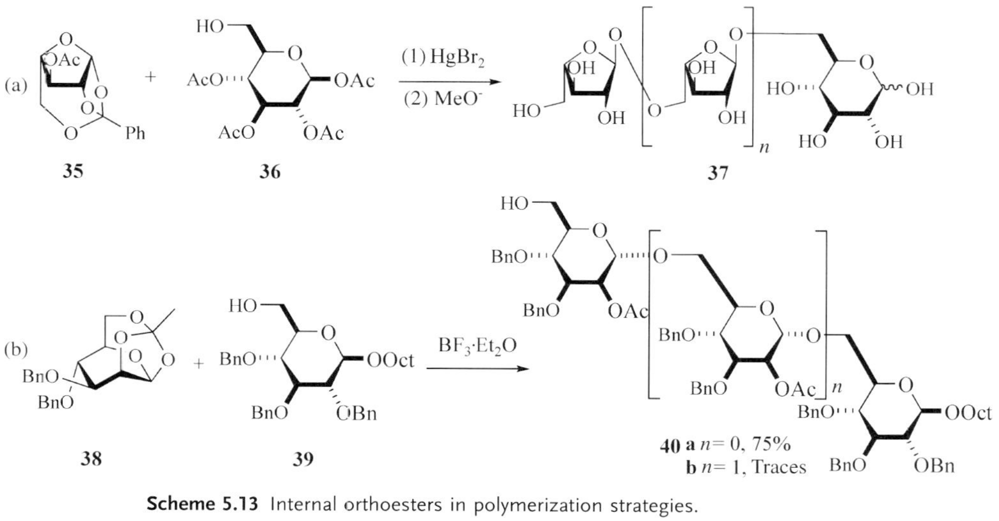

Scheme 5.13 Internal orthoesters in polymerization strategies.

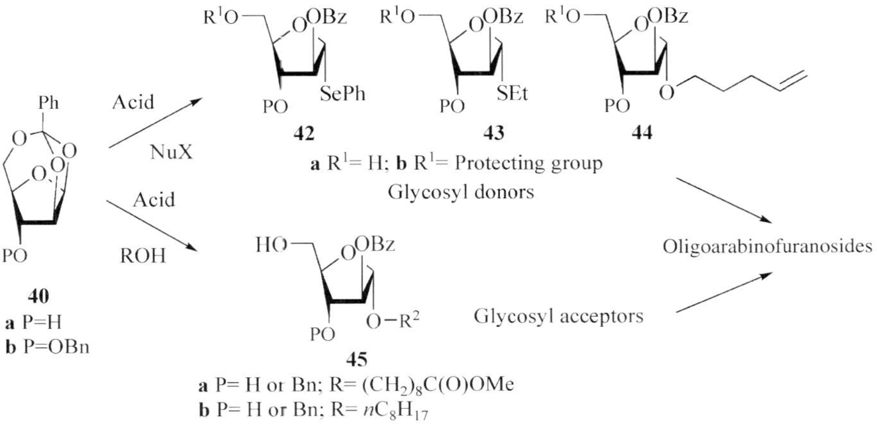

Scheme 5.14 Prandi's strategy to oligoarabinofuranosides from internal orthoester **41**.

Finally, Prandi and coworkers have described the use of D-arabinose 1,2,5-orthoesters, for example **41**, in a convergent–divergent strategy to oligoarabinofuranosides (Scheme 5.14) [51]. Acid-catalyzed opening of orthoesters **41** in the presence of selenophenol, thioethanol, 4-pentenol, followed by the protection of their 6-OH, gave raise to the corresponding selenyl- [52], thio- [53] and n-pentenyl- glycosyl donors [54] (**42b–44b**), whereas ring opening in the presence of alcohols furnished glycosyl acceptors **45**. Protecting-group manipulation and glycosyl assembly of these building blocks thus permitted the synthesis of oligoarabinofuranosides.

5.1.2.5 Miscellaneous Orthoesters

Kunz and Pfrengle reported the use of an oximate orthoester (**46**) in the $BF_3 \cdot Et_2O$-catalyzed glycosylation of some primary and secondary hydroxyl groups and phenols (70–80% yield) (Figure 5.1) [55]. A phosphite orthoester (**47**), isolated by Wong and coworkers as a side product in the attempted conversion of a glycosyl phosphite into a thioglycosyl derivative, reacted with thiocresol in the presence of TMSOTf to give the corresponding p-methylphenyl-1-thio-glycoside [56]. Griffith and Hindsgaul reported the formation of stable fluoro-orthoesters (**48**) in the reaction of 2,3,4,6-tetra-O-acetyl-α,β-D-glucopyranose with DAST. They could be rearranged to glycosyl fluorides in the presence of $BF_3 \cdot Et_2O$ [57].

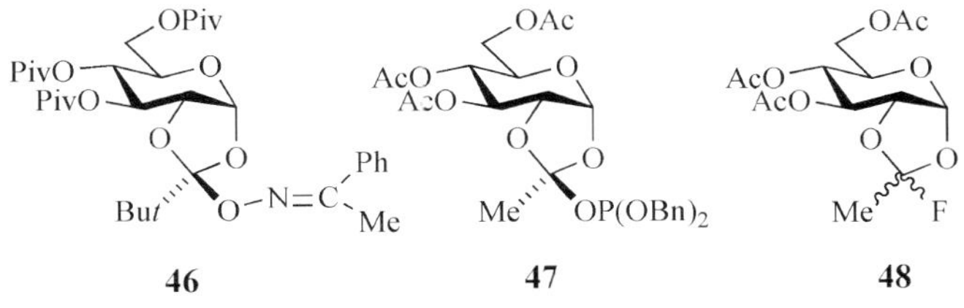

Figure 5.1 Miscellaneous orthoesters.

Scheme 5.15 Alternative strategies for efficient glycosylation with orthoesters.

5.1.3
Orthoester to Glycoside Rearrangement – The Two-Stage Glycosylation Method Revisited

The *two-stage glycosylation method* (**12** → **14** → **6**, Scheme 5.15) had been introduced to avoid problems related to the competitive incorporation of the acceptor (R^1OH) and the extruded alcohol (ROH) from **12**. In this context, more efficient glycosylation strategies (**4** → **14** → **6**, Scheme 5.15), which circumvent the competitive aspect, have been studied [58].

A remarkable example of this strategy came from Ogawa et al. [59]. They reported the regioselective preparation of bis-orthoester **51a** by the reaction of glycosyl chloride **55** with a partially stannylated [20] mannoside arising from **49** (Scheme 5.16). Rearrangement, after benzylation, of bis-orthoester **51b** took place in the presence of $HgBr_2$ at 120 °C to give trimannan **52** in 27% yield. Shortly after this work appeared, Ogawa et al. reported the value of TMSOTf, in replacing $HgBr_2$ as the reagent of choice for effecting orthoester to glycoside rearrangements [60].

Very recently, Kong has incorporated refinements to this approach and illustrated its potential with the regioselective synthesis of a large variety of oligosaccharides

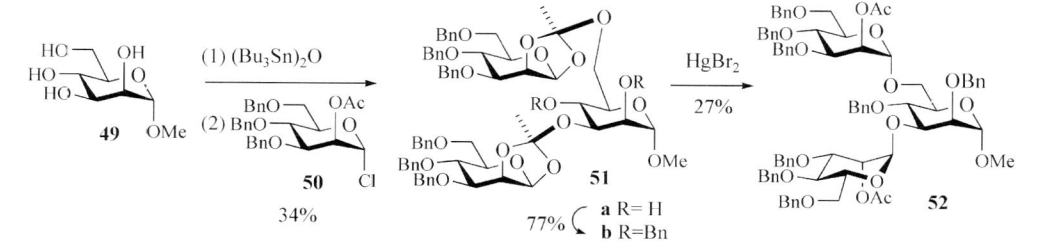

Scheme 5.16 Ogawa's synthesis of trisaccharide **52**.

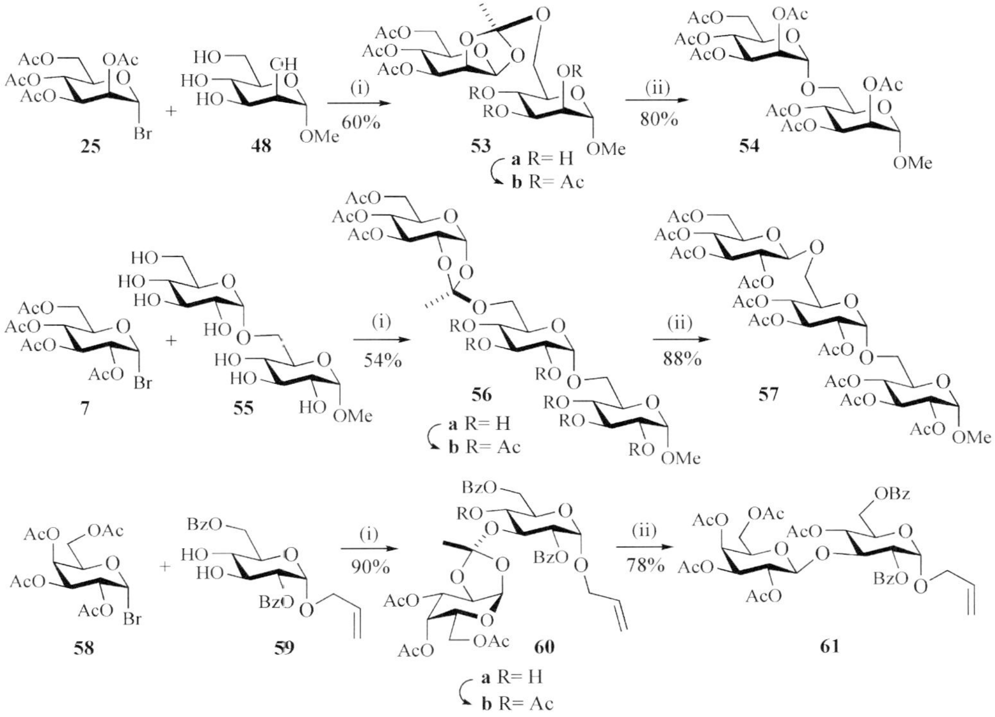

Scheme 5.17 Kong and coworkers approach to oligosaccharide synthesis. Reaction conditions: (i) 2,4-lutidine, AgOTf and molecular sieves; (ii) TMSOTf, 0 °C.

[61]. In his initial studies, he found that prior stannylation of the polyol acceptor was not a requisite for regioselective orthoester formation [62]. Indeed, a remarkable, see below, regioselectivity was found in the reaction of acetobromosugars and pyranosidic polyols leading to the formation of orthoesters (Scheme 5.17). Reaction of acetobromomannose (**25**), glucose (**7**) and galactose (**58**) with unprotected acceptors **48**, **55** and **59** in the presence of AgOTf and 2,6-lutidine [19] yielded mono-orthoesters **53a**, **56a** and **60a**, respectively, with complete regioselectivity. After the acetylation, the resulting orthoesters (**53b**, **56b**, **60b**) were transformed into the corresponding saccharides (**54**, **57**, **61**) by treatment with TMSOTf in CH$_2$Cl$_2$ at 0 °C.

Further application of this protocol, by Wang and Kong [63], led to the high-yielding synthesis of double glycosylated gluco- and mannoderivatives **64** and **67** by the reaction of bromoaldoses **25** and **7** with triols **62** and **65** (Scheme 5.18). Regioselective preference was found for the formation 1 → 6- (with acceptors containing unprotected 3,4,6-hydroxy groups) and 1 → 3- (with acceptors containing unprotected 2,3,4-hydroxy groups)-linked saccharides. This strategy was used to provide access to a series of 1 → 6- and 1 → 3-linked and 3,6-branched oligosaccharides such as **54**, **57**, **61**, **64** and **67** [64].

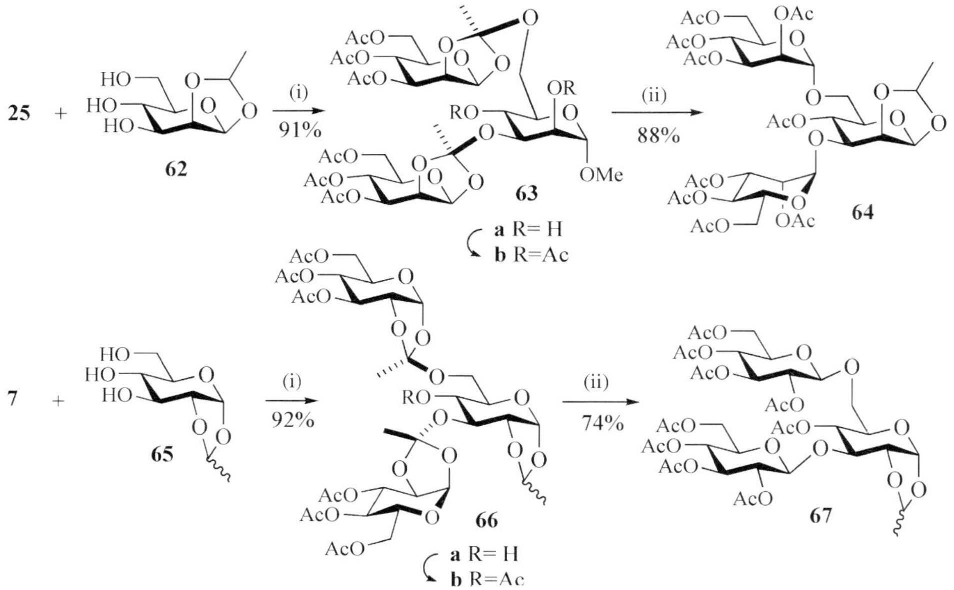

Scheme 5.18 Kong and coworkers one-pot method for saccharide synthesis. Reaction conditions: (i) 2,4-lutidine, AgOTf, molecular sieves; (ii) TMSOTf, 0 °C.

As the final regioselective outcome of the two-stage glycosylation has been ascribed to regioselection in orthoester formation (step 1, Scheme 5.19), one additional protection step (**14a** → **14b**, Scheme 5.20) was required to prevent hydroxyl scrambling during the final rearrangement step (**14b** → **6**, R = Ac), Scheme 5.19. Thus, the once two-stage method had to be converted into a three-step procedure if high regioselectivity was to be ensured. To avoid this, Kong and coworkers subjected hydroxylated orthoesters (e.g. **14a**, Scheme 5.19) to the action of TMSOTf and

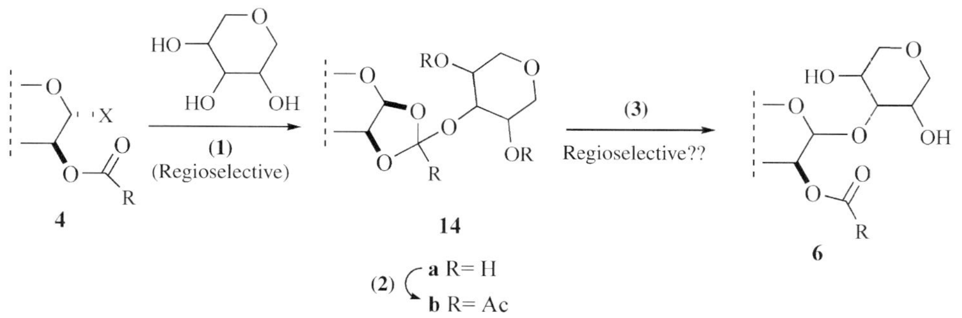

Scheme 5.19 Regioselectivity issue in the rearrangement of mixed orthoesters.

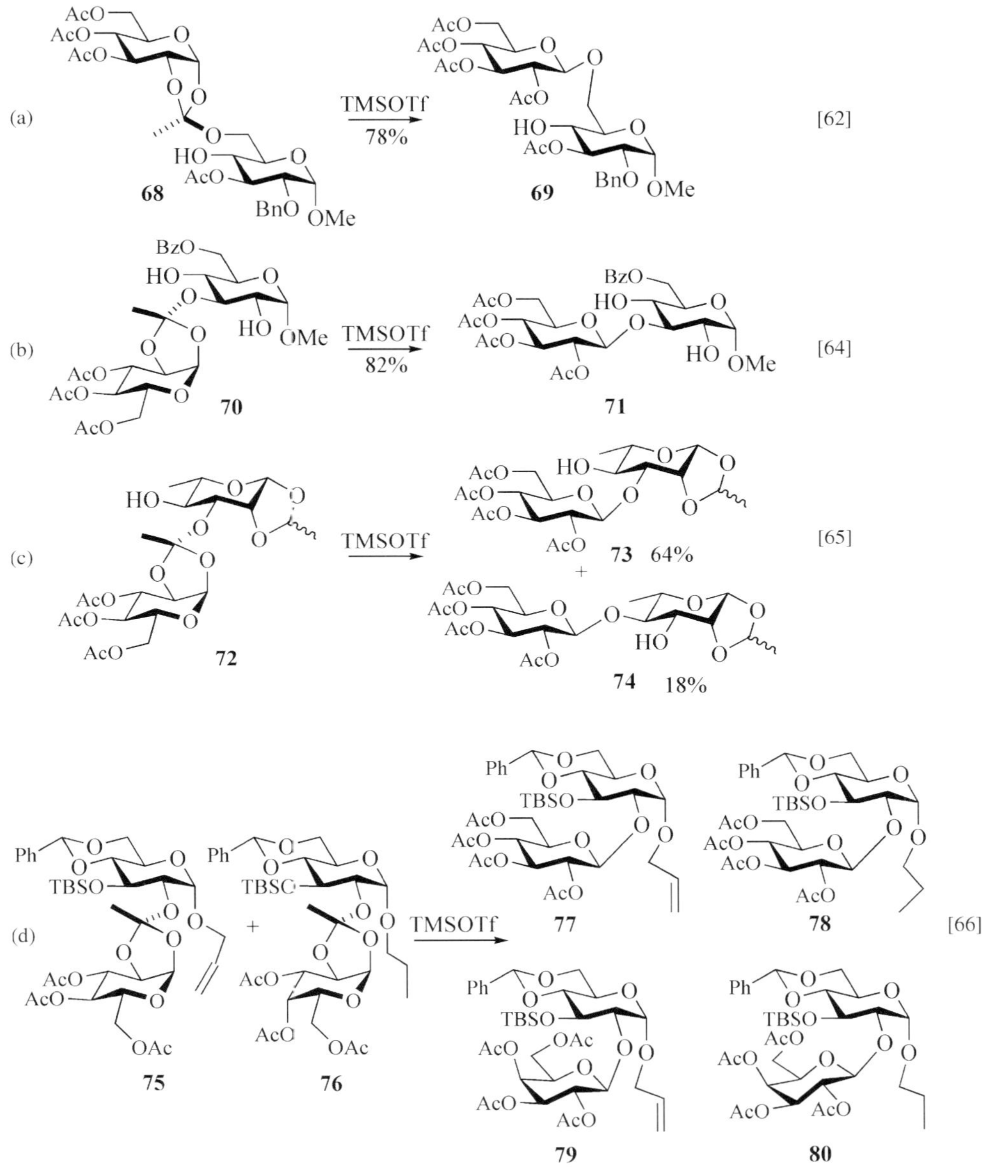

Scheme 5.20 Regioselectivity issue in the rearrangement of mixed orthoesters.

found that the rearrangement could sometimes be completely regioselective (Scheme 5.20a, b versus c) [62,64,65].

In an elegant experiment, Yu and coworkers showed that intermolecular *crossover* is also possible. Thus, the TMSOTf-catalyzed rearrangement of two structurally similar orthoesters **75** and **76** led to crossover products **77/78** and **79/80** (Scheme 5.20d) [66].

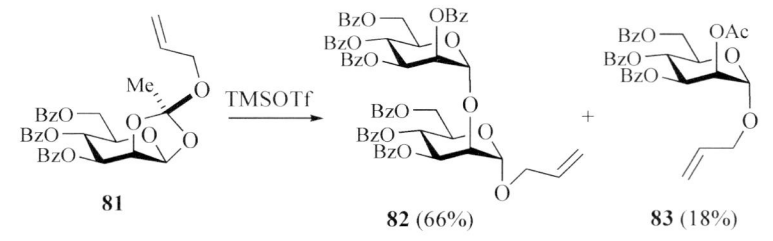

Scheme 5.21 Kong' approach to (1–2)-linked mannose oligosaccharides.

5.1.3.1 Self-Condensation of Mannose 1,2-Orthoesters: Ready Access to (1 → 2)-Linked Mannose Oligosaccharides

Although, as shown in Scheme 5.8, several reaction pathways are possible for a given orthoester (e.g. **12**), the transformation **12 → 17** is generally observed. Therefore, the presence of (usually minor) compounds such as **18** or **23**, arising from alternative reaction pathways (Scheme 5.8), is normally considered as the result of an undesired side reaction. However, recent studies by Kong and Zhu have revealed that the transformation **12 → 23** can be the main reaction course under appropriate conditions, and they have shown its usefulness for the synthesis of α-(1 → 2)-linked mannose disaccharides (Scheme 5.21) [67,68]. Thus, the treatment of allyl orthoester **81** with TMSOTf permitted the synthesis of α-(1 → 2)-linked disaccharide **82** in 66% yield, with the 'normal' rearranged allyl glycoside **83** also being obtained in 18% yield.

5.1.3.2 Rearrangement of Sugar–Sugar Orthoesters Leading to 1,2-cis-Glycosidic Linkages

The rearrangement of sugar–sugar orthoesters, as shown in Scheme 5.20, usually gives 1,2-*trans*-linked oligosaccharides. However, recent studies by Kong and Zhu have revealed that pure α-linked saccharides can be obtained by the glycosylation with glucosyl trichloroacetimidate donors with a C-2 ester capable of neighboring-group participation [69,70]. The authors assumed that the presence of 3-linked glucooligosaccharide–glucooligosaccharide orthoester intermediates, which might rearrange, gives rise to 1,2-*cis*-linked oligosaccharides.

5.1.4
n-Pentenyl-1,2-Orthoesters: Glycosyl Donors with Novel Implications

In 1988, a report from Fraser-Reid's laboratories drew attention to a novel anomeric leaving group [71]. *n*-Pentenyl glycosides (NPGs) were introduced as new derivatives that facilitated the chemospecific liberation of the anomeric center under *non*acidic conditions. The postulated mechanism for this transformation involved the participation of the exocyclic anomeric oxygen in the formation of a bromofuranylium ion (**86**), which triggered the formation of oxocarbenium ion **87** (Scheme 5.22a).

Although the first published *n*-pentenyl orthoester (NPOE) **89** appeared shortly after (Scheme 5.22b) [72], the initial link between NPGs and 1,2-orthoesters could be

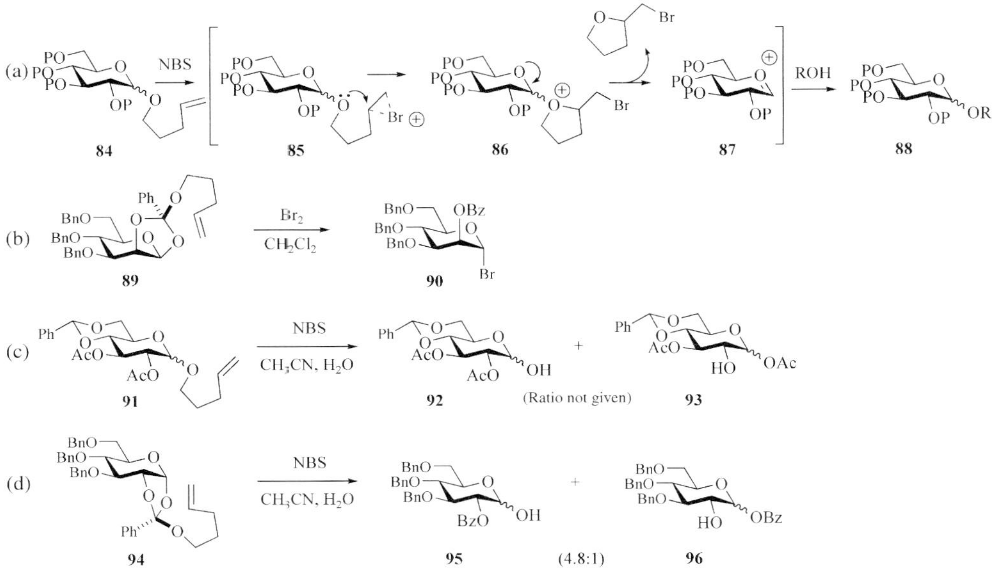

Scheme 5.22 n-Pentenyl glycosides and n-pentenyl orthoesters.

inferred from the formation of anomeric acetate **93** in the earlier contribution [71]. Thus, the formation of **93** as a side product in the oxidative hydrolysis (NBS, CH$_3$CN, H$_2$O) of **91** (Scheme 5.22c) could be rationalized through the rearrangement of either an orthoacid or an acyloxonium ion intermediate (see below). Indeed, a similar result (formation of **96**, Scheme 5.22d) was recently observed when *gluco*-NPOE **94** was treated under similar reaction conditions [73].

A rationale for the formation of **93** and **96** emerges from the seminal investigations of King [74], who established that bridging cations such as **100** (Scheme 5.23) scavenge water to generate unstable orthoacid intermediates (e.g. **101**). The latter undergo stereoelectronically driven [75] rearrangements favoring axially oriented esters on cyclohexyl scaffolds.

Disarmed glycosyl donors and orthoester analogs such as **97** and **99** (Scheme 5.23), upon appropriate treatment, give (potentially) equilibrating cations **98** and **100** and then products **101** and **102**. For substrates with alkoxy leaving groups (i.e. **97** and **99**, LVG = OAlkyl), this can be achieved by treatment with acidic reagents. However, for *n*-pentenyl analogs, activation can be carried out under nonacidic conditions, which brings about new possibilities. For example, it is possible to effect 'irreversible' *trans*-orthoesterification reactions that would be impossible under acidic conditions, as the work of the Kochetkov's school clearly shows [37]. Thus, the treatment of the *manno* and *gluco* NPOEs **89** and **105** with isopropanol under the agency of iodonium *sym*-collidinium triflate (IDCT) (Scheme 5.24) [73] led to the exchange of the alkoxy moieties in **104** and **107**, respectively.

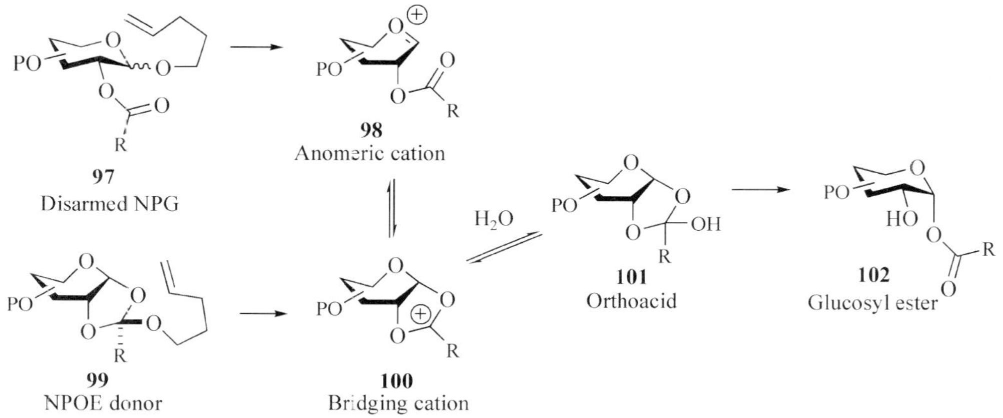

Scheme 5.23 Analogies between disarmed NPGs and NPOEs.

The first synthetic task assigned to an NPOE was the demanding glycosylation of an axial 2-OH group in the *pseudo*tetrasaccharide **109**, *en route* to the pseudopentasaccharide core of the protein membrane anchor found in *Trypanosoma brucei* that was obtained in 68% yield (Scheme 5.25) [76].

5.1.4.1 Divergent–Convergent Synthesis of Glycosylaminoglycan 120 from Glycosyl Donors and Acceptors Ensuing from NPOEs

The first extensive use of NPOEs as key intermediates in oligosaccharide synthesis was undertaken by Allen and Fraser-Reid with the synthesis of glycosylaminoglycan **120** (Scheme 5.26) [77]. Their approach benefited from the uniqueness of the NPOE → NPG transformation in combination with the sidetracking of NPGs [78]. In the synthetic strategy, NPOEs of D-galactose and D-xylose **111** and **112**, respectively, furnished all the required glycosyl donors and acceptors (Scheme 5.26a).

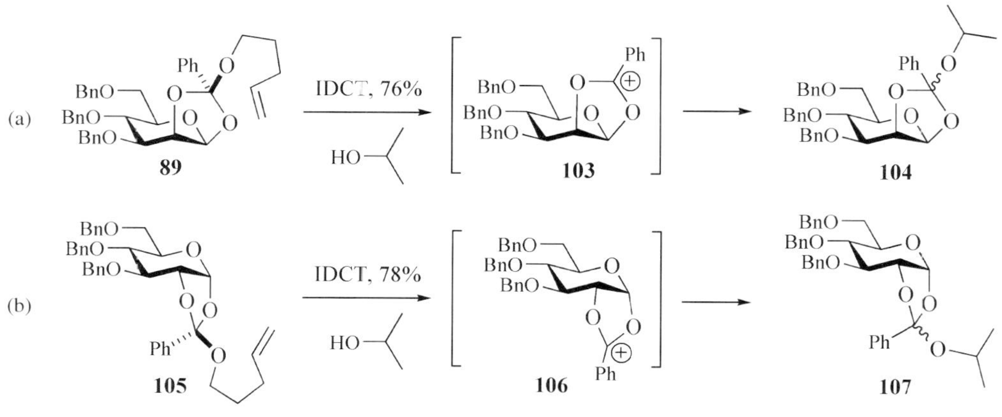

Scheme 5.24 Reaction of NPOEs promoted by IDCT.

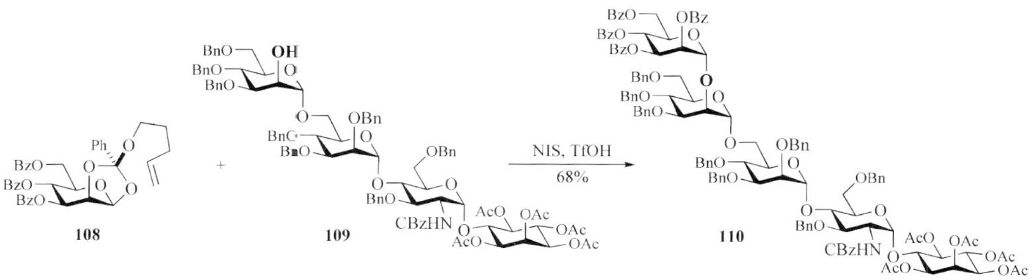

5.1 Orthoesters and Related Derivatives | 397

Scheme 5.25 First utilization of an NPOE in oligosaccharide synthesis.

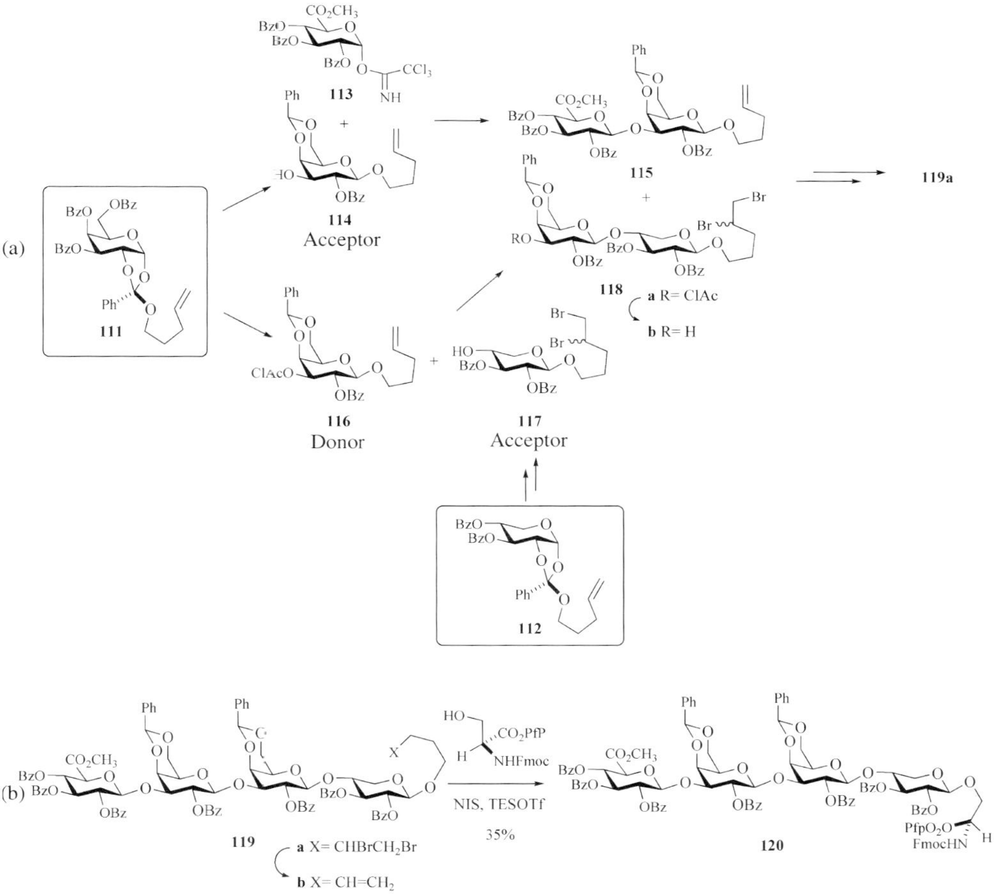

Scheme 5.26 Divergent–convergent use of NOPEs in oligosaccharide synthesis.

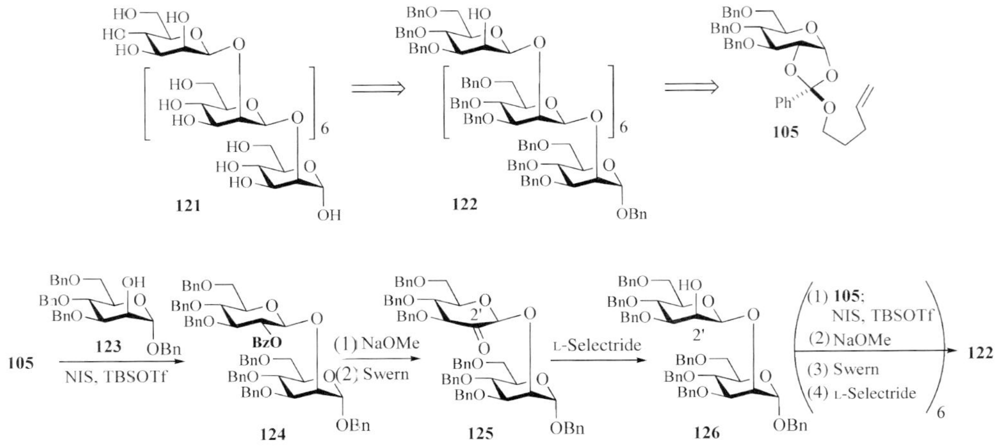

Scheme 5.27 NPOE-based strategy for preparation of β-mannosides.

Accordingly, orthoester **111** was transformed into NPGs **114** and **116**. The former functioned as an acceptor to the trichloroacetimidate **113** to yield **115**. The latter was then used as a donor to glycosylate sidetracked xylosyl acceptor **117**, prepared in turn from xylose orthoester **112**, to yield dibromopentanyl derivative **118**. The final glycosylation of **118** with donor **115** provided the sought tetrasaccharide **119a**, which was used in the final transformations: **119a** → **119b** → **120** (Scheme 5.26b).

5.1.4.2 From NPOEs to the 1,2-β-Linked Oligomannans of *Candida albicans*

Fraser-Reid and coworkers described an entirely NPOE-based protocol for the synthesis of the β-1,2-linked oligomannan components of *C. albicans*, **121** (Scheme 5.27) [79], using glucosyl orthoester **105** as the key building unit. The strategy benefited from two characteristics inherent to orthoesters: (a) the 1,2-*trans* selectivity in their glycosyl coupling and (b) the protecting-group differentiation at O-2 that takes place after glycosidation. Accordingly, the reaction of **105** with **123**, under the agency of NIS/TBSOTf exclusively furnished disaccharide **124**, bearing a strategically important 2′-*O*-benzoyl group. The manipulation of the latter through a sequence involving saponification, oxidation and stereoselective reduction yielded *manno*-disaccharide **126** via ulose **125**. The iteration of the process, up to six times, produced protected octasaccharide **122**, in yields ranging from 82 to 93% per cycle. Final deprotection of **122** (HCOOH, Pd/C, MeOH, room temperature) led to **121**.

5.1.4.3 From NPOEs to the Synthesis of a Malaria Candidate Glycosylphosphatidylinositol (GPI)

A congener of the GPI membrane anchor, present on the cell surface of the malaria pathogen *Plasmodium falciparum* **127**, was also tackled. The molecule is composed of one inositol moiety (**I**) and four monosaccharide units (**II–V**) [80]. The retrosynthesis identified a single NPOE retron **108** as the precursor for all glycan units (Scheme 5.28). Transformations of the latter afforded NPOE analogs **128**, **89**,

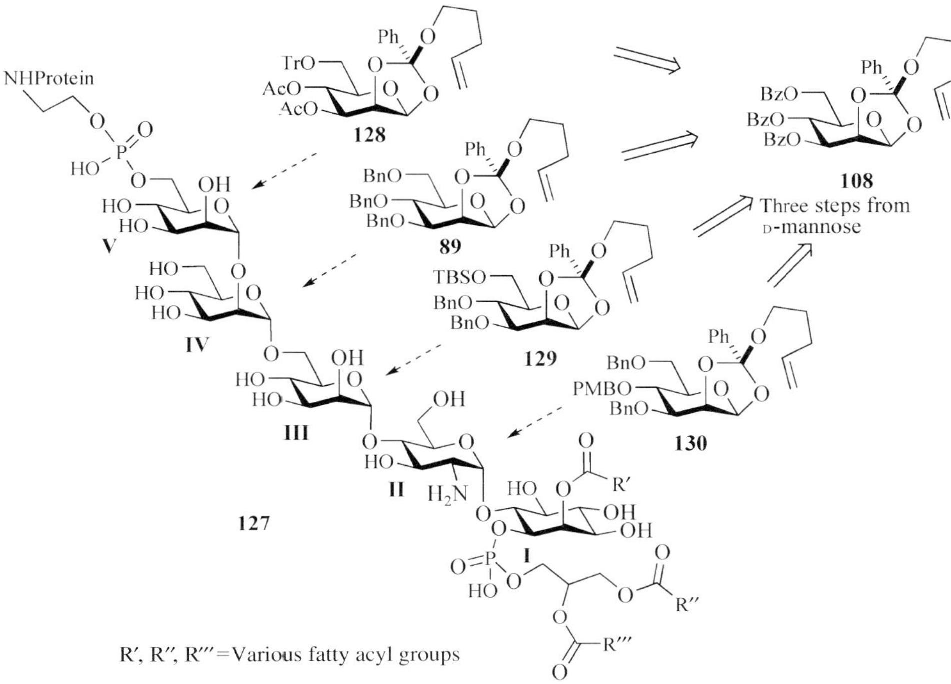

Scheme 5.28 Retrosynthesis of **127** from a single NPOE, **108**.

129 and **130**, which are correlated with glycan units **V–II**, respectively, as indicated in Scheme 5.28.

The synthetic protocol (Scheme 5.29) followed the guidelines shown in Scheme 5.28. Thus each NPOE **89**, **128–130** was specifically designed to allow the liberation of the required OH-group for further coupling or processing. The GPIs α-glucosaminide component (unit **II** in **127**, Scheme 5.28) requires special comment [81,82]. The glycosylation of inositol derivative **131** with NPOE **130** gave pseudodisaccharide **132a** in 98% yield, and the convenient 2-O-acyl group, after being replaced by a triflate, underwent azide displacement [83] to yield 2-azido-pseudodisaccharide **133**. Saccharide growth was then effected by sequential glycosylation with NPOEs **130**, **89** and **128**. All NPOE-mediated glycosyl couplings took place with excellent yields and paved the way to key intermediate **136**. Finally, the elaboration of **136** into **137** required the following: (1) attachment of the phosphoethanolamine complex [84] at unit **V**, (2) regioselective incorporation of the glyceryl chains [81] at the inositol moiety (**I**, Scheme 5.28) and (3) global debenzylation and azide reduction to isomeric **137a** and **137b** that differ by the acyl chains.

5.1.4.4 From NPOEs to the Preparation of Glycolipids for Multivalent Presentation

NPOEs have been employed in a straightforward route for the preparation of multivalent presentations of the trimannan array present at the distal end of

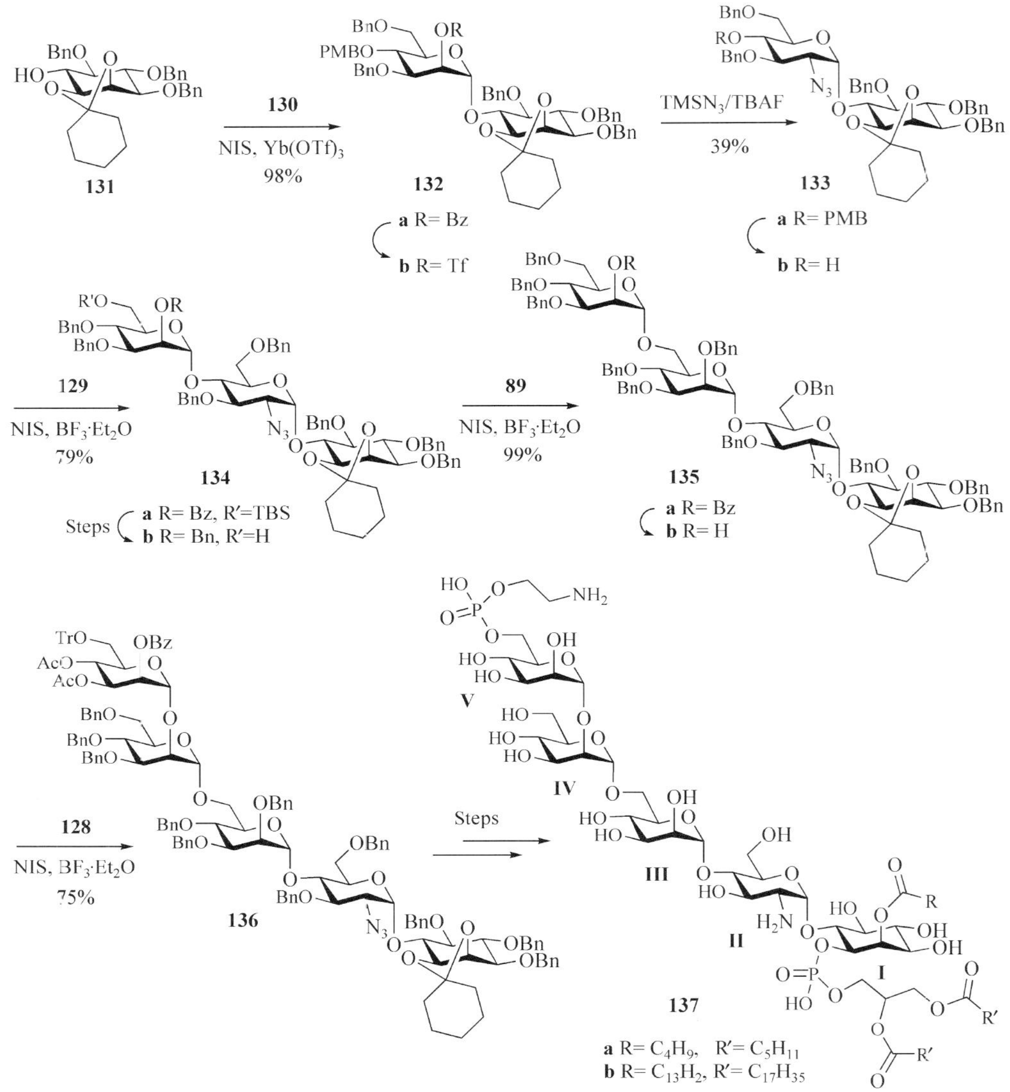

Scheme 5.29 Preparation of GPI **137** from NPOEs.

membrane-anchored GPIs (Scheme 5.30) [85]. The strategy for the preparation of mono- and divalent neoglycolipids relied on the unique characteristics of NPOEs for ensuring the following: (a) high-yielding glycosyl couplings, (b) rearrangement to NPGs, (c) chemoselective activation of NPOEs and (d) the potential of the *n*-pentenyl chain to be utilized itself in tethering processes [86].

The sole source progenitor was NPOE **108** by way of counterparts **89** and **129**. The synthesis started with the rearrangement of NPOE **89** to NPG **138** (TBSOTf, 92%) that

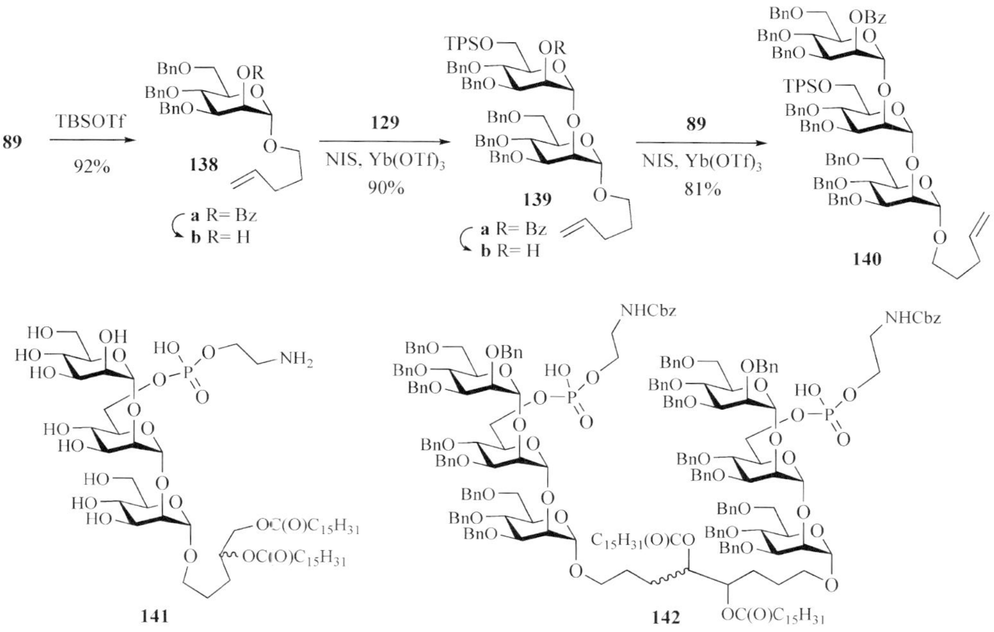

Scheme 5.30 Use of NPOEs for multivalent presentations.

was homologated to **139** and **140** by sequential de-*O*-benzoylation followed by the glycosylation with NPOE donors **129** and **89**, respectively. The exquisite chemoselective activation of NPOEs by NIS/Yb(OTf)$_3$ in the presence of NPGs [87] allowed the olefinic residue, in trisaccharide **140**, to remain unaltered for further processing. Notably, this chemoselective feature dispenses with the earlier 'sidetracking' strategy [77], where dibromination was required to neuter one of the pentenyl residues (e.g. Scheme 5.26a).

Hydroxylation of the terminal double bond in **140** followed by the esterification paved the way to phosphodiester **141**, whereas Grubbs' olefin metathesis followed by hydroxylation and esterification led to divalent phosphodiester **142**.

5.1.4.5 The Lipoarabinomannan Components of the Cell Wall Complex of *Mycobacterium tuberculosis*: NPOEs in Chemoselective, Regioselective and Three-Component Double Differential Glycosidations

The lipoarabinomannan (LAM) capsule is the major virulence factor of *M. tuberculosis* [88]. The multifaceted architecture, shown as **143** (Figure 5.2) by Turnbull *et al.* [89], anticipates the complexity of any synthetic scheme aimed to its conquest. However, a simple analysis of the molecule shows that the *mannan* and *arabino* segments possess 1,2-*trans* glycosidic linkages, which point toward a retrosynthesis in which NPOEs could be the donors of choice.

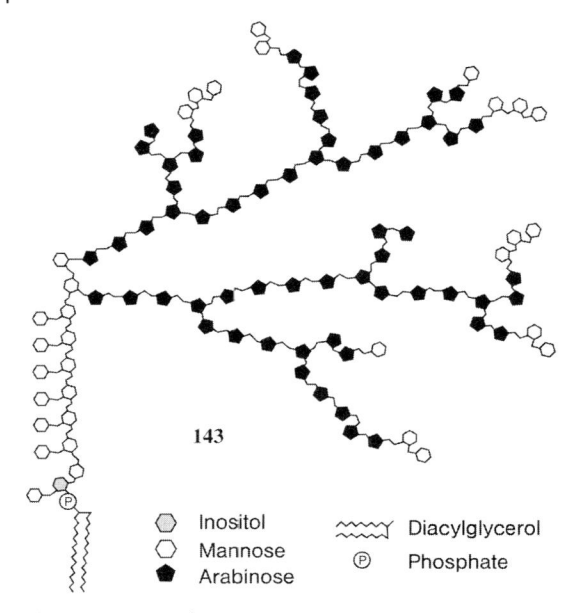

Figure 5.2 Lipoarabinomannan **143**.

The Inositol Core of Lipoarabinomannans: NPOE Donors in Chemo- and Regioselective Three-Component Double Differential Glycosidations Initial independent studies, by van Boom's [90] and Fraser-Reid's [91] groups, on the preparation of the pseudotrisaccharide core of LAM **144**, by mannosylation of inositol diol **145**, had made it clear that the choice of the glycosyl donor and/or the protecting group at O-1 was crucial for optimal results (Scheme 5.31). This observation is in keeping with the concept of matching/mismatching phenomena in glycosyl couplings [92,93].

Scheme 5.31 Retrosynthesis of **144**.

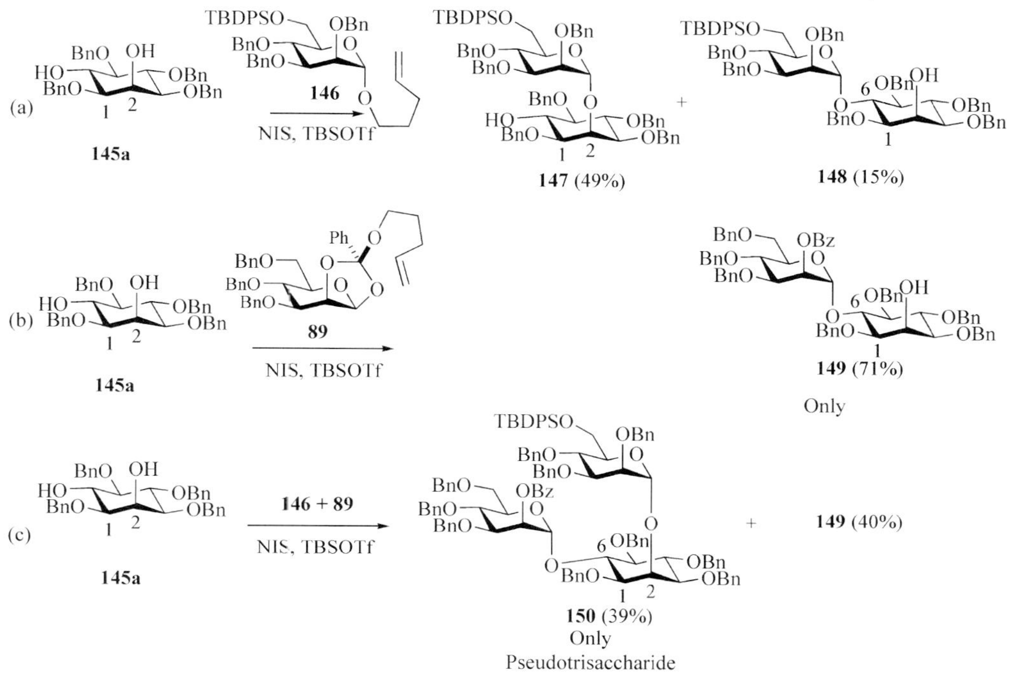

Scheme 5.32 Regioselective *in situ* three-component synthesis of **150**.

Subsequent studies by Anilkumar *et al.* [94] revealed that the glycosylation of a diol **145a** with NPG **146** was weakly regioselective, giving rise to a 3 : 1 mixture of pseudodisaccharides **147** and **148** arising from glycosylation at O-2 and O-6, respectively (Scheme 5.32a). By contrast, the glycosylation of **145a** with NPOE **89** furnished exclusively pseudodisaccharide **149** from glycosidation at O-6 (Scheme 5.32b). This different behavior, although not yet completely understood, permitted the authors to devise an *in situ*, site-selective protocol for double glycosidation of the inositol core of the LAM antigen **150** (Scheme 5.32c). The successful implementation of this strategy benefits not only from the exquisite regioselectivity of NPOE **89** but also from the possibility to chemoselectively activate NPOEs in the presence of armed NPGs. Notably, the more reactive NPOEs also display higher regioselectivity in glycosyl couplings than that achieved with NPGs.

The Lipomannan Component of LAM: Regioselective Couplings with NPOEs The architecture of compound **143** (Figure 5.2) indicated that the inositol phospholipids moiety is linked to the oligomannan portion of LAM. The retrosynthetic route to structure **163** again used NPOE **119** as the sole source for the *manno* components [95]. The glycosyl unit correlation in this approach is outlined in Scheme 5.33.

The synthesis was carried out by four successive glycosylations with NPOE **89**, which yielded a linear, α-linked 1,6-mannan chain, with benzoate esters strategically

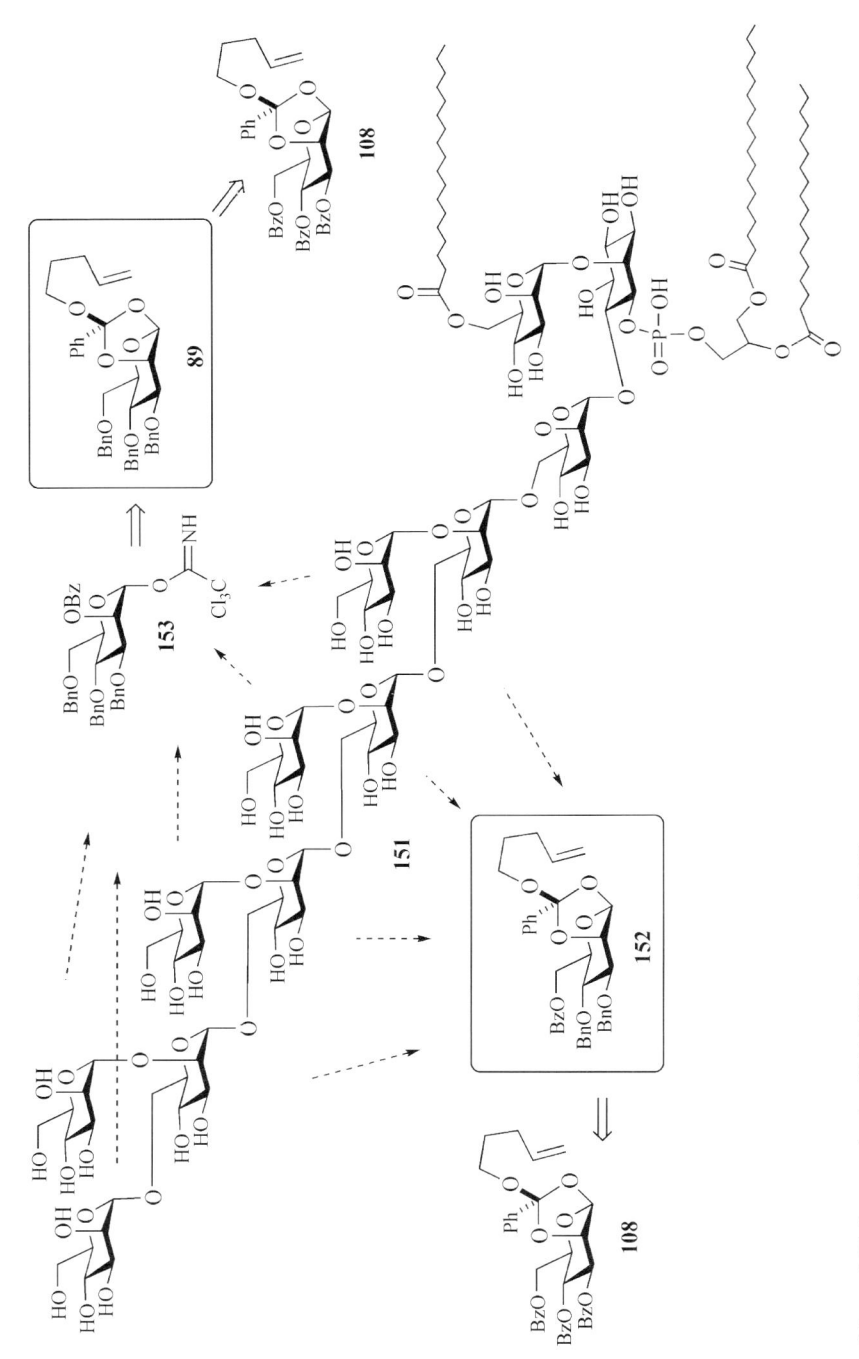

Scheme 5.33 Retrosynthesis of LAM **151** from two NPOE precursors.

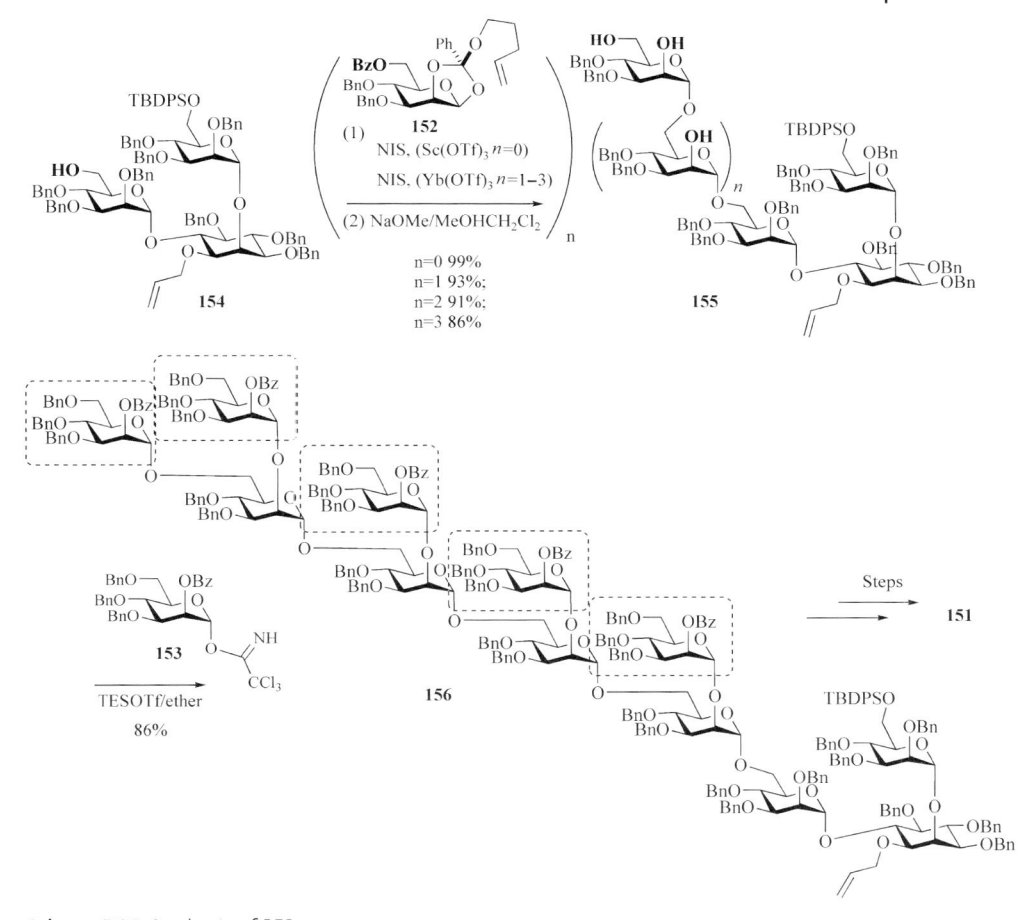

Scheme 5.34 Synthesis of **151**.

located at all O-2 sites (Scheme 5.34). Saponification of these, followed by exhaustive glycosylation with trichloroacetimidate **153**, readily prepared from NPOE **89**, achieved the one-pot incorporation of all five mannose branches. The synthesis was completed by the protecting-group manipulations that allowed the incorporation of stearoyl ester and phosphoglycerolipidation, using the previously credited procedure [81,82]. Finally, exhaustive debenzylation and purification led to **151**.

5.1.4.6 Relevance of NPOEs to the Regioselectivity in the Glycosylation of Primary Versus Secondary Hydroxyls

The conventional thinking in organic synthesis indicates that primary hydroxyls are more reactive than secondary hydroxyls [96]. Under this assumption, the glycosylation of triol **157** to furnish diol **158** by exclusive glycosylation at the primary-OH might seem to be routine laboratory experience (Scheme 5.35a). However, recent studies by López and coworkers have shown that the 'primary versus secondary' issue in regioselective glycosylation is not that simple and that the choice of the glycosyl donor plays a determinant role in eliciting regioselectivity [97].

Diol **159** was chosen as a model for probing regioselective glycosylation with armed and disarmed thioglycosides **160** and **163**, and NPOE **108**. The results obtained were not anticipated (Scheme 5.35b–d).

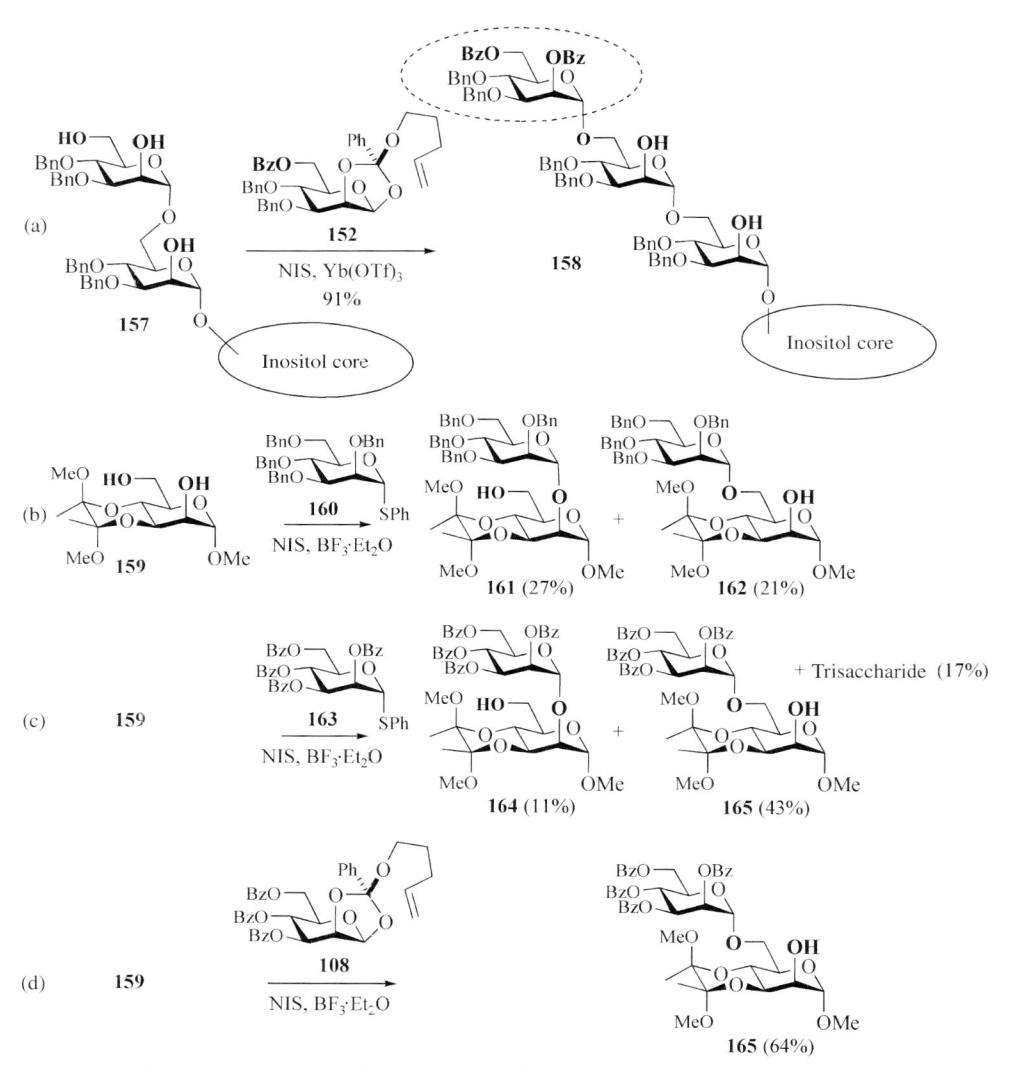

Scheme 5.35 The issue of regioselectivity in the glycosylation of primary versus secondary hydrosyls.

Remarkably, armed glycosyl donor **160** preferred the secondary-OH of **159** (Scheme 5.37b). Disarmed donor **163** and NPOE **108** both 'preferred' the primary-OH, but the NPOE furnished a single regioisomer **176** (Scheme 5.35c and d).

These results clearly demonstrate that (a) it is relevant (and importance) to match glycosyl donors and glycosyl acceptors to achieve regioselective couplings, (b) an absolute reactivity order cannot be ascribed to the glycosyl acceptor's hydroxyl groups and by corollary (c) NPOE regioselectivities (e.g. **157** → **158**, Scheme 5.35a and **154** → **155**, Scheme 5.34) cannot be regarded as 'trivial' or routine.

5.1.4.7 Iterative Regioselective Glycosylations of Unprotected Glycosyl Donors and Acceptors

In Scheme 5.35, the issue of regioselectivity is focused on polyol glycosyl acceptors. By contrast, examples in which the glycosyl donor itself has one or more free hydroxyl groups are rarely seen [98–100]. Encouraged by the regioselectivity displayed by NPOE donors, López *et al.* studied the regioselective couplings of NPOE diol **167** with polyol acceptors (Scheme 5.36) [101]. Reaction with methyl glucoside **166** furnished a single compound **168**, thus showing that self-coupling of **167** was not a competing reaction. Triol **169**, obtained by the deprotection of silyl ether **168** with TBAF, was next tested as a glycosyl acceptor. Again only one compound, **170** ($n = 1$), was obtained in the reaction of the NPOE donor diol **167** with triol **169**. The iteration of the process, up to three times, permitted the synthesis of pentasaccharide **170** ($n = 3$) with nine free-OH

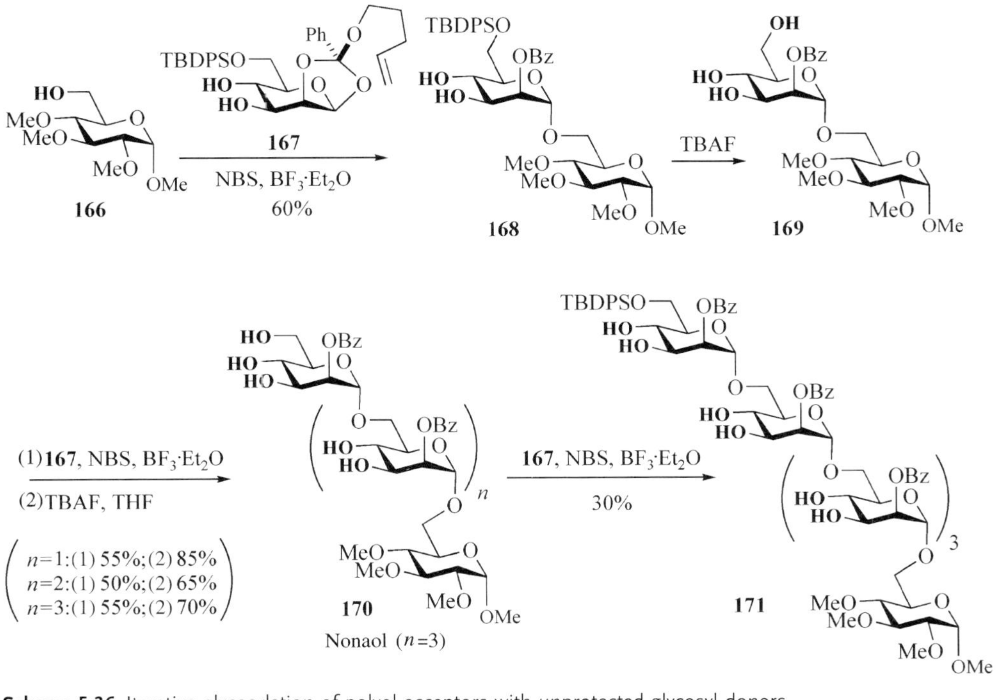

Scheme 5.36 Iterative glycosylation of polyol acceptors with unprotected glycosyl donors.

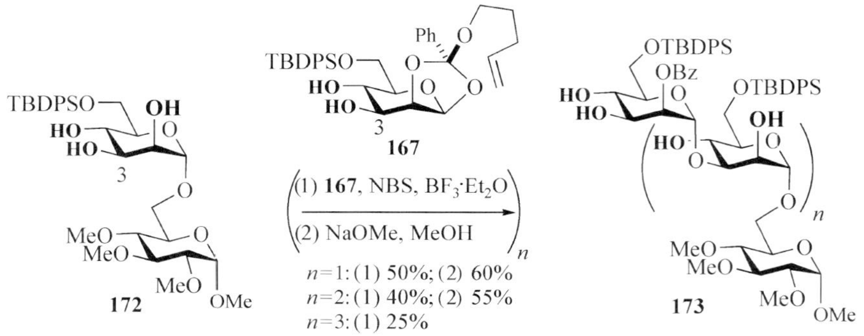

Scheme 5.37 Iterative glycosylation of secondary triols with NPOE diol **167**.

groups, resulting from the glycosylation of a heptaol acceptor with diol **167**. Glycosidation of **170** with **167** then gave hexasacharide **171** with 10 free-OH groups.

In none of the above cases were significant amounts of products from self-condensation of **167** seen.

The synthesis of a hexasaccharide by successful differentiation between one primary OH group and up to 10 secondary OH groups (eight in the acceptor plus two in the donor) prompted the authors to explore the feasibility of oligosaccharide synthesis through secondary versus secondary hydroxyl group selectivity. Accordingly, the glycosylation of triol **172** with NPOE **167** resulted in the formation of one single trisaccharide **173** ($n = 1$) in 50% yield by selective glycosylation at O-3, thus showing that discrimination was also possible among secondary OH groups (Scheme 5.37). The iteration of the protocol two more times resulted in the preparation of octaol pentasaccharide **173** ($n = 3$).

5.1.4.8 NPOEs of Furanoses: Key Intermediates in the Elaboration of the Arabino Fragment of LAM

The multibranched dodeca-arabinosaccharide **174** (Scheme 5.38) is a relevant part of the of LAM glycolipid of *M. tuberculosis*, illustrated in Figure 5.2. The structure features a repeated 3,5-branched motif, represented by the encircled units, along with linear catenated *arabino*-1 → 5 units. Lu and Fraser-Reid proposed two retrosynthetic plans (based on dissections **A** and **B**, Scheme 5.38), which employ as the sole source NPOE **175** [102,103]. In option **A**, a nona-arabinan donor would be delivered to the primary hydroxyl of a tri-arabinan acceptor, whereas in option **B** two identical tetra-arabinan donors would be delivered to the tetra-arabino acceptor.

In preliminary studies, Lu and Fraser-Reid had reported the efficient assembly of tetra-arabino derivative **178** by iterative couplings of donor **176** with NPG **177**, also accessible by TBDPSOTf-catalyzed rearrangement of **176** (Scheme 5.39a) [104]. They applied a similar protocol to the preparation of tetra-arabinan diol acceptor **182b** (Scheme 5.39b). In this case, the protected ethanolamine **179** [84] was used as the first acceptor for the iterative sequence with NPOE donors **176** to obtain **180**, and then with **181** to obtain **182a**, which on treatment with thiourea furnished 3,5-diol **182b**.

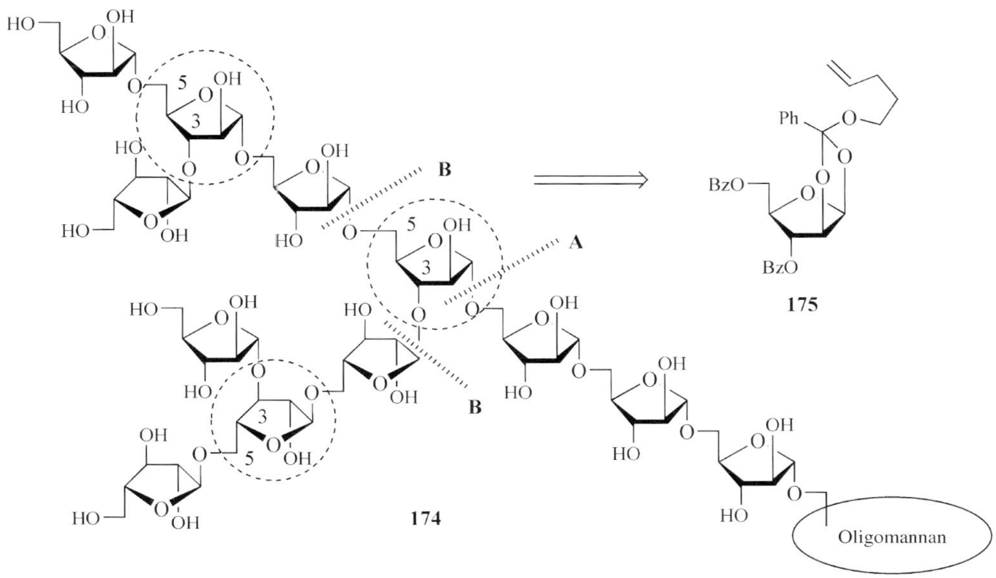

Scheme 5.38 Retrosynthesis of oligoarabinan **174** from NPOE **175**.

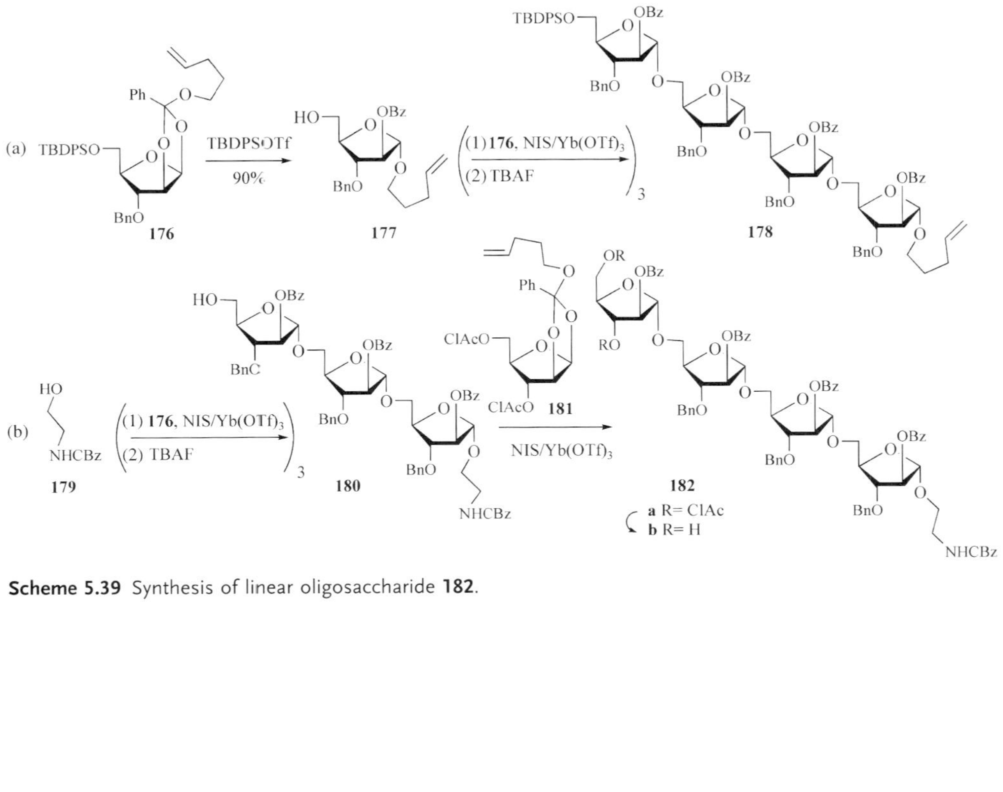

Scheme 5.39 Synthesis of linear oligosaccharide **182**.

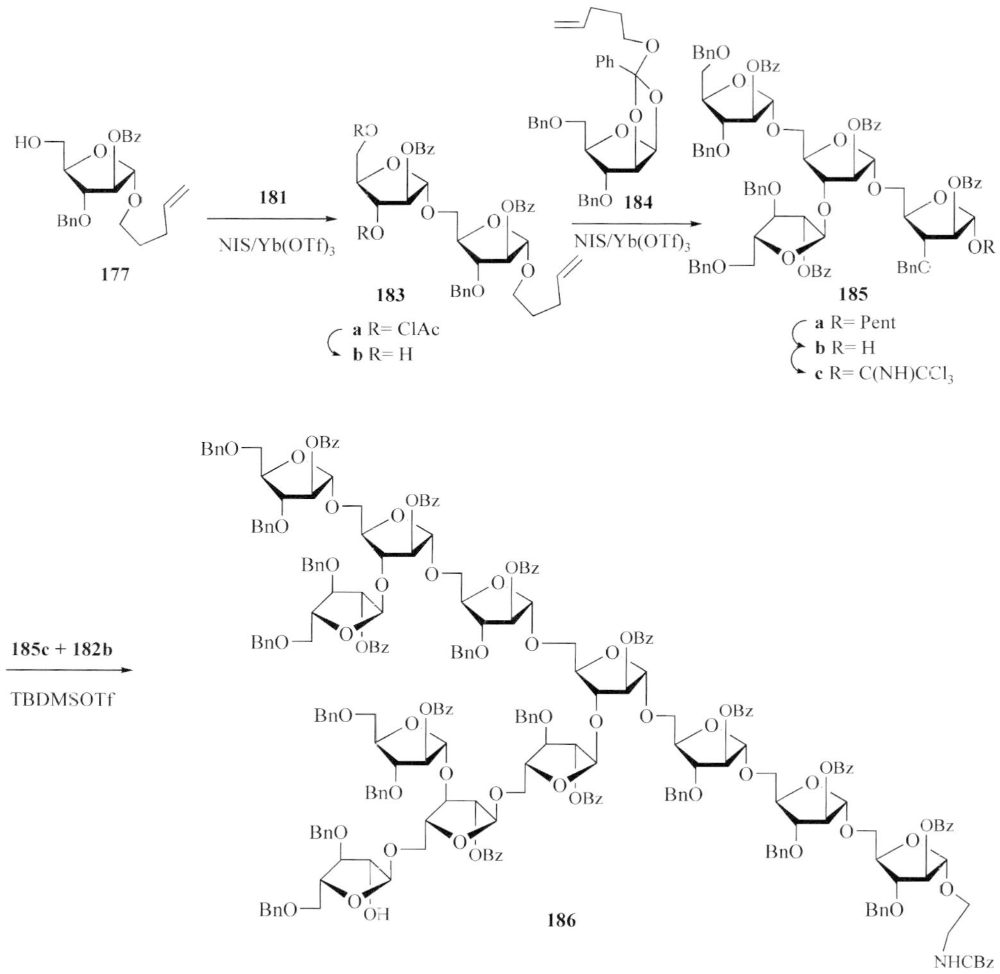

Scheme 5.40 Final assembly of arabinnan **186**.

The synthesis of the branched donor unit **185a** was carried out by using NPOE **184** to bis-glycosylate diol **183b** (Scheme 5.40), the latter having been easily prepared by the glycosylation of **177** with NPOE donor **181** followed by thiourea deprotection. The final step of the convergent synthesis (**B**, Scheme 5.38) was the glycosyl coupling of trichloroacetimidate **185c** with the tetrasaccharide acceptor **182b**.

5.1.5
Conclusions and Future Directions

In conclusion, NPOEs, which can be prepared and manipulated routinely, have excellent shelf life, but can yet be readily activated by the action of mild reagents.

Their value is enhanced by the facility with which they serve as progenitors to other donors, some of which are too reactive for prolonged storage, and also to other less readily prepared counterparts, such as NPGs, thioglycosides and glycosyl fluorides. These properties mean that NPOEs can be selectively, and even chemospecifically, targeted, thereby enabling one-pot sequences as well as *in situ* glycosylation(s). The recent extension to furanose systems is propitious in view of the generally greater reactivity *vis-a-vis* pyranoses. In this connection, the masked O-2 benzoate is extremely valuable, as it provides a nascent disarming device that confers stability on reactive species such as the trichloroacetimidate **185c**.

5.1.6
Typical Experimental Procedures

5.1.6.1 General Procedure for the Preparation of Orthoesters

The corresponding bromide (0.111 mol) was dissolved in dry CH_2Cl_2 (250 ml), and 2,6-lutidine (20 ml, 0.172 mmol), 4-pentenyl alcohol (14 ml, 0.14 mol) and tetra-*n*-butylammonium iodide (2.0 g, 5.4 mmol) were added. The resulting mixture was refluxed under argon for 24 h, during which TLC (hexane/ethyl acetate) showed complete disappearance of the starting material. After cooling to room temperature, water and diethyl ether were added, and the organic layer was washed with water and with brine and dried. After evaporation the residue was filtered through silica gel using hexanes/ethyl acetate (from 9:1 to 4:1) to effect partial purification of polar impurities.

5.1.6.2 General Procedure for Glycosidation with *n*-Pentenyl Orthoesters

General Glycosylation Procedure with NIS/BF$_3$·OEt$_2$ To a solution of glycosyl acceptor and NPOE (1.1 equiv) in dry CH_2Cl_2 were added 5 Å molecular sieves (1 mg mg^{-1} donor) and *N*-iodosuccinimide (1.5 equiv) at $-30\,°C$. After stirring the mixture for 5 min, BF$_3$·OEt$_2$ (0.045 mmol) was added. The reaction was monitored by TLC and after acceptor disappearance was observed, it was quenched with 10% aqueous sodium thiosulfate and saturated aqueous bicarbonate solutions. The reaction mixture was extracted with CH_2Cl_2 and dried over anhydrous sodium sulfate. The solvents were removed and the residue was purified by flash chromatography.

General Glycosylation Procedure with NIS/TBDMSOTf   The glycosyl acceptor and the NPOE (1.1 equiv) were dissolved in a small quantity of toluene, azeotroped to dryness and then dissolved in dry CH_2Cl_2 at $0\,°C$ under argon atmosphere. *N*-Iodosuccinimide (1.2 equiv) was added to the solution, and after stirring for 3 min, TBDMSOTf (0.25 equiv) was added. The reaction was quenched with 10% aqueous sodium thiosulfate and saturated sodium bicarbonate, extracted with CH_2Cl_2 and worked up as above. After work-up the mixture was purified by flash chromatography.

General Glycosylation Procedure with NIS/Yb(OTf)$_3$ The glycosyl acceptor and the NPOE (2.2 equiv) were dissolved separately in small amounts of toluene, and the solutions were evaporated to dryness and kept overnight under vacuum. The acceptor was dissolved in CH_2Cl_2, the solution cooled to $0\,°C$, NIS (2.5 equiv) was added, followed by the addition of $Yb(OTf)_3$ (0.3 equiv). After stirring for a few minutes, a methylene chloride solution of the NPOE was then added dropwise over 15 min. The reaction was quenched with 10% aqueous sodium thiosulfate and saturated aqueous bicarbonate, extracted with CH_2Cl_2 and dried. The resulting mixture was purified by flash chromatography.

References

1 Trost, B.M. (1983) *Science*, **219**, 245–250.
2 Michael, A.J. (1879) *American Chemical Journal*, **1**, 305.
3 Koenigs, W. and Knorr, E. (1901) *Chemische Berichte*, **34**, 957–981.
4 Kochetkov, N.K., Khorlin, A.J. and Bochkov, A.F. (1964) *Tetrahedron Letters*, **5**, 289–293.
5 Paulsen, H. (1990) *Angewandte Chemie (International Edition in English)*, **29**, 823–839.
6 Frush, H.L. and Isbell, H.S. (1941) *Journal of Research of the National Bureau of Standards*, **27**, 413.
7 Fraser-Reid, B., López, J.C., Gómez, A.M. and Uriel, C. (2004) *European Journal of Organic Chemistry*, 1387–1395.
8 Pacsu, E. (1945) *Advances in Carbohydrate Chemistry and Biochemistry*, **1**, 77–127.
9 Fischer, E., Bergmann, M. and Rabe, A. (1920) *Chemische Berichte*, **53**, 2362–2368.
10 Freudenberg, K. and Scholz, H. (1930) *Chemische Berichte*, **63**, 1969–1972.
11 Braun, E. (1930) *Chemische Berichte*, **63**, 1972–1974.
12 Bott, H.G., Haworth, W.N. and Hirst, E.L. (1930) *Journal of Chemical Society*, 1395–1405.
13 (a) Haworth, W.N., Hirst, E.H. and Stacey, M. (1931) *Journal of Chemical Society*, 2864–2872. (b) Fletcher, H.G., Jr and Ness, R.K. (1955) *Journal of the American Chemical Society*, **77**, 5337–5340. (c) Ness, R.K. and Fletcher, H.G., Jr (1954) *Journal of the American Chemical Society*, **76**, 1663–1667. (d) Lemieux, R.U. and Brice, C. (1955) *Canadian Journal of Chemistry*, **33**, 109–119. (e) Lemieux, R.U. and Cipera, J.D.T. (1956) *Canadian Journal of Chemistry*, **34**, 906–910.
14 Lemieux, R.U. and Morgan, A.R. (1965) *Canadian Journal of Chemistry*, **43**, 2198–2204.
15 (a) Lemieux, R.U., Hendricks, K.B., Stick, R.V. and James, K. (1975) *Journal of the American Chemical Society*, **97**, 4056–4062. (b) Lemieux, R.U. and Morgan, A.R. (1965) *Canadian Journal of Chemistry*, **43**, 2214–2221.
16 Kochetkov, N.K., Khorlin, A.J. and Bochkov, A.F. (1967) *Tetrahedron*, **23**, 693–707.
17 Zurabyan, S.E., Thkhomirov, M.M., Nesmeyanov, V.A. and Khorlin, A.Y. (1973) *Carbohydrate Research*, **26**, 117–123.
18 Hanessian, S. and Banoub, J. (1975) *Carbohydrate Research*, **44**, C14–C17.
19 Banoub, J. and Bundle, D.R. (1979) *Canadian Journal of Chemistry*, **57**, 2091–2097.
20 Ogawa, T. and Matsui, M. (1976) *Carbohydrate Research*, **51**, C13–C18.
21 Wulff, G. and Schmidt, W. (1977) *Carbohydrate Research*, **53**, 33–46.
22 Tsui, D.S. and Gorin, P.A.J. (1985) *Carbohydrate Research*, **144**, 137–147.
23 Matsuoka, K., Nishimura, S.-I. and Lee, Y.C. (1995) *Bulletin of the Chemical Society of Japan*, **68**, 1715–1720.

24 Banoub, J., Boullanger, P., Potier, M. and Descotes, G. (1986) *Tetrahedron Letters*, **27**, 4145–4148.
25 Shoda, S.-I., Moteki, M., Izumi, R. and Noguchi, M. (2004) *Tetrahedron Letters*, **45**, 8847–8848.
26 Radhakrishnan, K.V., Sajisha, V.S. and Chacko, J.M. (2005) *Synlett*, 997–999.
27 Ernst, B., De Mesmaeker, A., Wagner, B. and Winkler, T. (1990) *Tetrahedron Letters*, **31**, 6167–6170.
28 Adinolfi, M., Iadonisi, A., Ravida, A. and Schiattarella, M. (2003) *Tetrahedron Letters*, **44**, 7863–7866.
29 Sznaidman, M.L., Johnson, S.C., Crasto, C. and Hecht, S.M. (1995) *The Journal of Organic Chemistry*, **60**, 3942–3943.
30 Kochetkov, N.K., Bochkov, A.F., Sokolovskaya, T.A. and Snyatkova, V.J. (1971) *Carbohydrate Research*, **16**, 17–27.
31 Kochetkov, N.K., Malysheva, N.M., Torgov, V.I. and Klimov, E.M. (1977) *Carbohydrate Research*, **54**, 269–274.
32 Bochkov, A.F., Betanely, V.I. and Kochetkov, N.K. (1973) *Carbohydrate Research*, **30**, 418–419.
33 Bochkov, A.F. and Zaikov, G.E. (1979) *Chemistry of the O-Glycosidic Bond*, Pergamon, Oxford.
34 Garegg, P.J. and Kvarnstrom, I. (1977) *Acta Chemica Scandinavica Series B – Organic Chemistry and Biochemistry*, **31**, 509–513.
35 Garegg, P.J., Konradsson, P., Kvarnstrom, I., Norberg, T., Svensson, S.C.T. and Wigilius, B. (1985) *Acta Chemica Scandinavica Series B – Organic Chemistry and Biochemistry*, **39**, 569–577.
36 Bochkov, A.F. and Kochetkov, N.K. (1975) *Carbohydrate Research*, **39**, 355–357.
37 Kochetkov, N.K. (1987) *Tetrahedron*, **43**, 2389–2436.
38 Backinowsky, L.V., Abronina, P.I., Shashkov, A.S., Grachev, A.A., Kochetkov, N.K., Negopodiev, S.A. and Stoddart, J.F. (2002) *Chemistry – A European Journal*, **8**, 4412–4423.
39 (a) Tsvetkov, Y.U., Kitov, P.I., Backinowski, L.V. and Kochetkov, N.K. (1993) *Tetrahedron Letters*, **34**, 7977–7980. (b) Tsvetkov, Y.U., Kitov, P.I., Backinowski, L.V. and Kochetkov, N.K. (1996) *Journal of Carbohydrate Chemistry*, **15**, 1027–1050.
40 Demchenko, A. and Boons, G.-J. (1997) *Tetrahedron Letters*, **38**, 1629–1632.
41 Kochetkov, N.K., Backinovsky, L.V. and Tsvetkov, Y.E. (1977) *Tetrahedron Letters*, **18**, 3681–3684.
42 Backinovsky, L.V., Tsvetkov, Y.E., Balan, N.F., Byramova, N.E. and Kochetkov, N.E. (1980) *Carbohydrate Research*, **85**, 209–221.
43 Lemieux, R.U. and Morgan, A.R. (1965) *Canadian Journal of Chemistry*, **43**, 2190–2197.
44 Konradsson, P., Mootoo, D.R., McDevitt, R.E. and Fraser-Reid, B. (1990) *Journal of the Chemical Society: Chemical Communications*, 270–272.
45 Veeneman, G.H., van Leeuwen, S.H. and van Boom, J.H. (1990) *Tetrahedron Letters*, **31**, 1331–1334.
46 (a) Zuurmond, H.M., van der Marel, G.A. and van Boom, J.H. (1991) *Recueil Des Travaux Chimiques Des Pays-Bas (Journal of the Royal Netherlands Chemical Society)*, **110**, 301. (b) Zuurmond, H.M., van der Marel, G.A. and van Boom, J.H. (1993) *Recueil Des Travaux Chimiques Des Pays-Bas (Journal of the Royal Netherlands Chemical Society)*, **112**, 507.
47 Kartha, K.P.R., Karkkainen, T.S., Marsh, S.J. and Field, R.A. (2001) *Synlett*, 260–262.
48 Smoot, J.T., Pornsuriyasak, P. and Demchenko, A.V. (2005) *Angewandte Chemie (International Edition)*, **44**, 7123–7126.
49 Kochetkov, N.K., Bochkov, A.F. and Yazlovetsky, I.G. (1969) *Carbohydrate Research*, **9**, 49–60.
50 Hiranuma, S., Kanie, O. and Wong, C.-H. (1999) *Tetrahedron Letters*, **40**, 6423–6426.

51 (a) Sanchez, S., Bamhaoud, T. and Prandi, J. (2002) *European Journal of Organic Chemistry*, 3864–3873. (b) Sanchez, S., Bamhaoud, T. and Prandi, J. (2000) *Tetrahedron Letters*, **41**, 7447–7452. (c) Bamhaoud, T., Sanchez, S. and Prandi, J. (2000) *Chemical Communications*, 659–660.

52 Mehta, S. and Pinto, B.M. (1991) *Tetrahedron Letters*, **32**, 4435–4438.

53 (a) Ferrier, R.J., Hay, R.W. and Vethaviyasar, N. (1973) *Carbohydrate Research*, **27**, 55–61. (b) Ferrier, R.J. and Furneaux, R.H. (1976) *Carbohydrate Research*, **52**, 63–68.

54 Fraser-Reid, B., Konradsson, P., Mootoo, D.R. and Udodong, U. (1988) *Journal of the Chemical Society: Chemical Communications*, 823–825.

55 Kunz, H. and Pfrengle, W. (1986) *Journal of the Chemical Society: Chemical Communications*, 713–714.

56 Kondo, H., Aoki, S., Ichikawa, Y., Halcomb, R.L., Ritzen, H. and Wong, C.-H. (1994) *The Journal of Organic Chemistry*, **59**, 864–877.

57 Griffith, M.H.E. and Hindsgaul, O. (1991) *Carbohydrate Research*, **211**, 163–166.

58 Derevitskaya, V.A., Klimov, E.M. and Kochetkov, N.K. (1968) *Carbohydrate Research*, **7**, 7–11.

59 (a) Ogawa, T., Katano, K. and Matsui, M. (1978) *Carbohydrate Research*, **64**, C3–C9. (b) Ogawa, T., Katano, K., Sasajima, K. and Matsui, M. (1981) *Tetrahedron*, **37**, 2779–2786.

60 Ogawa, T., Beppu, K. and Nakabayashi, S. (1981) *Carbohydrate Research*, **93**, C6–C9.

61 Kong, F. (2003) *Current Organic Chemistry*, **7**, 841–865.

62 Wang, W. and Kong, F. (1998) *The Journal of Organic Chemistry*, **63**, 5744–5745.

63 Wang, W. and Kong, F. (1999) *The Journal of Organic Chemistry*, **64**, 5091–5095.

64 Wang, W. and Kong, F. (1999) *Angewandte Chemie (International Edition)*, **38**, 1247–1250.

65 Du, Y. and Kong, F. (1999) *Journal of Carbohydrate Chemistry*, **18**, 655–666.

66 Yang, Z., Lin, W. and Yu, B. (2000) *Carbohydrate Research*, **329**, 879–884.

67 Zhu, Y. and Kong, F. (2000) *Synlett*, 1783–1787.

68 Kong, F. (2007) *Carbohydrate Research*, **342**, 345–373.

69 Zeng, Y., Ning, J. and Kong, F. (2002) *Tetrahedron Letters*, **43**, 3729–3743.

70 Zeng, Y., Ning, J. and Kong, F. (2003) *Carbohydrate Research*, **338**, 307–311.

71 Mootoo, D.R., Date, V. and Fraser-Reid, B. (1998) *Journal of the American Chemical Society*, **110**, 2662–2663.

72 Konradsson, P. and Fraser-Reid, B. (1989) *Journal of the Chemical Society: Chemical Communications*, 1124–1125.

73 Mach, M., Schlueter, U., Mathew, F., Fraser-Reid, B. and Hazen, K.C. (2002) *Tetrahedron*, **58**, 7345–7354.

74 (a) King, J.F. and Allbutt, A.D. (1967) *Tetrahedron Letters*, **8**, 49–54. (b) King, J.F. and Allbutt, A.D. (1970) *Canadian Journal of Chemistry*, **48**, 1754–1769.

75 Deslongchamps, P. (1983) *Stereoelectronic Effects in Organic Chemistry*, Pergamon, New York, Chapter 3.

76 Madsen, R., Udodong, U.E., Roberts, C., Mootoo, D.R., Konradsson, P. and Fraser-Reid, B. (1995) *Journal of the American Chemical Society*, **117**, 1554–1565.

77 Allen, J.G. and Fraser-Reid, B. (1999) *Journal of the American Chemical Society*, **121**, 468–469.

78 Fraser-Reid, B., Wu, Z., Ottosson, H. and Udodong, U.E. (1990) *The Journal of Organic Chemistry*, **55**, 6068–6070.

79 Mathew, F., Mach, M., Hazen, K.C. and Fraser-Reid, B. (2003) *Synlett*, 1319–1322.

80 Lu, J., Jayaprakash, K.N., Schlueter, U. and Fraser-Reid, B. (2004) *Journal of the American Chemical Society*, **126**, 7540–7547.

81 Lu, J., Jayaprakash, K.N. and Fraser-Reid, B. (2004) *Tetrahedron Letters*, **45**, 879–882.

82 Anilkumar, G., Nair, L.G., Olsson, L., Daniels, J.K. and Fraser-Reid, B. (2000) *Tetrahedron Letters*, **41**, 7605–7608.

83 Soli, E.D., Manoso, A.E., Satterson, M.C., DeShong, P., Favor, D.A., Hirschmann, R. and Smith, A.B., III (1999) *The Journal of Organic Chemistry*, **64**, 3171–3177.

84 Campbell, A.S. and Fraser-Reid, B. (1994) *Bioorganic and Medicinal Chemistry*, **2**, 1209–1219.

85 Lu, J., Fraser-Reid, B. and Gowda, Ch. (2005) *Organic Letters*, **7**, 3841–3843.

86 Buskas, T., Soderberg, E., Konradsson, P. and Fraser-Reid, B. (2000) *The Journal of Organic Chemistry*, **65**, 958–963.

87 (a) Jayaprakash, K.N., Rachakrishnan, K.V. and Fraser-Reid, B. (2002) *Tetrahedron Letters*, **43**, 6953–6955. (b) Jayaprakash, K.N. and Fraser-Reid, B. (2004) *Synlett*, 301–305.

88 (a) Riviere, M., Moisand, A., López, A. and Puzo, G. (2004) *Journal of Molecular Biology*, **344**, 907–918. (b) Briken, V., Procelli, S.A., Besra, G.S. and Kremer, L. (2004) *Molecular Microbiology*, **53**, 391–403. (c) Brennan, P.J. (2003) *Tuberculosis*, **1**, 1. (d) Truemann, A., Xidong, X., McDonnell, L., Derrick, P.J., Ashcroft, A.E., Chatterjee, D. and Homans, S.W. (2002) *Journal of Molecular Biology*, **316**, 89–100. (e) Lowary, T.L. (2001) *Glycoscience: Chemistry and Biology*, vol. 3 (eds B. Fraser-Reid, K. Tatsuta and J. Thiem), Springer, Heidelberg, p. 2005. (f) Chatterjee, D. (1997) *Current Opinion in Chemical Biology*, **1**, 579–588.

89 Turnbull, W.B., Shimizu, K.H., Chatterjee, D., Homans S.W. and Truemann, A. (2004) *Angewandte Chemie (International Edition)*, **43**, 3918–3922.

90 Elie, C.J.J., Verduyn, R., Dreef, E.E., Brounts, D.M., van der Marel, G.A. and van Boom, J.H. (1990) *Tetrahedron*, **46**, 8243–8254.

91 Anilkumar, G., Gilbert, M.R. and Fraser-Reid, B. (2000) *Tetrahedron*, **56**, 1993–1997.

92 (a) Paulsen, H. (1982) *Angewandte Chemie (International Edition in English)*, **21**, 155–173. (b) Paulsen, H. (1984) *Selectivity a Goal for Synthetic Efficiency*, (eds W. Bartmann and B.M. Trost), Verlag Chemie, Weinheim, p169.

93 Spijker, N.M. and van Boeckel, C.A.A. (1991) *Angewandte Chemie (International Edition in English)*, **30**, 180–183.

94 Anilkumar, G., Lair, L.G. and Fraser-Reid, B. (2000) *Organic Letters*, **2**, 2587–2589.

95 Jayaprakash, K.N., Lu, J. and Fraser-Reid, B. (2005) *Angewandte Chemie (International Edition)*, **44**, 5894–5898.

96 (a) Sugihara, J.M. (1953) *Advances in Carbohydrate Chemistry and Biochemistry*, **8**, 1–44. (b) Haines, A.H. (1976) *Advances in Carbohydrate Chemistry and Biochemistry*, **33**, 11–110.

97 Uriel, C., Agocs, A., Gómez, A.M., López, J.C. and Fraser-Reid, B. (2005) *Organic Letters*, **7**, 4899–4902.

98 Boons, G.-J. and Zhu, T. (1997) *Synlett*, 809–811.

99 Hanessian, S. and Lou, B. (2000) *Chemical Reviews*, **100**, 4443–4463.

100 Plante, O.J., Palmacci, E.R., Andrade, R.B. and Seeberger, P.H. (2001) *Journal of the American Chemical Society*, **123**, 9545–9554.

101 López, J.C., Agocs, A., Uriel, C., Gomez, A.M. and Fraser-Reid, B. (2005) *Chemical Communications*, 5088–5090.

102 Lu, J. and Fraser-Reid, B. (2005) *Chemical Communications*, 862–864.

103 1,2-*cyano* alkylidene derivatives of furanoses (L-*arabino*, D-*galacto* and D-*gluco* series) were prepared and used (L-*arabino*) in glycosylation reactions of tritylated monosaccharides by Kochetkov and coworkers. Backinovsky, L.V., Negopod'ev, S.A., Shashkov, A.S. and Kochetkov, N.K. (1985) *Carbohydrate Research*, **138**, 41–54.

104 Lu, J. and Fraser-Reid, B. (2004) *Organic Letters*, **6**, 3051–3054.

5.2
Other Methods for Glycoside Synthesis: Dehydro and Anhydro Derivatives

David W. Gammon, Bert F. Sels

5.2.1
Introduction

Sugar derivatives containing a double bond, 'dehydro sugars', and those that feature an intramolecular anhydride, 'anhydro sugars', are versatile building blocks for the synthesis of glycosides and other natural products. Both classes of sugars have been utilized in the construction of an impressive range of complex carbohydrates and found application in emerging fields of solid phase and combinatorial synthesis of glycosides and sequential one-pot glycosylation strategies.

Of the unsaturated sugars, it is those with the double bond between C-1 and C-2 that can be utilized in glycosidic bond formation. These pyranoid or furanoid vinyl ethers (**I**, **II**, **IV** and **V**, Figure 5.3) are referred to as 'glycals' or more specifically *endo*-glycals to distinguish them from the *exo*-glycals bearing an exocyclic C-1–C-2 double bond (**III**, Figure 5.3). Their utility lies in the unique reactivity of the cyclic enol ether, with the ring oxygen influencing the regioselectivity of addition and rearrangement reactions, and the nature and orientation of ring substituents contributing to the overall reactivity and the stereochemical outcomes of reactions. They are glycosyl donors in their own right and are precursors to several other glycosyl donors, thereby serving as building blocks for the construction of a wide range of *O*-, *N*-, *S*- and *C*-glycosides. Their reactivity has been particularly exploited in the construction of

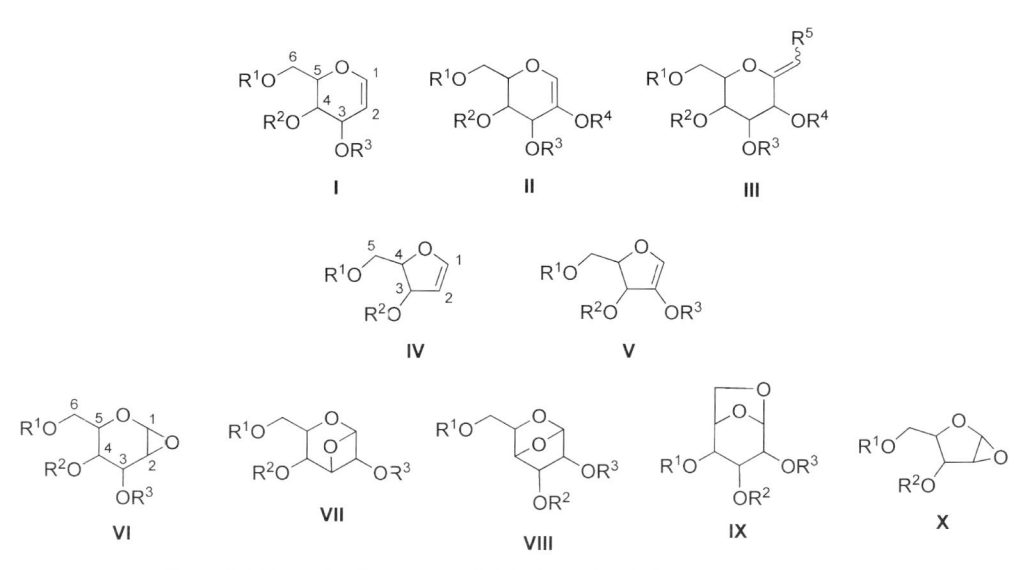

Figure 5.3 Generalized structures of dehydro and anhydro sugars.

5.2 Other Methods for Glycoside Synthesis: Dehydro and Anhydro Derivatives

2-deoxyglycosides, 2-amino-glycosides and in glycosidic bond formation where subsequent elaboration at C-2 in the glycosyl donor component is required.

Of the various anhydro sugar glycopyranoses (**VI–IX**, Figure 5.3), the 1,2-anhydro sugars (**VI**) and the 1,6-anhydro sugars (**IX**) are of most use in glycosidation reactions, the 1,3- and 1,4-anhydro sugars (**VII** and **VIII**) being more difficult to prepare and of insufficient stability to facilitate practical use. Similarly, the 1,2-anhydrofuranoses (**X**) are the most useful of the anhydrofuranoses. The 1,2-anhydro sugars are, indeed, derivable from the glycals, and their ease of preparation and excellent reactivity in glycosidation protocols contribute to the utility and versatility of this class of dehydro and anhydro sugars.

Comprehensive reviews of the chemistry of both the glycals [105–107] and anhydro sugars [108,109] have appeared in the recent literature. This chapter will attempt to summarize what is known about their preparation and reactivity in the formation of the glycosidic linkage and highlight recent examples to emphasize practical and strategic considerations in the choice of glycosyl donors.

5.2.2
Glycals in Glycoside Synthesis

5.2.2.1 Preparation of Glycals

Several protected glycals are available commercially, including the simple but important glycals 3,4,6-tri-O-acetyl-D-glucal (**1**) and 3,4,6-tri-O-acetyl-D-galactal (**2**), and these as well as others serve as precursors to a wide range of selectively protected glucals and galactals via strategies common in carbohydrate chemistry. The synthetic value of acetylated glucals has been enhanced by the availability of simple, one-pot procedures for the conversion of acyl- into alkyl-protected glucals [110].

Most of the common O-acetylated pyranoid glycals can be obtained using variations of the traditional Fischer–Zach method [111] of zinc-promoted elimination of glycosyl halides in acetic acid. The scope of this method has been extended with the development of efficient aprotic, nonacidic conditions for this reaction using zinc dust and 1-methyl imidazole in ethyl acetate [112] and other methods where the acetylated glycosyl halide is treated with catalytic amounts of Vitamin B_{12} (as source of Co(III)) together with zinc and ammonium chloride [113] or with THF solutions of the Ti(III) dimer $(Cp_2TiCl)_2$ [114]. These elimination methods are, however, limited to the readily available glucosyl and galactosyl halide precursors, and therefore are primary sources of D-glucal and D-galactal derivatives. The most attractive route to D-allal and D-gulal derivatives **7** and **8** (Scheme 5.41) is via oxidation and rearrangement of S-phenyl-2,3-dideoxy derivatives **3** and **4**, derived

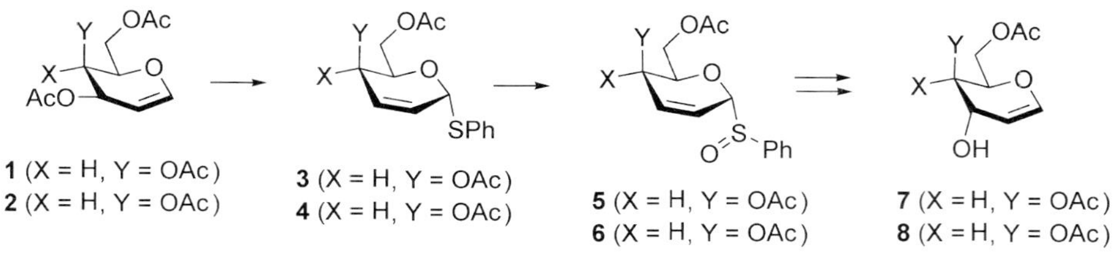

Scheme 5.41 Synthetic route to D-allal and D-gulal derivatives.

from acetylated D-glucal and D-galactal, respectively, using the Ferrier reaction on glycals [115].

A useful and promising addition to the methodologies for the synthesis of the whole range of glycals, including D-allal and D-gulal derivatives, is illustrated (Scheme 5.42) for the conversion of benzylated D-ribose into benzylated D-allal through an olefination–cyclization–elimination sequence [116]. Wittig–Horner olefination of protected ribose **9** gave alkenylated derivative **10**, which could be readily converted using NIS-activation of the olefin into 2-iodo-thioglycoside **11**, and this underwent smooth elimination to give benzylated allal **12**. The efficiencies of the individual steps in this sequence vary with the nature and orientation of the substituents, but it represents a useful route to rare glycals.

Glycals are also available from 2-deoxy sugars by acid- or base-induced eliminations of anomeric substituents. These methods are limited by the availability of the 2-deoxy sugars, for which the glycals themselves are the most obvious synthetic precursors. However, examples of these methods (Scheme 5.43) are in the direct preparation of tri-*O*-benzyl-D-glucal (**14**) from 2-deoxy-tri-*O*-benzyl-D-glucopyranose (**13**) via its 1-*O*-mesylate [117], and di-*O*-benzyl-D-ribal (**16**) from the phenylselenide **15** via oxidation to the selenoxide followed by elimination [118].

The latter elimination route to D-ribal is illustrative of the approach necessary for the preparation of the sensitive furanoid glycals, which are not accessible via the

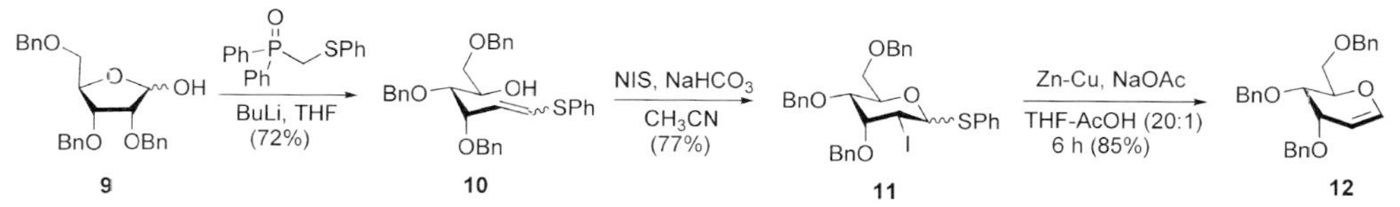

Scheme 5.42 Alternative synthesis of a D-allal derivative.

Scheme 5.43 Preparation of glycals via base-induced eliminations.

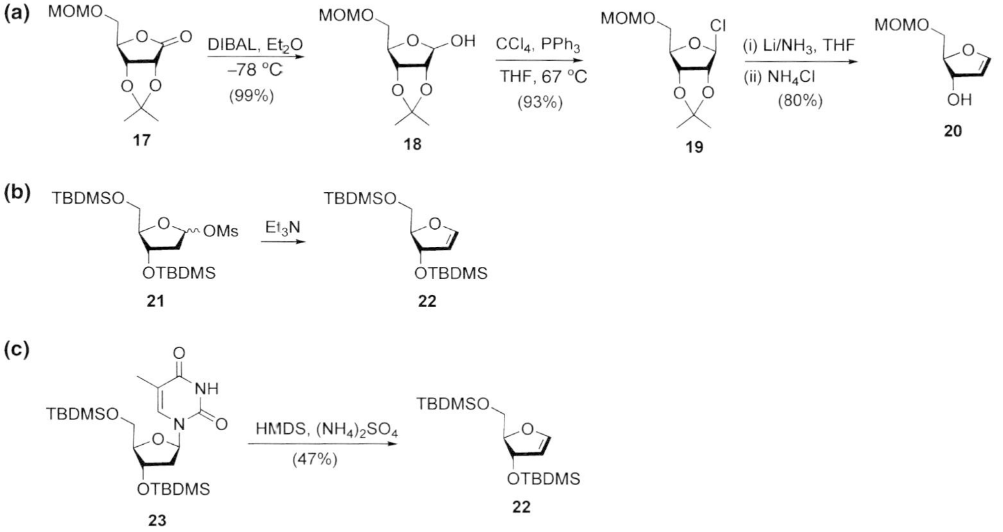

5.2 Other Methods for Glycoside Synthesis: Dehydro and Anhydro Derivatives | 419

Scheme 5.44 Preparations of D-ribal.

zinc-mediated reductive elimination routes to pyranoid glycals noted above. The first efficient preparation of a D-ribal [119] proceeded via DIBAL (diisobutyl aluminium hydride) reduction of ribonolactone **17** (pathway a, Scheme 5.44) to give ribose derivative **18**, followed by the formation of the ribosyl chloride **19** and then treatment of this with Li–NH$_3$ followed by careful quenching using NH$_4$Cl to give partially protected D-ribal **20**. A more practical preparation of D-ribal derivatives (pathway b, Scheme 5.44) involves treatment of 2-deoxyribosyl mesylate **21** with triethylamine, giving the silylated D-ribal **22** in good yield [120]. This can also be obtained by heating O-protected thymidine **23** in refluxing HMDS (hexamethyl disilazane) in the presence of ammonium sulfate under inert atmosphere [121].

The 2-substituted glycals, such as the 2-oxyglycals **II** and **V** in Figure 5.3, have traditionally enjoyed less attention than their 2-unsubstituted counterparts, presumably because of their limitations as glycosyl donors in glycosidation reactions. Earlier methods [122,123] for the preparation of the 2-oxyglycals involved acid- or base-induced eliminations of protected glycosyl halides, as illustrated more recently [124] in the preparation of 2-O-acetyl-3,4,6-tri-O-benzyl-D-glucal **27** (reaction a, Scheme 5.45). This was achieved by the conversion of tetra-O-acetyl glucosyl bromide **24** into orthoester **25**, exchange of acetyl for benzyl protecting groups to give **26**, and then heating in bromobenzene and pyridine to give **27** in good yield. A robust procedure of this kind, suitable for large-scale preparation of tetra-O-benzyl-D-glucal **29** (reaction b, Scheme 5.45), involves a sequential treatment of pentenyl glycoside **28** with bromine and DBU (1,8-diazabicyclo[5.4.0]undec-7-ene). These approaches are somewhat limited in terms of the range of protecting groups that are tolerated, but this is overcome by an attractive alternative that utilizes highly efficient *syn* elimination of 1,2-*trans* glycosyl sulfoxides [125] or glycosyl selenoxides

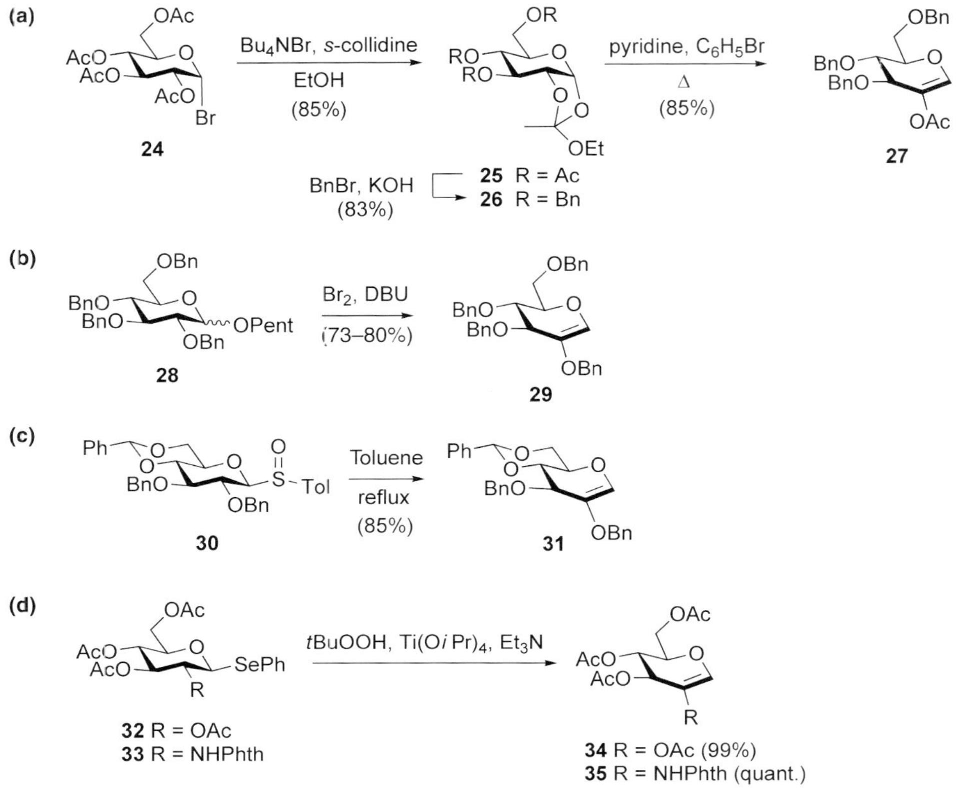

Scheme 5.45 Methods for preparing 2-substituted glycals.

[126], the latter generated *in situ* by the oxidation of the corresponding phenylselenides. Selected examples are shown in Scheme 5.45 (reactions c and d), illustrating that both 2-oxypyranoid glycals (**32** and **34**) as well as 2-amidoglycals (**35**) are available in this way.

5.2.2.2 Glycals as Glycosyl Donors

Glycals can be glycosylated not only directly via electrophilic addition, cyclo addition, nucleophilic addition and rearrangement reactions but also indirectly by conversion into a range of other glycosyl donors. One of the most important classes of these glycal-derived donors, the 1,2-anhydro sugars, is discussed in detail in Section 1.3.

Direct Synthesis of 2-Deoxyglcosides from Glycals Glycals are important donors in the preparation of glycosides of 2-deoxy sugars [106]. In polar addition reactions to the double bond, the regiochemical outcome is governed by the intermediacy of an oxocarbenium ion, which may be in equilibrium with a cyclic onium species across C-1–C-2, thus directing the incoming nucleophile to C-1. The stereochemistry is

5.2 Other Methods for Glycoside Synthesis: Dehydro and Anhydro Derivatives

Scheme 5.46 General mechanisms for polar additions to glycols.

influenced by a complex interplay of stereochemical and stereoelectronic factors including, *inter alia*, the nature and orientation of ring substituents, the nature of the electrophilic species, the solvent and the kinetic anomeric effect. These mechanisms and outcomes are summarized in Scheme 5.46 and have been comprehensively evaluated [127–129]. Representative examples of the synthetically useful addition protocols are discussed below.

The most direct route to 2-deoxyglycosides is by acid-promoted addition of alcohols to the glycals, although care must be taken to prevent acid-catalyzed rearrangements or deprotection. For example, treatment of *O*-acylated glycals in dichloromethane with alcohols and triphenylphosphine-hydrogen bromide [130], or with acid resins in acetonitrile containing anhydrous lithium bromide [131], mainly gives the thermodynamically more stable 2-deoxy-α-glycosides. This kind of direct glycosidation has found somewhat limited application because of the difficulties associated with the stereocontrol and acid sensitivity of glycals. However, the scope has been expanded by the recent report [132], wherein a Re(V)-oxo complex is used in catalytic amounts to activate a nucleophile for addition to glycals to give α-linked 2-deoxyglycosides in very high yields and selectivities (Scheme 5.47). For example,

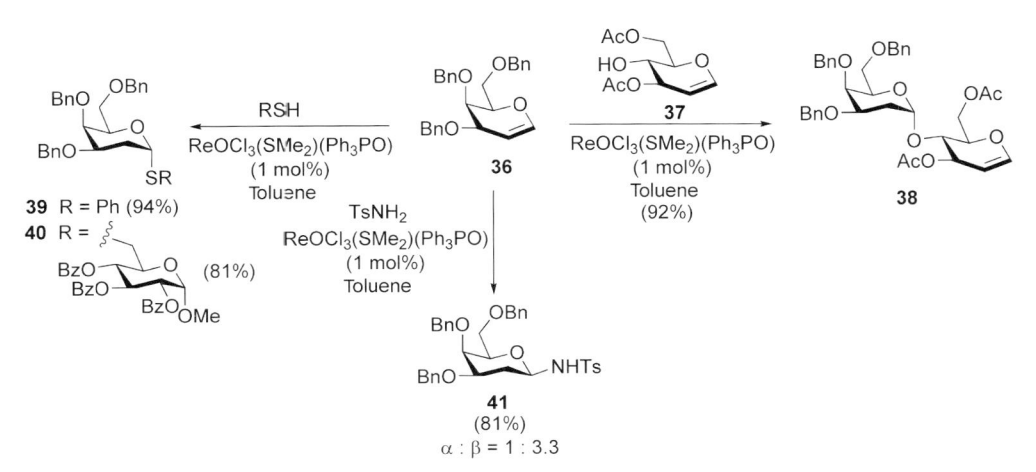

Scheme 5.47 Synthesis of 2-deoxy-α- and β-glycosides using Re(V)-oxo complex.

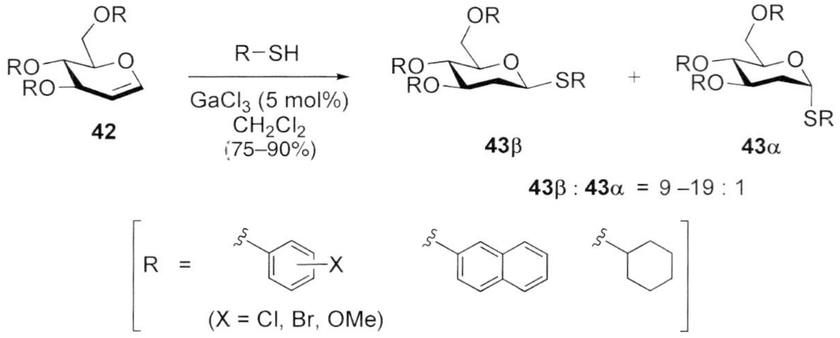

Scheme 5.48 GaCl$_3$-promoted formation of 2-deoxy-β-thioglycosides.

deactivated ('disarmed') glycal alcohol **37** could be coupled to benzylated galactal **36** to give α-linked disaccharide **38** in excellent yield. The methodology could also be applied to the preparation of 2-deoxythioglycosides such as **39** and **40** or N-glycosides **41**, although in the latter case the β-glycoside predominated.

In contrast to the α-selectivity in 2-deoxythioglycoside formation in the preceding example, excellent β-selectivity has been achieved [133] on treatment of protected glycals **42** with a range of thiophenols in the presence of GaCl$_3$ (Scheme 5.48).

Preparation of 2-deoxyglycosyl azides represents a useful alternative method for the construction of 2-deoxy-N-glycosides. Reddy [134] has recently described a method for direct conversion of glycals into 2-deoxyglycosyl azides using a TMS nitrate–TMS azide combination. Protected galactals gave only the α-galactosyl azides, whereas benzylated glucal (**14**) or its 3-deoxy analog gave mixtures of the α- and β-glucosyl azides.

An alternative and more stereocontrolled approach to 2-deoxyglycosides and C2-modified glycosides is the simultaneous introduction of glycosyl acceptor and a removable substituent at C-2. In this category, bromo- and iodoalkoxylation have emerged as the most useful reactions based on the early work of Lemieux [135,136] in which he introduced iodonium dicollidine perchlorate (IDCP) as the promoter, and subsequent reports on the use of N-bromo-succinimide [137] and N-iodo-succinimide (NIS) [138] as useful sources of the halonium species. These methods predominantly give the product of 1,2-*trans* addition, and subsequently the bromo- or iodo-substituents are reductively removed using reagent combinations such as Bu$_3$SnH and catalytic azobis(isobutyronitrile) (AIBN). The stereochemical outcome of the reactions is influenced by a range of factors, including the preferred conformation of the glycal ($^5H_4$ versus $^4H_5$, as illustrated in Figure 5.4), steric factors in both the glycal and nucleophile, the solvent and temperature.

With most protected D-glucals, the sequence is highly selective toward formation of α-2-deoxyglucosides because haloalkoxylation of glucals gives predominantly, and often exclusively, the 2-halo-2-deoxy-α-mannosides as a result of upper-face addition of the iodonium species and *trans*-1,2-addition. The two representative examples in Scheme 5.49 illustrate the utility of this methodology. In reaction (a), glucal **1** reacted cleanly with primary sugar alcohol **44** to give the α-linked disaccharide **45** in good yield [139], whereas reaction (b) illustrates the exploitation of the influence of

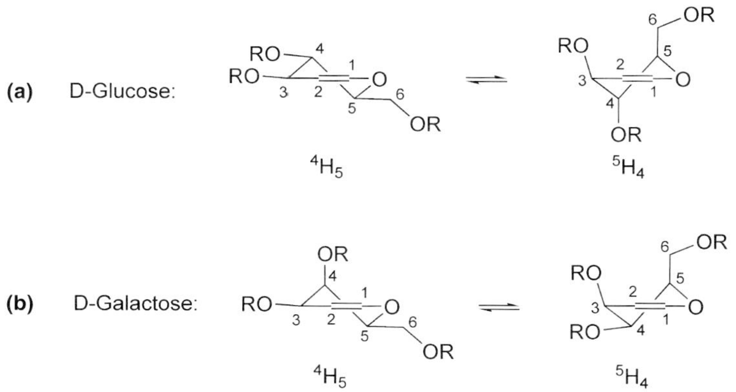

5.2 Other Methods for Glycoside Synthesis: Dehydro and Anhydro Derivatives

Figure 5.4 Conformational options in derivatives of D-glucose and D-galactose.

protecting groups in controlled, iterative assembly of oligosaccharides [140]. 'Armed' benzylated glucal **14** was combined with 'disarmed' benzoylated glucal **46** in the presence of the promoter IDCP to selectively give disaccharide **47**, which was in turn used in subsequent glycosylation of protected galactoside **48** to give trisaccharide **49**.

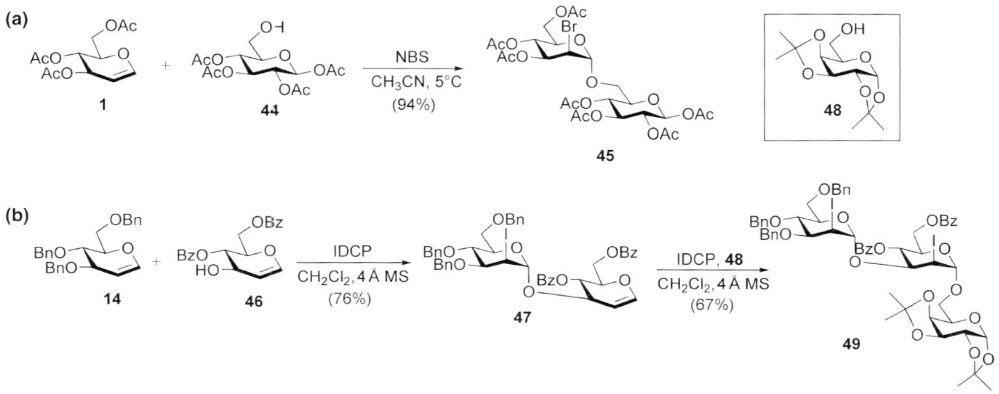

Scheme 5.49 Formation of 2-deoxyglycosides via haloalkoxylation of glycals.

An unusual application of the iterative iodoalkoxylation sequence was recently described in a synthetic route to oligo-(2,8)-3-deoxy-α-D-*manno*-2-octulosonic acid derivatives (Scheme 5.50) [141]. The C-1-substituted glycal **50** was treated with acyclic D-*manno*-2-octulosonic acid diol **51**, bearing alkyl sulfanyl and alkyl sulfoxide groups at C-2, in the presence of NIS and TfOH to give dimer **52**. The sequential activation with $(COCl)_2$ and AgOTf removed the sulfoxide and alkyl sulfanyl groups and regenerated the glycal ester **53**. The repetition of the sequence afforded an oligomer **54**, which upon deprotection and reductive removal of iodine gave the KDO oligomer **55**.

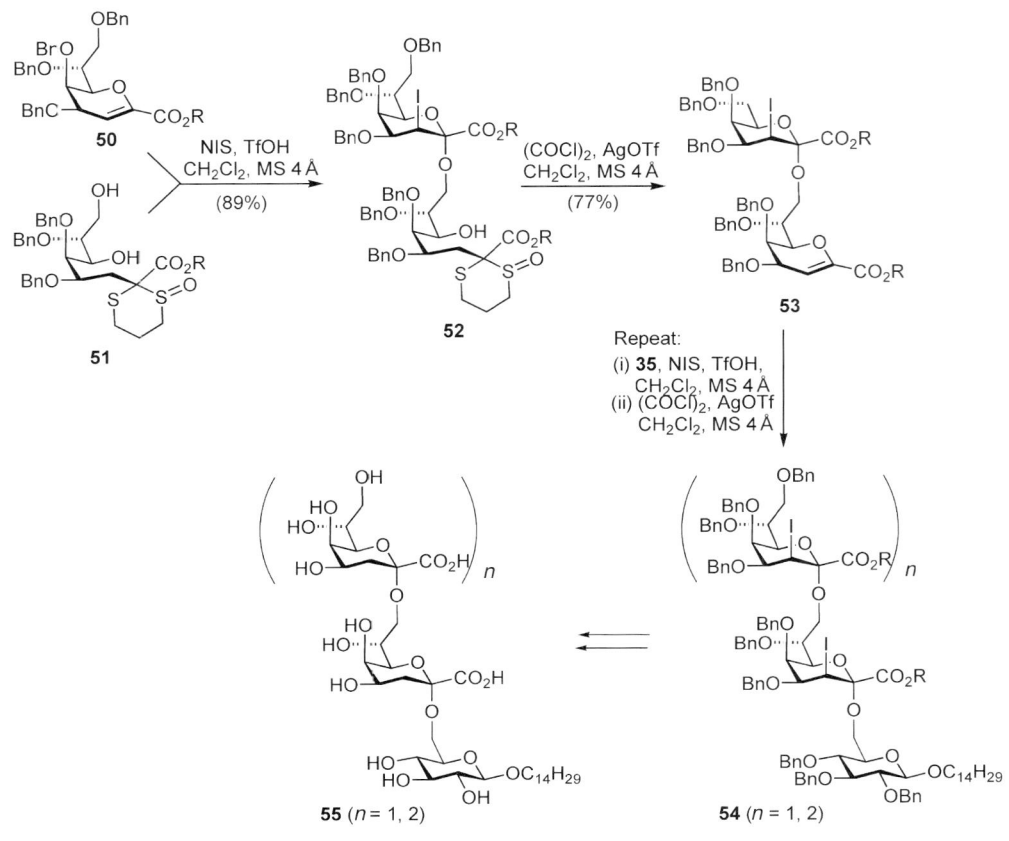

Scheme 5.50 Synthetic route to oligo-(2,8)-3-deoxy-α-D-*manno*-2-octulosonic acid derivatives via iterative iodoalkoxylation sequence.

Unusually-substituted deoxy sugars are found in a number of biologically important natural products, and an application of iodoglycosylation for the preparation of these compounds is shown in Scheme 5.51 [142]. Protected glucals **1** or **14** were selectively converted into 2-deoxy-2-iodo-α-mannopyranosyl glycosides **56** or **57**. The

Scheme 5.51 Method for preparation of 2-deoxy-2-C-alkylated glucosides.

allyl group was then transformed by reductive ozonolysis to the aldehydes **58** or **59** and these underwent intramolecular radical cyclization on treatment with Bu$_3$SnH and AIBN to eventually give the C-2 branched glucosides **60** and **61**.

In brief, a wide range of 2-deoxyglycosides are available from glucals, either directly using proton or Lewis acids together with nucleophiles, or by haloalkoxylation with subsequent reductive removal of the halide at C-2.

Conversion of Glycals into Other Glycosyl Donors The versatility of glycals in glycoside synthesis lies not only in their direct use as glycosyl donors but also in the ease with which they can be converted into a wide range of other glycosyl donors. Apart from the available variation in the anomeric substituent, there is also a potential for installing new substituents at C-2, which can be selected according to stereodirecting requirements in the subsequent glycosylation steps, or the need to elaborate the structures further at this key position in the newly formed glycoside. These considerations are particularly important in addressing the problem of stereoselective synthesis of 2-deoxy-β-glycosides.

The methods can be divided into three categories: those directly leading to glycosyl acetates, halides or other glycosyl donors, mostly with simultaneous introduction of a removable substituent at C-2; a second similar category resulting in glycosyl donors bearing a protected amino function at C-2; finally, those methods involving conversion of glycals into 1,2-anhydro sugars.

The glycals are easily converted into the 1,2-dihalo-derivatives, which in principle can act as glycosyl donors. However, these derivatives have not found wide application in glycoside synthesis, mainly because of the low facial selectivity in the initial addition of the electrophilic species [143–145]. In an example of a successful application, 2-deoxy-2-bromo-α-D-glucopyranosyl bromide [146] has been shown to give predominantly the 2-deoxy-β-D-glucopyranosides in silver-triflate-promoted reactions with alcohols.

However, the 2-bromo- and 2-iodo glycosyl acetates are more synthetically useful glycosyl donors. These are available either through formation of halohydrins from glycals using a source of halonium ions in the presence of water [147], followed by acetylation, or by direct haloacetoxylation of the glycals [148]. The halohydrin route has been the subject of renewed interest in the exploration of alternative catalytic routes to generating the required bromonium or iodonium species [149,150]. Although the generation of these species is efficient, the diastereoselectivity in the subsequent halohydrin formation is somewhat limited, with the 2-halo-α-mannopyranoses predominating.

The direct route from glycals to 2-halo-glycosyl acetates using a halonium ion source in the presence of acetic acid is more useful. A number of variations have been described, notably that of Roush [148] in which 2-deoxy-2-iodo-α-mannoanosyl acetates are formed with high α-selectivity using cerium(IV) ammonium nitrate and sodium iodide in acetonitrile. Comparable results [151,152] were obtained using a reagent mix of NH$_4$I/50% H$_2$O$_2$/AcOH/Ac$_2$O, wherein the iodonium species is generated by oxidation of the simple iodide, with reactions proceeding at low temperatures and in short reaction times with high yields and stereoselectivities.

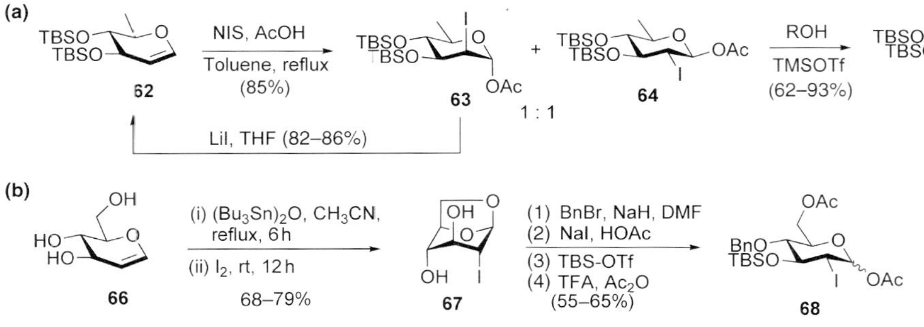

Scheme 5.52 Preparation of 2-deoxy-2-iodo-β-glycosyl acetates and their use in formation of 2-deoxy-β-glycosides.

As glycosyl donors, 2-deoxy-2-iodo-α-mannopyranosyl acetates lead predominantly to the formation of 2-deoxy-α-glycosides, whereas 2-deoxy-2-iodo-β-glucopyranosyl acetates lead to 2-deoxy-β-glycosides [153,154]. Thus, for example the 2-iodoglucosyl acetate **64** (pathway a, Scheme 5.52) was treated with a range of alcohols in the presence of TMSOTf or TBSOTf as the promoter to give the 2-deoxy-β-glycosides **65** after the reductive removal of 2-iodo substituent [153]. The required 2-deoxy-2-iodo-glucosyl acetates can be prepared by iodoacetoxylation of TBS-protected glucal **62**, where the undesired mannosyl acetate **63** is separated from **62** and recycled to the starting glucal. Alternatively (pathway b, Scheme 5.52), they can be selectively obtained from 1,6-anhydro-2-deoxy-2-iodo-D-glucose **67**, itself prepared from the D-glucal **66** using stannylene chemistry [155], in a sequence involving initial simultaneous benzylation at O-4 and formation of the β-2,3-epoxide, followed by reintroduction of the iodine at C-2, protection of O-4 and finally acetolysis to give 2-iodo-β-glucoside **68**.

The addition of phenyl sulfenyl choride and phenyl selenenyl chloride to glycals has been investigated, which provides another entry to the 2-deoxy-β-glucosides. As summarized by Roush *et al.* [156] (Scheme 5.53), the method gives the best selectiv-

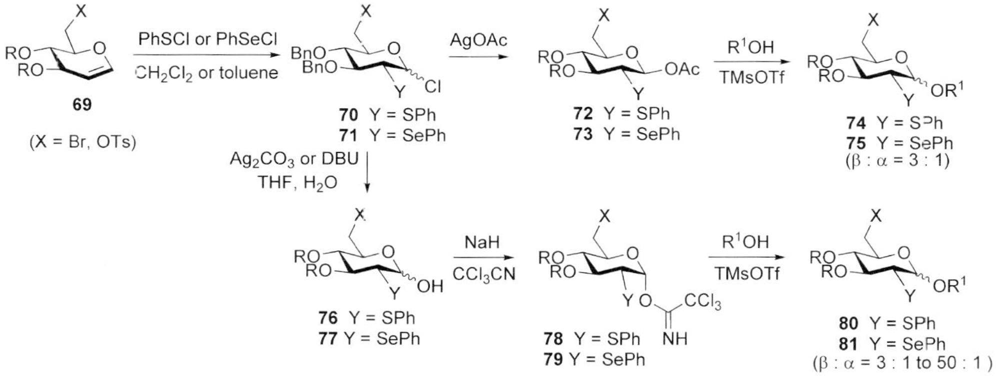

Scheme 5.53 Preparation of glycosyl donors from glycals using PhSCl and PhSeCl.

ities with glucals bearing an electron-withdrawing substituent at C-6 (X = Br or OTs in Scheme 5.53). In these cases, the 2-sulfenyl- or 2-selenenyl glucosyl chlorides **70** and **71** are formed selectively from glucals **69** on treatment with phenyl sulfenyl chloride or phenyl selenenyl chloride, respectively, and these can be transformed to glycosyl acetates (**72** and **73**) or trichloroacetimidates (**78** and **79**) via the corresponding hemiacetals **76** and **77**. Subsequently, β-selective glycosidation is achieved in glycosylations of a range of alcohols, with highest selectivity being achieved with the least sterically demanding alcohols.

Novel entry to the biologically important 2-fluoro derivatives has been recently described (Scheme 5.54) [157] in the context of a synthesis of 2′-fluoro-2′-deoxy-α-D-galactopyranosylceramide **88**, an immunomodulatory galactoglycosphingolipid. Selectfluor® was used to convert benzylated D-galactal (**36**) into 2-fluorogalactose **82**. This was followed either by acetylation to give the 2-fluorogalactosyl acetate **83** or treatment with diethylaminosulfur trifluoride (DAST) to give 2-fluorogalactosyl fluoride **84**. Both were then used as glycosyl donors in the preparation of glycoside **87αβ**, albeit with modest yields and selectivities, *en route* to the desired compound **88**.

Finally, S-(2,6-dideoxyglycosyl)phosphorodithioates **90** were prepared from 6-deoxy-L-glycals **89** [158] and transformed to either α- or β-2-deoxyglycosides **91** and **92** depending on the conditions used in adding the alcohol (Scheme 5.55).

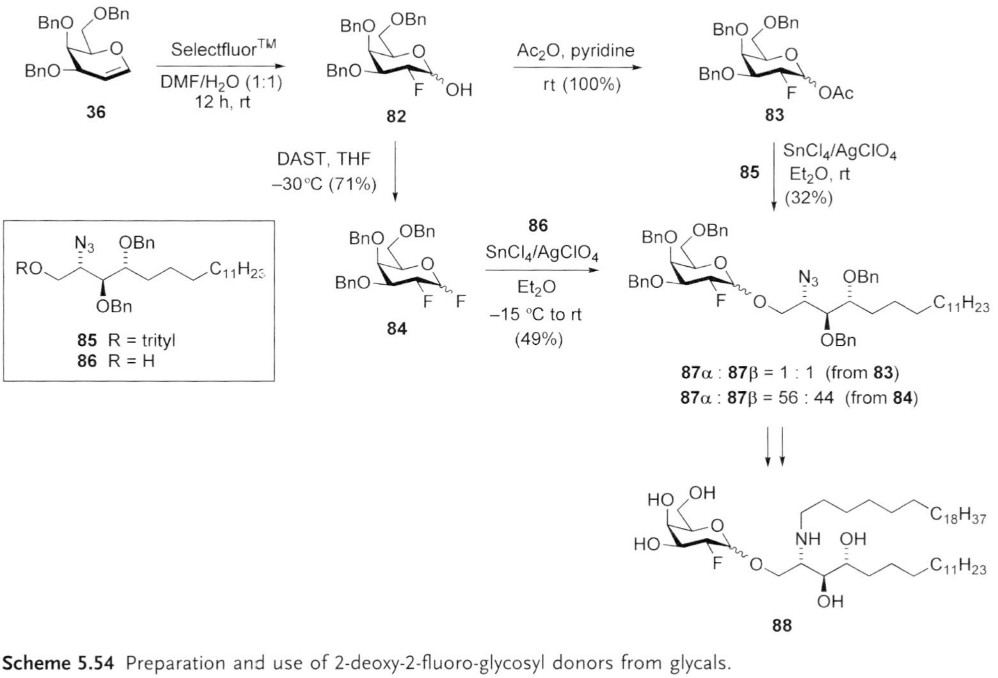

Scheme 5.54 Preparation and use of 2-deoxy-2-fluoro-glycosyl donors from glycals.

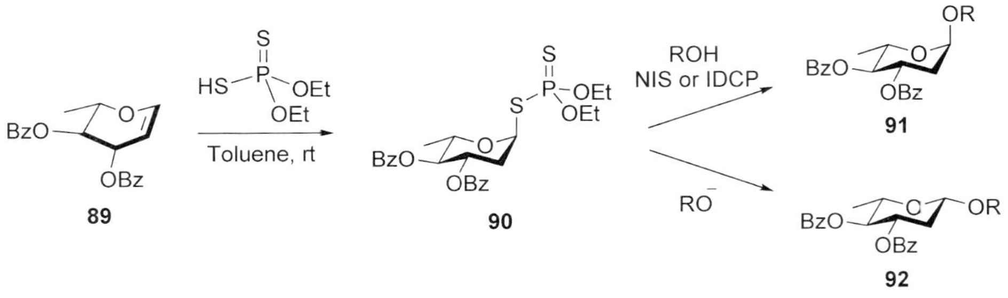

Scheme 5.55 Conversion of glycals to phosphorodithioates as glycosyl donors.

A wide range of glycosyl donors is, thus, directly available from glycals, allowing for careful tuning of reactivity depending on requirements for the synthesis of target 2-deoxyglycosides.

Synthesis of 2-Amino-2-Deoxyglycosides from Glycals Glycals are a useful source for preparation of 2-amino-2-deoxyglycosides as summarized in Scheme 5.56, with selectivity dependent on the structure and substituents in the starting glycals. A useful comparison of the glycal approach to this important class of sugars is provided in a recent review [159].

In the first of these approaches from glycals (pathway a), an azidonitration reaction using sodium azide and cerium(IV) ammonium nitrate is used to convert galactals and glucals into 2-azido-2-deoxyglycosyl nitrates [160,161]. The best regioselectivity is obtained in the case of D-galactals **2** that predominantly give the 2-azido-D-galactosyl nitrates **93**, whereas the D-glucals give mixtures of 2-azido-*gluco-* and *manno-*adducts. Hydrolysis of **93** gives 2-azido-D-galactose **94** that can be converted into other donors to form acetamido glycosides **95** after the reduction of the azide and acetylation.

A second approach (pathway b) involves the addition of iodoazide to glucal **1** to give 2-iodoglycosyl azides **96**, predominantly in the α-*manno-*configuration [162]. These can be converted into 2-deoxy-2-acetamido-β-glucosides **97** by treatment with PPh₃ and an alcohol, followed by subsequent protection of the amino moiety.

Pathway (c) illustrates the third approach, the 'azaglycosidation' route pioneered and developed by Danishefsky and coworkers [163,164]. Thus, for example treatment of benzylated glucal **14** with IDCP and benzenesulfonamide leads to the formation of 2-iodo-1-benzenesulfonamido-α-mannoside **98**, which can be induced to rearrange via aziridinium intermediates in the presence of alcohols and bases such as lithium tetramethylpiperidide (LTMP) to give 2-deoxy-2-sulfonamido-β-glucoside **99**. The iodobenzenesulfonamido derivatives can also be converted into the alternative thioethyl glycosyl donors by substituting ethane thiol for the alcohol. An example of the efficiency and versatility of this method is its use in the preparation of complex oligosaccharide fragments of Lewis-Y and KH-1 tumor-associated carbohydrate antigens (Scheme 5.57) [165]. 2-Iodo-sulfonamido glycoside **107**, derived from the corresponding glycal-terminated tetrasaccharide, was treated with tributylstannyl ether **108** in the presence of AgBF₄ in a stereoselective coupling to give protected hexasaccharide fragment **109** in an acceptable yield.

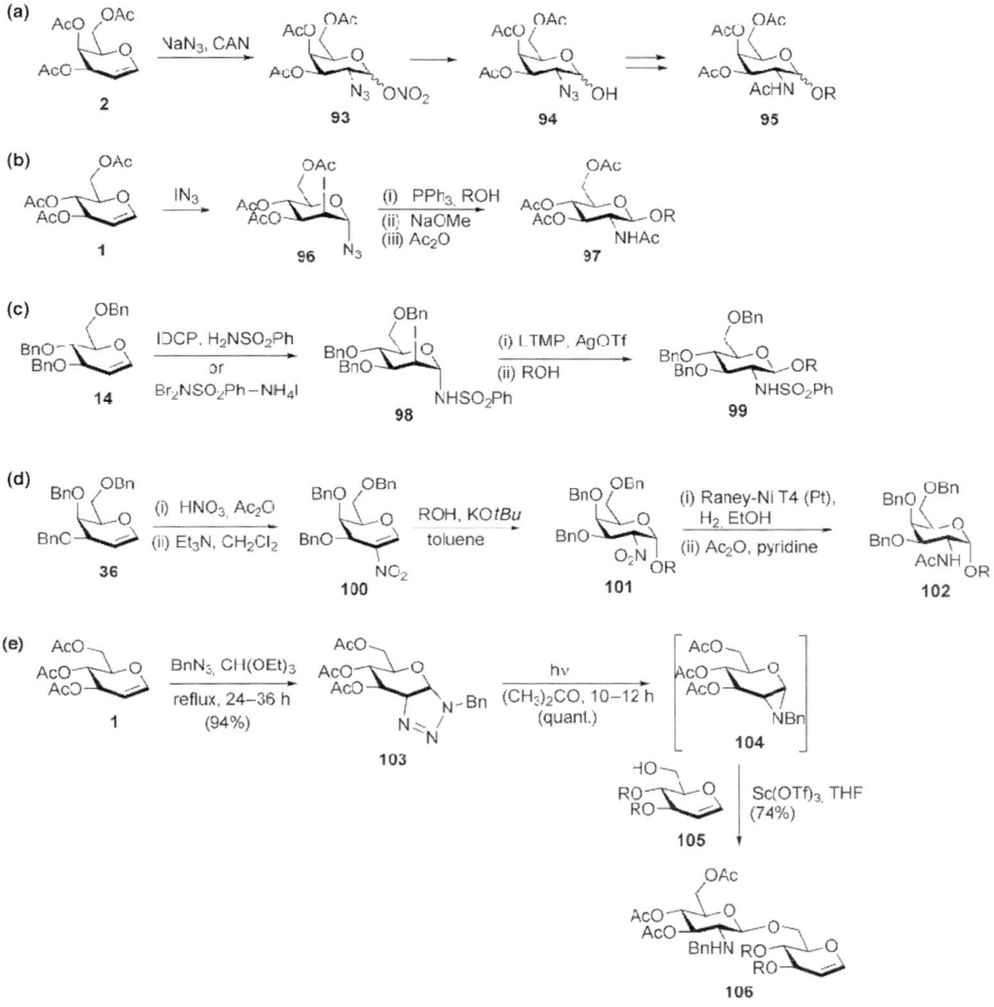

Scheme 5.56 Routes to 2-amino-2-deoxy-glycosides from glycals.

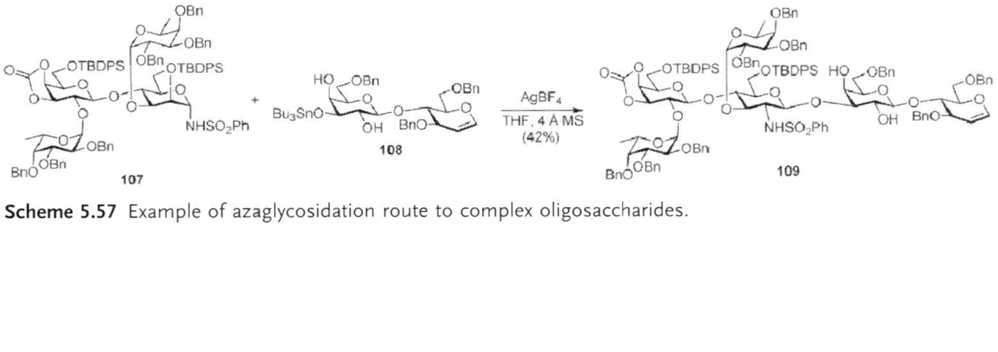

Scheme 5.57 Example of azaglycosidation route to complex oligosaccharides.

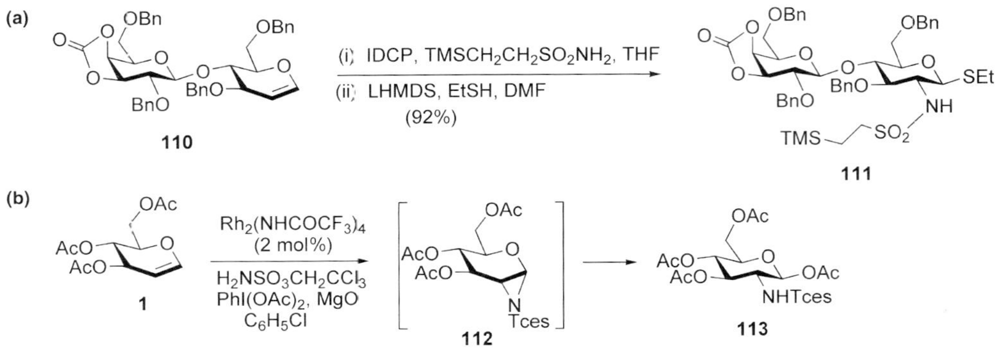

Scheme 5.58 Variations on azaglycosidation.

Two variations of this method are shown in Scheme 5.58. In reaction (a), the benzenesulfonamide is replaced by (2-trimethylsilyl)ethylsulfonamide, which is subsequently removed using cesium fluoride in DMF. The key thioglycoside building block **111**, required in the synthesis of the fucosylated biantennary N-glycan of erythropoietin, [166] was prepared from glycal **110** by a sequence involving the formation of the 2-iodo-1-(2-trimethylsilyl)ethylsulfonamide followed by rearrangement induced by ethanethiolate formed by the treatment of ethane thiol with lithium hexamethyldisilazide (LHMDS). In a more recent variant (reaction b) [167], tri-O-acetyl-D-glucal **1** is treated with a catalytic amount of $Rh_2(COCF_3)_4$, together with 2,2,2-trichloroethylsulfamate, $PhI(OAc)_2$ and MgO, to give the 2-trichloroethylsulfonamido-β-glucosyl acetate **113** via the aziridine **112**, with the trichloroethylsulfonamide (Tces) protecting group removable with Zn under relatively mild reducing conditions. The impressive scope of this methodology was recently illustrated in the synthesis of β-2-acetamidoglucosyl linkages in a complex β-N-linked glycopeptide **114** containing the H-type 2 blood group determinants [168].

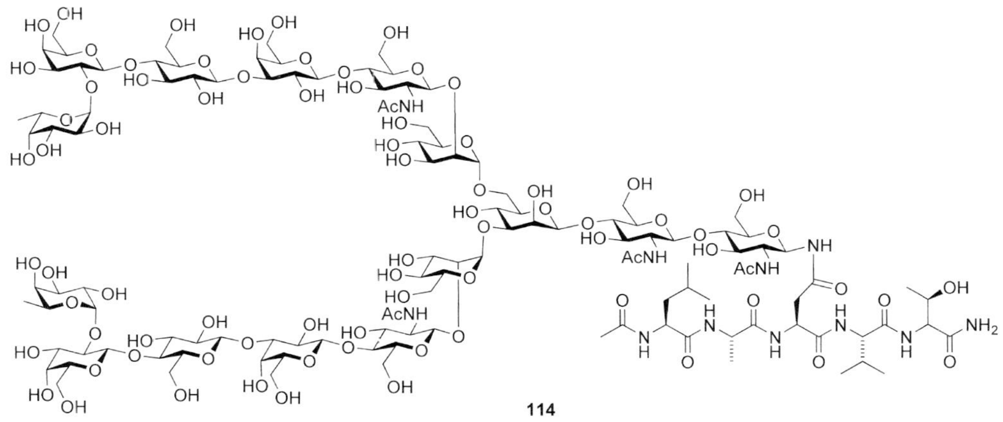

5.2 Other Methods for Glycoside Synthesis: Dehydro and Anhydro Derivatives

The fourth method (pathway d, Scheme 5.56) is particularly efficient in accessing 2-acetamido-2-deoxy-α-galactosides via 2-nitrogalactals, with lower selectivities observed using the corresponding 2-nitroglucals [169–175]. For example, 2-nitrogalactal **100** is formed from benzylated galactal **36** on treatment with nitric acid and acetic anhydride, and this is followed by Michael-type addition of alkoxide to form 2-nitro-α-galactosides **101** with subsequent reduction of the nitro group and acetylation giving 2-acetamido-2-deoxy-α-galactoside **102**.

Finally (pathway e, Scheme 5.56), triazoline **103** formed by cyclo addition of azide to glycal **1** can be photolytically converted into a 1,2-aziridine intermediate **104**, from which 2-benzylamino-2-deoxy-β-glucosides can be formed on addition of an alcohol and catalytic scandium triflate [176].

These methods complement other approaches [159] to the preparation of 2-deoxy-2-aminoglycosides and serve to further emphasize the rich diversity of products available from the glycals.

2,3-Dehydro-Glycosides via Ferrier Reactions of Glycals As noted earlier, glycals react with protic acids in the presence of nucleophiles to give 2-deoxyglycosides. However, the reaction with Lewis acids generally induces loss of the allylic substituent to give 2,3-dehydroglycosides. This reaction, fully investigated and described by Ferrier and others [105,177,178], has continued to attract attention for the stereoselective formation of O-, N-, S- and C-glycosides of interest in the synthesis of a variety of complex natural products. The general reaction is illustrated in Scheme 5.59, where acetylated glucal **1** reacts with nucleophiles in the presence of Lewis acids such as BF_3 to give 2,3-dehydroglycosides **115**, predominantly with anomeric α-configuration, and glycal **116**, a possible alternative product, given the fact that the intermediate in this reaction is an allylic carbenium ion.

Priebe and coworkers [107,178] have attempted to rationalize the product distribution in terms of Pearson's theory of hard and soft acids and bases (HSAB) [179], concluding as a broad generalization that soft bases (S-, N- and C-nucleophiles) form bonds at the softer C-3 electrophilic center, whereas hard bases (O-based nucleophiles) react preferentially at the harder C-1 center to give glycosides. They acknowledge that other factors may overrule this interpretation, such as when C-nucleophiles give kinetic C-1-alkylated products whose formation is not reversible.

The 2,3-dehydroglycosides continue to be of interest in view of the potential for further functionalization of the double bond toward unusual sugars or other natural products. The efficiency and selectivity depends on the catalyst and the conditions

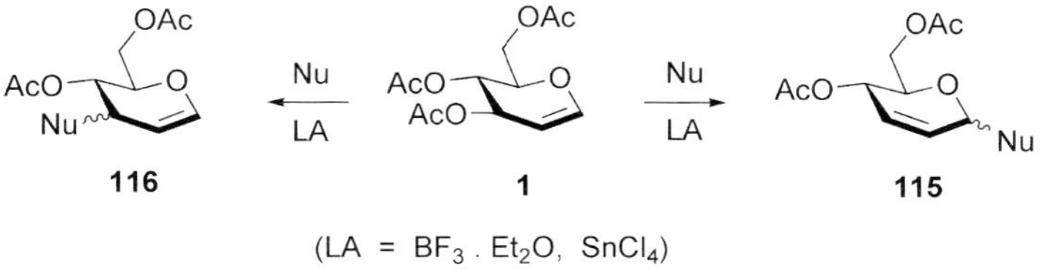

Scheme 5.59 Ferrier reaction of glycals. (LA = $BF_3 \cdot Et_2O$, $SnCl_4$)

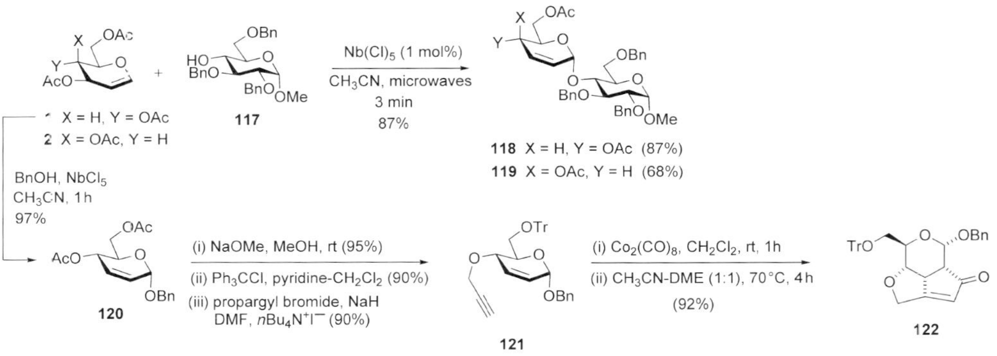

Scheme 5.60 Use of catalytic NbCl$_5$ in microwave-assisted glycosylations of glycals.

for the reaction, and considerable effort has been expended recently on the evaluation of a range of homogeneous and heterogeneous catalysts. This is summarized in a recent publication of Hotha and Tripathi [180] in which the potential of these investigations is exemplified by the use of catalytic NbCl$_5$ under microwave irradiation in acetonitrile to achieve remarkably quick and efficient glycosylation of glycals using a range of alcohols. Thus, for example when acetylated glucal **1** or galactal **2** were treated with sugar alcohol **117** under these conditions (Scheme 5.60), the 2,3-dehydro-α-glycosides **118** and **119** were produced cleanly within 3 min. The usefulness of unsaturated glycosides like these as substrates for 'diversity oriented synthesis' of tricyclic molecules was then demonstrated by preparing benzyl-2,3-dehydro-α-glucoside **120**, converting this into the 4-O-propargyl derivative **121** and carrying out an intramolecular Pauson–Khand reaction on this enyne system to give tricyclic product **122** [181].

The Ferrier reaction of glucals proceeds in most cases with high α-selectivity, but a recent example involving Pd-catalyzed glycosylation of glycals (Scheme 5.61) [182] illustrates that under appropriate conditions high β-selectivity can be achieved. Addition of acceptor **124** to glycal **123** in the presence of catalytic Pd(OAc)$_2$ and other additives gave a good yield of β-linked 2,3-dehydroglycoside **125**, with the finding that α-/β-selectivity was dependent on the choice of ligand. Thus, di(*tert*-butyl)-2-biphenylphosphine (DTBBP) gave almost exclusively β-selectivity, whereas the use of P(OMe)$_3$ gave variable outcomes.

A concluding example illustrates an interesting reaction whose outcome is analogous to the Ferrier rearrangement but involves a conjugate addition of alcohols to

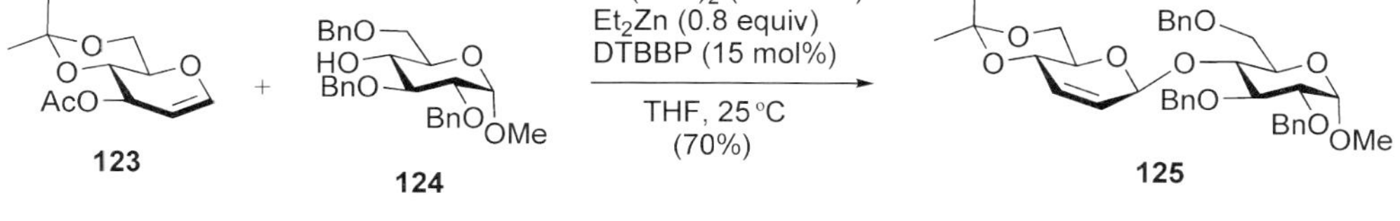

Scheme 5.61 Pd-catalyzed β-selective glycosylation of glycals.

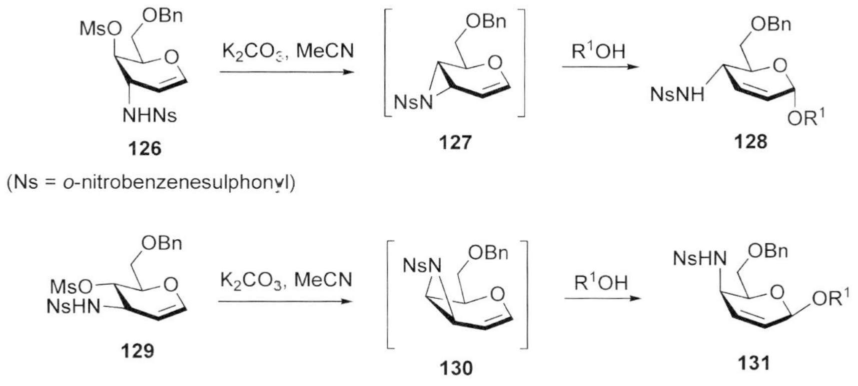

Scheme 5.62 Conjugate addition to allylic aziridines in glycals.

allyl aziridines [183]. Glycals **126** and **129** (Scheme 5.62), having vicinal *trans*-oriented mesyloxy and nosylamino groups, were treated with base to form the allyl aziridines **127** and **130**, respectively, and these then reacted smoothly with a range of alcohols in highly regio- and stereoselective manner to give 2,3-dehydro-4-nosylamino-α-glycoside **128** or 2,3-dehydro-4-nosylamino-β-glycoside **131**. The authors rationalize the observed regio- and stereoselectivities as arising from the hydrogen bonding of the alcohols with the nitrogen in the aziridine intermediates **127** and **128**, with the resultant face-selective attack at the anomeric carbon, migration of the double bond and opening of the C(3)-N bond. The nosyl (*o*-nitrobenzensulfonyl) group is easily removed by treatment with thiophenol and potassium carbonate to give the corresponding amines, and this, thus, appears to provide an excellent route to 4-aminoglycosides.

These examples and many others illustrate the ongoing fascination with the Ferrier rearrangement. The growing catalog of catalysts and subtle variations in outcomes that are associated with these serve to enhance the practical utility of this reaction in glycoside and natural product synthesis.

Glycals in Solid-Phase Synthesis of Glycosides The role of glycals in solid-phase synthesis of oligosaccharides has been recently reviewed [184–186], and the broad strategy of this application of glycals is briefly highlighted here. Bearing in mind the precautions and limitations associated with resin-bound saccharides, the glycals have found particularly effective use in 'donor-bound' strategies and 'bidirectional' strategies [186]. In the donor-bound strategy, the partially protected glycals are linked via a free hydroxyl group to a resin (compound **132**, Scheme 5.63) and then efficiently converted using standard glycal chemistry into a range of polymer-supported glycosyl donors, including 1,2-anhydro glycoses (**133**), thioethyl glycosides (**135, 136**) and 2-phenylsulfonylamido-thioglycosides (**139**), to mention a few. These are then efficiently converted into resin-linked glycosides such as **134**, **137** and **140**, utilizing the high efficiency and stereoselectivity of the coupling reactions. The assembled oligosaccharide is then cleaved from the resin using selective reactions that are

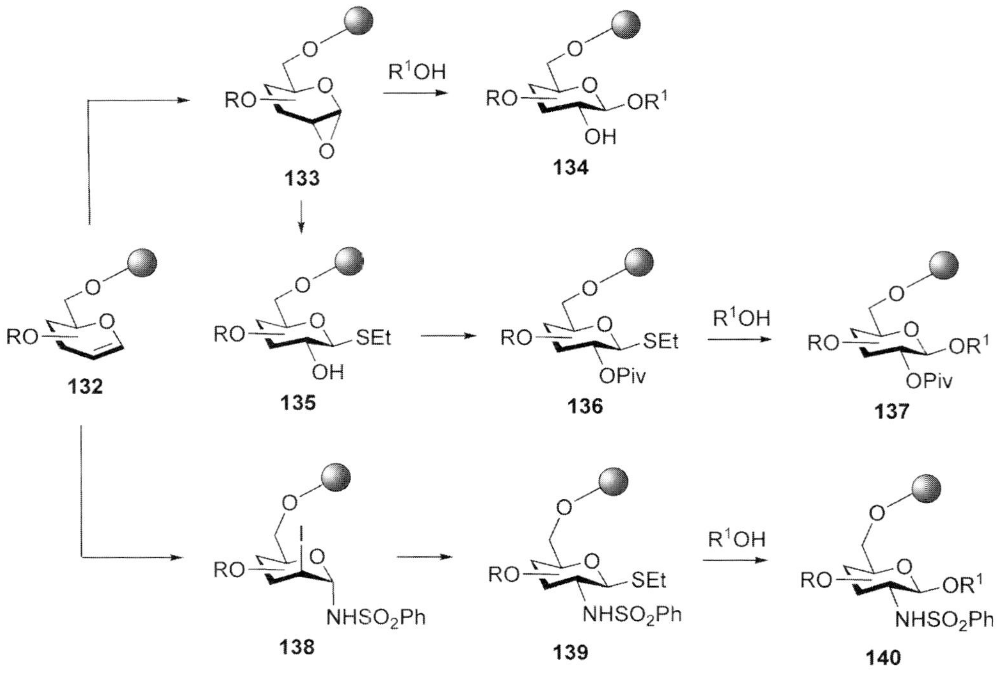

Scheme 5.63 Use of glycals in solid-phase synthesis of glycosides.

compatible with sugar functionalities. The bidirectional strategy arises when polymer-bound glycals can act as either donors or acceptors in glycoside synthesis, by virtue of selectively removable protecting groups, with the potential to assemble branched glycans. Despite some shortcomings of these approaches, they have been used to good effect in the preparation of a range of complex oligosaccharides.

Conclusion and Outlook The foregoing discussion has illustrated the impressive range of structures available from glycals. Current challenges in oligosaccharide chemistry include not only the synthesis of naturally occurring compounds in all their complexity but also the assembly of analogs of these as probes of their biological activity. It is thus pertinent to conclude this section with a selected recent example in which many facets of glycal chemistry were incorporated into the assembly of analogs of a complex oligosaccharide.

Awad et al. have recently described the synthesis of C-disaccharide analogs of the T-epitope (β-D-Galp(1–3)-α-D-GalpNac). Their approach to these structures features formation of glycals and manipulation of glycals toward glycosides and 2-aminoglycosides and illustrates the scope and some limitations of the use of glycals in synthesis. As shown in Scheme 5.64, C-disaccharide **141**, with the 1,6-anhydro unit arising from levoglucosenone, was converted into the anomeric acetate of 2-deoxy-D-galactose containing C-disaccharide **142**, which underwent efficient elimination to glycal **143** upon

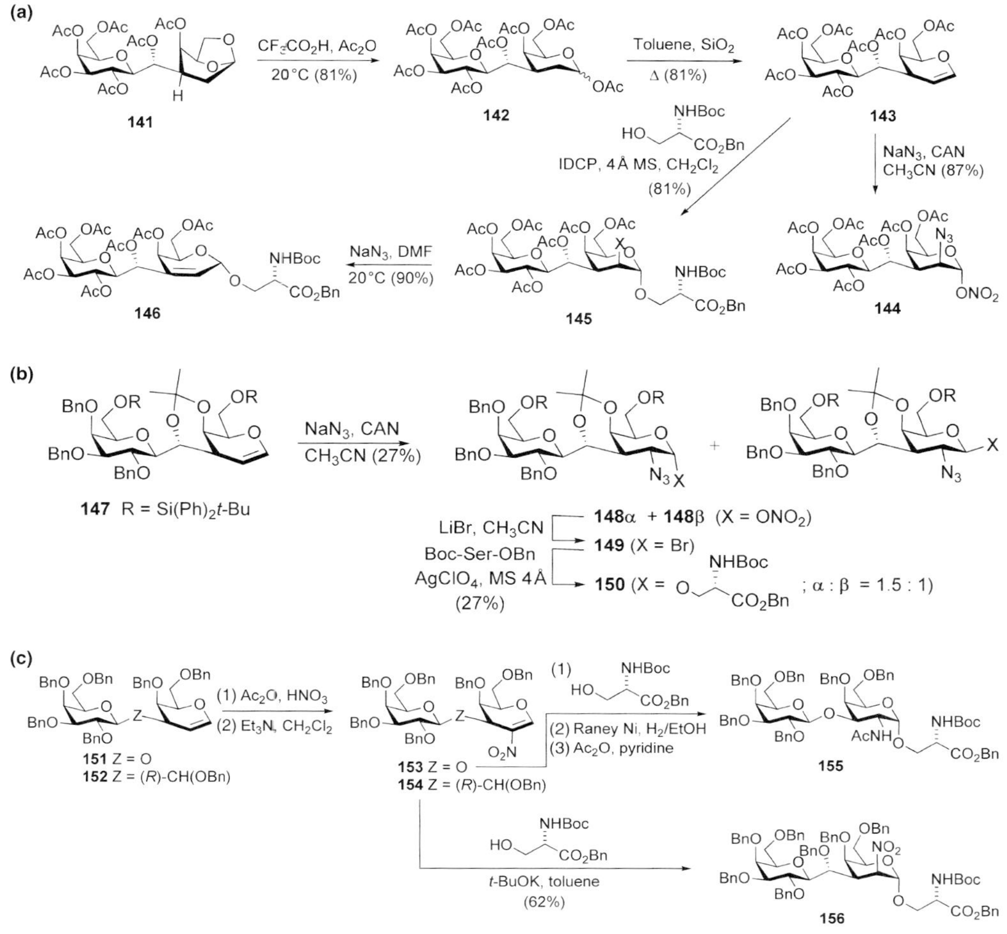

5.2 Other Methods for Glycoside Synthesis: Dehydro and Anhydro Derivatives | 435

Scheme 5.64 Use of glycal methodology in the synthesis of complex C-glycosides.

heating in toluene over silica. Iodoalkoxylation using IDCP and a serine derivative gave the glycopeptide **145**. However, on treatment of **145** with sodium azide, in an attempt to generate the equatorial 2-azide, an elimination took place to give the 2,3-dehydro derivative **146**. In contrast to the earlier results on analagous O-linked disaccharides and with expectations from the original Lemieux work, the attempted Lemieux azidonitration of **143** unexpectedly gave the *talo*-2-azido pseudodisaccharide **144**, thus emphasizing the dependence of the stereochemical outcome on substituents and conformational mobility. Azidonitration from the α-face was then engineered by preparing the conformationally restrained derivative **147** that was then shown to give azidonitrates **148α** and **148β**, bearing the equatorial azido group, in moderate yield. Conversion into the glycosyl bromide **149** and Königs–Knorr glycosidation gave the desired glycosides **150** in an α : β ratio of 1.5 : 1.

The importance of substituent and conformational effects in glycal additions was also demonstrated in attempting Michael addition of an alkoxide to the O-linked 2-nitrogalactal **153** and its C-linked analog **154**. The α-*talo*-isomer **156** was obtained from **154** in contrast to the result of Michael addition to the analogous O-linked disaccharide **153**, which eventually gave the 2-acetamido-α-galactoside-terminated disaccharide **155**.

5.2.3
Anhydro Sugars as Glycosyl Donors

5.2.3.1 1,2-Anhydro Sugars

Since the first description of 3,4,6-tri-O-acetyl-1,2-anhydro-α-D-glycopyranose, 'Brigl's anhydride' **157** [187,188], compounds of this type have gained prominence [109] as glycosyl donors, particularly through the contribution of Danishefsky and coworkers [189–192]. Attention will be focused here on the most widely applicable methods for preparation of these derivatives, as well as on promising new methodologies and the scope for efficient and stereoselective glycoside formation using the 1,2-anhydro sugars.

157

Synthetic Routes to 1,2-Anhydro Sugars from Glycals Involving a Single-Step Oxirane Formation The potential for the use of 1,2-anhydro sugars as glycosyl donors was only fully realized in the late 1980s and early 1990s with the development by Danishefsky's group of efficient procedures for the formation of these from glycals using Murray's reagent dimethyldioxirane (DMDO) [193], and their use in controlled glycosidations with appropriate Lewis acids. As a reflection of this method of preparation, the 1,2-anhydro sugars are often referred to as 'glycal epoxides'. These developments and applications thereof are elegantly summarized in the literature [194], and Scheme 5.65 summarizes the efficiencies and stereochemical outcomes of dioxirane-mediated glycal epoxide formation. Protected glucal **14** and galactal **159** predominantly give α-1,2-anhydroglycopyranoses **158** and **160**, whereas oxidation of D-allal derivative **161** predominantly gives β-epoxide **162**, and stereoselectivity is lost in the oxidation of D-gulal derivative **163**, presumably because of the competing steric influences on either face.

The oxidation is carried out with DMDO solutions prepared from Oxone and acetone [193], and the epoxide products are obtained in virtually quantitative yields and often used in subsequent steps after simple work-up and little or no further purification. A simplified procedure for high-yielding multigram-scale epoxidation of glucal and galactal has been recently described [195], where DMDO is generated *in situ* using Oxone and acetone in a biphasic system (CH_2Cl_2–aqueous $NaHCO_3$).

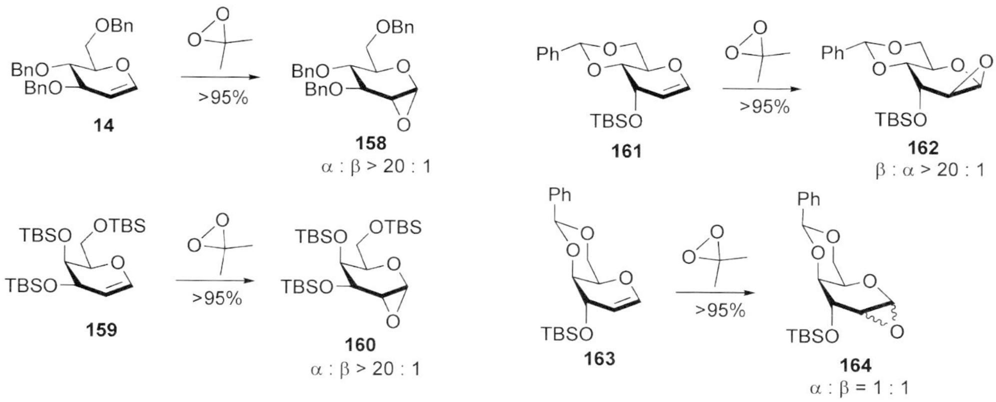

Scheme 5.65 Summary of reaction outcomes in DMDO epoxidation of glycals.

Several alternative methods for converting glycals into epoxides have been described. For example, Di Bussolo et al. [196] have described a one-pot method for oxidative glycosylation using the combination of triflic anhydride and diphenylsulfoxide to effect the formation of 1,2-anhydro-α-glucose from protected glucals. Subsequent addition of zinc chloride and an acceptor alcohol leads to the formation of β-glucosides in acceptable yields. In one interesting exception to the observed β-selectivity, glycosylation of the hindered 3-OH of an otherwise protected glycosyl acceptor using zinc chloride was not successful, but yielded the α-linked disaccharide when the stronger Lewis acid, Sc(OTf)$_3$, was used.

The use of other peroxides in the epoxidation of glycals is limited by selectivities that are often inferior to those achieved with DMDO. One notable exception is the use of the mCPBA (m-chloroperbenzoic acid)/KF combination and its recent successful application in one-pot epoxidation alcoholysis (Scheme 5.66) [197]. This involved treatment of benzylated D-galactal **36** in dichloromethane/methanol with a mixture of mCPBA and KF (2 : 1) in anhydrous dichloromethane to give methyl-2-hydroxygalactoside **165**.

Other Synthetic Routes to 1,2-Anhydro Sugars The most common alternative preparation of 1,2-anhydro sugars proceeds via 1,2-trans-glycosyl halides having the 2-OH free or latent, such that on treatment with a base an intramolecular displacement of the halide takes place to give the 1,2-oxiranes [198]. Examples (Scheme 5.67) include the formation of 1,2-anhydro-tri-O-benzyl-D-glucose **158** from β-glucosyl fluoride **166** (pathway a). The corresponding β-epoxide **169**, not readily available

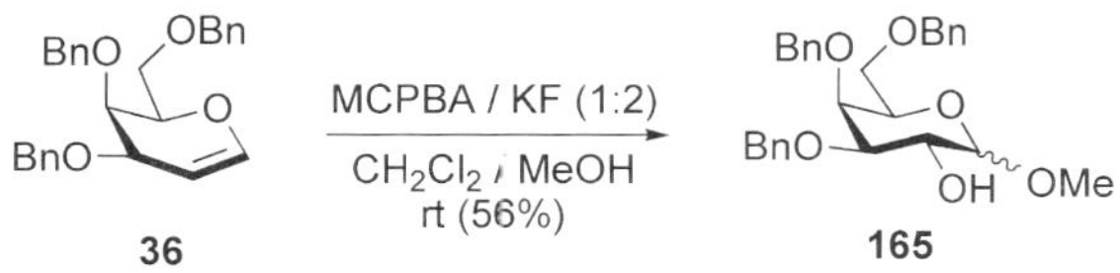

Scheme 5.66 Use of mCPBA/KF in the preparation of 1,2-anhydro sugars.

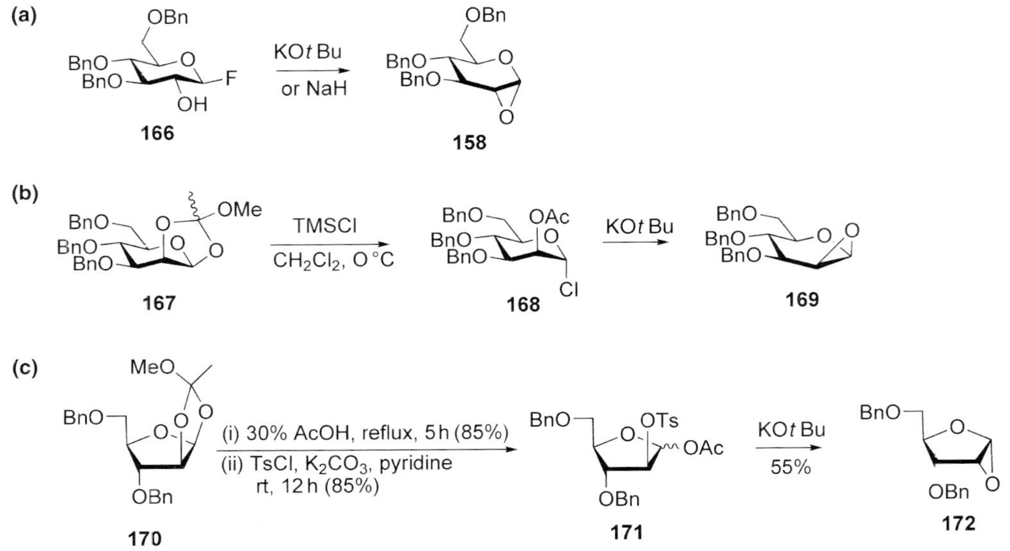

Scheme 5.67 Alternative routes to 1,2-anhydro sugars.

by DMDO oxidation, has been prepared [199–201] by treating the precursor 2-*O*-acetyl-α-D-mannosyl chloride **168** with potassium *tert*-butoxide in THF (pathway b). In a related approach [202], the sensitive D-ribofuranosyl epoxide **172** [203] could be prepared from 2-*O*-tosyl-D-ribofuranosyl acetate **171**, obtained in turn from the orthoester **170** (pathway c). This and similar methods for preparing 1,2-anhydropyranose sugars [204] proceed via prior installation of a leaving group such as a tosylate at C-2, and subsequent intramolecular attack by an alkoxide generated at C-1. Interestingly, the more direct approach to preparing 2-bromoglycoses from glycals, followed by sequential base-mediated 1,2-epoxide formation and trapping with thiolates or alkoxides, suffers from poor yields and selectivities in the final steps and has not found practical application [147].

1,2-Anhydro Sugars in the Synthesis of *O*-, *N*- and *S*-Glycosides The tremendous versatility of 1,2-anhydro sugars in the synthesis of glycosides is illustrated in Scheme 5.68, and there are now numerous examples of their use in the preparation of a wide array of complex oligosaccharides, glycoconjugates and other glycosides [190,194]. Glycal epoxides such as **173** can be used directly as glycosyl donors, with dual advantages of stereocontrol in glycoside formation and the simultaneous generation of glycosides with a free 2-OH, for further modification or use as a glycosyl acceptor in iterative glycosylation processes. Perhaps, the most frequent use of this methodology is in glycosylation reactions to generate β-*O*-glucosides [192,205] such as **175** and **178** and the corresponding β-thioglucosides [205–208] **176**, **177** and **182**, using conditions that ensure controlled, concerted opening of the epoxide. However, α-glucosides are also accessible [191] in a direct manner using alternative Lewis acids

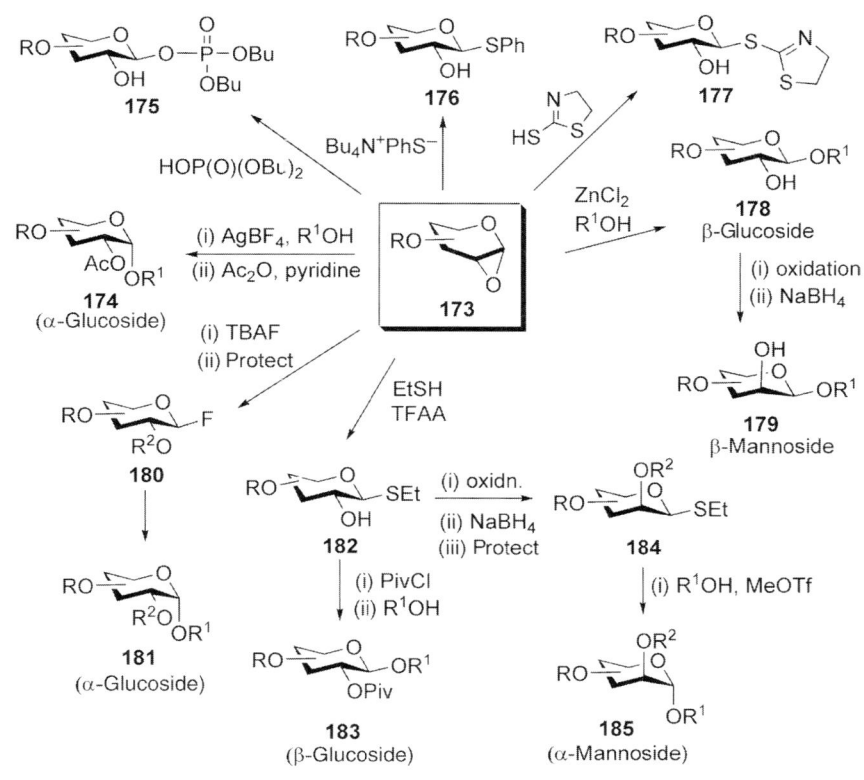

Scheme 5.68 1,2-Anhydro sugars as glycosyl donors.

such as AgBF$_4$ in the formation of **174**, or indirectly via β-glycosyl fluoride **180**, which after introduction of a suitable nonparticipating protecting group can be stereoselectively glycosylated to give **181**.

Access to β-mannosides [209] is illustrated by the preparation of **179** from β-glucoside **178** by oxidation of the equatorial 2-OH followed by stereoselective reduction to give the axial alcohol an efficient indirect route to the α-mannosides [206] utilizes the β-thioglucoside **182**, readily obtained from epoxide **173**, proceeding via an oxidation–reduction protection sequence to give β-thiomannoside glycosyl donor **184**, from which α-mannoside **185** can be stereoselectively prepared.

As noted above, the possibility of converting 1,2-anhydro sugars into a range of glycosyl donors significantly enhances their scope in glycosidation reactions. For example, thiazolinyl glycosides **177** represent a recently introduced attractive alternative [208], as these thioglycosides are easily activated by AgOTf in the presence of alcohols to form O-glycosides. In addition, the formation and use of glycosylphosphates [210] such as **175** continues to be explored, as illustrated in recent investigations [211,212] of an alternative approach to their preparation involving treatment of benzylated glucal **14** with catalytic methyltrioxorhenium (MTO) and urea-hydroperoxide (UHP) in the presence of dibutylphosphate to give mixtures of gluco- and mannosylphosphates, with the *gluco*-isomers favored. The efficiency of this process was found to be crucially dependent on addition of the ionic liquid dimethylimidazolium tetrafluoroborate and the accelerating ligand imidazole in dichloromethane or toluene. This example highlights the fact that the search for efficient catalytic routes to the sugar epoxides and other derivatives may well drive further efforts in this arena.

The use of 1,2-anhydro sugars in the preparation of N-glycosides is illustrated (pathway a, Scheme 5.69) by the glycosylation of protected bisindole **186** by deprotonation using NaH, then addition of 3 equiv of epoxide **187** to give glycoside **188**, a precursor of the anticancer compound rebeccamycin [213]. The scope is widened to provide routes to glycosyl amines and glycosyl amides by the development of methodology for the formation of glycosyl azides from 1,2-anhydro sugars. Lee *et al.* [214] have shown (pathway b, Scheme 5.69) that the treatment of α-1,2-anhydrides **190** of D-glucose and D-galactose, bearing a variety of protecting groups, with lithium azidohydridodiisobutylaluminate (DIBAH–LiN$_3$) in THF, exclusively gives the β-azidoglycosides **191**. The corresponding β-1,2-anhydrides **193** derived from the protected D-allal **192** were shown to give α-altrosyl azides **194** as the major products (pathway c, Scheme 5.69).

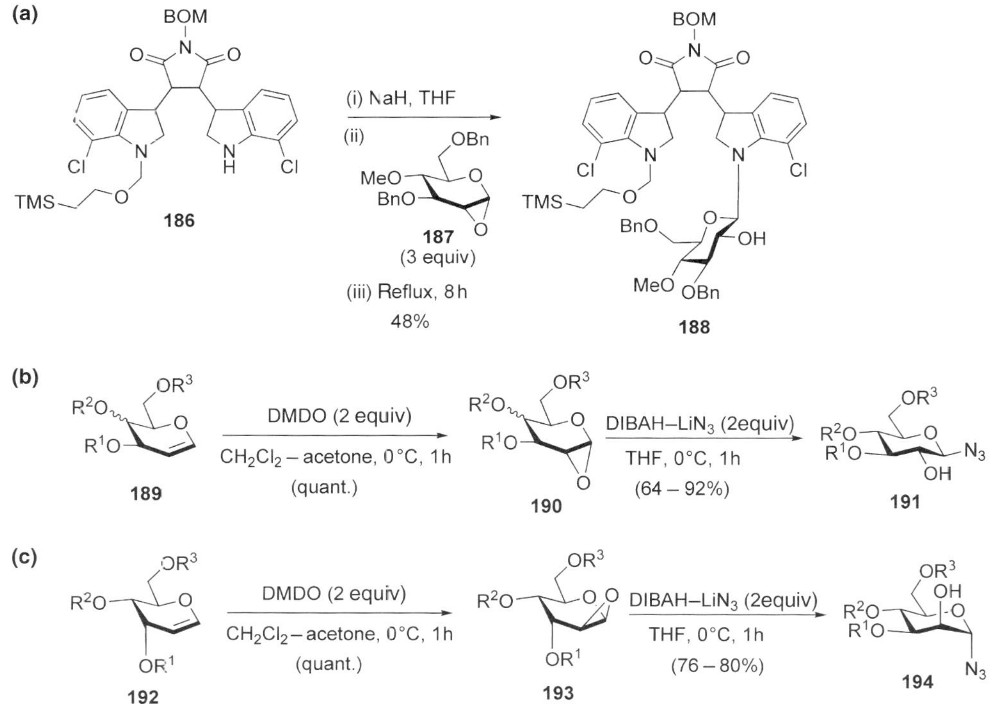

Scheme 5.69 Synthesis of N-glycosides from 1,2-anhydro sugars.

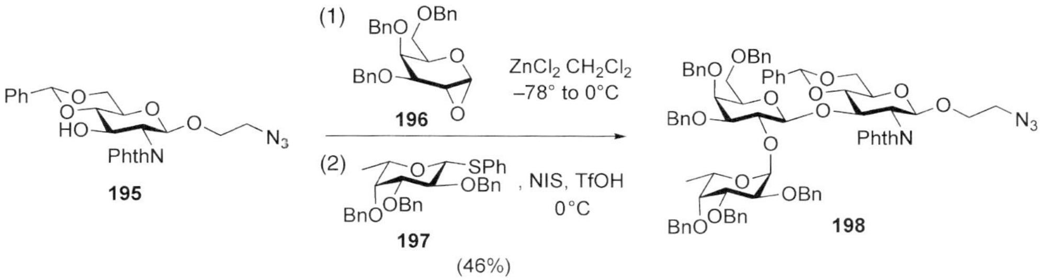

Scheme 5.70 Sequential, one-pot glycosylation using 1,2-anhydro sugars.

Examples and Future Outlook From the foregoing discussion, it is evident that 1,2-anhydro sugars are readily prepared and are highly versatile glycosyl donors. Further developments in their *in situ* generation and in the management and control of glycosylation sequences will enhance their usefulness in rapid generation of complex, diverse structures. A concluding example that illustrates their potential is the account of a sequential one-pot glycosylation (Scheme 5.70) in which benzylated galactal epoxide **196** was combined with acceptor **195** in the presence of the promoter zinc chloride, followed immediately by the addition of NIS and the thioglycoside donor **197** to give the trisaccharide **198** in 46% yield based on epoxide donor **196**.

5.2.3.2 1,6-Anhydro Sugars as Glycosyl Donors

The 1,6-anhydro sugars are another member of the class of anhydro sugars that are of significant use in the preparation of glycosyl donors and in glycoside bond formation. The chemistry of these derivatives has recently been comprehensively reviewed [109], and it will only be necessary here to summarize the key features of these sugars and highlight their role in glycoside synthesis.

The significance of these bicyclic 1,6-anhydro sugars is illustrated by considering a common member of this family, 1,6-anhydro-β-D-glucopyranose **199** (levoglucosan). It exists predominantly in the $^1C_4$ conformation, and the availability of O-2, O-3 and O-4 in inverted orientations relative to glucopyranose itself allows for unique manipulations and generation of a range of selectively protected or substituted derivatives **200** (Scheme 5.71). A range of methods is available for the synthesis of **199**, including pyrolysis of cellulose, which is practical for large-scale

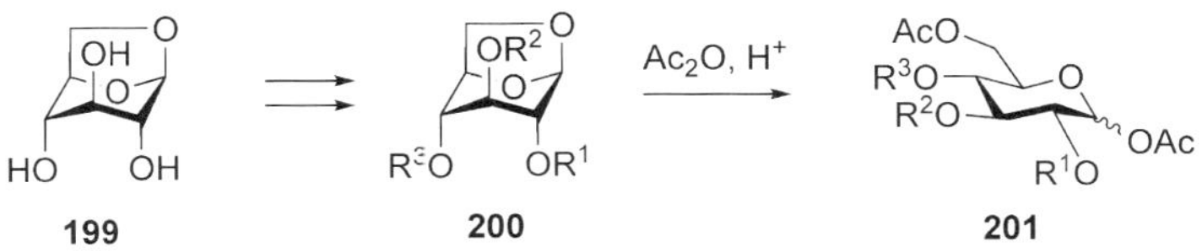

Scheme 5.71 Structure of 1,6-anhydro glucose, and its use in the synthesis of selectively protected glycosyl donors.

preparations. In addition to this *gluco* version, the *allo, altro, manno, gulo, ido, galacto* and *talo* anhydropyranoses are also known. Differentially protected glucosyl acetate **201** is easily regenerated by a number of methods, of which the most common and efficient is acetolysis using acetic anhydride with catalysts such as trifluoroacetic acid, triethylsilyl trifluoromethanesulfonate or scandium triflate, and 1,6-anhydrosugar derivatives can also be directly *S*-glycosylated [215] by treatment with trimethylsilyl phenylsulfide (TMSSPh) and ZnI_2. The fact that the 1,6-anhydro sugars are readily converted into glycosyl donors thus enhances their utility as building blocks for the synthesis of oligosaccharides.

A few selected examples from the recent literature illustrate the strategic importance of these sugar derivatives. In a recent synthesis of a key trisaccharide intermediate for the preparation of inner core structures of *Haemophilus* and *Neisseria* lipopolysaccharides [216], heptoses donors were assembled via 1,6-anhydro derivatives. Mannoside **202** (Scheme 5.72) was oxidized at C-6 and then reacted with vinyl magnesium bromide to give the chain-extended sugar **203**. This was converted into 1,6-anhydro sugar **204** on treatment with ferric chloride, and the heptose derivative **205** was then obtained by the oxidation of the olefin followed by further manipulation of protecting groups. Glycosylation with glycosyl bromide **206**, followed by the hydrolysis of the isopropylidene group and selective benzylation, gave glycosyl acceptor **207**, which could be glycosylated with donor **208** to give, after acetolysis, the protected heptose-containing trisaccharide **209** in a form in which it can be readily converted into other suitable glycosyl donors.

As a further illustration of the scope for use of 1,6-anhydro sugars in the preparation of uncommon sugars, the rare L-idose derivative **210** has been prepared recently and used in the preparation of oligosaccharides containing iduronic acid

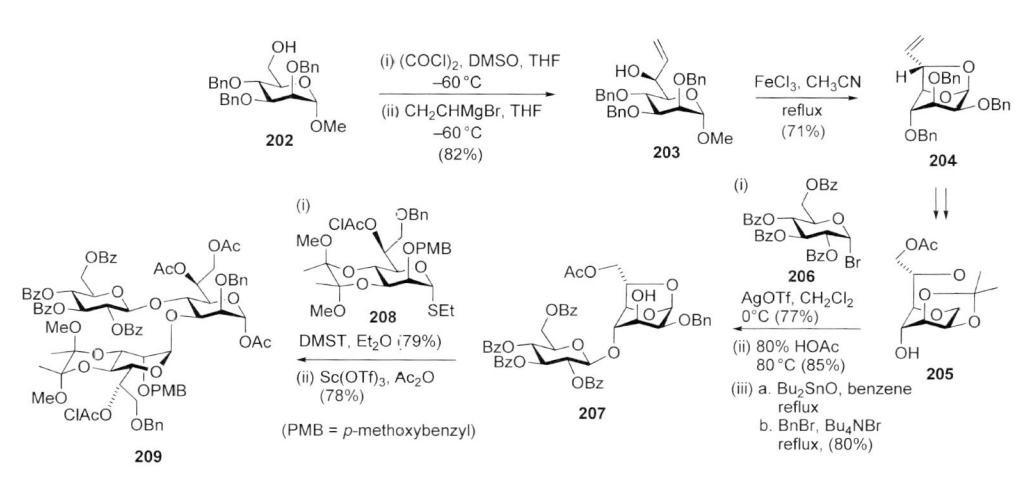

Scheme 5.72 Preparation of modified 1,6-anhydromannosides and their use in the synthesis of heptose-containing oligosaccharide.

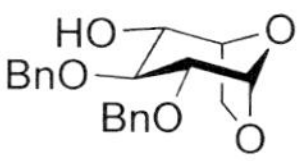

derivatives [217]. The 6-*exo*-bromo-derivatives such as **211** are also readily prepared [218,219] in a highly regio- and stereoselective radical bromination of acetylated 1,6-anhydroglycoses, providing a template for stereoselective preparation of C-6 alkylated sugars [220].

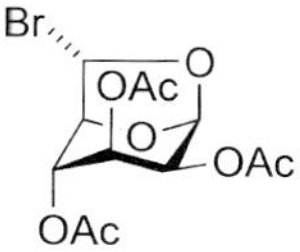

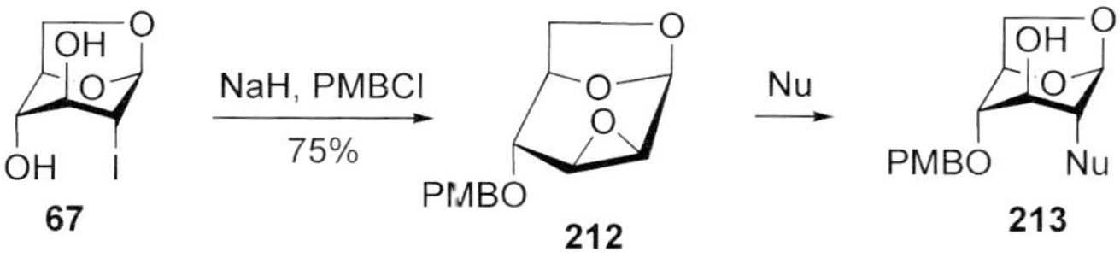

(PMB = *p*-methoxybenzyl)   (Nu = N_3^-, PhthN$^-$, AllO$^-$, BzO$^-$)

Scheme 5.73 Selective preparation of 2-substituted glucosides via 1,6-anhydro-D-glucose derivatives.

The final example (Scheme 5.73) illustrates the efficient preparation of epoxide **212** from 2-iodo-1,6-anhydro-D-glucose **67** (see Scheme 5.5) [221,222]. This "Černý epoxide" [109] is a useful and versatile precursor to a range of 2-substituted glucose derivatives **213**, which can be elaborated further to suitable glycosyl donors.

In the 1970s and 1980s considerable effort was directed at establishing the application of the 1,6-anhydro sugars in the preparation of stereoregular polysaccharides. Although this effort appears to have dissipated, a recent report [223] described the ring-opening polymerization and copolymerization of a benzylated 1,6-anhydro-3-azido-3-deoxy-β-D-allopyranose in the preparation of aminopolysaccharides containing 1,6-α-allopyranosidic linkages.

5.2.4
Conclusion

This overview has served to emphasize the versatility of dehydro and anhydro sugars in the preparation of a wide range of glycosidic linkages. They are, in general, easily accessible from readily available starting materials, and the principles governing their reactivities are now sufficiently understood to enable rational planning of synthetic strategies toward complex targets. They have also been well adapted to the rigorous requirements of modern-day synthetic chemistry, with its renewed emphasis on atom efficiency and low environmental impact without sacrificing stereo- and regiocontrol. The need for methodologies that permit rapid assembly of complex arrays of well-defined structures for biological evaluation, coupled with the search for new and better chemical technologies such as

5.2.5
General Experimental Procedures

5.2.5.1 General Method for the Preparation of 2-Deoxy-2-Iodoglycosides from Glycals

Powdered 4-Å molecular sieves (200 mg) was added to a solution of glycal (0.5 mmol) and glycosyl acceptor (0.55 mmol, 1.1 equiv) in dry CH_2Cl_2 (0.04 M in glycal). The resulting mixture was stirred at room temperature for 30 min and then IDCP was added as a solid. When TLC analysis indicated the completion of the reaction (typically, 1–2 h), the mixture was filtered and the solid washed with CH_2Cl_2. The combined filtrate was washed with 10% aqueous $Na_2S_2O_3$, dried over $MgSO_4$ and concentrated. Chromatography of the residue on silica gel (gradient hexanes-ethyl acetate) provided the coupled product (60–85%).

5.2.5.2 Preparation of 1,2-Anhydro-tri-O-Benzyl-α-D-Glucose and General Method for Its Use as a Glycosyl Donor in the Formation of β-Glycosides

Benzylated glucal (0.5 mmol) was dissolved in dry CH_2Cl_2 (2 ml) and cooled to 0 °C in a nitrogen atmosphere. DMDO (20 ml of a 0.03 M solution in acetone, 0.6 mmol) was added dropwise and the solution was stirred at 0 °C for 15 min. The α-1,2-anhydro sugar was concentrated to dryness by passing a stream of nitrogen over the reaction mixture and placing it under vacuum for 1 h. A solution of the glycosyl acceptor (0.75 mmol, 1.5 equiv) in dry THF (1.5 ml) was then added to the 1,2-anhydro sugar and the temperature reduced to −78 °C. $ZnCl_2$ (1.5 equiv of a 1 M solution in diethyl ether) was added dropwise to the reaction mixture and the reaction allowed to slowly warm to 25 °C with stirring overnight. The reaction was quenched by the evaporation of the solvent and the resulting residue purified by chromatography on silica (gradient hexanes-ethyl acetate) to give the β-glycosides (42–80%).

5.2.5.3 General Method for the Preparation of 2-Deoxy-2-Iodoglycosylbenzenesulfonamides from Glycals and Its Use as Glycosyl Donors in the Synthesis of 2-Benzenesulfonamido-2-Deoxy-β-Glycosides

(i) Glycal (0.5 mmol), benzene sulfonamide (1.25 mmol, 2.5 equiv) and 4-Å molecular sieves (0.6 g) were suspended in dry CH_2Cl_2 (10 ml), and the resulting suspension was stirred at room temperature for 10 min. The suspension was then cooled to 0 °C and treated with IDCP (1 mmol, 2 equiv). The reaction mixture was stirred in the absence of light at 0 °C for 1 h, then the reaction mixture was diluted with EtOAc (25 ml) and filtered through a pad of silica gel. The clear yellow filtrate was washed successively with saturated $Na_2S_2O_3$ (aq) (3×70 ml), saturated $CuSO_4$ (aq) (3×70 ml) and brine (3×70 ml), and dried over Na_2SO_4. The resulting crude product was purified by column chromatography (gradient ethyl acetate–hexanes) to yield the iodosulfonamide (90–99%). (ii) A solution of the iodosulfonamide

(0.5 mmol) and glycosyl acceptor (0.65 mmol, 1.3 equiv) in THF (0.15 M in iodosulfonamide) was cooled to −78 °C and a solution of lithium tetramethylpiperidide(LTMP) (1 M in THF, 1.1 ml, 2.2 equiv) was added dropwise, followed by dropwise addition of silver trifluoromethanesulfonate (1 M in THF, 0.7 ml, 1.4 equiv) after 10 min. Ensuring the exclusion of light, the reaction was allowed to slowly warm to room temperature. The reaction was monitored by TLC, and after several hours (5–15 h) solid NH_4Cl (several equiv) was added with stirring. Thereafter, the suspension was filtered and the filtrate concentrated to dryness. Flash chromatography of the resulting residue on silica (gradient hexanes-ethyl acetate) provided the coupled product (23–64%).

References

105 Ferrier, R.J. and Hoberg, J.O. (2003) *Advances in Carbohydrate Chemistry and Biochemistry*, **58**, 55–119.

106 Veyrières, A. (2000) *Carbohydrates in Chemistry and Biology: A Comprehensive Handbook*, vol. 1 (eds B. Ernst, G.W. Hart and P. Sinay), Wiley-VCH Verlag GmbH, Weinheim, New York, pp. 367–405.

107 Priebe, W. and Grynkiewicz, G. (2001) *Glycoscience – Chemistry and Chemical Biology*, vol. 1 (eds B.O. Fraser-Reid, K. Tatsuta and J. Thiem), Springer, pp. 749–783.

108 Jarosz, S. (2001) *Glycoscience - Chemistry and Chemical Biology*, vol. 1 (eds B.O. Fraser-Reid, K. Tatsuta and J. Thiem), Springer, pp. 291–304.

109 Černý, M. (2003) *Advances in Carbohydrate Chemistry and Biochemistry*, **58**, 121–198.

110 Madhusudan, S.K., Agnihotri, G., Negi, D.S. and Misra, A.K. (2005) *Carbohydrate Research*, **340**, 1373.

111 Fischer, E. and Zach, K. (1913) *Sitzungsberichte der Königlich Preussischen Akademie der Wissenschaften*, **16**, 311–317.

112 Somsák, L. and Németh, I. (1993) *Journal of Carbohydrate Chemistry*, **12**, 679–684.

113 Forbes, C.L. and Franck, R.W. (1999) *Journal of Organic Chemistry*, **64**, 1424–1425.

114 Spencer, R.P. and Schwartz, J. (2000) *Tetrahedron*, **56**, 2103–2112.

115 Wittman, M.D., Halcomb, R.L., Danishefsky, S.J., Golik, J. and Vyas, D. (1990) *Journal of Organic Chemistry*, **55**, 1979–1981.

116 Boutureira, O., Rodriquez, M.A., Matheu, M.I., Diaz, Y. and Castillon, S. (2006) *Organic Letters*, **8**, 673–675.

117 Charette, A.B. and Cote, B. (1993) *Journal of Organic Chemistry*, **58**, 933–936.

118 Kassou, M. and Castillon, S. (1994) *Tetrahedron Letters*, **35**, 5513–5516.

119 Ireland, R.E., Thaisrivongs, S., Vanier, N. and Wilcox, C.S. (1980) *Journal of Organic Chemistry*, **45**, 48–61.

120 Walker, J.A., Chen, J.J., Wise, D.S. and Townsend, L.B. (1996) *Journal of Organic Chemistry*, **61**, 2219–2221.

121 Cameron, M.A., Cush, S.B. and Hammer, R.P. (1997) *Journal of Organic Chemistry*, **62**, 9065–9069.

122 Rao, D.R. and Lerner, L.M. (1972) *Carbohydrate Research*, **22**, 345–350.

123 Pravdic, N., Franjic-Mihalic, I. and Danilov, B. (1975) *Carbohydrate Research*, **45**, 302–306.

124 Lichtenthaler, F.W. and Schneider-Adams, T. (1994) *Journal of Organic Chemistry*, **59**, 6728–6734.

125 Liu, J., Huang, C.-Y. and Wong, C.-H. (2002) *Tetrahedron Letters*, **43**, 3447–3448.

126 Chambers, D.J., Evans, G.R. and Fairbanks, A.J. (2004) *Tetrahedron*, **60**, 8411–8419.

127 Kirby, A.J. (1983) *The Anomeric Effect and Related Stereoelectronic Effects at Oxygen*, Springer Verlag, New York.

128 Deslongchamps, P. (1983) *Stereoelectronic Effects in Organic Chemistry*, Pergamon Press, Oxford.

129 Franck, R.W. (1995) *Conformational Analysis and Stereochemistry of Six-Membered Rings*, VCH Publishers, Inc., New York.

130 Bolitt, V., Mioskowski, C., Lee, S.G. and Falck, J.R. (1990) *Journal of Organic Chemistry*, **55**, 5812–5813.

131 Sabesan, S. and Neira, S. (1991) *Journal of Organic Chemistry*, **56**, 5468–5472.

132 Sherry, B.D., Loy, R.N. and Toste, F.D. (2004) *Journal of the American Chemical Society*, **126**, 4510–4511.

133 Yadav, J.S., Reddy, B.V.S., Bhasker, E.V., Raghavendra, S. and Narsaiah, A.V. (2007) *Tetrahedron Letters*, **48**, 677–680.

134 Reddy, B.G., Madhusudanan, K.P. and Vankar, Y.D. (2004) *Journal of Organic Chemistry*, **69**.

135 Lemieux, R.U. and Levine, S. (1964) *Canadian Journal of Chemistry*, **42**, 1473–1480.

136 Lemieux, R.U. and Morgan, A.R. (1965) *Canadian Journal of Chemistry*, **43**, 2190–2198.

137 Tatsuta, K., Fujimoto, K. and Kinoshita, M. (1977) *Carbohydrate Research*, **54**, 85–104.

138 Thiem, J., Karl, H. and Schwentner, J. (1978) *Synthesis*, 696–698.

139 Tatsuta, K., Fujimoto, K., Kinoshita, M. and Umezawa, S. (1977) *Carbohydrate Research*, **54**, 85–104.

140 Friesen, R.W. and Danishefsky, S.J. (1989) *Journal of the American Chemical Society*, **111**, 6656–6660.

141 Tanaka, H., Takahashi, D. and Takahashi, T. (2006) *Angewandte Chemie (International Edition)*, **45**, 770–773.

142 Choe, S.W.T. and Jung, M.E. (2000) *Carbohydrate Research*, **329**, 731–744.

143 Boullanger, P. and Descotes, G. (1976) *Carbohydrate Research*, **51**, 55–63.

144 Horton, D., Priebe, W. and Varela, O. (1986) *Journal of Organic Chemistry*, **51**, 3479–3485.

145 Bellucci, G., Chiappe, C., D'Andrea, F. and Lo Moro, G. (1997) *Tetrahedron*, **53**, 3417–3424.

146 Bock, K., Pedersen, C. and Thiem, J. (1979) *Carbohydrate Research*, **73**, 85–91.

147 Marzabadi, C.H. and Spilling, C.D. (1993) *Journal of Organic Chemistry*, **58**, 3761–3766.

148 Roush, W.R., Narayan, S., Bennett, C.E. and Briner, K. (1999) *Organic Letters*, **1**, 895–897.

149 Barluenga, J., Marco-Arias, M., Gonzalez-Bobes, F., Ballesteros, A. and Gonzalez, J.M. (2004) *Chemistry – A European Journal*, **10**, 1677–1682.

150 Sels, B., Levecque, P., Brosius, R., De Vos, D., Jacobs, P., Gammon, D.W. and Kinfe, H.H. (2005) *Advanced Synthesis and Catalysis*, **347**, 93–104.

151 Gammon, D.W., Kinfe, H.H., De Vos, D.E., Jacobs, P.A. and Sels, B.F. (2004) *Tetrahedron Letters*, **45**, 9533–9536.

152 Gammon, D.W., Kinfe, H.H., De Vos, D.E., Jacobs, P.A. and Sels, B.F. (2007) *Journal of Carbohydrate Chemistry*, **26**, 141–157.

153 Roush, W.R. and Bennett, C.E. (1999) *Journal of the American Chemical Society*, **121**, 3541–3542.

154 Chong, P.Y. and Roush, W.R. (2002) *Organic Letters*, **4**, 4523–4526.

155 Leteux, C., Veyrieres, A. and Robert, F. (1993) *Carbohydrate Research*, **242**, 119–130.

156 Roush, W.R., Sebesta, D.P. and James, R.A. (1997) *Tetrahedron*, **53**, 8837–8852.

157 Barbieri, L., Costantino, V., Fattorusso, E., Mangoni, A., Basilico, N., Mondani, M. and Taramelli, D. (2005) *European Journal of Organic Chemistry*, 3279–3285.

158 Borowiecka, J., Lipka, P. and Michalska, M. (1988) *Tetrahedron*, **44**, 2067–2076.

159 Bongat, A.F.G. and Demchenko, A.V. (2007) *Carbohydrate Research*, **342**, 374–406.

160 Lemieux, R.U. and Ratcliffe, R.M. (1979) *Canadian Journal of Chemistry*, **57**, 1244–1251.

161 Paulsen, H. and Lorentzen, J.p. (1984) *Carbohydrate Research*, **133**, C1–C4.

162 Lafont, D., Guilloux, P. and Descotes, G. (1989) *Carbohydrate Research*, **193**, 61–73.

163 Griffith, D.A. and Danishefsky, S.J. (1990) *Journal of the American Chemical Society*, **112**, 5811–5819.

164 Danishefsky, S.J., Koseki, K., Griffith, D.A., Gervay, J., Peterson, J.M., McDonald, F.E. and Oriyama, T. (1992) *Journal of the American Chemical Society*, **114**, 8331–8333.

165 Spassova, M.K., Bornmann, W.G., Ragupathi, G., Sukenick, G., Livingston, P.O. and Danishefsky, S.J. (2005) *Journal of Organic Chemistry*, **70**, 3383–3395.

166 Wu, B., Hua, Z., Warren, J.D., Ranganathan, K., Wan, Q., Chen, G., Tan, Z., Chen, J., Endo, A. and Danishefsky, S.J. (2006) *Tetrahedron Letters*, **47**, 5577–5579.

167 Guthikonda, K., Wehn, P.M., Caliando, B.J. and DuBois, J. (2006) *Tetrahedron*, **62**, 11331–11342.

168 Wang, Z.-G., Zhang, X., Visser, M., Live, D., Zatorski, A., Iserloh, U., Lloyd, K.O. and Danishefsky, S.J. (2001) *Angewandte Chemie (International Edition)*, **40**, 1728–1732.

169 Lemieux, R.U., Nagabushan, T.L. and O'Neill, I.K. (1964) *Tetrahedron Letters*, **5**, 1909–1916.

170 Lemieux, R.U., Nagabushan, T.L. and O'Neill, I.K. (1968) *Canadian Journal of Chemistry*, **46**, 413–418.

171 Das, J. and Schmidt, R.R. (1998) *European Journal of Organic Chemistry*, 1609–1613.

172 Winterfeld, G.A. and Schmidt, R.R. (2001) *Angewandte Chemie (International Edition)*, **40**, 2654–2657.

173 Barroca, N. and Schmidt, R.R. (2004) *Organic Letters*, **6**, 1551–1554.

174 Khodair, A.I., Pachamuthu, K. and Schmidt, R.R. (2004) *Synthesis*, 53–58.

175 Geiger, J., Reddy, B.G., Winterfeld, G.A., Weber, R., Przybylski, M. and Schmidt, R.R. (2007) *Journal of Organic Chemistry*, **72**, 4367–4377.

176 Dahl, R.S. and Finney, N.S. (2004) *Journal of the American Chemical Society*, **126**, 8356–8357.

177 Ferrier, R.J. (2001) *Topics in Current Chemistry*, vol. 215 (ed. A.E. Stütz), Springer-Verlag, Berlin, pp. 153–175.

178 Priebe, W. and Zamojski, A. (1980) *Tetrahedron*, **36**, 287–297.

179 Pearson, R.G. (1963) *Journal of the American Chemical Society*, **85**, 3533–3539.

180 Hotha, S. and Tripathi, A. (2005) *Tetrahedron Letters*, **46**, 4555–4558.

181 Hotha, S. and Tripathi, A. (2005) *Journal of Combinatorial Chemistry*, **7**, 968–976.

182 Kim, H., Men, H. and Lee, C. (2004) *Journal of the American Chemical Society*, **126**, 1336–1337.

183 Di Bussolo, V., Romano, M.R., Pineschi, M. and Crotti, P. (2007) *Tetrahedron*, **63**, 2482–2489.

184 Seeberger, P.H. and Danishefsky, S.J. (1998) *Accounts of Chemical Research*, **31**, 685–695.

185 Seeberger, P.H. and Haase, W.-C. (2000) *Chemical Reviews*, **100**, 4349–4394.

186 Seeberger, P.H. (2003) *Carbohydrate-Based Drug Discovery*, vol. 1 (ed. C.-H. Wong) Wiley-VCH Verlag GmbH, Weinheim, Germany, pp. 103–127.

187 Brigl, P. (1922) *Hoppe-Seyler's Zeitschrift fur Physiologische Chemie*, **122**, 245–262.

188 Lemieux, R.U. and Howard, J. (1963) *Methods in Carbohydrate Chemistry* v.2, 400–402.

189 Halcomb, R.L. and Danishefsky, S.J. (1989) *Journal of the American Chemical Society*, **111**, 6661–6666.

190 Bilodeau, M.T. and Danishefsky, S.J. (1996) *Modern Methods in Carbohydrate Synthesis*, (eds S. Khan and R.A. O'Neill), Harwood Academic Publishers GmbH, Amsterdam, pp. 171–193.

191 Liu, K.K.C. and Danishefsky, S.J. (1994) *Journal of Organic Chemistry*, **59**, 1895–1897.

192 Gervay, J. and Danishefsky, S. (1991) *Journal of Organic Chemistry*, **56**, 5448–5451.

193 Murray, R.W. and Jeyaraman, R. (1985) *Journal of Organic Chemistry*, **50**, 2847–2853.

194 Williams, L.J., Garbaccio, R.M. and Danishefsky, S.J. (2000) *Carbohydrates in Chemistry and Biology: A Comprehensive Handbook*, vol. 1 (eds B. Ernst, G. W. Hart and P. Sinay), Wiley-VCH Verlag GmbH, Weinheim, New York, pp. 61–92.

195 Cheshev, P., Marra, A. and Dondoni, A. (2006) *Carbohydrate Research*, **341**, 2714–2716.

196 Di Bussolo, V., Kim, Y.-J. and Gin, D.Y. (1998) *Journal of the American Chemical Society*, **120**, 13515–13516.

197 Bellucci, G., Catelani, G., Chiappe, C. and D'Andrea, F. (1994) *Tetrahedron Letters*, **35**, 8433–8436.

198 Yamaguchi, H. and Schuerch, C. (1980) *Carbohydrate Research*, **81**, 192–195.

199 Sondheimer, S.J., Yamaguchi, H. and Schuerch, C. (1979) *Carbohydrate Research*, **74**, 327–332.

200 Du, Y. and Kong, F. (1995) *Journal of Carbohydrate Chemistry*, **14**, 341–352.

201 Ding, X. and Kong, F. (1999) *Journal of Carbohydrate Chemistry*, **18**, 775–787.

202 Ning, J., Xing, Y. and Kong, F. (2003) *Carbohydrate Research*, **338**, 55–60.

203 Chow, K. and Danishefsky, S. (1990) *Journal of Organic Chemistry*, **55**, 4211–4214.

204 Wu, E. and Wu, Q. (1993) *Carbohydrate Research*, **250**, 327–333.

205 Gordon, D.M. and Danishefsky, S.J. (1990) *Carbohydrate Research*, **206**, 361–366.

206 Seeberger, P.H., Eckhardt, M., Gutteridge, C.E. and Danishefsky, S.J. (1997) *Journal of the American Chemical Society*, **119**, 10064–10072.

207 Kochetkov, N.K., Malysheva, N.N., Demchenko, A.V., Kclotyrkina, N.G. and Klimov, E.M. (1991) *Bioorganicheskaya Khimiya*, **17**, 1655–1659.

208 Pornsuriyasak, P. and Demchenko, A.V. (2006) *Chemistry – A European Journal*, **12**, 6630–6646.

209 Liu, K.K.C. and Danishefsky, S.J. (1994) *Journal of Organic Chemistry*, **59**, 1892–1894.

210 Plante, O.J., Palmacci, E.R., Andrade, R. B. and Seeberger, P.H. (2001) *Journal of the American Chemical Society*, **123**, 9545–9554.

211 Soldaini, G., Cardona, F. and Goti, A. (2003) *Tetrahedron Letters*, **44**, 5589–5592.

212 Soldaini, G., Cardona, F. and Goti, A. (2005) *Organic Letters*, **7**, 725–728.

213 Gallant, M., Link, J.T. and Danishefsky, S. J. (1993) *Journal of Organic Chemistry*, **58**, 343–349.

214 Lee, G.S., Min, H.K. and Chung, B.Y. (1999) *Tetrahedron Letters*, **40**, 543–544.

215 Demchenko, A.V., Wolfert, M.A., Santhanam, B., Moore, J.N. and Boons, G.J. (2003) *Journal of the American Chemical Society*, **125**, 6103–6112.

216 Segerstedt, E., Mannerstedt, K., Johansson, M. and Oscarson, S. (2004) *Journal of Carbohydrate Chemistry*, **23**, 443–452.

217 Lu, L.D., Shie, C.R., Kulkarni, S.S., Pan, G.R., Lu, X.A. and Hung, S.C. (2006) *Organic Letters*, **8**, 5995–5998.

218 Ferrier, R.J. and Furneaux, R.H. (1980) *Australian Journal of Chemistry*, **33**, 1025–1036.

219 Somsák, L. and Ferrier, R.J. (1991) *Advances in Carbohydrate Chemistry and Biochemistry*, **49**, 37–92.

220 Nishikawa, T., Mishima, Y., Ohyabu, N. and Isobe, M. (2004) *Tetrahedron Letters*, **45**, 175–178.

221 Arndt, S. and Hsieh-Wilson, L.C. (2003) *Organic Letters*, **5**, 4179–4182.

222 Krohn, K., Gehle, D. and Flörke, U. (2005) *European Journal of Organic Chemistry*, 4557–4562.

223 Hattori, K., Yoshida, T. and Uryu, T. (1997) *Macromolecular Chemistry and Physics*, **198**, 29–39.

5.3
Miscellaneous Glycosyl Donors

Kazunobu Toshima

5.3.1
Introduction

This chapter describes the preparations and chemical glycosylation reactions of miscellaneous glycosyl donors and some of their applications to the synthesis of natural products. For a survey on the general current methodological advances, miscellaneous glycosyl donors are classified into 18 groups on the basis of the type of anomeric functional group and their activating methods: (1) 1-*O*-silyl glycoside, (2) diazirine, (3) telluroglycoside, (4) carbamate, (5) 2-iodosulfonamide, (6) *N*-glycosyl triazole, (7) *N*-glycosyl tetrazole, (8) *N*-glycosyl amide, (9) DNA and RNA nucleosides, (10) oxazoline, (11) oxathiine, (12) 1,6-lactone, (13) sulfate, (14) 1,2-cyclic sulfite, (15) 1,2-cyclopropane, (16) 1,2-*O*-stannylene acetal, (17) 6-acetyl-2*H*-pyran-3(6*H*)-one and (18) *exo*-methylene. Furthermore, the stereochemical aspects of the glycosidic bonds formed by the glycosylations are also discussed in this chapter.

5.3.2
1-*O*-Silyl Glycoside

In the use of 1-*O*-silyl glycoside as a glycosyl donor, trimethylsilyl (TMS) and *t*-butyldimethylsilyl (TBS) groups were more commonly used (Table 5.1). These donors were prepared from the corresponding hemiacetals using an appropriate silylating agent such as a chlorotrialkylsilane and a base. Tietze *et al.* [224] introduced the glycosylation reaction of 1-*O*-trimethylsilyl glycoside with aryltrimethylsilyl ethers in the presence of a catalytic amount of TMSOTf as a Lewis acid. In a similar manner, Nashed and Glaudemans [225] used a 6-*O*-*t*-butyldiphenylsilyl-protected sugar as a glycosyl acceptor (Scheme 5.74). Cai and coworkers applied the method for the synthesis of alkyl glycosides from 1-*O*-trimethylsilyl glycoside using $BF_3 \cdot Et_2O$ instead of TMSOTf as an activator [226]. On the contrary, Mukaiyama and his coworkers developed stereoselective glycosylation reactions using 1-*O*-trimethylsilyl furanosides and pyranosides. According to this method, 1,2-*trans* ribofuranosides were predominantly synthesized by the glycosylation of 1-*O*-trimethylsilyl ribofuranose and trimethylsilyl ethers as glycosyl acceptors in the presence of a catalytic amount of TMSOTf and $Ph_2Sn=S$ as an additive. Interestingly, predominantly 1,2-*cis* ribofuranosides and 1,2-*cis* glucopyranosides were stereoselectively prepared by using $LiClO_4$ additive to the above reaction conditions (Scheme 5.75) [227]. Furthermore, Mukaiyama *et al.* introduced the glycosylation of 1-*O*-trimethylsilyl arabinofuranose and trimethylsilyl ethers using the combined catalyst, TMSOTf-[1,2-benzenediolato(2−)-*O*,*O*′]oxotitanium, to furnish the corresponding 1,2-*cis* arabinofuranosides (Scheme 5.76) [228]. On the contrary, the 1-*O*-*t*-butyldimethylsilyl glycosyl donor was used for the synthesis of 2-deoxy glycosides by Priebe *et al.* [229], and it

Table 5.1 Glycosylation of 1-O-silylglycoside.

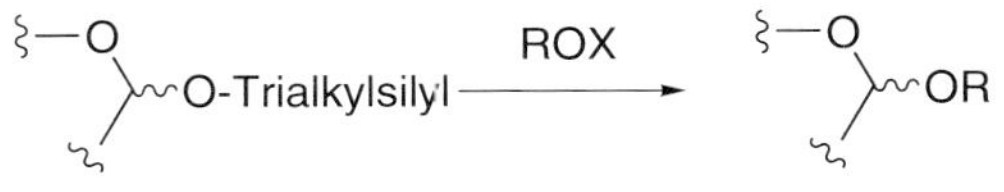

R=aryl, alkyl, sugar residue

Trialkylsilyl	Activator	X	References
TMS	TMSOTf (cat.)	TMS or TBS	[224,225]
	BF$_3$·Et$_2$O	H	[226]
	TMSOTf (cat.)−Ph$_2$Sn=S	TMS	[227]
	TMSOTf (cat.)−Ph$_2$Sn=S−LiClO$_4$	TMS	[228]
	TMSOTf (cat.) 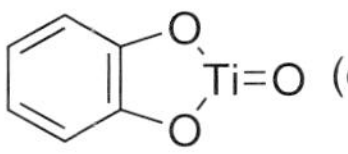 catecholato-Ti=O (cat.)	TMS	[228]
TBS	TMSOTf (cat.)	H	[229,230]

was employed in the anthracycline oligosaccharide synthesis by Kolar and Kneissl (Scheme 5.77) [230].

5.3.3
Diazirine

Vasella *et al.* introduced an approach to glycoside synthesis using the glycosylidene carbene generated from the diazirine sugar as a novel type of glycosyl donor. The

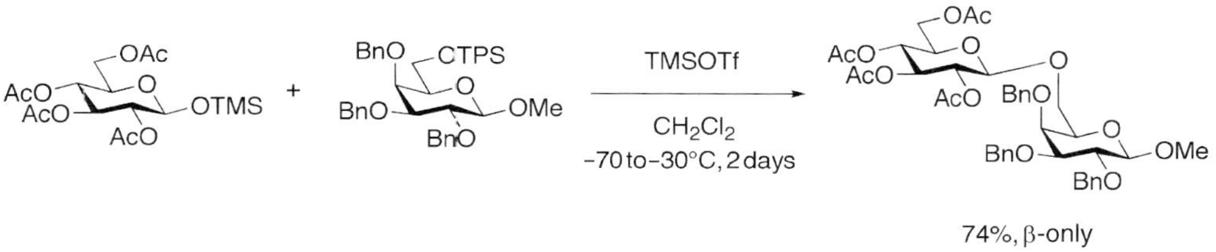

Scheme 5.74 Ref. [225].

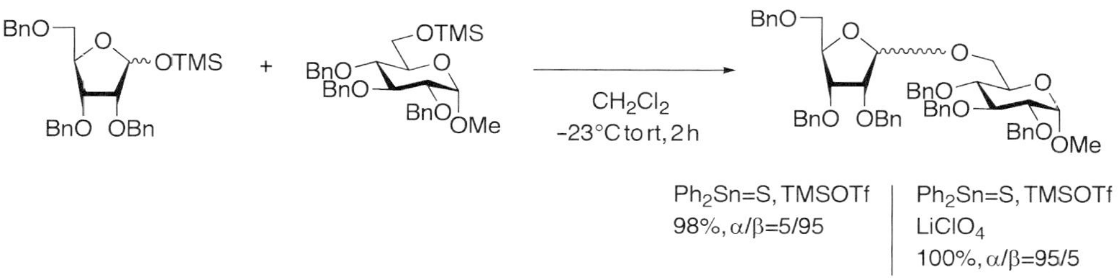

74%, β-only

Scheme 5.75 Ref. [227].

Ph$_2$Sn=S, TMSOTf
98%, α/β=5/95

Ph$_2$Sn=S, TMSOTf
LiClO$_4$
100%, α/β=95/5

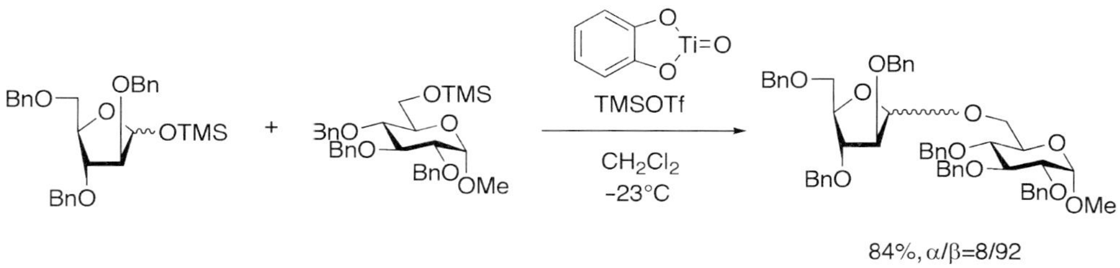

Scheme 5.76 Ref. [228].

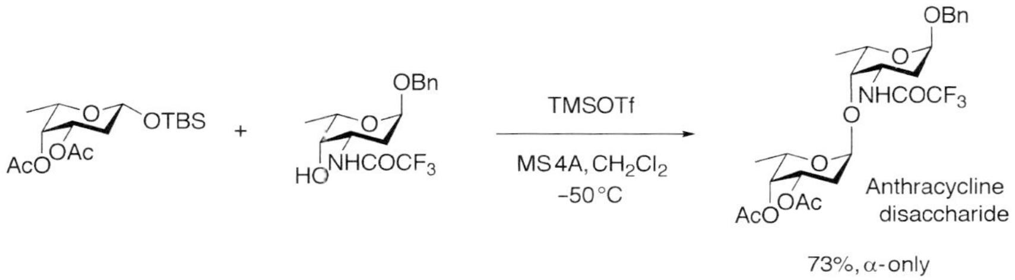

Scheme 5.77 Ref. 230a.

glycosylidene diazirines were prepared by I$_2$-mediated oxidation of the corresponding diaziridines, which were in turn formed from [(glycosylidene)amino]methanesulfonates with a saturated solution of NH$_3$ in MeOH (Scheme 5.78) [231]. The glycosylidene diazirine reacted with alcohols via the glycosylidene carbene in the absence of any additives under thermal and/or photolytic conditions to give the corresponding glycosides.

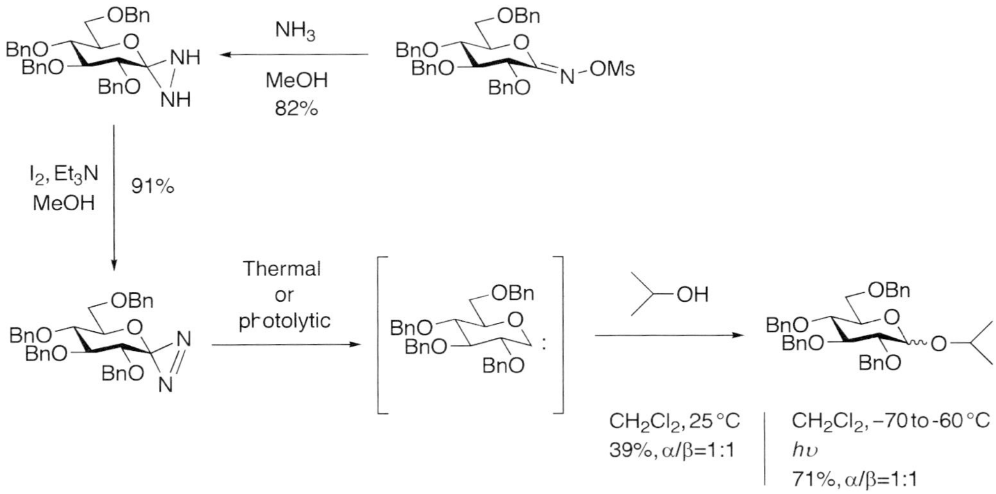

Scheme 5.78 Ref. [231].

5.3.4
Telluroglycoside

Barton and Ramesh first introduced the telluroglycoside as a glycosyl donor for C-glycosylation by means of a radical reaction [232]. Later, Yamago and Yoshida and coworkers reported O-glycosylations using aryl telluroglycosides [233]. Aryl telluroglycosides were prepared by the reaction of the corresponding bromoglycosides with diaryl ditelluride in the presence of $NaBH_4$. The activation methods including NBS, NIS, NIS-TMSOTf or by electrochemical activation, all of which were also used for the activation of thio- and selenoglycosides, were demonstrated for telluroglycosides (Scheme 5.79).

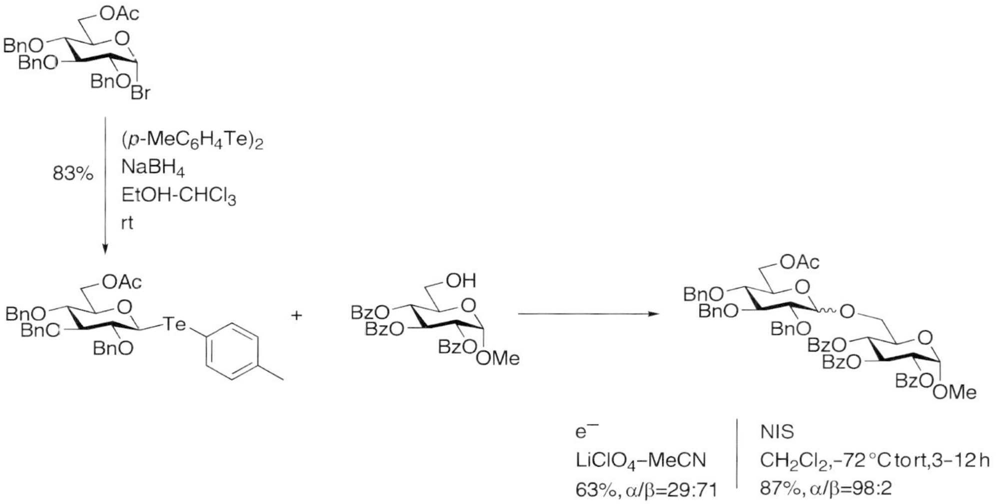

Scheme 5.79 Ref. [233].

5.3.5
Carbamate

Kunz and Zimmer introduced glycosyl N-allyl carbamates as glycosyl donors [234]. The glycosyl donors were prepared from the corresponding hemiacetals by the reaction with allyl isocyanate in the presence of $^iPr_2NEt$ or their 1-O-acyl derivatives after treatment with hydrazine acetate. The glycosylation of glycosyl N-allyl carbamates is based on an electrophile-induced lactonization of anomeric alkenoic esters, and is somewhat similar to the Fraser-Reid method [235] for the activation of pentenyl glycosides. Thus, soft electrophiles such as dimethyl methylthiosulfonium trifluoromethansulfonate (DMTST), di-(sym-collidine) iodonium perchlorate (IDCP) or methyl bis-methylthiosulfonium hexachloroantimonate (TMTSB) were used as the activator (Scheme 5.80). Along similar lines, Kiessling introduced glycosyl N-sulfonylcarbamates as new glycosyl donors with tunable reactivity [236]. The donors were synthesized by the reaction of

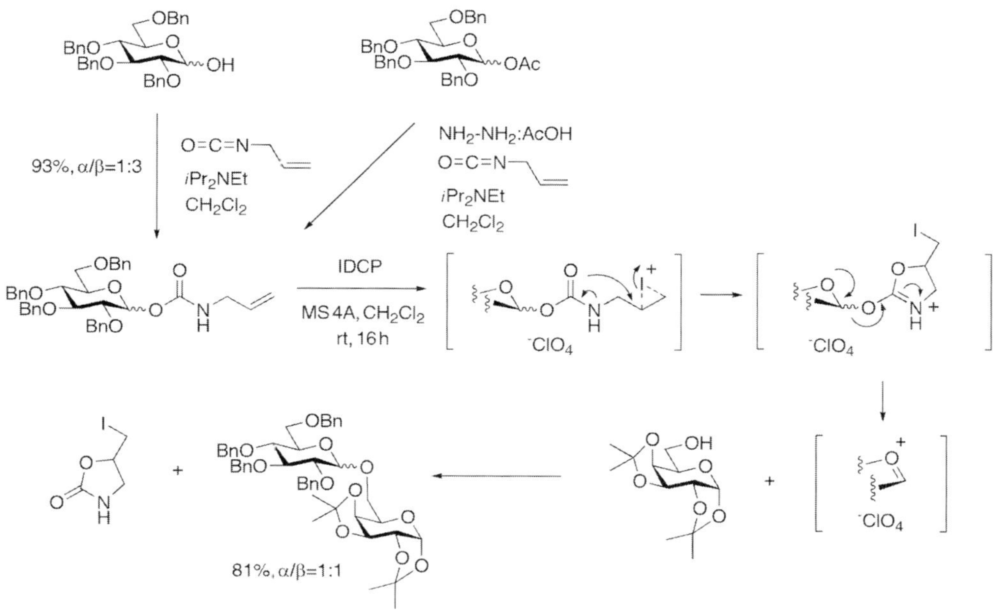

Scheme 5.80 Ref. [234].

the corresponding aldoses and *N*-sulfonyl isocyanates with 1,4-diazabicyclo[2.2.2]
octane (DABCO), and activated by several protic and Lewis acids such as
TfOH, BF$_3$·Et$_2$O, Yb(OTf)$_3$ or TMSOTf (Scheme 5.81). Furthermore, it was found
that the reactivity of glycosyl *N*-sulfonylcarbamates toward TMSOTf could
be tuned with the variation of the alkyl substituent on the nitrogen as shown
in Table 5.2.

5.3.6
2-Iodosulfonamide

2-Iodosulfonamides were employed as glycosyl donors by Danishefsky *et al.* for the
construction of 2-aminoglycosides [237]. Reaction of 2-(trimethylsilyl)ethanesulfo-
namide (SESNH$_2$) and glycal with IDCP provided *trans*-2-iodosulfonamide, which
was activated by silver catalyst (AgOTf, AgBF$_4$) and then coupled with alcohols to
furnish 1,2-*trans*-2-aminoglycosides. This method was successfully applied to the
syntheses of sialyl-Lewis X antigen oligosaccharides and chitinase inhibitors
(Scheme 5.82).

5.3.7
N-Glycosyl Triazole

Kunz *et al.* demonstrated the glycosylation using *N*-glycosyl triazoles as glycosyl
donors [238]. These glycosyl donors were prepared from the corresponding glycosyl

Scheme 5.81 Ref. [236].

azides by the reaction with alkynes. The glycosylation of N-glycosyl triazoles with alcohols in the presence of TMSOTf produced the corresponding glycosides (Scheme 5.83).

Table 5.2 Glycosylation using different glycosyl sulfonylcarbamates [236].

R	Equiv. of TMSOTf	Yield	α/β ratio
H	1.1	82	6/1
Me	0.1	84	1.7/1
CH$_2$CN	0.1	89	1.3/1
⇘	0.1	0	—

5.3.8
N-Glycosyl Tetrazole

Besides N-glycosyl triazoles, N-glycosyl tetrazoles were introduced by Sulikowski and coworkers [239]. N-Glycosyl tetrazoles were formed via phosphitylation of

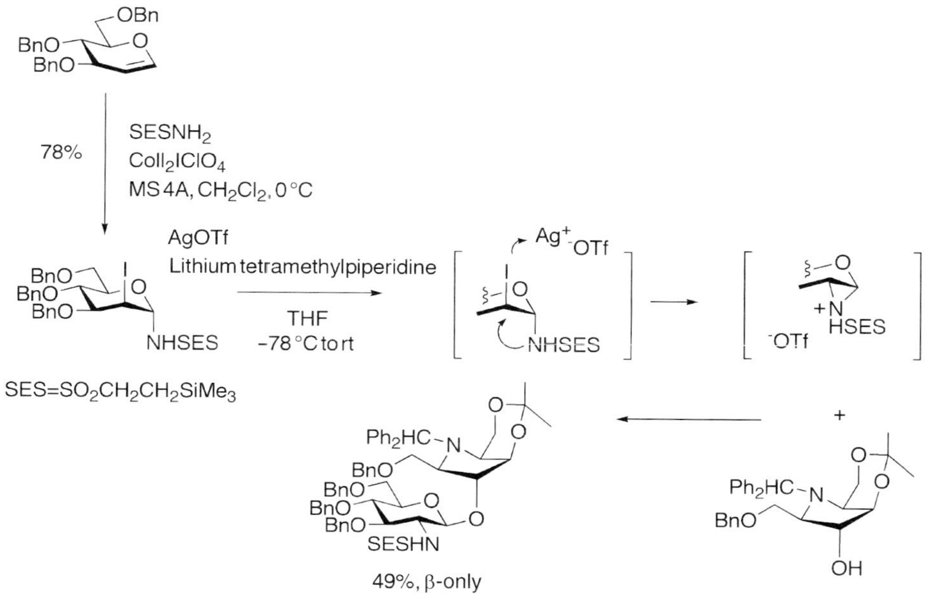

Scheme 5.82 Ref. [237].

2-deoxy sugars, and these worked as glycosyl donors when activated by ZnCl$_2$ or Me$_3$OBF$_4$ furnishing α-glycosides selectively (Scheme 5.84). The method was successfully employed in the syntheses of the deoxyoligosaccharides of the antibiotics PI-080 and landomycin A (Scheme 5.85) [239b,c].

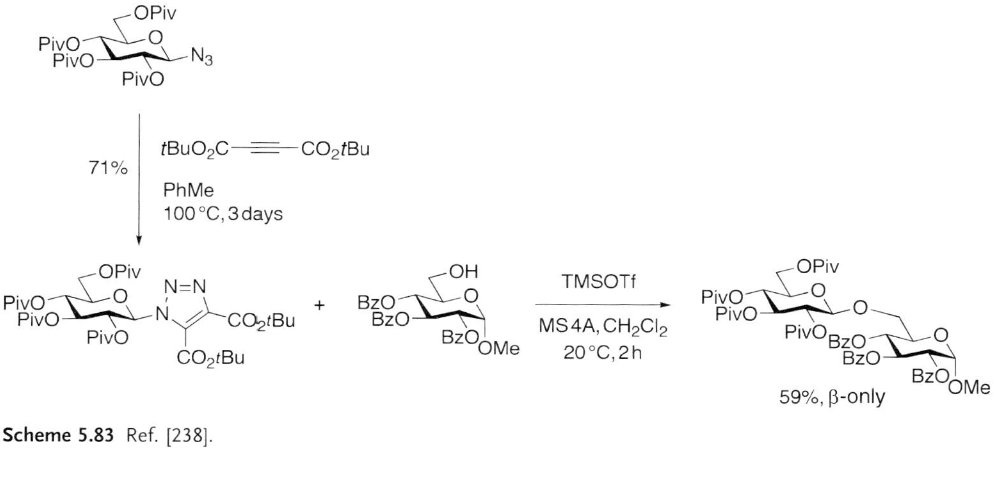

Scheme 5.83 Ref. [238].

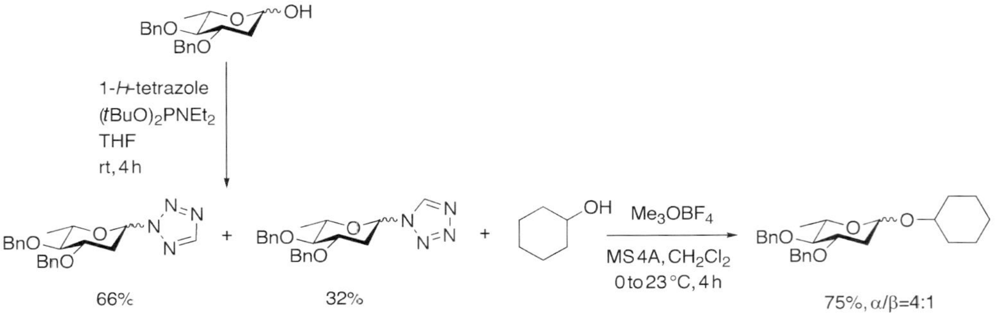

Scheme 5.84 Ref. [239].

5.3.9
N-Glycosyl Amide

The use of N-glycosyl amides as glycosyl donors was reported by Pleuss and Kunz [240]. These amides were activated by Ph_3P and CBr_4 to produce bromo-N-imidates, which were spontaneously converted into the corresponding bromide concomitant with releasing nitrile, and then coupled with alcohols by activation with AgOTf (Scheme 5.86).

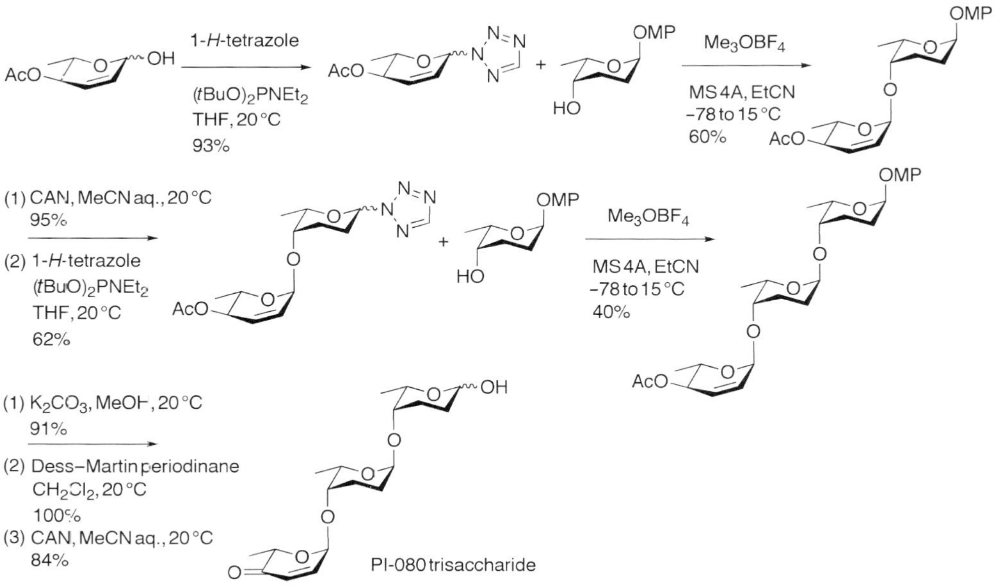

Scheme 5.85 Ref. [239b].

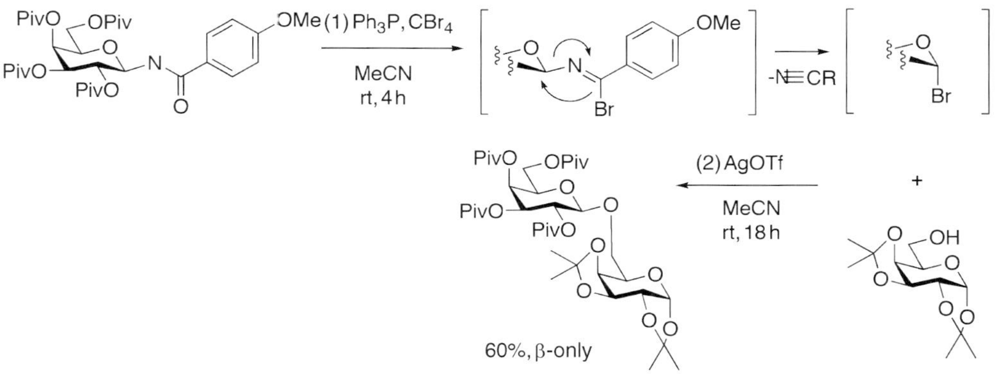

Scheme 5.86 Ref. [240].

5.3.10
DNA and RNA Nucleosides

Toshima and coworkers developed a novel glycosylation method using DNA bases as leaving groups based on one of the most typical DNA sequence protocols, Maxam–Gilbert method [241]. Thus, the simple and practical synthesis of alkyl glycosides by protic and alkylative glycosylations using natural starting compounds, DNA and RNA nucleosides, was realized (Scheme 5.87). In this study, it was found that purine bases (adenine and guanine) worked as good leaving groups, whereas the pyrimidine bases (cytosine and thymine) had no such ability under MeOTf- or TfOH-promoted glycosylation conditions. Importantly, the chemoselectivity of this glycosylation method was exploited for the preparation of modified DNA oligomers from naturally occurring DNA fragments (Scheme 5.88).

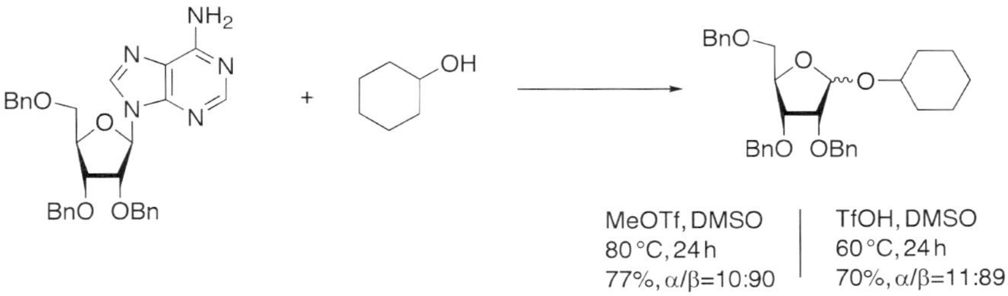

Scheme 5.87 Ref. [241].

5.3.11
Oxazoline

Oxazoline glycosyl donors are generally prepared from the corresponding 2-acylamido-2-deoxyglycosides by the activation of the C-1 leaving group using an

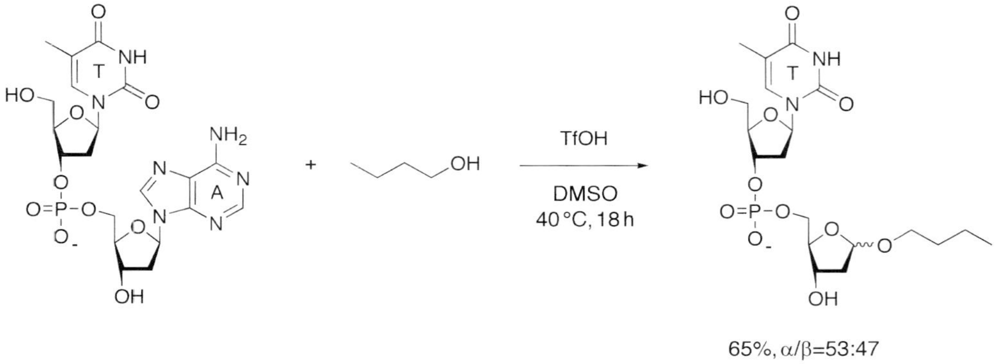

Scheme 5.88 Ref. [241].

appropriate activator (Scheme 5.89). Oxazolines can then be activated for glycosylation with Yb(OTf)$_3$ [247] (Scheme 5.90), p-TsOH [242], FeCl$_3$ [243], TMSOTf [60], camphorsulfonic acid (CSA) [244], CuCl$_2$ [245], pyridinium triflate with microwave irradiation [246] and PPTS [246]. Instead of the most commonly used methyloxazolines, trichloro-oxazolines, which could be activated under milder conditions with TMSOTf, were also introduced [248]. The advantage of using oxazolines as glycosyl donors in glycosylation reactions is the direct and stereoselective access to the *trans*-2-aminoglycosides.

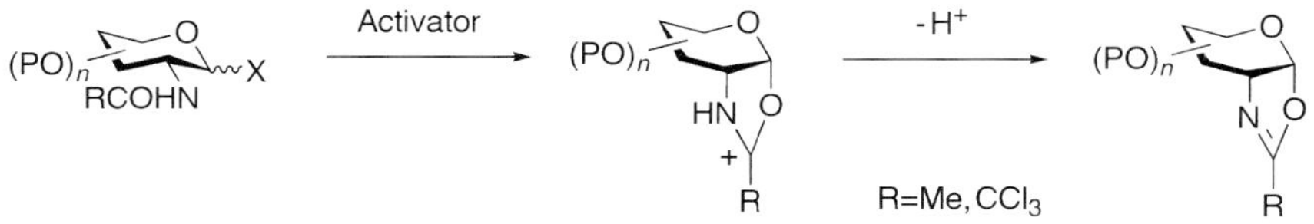

Scheme 5.89

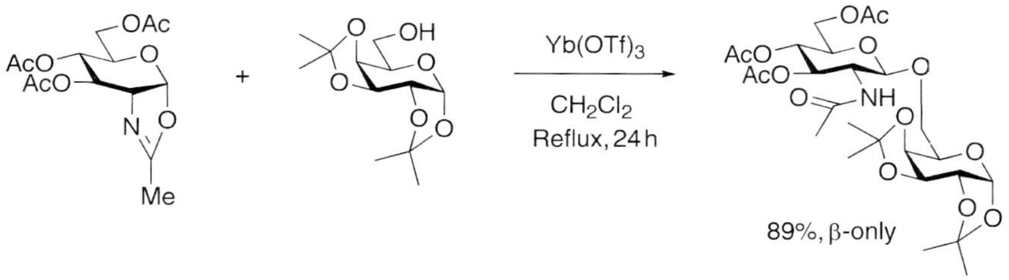

Scheme 5.90 Ref. [247].

5.3.12
Oxathiine

Franck, Marzabadi, Capozzi and Nativi introduced bicyclic glycosyl donors, oxathiines [249]. These glycosyl donors were prepared via Diels–Alder cycloaddition

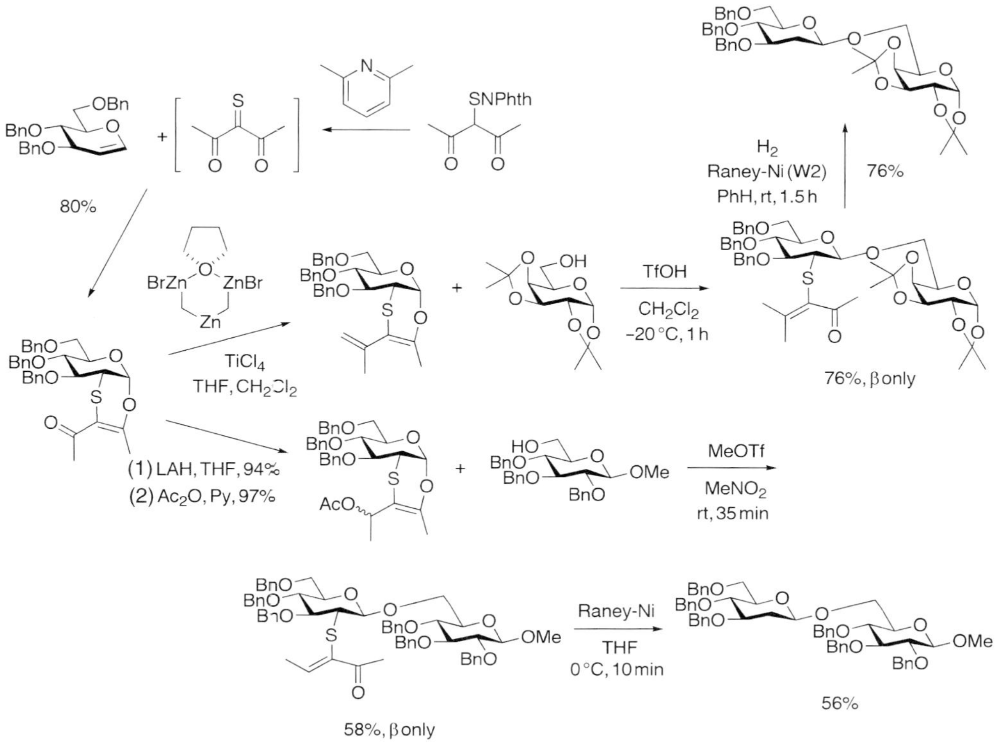

Scheme 5.91 Ref. [249].

reactions between glycals and 3-thioxopentane-2,4-dione (Scheme 5.91). Treatment of the cycloadducts with Nysted reagent in the presence of TiCl$_4$ gave the vinyl glycosides as glycosyl donors, which could be activated by TfOH and then coupled with alcohols to furnish β-glycosides. These products were converted into the corresponding 2-deoxy-β-glycosides by desulfurization using Raney-Ni. On the contrary, the allyl acetates as glycosyl donors were prepared from the cycloadducts by reduction using lithium aluminum hydride (LAH) followed by acetylation. The glycosylations of alcohols with the allyl acetates using MeOTf proceeded to afford β-glycosides, which were also converted into the corresponding 2-deoxy-β-glycosides by hydrogenolysis using Raney-Ni. It was also found that the obtained β-glycosides could be transformed into the corresponding α-anomer by anomerization in the same reaction medium; the isomerization was presumably induced by acid catalysis.

5.3.13
1,6-Lactone

1,6-Lactone derivatives derived from glucuronic acid were used as glycosyl donors by Murphy and coworkers [250]. These glycosyl donors were found to react with silyl

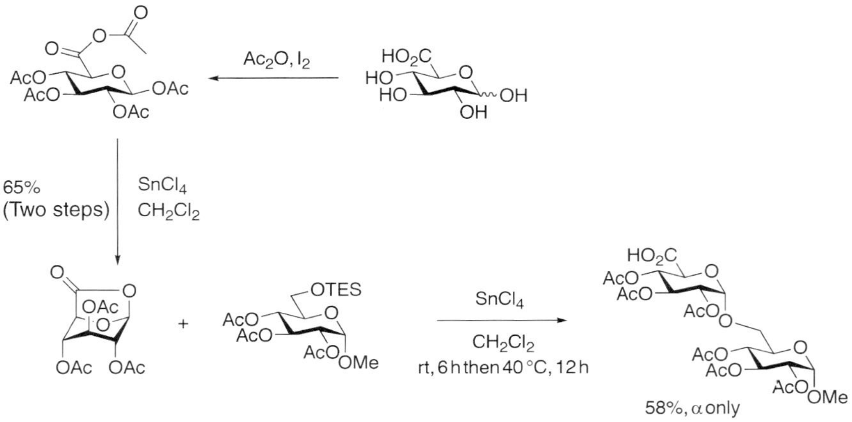

Scheme 5.92 Ref. [250].

ethers in the presence of SnCl$_4$ to stereoselectively give the corresponding α-glucuronides (Scheme 5.92). In contrast, the donor possessing an iodo-substituent at the C-2 position afforded the corresponding β-glucuronides through the iodonium intermediate.

5.3.14
Sulfate

Russo and coworkers introduced glycosyl sulfates as glycosyl donors [251]. These glycosyl donors were synthesized by the treatment of the corresponding aldoses with SO$_3$·NMe$_3$ complex and could be stored for several weeks as the triethylammonium salts. The glycosylations using BF$_3$·Et$_2$O or TMSOTf gave glycosides in fair to good yields (Scheme 5.93).

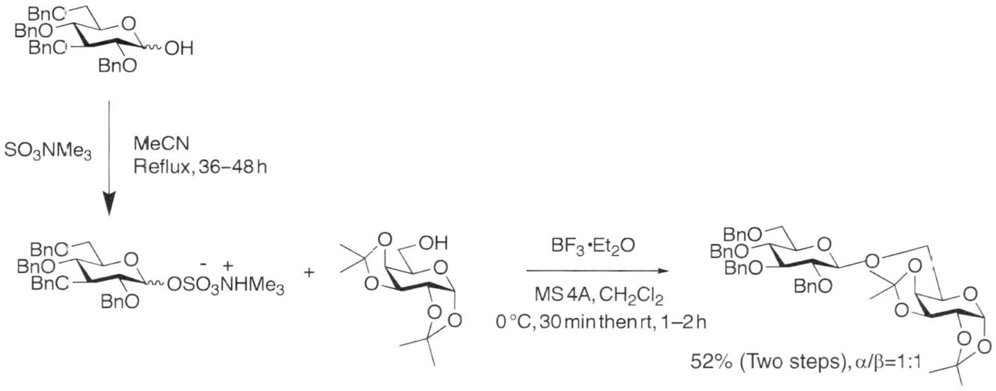

Scheme 5.93 Ref. [251].

5.3.15
1,2-Cyclic Sulfite

Beaupere and coworkers [252] and Sanders and Kiessling [253] independently reported the glycosylations of 1,2-cyclic sulfites as glycosyl donors, which were prepared from 1,2-dihydroxy sugars by the reaction using thionyldiimidazole (SO(Im)$_2$). These glycosyl donors reacted with phenoxide ions to yield the corresponding β-aryl glycosides (Scheme 5.94), and also coupled with alcohols by the activation using Yb(OTf)$_3$ or Ho(OTf)$_3$ to afford the corresponding β-glycosides (Scheme 5.95).

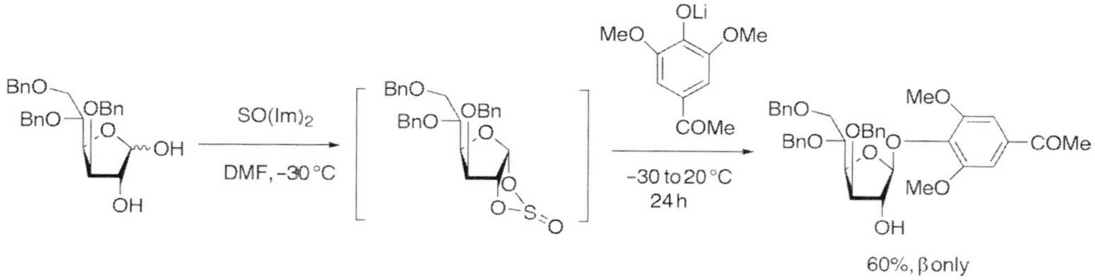

Scheme 5.94 Ref. [252].

5.3.16
1,2-Cyclopropane

Madsen et al. reported platinum(II)-mediated ring-opening glycosylation of 1,2-cyclopropanated sugars, which were prepared from the corresponding glycals by Simmons–Smith cyclopropanation reaction using CH$_2$I$_2$, Zn and CuCl. These glycosyl donors were found to react with alcohols by activation using Zeise's dimer [Pt(C$_2$H$_4$)Cl$_2$] to yield the corresponding 2-C-methyl glycosides [254] (Scheme 5.96).

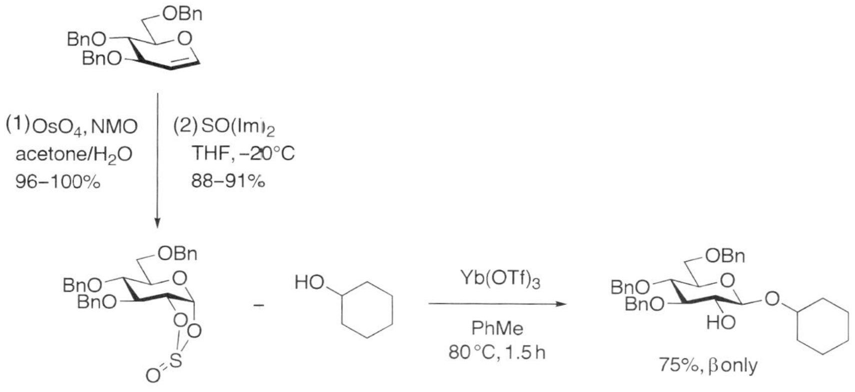

Scheme 5.95 Ref. [253].

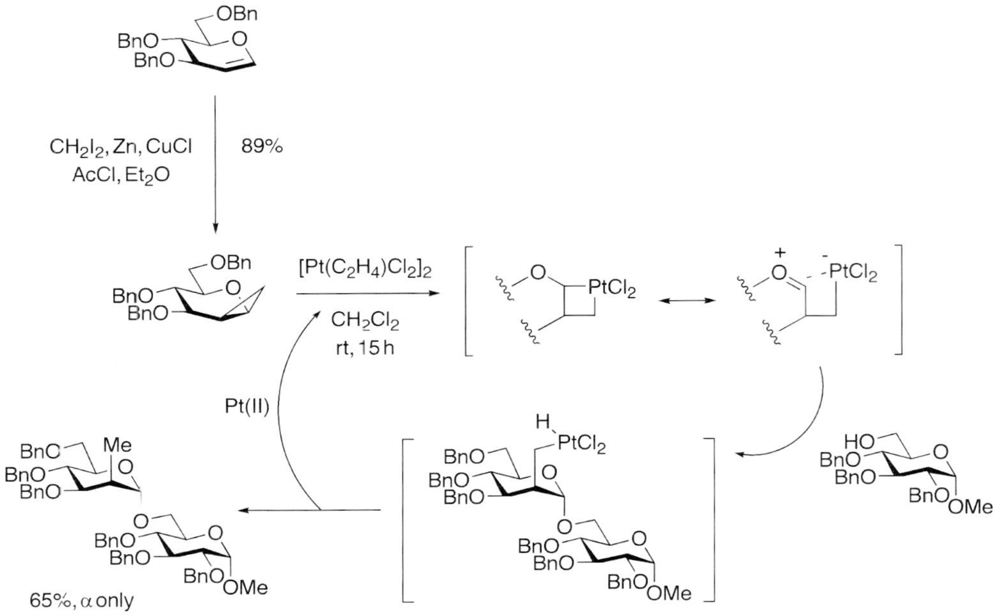

Scheme 5.96 Ref. [254].

5.3.17
1,2-O-Stannylene Acetal

Srivastava and Schuerch [255] and Desinges *et al.* [256] demonstrated the potential utility of glycosyl 1,2-O-stannylene acetals, and Hodosi and Kovác [257] established this method involving alkylation of a 1,2-O-stannylene acetal with a triflate derivative as an aglycon (Scheme 5.97). This glycosyl donor was prepared by the

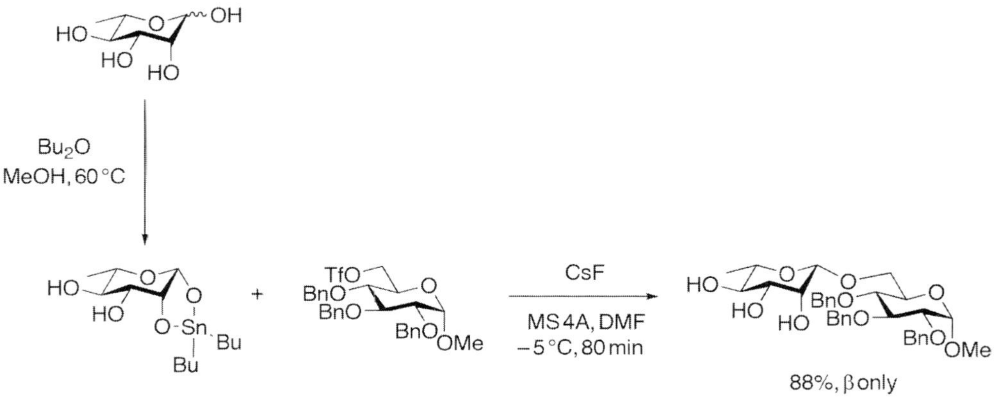

Scheme 5.97 Ref. [257].

reaction of the corresponding 1,2-dihydroxy sugar with dibutyltin oxide, and the glycosylation was shown to be useful for constructing 1,2-*cis*-glycosidic linkages such as β-mannopyranosyl or β-rhamnopyranosyl linkages. Importantly, the glycosylation proceeded with the retention of anomeric configuration in the glycosyl donor.

5.3.18
6-Acyl-2H-Pyran-3(6H)-One

Feringa and coworkers [258] and O'Doherty *et al.* [259] independently reported palladium-catalyzed glycosylations of 2-substituted 6-acyl-2*H*-pyran-3(6*H*)-one derivatives and alcohols (Scheme 5.98). This reaction presumably involves electrophilic Pd π-allyl complex intermediate, which was generated by the reaction of 2-substituted 6-acyl-2*H*-pyran-3(6*H*)-one and Pd(0)/PPh$_3$. It is noteworthy that 2-substituted 6-acyl-2*H*-pyran-3(6*H*)-one derivatives were stereoselectively converted into 2-substituted 6-alkoxy-2*H*-pyran-3(6*H*)-one derivatives with complete retention of configuration by this reaction. A two-step reduction/oxidation manipulation after the glycosylation can install new stereocenters in the obtained glycosides.

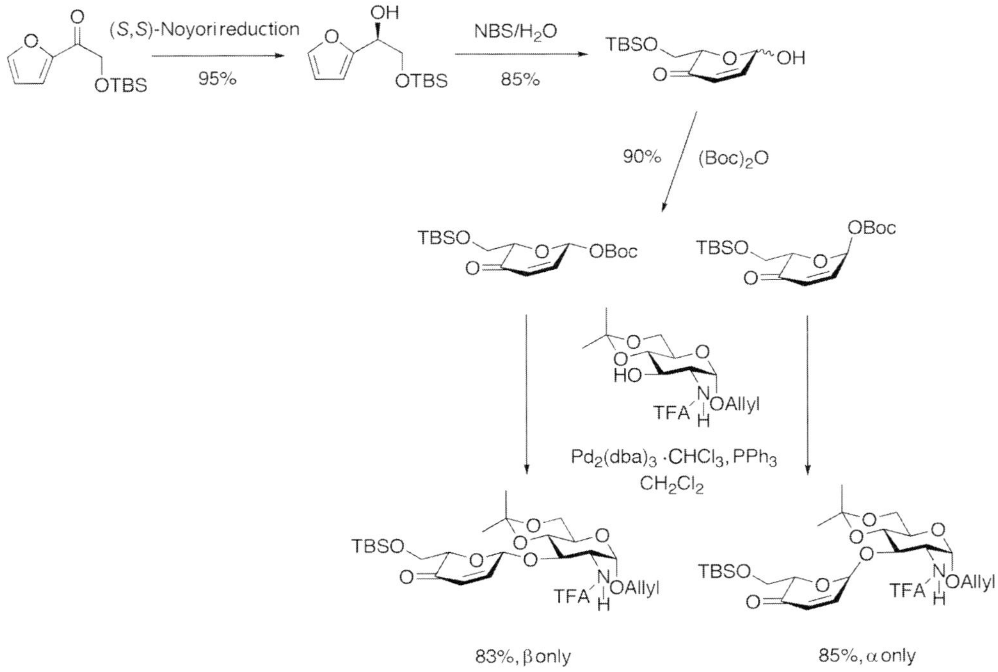

Scheme 5.98 Ref. [259].

5.3.19
exo-Methylene

The use of 1-methylene sugars as glycosyl donors was first demonstrated by van Boom et al. [260]. These glycosyl donors were prepared from the corresponding sugar lactones with Tebbe's reagent, and could be converted into α-linked ketopyranosides by the reactions with alcohols in the presence of IDCP (Scheme 5.99). Similar strategies using protic and Lewis acids, such as TfOH, TMSOTf, MsOH, 10-CSA, BCl$_3$ and HBr·PPh$_3$, were also reported by Ikegami et al. [261] and Bravo et al. [262]. On the contrary, Lin et al. reported a glycosylation of 1-methylene sugars using BF$_3$·Et$_2$O, which was promoted by Ferrier-type rearrangement [263] (Scheme 5.100).

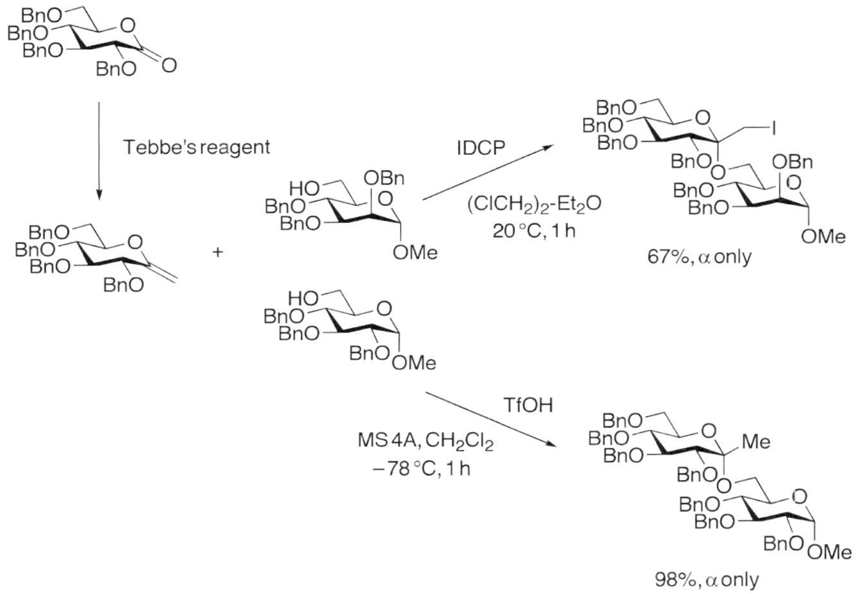

Scheme 5.99 Refs. [260,261].

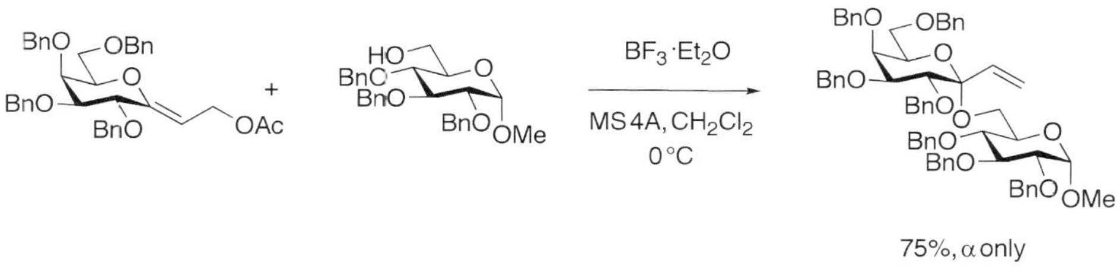

Scheme 5.100 Ref. [263].

5.3.20
Concluding Remarks

The glycosylation methodology has made tremendous progress in the past three decades and has been successfully applied to the synthesis of glycomolecules. However, a universal method for chemical glycosylation has not yet appeared from the point of view of chemical yield and stereoselectivity. Therefore, we always ask as to which method is the most suitable for the synthesis. Does a single powerful method in the glycosylation area really exist? In the future, two alternative ways may be determined to be efficient for glycosylation reactions: one is the development of a more general method, and another is the creation of a special method that is peculiar to each type of sugar or linkage, considering the features of each sugar structure. Furthermore, a major breakthrough may be needed for synthesizing any given glycomolecule by fully controlled chemistry. Because carbohydrates are indispensable substances in our life activities, the study of carbohydrate chemistry will continue for a long time.

5.3.21
Typical Experimental Procedure

5.3.21.1 General Procedure for the Preparation of Diazirines from Glycosyl Sulfonates
After the treatment of a sulfonate (16.0 mmol) with a saturated NH_3 solution in MeOH (180 ml) for 36 h, until the starting material had disappeared, according to TLC, half of the solvent was distilled. Keeping the remaining solution at $-25\,°C$ afforded the corresponding crystalline diaziridine. The solution of the oily diaziridine was evaporated, the residue was taken up in diethyl ether and the precipitated NH_4OSO_2Me was filtered to give the crude diaziridine. A solution of I_2 (1.81 mmol) in dry MeOH (9 ml) at $-25\,°C$ was added dropwise over 15 min to a mixture of diaziridine (1.81 mmol) and triethylamine (28.7 mmol) in dry MeOH (50 ml). The precipitated crystalline glycosyl diazirine was filtered. A second crop of crystals was obtained by the concentration of the mother liquor and crystallization at $-25\,°C$.

5.3.21.2 General Procedure for the Glycosylation of Diazirines
Under thermal conditions, reaction of glycosyl diazirine (0.18 mmol) and glycosyl acceptor (0.19 mmol) in dry dichloromethane (1.5 ml) for 3 h at $25\,°C$ yielded, after column chromatography, the corresponding glycoside. Under photolytic conditions, irradiation of a solution of glycosyl diazirine (0.087 mmol) and glycosyl acceptor (0.13 mmol) in dry dichloromethane (0.75 ml) with an HPK-125-Philips high-pressure Hg lamp (Solidex filter) at $-65\,°C$ for 20 min gave, after column chromatography, the corresponding glycoside.

5.3.21.3 General Procedure for the Preparation of Glycosyl Sulfonylcarbamates from Hemiacetals
A hemiacetal (1.0 mmol) and DABCO (10 mmol) were azeotroped thrice with dry benzene. The resultant residue was dissolved in dry toluene (10 ml) and cooled to

0 °C, and freshly distilled TsNCO (1.1 mmol) dissolved in dry toluene (5 ml) was added dropwise over 2 h. Upon completion, the reaction was warmed to room temperature and quenched by pouring 10% citric acid into it (50 ml). The resulting mixture was extracted with Et_2O (50 ml) and washed with sat. $NaHCO_3$ (aq) (25 ml) and brine (25 ml). The organic phase was dried over $MgSO_4$ and concentrated. The purification of the residue by column chromatography on silica gel gave the corresponding glycosyl sulfonylcarbamate.

5.3.21.4 General Procedure for the Glycosylation of Glycosyl Sulfonylcarbamates

Glycosyl sulfonylcarbamate (0.10 mmol) and glycosyl acceptor (0.15 mmol) were azeotroped thrice with dry toluene, and the resulting residue was dissolved in dry Et_2O (1.0 ml) under nitrogen atmosphere. TMSOTf (0.11 mmol) was then added dropwise and the reaction was stirred for 1.5 h. The reaction was quenched by the addition of solid $NaHCO_3$, and the Et_2O was removed under reduced pressure. Purification of the residue by column chromatography on silica gel gave the corresponding glycoside.

5.3.21.5 General Procedure for the Preparation of 1,2-O-Stannyl Acetals from Hemiacetals and the Glycosylation

A mixture of a hemiacetal (3.92 mmol) and dibutyltin oxide (3.52 mmol) in dry MeOH (25 ml) was stirred at 60 °C until a clear solution was obtained (~1.5 h). CsF (4.7 mmol) and toluene (5 ml) were added, and the mixture was concentrated. After having been kept at 50 °C and 0.2 Torr for 2 h to assure dryness, the residue was dissolved in DMF (5 ml), 4-Å molecular sieves (0.5 g) were added and the solution was cooled to −5 °C. After the addition of a triflated glycosyl donor (0.79 mmol), the mixture was stirred vigorously at −5 °C for 80 min and concentrated. The residue was triturated with acetonitrile, the resulting suspension was filtered through a pad of Celite, solids were washed with acetonitrile and the combined filtrate was concentrated. Purification of the residue by column chromatography on silica gel gave the corresponding glycoside.

5.3.21.6 General Procedure for the Preparation of 6-Acyl-2H-Pyran-3(6H)-Ones from 1-(2′-Furyl)-2-*tert*-Butyldimethylsilanyloxyethan-1-Ols

1-(2′-Furyl)-2-*tert*-butyldimethylsilanyloxyethan-1-ol (6.97 mmol), THF (12 ml), and H_2O (3 ml) were added to a round-bottom flask and cooled to 0 °C. Solid $NaHCO_3$ (13.9 mmol), $NaOAc·3H_2O$ (6.98 mmol) and NBS (6.97 mmol) were added to the solution and the mixture was stirred for 1 h at 0 °C. The reaction was quenched with saturated $NaHCO_3$ (aq) (15 ml), extracted with Et_2O (25 ml × 3), dried over Na_2SO_4 and concentrated under reduced pressure. Purification of the residue by column chromatography on silica gel gave 6-hydroxy-2-*tert*-butyldimethylsilanyloxymethyl-2H-pyran-3-(6H)-one. This compound (10 mmol) was dissolved in CH_2Cl_2 (8 ml), and the solution was cooled to −78 °C. A CH_2Cl_2 (2 ml) solution of $(Boc)_2O$ (12 mmol) and a catalytic amount of DMAP (1 μmol) was added to the reaction mixture. After stirring for 1 h at −78 °C, the reaction was quenched with saturated $NaHCO_3$ (aq) (50 ml), extracted with Et_2O (50 ml × 3), dried over Na_2SO_4 and concentrated under reduced pressure. The purification of the residue by column chromatography on silica gel gave carbonic acid *tert*-butyl ester 6-(*tert*-butyl-dimethylsilanyloxymethyl)-5-oxo-5,6-dihydro-2*H*-pyran-2-yl ester.

5.3.21.7 General Procedure for the Glycosylation of 6-Acyl-2*H*-Pyran-3(6*H*)-Ones

A CH_2Cl_2 (0.3 ml) solution of carbonic acid *tert*-butyl ester 6-(*tert*-butyl-dimethylsilanyloxymethyl)-5-oxo-5,6-dihydro-2*H*-pyran-2-yl ester (0.279 mmol) and glycosyl acceptor (0.558 mmol) was cooled to 0 °C. A CH_2Cl_2 (0.3 ml) solution of $Pd_2(DBA)_3 \cdot CHCl_3$ (2.5 mol%) and PPh_3 (10 mol%) was added to the reaction mixture at 0 °C. The reaction mixture was stirred at 0 °C for 2 h. The reaction mixture was quenched with saturated $NaHCO_3$ (aq) (5 ml), extracted with Et_2O (5 ml × 3), dried over Na_2SO_4 and concentrated under reduced pressure. Purification of the residue by column chromatography on silica gel gave the corresponding glycoside.

References

224 Tietze, L.-F., Fischer, R. and Guder, H.-J. (1982) *Tetrahedron Letters*, **23**, 4661–4664.

225 Nashed, E.M. and Glaudemans, C.P.J. (1989) *The Journal of Organic Chemistry*, **54**, 6116–6118.

226 Qiu, D.-X., Wang, Y.-F. and Cai, M.-S. (1989) *Synthesis Communications*, **19**, 3453–3456.

227 Mukaiyama, T. and Matsubara, K. (1992) *Chemistry Letters*, 1041–1044.

228 Mukaiyama, T., Yamada, M., Suda, S., Yokomizo, Y. and Kobayashi, S. (1992) *Chemistry Letters*, 1401–1404.

229 Priebe, W., Grynkiewicz G. and Neamati, N. (1991) *Tetrahedron Letters*, **32**, 2079–2082.

230 (a) Kolar, C. and Kneissl, G. (1990) *Angewandte Chemie (International Edition in English)*, **29**, 809–811. (b) Kolar, C., Kneissl, G., Knödler, U. and Dehmel, K. (1991) *Carbohydrate Research*, **209**, 89–100.

231 (a) Briner, K. and Vasella, A. (1989) *Helvetica Chimica Acta*, **72**, 1371–1382. (b) Briner, K. and Vasella, A. (1990) *Helvetica Chimica Acta*, **73**, 1764–1778. (c) Vasella, A. (1991) *Pure and Applied Chemistry*, **63**, 507–518. (d) Briner, K. and Vasella, A. (1992) *Helvetica Chimica Acta*, **75**, 621–635.

232 Barton, D.H. and Ramesh, M. (1990) *Journal of the American Chemical Society*, **112**, 891–892.

233 (a) Yamago, S., Kokubo, K., Matsuda, S. and Yoshima, J. (1996) *Synlett*, 929–930. (b) Yamago, S., Kokubo, K. and Yoshida, J. (1997) *Chemistry Letters*, 111–112. (c) Yamago, S., Kokubo, K., Murakami, H., Mino, Y., Hara, O. and Yoshida, J. (1998) *Tetrahedron Letters*, **39**, 7905–7908. (d) Yamago, S., Kokubo, K., Hara, O., Matsuda, S. and Yoshida, J. (2002) *The Journal of Organic Chemistry*, **67**, 8584–8592.

234 Kunz, H. and Zimmer, J. (1993) *Tetrahedron Letters*, **34**, 2907–2910.

235 Fraser-Reid, B., Konradsson, P., Mootoo, D.R. and Udodong, U. (1988) *Journal of the Chemical Society: Chemical Communications*, 823–825.

236 Hinklin, R.J. and Kiessling, L.L. (2001) *Journal of the American Chemical Society*, **123**, 3379–3380.

237 Griffith, D.A. and Danishefsky, S.J. (1996) *Journal of the American Chemical Society*, **118**, 9526–9538.

238 (a) Bröder, W. and Kunz, H. (1990) *Synlett*, 251–252. (b) Bröder, W. and Kunz, H. (1993) *Carbohydrate Research*, **249**, 221–241. (c) Peto, C., Batta, G.,

Gyorgydeak, Z. and Sztaricskai, F. (1996) *Journal of Carbohydrate Chemistry*, **15**, 465–483.
239 (a) Falahatpisheh, N. and Sulikowski, G.A. (1994) *Synlett*, 672–674 (b) Sobti, A., Kim, K. and Sulikowski, G.A. (1996) *The Journal of Organic Chemistry*, **61**, 6–7. (c) Guo, Y. and Sulikowski, G.A. (1998) *Journal of the American Chemical Society*, **120**, 1392–1397.
240 Pleuss, N. and Kunz, H. (2003) *Angewandte Chemie (International Edition)*, **42**, 3174–3176.
241 Nishikubo, Y., Sasaki, K., Matsumura, S. and Toshima, K. (2006) *Tetrahedron Letters*, **47**, 1041–1045.
242 (a) Zurabyan, S.E., Volosyuk, T.P. and Khorlin, A.J. (1969) *Carbohydrate Research*, **9**, 215–220. (b) Zurabyan, S.E., Antonenko, S.T. and Khorlin, A.Y. (1970) *Carbohydrate Research*, **15**, 21–27. (c) Warren, C. and Jeanloz, R.W. (1977) *Carbohydrate Research*, **53**, 67–84. (d) Kobayashi, S., Warren, C.D. and Jeanloz, R.W. (1986) *Carbohydrate Research*, **150**, C7–C10.
243 (a) Kiso, M. and Anderson, L. (1979) *Carbohydrate Research*, **72**, C12–C14. (b) Kiso, M. and Anderson, L. (1979) *Carbohydrate Research*, **72**, C15–C17.
244 (a) Nishimura, S., Matsuoka, K. and Lee, Y.C. (1994) *Tetrahedron Letters*, **35**, 5657–5660. (b) Matsuoka, K., Ohtawa, T., Hinou, H., Koyama, T., Esumi, Y., Nishimura, S., Hatano, K. and Terunuma, D. (2003) *Tetrahedron Letters*, **44**, 3617–3620.
245 Wittmann, V. and Lennartz, D. (2002) *European Journal of Organic Chemistry*, 1363–1367.
246 Mohan, H., Gemma, E., Ruda, K. and Oscarson, S. (2003) *Synlett*, 1255–1256.
247 Crasto, C.F. and Jones, G.B. (2004) *Tetrahedron Letters*, **45**, 4891–4894.
248 Donohoe, T.J., Logan, J.G. and Laffan, D.D.P. (2003) *Organic Letters*, **5**, 4995–4998.
249 (a) Marzabadi, C.H. and Franck, R.W. (1996) *Chemical Communications*, 2651–2652. (b) Capozzi, G., Mannocci, F., Menichetti, S. and Nativi, C. (1997) *Chemical Communications*, 2291–2292. (c) Franck, R.W. and Marzabadi, C.H. (1998) *The Journal of Organic Chemistry*, **63**, 2197–2208. (d) Dios, A., Nativi, C., Capozzi, G. and Franck, R.W. (1999) *European Journal of Organic Chemistry*, 1869–1874. (f) Bartolozzi, A., Capozzi, G., Menichetti, S. and Nativi, C. (2001) *European Journal of Organic Chemistry*, 2083–2090
250 Poláková, M., Pitt, N., Tosin, M. and Murphy, P.V. (2004) *Angewandte Chemie (International Edition)*, **43**, 2518–2521.
251 Cipolla, L., Lay, L., Nicotra, F., Panza, L. and Russo, G. (1994) *Tetrahedron Letters*, **35**, 8669–8670.
252 Aouad, M.E.A., Meslouti, A.E., Uzan, R. and Beaupere, D. (1994) *Tetrahedron Letters*, **35**, 6279–6282.
253 Sanders, W.J. and Kiessling, L.L. (1994) *Tetrahedron Letters*, **35**, 7335–7338.
254 Beyer, J. and Madsen, R. (1998) *Journal of the American Chemical Society*, **120**, 12137–12138.
255 Srivastava, V.K. and Schuerch, C. (1979) *Tetrahedron Letters*, **20**, 3269–3272.
256 Desinges, A., Olesker, A. and Lukacs, G. (1984) *Carbohydrate Research*, **126**, C6–C8.
257 (a) Hodosi, G. and Kovác, P. (1997) *Journal of the American Chemical Society*, **119**, 2335–2336. (b) Hodosi, G. and Kovác, P. (1998) *Carbohydrate Research*, **308**, 63–75.
258 Comely, A.C., Eelkema, R., Minnaard, A.J. and Feringa, B.L. (2003) *Journal of the American Chemical Society*, **125**, 8714–8715.
259 (a) Babu, R.S. and O'Doherty, G.A. (2003) *Journal of the American Chemical Society*, **125**, 12406–12407. (b) Babu, R.S., Zhou, M. and O'Doherty, G.A. (2004) *Journal of the American Chemical Society*, **126**, 3428–3429.
260 (a) Noort, D., Veeneman, G.H., Boons, G.-J.P.H., van der Marel, G., Mulder, G.J. and van Boom, J.H. (1990) *Synlett*, 205–206. (b) Heskamp, B.M., Noort, D.,

van der Marel, G.A. and van Boom, J.H. (1992) *Synlett*, 713–715. (c) Heskamp, B.M., Veeneman, G.H., van der Marel, G.A., van Boeckel, C.A.A. and van Boom, J.H. (1995) *Tetrahedron*, **51**, 5657–5670.

261 (a) Li, X., Ohtake, H., Takahashi, H. and Ikegami, S. (2001) *Tetrahedron*, **57**, 4283–4295. (b) Li, X., Ohtake, H., Takahashi, H. and Ikegami, S. (2001) *Tetrahedron*, **57**, 4297–4309. (c) Namme, R., Mitsugi, T., Takahashi, H. and Ikegami, S. (2005) *Tetrahedron Letters*, **46**, 3033–3036.

262 (a) Colinas, P.A., Lieberknecht, A. and Bravo, R.D. (2002) *Tetrahedron Letters*, **43**, 9065–9068. (b) Colinas, P.A., Ponzinibbio, A., Lieberknecht, A. and Bravo, R.D. (2003) *Tetrahedron Letters*, **44**, 7985–7988.

263 (a) Lin, H.-C., Yang, W.-B., Gu, Y.-F., Chen, C.-Y., Wu, C.-Y. and Lin, C.-H. (2003) *Organic Letters*, **5**, 1087–1089. (b) Lin, H.-C., Du, W.-P., Chang, C.-C. and Lin, C.-H. (2005) *Tetrahedron Letters*, **46**, 5071–5076.

5.4
The Twenty First Century View of Chemical *O*-Glycosylation

Thomas Ziegler

5.4.1
Indirect and Special Methods

Indirect and special methods for *O*-glycosidic bond formation are usually only applied to special cases where common glycosylation methods via glycosyl donors or anomeric *O*-alkylation, as outlined in the previous chapters, fail or appear to be impractical. Nevertheless, some of these indirect and rather special methods provide for the efficient preparation of glycosides and saccharides that are otherwise difficult to obtain. Special methods for glycoside synthesis also give more insight into the intriguing and sometimes puzzling effects reigning stereo- and regioselectivity and reactivity of glycosylation reactions. This chapter summarizes some of these indirect and special methods and discusses their applicability for glycoside and oligosaccharide synthesis.

5.4.1.1 Intramolecular *O*-Glycosylation

Intramolecular glycosylations can be regarded as a biomimetic variant of *O*-glycosidic bond formation as they resemble enzymatic glycoside synthesis where glycosyl donor and acceptor are bound to the active site of an enzyme, and thus the *O*-glycosidic bond forms intramolecularly. Three different approaches or concepts to achieve the intramolecularization of *O*-glycosidic bond formation can be envisaged and had been studied over the past years (Scheme 5.101) [264,265].

In the following chapters, these three concepts will be discussed and the currently available literature will be summarized.

5.4.1.2 Leaving-Group-Based Concept

In the 'leaving-group-based concept', the glycosyl acceptor (nucleophile) is attached to the leaving group of the glycosyl donor. Upon activation, the leaving group is

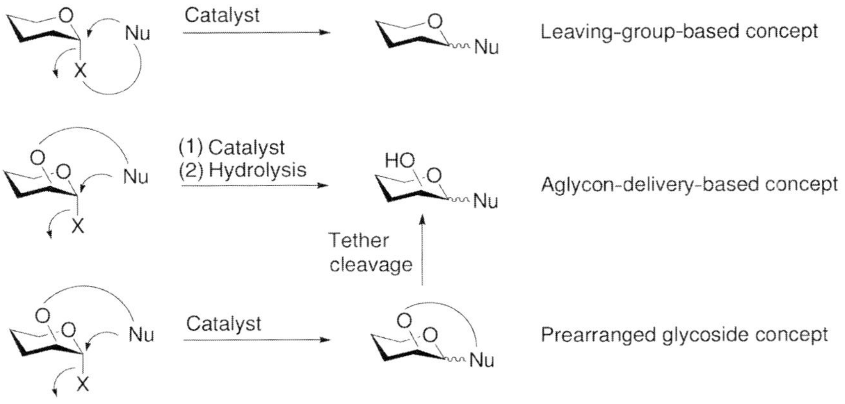

Scheme 5.101 Concepts for intramolecular glycosylation. X = leaving group, Nu = glycosyl acceptor.

released and the glycosidic bond forms. Glycosyl carbonates have been used for this purpose (Table 5.3, entries 1–8) [266–269]. Activation can be induced either by simple heating or, preferentially, by Lewis acid catalysis. Stereoselectivities for these leaving-group-based glycosylations were similar to those obtained for the corresponding intermolecular glycosylations, suggesting that glycosylations via anomeric carbonates rather proceed intermolecularly. Indeed, it could be shown from competitive experiments that such glycosylations proceed, at least in part, intermolecularly as well [266]. Other examples for leaving-group-based concept glycosylations are glycosyl hexynoates that can get activated by converting the alkyne moiety into a dicobalthexacarbonyl complex (Table 5.3, entries 9–13) [270]. Stereoselectivities are high, provided a participating neighboring group is present in the donor moiety. Yet another approach for leaving-group-based concept glycosylations uses benzo-annelated pentenyl-type tethers that, however, were also shown by competitive experiments to proceed intermolecularly (Table 5.3, entries 14–15) [271]. Similarly, several thio-linked glycosides have been used as well (Table 5.3, entries 16–18) [272]. However, stereoselectivities were similarly unsatisfactory here. To this point, it remains unclear as to which extend this concept can be regarded as a truly intramolecular glycosylation and whether further experiments are necessary.

Recently, Jensen and coworkers presented yet another protocol for leaving-group-based glycosylation using an intramolecular S_N2 reaction without the need of any activation of the leaving group [273] (Scheme 5.102). Glycosyl donor and acceptor are first tethered via dinitrofluorobenzoic acid. Next, glycosylation is effected either by simple heating of the tethered glycoside in nitromethane or, in the alternative, by activation with a Lewis acid. The anomeric selectivity is usually low in this approach.

Intramolecular Aglycon Delivery (IAD) Concept The 'intramolecular aglycon-delivery concept' for intramolecular glycosylations has gained significant applications for oligosaccharide synthesis. Originally developed for the highly stereoselective synthesis of β-mannosidic linkages, still an imminent problem in saccharide synthesis

Table 5.3 Leaving-group-based concept glycosylations.

Entry	Starting material	Product	Activation	Yield %	α : β	References
1			170 °C	46	β only	[267]
2			TMSOTf, CH$_2$Cl$_2$ TMSOTf, toluene TBDMSOTf, CH$_2$Cl$_2$ TBDMSOTf, toluene	81 85 75 73	41:59 32:68 32:68 19:81	[268]
4			TMSOTf, toluene TBDMSOTf, toluene	79 85	32:68 28:72	[268]
5			TMSOTf, toluene TBDMSOTf, toluene	67 67	42:58 36:64	[268]
6			TMSOTf, toluene TBDMSOTf, toluene	69 72	42:58 34:66	[268]
7			TMSOTf, CH$_2$Cl$_2$	63	1:99	[269]
8			TMSOTf, CH$_2$Cl$_2$	72	1:99	[269]
9			(1) Co$_2$(CO)$_8$, Et$_2$O (2) TMSOTf, CH$_2$Cl$_2$	60	46:54	[270]
10			(1) Co$_2$(CO)$_8$, Et$_2$O (2) TMSOTf, CH$_2$Cl$_2$	77	42:58	[270]
11			(1) Co$_2$(CO)$_8$, Et$_2$O (2) TMSOTf, CH$_2$Cl$_2$	65	94:4	[270]
12			(1) Co$_2$(CO)$_8$, Et$_2$O (2) TMSOTf, CH$_2$Cl$_2$	37	1:99	[270]

(*continued*)

Table 5.3 (Continued)

Entry	Starting material	Product	Activation	Yield %	α : β	References
13	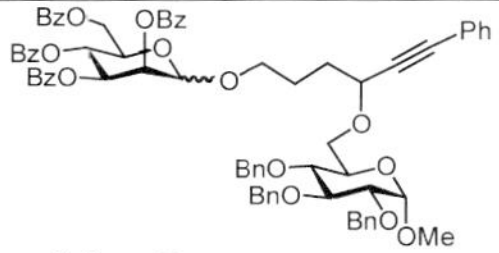	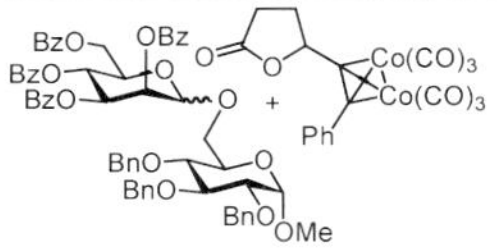	(1) Co$_2$(CO)$_8$, Et$_2$O (2) TMSOTf, CH$_2$Cl$_2$	43	99:1	[270]
14	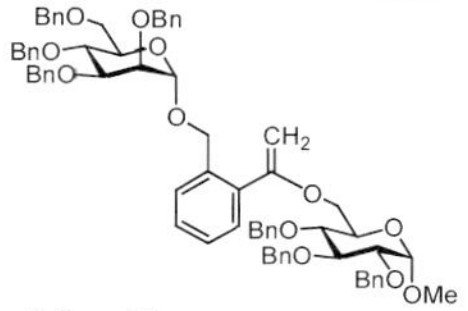	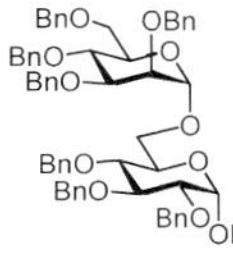	PhSeCl, AgOTf, toluene	80	α only	[271]
15	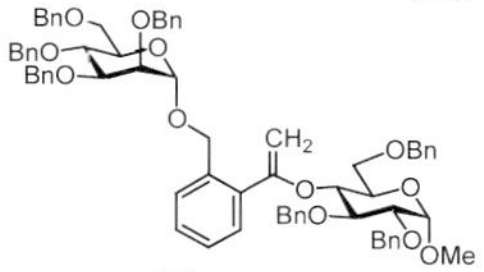	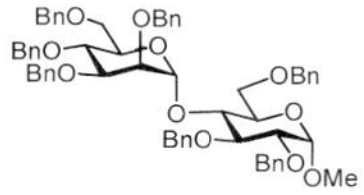	PhSeCl, AgOTf, toluene	80	α only	[271]
16	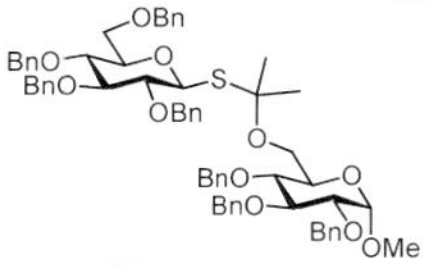	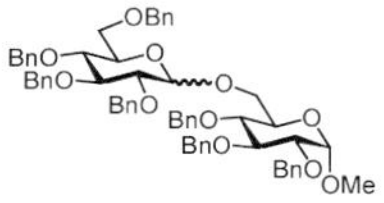	DMTST, CH$_2$Cl$_2$	27	50:50	[272]
17	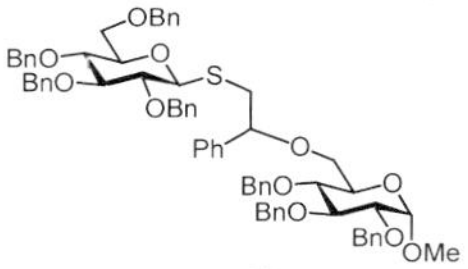	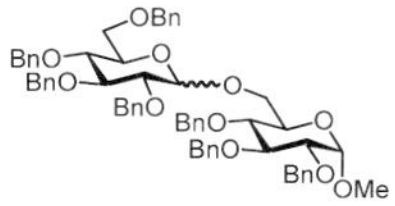	DMTST, CH$_2$Cl$_2$	51	40:60	[272]
18	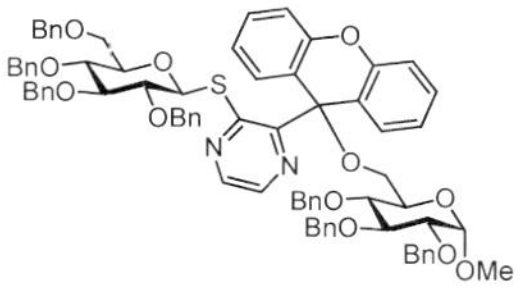	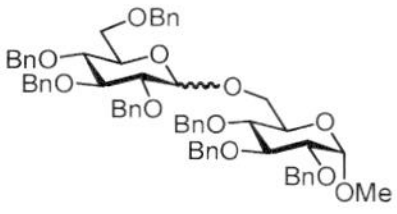	AgOTf, CH$_2$Cl$_2$	75	60:40	[272]

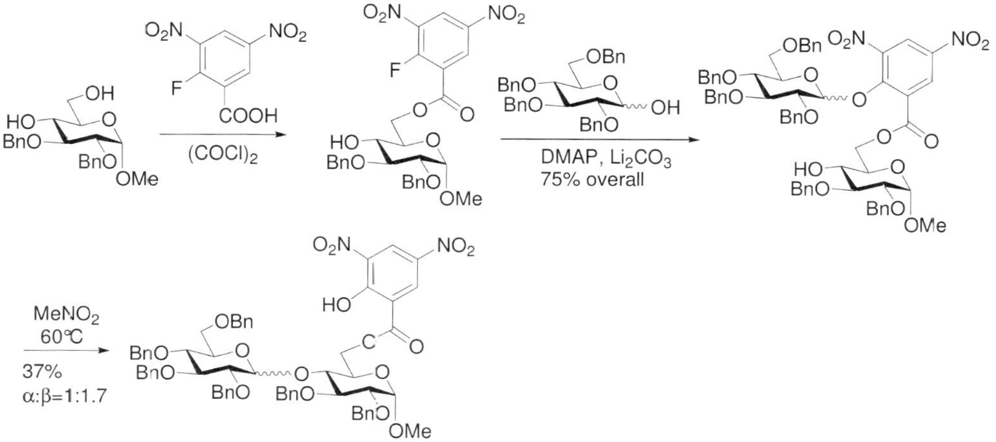

Scheme 5.102

and later extended to other glycoside syntheses, two approaches via acetal tethers and silylene tethers have been grown into useful methods for the efficient construction of oligosaccharides.

IAD via Acetal Tethering The first example for this concept was presented by Barresi and Hindsgaul in 1991 and later extended to other examples [274,275]. Starting from ethyl 3,4,6-tri-O-benzyl-2-O-(2-propenyl)-1-thio-α-D-mannopyranoside, which is easily accessible from the corresponding 2-O-acetyl derivative with Tebbe reagent, acid-catalyzed acetalization with suitable glycosyl acceptors affords the tethered glycosides in medium yield. Next, activation of the donor with NIS exclusively gives the corresponding β-linked disaccharides [275]. A major drawback, however, are rather low yields when this concept is applied to the synthesis of more complex oligosaccharides, where both the acetalization step and the IAD glycosylation proceed in less than 30% yield owing to side reactions [276] (Scheme 5.103).

Recently, Fairbanks and coworkers published a highly efficient protocol that circumvents the above-mentioned problems and allows for the efficient preparation of complex oligosaccharides using the IAD concept [277–279]. For example, the acetal tether is efficiently established by the reaction of a 2-O-(1-propenyl)-mannoside, which is generated from the corresponding 2-O-allyl-mannoside, with the glycosyl acceptor and N-iodosuccinimide (NIS), affording the tethered glycosides in high yield. NIS-induced activation of the tethered glycosides then results in a clean intramolecular glycosylation to give the β-mannosides in high yield as well [278] (Scheme 5.104). Another advantage of this protocol is that acetalization and intramolecular glycosylation can be performed with the same reagents. The solvent

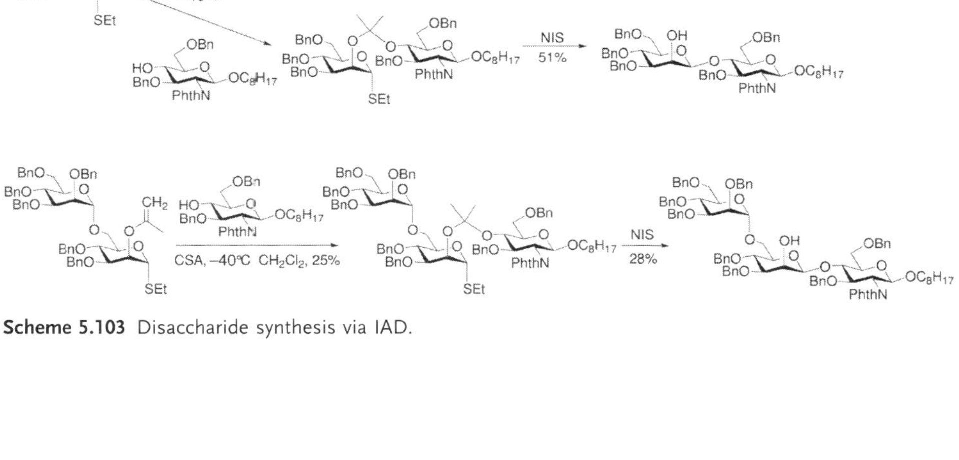

Scheme 5.103 Disaccharide synthesis via IAD.

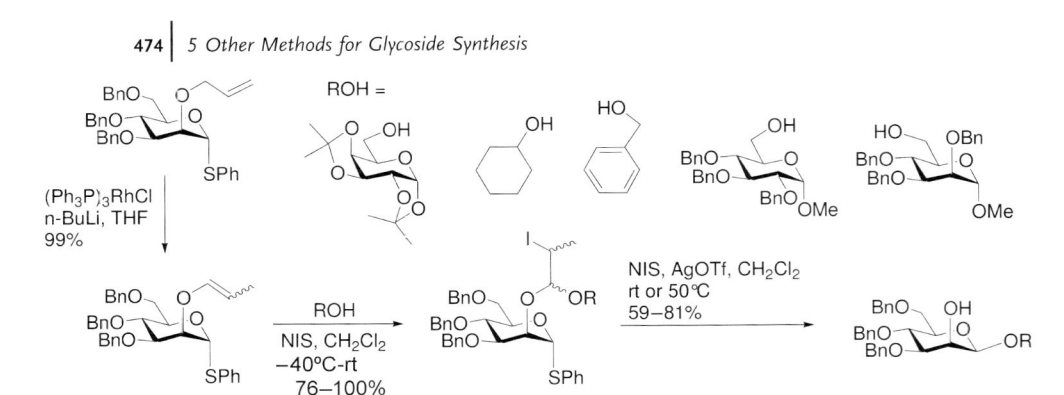

Scheme 5.104 Disaccharide synthesis via Fairbanks' IAD concept.

solely needs to be changed [280]. Table 5.4 summarizes the examples of this promising approach published so far [280–283].

Yet another improvement of the IAD concept was presented by Ito and Ogawa [284]. Substituting the aliphatic tether used by Hindsgaul and Fairbanks by an aromatic tether circumvents the difficulties encountered during acetal formation. Ogawa's *p*-methoxyphenylmethyl tether can be generated from a *p*-methoxyphenyl group (PMP) by oxidation with DDQ in the presence of the respective glycosyl acceptor. Thus, tethered glycosides are very conveniently accessible in high yield. Another significant improvement is the fact that the intermediate tethered glycosides do not need to get isolated but, instead, can directly get converted into the corresponding saccharides (Scheme 5.105) [284].

Another advantage of the IAD concept via *p*-methoxybenzylidene acetals lies in the compatibility of the oxidative tethering procedure with 1-thio-glycosyl donors [285,286]. Thus, highly flexible strategies for oligosaccharide synthesis can be realized [287–289]. For example, fragments related to the core region of Asn-linked glycans have been prepared efficiently this way [285–291] (Scheme 5.106).

The IAD concept via *p*-methoxybenzylidene acetals was also shown to be suitable for polymer-supported syntheses of disaccharides (Scheme 5.107) [292]. A suitable *p*-allyloxybenzyl group at position 2 of a 1-thio-mannosyl donor is first converted into a PEG-modified benzyl group that allows for the convenient isolation of the intermediate tethered glycosides.

The synthesis of β-D-fructofuranosides is yet another useful application of this concept [293–295]. The latter 1,2-*cis*-glycosidic linkage is as difficult to establish as in the case of β-mannosides. In an elegant synthesis of α-D-fucofuranose-containing disaccharides, Plusquellec and coworkers used the IAD concept via *p*-methoxybenzylidene acetals in combination with a glycosylation protocol via pentenyl glycosides. Here, the intermediate NIS-adduct could be isolated (Scheme 5.108) [295].

IAD via Silylene Tethering IAD through silylene-tethered glycosides was introduced by Bols [296–301] and Stork [302,303]. The silylene tether is usually established by

Table 5.4 Examples applying the IAD concept.

Entry	Starting materials	Product	Conditions	Yield[a] %	References
1			NIS, AgOTf, CH$_2$Cl$_2$	72	[278]
2			I$_2$, AgOTf, CH$_2$Cl$_2$ −78°C–rt, 90% Then I$_2$, AgOTf, MeCN −20°C–rt	51	[280]
4			I$_2$, AgOTf, CH$_2$Cl$_2$ −78°C–rt, 86% then I$_2$, AgOTf, MeCN −20°C–rt	65	[280]
5			I$_2$, AgOTf, CH$_2$Cl$_2$ −78°C–rt, 81% then I$_2$, AgOTf, MeCN −20°C–rt	78	[280]
6			I$_2$, AgOTf, CH$_2$Cl$_2$ −78°C–rt, 79% then I$_2$, AgOTf, MeCN −20°C–rt	79	[280]
7			I$_2$, AgOTf, CH$_2$Cl$_2$ −78°C–rt, 71% then I$_2$, AgOTf, MeCN −20°C–rt	55	[280]
8			I$_2$, AgOTf, CH$_2$Cl$_2$ −78°C–rt, 75% then I$_2$, AgOTf, MeCN −20°C–rt	63	[280]
9			NIS, AgOTf, CH$_2$Cl$_2$, rt	70	[281]
10			NIS, AgOTf, CH$_2$Cl$_2$, rt	39	[281]

(continued)

476 | *5 Other Methods for Glycoside Synthesis*

Table 5.4 (*Continued*)

Entry	Starting materials	Product	Conditions	Yield[a] %	References
11	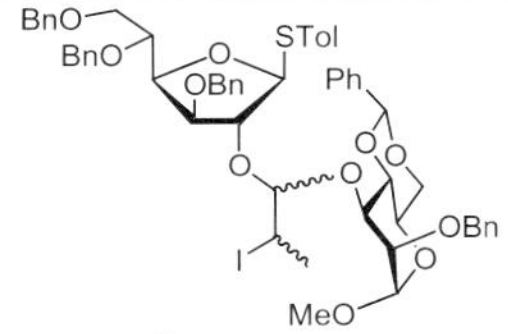	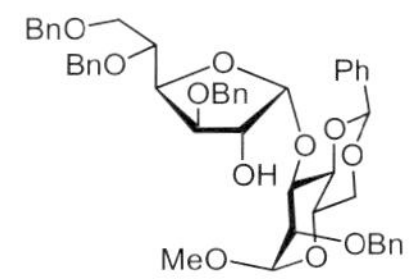	MeOTf, Me$_2$S$_2$, CH$_2$Cl$_2$,	32	[281]
12	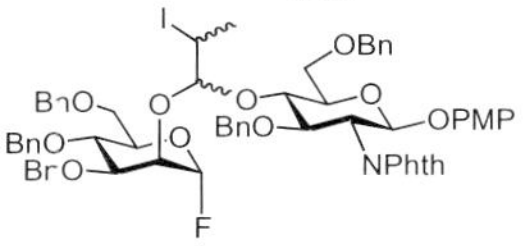	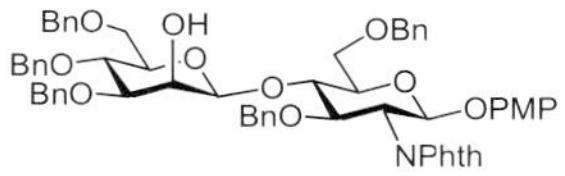	SnCl$_2$, AgOTf, CH$_2$Cl$_2$, 50 °C	54	[282]

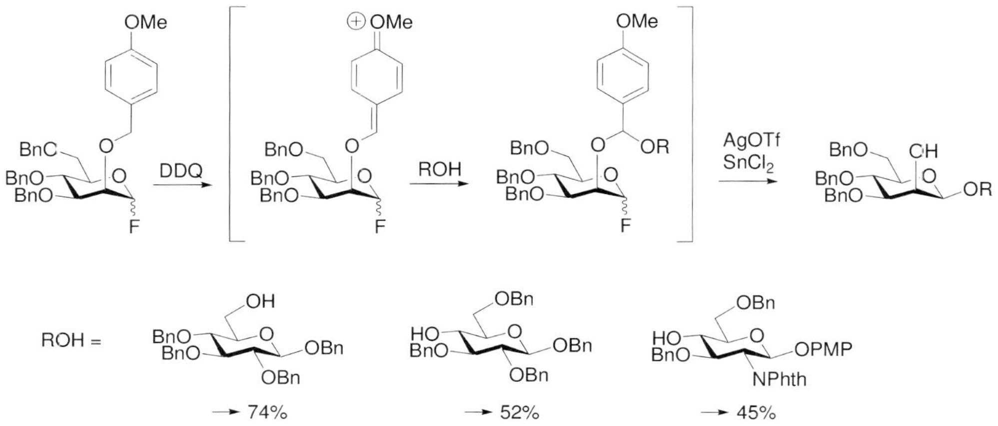

Scheme 5.105 IAD via *p*-methoxybenzylidene acetals.

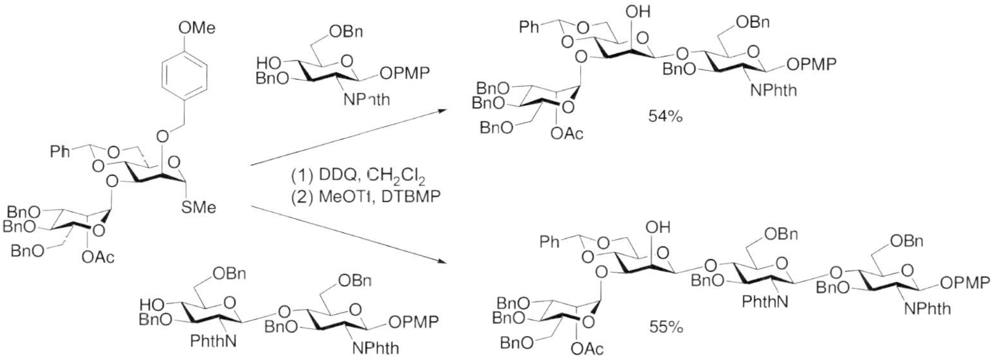

Scheme 5.106

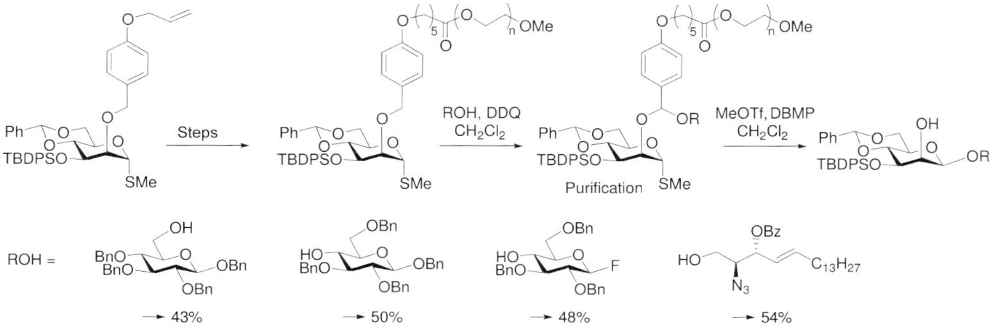

Scheme 5.107 Polymer-supported synthesis via IAD.

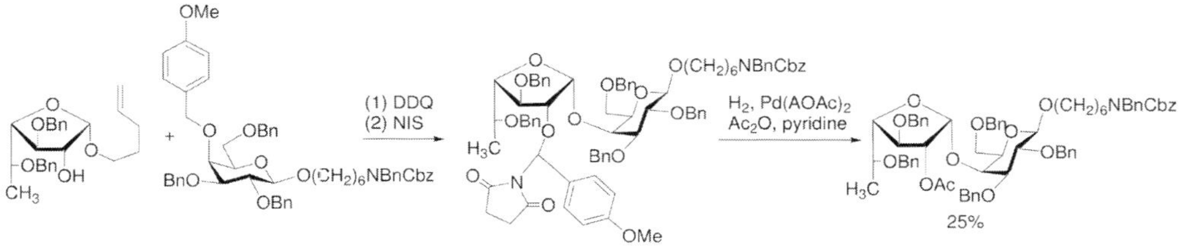

Scheme 5.108 Synthesis of furanosides via IAD concept.

reacting a suitable 1-thio-glycoside with dichlorodimethyl silane, followed by the corresponding glycosyl acceptor. Next, glycosylation can be effected either by NIS, resulting in intramolecular glycosylation with concomitant removal of the silylene tether, or by first converting the thio group into a sulfoxide and initiating glycosylation with triflic anhydride (Scheme 5.109) [303]. The literature up to 1995 was reviewed [304,305], and Table 5.5 summarizes the glycosylations using this approach.

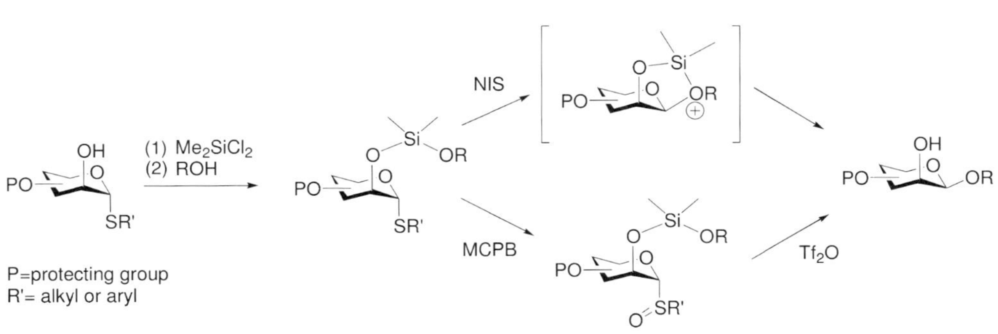

Scheme 5.109 IAD concept via silylene tethers.

Table 5.5 IAD concept via silylene tethers.

Entry	Starting materials	Product	Conditions	Yield[a] %	References
1	R= n-octyl, cyclohexyl, t-butyl, phenyl		NIS, TfOH, CH$_2$Cl$_2$, rt	59, 62, 61, 72	[296]
2			NIS, MeNO$_2$, 100 °C	74	[297]
3			NIS, MeNO$_2$, 100 °C	85	[297]
4			NIS, MeNO$_2$, 100 °C	39	[297]
5			NIS, MeNO$_2$, 25 °C	22	[297]
6			NIS, MeNO$_2$, 25 °C	45	[300]
7			NIS, MeNO$_2$, 25 °C, glycoside:anhydro glycose = 2:7	70	[300]
8			NIS, MeNO$_2$, 25 °C	63	[300]
9	ROH = MeOH		Tf$_2$O, DTBP, CH$_2$Cl$_2$, 78 °C–25 °C	—[a]	[302]
10			Tf$_2$O, DTBP, CH$_2$Cl$_2$, 100 °C–25 °C	92	[303]
11			Tf$_2$O, DTBP, CH$_2$Cl$_2$, 100 °C–25 °C	65	[303]
12			Tf$_2$O, DTBP, CH$_2$Cl$_2$, 100 °C–25 °C	82	[303]
13			Tf$_2$O, DTBP, CH$_2$Cl$_2$, 100 °C –25 °C	12	[303]

Table 5.5 (Continued)

Entry	Starting materials	Product	Conditions	Yield[a] %	References
14	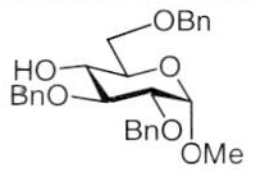 [structure: HO, BnO, BnO, OBn, OMe]		Tf$_2$O, DTBP, CH$_2$Cl$_2$, 100 °C–25 °C	48	[303]
15	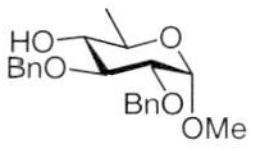 [structure: HO, BnO, BnO, OMe] and [structure: HO, BnO, OBn, OC$_8$H$_{17}$, PhthN]		Tf$_2$O, DTBP, CH$_2$Cl$_2$, 100 °C–25 °C	54	[303]

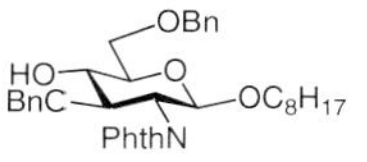

5.4.1.3 Prearranged Glycoside Concept

This concept uses stable tethers that link glycosyl donor and acceptor at positions that are not involved in the glycosylation step. This concept is sometimes referred to as 'remote glycosylation'. To this extend, this concept for intramolecular glycosylation resembles best enzymatic glycosylations, where glycosyl donor and acceptor are also bound to the active site of an enzyme through hydroxyls that are not primarily involved in the formation of the glycosidic bond. In general, the concept of O-glycosidic bond formation via prearranged glycosides allows for both stereoselective and regioselective glycosylations depending on the nature of the tether and the positions through which glycosyl donor and acceptor are linked (Scheme 5.110).

As various parameters concerning the nature of the tether (length, torsional rigidity, etc.) and the positions in the glycosyl donor and acceptor to which it is attached govern the outcome of intramolecular glycosylations applying this concept, a vast number of examples have been investigated so far. Tables 5.6A–5.6C

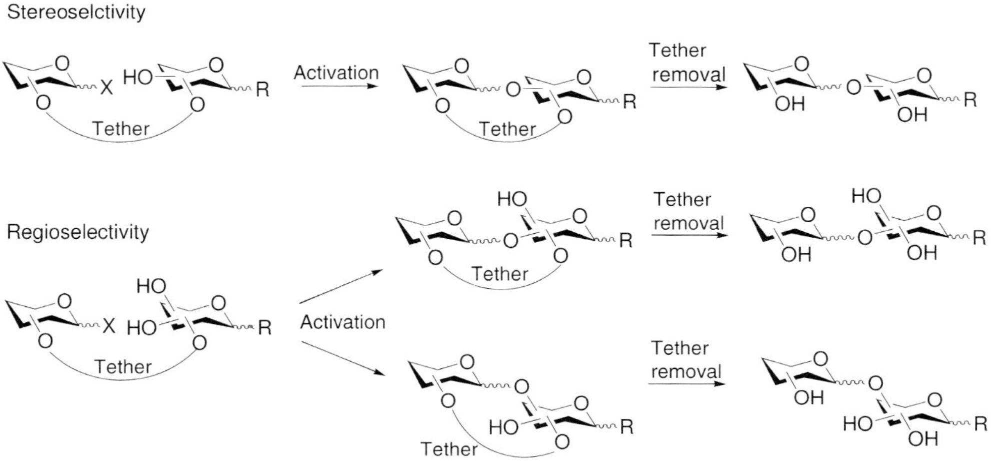

Scheme 5.110 Prearranged glycoside concept.

Table 5.6A Intramolecular glycosylations via prearranged glycosides – glucose donors.

Entry	Starting material	Product	Activation	Yield %	α:β	References
1	X = -CO-(CH$_2$)$_3$-CO- -CO-(CH$_2$)$_2$-CO- phthaloyl -Si(t-Bu)$_2$-	X = -CO-(CH$_2$)$_3$-CO- -CO-(CH$_2$)$_2$-CO- phthaloyl -Si(t-Bu)$_2$-	PhIO, TMSOTf PhIO, TMSOTf PhIO, TMSOTf PhIO, TMSOTf	37 67 86 77	89:11 93:7 99:1 3:97	[306] [306] [306] [306]
2	R = OBz NPhth		NIS, TMSOTf NIS, TMSOTf	80 75	α only α only	[307] [307]
4			NIS, TMSOTf	84	β only	[308]
5			NIS, TMSOTf	65	β only	[309]
6			NIS, TMSOTf	72	β only	[309]
7			NIS, TMSOTf	81	β only	[309]
8			NIS, TMSOTf	9	α only	[309]

5.4 The Twenty First Century View of Chemical O-Glycosylation

Table 5.6A (Continued)

Entry	Starting material	Product	Activation	Yield %	α : β	References
9	R = 2,3,4,6-tetra-O-benzyl-α-D-glucosyl 2,3,4,6-tetra-O-benzyl-β-D-glucosyl 2,3,4,6-tetra-O-benzyl-β-D-galactosyl		NIS, TMSOTf NIS, TMSOTf NIS, TMSOTf	64 65 65	α only α only α only	[310] [310] [310]

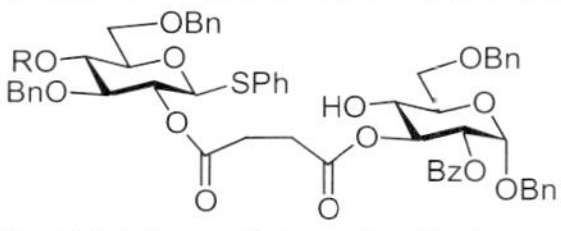

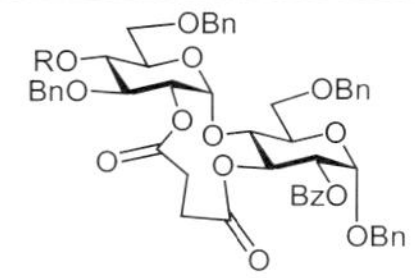

Table 5.6B Intramolecular glycosylations via prearranged glycosides – glucose donors.

Entry	Starting material	Product	Activation	Yield %	α : β	References
10	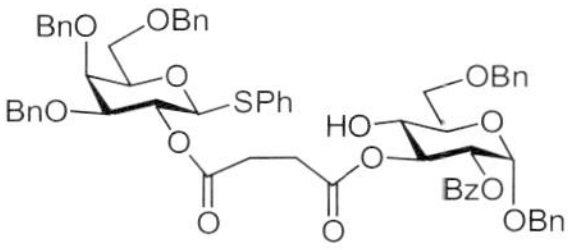	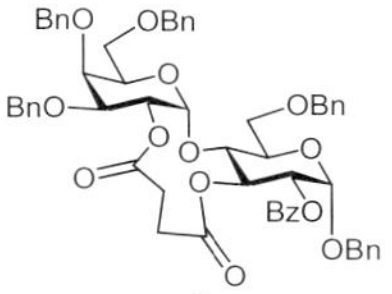	NIS, MeCN	51	α only	[311]
11	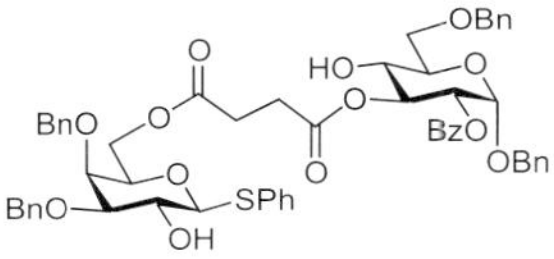	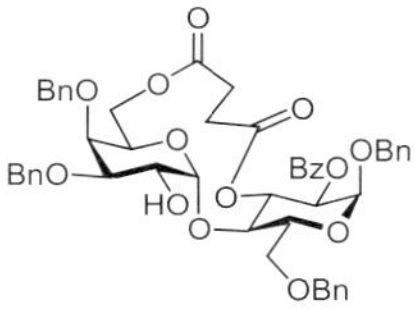	NIS, MeCN	69	α only	[311]

Table 5.6C Intramolecular glycosylations via prearranged glycosides – glucose donors.

Entry	Starting material	Product	Activation	Yield %	α : β	References
12			NIS, TMSOTf, MeCN NIS, TMSOTf, CH$_2$Cl$_2$	74 75	18:82 71:29	[312] [313]
13	X = phthaloyl malonyl	X = phthaloyl malonyl	NIS, TMSOTf, MeCN NIS, TMSOTf, MeCN	68 64	50:50 46:54	[313] [313]
14	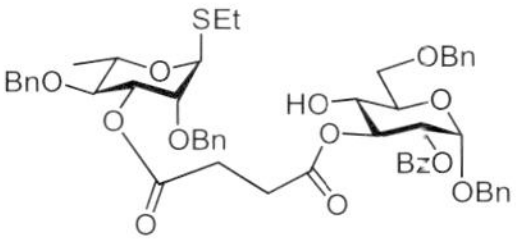	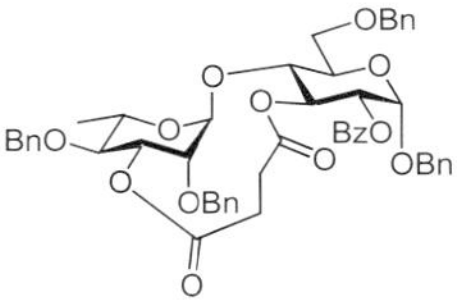	NIS, TMSOTf, MeCN	48	α only	[313]
15	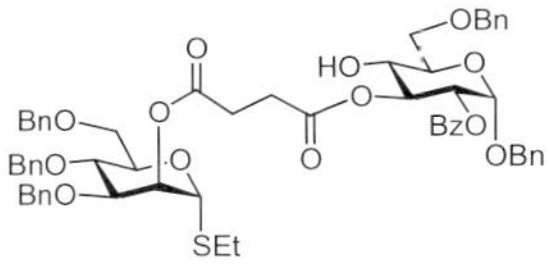	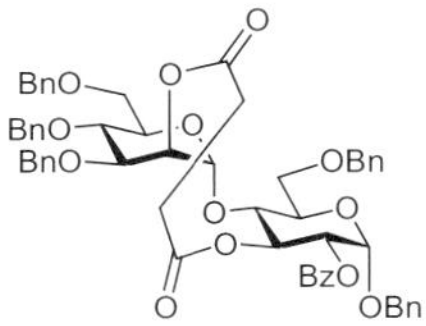	NIS, TMSOTf, MeCN	54	α only	[314]

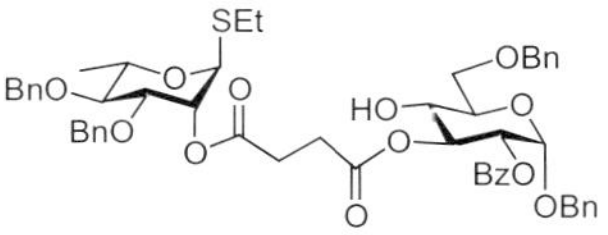

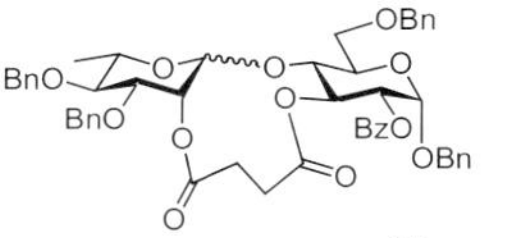

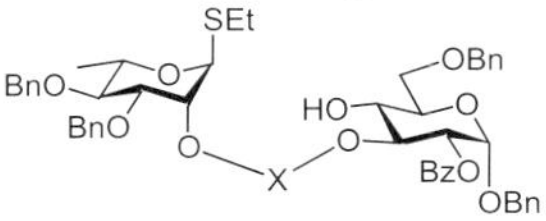

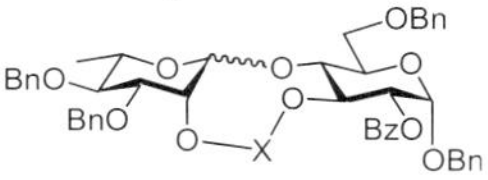

(continued)

Table 5.6C (Continued)

Entry	Starting material	Product	Activation	Yield %	α : β	References
16	(structure)	(structure)	NIS, TMSOTf, MeCN	72	23:77	[314]
17	(structure)	(structure)	NIS, TMSOTf, MeCN MeOTf, MeCN	66 64	β only α only	[315] [316]
18	X = Carbonyl		NIS MeOTf (1) MCPBA (2) Tf₂O	— — 79%	— — α only	[317]
	Oxalyl		NIS MeOTf (1) MCPBA (2) Tf₂O	— — 75%	— — α only	[317] [317] [317]
	Malonyl		NIS MeOTf	51% 52%	β only β only	[317]
	Succinyl		NIS MeOTf	70% 70%	— 61:39 39:61	[317]
	Phthaloyl		NIS MeOTf	69% 63%	89:11 86:11	[317]

summarizes the most significant examples with respect to anomeric selectivity and yield of the glycosylation step. It must be noted that it is still rather difficult to predict the outcome of such intramolecular glycosylations via prearranged glycosides though.Tables 5.6A–5.6C.

As can be seen from the data in Tables 5.6A–5.6C, the outcome of the intramolecular glycosylations via prearranged glycosides is somehow confusing. Especially in the manno series, the stereoselectivity of the glycosylation is governed not only by the tether and the position to which it is linked but also by the activation procedure of the donor moiety (Tables 5.6A–5.6C, entries 15–18). In one case, the anomeric selectivity was even inverted depending on the activation procedure [315,317]. Another efficient β-mannosylation uses more rigid m-xylylene tethers [318] (Scheme 5.111).

One aspect, besides the size of the ring that forms upon intramolecular glycosylation, governing the stereoselective outcome using this concept has been shown to be

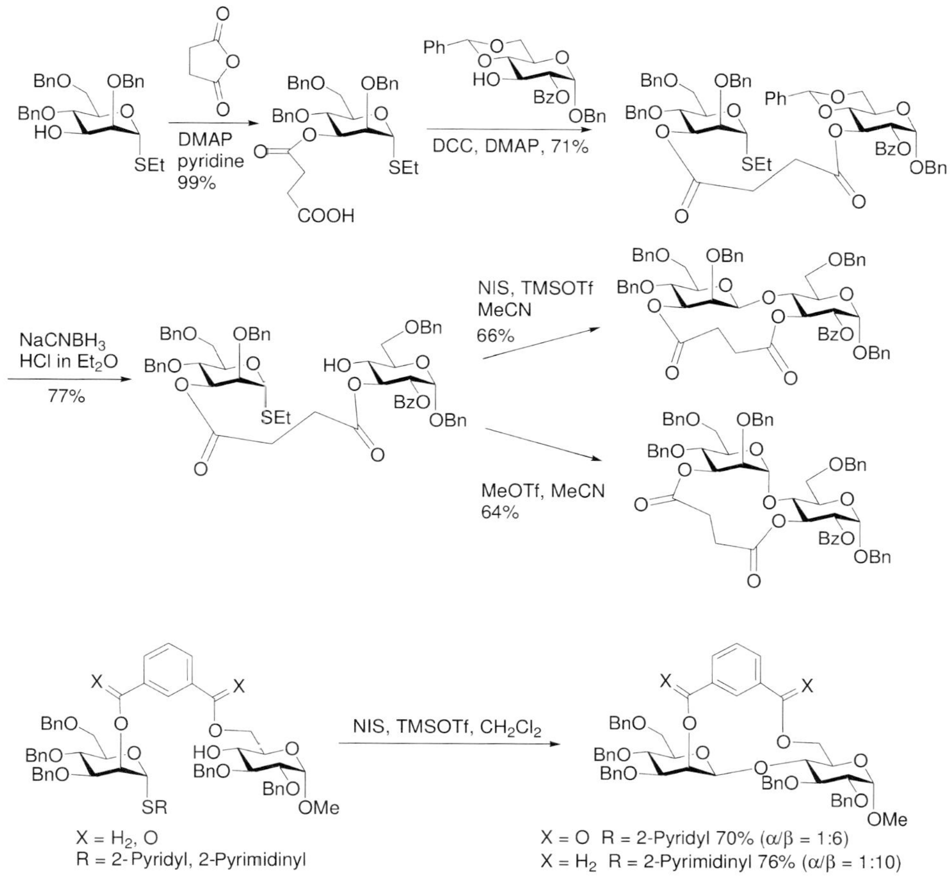

Scheme 5.111 Mannosylation via prearranged glycosides.

the site where the tether is attached to the donor and the configuration of the positions in the acceptor where the tether is bound and where the glycosylation occurs (1,2- versus 1,3-diols, and D/L-*threo* or D/L-*erythro*) [308,309]. It was also shown that a double asymmetric induction influences the anomeric selectivity of the intramolecular glycosylation. For example, this was shown for the pair D/L-mannose donor and D/L-glucose acceptor prearranged by a succinyl tether via positions 6 of the mannose and position 3 of the glucose moieties for a 1,4-glycosylation (Scheme 5.112). Here, no significant change in the stereoselectivity of the intramolecular glycosylation step was observed upon changing the topographic properties of the pairs whereas an inversion was found upon changing the geometric properties [319].

The regioselectivity of glycosylations via prearranged glycosides is even more intriguing. Valverde *et al.* have reported several examples that are summarized in Table 5.7 [320–324]. Although it has been shown that the tethering has a dramatic influence on the regioselectivity compared to the corresponding intermolecular

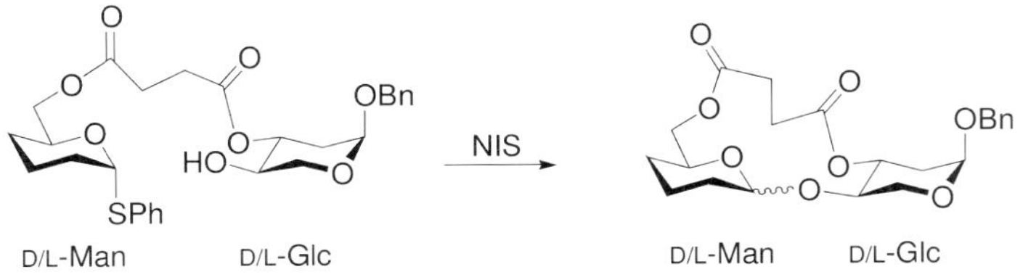

Scheme 5.112 Double asymmetric induction.

Donor	Acceptor	Yield%	α/β Ratio
L-Man	L-Man	76	65:35
D-Man	D-Man	70	68:32
L-Man	L-Man	75	44:56
D-Man	D-Man	78	30:70

glycosylations [324], more examples are still needed to deduct more general rules for predicting selectivity.

Yet another approach uses peptides as tethers for intramolecular glycosylations via prearranged glycosides (Scheme 5.113) [326,327]. The regio- and anomeric selectivity of the intramolecular glycosylation depends on the amino acid sequence of the peptide, which links glycosyl donor and acceptor.

Despite the still virulent difficulties for predicting the outcome of intramolecular glycosylations via prearranged glycosides, several examples of complex oligosaccharide syntheses using this approach had been published. Müller and Schmidt demonstrated the usefulness of the 'rigid spacer concept' via *m*-xylylene tethers with a stepwise synthesis of maltotriose by applying two consecutive intramolecular glycosylation steps [328] (Scheme 5.114).

An efficient synthesis of the pyruvated repeating unit of the *exo*-polysaccharide of *Streptococcus pneumoniae* containing a β-linked L-rhamnosyl moiety was also

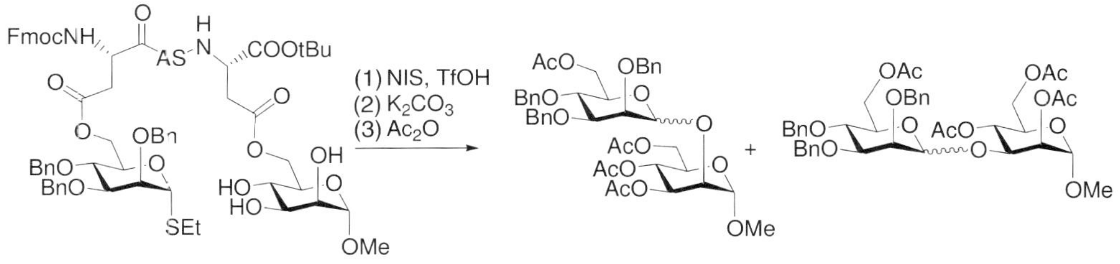

AS	α(1,2)%	β(1,2)%	α(1,3)%	β(1,3)%
—	—	13%	11	23
Gly	—	—	21	20
Phe	—	—	13	18
Pro	14%	16%	—	19
GlyGly	—	—	—	56

Scheme 5.113

5.4 The Twenty First Century View of Chemical O-Glycosylation | 485

Table 5.7 Regioselectivity in intramolecular glycosylations via prearranged glycosides.

Entry	Starting material	Product	Activation	Yield %	Linkage	α:β	References
1	X = phthaloyl, succinyl		NIS, TfOH, 0°C–25°C	80 65	1,3 1,3	— —	[320] [320]
2	X = phthaloyl, Iso-phthaloyl		NIS, TfOH, CH$_2$Cl$_2$, 0°C	72 47	1,3 1,3	58:42 α only	[323] [323]
3	X = phthaloyl, Iso-phthaloyl		NIS, TfOH, CH$_2$Cl$_2$, 0°C	72 47	1,3 1,3	72:28 α only	[323] [323]
4			NIS, TfOH, CH$_2$Cl$_2$, 0°C	76	1,3	α only	[323]
5			NIS, TfOH, CH$_2$Cl$_2$, 0°C	72	1,4	α only	[323]
6			NIS, AgOTf, CH$_2$Cl$_2$	48	1,6	α only	[322]
7			NIS 25°C	25 75	1,3 1,2	α only α only	[324] [324]

(*continued*)

Table 5.7 (Continued)

Entry	Starting material	Product	Activation	Yield %	Linkage	α:β	References
8	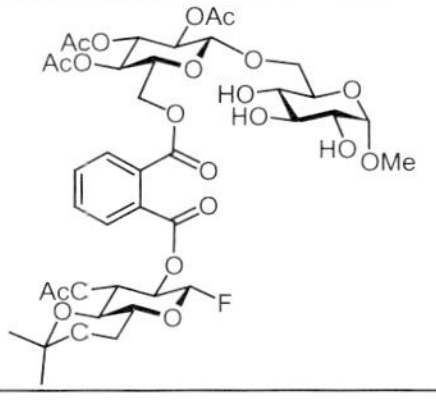	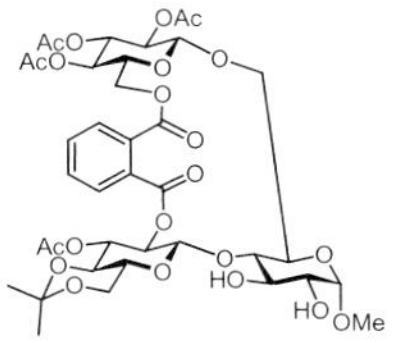	Cp₂HfCl₂, AgOTf	37	1,4	β only	[325]

established through intramolecular glycosylation (Scheme 5.115) [329] Here, starting from monosaccharide precursors, prearranged succinyl-tethered L-Rha-Glc-glycosides were first prepared by a standard condensation/reduction sequence, and their intramolecular rhamnosylation was studied. Next, the β-linked disaccharide building block obtained in 55% yield was converted into the desired tetrasaccharide 5-aminopentyl glycoside in one step. The latter may serve as an antigen for synthetic vaccines against *S. pneumoniae* infection.

Intramolecular glycosylations via prearranged glycosides can also be combined with common strategies usually applied for the synthesis of complex oligosaccharides. For example, prearranged glycosides can be combined with the strategy of armed and disarmed glycosyl donors as outlined in Scheme 5.116 [330]. Once again, in a standard condensation/reduction sequence, an armed ethyl 1-thio-mannosyl

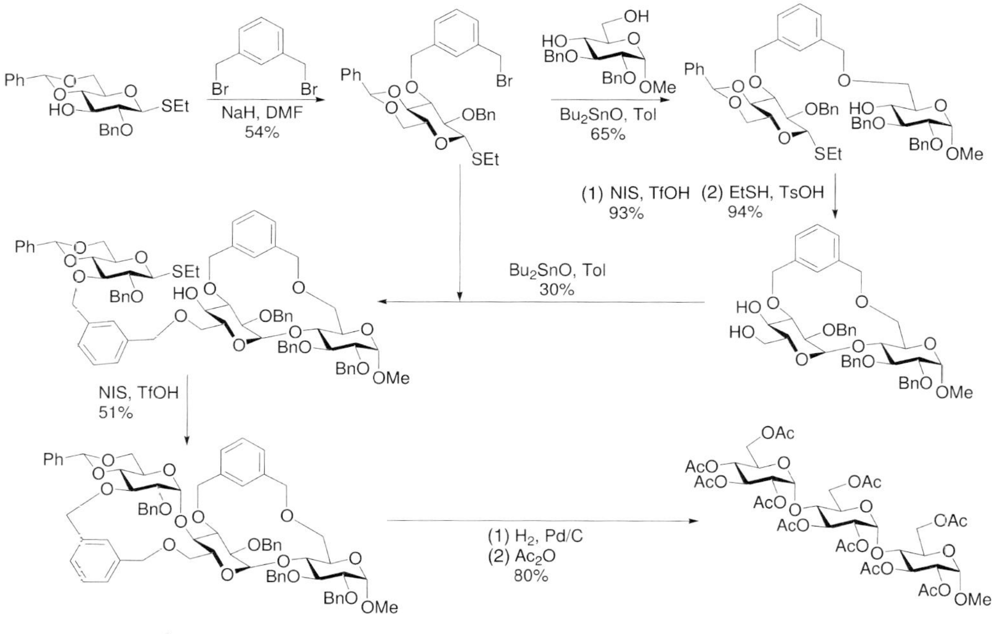

Scheme 5.114

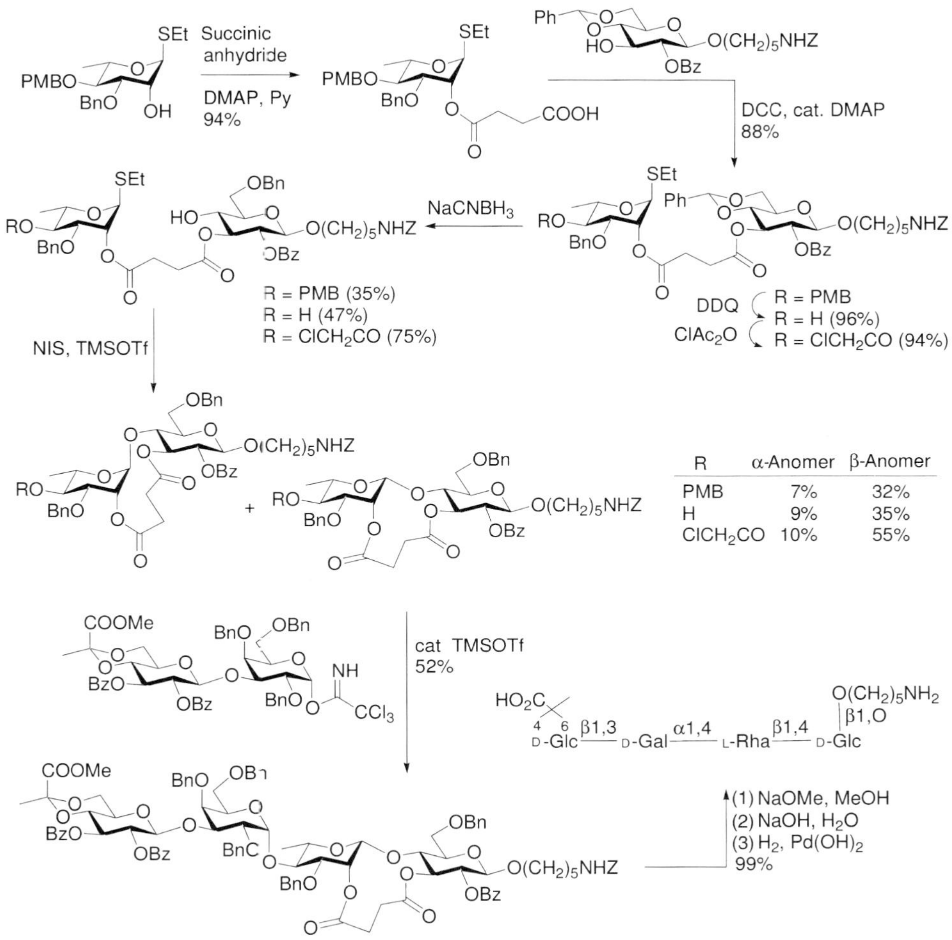

Scheme 5.115

donor was succinyl tethered with a disarmed phenyl 1-thio-glucosyl donor that also functions as an acceptor for the intramolecular glycosylation step. Activation of the mannose donor in the prearranged glycoside with methyl triflate gave the β-linked disaccharide in 69% yield, which upon activation with N-iodosuccinimide was coupled to disaccharide acceptor in 64% yield. Thus, a tetrasaccharide 5-aminopentyl glycoside related to the exopolysaccharide of *Arthrobacter sp.* could be prepared in just a few steps.

Another significant improvement of the prearranged glycoside concept are non-symmetric tethers that allow flexible oligosaccharide syntheses [331]. As shown in Scheme 5.117, brommethyl benzoates are well suited for that purpose. Alkylation of either the glycosyl donor or the glycosyl acceptor followed by condensation with the respective counterpart donor or acceptor results in tethered glycosides that after intramolecular glycosylation and saponification of the tether afford disaccharides

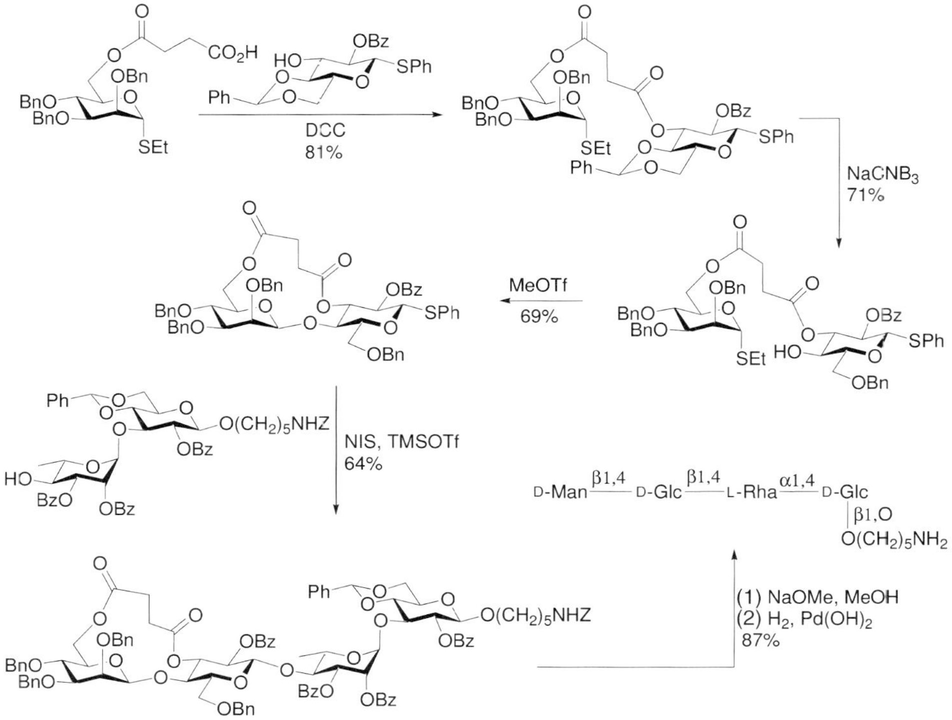

Scheme 5.116

suitable for further chain elongation at different positions. Nonsymmetric tethers have also been applied for iterative intramolecular glycosylations [332].

The strategy via nonsymmetrical tethers has also been applied to the synthesis of pentasaccharides related to the repeating unit of *Shigella sp.* (Scheme 5.118) [333]. Here, *o*-brommethyl benzoate was used as nonsymmetrical tether for the synthesis of a Glcα(1,4)GlcNAc disaccharide donor that was used for the preparation of the pentasaccharides either directly or after regioselective opening of the tether with concomitant chain extension at position 3 of the GlcNAc moiety.

5.4.2
Other Indirect and Special Methods

There are several other indirect and special methods that have not found any broader application for glycoside and saccharide synthesis yet but that might be useful for certain cases or might develop into more common procedures in the future. Such other indirect and special methods are discussed in this chapter.

5.4.2.1 [4 + 2] Cycloadditions of Glycals

When irradiated with light, azidodicarboxylates react with glycals as dienophiles to give Diels–Alder adducts via formal [4 + 2] cycloadditions [334]. Upon treatment of

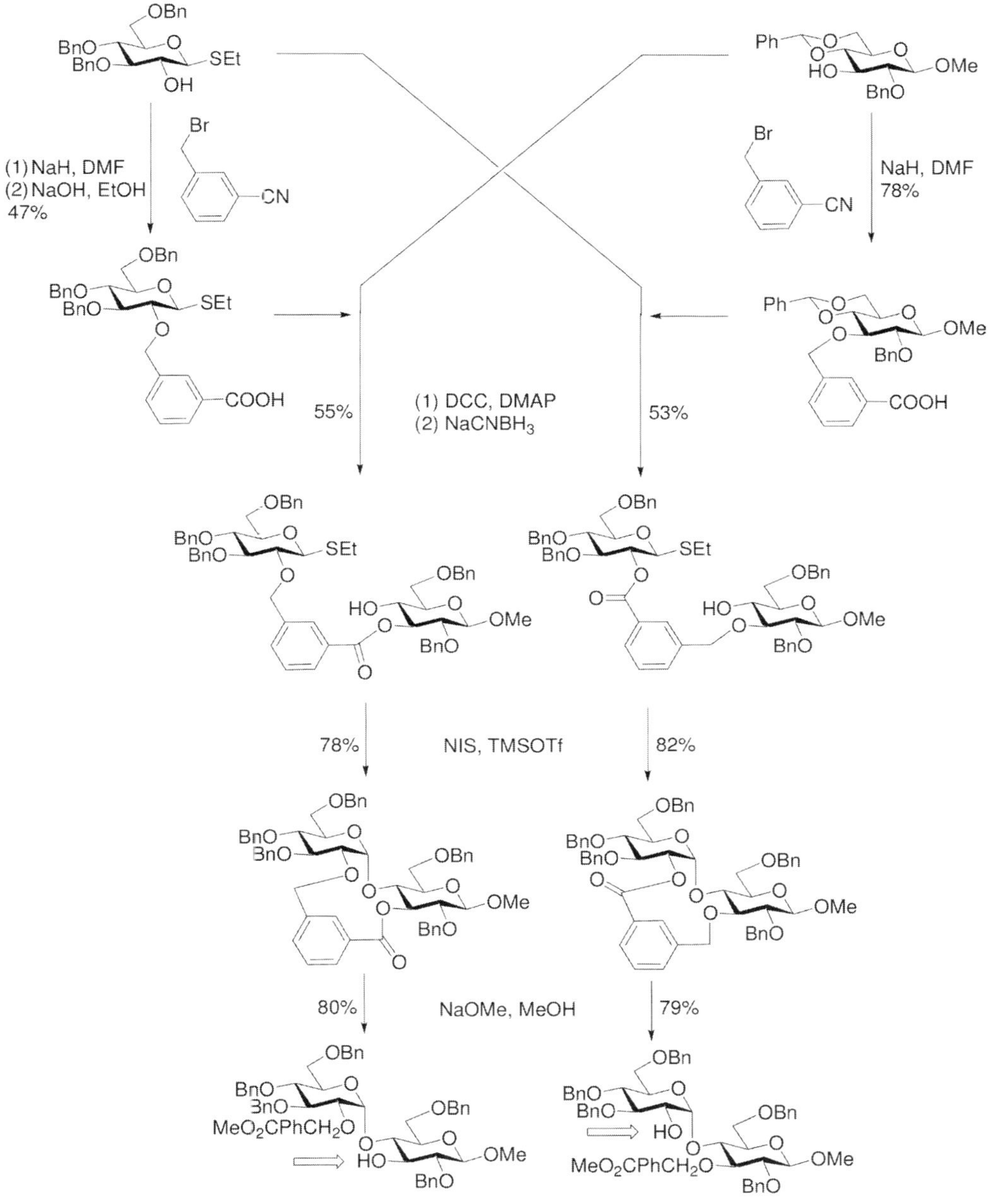
OBn
BnO
BnO
O
SEt
OH
Br
CN
(1) NaH, DMF
(2) NaOH, EtOH
47%
Ph
O
O
HO
BnO
OMe
NaH, DMF
78%
COOH
55%
(1) DCC, DMAP
(2) NaCNBH3
53%
78%
NIS, TMSOTf
82%
80%
NaOMe, MeOH
79%
MeO2CPhCH2O

Scheme 5.117

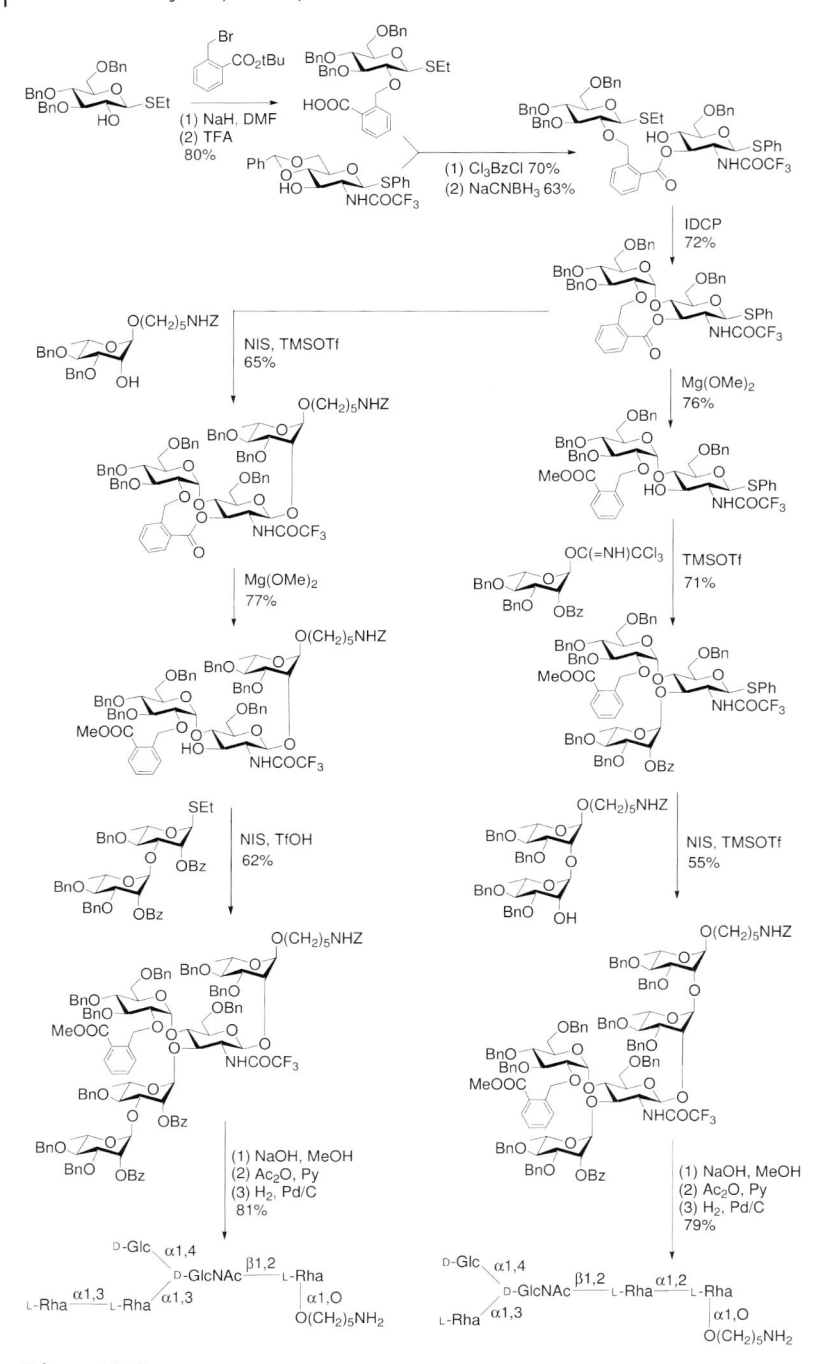

Scheme 5.118

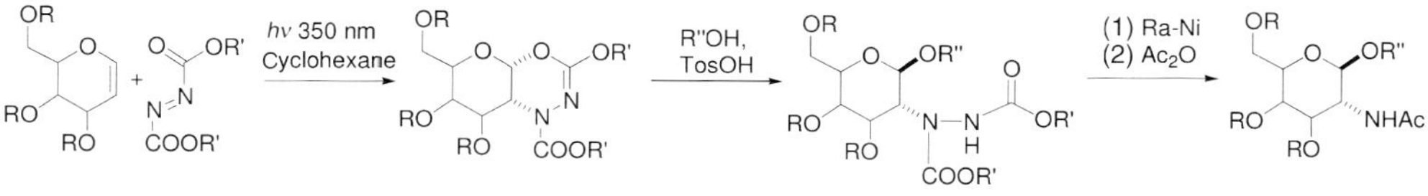

Scheme 5.119

these initial Diels–Alder adducts with alcohols under acidic conditions, 2-hydrazino-glycosides are formed stereoselectively that can be reduced to 2-amino-glycosides (Scheme 5.119). The initial cycloadducts are *cis*-configurated and are opened with alcohols by a S_N2 mechanism giving exclusively 1,2-*trans*-configurated 1-aminoglycosides.

Table 5.8 summarizes the glycals that had been converted into 2-aminoglycosides so far. In general, photolytic *cis/trans* isomerization of *trans*-azodicarboxylates is necessary to accomplish good yields of the initial cycloadducts [335–337]. However, bis-trichloroethyl azodicarboxylate also reacts under thermal conditions and also gives a more reactive cycloaddition product [338,339].

Table 5.8 [2 + 4] Cycloadditions according to Scheme 5.19.

Entry	Glycal	[4+2] Adduct	Yield %	Alcohol	Product	Yield %	References
1	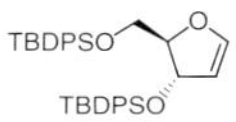	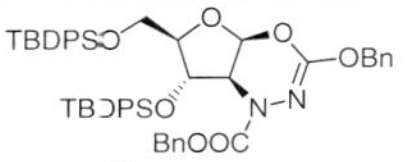	70	MeOH	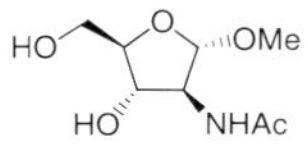	64	[335]
2	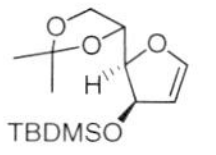	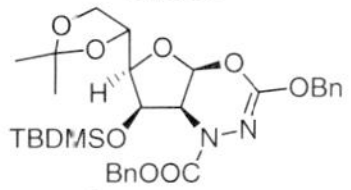	73	MeOH	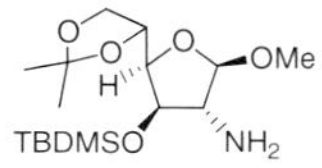	84	[335]
3	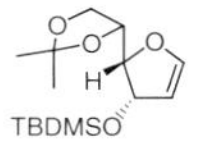	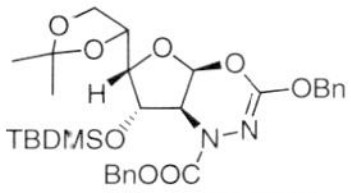	80	MeOH	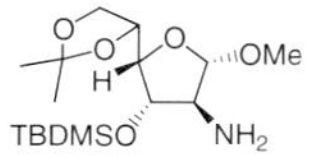	78	[335]
4		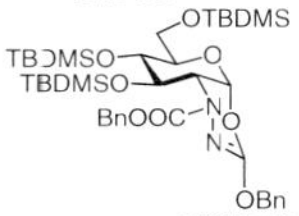	71	MeOH	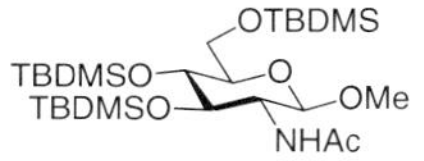	63	[335]
5		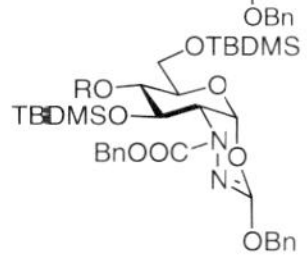	71	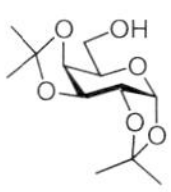	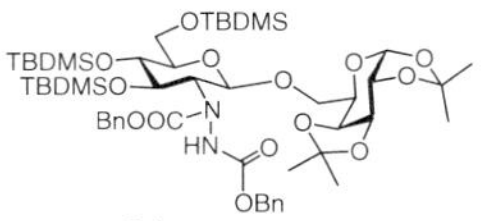	95	[336]
6	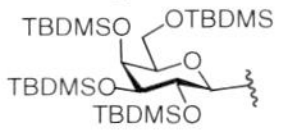	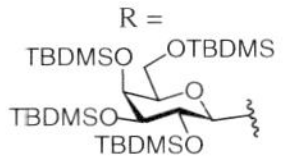	92	MeOH	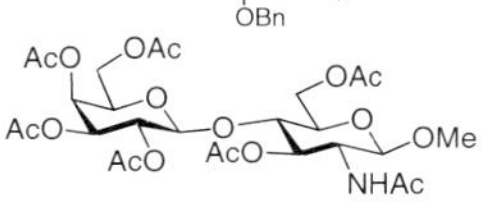	48	[339]

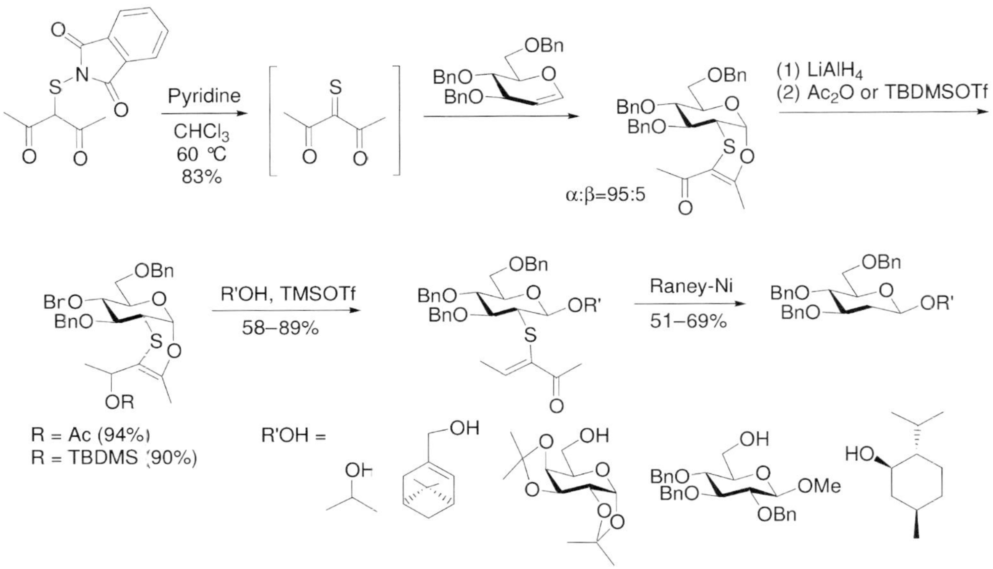

Scheme 5.120

Yet another hetero-dien for cycloaddition reactions of glycals that had been used for the preparation of 2-deoxyglycosides is 2,4-dioxo-3-thioxo-pentane, generated *in situ* from 3-thiophthalimido-pentane-2,4-dion. The cycloaddition occurs with high 1,2-*cis* selectivity, giving thioxins [340,341]. The latter can be reacted with alcohols under Lewis acid catalysis to afford 2-thio-glycosides that finally give 2-deoxy-glycosides upon desulfuration with Raney-Ni. [342] (Scheme 5.120). Yields are medium to high, but more examples need to be investigated to judge the broad applicability of this approach. At least this special method is a considerable alternative for the preparation of β-2-deoxy-glycosides that are otherwise difficult to obtain.

5.4.2.2 1,2-Cyclopropanated Sugars

1,2-Cyclopropanated sugars, easily available from glycols [343,344] react with alcohols to give 2-*C*-branched glycosides. As both diastereomeric cyclopropanated sugars can be obtained, this special method for the preparation of glycosides is highly flexible and affords a wide variety of 2-*C*-branched glycosides. For example, tri-*O*-benzyl-D-glycal affords the β-manno-configurated 1,2-cyclopropane derivative upon Simmons–Smith cyclopropanation, whereas the α-gluco-configurated counterpart is obtained through cyclopropanation with dichlorocarbene followed by dehalogenation. Upon methanolysis, the dichlorocyclopropane sugar reacts under ring opening to give an oxepine derivative whereas the cyclopropane counterparts afford the corresponding 2-bromomethyl mannosides and glycosides when treated with NBS and methanol [344]. 2-*C*-Methyl glycosides can be obtained by Pt-catalyzed ring opening of the cyclopropane ring [254,345] (Scheme 5.121).

Similarly, *tri-O*-benzyl-glucal affords the corresponding manno-configurated methoxycarbonyl cyclopropanes upon treatment with methyl diazoacetate under

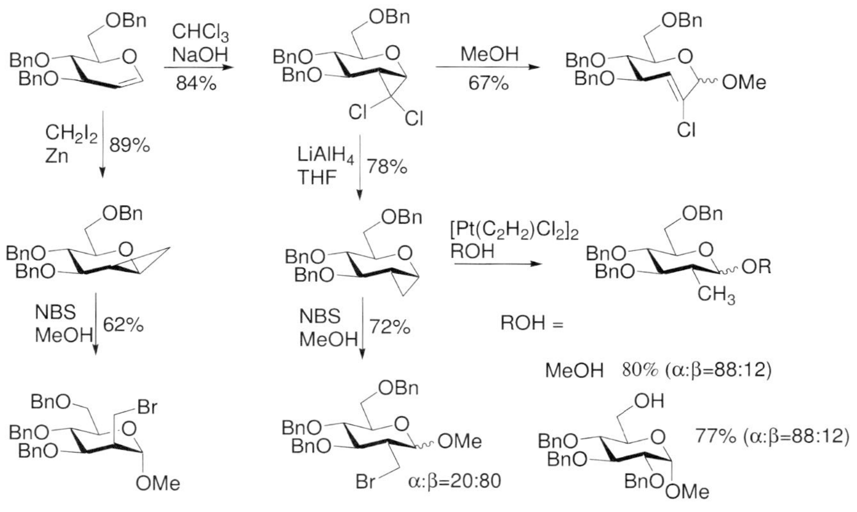

Scheme 5.121

rhodium acetate catalysis [346]. The latter can get opened at the cyclopropane ring with NIS and methanol, and the formed intermediates can be converted into 2-C-aminomethyl-glycosides.

Yet another indirect method for the preparation of O-glycosides makes use of intermediate cyclopropanated sugars. Treatment of benzyl-protected 2-(C-2-O-mesyl-α-D-mannosyl)acetaldehyde with alcohols and base results in the intermediate formation of cyclopropanated sugar that undergoes immediate ring opening to afford the corresponding 2-C-formylmethyl glycosides [347,348] (Scheme 5.122). The method is also suitable for the preparation of 1-thio-glycosides and glycosyl azides.

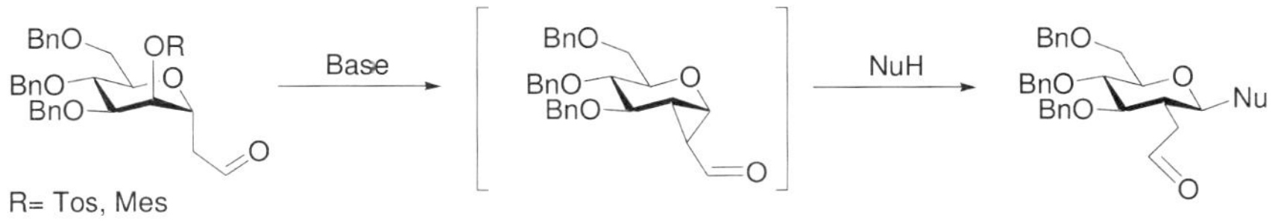

R= Tos, Mes

R	Base	NuH	Yield%
Mes	Et$_3$N	MeOH	72
Mes	Et$_3$N	EtOH	71
Tos	Et$_3$N	H$_2$C=CH-CH$_2$OH	76
Tos	K$_2$CO$_3$	H$_2$C=CH-(CH$_2$)$_4$OH	78
Tos	K$_2$CO$_3$	Ph-CH$_2$OH	75
Tos	K$_2$CO$_3$	Ph-OH	62
Tos	K$_2$CO$_3$	Ph-SH	86
Tos	K$_2$CO$_3$	p-MeOPh-SH	85
Tos	K$_2$CO$_3$	p-ClPh-SH	83
Tos	Et$_3$N	NaN$_3$/MeOH	52

Scheme 5.122

References

264 Jung, K.-H., Müller, M. and Schmid, R.R. (2000) *Chemical Reviews*, **100**, 4423–4442.
265 Demchenko, A.V. (2003) *Current Organic Chemistry*, **7**, 35–79.
266 Scheffler, G. and Schmid, R.R. (1997) *Tetrahedron Letters*, **38**, 2943–2946.
267 Inaba, S., Yamada, M., Yoshino, T. and Ishido, Y. (1973) *Journal of the American Chemical Society*, **95**, 2062–2063.
268 Iimori, T., Shibazaki. T. and Ikegami, S. (1996) *Tetrahedron Letters*, **37**, 2267–2270.
269 Azumaya, I., Niwa, T., Kotani, M., Iimori, T. and Ikegami, S. (1999) *Tetrahedron Letters*, **40**, 4683–4686.
270 Mukai, C., Itoh, T. and Hanaoka, M. (1997) *Tetrahedron Letters*, **38**, 4595–4598.
271 Scheffler, G. and Schmidt, R.R. (1999) *The Journal of Organic Chemistry*, **64**, 1319–1325.
272 Scheffler, G., Behrendt, M.E. and Schmidt, R.R. (2000) *European Journal of Organic Chemistry*, 3527–3539.
273 Laursen, J.B., Petersen, L. and Jensen, K.J. (2001) *Organic Letters*, **3**, 687–690.
274 Barresi, F. and Hindsgaul, O. (1991) *Journal of the American Chemical Society*, **113**, 9376–9377.
275 Barresi, F. and Hindsgaul, O. (1992) *Synlett*, 759–761.
276 Barresi, F. and Hindsgaul, O. (1994) *Canadian Journal of Chemistry*, **72**, 1447–1465.
277 Ennis, S.C., Fairbanks, A.J., Tennant-Eyles, R.J. and Yates, H.S. (1999) *Synlett*, 1387–1390.
278 Seward, C.M.P., Cumpstey, I., Aloui, M., Ennis, S.C., Redgrave, A.J. and Fairbanks, A.J. (2000) *Chemical Communications*, 1409–1410.
279 Fairbanks, A.J. (2003) *Synlett*, 1945–1958.
280 Chayajarus, K., Chambers, D.J., Chughtai, M.J. and Fairbanks, A.J. (2004) *Organic Letters*, **6**, 3797–3800.
281 Cumpstey, I., Fairbanks, A.J. and Redgrave, A.J. (2004) *Tetrahedron*, **60**, 9061–9074.
282 Cumpstey, I., Chayajarus, K., Fairbanks, A.J., Redgrave, A.J. and Seward, C.M.P. (2004) *Tetrahedron: Asymmetry*, **15**, 3207–3221.
283 Attolino, E., Cumpstey, I. and Fairbanks, A.J. (2006) *Carbohydrate Research*, **341**, 1609–1618.
284 Ito, Y. and Ogawa, T. (1994) *Angewandte Chemie (International Edition in English)*, **33**, 1765–1768.
285 Ito, Y., Ohnishi, Y., Ogawa, T. and Nakahara, Y. (1998) *Synlett*, 1102–1104.
286 Lergenmüller, M., Nukada, T., Kuramochi, K., Dan, A., Ogawa, T. and Ito, Y. (1999) *European Journal of Organic Chemistry*, 1367–1376.
287 Dan, A., Ito, Y. and Ogawa, T. (1995) *The Journal of Organic Chemistry*, **60**, 4680–4681.
288 Dan, A., Ito, Y. and Ogawa, T. (1996) *Carbohydrate Letters*, **1**, 469–474.
289 Dan, A., Ito, Y. and Ogawa, T. (1995) *Tetrahedron Letters*, **36**, 7487–7490.
290 Dan, A., Lergenmüller, M., Amano, M., Nakahara, Y., Ogawa, T. and Ito, Y. (1998) *Chemistry – A European Journal*, **4**, 2182–2190.
291 Matsuo, I., Totani, K., Tatami, A. and Ito, Y. (2006) *Tetrahedron*, **62**, 8262–8277.
292 Ito, Y. and Ogawa, T. (1997) *Journal of the American Chemical Society*, **119**, 5562–5566.
293 Krog-Jensen, C. and Oscarson, S. (1996) *The Journal of Organic Chemistry*, **61**, 4512–4513.
294 Krog-Jensen, C. and Oscarson, S. (1998) *The Journal of Organic Chemistry*, **63**, 1780–1784.
295 Gelin, M., Ferrieres, V., Leveuvre, M. and Plusquellec, D. (2003) *European Journal of Organic Chemistry*, 1285–1293.
296 Bols, M. (1992) *Chemical Communications*, 913–914.
297 Bols, M. (1993) *Chemical Communications*, 791–792.
298 Bols, M. (1993) *Acta Chemica Scandinavica*, **47**, 829–834.

299 Bols, M. (1993) *Tetrahedron*, **49**, 10049–10060.
300 Bols, M. and Hansen, H. (1994) *Chemistry Letters*, 1049–1052.
301 Bols, M. (1996) *Acta Chemica Scandinavica*, 931–937.
302 Stork, G. and Kim, G. (1992) *Journal of the American Chemical Society*, **114**, 1087–1088.
303 Stork, G. and La Clair, J.J. (1996) *Journal of the American Chemical Society*, **118**, 247–248.
304 Bols, M. and Skrydstrup, T. (1995) *Chemical Reviews*, **95**, 1253–1277.
305 Madsen, J. and Bols, M. (2000) *Carbohydrates in Chemistry and Biology*, vol. 1 (eds B. Ernst, G.W. Hart and P. Sinay), Wiley-VCH Verlag GmbH, Germany, 449–466.
306 Wakao, M., Fukase, K. and Kusumoto, S. (1999) *Synlett*, 1911–1914.
307 Ziegler, Th., Ritter, A. and Hürttlen, J. (1997) *Tetrahedron Letters*, **38**, 3715–3718.
308 Huchel, U. and Schmidt, R.R. (1998) *Tetrahedron Letters*, **39**, 7693–7697.
309 Müller, M., Huchel, U., Geyer, A. and Schmidt, R.R. (1999) *The Journal of Organic Chemistry*, **64**, 6190–6210.
310 Lemanski, G., Lindenberg, T., Fakhrnabavi, H. and Ziegler, Th. (2000) *Journal of Carbohydrate Chemistry*, **19**, 727–745.
311 Ziegler, Th., Dettmann, R. and Zettl, U. (1999) *Journal of Carbohydrate Chemistry*, **18**, 1079–1095.
312 Ziegler, Th. and Lau, R. (1995) *Tetrahedron Letters*, **36**, 1417–1420.
313 Lau, R., Schüle, G., Schwaneberg, U. and Ziegler, Th. (1995) *Liebigs Annalen der Chemie*, 1745–1754.
314 Ziegler, Th., Lemanski, G. and Rakoczy, A. (1995) *Tetrahedron Letters*, **36**, 8973–8976.
315 Ziegler, Th. and Lemanski, G. (1998) *Angewandte Chemie (International Edition in English)*, **37**, 3129–3132.
316 Lemanski, G. and Ziegler, Th. (2000) *Helvetica Chimica Acta*, **83**, 2655–2675.
317 Lemanski, G. and Ziegler, Th. (2000) *Tetrahedron*, **56**, 563–579.
318 Abdel-Rahman, A.A.-H., ElAshry, E.S.H. and Schmidt, R.R. (2002) *Carbohydrate Research*, **337**, 195–206.
319 Ziegler, Th. and Lemanski, G. (1998) *European Journal of Organic Chemistry*, 163–170.
320 Valverde, S., Gomez, A.M., Hernandez, A., Herradon, B. and Lopez, J.C. (1995) *Chemical Communications*, 2005–2006.
321 Valverde, S., Gomez, A.M., Lopez, J.C. and Herradon, B. (1996) *Tetrahedron Letters*, **37**, 1105–1108.
322 Valverde, S., Garcia, M., Gomez, A.M. and Lopez, J.C. (2000) *Chemical Communications*, 813–814.
323 Valverde, S., Garcia, M., Gomez, A.M. and Lopez, J.C. (2000) *Synlett*, 22–26.
324 Cid, M.B., Valverde, S., Lopez, J.C., Gomez, A.M. and Garcia, M. (2005) *Synlett*, 1095–1100.
325 Yamada, H., Imamura, K. and Takahashi, T. (1997) *Tetrahedron Letters*, **38**, 391–394.
326 Tennant-Eyles, R.J., Davis, B.G. and Fairbanks, A.J. (1999) *Chemical Communications*, 1037–1038.
327 Tennant-Eyles, R.J., Davis, B.G. and Fairbanks, A.J. (2000) *Tetrahedron: Asymmetry*, **11**, 231–243.
328 Müller, M. and Schmidt, R.R. (2001) *European Journal of Organic Chemistry*, 2055–2066.
329 Schüle, G. and Ziegler, Th. (1996) *Liebigs Annalen der Chemie*, 1599–1607.
330 Lemanski, G. and Ziegler, Th. (2000) *European Journal of Organic Chemistry*, 181–186.
331 Ziegler, Th., Lemanski, G. and Hürttlen, J. (2001) *Tetrahedron Letters*, **42**, 569–572.
332 Paul, S., Müller, M. and Schmidt, R.R. (2003) *European Journal of Organic Chemistry*, 128–137.
333 Lemanski, G. and Ziegler, Th. (2006) *European Journal of Organic Chemistry*, 2618–2630.
334 Leblanc, Y. and Labelle, M. (1992) Cycloaddition reactions in carbohydrate

335 Fitzsimmonis, B.J., Leblanc, Y. and Rokach, J. (1987) *Journal of the American Chemical Society*, **109**, 285–286.

336 Fitzsimmonis, B.J., Leblanc, Y., Chan, N. and Rokach, J. (1988) *Journal of the American Chemical Society*, **110**, 5229–5231.

337 Leblanc, Y., Fitzsimmonis, B.J., Springer, J.P. and Rokach, J. (1989) *Journal of the American Chemical Society*, **111**, 2995–3000.

338 Leblanc, Y. and Fitzsimmonis, B.J. (1989) *Tetrahedron Letters*, **30**, 2889–2892.

339 Toepfer, A. and Schmidt, R.R. (1993) *Carbohydrate Research*, **247**, 159–164.

340 Capozzi, G., Dios, A., Franck, R.W., Geer, A., Marzabodi, C., Menichetti, S., Nativi, C. and Tamarez, M. (1996) *Angewandte Chemie (International Edition in English)*, **35**, 777–779.

341 Capozzi, G., Franck, R.W., Mattioli, M., Menichetti, S., Nativi, C. and Valle, G. (1995) *The Journal of Organic Chemistry*, **60**, 6416–6424.

342 Bartolozzi, A., Capozzi, G., Menichetti, S. and Nativi, C. (2001) *European Journal of Organic Chemistry*, 2083–2090.

343 Murali, R., Ramana, C.V. and Nagarajan, M. (1995) *Chemical Communications*, 217–218.

344 Ramana, C.V., Murali, R. and Nagarajan, M. (1997) *The Journal of Organic Chemistry*, **62**, 7694–7703.

345 Beyewr, J., Skaanderup, P.R. and Madsen, R. (2000) *Journal of the American Chemical Society*, **122**, 9575–9583.

346 Sridhar, P.R., Ashlau, K.C. and Chandrasekaran, S. (2004) *Organic Letters*, **6**, 1777–1779.

347 Shao, H., Ekthawatchai, S., Wu, S.-H. and Zou, W. (2004) *Organic Letters*, **6**, 3497–3499.

348 Shao, H., Ekthawatchai, S., Chen, C.-S., Wu, S.-H. and Zou, W. (2005) *The Journal of Organic Chemistry*, **70**, 4726–4734.

chemistry. ACS Symposium Series No. 494 (ed. R.M. Giuliano), 81–96.

Index

a

N-acetyllactosamine residue 151
N-acetylneuraminic acid derivative 38, 274
N-acetylneuraminic acid residues 245
acidic glycosyl acceptors
– carboxylic acids 103
– hydroxyphthalimides 103
– imides 103
– phenols 103
acylated bromosugars 76
alcohol acceptor 102
aliphatic alcohol nucleophile 105
allyl sulfoxide–allyl sulfenate equilibrium 304
alkoxide Michael-type addition 431
alkyl glycosides 187
alkyl iodides (bromides) 262
alkyl thioglycosides 261
angucycline antibiotic family 159
anhydro sugars 416
anomeric carbon–chalcogen bond 361
anomerization process 267
anomeric bromide intermediate 106
anomeric esters 116
– glycosylation 116
anomeric hydrogenolysis 96
anomeric hydroxyl group 178
anomeric oxophosphonium intermediate 106, 107, 109, 122
anomeric oxotitanium moiety 114
anomeric oxotitanium species 114
anomeric pivaloate donors 122
anomeric p-nitrobenzenesulfonate 108
anomeric p-nitrobenzoate donors 122
anomeric O-trichloroacetimidates 144
– preparation 144
anomeric triflate intermediate 108
Appel method 107
Appel-type glycosylation
– representative procedure 129

arabinopyranose donor 102
armed–disarmed glycosylation
– approach 275
– methodology 246
– strategy 202
O-aryl disaccharides 117
aryl thioglycosides 261, 262
aryl trimethylsilyl ether 187
aureolic acid antibiotics 120
2-azido-2-deoxyglycopyranosyl diethyl 233
2-azido-2-deoxyglycosyl donors 242
2-azido lactosyl α-chloride 63

b

backbone amide linker (BAL) 194
benzenesulfenyl triflate promoter systems 310
benzothiazoyl vinyl sulfone 69
benzotrifluoride solvent 100
benzoyl-protected acceptor diol 216
benzylated thioglycosyl donor 288
benzyl-protected glycosyl diphenyl phosphates 228
bleomycin derivatives 248
boron trifluoride diethyl etherate 119, 261
bromofuranylium ion 394
tert-butyldimethylsilyl (TBDMS) groups 104, 187
tert-butyl hydroperoxide 227
tert-butyldimethylsilyl glycosyl donors 188

c

camphor sulfonic acid (CSA) 458
carbamate glycosyl donors 125
carbohydrate orthoesters 382
carbon electrophiles
– hemiacetal activation 111
carbon-sulfur bond 292
Chagas disease 158, 372
chemoselective condensation sequence 288

Handbook of Chemical Glycosylation: Advances in Stereoselectivity and Therapeutic Relevance.
Edited by Alexei V. Demchenko.
Copyright © 2008 WILEY-VCH Verlag GmbH & Co. KGaA. All rights reserved.
ISBN: 978-3-527-31780-6

chemoselective glycosylations   278
– method   274
– technology   290
chitinase inhibitors   453
cholestanyl riboside   106, 115
cholesteryl glucoside   107
chondroitin sulfates (CS)   154
Conandron ramoidioides   168
coupling–deacetylation sequence   83
contact ion pair (CIP)   307
cyanoethylidene derivatives   385

d

2-deoxy-2-iodoglycosides
– preparation   444
– protocol   109
dibenzothiophene-5-oxide (DBTO)   225
dibenzyl phosphates   233, 244
dicyclohexyl carbodiimide   112
diethylaminosulfur trifluoride (DAST)   31, 115
diethyl ether solvent   116
diethyl phosphorochloridite   251
diglucosyl dehydrodigalloyl diester   168
dimethyl methylthiosulfonium trifluoromethansulfonate (DMTST)   125, 452
N,N-dimethyl 4-aminopyridine (DMAP)   114
N,N-dimethylformamide (DMF) solvent   107
diphenyldichlorosilane   101
diphenyl sulfoxide   109
disaccharide fluoride donor   45
disarmed glycosyl iodides   82
DNA sequence protocols   457

e

electrophilic promoters   339
electrophilic selenylating agents   367
enzymatic transglycosylation   185
ester-directed couplings   315
exo-anomeric effect   304

f

Ferrier reactions   306, 418
Fischer and Koenigs–Knorr approaches   17
Fischer glycosylation   96, 100
Fischer–Zach method   417
fluorobenzoyl groups   70
fluorosulfonium ion   32
N-Fmoc-serine acceptor   119
Fraser-Reid's laboratories   394
l-fucose glycopeptides   119
l-fucosyl fluoride derivative   38
furanose hemiacetal donors   98, 114

g

galactofuranosyl-containing oligosaccharides   372
– synthesis of   372
galactosyl trichloroacetimidate donor   161
Globo H antigen   177
glycal-based glycosyl acceptors   19
glycal-derived donors   420
glycal-terminated tetrasaccharide   428
glycosidic bond   16
– stereoselective synthesis   16
glycosidic linkage   16, 21
glycosphingolipids (GSLs)   160
glycosyl acceptor alcohol   41, 96
glycosyl amino acids   163
glycosylation   4, 21
– mechanism   4
– process   1, 6, 9, 10, 11, 367
– reactions   18, 314, 370
– sequencing   16
glycosyl bromide   66, 98, 107, 121, 203, 442
– preparation   66
– reactivity patterns   68
glycosyl fluoride synthesis   31
– from glycosyl esters   33
– from hemiacetals   32
– from S-glycosides   35
glycosyl dithiocarbonates (xanthates)   329, 350
glycosyl donor   272, 329, 341, 367, 382
– electrophilic activation   272
– self-condensation   289
glycosyl iodides   367, 383
– *in situ* generation of   383
glycosyl phenyl sulfones   316
glycosylphosphatidylinositol (GPI)   398
glycosyl phosphorodithioates   357
– glycosidation of   357
glycosyl sulfonylcarbamates   465, 466
glycosyl thiocyanates   356
– synthesis   356
glycosyl thioimidates (heteroaryl thioglycosides)   335
glycosyltransferase inhibitors   69
glycosyl trichloroacetimidates   143
glucogalactosyl hydroxylysine   163
glucopyranose acetate donors   120
glucosamine–glycerophospholipid conjugates   161
Grignard reagents   263
Grubbs catalyst   173

h

halogen-substituted acetate donors   122
hard and soft acids and bases (HSAB)   431

hederagenin saponins 166
Helferich glycosylation 122
Helicobacter pylori strains 215
hemiacetal acylation 114
hemiacetal hydroxyl groups 96
heparin tetrasaccharides 153
heptosyl trichloroacetimidate 159
hetaryl onium salts 111
heteroaryl glycosides 190
high-yielding multigram-scale epoxidation 436
hordenine glycoalkaloid 117
Hunsen's procedures 66
hydroxyl perbenzylation 96
hydroxyl-protecting groups 32
hydroxymethyl-benzyl benzoate spacer–linker system 171

i

iodide 367
– *in situ* oxidation 367
iodobenzenesulfonamido derivatives 428
iodonium dicollidine perchlorate (IDCP) 11, 264, 387, 422
– mediated chemoselective glycosylation 274
iodonium *sym*-collidinium triflate (IDCT) 395
N-iodo-succinimide-triflic acid (NIS-TfOH) 264
inositolphosphoglycans (IPGs) 163
in situ anomerization method 80
intermediate sulfoxide moiety 33
intramolecular aglycon delivery (IAD) 270, 319, 470
intramolecular glycosylation 477, 488
intramolecular hydrogen-bond 7
iodonium dicollidine perchlcrate (IDCP) 195, 201
iodonium di-*sym*-collidine perchlorate 123
IR spectroscopy 112
isoflavone glucuronide 178
isopropenyl glycoside 107, 194
iterative iodoalkoxylation sequence 423

k

kalopanaxsaponin A 166
Koenigs–Knorr glycosylation 53, 59, 200
Koenigs–Knorr method 381
Koenigs–Knorr reaction 382

l

labor-intensive multiglycosylation sequences 95
latent–active glycosylation method 198
Lawesson's reagent 115, 127

Leloir pathway 223
Lemieux azidonitration 435
Lemieux's *in situ* anomerization method 235
Lewis acid catalysis 62, 98, 438
Lewis acid promoted glycosylation 129, 131
– representative procedure 129, 131
lipoarabinomannan (LAM) capsule 401
liquid-phase technique 174
lithium aluminum hydride (LAH) 459
lithium hexamethyldisilazide (LHMDS) 428, 430
lithium perchlorate 100, 114
– role 100
lithium tetramethylpiperidide (LTMP) 428
low-temperature NMR spectroscopy 307

m

magnesium monoperoxyphthalate (MMPP) 264, 305
mannodendrimers 385
mannofuranose hemiacetal donor 103–104
mannose 1,2-orthoesters 394
mannosyl diethyl phosphate 241
Maxam–Gilbert method 457
mechanistic pathways 102
medium-pressure column chromatography 55
Merrifield resin 171
metal-salt-based promoters 340
methanesulfonic acid 97
methanesulfonyl chloride 108
p-methoxybenzylidene acetals 474
p-methoxyphenyl group (PMP) 474
methyl glycoside 108
methyltrioxorhenium (MTO) 439
Mitsunobu protocol 103
mono neoglycolipids 400
monosaccharide acceptor 105
montmorillonite K-10 (MK-10) catalyst 241
morphine-6-glucuronide 81
multibranched dodeca-arabinosaccharide 408
multivalent presentations 399
Mycobacterium avium 209

n

Neisserial lipooligosaccharides 159
neomycin mimetics 171
nosyl chloride promoted glycosylation
– representative procedure 130
novel capping reagents 173
novel lactamized gangliosides 161
NPOE-mediated glycosyl couplings 399
nucleic acids 381

nucleophilic acceptor 2, 96
nucleophilic anionic substitutions 77
nucleoside diphosphates (NDPs) 223
2-naphthylmethyl group (NAP) 153
para-nitrobenzenesulfonyl chloride 108

o

oligo-ethylene glycol 161
oligosaccharides 100, 146
– aminosugar-containing 149
– solid-phase synthesis 433
one-electron transfer reaction 266
one-pot glycosylation 90, 247, 441
– procedure 90
– strategy 50
oxacarbenium ion 5, 6, 231, 237, 269, 316, 420
– destabilizing 9
oxazoline glycosyl donors 457
oxophosphonium intermediate 106

p

palladium-catalyzed glycosylations 463
pentaacetyl galactose 119
pentasaccharide trichloroacetimidate 163
n-pentenyl orthoesters 411
– glycosidation 411
perbenzoyl glycosyl donors 120
phase transfer catalysis 62
N-phenyl diethyl phosphorimidates 239
phenyl selenenyl chloride 426
phenylsulfanyl moiety 156
phenyl sulfenyl choride 426
N-phenyl trifluoroacetimidates (PTFA) 174
phosphine oxide nucleophilic catalysis 122
phosphonium dibromide salt 106
phosphonium halide 106
phosphorothioate anion 350
phosphorus electrophiles
– hemiacetal activation 103
phosphorylated pseudohexasaccharide 163
N-phthaloyl glycosyl acetate donors 120
Phytophthora megasperma 264
phytoalexin elicitor glucohexaose 146
podophyllum lignan 239
polycondensation reaction 386
polymer-supported glycosyl acceptors 321
polymer-supported synthesis 321
polyol glycosyl acceptors 407
– regioselectivity 407
polymer-bound triphenylphosphine–iodine complex 75
promoter iodonium ion 367
pyranose hemiacetals donors 98, 104

r

radical-cation-based activation 369
real-time reaction-monitoring method 173
reduction sequence 486
resin-linked glycosides 433
rhamnopyranose acceptor 110
ribofuranose acetate donor 122
ribofuranose nucleosides 117
ribonucleoside synthesis 103
ribosyl cholesterol 115
regioselective couplings 382, 403
– privileged donors 382
regioselective saccharide 381
reversible trans-orthoes-terification step 384
Rhodotorula mucilaginosa 269

s

Schmidt glycosidation
– protocol 163
– reaction 143, 145
Scrapie prion protein 162
selenium-containing sugars 361
selenoglycoside building blocks 375
sensitive furanoid glycols 418
– preparation 418
short-column chromatography 178
sialyl xanthate 329
silica gel flash column chromatography 219
silicon electrophiles 100
– hemiacetal activation 100
silver fluoride
– fluorination with 53
silylated glycosyl acceptor
– glycosylation 55
silylated serine acceptor 115
silyl ether glycosyl acceptor 114
silyl halide promoters 100, 102
silyl sulfonates 100
single-electron transfer (SET) process 361
solid-phase oligosaccharide synthesis 121, 171
solution-phase Helferich glycosylation 117
solvent-separated ion pair (SSIP) 269
Staudinger-type reactions 233
stereochemical interactions 11
stereoselective saccharide 381
Streptococcus pneumoniae 484
sulfamidogalacto donor 9
sulfonyl chloride hemiacetal activating agents 108
sulfoxide-based glycosylation reactions 310
sulfoxide-mediated intramolecular aglycone 320
sulfur electrophiles
– hemiacetal activation 107

t

tethered glycosides 473, 474
- NIS-induced activation 473
tetrabutylammonium bromide 102, 107
tetrabutyl ammonium hydrogen sulfate (TBAHS) 68, 331
tetrabutylammonium perchlorate 116
tetramethylguanidium azide (TMGA) 78
tetramethyl urea (TMU) 106, 107
tetra-TMS protected fucose donor 102
thallous ethoxide 251
thioglycoside-type linker 321
thio-linked oligosaccharides 262
- synthesis 262
thioorthoesters 387
thiophilic promoters 339
2-thiopyridylcarbonate (TOPCAT) donors 124
tin-sulfide reagents 114
TMS triflate promoted glycosylation 132, 240
- representative procedure 132
toluenesulfonyl chloride 108
trans–trans-linked oligosaccharides 341
trichloroethylsulfonamide (Tces) 430
trimethylsilyl triflate (TMSOTf) 261
trehalose-type disaccharides 98
trialkylsilyl sulfonates 102
trichloroacetic anhydride 114
trichloroacetimidate chemistry 155
trichloromethyl carbonate donors 127
trifluoroacetimidate method 144, 174
- activation 174
- preparation 174
trifluoromethanesulfonic (triflic) anhydride 105

trifluorozincbromide reagent 34
trityl salt promoted glycosylation 132
- representative procedure 132
trityl-butyldimethylsilyl protective group 104
trivalent lanthanide triflates 145
Trypanosoma brucei 204
two-stage activation procedure 42
two-direction glycosylation strategy 289
two-stage glycosylation method 390
two-step glycosylation strategy 279

u

urea-hydroperoxide (UHP) 439
1-*O*-unprotected sugars
- phosphorylation of 224, 232
2-*O*-unprotected β-glycosyl dibutyl phosphates 229

v

vicinal dibromide G 203
vinyl glycosides 186, 194, 218

w

Waldman's method 76
Wilkinson's catalyst 196
Wittig–Horner olefination 418

x

xylopyranose donor 102
zirconium(IV) chloride 262

y

Yamaguchi lactonization 168